2800

Understanding Dying, Death, and Bereavement

Understanding Dying, Death, and Bereavement

EIGHTH EDITION

Michael R. Leming
St. Olaf College

George E. Dickinson
College of Charleston

CENGAGE
Learning®

Australia • Brazil • Japan • Korea • Mexico • Singapore • Spain • United Kingdom • United States

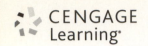
CENGAGE
Learning·

Understanding Dying, Death, and Bereavement, Eighth Edition
Michael R. Leming and George E. Dickinson

Product Director: Marta Lee-Perriard

Content Developer: Lori Bradshaw

Product Assistant: Julia Catalano

Media Developer: John Chell

Marketing Manager: Kara Kindstrom

Content Project Manager:
Ruth Sakata Corley

Art Director: Caryl Gorska

Manufacturing Planner: Judy Inouye

Production and Composition: MPS Limited

Text and Photo Researcher: PMG

Copy Editor: Bill Clark

Cover Designer: Lee Friedman

Cover Image: LuckyPix/Corbis

Library of Congress Control Number: 2014941993

ISBN-13: 978-1-305-09449-9

ISBN-10: 1-305-09449-2

Cengage Learning
200 First Stamford Place, 4th Floor
Stamford, CT 06902
USA

Cengage Learning is a leading provider of customized learning solutions with office locations around the globe, including Singapore, the United Kingdom, Australia, Mexico, Brazil, and Japan. Locate your local office at **www.cengage.com/global**.

Cengage Learning products are represented in Canada by Nelson Education, Ltd.

To learn more about Cengage Learning Solutions, visit **www.cengage.com**.

Purchase any of our products at your local college store or at our preferred online store **www.cengagebrain.com**.

Printed in the United States of America
1 2 3 4 5 6 7 18 17 16 15 14

CONTENTS

ABOUT THE AUTHORS

Michael R. Leming is professor emeritus of sociology and anthropology at St. Olaf College in Minnesota and co-director, Spring Semester in Thailand. He holds degrees from Westmont College (B.A.), Marquette University (M.A.), and the University of Utah (Ph.D.), and he has done additional graduate study at the University of California at Santa Barbara.

He is co-editor (with George E. Dickinson) of *Annual Editions: Dying, Death and Bereavement*, 14th ed. (McGraw-Hill, 1993, 1994, 1995, 1997, 2000, 2002, 2004, 2005, 2007, 2008, 2010, 2011, 2012, 2014) and co-author of *Understanding Families: Diversity, Continuity, and Change* (Allyn & Bacon, 1990; Harcourt Brace, 1995). He is also co-editor (with Raymond DeVries and Brendan Furnish) of *The Sociological Perspective: A Value-Committed Introduction* (Zondervan, 1989; Wipf & Stock Publishers, 2009). In 1995 he produced a documentary film entitled *The Karen of Musikhee: Rabbits in the Mouth of the Crocodile.* His most recent film project is a documentary film on the Karen produced by the BBC, for which he was the chief research consultant.

Dr. Leming was the founder and former director of the St. Olaf College Social Research Center, former member of the board of directors of the Minnesota Coalition on Terminal Care and the Northfield AIDS Response, and has served as a hospice educator, volunteer, and grief counselor. He has been teaching courses on death and dying for over 35 years. For the past 14 years he has directed The Spring Semester in Thailand program (http://www.springsemesterinthailand.com), which is affiliated with Chiang Mai University, and he lives in Thailand during Minnesota's coldest months.

George E. Dickinson is professor of sociology at the College of Charleston in South Carolina. He holds degrees from Baylor University (B.A. in biology and M.A. in sociology) and Louisiana State University (Ph.D. in sociology, minor in anthropology). He has completed postdoctoral studies in gerontology at Pennsylvania State University, thanatology at the University of Kentucky School of Medicine, and medical sociology at the University of Connecticut. He was visiting research fellow in palliative medicine at the University of Sheffield's School of Medicine in England in 1999, the International Observatory on End of Life Care in the Institute for Health Research at Lancaster University in England in 2006, and the University of Bristol School of Veterinary Science's Department of Animal Behavior and Welfare in England in 2013.

He has published more than 90 articles in professional journals and has co-authored other books with Michael Leming, as noted above. Like Dr. Leming, he has been teaching courses on death and dying for nearly 40 years and has been actively involved as a hospice educator. Dr. Dickinson is on the international editorial board of *Mortality* in the United Kingdom and on the editorial review board of the *American Journal of Hospice & Palliative Medicine*. He received the South Carolina Governor's Distinguished Professor Awards in 2003 and 2008 and was the recipient of the 2009 Death Educator Award from the Association of Death Education and Counseling. At the College of Charleston he was awarded the Distinguished Teacher-Scholar Award in 2002 and the Distinguished Research Award in 2008.

PREFACE

WHY DID WE WRITE A BOOK ON DEATH AND DYING?

Back in the early 1970s, following a Nobel Conference at Gustavus Adolphus College in St. Peter, Minnesota, entitled "The End of Life," George Dickinson's cultural anthropology students began writing term papers on how different cultures dealt with dying, death, and bereavement. About this same time one of his former students, then in his third year of medical school, stopped by the office for a visit. Dickinson naïvely asked about the student's death and dying course in medical school. With a somewhat bewildered look on his face and a shrug of his shoulders, he replied that they did not have anything like that. Would not one think that medical schools offered a course on dying and death? Both the students' written papers and the encounter with the medical student sparked an interest in Dickinson's research and teaching in the area of death and dying.

Soon thereafter, Michael Leming, a friend and former colleague who had written his Ph.D. dissertation about terminally ill cancer patients, contacted Dickinson about co-authoring a textbook on dying, death, and bereavement. Both were then teaching death and dying courses. There were very few texts available at the time. Leming and Dickinson became frustrated over a lack of reading materials appropriate for students. Thus, the conception of this textbook back in the early 1980s came from the experience of having limited classroom materials and from the interest and enthusiasm of students in the topic of thanatology.

WHAT DO WE HOPE TO ACCOMPLISH IN THIS BOOK?

It has been over 30 years since we began work on the first edition of *Understanding Dying, Death, and Bereavement*. Our goal in the eighth edition, as in the previous editions, has been to create a book that is both informative and practical, yet theoretical, and a book that was reader-friendly to the students. We visualized a humanistic text that was cross-cultural, multidisciplinary in orientation, and inclusive of the major foci of the interdisciplinary subject of social thanatology. Indeed, many changes have occurred in end-of-life issues since the first edition. The topic of dying, death, and bereavement has come out of the closet and is less a taboo topic than it was in the early 1980s. Though thanatology is not yet a household word, it is certainly better known than it was a quarter century ago. End-of-life issues are more "in" today, especially in the context of highly publicized and all-too-frequent school shootings, terrorist attacks in various parts of the world, and the much-debated U.S. Affordable Care Act of the 21st century, which was, at one juncture, falsely accused of including a measure to euthanize the elderly. Additionally, body disposition options are changing in the 21st century, as environmentally correct ways of disposal are evolving. The current edition reflects these changes. Specifically, we have the following objectives for this book:

1. To sensitize students to the subject of dying, death, and bereavement.
2. To aid students in adjusting to the death of a significant other.
3. To help individuals examine their own feelings and reactions to death and grieving.
4. To make readers aware of different cultural groups' death and bereavement customs in America and internationally.

The eighth edition of *Understanding Dying, Death, and Bereavement* will equip the reader with the necessary information to both understand and cope with the social aspects of dying, death, and bereavement. Because we have each taught courses on thanatology for some 40 years, we are convinced that every student carries both academic and personal agendas when approaching the subject. Although every student reading this book may not have family crises such as divorce and violence, every individual will eventually have to deal with deaths in her or his family (if they have not already done so). Having been exposed to the material in this text, we hope that each student will be in a better position to cope with dying and death. One does not "get over" the death of a significant other, like getting over the flu or a bad cold, but one must learn to live with the fact that the loved one is indeed dead and will never again physically be with us. One of our more satisfying experiences in academia is the occasional e-mail, letter, or phone call from a former student who shares with us the usefulness of this book in his or her dealing with the death of a significant other. For some, it is the benefit to themselves; and others relate how they were a good information source and support for other family members.

This textbook will make a significant contribution to your class because it is a proven text informed by some 80 years of combined experience of teaching and researching on this topic. *Understanding Dying, Death, and Bereavement,* Eighth Edition, is comprehensive and covers the wide range of topics in social thanatology. It is scholarly and academically sound, and it is practical for students because it addresses personal issues relating to an individual's ability to cope with the social and psychological processes of dying, death, and bereavement. The book has a strong cross-cultural emphasis that allows one to understand both the universality of death, dying, and bereavement and also the incredible diversity and similarity of social customs relating to this experience. This text appeals to a vast audience, not only because of its wide adaptability on college and university campuses, but also because of its practical implications for all persons. Although intended primarily for under-graduate students in sociology, psychology, anthropology, nursing, social work, kinesiology, religion, gerontology, health science, family studies, public health, philosophy, and education courses, it is also appropriate for professional courses in medicine, nursing, mortuary science, social work, child life, and personal and pastoral counseling. We have been somewhat surprised at the variety of disciplines and professional schools where earlier editions of this book have been used, including one large business school where the focus was on consumer efforts about dying and death.

WHAT'S NEW ABOUT THIS TEXT IN THE 21ST CENTURY?

We have both added and removed boxed inserts in the chapters but have maintained the four box categories: (1) **Practical Matters** boxes basically offer practical advice; (2) **Listening to the Voices** boxes consist of excerpted material from people writing about their own experiences with dying and death; (3) **Words of Wisdom** boxes contain excerpted materials—poems, literature, and other words of wisdom; and (4) **Death Across Cultures** boxes examine cross-cultural examples of death practices and beliefs. In addition to the boxes, chapter **conclusions, summaries**, and **discussion questions** at the end of each chapter serve as study aids. A **glossary** also appears at the end of each chapter with definitions of words in bold print in the chapter. Additionally, **suggested readings** are given at the end of each chapter for additional reading if desired. New tables have been added in several chapters at the request of reviewers. Some new boxes include child life specialists, death and the musical requiem, lessons for the end of life, resistance to pain medication across cultures, U.K. and U.S. prison hospice programs, animal hospices, veterinarians and end-of-life issues, a guide to survival following a suicide of a friend or family member, a letter to one's dad some ten years after his suicide, a suicide risk assessment, suicide rates in the military, bullying and suicide, capital punishment, a letter on why one should always go to a funeral, a story of Hmong funeral traditions and why one needs to return home, an explanation of the high cost of saying goodbye, and a box on diversity in death—body disposition and memorialization.

New material in the eighth edition includes updates of statistical material throughout the text and new information on various topics since the 2010 edition, including new sections on homicide and accidental deaths, personal death trends in body disposition and memorialization, and pet loss.

As in previous editions, pictures and cartoons are scattered throughout the book. Numerous new pictures and cartoons have been added to this edition, making it the most comprehensive of all of our books on the topic.

SUPPLEMENTS

An online test bank and online instructor's manual are available for instructors. The test bank contains 25 to 30 true/false questions, 25 to 30 multiple-choice questions, and 10 to 15 essay questions for every chapter, to save time creating tests. Answers to the true/false and multiple-choice questions are provided. The instructor's manual contains detailed chapter outlines and sample syllabi, which list suggested readings and class projects.

ACKNOWLEDGMENTS

We gratefully acknowledge those who reviewed previous editions and made valuable suggestions, which are incorporated in this edition.

We are indebted to Adansi Amankwaa, Albany State University; M. Terry Andrews, Mount Wachusett Community College; Angela Andrus, Fullerton College; Catherine Bacus, Chaffey College; Stephanie E. Afful, Missouri Baptist University; John Baugher, University of Southern Maine; Meredith McGuire, Trinity University; Carol Nowak, State University of New York at Buffalo; Paul Rosenblatt, University of Minnesota; Sylvia Zaki, Rhode Island State College; Dale Lund, University of Utah; Tillman Rodabough, Baylor University; John R. Earle, Wake Forest University; Clifton D. Bryant, Virginia Polytechnic Institute and State University; Dennis E. Ferrara, Mott Community College; Lori Henderson, Southeastern Community College, Burlington; Jennifer Keene, University of Nevada, Las Vegas; Patricia Kolar, University of Pittsburgh; Leslie A. Muray, Lansing Community College; Kelly Niles-Yokem, York College of Pennsylvania; Mari Plikhun, University of Evansville; Rebecca Reviere, Howard University; Julia Tang, Mount St. Mary's College; Alban L. Wheeler, Morehead State University; Martha Shwayder, Metropolitan State College of Denver; James F. Paul, Kankakee Community College; Catherine Wright, Mitchell College; Dolores Mysliwec, Glenville State College; and Gerry R. Cox, University of Wisconsin–La Crosse.

As in the past, we acknowledge those we have encountered in life. We have learned so much from significant others and our students who have shared their experiences with dying and death. Our own lives have indeed gained from these experiences. We hope our readers will experience an appreciation for life as they begin to understand, both intellectually and emotionally, the social-psychological processes of dying, death, and bereavement. This book is about dying and death, but it is really about living and life. May your reading of this book help to make your life more meaningful.

Michael R. Leming and
George E. Dickinson

The symbols of death say what life is and those of life define what death must be.
The meanings of our fate are forever what we make them.
—**Lloyd Warner, The Living and the Dead**

The day we die the wind comes down to take away our footprints.
—**Song of the Southern Bushmen**

Studying Dying, Death, and Bereavement

CHAPTER I

You are enrolled in a course in death and dying. Note the reactions of others when you inform them of this. They are likely okay with your saying you are taking such courses as Principles of Biology, English Literature, Introduction to Philosophy, History of the United States, or Calculus, but *Death and Dying*? Being informed of this course may raise some eyebrows. That sounds weird and certainly morbid, or so some seem to think. We often experience startled reactions when we tell that we teach a course entitled "Death and Dying." We wonder what images pop into individuals' minds when so informed. One woman, when introduced to George Dickinson and told that he taught a course on death and dying, drew back and in a rather startled voice said, "How could you do such a thing?" She must have imagined Dracula team-teaching and Frankenstein working as a lab assistant, along with the Grim Reaper. She then said, "What do you do in a course *like that*?" After giving her a quick synopsis of the course, she seemingly breathed a sigh of relief and said, "Oh, that doesn't sound so bad."

Imagine the responses funeral home personnel must endure if we receive such startled looks from just teaching about dying, death, and bereavement, and not actually dealing with dead human remains. Stephanie Schim and colleagues (2007) in various health care professions note that responding honestly at a party to the question of "What is your area of study?" with "Oh, I'm working with death and dying" is a real conversation terminator. From reactions that many individuals have to death-related issues, it appears that death discussions are considered in bad taste and something to be avoided. **Road Scholar** programs (formerly called Elderhostel), for example, can present any topic of interest to older persons *except dying and death*. Likewise, at least according to a survey in the late 20th century, U.S. high school physical education and health teachers spend less time teaching death and dying than any of the 21 health topics, and the topic of dying and death is less likely to be required in the 50 states than are any of the other 21 health topics.

Dying in the United States occurs offstage, away from the arena of familiar surroundings of kin and friends, with 80 percent of deaths occurring in institutional settings—hospitals and nursing homes. Back in 1949, deaths in institutional settings accounted for 50 percent of U.S. deaths; thus there has been a rapid rise in away-from-home deaths. Grandfather seldom dies at home today, where he spent most of his life; rather his death likely occurs in an impersonal institutional setting. Over the course of U.S. history, death has migrated "from a position of prominence to one of near invisibility" (Jones, 2008, p. 2). The removal of death from the usual setting prompted Dumont and Foss (1972, p. 2) to raise the question: "How is the modern American able to cope with her/his own death when the deaths experienced are infrequent, highly impersonal, and viewed as virtually abnormal?" For some time now, modern-day Americans have been as repressed about death as the Victorians were about sex. We have become rather adept as a society at avoidance tactics, at removing sickness and dying from everyday life (Thomson, 2008).

Death and dying is a topic to which college students should be able to relate easily—the dead body and its management are mysterious, yet most students likely have aged relatives at or near the end of their lives, notes British sociologist Tony Walter (2008). Death and sexuality are exotic, yet familiar. Our bodies, being living organisms, eventually deteriorate and die, yet the death of a human body is also inherently social. How we understand dying and death and how we explain our reactions to death are major themes throughout this book. We will look at how academics approach the study of death. A medical doctor will focus on the biological aspects of dying and may rely on an autopsy to definitely determine the cause(s) of a particular death. There are also social, psychological, and spiritual aspects of dying and death as well. Dying and death can be examined from a developmental perspective, viewing death at all stages of the life cycle. In the end, however, what is important to many individuals is the meaning of death. We feel, therefore, that an examination of the meaning of death and death-related behaviors is one of the most important ways to approach the study of dying, death, and bereavement.

CURRENT INTEREST IN DEATH AND DYING

Many Americans express death anxiety, yet many also have an obsessive fascination with death, dying, and the dead. This paradox is apparent in our popular culture, as television programs, movies, songs, and the print media are fraught with thanatological content (Durkin, 2003). The fascination has been fueled by various legal bodies and by legal gymnastics surrounding several prolonged and much-observed deaths of individuals. Though we have a tendency in general to avoid the topic of death, Americans are finally willing to acknowledge that death is part of life and seem to want to talk about it (Foderaro, 1994). As Muriel Gillick (2000) noted, "Death happens 55 million times each year throughout the world and 2.3 million times annually in the United States. Of the tens of billions of people that have ever existed, everyone born before 1880 has died, and nearly everyone currently alive today will perish in this century." Today's **thanatology** (the study of dying, death, and bereavement) student is bombarded by pressing issues of the day that involve death and death-related matters: memories of September 2001; terrorist attacks on the United States and threats of future attacks; the tsunami caused by the earthquake off the West Coast of Northern Sumatra in 2004; Hurricane Katrina hitting the Gulf Coast in 2005; the shootings at Virginia Tech in 2007; the war in Afghanistan; the shootings at Fort Hood, Texas in 2009; the earthquake in Haiti in 2010; the "perfect storm" Hurricane Sandy in New England in 2012; the shootings at the elementary school in Newtown, Connecticut in 2012; the typhoon in the Philippines in 2013; the bombings at the Boston Marathon in 2013; the Malaysia Airlines Boeing Flight 370 disappearance with 239 individuals on board in 2014: the worldwide AIDS crisis; the prolongation of dying from cancer; the growing incidence of chronic illnesses with uncertain courses; murder; ecological disasters; fetal transplants; cloning; and abortion. The topic of death is alive and well in today's contemporary society. Let's look at some of the reasons why dying and death as topics of discussion have come into their own in recent years.

LISTENING TO THE VOICES | DEATH ON THE FARM

As one brought up exposed to a farm environment, George Dickinson remembers on one occasion helping his father dispose of the body of a polled Hereford bull. The bull had been dead for about a day before he was found in the woods. The buzzards circling overhead were a good indication of where the lost bull would be. When he was actually found, about a dozen buzzards were already there. Just shooing those aggressive vultures away from the death scene was indeed a grim portrayal of death. Since the bull weighed approximately 2,000 pounds, Dad decided against earth burial and we then proceeded to cremate the animal. After soaking him in kerosene and putting numerous pieces of timber over the corpse (not completely unlike a pyre, as discussed in Chapter 10), we set him on fire. The corpse was still smoking several days later, though the buzzards had long since gone away. On other occasions, we would have to "put down" a cow and thus practice euthanasia (to be discussed in Chapter 8). At other times a calf would be born dead or die soon after birth. Exposure to births and deaths is commonplace in such a setting.

WHY THE INCREASED INTEREST?

Though death has been around since the beginning of humankind, in recent years a near fascination with death has evolved. Such an increased interest in thanatology is due to several reasons: An aura of mystery surrounding death perhaps brought on in part because of lingering deaths due to chronic illnesses rather than the acute illnesses of an earlier day, terrorism, an interest in ethical issues concerning death and dying, and increased media coverage of deaths, especially violent deaths.

THE MYSTERY OF DEATH   An aura of mystery developed. One is taken to the hospital or nursing home and is next seen dead. Thus, a child may begin to wonder what is going on with this thing called death that takes people away, not to return. As one little boy said in writing a letter to God, "Dear God, What is it like when you die? I don't want to do it. I just want to know." Often small children are not allowed to go into certain parts of a hospital; thus, the taboo nature of a hospital setting makes one wonder what is going on in there—"I just want to know." Hospitals do not allow such entry in part because of contagious diseases to which small children would be more vulnerable than adults. Such rejection is not unlike Charles Schulz's saying "no dogs allowed" as Snoopy tries to enter a forbidden area. "No children allowed" is indeed the strong message of rejection here. To say "no" to a child often increases his or her curiosity tremendously. "Why is this place such that I cannot enter?" the child may ask.

A desire to examine this mysterious thing called death has contributed to a growing interest in thanatology. Our society has done little to achieve formal **socialization** of its members to deal with death on personal and emotional levels. Even though the hospital may not allow children to enter certain areas or at particular times, parents have often tried to shield their innocent children from death scenes. Medical and theology schools have not had significant curricular offerings to prepare their students for this death-related work. Overall, our socialization to dying and death situations has been unsystematic and ineffective.

Today's lack of familiarity with death may be in part due to fewer individuals being raised on farms than was the case in the early 20th century. Being brought up in a rural environment gave one direct exposure to birth and death as everyday events. Children were surrounded by the alpha and omega of the life cycle. Kittens, puppies, piglets, lambs, calves, chicks, and colts were born—and also died. Thus, it was commonplace to make observations of death and to deal with these situations accordingly.

Death can be very visible down on the farm. With less than 10 percent of the U.S. population engaged in farming today, birth and death scenes have largely been removed from the personal observations of most individuals. The mystery of these events at the beginning and at the end of life is at least somewhat addressed with farm dwellers—less likely the case with urbanites.

TERRORISM   In the post 9/11 era, the ever-present threat of terrorist attacks has injected the unpredictable nature of mortal danger and mass destruction into our collective awareness. Victory in the war on terror depends in part on our ability to live with death threats (Wong & Tomer, 2011). Such fear of an invasion had not presently occurred in the lifetime of most living Americans. A stunned nation watched on television that day, and the days that followed, absolutely in disbelief that such an attack had occurred. Many television viewers saw the second plane hit New York's World Trade Center. In one of the more horrific images, the dead and the doomed plummeted from the skyscrapers. A nightmare of this magnitude was unknown to the majority of U.S. citizens. Not since the attack on Pearl Harbor had such a surprise occurrence taken place, yet this time it hit the U.S. mainland. The sense of security and self-confidence that Americans take as their birthright suffered a grievous blow, from which recovery will take a long time, observed Washington, DC reporter R. W. Apple, Jr. (2001). These deaths were sudden and violent, thus more intense than lingering or expected demises. More guilt may occur as a result because there is no opportunity to say goodbye, express feelings, or make amends (Dickinson, 2011a).

In the aftermath of September 11, 2001, more than 60 percent of Americans who were asked about their emotions responded that their personal sense of security had been shaken, compared with 24 percent in the preceding year; 54 percent feared that they or a member of their family would become a victim of a future terrorist attack (Jonas & Fischer, 2006). Regarding children, in counseling after the 9/11 attack, N. B. Webb (2002) observed that the overwhelming magnitude of this tragedy affected even those children who did not suffer personal losses. Terrorist attacks induce fear, anxiety, and concern about death and seem to be a natural reminder of mortality.

Ken Doka (2003) stated that one of the most significant elements of public tragedy involves the degree to which it is perceived as being caused by humans, particularly in an intentional way. Assigning blame, Doka notes, provides a target for anger, which may help promote an illusion of safety and control. Regarding adolescents' reactions to the terrorist attacks, some good may come out of the bad. For instance, it is suggested in the midst of stress-related disorders that many active attempts at coping and adapting to difficulty may be seen. Following the 9/11 attacks, there was the potential to seriously compromise adolescents' sense of fairness and justice, yet their increased capacity to think in abstractions and ability to problem solve leaves the potential for adolescents to cope by constructing a sense of order and justice even in the face of senseless acts (Noppe, Noppe, & Bartell, 2006). Such positive approaches to negative circumstances may enhance resilience by promoting greater flexibility in thinking and openness to new information.

Religiousness is generally associated with lower death anxiety, thus religious belief plays a protective role in terror management (Jonas & Fischer, 2006). For example, 84 percent of individuals asked if they had said special prayers in response to the September 11 terrorist attack answered in the affirmative (Smith, Rasinski, & Toce, 2001). Additionally, following the 9/11 attack, the highest level of church attendance since the 1950s was observed in America, with similar spikes in Canada, England, and Australia (Pyszczynski, Solomon, & Greenberg, 2003). Terror management studies have suggested that the reminder of mortality leads people to defend their religious faith (Jonas & Fischer, 2006).

Terror management theory (TMT) suggests that people adhere to cultural worldviews and beliefs in order to suppress death and mortality-related thoughts. TMT says that individuals combat the terror of their mortality with the same cognitive abilities that cause this terror to arise, by developing "death-denying cultural belief systems" (Goldenberg et al., 2000). TMT could explain why many people have trouble interacting with people from different cultures.

Studies by Eva Jonas and Peter Fischer (2006) suggest that those intrinsically vested in their religion (those for whom religion serves as a framework for life by providing both meaning and value) derive terror management benefits from religious beliefs. Religion is unique among meaning systems because it equips individuals to respond to situations in which they come face-to-face with the limits of human control and power and are confronted with their finitude (Smith, Pargament, Brant, & Oliver, 2000).

The September 11 attack contributed to an omnipresent feeling of compassion and a yearning to believe something redemptive could come out of horrific tragedy, noted New York reporter N. R. Kleinfield (2002). Altruism, patriotism, and a sense of unity followed this event, as is often true following a disaster. A sharing of grief, even with strangers, seems to console; it helps to know that others care. The 9/11 dead were like us, and identification with and connections to them happened through the media. Grief shared is grief relieved (Dickinson, 2011a). This event helped many individuals realize just how precious life is and that we should be thankful for what we have. One wakes up in the morning not knowing what the day may bring. An outcome of 9/11 is that individuals today are more aware that the beginning of a bright sunshine-filled day could end up in tragedy.

The events of September 11, 2001, have caused our society to become more paranoid. We tend to "look over our shoulder" more often. We are more suspicious of certain individuals whom we fear might harm us. Airports, government buildings in the United States and abroad, and any facility where large crowds gather have more surveillance and personnel to guard the premises. The possibility of death from terrorism is all around us today, and, unfortunately, it is not likely to go away anytime soon.

ETHICAL ISSUES Individuals are living longer today in part because of medical break-throughs such as life-support equipment, organ transplants, penicillin and other miracle drugs, clean water supplies, sanitation and other public health measures, healthier diets, more exercise, and improved personal habits (e.g., stopping smoking). We do have a problem with obesity, however, which counters the positive aspects of 21st-century living. Such prolongation of life has raised ethical issues dating back to the 1970s (e.g., Karen Ann Quinlan and Baby Jane Doe) that involve the right to die, causing a furor in philosophy, law, and medicine. The case of Terri Schiavo in Florida in 2005 brought this end-of-life issue to the fore-front. With the media highly publicizing these cases, the public was alerted to moral and legal questions on death not previously posed. Whether or not to "pull the plug" and disconnect life-supporting equipment present questions for which ready answers are not found. An elderly woman summed up this dilemma when she spoke to Jinny Tesik of Compassion in Dying (Sturgill, 1995), an organization providing education regarding terminal care: "We used to be afraid to go to the hospital because that's where you went to die; now we're afraid to go because that's where they won't let us die."

The whole issue of when death occurs evolves from these medical developments. The question of who determines when one is alive or dead has been addressed by physicians, lawyers, philosophers, and theologians. These questions, along with the controversy over abortion rights, were a few of the significant ethical issues of the 1970s that provided an open forum for discussion and debate concerning the topic of dying and death. Specific definitions of death were not as necessary, prior to the coming of these 20th-century medical breakthroughs.

POPULAR CULTURE Having grown up in Texas, George Dickinson has fond memories of spending the night with his grandparents on Fridays and going to the movie with his grandfather. The movie was always a Western with basically the same theme—the "good guys" (the cowboys) wore white hats, and the "bad guys" wore black hats. Though toward the end of each movie it appeared that the good guys were going to be wiped out, military reinforcements would always come to the rescue just at the last minute by fighting fair and because the good guys were morally superior. The guys in the white hats would be rescued, and the bad guys would receive their just reward of death. Similarly, recent research (Yokota & Thompson, 2000) of G-rated animated feature films released in theaters between 1937 and 1999 revealed that characters portrayed as bad were much more likely to die of injury than other characters.

In the early 1970s, however, Hollywood began to additionally produce films revolving around the theme of death in which the good guy (the box office star) died. One of the first of these movies to deal with death was *Love Story,* in which one of the two main characters is dying throughout much of the film and eventually does die during the film. Then Tom Hanks dies of AIDS in *Philadelphia Story,* and Susan Sarandon dies of cancer in *Stepmother.* Other movies provide a realistic portrayal of historical events involving mass deaths, such as *Schindler's List* and *Saving Private Ryan.* Twenty-first-century movies include (1) *Igby Goes Down, a* portrayal of two brothers helping their terminally ill mother die with the aid of drugs and a plastic bag, (2) *The*

Hours, with an undercurrent theme of the reasons for suicide, an attempted suicide, and a rational suicide, (3) *The Event,* about a series of unexplained deaths that occur among the gay community in New York City, (4) *The Sea Inside,* about the death of a sailor who became a quadriplegic after injuries in a diving accident, (5) *Million Dollar Baby,* with an underlying theme of assisted suicide, and (6) and *You Don't Know Jack,* a 2010 HBO TV movie about Jack Kevorkian, a medical doctor who helped over 100 individuals die from physician-assisted suicide. Almost an obsession and certainly a fascination with death seem to occur in today's society—especially on the screen in a somewhat imaginative world.

Likewise, many situation comedies on prime-time television in the 1970s, 1980s, 1990s, and now the 21st century have addressed the topic of death. Some have viewed death in a serious vein, whereas others have taken a humorous approach. Among the first situation comedies to talk about death was *The Cosby Show*, starring Bill Cosby. This popular late-1980s and early-1990s program showed an episode in which a goldfish had died. Bill Cosby gathered the family around the toilet bowl in the bathroom and insisted that they have a proper funeral for the deceased fish. One read scripture, one said a eulogy, and then another flushed the fish out into its eternal resting place in the sewer. The episode made a serious effort to explore some issues involving the death of pets, even though one might not experience the death of a goldfish in the same way as the death of a dog.

In addition, *Sesame Street* devoted a 15-minute segment on the death of Mr. Hooper, a few days after his death. However, rather than explore the feelings of loss, the characters focused on all the good qualities of Mr. Hooper. Only in the end did Big Bird, who seemingly was having a more difficult time with the death of his friend, say, "We're going to miss you, Mr. Looper," to which the entire cast said in mass, "That's Hooper, Big Bird, Mr. Hooper." We do miss individuals when they die. Big Bird's comment was a most realistic way of expressing his feelings about the death of a friend on a popular children's television program. Although these two programs were both rather serious, we often respond to both death and sex as topics about which we joke. They are uneasy topics; thus, to laugh about them is a way to cope. More recently from *Sesame Street* in 2010, Katie Couric appeared with Elmo in a special in which they dealt with families coping with the death of a parent.

The sudden, untimely death of John Ritter, just a week before the beginning of the 2003 fall season, posed a dilemma for the producers of his popular show *Eight Simple Rules for Dating My Teenage Daughter*. After much discussion, the producers decided to do a tribute to John Ritter and then to air the first three programs that were previously taped. After the airing of these three programs, the network chose to work John Ritter's sudden death in September into the plot of a special one-hour episode in November. That program opened with Ritter's wife receiving a telephone call about his dying of a heart attack in the supermarket, leaving her and three teenage children to deal with their shock and loss. Thus, the network chose to deal with the death of the primary actor in a real way on the show.

Some of the earliest television programs to discuss death appeared in the early 1970s. *Living with Death* presented various death-related situations observed through the eyes of a CBS reporter. ABC's *The Right to Die* addressed moral questions of mercy killing and suicide. The National Endowment for the Humanities sponsored a program entitled *Dying*. For two hours this program very sensitively portrayed four cancer patients, ranging in age from their late 20s to early 70s. Each died during the course of the filming. A PBS documentary in 1979 showed the last three years of Joan Robinson's life. This film revealed the experience of a woman and her husband as they tried to live with her cancer of the breast and uterus. In the fall of 2000, *On Our Own Terms: Moyers on Dying*, a four-part, six-hour series, explored issues related to death and dying, including candid conversations with people dying in their homes and in hospitals. This series was followed by a companion show, *With Eyes Open*, a four-part PBS series of half-hour interviews examining grief, mortality, caregiving, and the afterlife. HBO's comedy-drama *Six Feet Under* came onto television screens at the beginning of the 21st century and depicted life in a funeral home. Such a weekly encounter gave viewers an idea as to what goes on behind the scenes in a funeral home, thus adding to the audience's awareness of death. *Six Feet Under* received critical acclaim and was praised for being so frank about death and its effects (Harper, 2009). *Death and the Civil War* aired on PBS in 2012 and gave a vivid depiction of death on the battlefield and discussed how death in that time period contributed to changes in the way death is viewed in the United States. The 2014 HBO show *Girls* twice addressed the issue of grief and death, with the show highlighting how the Internet has made grief more casual and public, and another episode with the sudden death of the editor on the show (Seligson, 2014).

Violence, including death, is highly revealed on television today. For example, a study by the Parents Television Council (2011) found that 2002 depictions of violence in television programs were 41 percent more frequent during the 8 p.m. Family Hour, and 134 percent more frequent during the 9 p.m. hour than in 1998. Television violence has become more graphic over time, with frequent use of guns or other weapons, more depictions of blood in violent scenes, and more on-scene killings and depictions of death in 2002 than in 1998. UPN and Fox had the highest rates for violence during the Family Hour in 2012.

Two movies with military themes, *Black Hawk Down* and *We Were Soldiers*, showed U.S. military attempts to recover fallen soldiers in Somalia and Vietnam. In that same theme, *Platoon* gave a rather realistic view of the Vietnam War. Such movies bring to life on the screen what is happening in military encounters in various places in the world even today. Death is alive and well in contemporary U.S. movies. The media are bringing death into movie theaters and right into our homes. By these occasional reminders of death, our interest in death is probably increased.

The topic of dying and death maintained a high profile throughout the 1990s, in part because of Dr. Jack Kevorkian. His "suicide machine" to terminate the life of a terminally ill patient with Alzheimer's disease in Michigan in 1990 drew headlines in the media. Subsequent suicide deaths assisted by Dr. Kevorkian through the

1990s fanned the fire of this controversy. Kevorkian's actions peaked with the video showing on *60 Minutes* in 1998, after which he ended up in prison. Much was aired about the whole controversy of CBS's showing this on television. This "shot" was heard (and seen) around the world. The various trials of Dr. Kevorkian brought extensive coverage in the media. What more drama could the media request? Although some individuals might have viewed Kevorkian more or less as a serial killer, he continued to be allowed to practice his physician-assisted suicide in defiance of the law. His goal was to change the law and make this behavior legal. In standing up for that which he felt to be right and at the risk of his own life, Kevorkian fought on and gained many supporters. After all, here was a man who would risk life and reputation (no matter how obnoxious he seemed to many) to prove his point. The media loved it and gave him the publicity he desired.

In addition to Dr. Kevorkian's actions, the issue of physician-assisted suicide made headlines when voters in Washington, California, and Oregon went to the polls in the early 1990s to decide the legality of this issue. Although the final vote counts in Washington and California were close, such action was not approved. On the other hand, the voters of Oregon successfully approved physician-assisted suicide in 1994 (and again in 1997). The Oregon voting was followed by court action and ultimately action by the Supreme Court (see Chapter 8). The pros and cons of assisted deaths were debated on talk shows on both radio and television. Individuals tend to have strong opinions regarding topics such as abortion and assisted deaths. Therefore, such controversial topics solicit heavy media coverage. Then in 2008 the voters of Washington returned to the polls and this time approved physician-assisted suicide. In 2010 the Supreme Court of Montana, not the voters, ruled that state law protects doctors from prosecution for helping terminally ill patients die. The Montana Supreme Court, in taking such action, side-stepped the larger landmark issue of whether physician-assisted suicide is a right guaranteed under the state's constitution (Enck, 2010).

Peter Nardi (1990) analyzed how the media handles AIDS and obituaries. He noted that one of the main issues raised by obituary reporting is personal privacy versus journalistic ethics. Unlike other diseases often unreported in earlier generations such as tuberculosis and cancer, AIDS raises questions of both medical and sexual ethics. The dilemmas faced by the media in AIDS-related deaths draw attention to the continuing stigma attached to AIDS. Sociologists and other social scientists have the task of identifying the circumstances and social-psychological processes underlying these negative attitudes.

The media have been a positive force in bringing the topic of AIDS out of the closet. For example, when a celebrity or sports hero is infected with human immunodeficiency virus (HIV), the media have contributed to *AIDS* and *HIV* becoming household words (Lemelle, Harrington, & Leblanc, 2000). Back in 1991, when NBA star Magic Johnson publicly announced that he was HIV-positive, this hero for many, a heterosexual athlete, produced new ways to respond to AIDS from both the media and the public. Today, the public is more accepting of the identification of HIV infection and the testing procedures for detecting HIV antibodies. Yet, these discoveries were scientific breakthroughs and altered life for many, noted Lemelle and colleagues.

Historically, thanatological themes have been present in numerous musical styles from operas and classical music to folk songs about serial killers dating back into the 19th century (Schecter & Everitt, 1997). Music popular with young people often has a morbid element that highlights death in a catastrophic and destructive manner such as in heavy metal and rap. Ice Cube's "Death Certificate" and "Lethal Injection" rode high on the charts in the 1990s. Some heavy metal bands have death-associated names: Megadeth, Anthrax, Slayer, and Grim Reaper. Examples of their song titles include "Suicide Solution," "Highway to Hell," and "Psycho Killer." One of the more successful rap recording companies is named Death Row Records. Rap song titles include "Murder Was the Case," "Cop Killer," "Sex, Money, and Murder," and "Natural Born Killers" (Durkin, 2003). Lyrics of some of the songs have such wording as "Baby join me in death, this life ain't worth living," with these lyrics being repeated over and over by the Finnish group HIM. Yet there is Bob Dylan's "Death is not the end," which tries to give hope to the listener, with various lines such as "and you just can't find a friend" or "you don't know what's around the road" sang just before "death is not the end."

Reports of dying and death are found in daily newspapers through death notices, sometimes accompanied by a photo of the deceased individual. Newspapers also highlight sensational deaths or catastrophic events. Printed matter such as novels and crime books are often filled with deaths, both violent and natural. Weekly magazines typically have a section referred to as "transitions" where they denote deaths of celebrities. In addition to media interest with death and dying is the rise of spontaneous shrines to victims of road accidents, a reminder to travelers of a death at that roadside spot.

George Dickinson

Roadside memorials are often placed near the spot where a highway fatality occurred. Typically the memorial has a religious significance, such as the crosses seen here. Such memorials are found in many countries around the world.

DEATH EDUCATION

The topic of dying and death did not come into its own in recent times until the 1970s. Certainly death, like sex, was not a new event, yet was something joked about but rarely discussed openly. Sex, as a subject of discussion, came "out of the closet" in the 1960s, followed by death in the 1970s.

THANATOLOGY CLASSES If the mission of colleges and universities is, in part, to prepare students for personal challenges, then undergraduate education is an excellent place to include death and dying education, as this is something which all students will definitely face (Ratner & Song, 2002). Some of the goals of death education are to promote discourse about death, explain the developmental processes of death understanding, integrate the dying with the living, heighten sensitivity about cultural variations in dying, death and grief, and appreciate the universal and individual course of the grief experience (Cupit, Sofka, & Gilbert, 2012). Death and dying offerings began to flourish back in the 1970s. Though a count as to the number of offerings in different departments in the United States is not available, Lizabeth Eckerd (2009) found in a recent survey of psychology departments in nine midwestern states that approximately 20 percent had offered a course in death, dying, and bereavement within the last five years. Overall, death and dying courses in the United States tend to be offered in psychology, sociology, social work, religious studies, philosophy, human services, and health sciences departments and have increasingly become legitimate topics of study over the past decade (Cupit, Sofka, & Gilbert, 2012).

In elementary and secondary school curricula, death and dying is on the "approval list" in many states, yet such an option does not always have a high priority. When we ask students in our death and dying classes if they had any death and dying orientation in public schools, very few answer in the affirmative. Students who attended a parochial school are more likely to have had such instruction. Hannelore Wass, a pioneer in death education in the United States, observed that even though death education and crisis intervention programs exist, they are available to only a minority of the 48 million students attending public school in the United States (2004).

Some outcomes of death education have demonstrated greater cognitive understanding and behavioral changes for students exposed to such education (Rocco, Jaclyn, & Heather, 2007). There has also shown to be a decrease in fear, less concern about dying, and a decline in death anxiety for those with death education, resulting in affective changes and changes of knowledge and behavior about death. Most schools have established protocols for crisis intervention, although preventative education through the study of death and dying remains controversial. Yet, death education could be part of a student's cultural education and could promote life-affirming and constructive attitudes about behavior toward self and others, note Rocco and colleagues. A better understanding of life, respect, empathy, and compassion all contribute to a higher quality of life.

Fergus Bordewich (1988) stated a quarter of a century ago that death was less likely than sex to be found as a subject in the curricula of American public schools. If death was acknowledged at all, it was usually discussed within the context of literary classics. In recent years, courses treating dying and death far more explicitly have begun to appear in schools across the country. Many schools have

"I think you'll be interested in the next patient. He's ninety two years old and accompanied by his parents."

Campbell, Martha/CSL, CartoonStock Ltd.

blended some of the philosophies and techniques of death education into health, social studies, and literature courses. Others have introduced suicide-prevention programs. With various shootings in high schools around the United States, interest in thanatology has increased significantly since the mid-1990s.

An argument for death education, noted by Bordewich said:

> An underlying, but seldom spoken, assumption of much of the death-education movement is that Americans handle death and dying poorly and that we ought to be doing better at it. As in the case of many other problems, many Americans believe that education can initiate change. Change is evident, and death education will play as important a part in changing attitudes toward death as sex education played in changing attitudes toward sex information and wider acceptance of various sexual practices. (1988, p. 31)

Edward Ratner and John Song (2002) observed that not enough is being done with death education. They noted that the most common approach to death education is to create an elective course for students interested in studying thanatology or in providing health care to dying patients. Such electives, they argue, reach very few students and often treat dying as a separate field of study, rather than as a part of everyone's future. Instead, colleges and universities should require students to take a general course on dying or integrate material on death into the curricula of as many disciplines as possible—or do both, they suggested. Furthermore, Ratner and Song, faculty members in the Center for Bioethics at the University of

Minnesota, argued that every undergraduate major and postgraduate field of study could include information on death.

Death education provides an opportunity to familiarize students and professionals with the needs and issues surrounding dying and death. One of the goals of courses on dying and death for all age groups is to increase knowledge about death and about the professions involved with death—funeral directors, medical personnel, and governmental organizations. Other goals are to help students learn to cope with the deaths of significant others, to deal with their own mortality, to be more sensitive to the needs of others, and to become more aware of different cultural groups' death and bereavement customs. A more abstract goal is to understand the social and ethical issues concerning death as well as the value judgments involved with these issues. As some students with personal problems may enroll in a death and dying course, instructors should be alert to this and not want to heighten their anxieties regarding unresolved losses. Instructors should look for signs displayed by at-risk students in death education classes.

The emotional conflicts related to dying and death are particularly acute for primary care health professionals. Although limited emphasis has historically been placed on death education in schools for health professionals in the United States, more recent studies of death education offerings in nursing, medicine, pharmacy, dentistry, child life specialty, and social work are somewhat encouraging (Dickinson, Sumner, & Frederick, 1992; Dickinson, 2007a; Parvin & Dickinson, 2010; Sirmons, Dickinson, & Burkett, 2010; Dickinson, 2011b; Dickinson, 2012a; Dickinson, 2012b).

As more emphasis is placed on relating to terminally ill patients and their families in the health professions, we would hope that more positive attitudes toward treatment of the dying patient will emerge when these students become practitioners. Helping students deal with their own anxieties about death at the time of actually facing terminally ill patients would seem to be an appropriate time for intervention. In the end, the young professional, the patient, and the patient's family all should benefit from this emphasis on death education.

THANATOLOGY PUBLICATIONS In the 1970s Michel Vovelle (1976) published an article entitled "The Rediscovery of Death," in which he documented the sudden flurry of publications on the subject of death that appeared in the Western world from the 1950s on. Geoffrey Gorer's 1955 essay on "The Pornography of Death" seemed to open the door for publications on the subject of death. Gorer argued that death had replaced sex as contemporary society's major taboo topic. With death in the community becoming rarer and with individuals actually seeing fewer corpses, a relatively realistic view of death had been replaced by a voyeuristic, adolescent preoccupation with it, observed Gorer. Vovelle argued that the focus of attention represented by these texts amounted to nothing less than the "displacement of a deeply seated taboo on a subject that had lain hidden in the shadows of the western psyche since some unspecified point in the 19th century" (Prior, 1989, p. 4).

One of the early texts for thanatology, published in 1959, was an anthology by psychologist Herman Feifel entitled *The Meaning of Death*. Feifel's work was an interdisciplinary attempt to restore death to cultural consciousness. In 1963, Jessica Mitford's *The American Way of Death* was very critical of the funeral industry. Elisabeth Kübler-Ross's *On Death and Dying*, published in 1969, advised

Tennessee Tech University Department of Archives

Though some early parish registers did not list the deaths of babies, this early 20th-century portrait of an Appalachian family reveals the importance of every member of the family, even the deceased one. This was obviously a sad time for the family, yet the somber expressions on the members' faces were typical of photos from this period.

LISTENING TO THE VOICES | STILL LIFE

Take photographs of a dead child? No way. To me, it was creepy, exploitative, and completely out of the question. I could not stop envisioning scenes typical of forensic crime lab dramas. Gray-hued cadavers placed on shiny tables in a windowless, disinfected room.

I was already a hormonal mess, sleep-deprived, and completely traumatized by what was about to happen. First of all, this is not at all how I had foreseen my first childbirth experience. I was supposed to be at least eight months along with a lost mucous plug or ruptured membranes. I was supposed to be fat with rosy cheeks (like Mrs. Claus, only with anxiety and contractions).

My husband and I had never been parents before, and now we were about to meet a child we'd never change, feed, or soothe. Our pastor told us that our pain was that of mourning our dashed hopes and anticipated joys. I just wanted this stillbirth nightmare to be over so I could go home and scream at the top of my lungs and pack away the crib and blankets. I wanted to hide in my bedroom and reflect upon why I was not meant to be a mother. I even felt like a disappointment to the labor and delivery staff in that I could not produce what so many thousands before me had.

My arrival at the hospital was like that of any normal expectant mother for a scheduled induction. I wore house socks and held my husband's quivering hand. We entered the business office and received identification bracelets. We waited in a stark white room with a wall clock and a barrage of television infomercials declaring the merits of Magic Bullets, OxiClean, and bareMinerals. When the doctor finally delivered her, I was afraid to look. After all, she was arriving so early, with so many internal abnormalities of which we were already aware, that I thought she might appear alien-like. Nevertheless, when my very first flesh and blood production was placed lifeless in my arms, she looked like a sleeping cherub. With the pink tone of life slowly fading from her face, she appeared strangely content with her unfortunate fate, as if she had maturely accepted it long before we had. It was difficult knowing that for so long (six months to be exact), I had essentially been her life support (her ventilator, if you will). According to the doctors, I wasn't doing her any favors. I held her and stared at her for a very long time. I talked to her. My husband rocked her. We most definitely did not want the video camera. But several family members were really pushing the photograph issue.

"That way you'll never look back and say, 'I wish I had.'"

"That way you will know that you did every thing you could to honor her."

But I was afraid to take her picture. The mere suggestion felt like an invasion of privacy. Would we not be disrespecting the deceased? I began contemplating the meaning of a "snapshot." What purpose does it serve? I suppose it's how we, as humans, attempt to hard-wire a memory. Among other things, we also use photos for evidence, protection, justification, and art. Did I wish to place on a back burner in my mind the physical and emotional despair of this day? Or did I need to prove to myself what a glorious part in the circle of life I had played? Granted, my most significant role in the universe thus far would be short-lived. However, I had indeed become a parent. I was the only female who would ever nurture this one-pound, 15-ounce being. Even though life had failed her, I was still a proud mother. I describe the experience as the single best and worst day of my life. My child had not lived, but I had met her. That was enough.

And we *did* take photographs—moments that to this day have only been witnessed by my husband and very few family members. I even made a small keepsake album that I keep hidden away. From time to time, I will glance at these sacred images even today, six years later. Looking back now, I don't know why I was so opposed to taking these photographs. No, they did not include Santa, her teammates, pets, or birthday cakes, but they are ours to touch and stare at when we need proof of that wonderful nightmare. I can marvel at how her features resemble those of her younger sisters, who are still too young to comprehend her tragic fate.

I now do not know how I would cope *without* those photographs. I shudder to think that my fear of the unknown nearly destroyed my firstborn's opportunity to achieve a physical permanence within her mother's life. And that would have been *my* greatest tragedy.

Source: H. Philpot. (January 2010). Still Life. *Skirt*, p. 24. Heather Philpot is a wife, mother, and freelance writer who lives in Travelers Rest, South Carolina.

Americans that they can play a significant role in the lives of the dying. Ernest Becker's *The Denial of Death*, published in 1973, argued that denying death is commonplace in our society. Both Kübler-Ross's and Becker's books became best-sellers. It was Kübler-Ross, a physician, who was a real catalyst in making the medical profession realize that terminally ill patients are more than the cancer patient in Room 713 and are warm, living human beings who have personal needs. Both of these books alerted the public to the issue of dying in America.

A popular book discussing dying that was on the bestseller list from 1998 through 2001 is Mitch Albom's *Tuesdays with Morrie* (1997), in which the author writes about his former college professor who is dying of amyotrophic lateral sclerosis (ALS). Sandra Gilbert's *Death's Door: Modern Dying and the Ways We Grieve* (2006) combines literary and cultural criticism with the intimacy of memoir. From a physician's perspective, Christine Montross's *Body of Work: Meditations on Mortality from the Human Anatomy Lab* (2007) gives a glimpse into the day-to-day life of a medical student dealing with dying and with dead human bodies. Shelly Kagan's *Death* (2012) invites the reader to take a fresh look at one of the central features of the human condition—the fact that we will die. Karla Erickson's *How We Die Now: Intimacy and the Work of Dying* (2013) is about what can be learned from those who work with the dying. Candi Cann's *Virtual Afterlives: Grieving the Dead in the Twenty-First Century* (2014) investigates emerging popular bereavement traditions.

Two professional journals on thanatology emerged in the United States in the 1970s: *Death Studies* (formerly *Death Education*) and *Omega: The Journal of Death and Dying*. In 1996 in England, *Mortality* evolved as a thanatology journal. Whereas the U.S. thanatology journals are psychology-oriented, *Mortality* has more of a mix of disciplines, with an orientation toward sociology. Other related journals in the United States and United Kingdom are *The Hospice Journal*, *Progress in Palliative Care*, *Palliative Medicine*, *Social Science and Medicine*, *American Journal of Hospice and Palliative Medicine*, *Suicide and Life-Threatening Behavior*, and *Illness, Crisis & Loss*. The number of articles on death also expanded considerably in journals of education, family, medicine, health, nursing, psychology, social work, and sociology.

Death education should not only prove useful in coping with dying and death situations, but also should actually improve the quality of our living. As Elisabeth Kübler-Ross noted, relating to the dying does not depress her, but rather it makes her both appreciative of each day of life and thankful each morning that she awakes with the potential of another day. Learning more about dying and death should provoke one to strive to make each day count in a positive way. Educating oneself about death will tend to make one "look for the good in others and dwell on it," as the late author Alex Haley suggested on his letterhead stationery.

MORTALITY STATISTICS

Ask an individual how he or she wishes to die. With the exception of the comical reply, "When I am 92 and at the hands of a jealous lover," most people will respond, "When I am very old, at home, unexpectedly, in my own bed, while sleeping—and with my full mental and physical capabilities." Unfortunately for most of us, we will not die as we would like. For some, this fact may be a source of apprehension and anxiety.

DEATH ETIOLOGY AND LIFE EXPECTANCY

Mortality rates have been declining in most developed countries since the mid-1800s. There has been a steady upward trend in the highest life expectancy at birth with an average increase of about three months per year since 1840 (Crimmins & Beltran-Sanchez, 2010). Recent increases in life expectancy have depended on increasing survival among the older adult population, however. A study using the World Health Organization's mortality data from 1955 to 2004 for 50 countries (Viner et al., 2011) revealed that in the 1950s, mortality in the one-to-four age group far exceeded that of all other age groups in all regions studied. In the 50 years leading up to 2004, however, death rates in children aged one to nine fell by 80 to 93 percent, mostly due to reductions in deaths from infectious diseases. By contrast, declines in death rates in those aged 15 to 24 years were only about half that of children, mainly because of increases in injury-related deaths, particularly in young men. For example, by the start of the 21st century, injuries (e.g., car crashes and street or gang violence) were responsible for 70 to 75 percent of all deaths in males aged 10 to 24 in all the regions studied. By 2004, suicide and violence were responsible for between one-fourth and one-third of deaths in males aged 10 to 24 years, and death rates in males between ages 15 and 24 are now two to three times higher than in boys aged one to four.

Whereas in the 19th century the mortality rate was particularly high among children and continued at a high level throughout adult life, the incidence of death is now heavily concentrated among the elderly. Thus, average life expectancy is now much longer. For most individuals, death approaches slowly over years of gradual decline. For example, the number of years individuals will live with conditions like vision or hearing loss and mental health issues such as depression is also increasing (Cheng, 2012). As noted in Table 1.1, the cause of death (**etiology**) for nearly half of us will be from one of two chronic diseases—heart disease or cancer. With these **chronic diseases**, deaths are usually prolonged and are not sudden and unexpected, as most people might desire. In the early 1900s, infectious diseases accounted for more deaths, whereas today cardiovascular diseases and cancers account for over half of deaths in the United States. While chronic diseases are killing more people nearly everywhere, the overall trend is the opposite in Africa, however, where illnesses like AIDS, malaria, and tuberculosis are still major threats (Cheng, 2012).

Cardiovascular disease has been the leading cause of death in the United States since the 1950s. Likewise, heart disease is thought to be the largest cause of death worldwide with estimates of 16.7 million lives annually, with cancer accounting for some 7.9 million deaths (Howarth, 2009). Yet in the United States, mortality rates from cardiovascular disease declined between 1970 and 2000, accounting for much of the increase in life expectancy at birth during this period (Crimmins & Beltran-Sanchez, 2010). The reduction of high cholesterol and hypertension, largely resulting from prescription drugs, has likely contributed to this decline. The growing problem of obesity, however, is projected to increase in the future, contributing to cardiovascular disease and diabetes. Additionally, mortality rates from a number of cancers, the second-leading cause of death in the United States, have decreased in recent years. Leading causes of cancer deaths are from lung, breast, prostate, and colorectal cancer.

In the United States, cigarette consumption remains the single most preventable cause of sickness and premature death (Anderson, 1996). Although smoking rates have declined since the 1960s, rates have leveled off in the past decade. About

TABLE 1.1	LEADING CAUSES OF DEATH IN THE UNITED STATES, 1900 AND 2010 (IN DEATH RATES PER 100,000 POPULATION)

Causes of Death	Death Rates Per 100,000 Population
1900*	
1. Pneumonia	191.9
2. Consumption (tuberculosis)	190.5
3. Heart disease	134.0
4. Diarrheal diseases	85.1
5. Kidney diseases	83.7
6. All accidents	72.3
7. Apoplexy (stroke)	66.6
8. Cancer	60.0
9. Old age	54.0
10. Bronchitis	48.3
2010	
1. Heart disease	192.9
2. Cancer	185.9
3. Chronic lower respiratory diseases (lungs)	44.6
4. Stroke (cerebrovascular diseases)	41.8
5. Accidents (unintentional injuries)	38.2
6. Alzheimer's disease	27.0
7. Diabetes	22.3
8. Kidney diseases	16.3
9. Influenza and pneumonia	16.2
10. Suicide	12.2

*These data are limited to the registration area, which included 10 registration states and all cities having at least 8,000 inhabitants. In 1900 this comprised 38 percent of the entire population of the continental United States. Since accidents were not reported in the 1900 census, this rate was taken from Lerner (1970).

Sources: *Abstract of the Twelfth Census of the United States, 1900.* Table 93. Washington, DC: U.S. Government Printing Office, 1902; National Center for Health Statistics, *National Vital Statistics Reports,* 2010, 60(4), Centers for Disease Control and Prevention. Atlanta, GA, January 11, 2012.

21 percent of U.S. adults and nearly 20 percent of high school students smoke cigarettes (Salahi, 2011). Smoking is responsible for the deaths of more people each year than alcohol, AIDS, car accidents, murders, suicides, and illegal drugs combined. Overall, smoking kills 434,000 Americans each year. Perhaps even more alarming, over a three-week period, second-hand smoke kills an equal number of innocent Americans to those killed in the World Trade Center attacks of September 11 (Snell, 2005). American women did not begin smoking in large numbers until after World War II, and their rates of lung cancer are now matching those of men (Cockerham, 2012). Worldwide, cigarette smoking has increased significantly in recent years and is now responsible on a global basis for about 5 percent (a conservative estimate) of all deaths (approximately 20 percent of all deaths in the United States) (Anderson, 1996).

In the United Kingdom, a ban was imposed on public smoking in July of 2007 (Leake, 2009). The ban on smoking has caused a fall in heart attack rates of about 10 percent within a year. In Scotland, where the public ban on smoking was introduced a year earlier, heart attack rates have fallen by about 14 percent because of the ban. The success of the smoking ban is emerging as one of the most significant improvements in public heath that Britain has seen. There are now movements in England to ban smoking in cars where children are present and at home in front of children. Each year in the United Kingdom 114,000 die of smoking-related diseases.

In June of 2011 the U.S. Food and Drug Administration unveiled the final nine graphics to appear on cigarette packs starting October 22, 2012. These include images of a man with rotting teeth and smoking from a tracheotomy hole with one-line facts like "cigarettes cause cancer" (Salahi, 2011). Graphic health warnings displayed in other countries seemingly work better than text warnings to motivate smokers to quit. For example, images used on cigarette packs in countries such as Canada are so disturbing that some smokers buy covers for their cigarette packs to block out the images. Though the United States was the first country to require health warnings on tobacco products, it is now playing catch-up to more than 30 countries that already require large, graphic cigarette warnings.

Yet despite lower death rates with bans on cigarette smoking, cigarette smoking is becoming more popular in movies today, as it was recently reported that there are as many people smoking on screen today as there were in 1950 (Rifkind, 2006). In the 1970s and 1980s, films featured or made reference to smoking an average of eight times; by the late 1990s, this number had risen to 25 times. Some 85 percent of the 250 top-grossing films in the 1990s featured smoking. Smoking is often considered to be cool, sexy, and glamorous in movies. What kind of image is this presenting for movie-goers in the 21st century?

Life expectancy has increased considerably in the United States since 1900, when it was 47 years at birth, compared to 78 in 2011. A hundred years ago, only 4 percent of the U.S. population was over the age of 65; today, that figure is 13 percent and is projected by 2050 to constitute 20 percent of the American population. Since 1950 life expectancy for males in the United States has increased from 66 years to 75 years, whereas females' life expectancy during this same period has increased from 71 to 80 years. Whites live nearly seven years longer than blacks in the United States. According to a study reported in the *New York Times* ("Is Life Expectancy Now Stretched to Its Limit?", 1990), however, there may be a limit to how long our bodies can hold out. The study concluded that science and medicine have pushed human life expectancy to its natural limit of about 85 years. The researchers note that even if a cure were found for most fatal diseases such as heart disease and cancer, the natural degeneration of the body puts a cap of about 85 years on the average life span.

White Americans living to age 80 and older, however, can expect to keep on living longer than octogenarians in other industrialized countries. The United States, which ranks 25th among nations in life expectancy at birth, is having its average life expectancy lowered by a high **infant mortality rate** and higher death rates until middle age (National Center for Health Statistics, 2009). For example, infant mortality rates in Japan and Sweden (2.6 and 2.8, respectively) are much lower than those in the United States (6.7). Though the reason is not known for sure, demographers suggest that the explanation why the oldest Americans are so

These cigarette cartons in England reveal the dangers of smoking. The smoker is warned as to his or her behavior prior to lighting up. But then the smoker may think, "That won't happen to me."

These white crosses in a cemetery in Zermatt, Switzerland depict the deaths of infants. This entire section of the cemetery has only graves of infants with the year of birth and death, and a name. Note the flowers growing within the enclosed area of the small graves.

long-lived is because their universal health insurance through Medicare may provide better care than old people receive elsewhere. Another factor might be the effect of education. A more educated society means a higher standard of living, which ultimately contributes to longevity.

Though medical and scientific breakthroughs have obviously contributed to an increased life expectancy, the American public has also strived to change its lifestyle and improve its health in recent years. Priorities for health promotion are smoking cessation, physical exercise, nutrition and weight control, stress management, safe sex, and appropriate use of alcohol and other drugs. Although one cannot change his or her heredity, one can have an impact on life expectancy through exercise, nutrition, and patterns of living.

GENDER DIFFERENCES IN MORTALITY RATES

Beatrice Gottlieb in *The Family in the Western World: From the Black Death to the Industrial Age* (1993) observed that in spite of childbirth always being risky for women—riskier in the past than now—men do not as a rule outlive women, nor did they in the past. The only exceptions worldwide are in southern Asia in countries such as Bangladesh and Nepal where men outlive women by a slight margin (Cockerham, 2012). Nutritional deprivation and lessened access to medical care are among the possible reasons for this reversal of the usual female superiority in life expectancy. Outside of southern Asia, women have a definite advantage over men in longevity.

In modern Western countries life expectancy is longer for women than for men, but this is not a new development. Although female infants initially are almost always outnumbered by male infants, a more equal balance is reached now, and also in the past, because male infants had a higher death rate. Indeed, the danger of childbirth was an important experience of families in the past; yet whatever is said about women dying in childbirth has to be put into this context of overall female survival (Gottlieb, 1993).

In the 19th-century United States, however, women typically died younger than men (Freund & McGuire, 1995). Reasons included women incurring risks in pregnancy and childbirth and women often getting what food was left after men and children ate their share. However, by the 1920s the gender patterns in mortality rates in the United States had changed, and women generally were living longer than men, as had seemingly been the case in the rest of the Western world. Today, mortality rates for the two leading causes of death in the United States, heart disease and cancer, are higher for men than for women (Weiss & Lonnquist, 2009). Higher death rates from heart disease in men could be attributed to a higher rate of smoking and a harder-driving personality; also, sex hormones secreted by women's ovaries may provide some protection against heart disease. Higher cancer rates in men are probably attributed again to smoking cigarettes, drinking alcohol more excessively, and being exposed to cancer-causing agents in the workplace. Indeed, men are more likely than women to die from almost all of the most common fatal diseases.

Biological advantages contributing to greater longevity for women may come from hormonal differences. Also, the monthly menstrual cycle of women reduces their iron count; thus women are less likely than men to build up a surplus of iron in their bodies. In addition, the conception ratio, projected to be higher than 120 males per 100 females, favors males in the United States, yet the **sex ratio** at birth is down to 105. Thus, male embryos and fetuses are the "weaker" of the sexes and die off more quickly. In the teens the sex ratio levels off, and after age 80, the ratio is less than 50 males per 100 females. There is evidence that female life expectancy is also higher among many other animal species (Sagan, 1987). Perhaps females simply have better-built bodies. Their bodies have the potential of carrying and supporting an embryo or fetus. Thus, females are the Porsche or Lamborghini model, whereas males are the more thrown-together model.

Regarding cultural differences between the sexes, conceivably females watch their diet more carefully than do males due to their traditional knowledge about food and a special cultural emphasis on weight maintenance. With more current emphasis on diet and exercise for both males and females, however, this difference may diminish.

Anthropologist Ashley Montagu (1968) suggests that women have a superior use of emotions because they are more likely to cry than men. Because it is not macho to cry, men generally refrain from such behavior, resulting in more psychosomatic disorders such as peptic ulcers. Montagu asks, "Is this a superior use of emotions?" Perhaps being freer to express themselves through the release of emotional feelings contributes to a decrease of stress for women.

Males have historically been involved in more risk-taking activities, as mentioned earlier, through masculine behavior such as smoking (the rugged Marlboro man, for example) and drinking. Males have also been more inclined to drive fast cars, live the James Dean devil-may-care life, and participate in violent sports. Accidents cause more deaths among males than among females (Cockerham, 2012). Males also have had more dangerous jobs, such as coal mining. Males have been in higher stress-producing positions, such as CEOs and other high-ranking administrative positions, than have females, though this is changing in the 21st century.

Despite the fact that men die earlier and have more life-threatening illnesses, women have higher **morbidity** (illness) rates than men (Freund & McGuire, 1995). Chronic illnesses are more prevalent among women than men, but they are less severe and life threatening. When sex differences are considered, an inverse relationship appears to exist between mortality and morbidity (Cockerham, 2012). Women may be sick more often but live longer. Men may be sick less often but die sooner. Women report more episodes of illness and more contact with physicians. Perhaps women are more willing than macho men to report that they are sick. Because illness indicates weakness, men are less likely to visit a physician or admit that they are anything less than healthy. A recent **longitudinal study** of 1,000 middle-age men in Wisconsin (Springer & Mouzon, 2009), for example, revealed that men who strongly endorse "old-school" notions of masculinity were half as likely as other men to seek preventive health care, such as an annual physical exam. Even men with a high level of education were less likely to seek preventive health care if they adhered to the ideal of the macho man. Such evidence contributes to the explanation of the gender longevity gap with women outliving men.

As gender roles continue to change, as females are found in greater numbers and in a greater variety of nontraditional occupations, as males share more in domestic tasks, and as the sexes come together in more unisex behaviors (more women smoking and drinking, having more stressful jobs, and living the "fast" life), stresses and strains of life should be more equally distributed between men and women. The argument of biology versus culture as influencing life expectancy by sex can then be better addressed.

APPROACHES TO THE STUDY OF DYING AND DEATH

There are several ways to approach the study of dying and death. The two primary areas of study are the natural sciences and the social sciences. Natural sciences include the biological approach. Social sciences include the sociological approach. The anthropological and psychological approaches mainly fall within the social sciences arena—certainly the case for studying death and dying—though they have sub-areas that lean toward the natural sciences. In addition, the humanities may address dying and death through literature, music, history, and philosophy.

In following the scientific method for research purposes, all the approaches use the same empirical methodology—the specific ways of gathering data simply vary. Nonetheless, the methodology is based on observation and reasoning, not on supernatural revelation, intuition, appeals to authority, or personal speculation.

The theoretical approach of this book is largely sociological, though somewhat slanted toward social psychology (the discipline that bridges sociology and psychology). We will take an anthropological approach at times, particularly when we discuss different cultures. The psychological orientation is woven into the social psychology bent of the book. A philosophical approach is another way to approach the study of dying and death. The biological approach is evident when medical issues pertaining to the human body (structure and function) are discussed. Medical issues regarding causes of disease, for example, are biological (genetic causes), yet also social (environmental causes).

THE BIOLOGICAL APPROACH

Biology is the study of life. Though this is a book on dying and death, it is really about living and life. Thus, the biological approach indeed has merit in a book like this. The process of dying is primarily a biological process—something that the body does to the person. In this book, however, we are more concerned with what people *do* with that process. For example, biologists and medical personnel (or anyone else, including social scientists) respond to the *meaning* of the biology rather than to the biology per se. The physician's decisions are made on the basis of what the biological condition means to that physician. Making a medical diagnosis is the process by which the physician decides the meaning of the biological factors. This diagnosis, then, represents the process of transposing biological factors into meaning factors.

A recurring dramatic illustration of the fact that the behavior of the physician and others stems from the meaning rather than from the biology per se can be seen each time the news media report that a corpse in the morgue has come back to life after being wrongly pronounced dead by the experts. The physician's belief that a body is dead does not guarantee that it is. Behavior follows from the meaning, not from the biological factors per se—a *living* body was sent to the morgue.

A later discussion of dying in the American health care system (see Chapter 7) will focus on the medical model. The medical model suggests that when sick, we go to a physician to be made well. Yet the dying patient in the later stages of the illness cannot usually be made well. The dying patient then becomes the "deviant" within the medical system—the body is beyond healing. Chapters 5, 6, and 7 will discuss the dying process and will bring in biological issues (Chapter 8 is about biomedical issues) but will also bring out relationships between physicians and patients and families (more of a social science orientation). Indeed, the biological approach is about life, yet life includes dying and death.

THE PSYCHOLOGICAL APPROACH

A psychological approach to dying examines, among other issues, the experiences of pain (see Chapter 6), death anxiety (discussed in Chapter 2), and emotional stages in dying (discussed in Chapter 5). Research on death anxiety has been almost exclusively conducted by psychologists. Death anxiety (fear) might involve the unknown, the nature of one's identity, fear of growing old (**gerontophobia**), and immortality, and continues across the life span. The psychological approach looks at dying from a developmental perspective (sometimes called life stages) and examines attitudes toward dying and death from the cradle to the grave (the womb to the tomb). Chapter 3 in this book takes such a developmental perspective; thus, psychology plays a particularly large role in looking at death through the eyes of children, adolescents, and adults (younger, middle-aged, and older).

A psychological approach to dying and death looks at death denial in different cultures, examines the influence of mass media desensitization, and studies death denial in both patients and physicians. As noted in Chapter 5, a very common reaction to the announcement of a terminal prognosis or the death of someone is denial (shock and disbelief). This kind of reaction is a type of defense mechanism: "I must have heard incorrectly." Psychology also analyzes emotions around dying and death and thoroughly studies the process of grief both before and after the death of a significant other, as discussed in Chapters 13 and 14.

The psychoanalytic perspective comes out of psychology and bases much of its argument on the unconscious portion of the mind. This approach looks back into the patient's past to help explain current behavior. Death fear is often taken to be the result of repressed guilt, traced to unresolved earlier experiences. Such a perspective has an applied function and can be useful in helping one to cope with dying and death.

The Philosophical Approach

Socrates said that the true philosopher welcomes death. Death is not an end, but a transition. We human beings are unique among the inhabitants of the earth in being creatures both of emotion and of reason (Rosenberg, 1983). As feeling beings, it is fitting for us to be touched and moved by the death of someone close to us, as through death that individual is lost to us irredeemably and forever. Thus, we should seek some comfort and solace in the face of such a loss. It is also fitting for us, as thinking beings, not to find our comfort and our solace only in myth and in muddle and in self-delusion. Rather, we should search out our consolations from the standpoint of a clear and reasoned understanding of the truth, which only rational beings such as we humans could ever achieve.

An **existentialist philosophical** approach to death is a rather practical one and suggests that we must all face death. We can face death. Death becomes an extraordinary event to each of us. No matter how surrounded by others we may be at the time, we will face death alone. Not unlike the little engine that could, "I think I can, I think I can, I think I can," and "I did." We can "do it."

Existential issues could be defined as including philosophical, psychological, and religious aspects of all kinds and should be seen as an overall concept. Existential means being in space and time, grounded in existence, or the experience of existence. For example, for some women with breast cancer, spiritual/existential issues are represented as faith in God, and faith might be seen as a source of strength in coping with a changed life situation (Vargens, 2012).

An offshoot of the existentialist philosophical perspective is **phenomenology**. This aspect of philosophy studies "the thing itself"—the phenomenon. The phenomenological study of dying would, for example, try to discover what dying is to those experiencing it. It studies the phenomenon directly and how it is constituted. If one were studying near-death experiences, who better could discuss these than persons having had near-death experiences? Or if one wished to study suicide and wanted to know what it is like, who better to ask than individuals who have attempted to take their own lives? What was the person thinking when trying to commit suicide? Study the phenomenon—the suicide attempt, the person who had the near-death experience. Obtain a first-hand account straight from the "horse's mouth."

The phenomenologist most fundamentally attempts to describe the objects of perception or those held in consciousness (Charmaz, 1980). The commonsense meanings of individuals and their construction become the special interest of sociologists who take a phenomenological approach in their work. This view of human nature assumes that individuals are capable of reflective thought and are not simply products of social forces. The phenomenological method assumes that the researcher may not share this experience or understand its rationality. The phenomenological sociologist must understand the logic of the actor's experiences from her or his particular perspective. Sociologist Kathy Charmaz observes that descriptive terms and concepts used to define and describe the study of death need to be systematically clarified, if using a phenomenological approach.

THE ANTHROPOLOGICAL APPROACH

Anthropologists, particularly cultural anthropologists, study rituals through which people deal with death and hence celebrate life. As anthropologists Huntington and Metcalf (1992) observe in *Celebrations of Death: The Anthropology of Mortuary Ritual*, the study of death rituals is a positive endeavor. In all societies, the issue of death throws into relief the most important cultural values by which people live their lives and evaluate their experiences. Life becomes transparent against the background of death—fundamental social and cultural issues are revealed. For example, death for Native Americans is very much a part of life. As we shall see, this is a very different view than that held by many Americans.

Chapter 10 relies heavily on the research of cultural anthropologists in looking at death rituals in different cultures. Cultural anthropologists also look at emotional responses in different cultures, how different groups prepare the body for final disposition, and how they dispose of the body. Whereas in the United States we primarily bury a body in the ground, other cultures might place the dead person in a tree, and others are likely to cremate the body. Cultural anthropologists study these kinds of rituals.

Another subfield of anthropology, physical (biological) anthropology, however, leans in the direction of the biological sciences. Such a specialist, for example, might work with bone identifications (from humans and other animals) both from the past and present. This specialty often assists in identifying the remains of a human body and helps to determine age, sex, and other physical characteristics. Thus, anthropology (the study of humankind) has subfields in the social and the natural sciences.

THE SOCIOLOGICAL APPROACH

The sociological approach to the study of dying and death generally includes four theories: structural-functional theory, conflict theory, social exchange theory, and symbolic interaction theory. Social exchange and symbolic interactionism differ from structural functionalism and conflict theory on two crucial points. The first is that social exchange and symbolic interactionism examine the individual in society, whereas structural functionalism and conflict theory examine social facts, institutions, and forces. Second, interactionists and exchange theorists contend that the essential feature of society is its subjective character, in contrast to the focus on the objective approach taken by the other two.

In the view of interactionists and exchange theorists, social facts do not have any inherent meaning other than what humans attribute to them. W. I. Thomas (1923) argued that if people define situations as real, the situations will be real in their consequences. Robert Merton (1968) took Thomas's idea a step further and argued that individuals *act* on their perceptions of the situation; thus, the prediction of the situation comes true. This is known as the **self-fulfilling prophecy**. Max Weber (1966) defined social action as human behavior to which the acting individual attaches subjective meaning and which takes into account the behavior of others.

Researchers using a social exchange or symbolic interaction approach are often required to develop empathy for the subjects they study. At times this will require the researcher to enter the subjective world of the subject by participating in that person's life experience. A Native American proverb encourages us not to judge the behavior of others until we have "walked a mile in their moccasins." In this context, social behavior is explained from the perspective of the subjective meanings of the actors' intentions for their behavior.

Structural-functional theory assumes that society is in a state of equilibrium and that the various social institutions (e.g., family, religion, economy, and politics) function on behalf of each other. This approach might be used to explain how a particular death ritual maintains social structure. The conflict perspective assumes that society is in a state of disequilibrium (imbalance), and tends to focus on inequalities in society. This approach would examine, for example, the unequal access to medical care for the terminally ill. Let's look more carefully at these four sociological approaches to studying dying and death.

STRUCTURAL-FUNCTIONAL THEORY   French sociologist Emile Durkheim's work served as the primary foundation for theoretical frameworks relating to group actions and societal structures. Durkheim defined sociology as the study of social facts, which are external to the individual. Language, religion, and money-exchange systems, for example, are all social facts that were in the world before we were born and will be here after we die. Social facts are constraining—they place limitations on what we can do (rules and regulations that we must follow).

Durkheim (1964) believed that society is a social system composed of parts which, without losing their identity and individuality, constitute a whole that transcends its parts. From this point of view, social groups or collectivities (e.g., a particular nuclear family) cannot be reduced to merely a collection of individuals, and social phenomena have a reality of their own that transcends the constituting parts. This perspective, therefore, studies group-related phenomena (e.g., death rituals, structures that provide care for dying patients, and professional groups of funeral functionaries) rather than behaviors of particular individuals.

Group behavior provides strength to the individuals involved and shows "togetherness" to outsiders. After the death of a police officer in the line of duty, for example, hundreds of police officers from around the area will attend the funeral and ride in the funeral procession. As a large group, in essence, they are saying to the world, "Though one of our members is gone and a void occurs within our ranks, we are banding together to cover for the dismemberment within the police family. We remain strong and ready to function to protect society." The message is loud and clear. And in their behavior, they are consoling each other. With the deaths of Nelson Mandela, Michael Jackson and Princess Diana, many individuals brought flowers to central places in countries around the world. Being together and "sharing grief" is helpful. Being in the presence of other individuals is indeed consoling.

As stated earlier, the structural functionalists believe that society is in a state of balance, of equilibrium. Death itself is functional to the overall "balance" of a society. As Charlotte Perkins Gilman so wisely put it, "Death? Why this fuss about death? Use your imagination and try to visualize a world without death. Death is the essential condition of life, not an evil" (Gilman, 1935).

Death is a normal aspect of society. Deaths create jobs—the funeral industry, cemeteries, and estate lawyers, for example. Deaths open up jobs within the structure of organizations—someone "dies off" and must be replaced; thus, death contributes to a smooth-running social system. The structure of society is such that deaths are expected to occur—they help to give equilibrium.

Structural functionalists view society as a social system of interacting parts in which death-related behavior is analyzed from two perspectives:

1. How do death-related meaning systems and death institutions contribute to the maintenance of the larger social system?
2. In what ways are death-related meaning systems and death institutions affected by their relationships to the larger social system?

Structural functionalists are interested in positive (**eufunctional**) and negative (**dysfunctional**) results of social interaction as well as the intended (**manifest**) and unintended (**latent**) consequences of death-related behavior. When family members commit themselves to caring for a dying family member, their behavior can be eufunctional for emotional ties within the family; however, the behavior can be dysfunctional (especially if family members must interrupt their employment) for the security of family financial resources. A manifest function of attending a funeral is to support the bereaved, as family members attempt to adapt to the loss of a loved one, but a latent function is to strengthen the relationships that exist within social groups (a funeral becomes a family reunion).

If structural-functional theorists were interested in the funeral rites and rituals, they might investigate one or more of the following research questions:

1. How do funerals help to celebrate and maintain society's most salient social values?
2. How do funerals help to promote relationships within kinship groups (grandparents, parents, children, aunts and uncles, cousins, brothers, sisters)?
3. How do funerals contribute to and/or affect the relationships between bereaved families and the larger society?
4. How do funerals facilitate the grieving process as one mourns the death of a loved one?
5. How do death-related rituals help return bereaved persons to their normal social responsibilities?
6. How do differing methods of body disposition and funeral rituals socially differentiate families regarding social status?

Utilizing a structural-functional perspective, sociologist Kathy Charmaz (1975) described the strategies used by coroner's deputies in maintaining the routine character of their work. The coroner's deputies attempt to get surviving family members to take over the responsibility for the care of the dead body and the financial obligations related to final disposition and to minimize personal involvement of the coroner's deputies. In Charmaz's description we can observe that if each party

Some two decades ago, following the unexpected death of my father, I* returned at the age of 50 to my native Texas to attend his funeral. I had moved away from home after graduation from high school, only to return once or twice annually to visit.

At Dad's funeral, I had an opportunity to reflect on changes in the family over time. At the end of the funeral service, various individuals filed by the open casket: kin, **fictive kin**, long-time friends, and others unknown to me. Cousins I had not seen in years (decades!) came from far and wide. They were all grown up now and showing signs of the aging process (not unlike myself). I recalled having played with one set of cousins as little children decades ago when we would visit our grandmother in another part of Texas, some 400 miles away. Seeing them once a year for a few years was all the contact we had. Then, there they were at my dad's funeral. Funerals become a family reunion, a latent function (hidden dimension) of a funeral.

Several great aunts (the uncles had all died), some of whom I had not seen for several years, seemed suddenly so old and frail (and indeed they were!). I recalled them as adults in their 40s and 50s when I was a small child. My, but they had changed with the passing of time. Former grade school and high school classmates walked by the open casket. They, too, showed what the passing of time can do to one's body (again, I only needed to look in a mirror).

It is within the family setting (including fictive kin) that one can observe these changes over time. These were individuals I had known all my life, both as a small child and now as a middle-aged adult. The dismemberment of the family through deaths left "gaps" from persons who previously played major roles in my life. My grandparents were now all dead. In the logical order of events in the family life course, it was time for my parents to die.

Babies are born, with most growing to old age, producing changes in the number, structure, and roles of families. Offspring of relatives that I had never before seen appeared at the funeral. They were born after I grew up and left the community. I was meeting **extended family** members for the first time. Changes had occurred in our extended family over the decades—changes which were obvious to me on this sad occasion.

The composition of the family life course produces changes as the clock ticks away. There is a time and a place for all things. There is a time to be born and a time to die. The family network is a setting in which these changing times are observed, as painful as it may sometimes be.

*George E. Dickinson

performs his or her socially prescribed role (or social function), the social system runs efficiently and social equilibrium is maintained.

Strategic control of the encounter between the coroner's deputies and the relatives of the deceased is enhanced by making the announcement in person. Part of the self-protection strategy is to remain polite and sincere, yet authoritative. The deputies must create the kind of ambiance wherein the announcement of the death is effective and believable. Unlike physicians in giving "bad news," the coroner's deputies lack their prestige and also do not have a relationship with the person. Deputies also lack the structural supports provided by the hospital situation; thus, they must devise tactics to get their work done without incident. Typically, their objectives are to announce quickly and then to turn to the responsibility of the body and its subsequent disposal. Deputies feel that they get a better response when they successively lead the relative into questioning them.

Charmaz noted that the deputies usually avoid the word *dead*, because it is such a harsh word, and use euphemisms instead. They try to manipulate the situation so that the relative actually uses the word *dead*, thus making the death more "real." The deputies then reaffirm the statements and elaborate on them. When the strategies work, the transition from perceiving one's relative from alive to dead can be made rapidly. If the encounter is successful, the relative is likely to express appreciation for the deputies' sensitivity. Relatives ask about the circumstances of dying. The deputies give the information they have and can release and then turn to the funeral and burial arrangements. Thus, the deputies have played the role of officials who cut through the survivor's grief and shock by pointing to the work that has to be done and have strategically structured the situation in ways that foster the relatives' acceptance of their directives.

CONFLICT THEORY   Whereas structural-functional theory focuses on the issues of societal maintenance and social equilibrium, conflict theory focuses on issues related to social change and disequilibrium. Conflict theorists focus on competition, conflict, and dissension resulting from individuals and groups competing for limited societal resources.

In emphasizing social competition for limited resources, conflict theorists interested in death-related behavior would point out the inequality in the availability and quality of medical care and the differential death rates. For example, the poor are deprived of optimal care, in general, and of life-saving procedures, in particular (Weaver & Rivello, 2006–2007). Access to resources that might prolong life or postpone death depends not only on the absolute level available but also on their distribution across the population. Societies where resources are unequally distributed often experience higher levels of morbidity and mortality than those where resources are more equally accessible. Allen Kellehear argued that those who control life are the same as those who control death. According to Kellehear (1990) the power elite in any society determines one's life and death chances. Society determines not only what types of death will occur but also when deaths occur and to whom, and the chances, if any, of doing anything about it.

As will be discussed in Chapters 6 and 8, organ transplantations are an example of the conflict perspective. There simply are more recipients waiting for organ and tissue donations than there are donors. Thus, the issue becomes one of deciding which individuals receive the organs and which ones do not. There is competition for this limited supply of resources. Another example of conflict theory is health insurance. Some individuals have more limited choices as to their medical care options. The Affordable Care Act is beginning to address this issue in 2014, however. If conflict theorists were interested in funeral rites and rituals, they might investigate one or more of the following research questions:

1. What are the dysfunctional consequences of attending funerals?
2. What role conflicts and family disputes arise as a result of planning a funeral for family members?
3. How does not attending a funeral create conflicts between adults in neighborhood, friendship, and occupational groups?
4. What are the problems created by the presence of children at funerals?
5. How do particular family relationships contribute to increased competition for status among family members as they participate in the funeral of a family member?
6. How do methods of planning a funeral, and the related expenditures, contribute to increased family conflict and competition for scarce financial resources within the family?

7. In what ways might members of the clergy and funeral directors engage in social conflict because of the fact that family members allocate to each functionary differing amounts of authority and social rewards (money) for conducting the funeral?

8. How is it possible for unethical funeral industry personnel to exploit bereaved survivors during a time of emotional distress?

9. How does the death of a parent create sibling rivalry among the children, and how does the death of a child create marital problems between the parents?

In distributing property at death through a will or inheritance, if all beneficiaries want fair treatment, conflict may occur because beneficiaries have different perceptions of what is fair. Thus, conflict theory applies to this "fairness" issue after death (Titus, Rosenblatt, & Anderson, 1979). Even if the will seems fair, to some individual recipients a particular item, such as a rocking chair or antique

Many sociologists trace the intellectual roots of conflict theory to Karl Marx. Marx is buried at Highgate Cemetery in London, England. The epitaph on his tomb reads: "Workers of all lands unite. The philosophers have only interpreted the world in various ways; the point is to change it."

clock, might be personally more valuable. Individuals simply have different perceptions of what is fair. Fairness can mean that something is divided equally, but fairness also takes into account various principles of deservingness or right. Because there are so many possible interpretations of what is fair or what is equal and because people often seek fairness or equality, a dispute may not be resolved easily. Disputes over inheritance may be one of the major reasons for adult siblings to break off relationships with each other. In some cases the inheritance dispute may be the final battle between competitive siblings, and in that sense it resembles the "last straw" reported in breakups in close relationships.

SOCIAL EXCHANGE THEORY   Two traditions are followed by social exchange theorists. The first tradition is consistent with principles of behavioral psychology and stresses psychological reductionism and behavioral reinforcement techniques. The second tradition has been influenced by the work of Peter Blau (1964) and is committed to many of the assumptions held by symbolic interaction theorists. Social exchange theories of this type contend that human behavior involves a subjective and interpretative interaction with others that attempts to exchange symbolic and nonsymbolic rewards. It is important that such social exchange involves reciprocity so that each interacting individual receives something perceived as equivalent to what is given.

From this perspective, individuals will continue to participate in social situations as long as they perceive that they derive equal benefits from their participation. For example, the social exchange theorist would contend that individuals will attend funerals (even though they tend to feel uncomfortable in such situations and find viewing the body, if on display, as distasteful and anxiety-producing) because they perceive social benefit in being supportive of bereaved friends. An individual would want (or expect) a friend to attend the funeral of someone close to them, thus reciprocity occurs here—do unto others as you would have them do unto you.

If social exchange theorists were interested in funeral rites and rituals, they might investigate one or more of the following research questions:

1. Why do some individuals attend funerals—what are the social rewards acquired by attending?
2. Why do some individuals not attend funerals—what are the social punishments or sanctions acquired by not attending?
3. What are the social and personal costs and benefits for families when they provide wakes, funerals, and other death-related rites of passage?
4. What are the social and personal costs for families when they do not provide wakes, funerals, and other death-related rites of passage?
5. Why would the average American family spend thousands of dollars to bury its dead when it could accomplish the same purpose at a fraction of the cost?

SYMBOLIC INTERACTION THEORY   The foundation of symbolic interaction theory is that **symbols** (meanings) are a basic component of human behavior. People interact with each other based on their understanding of the meanings of social situations and their perceptions of what others expect of them within these situations.

Stressing the symbolic nature of social interaction, Jonathan Turner said: "Symbols are the medium of our adjustment to the environment, of our interaction with others, of our interpretation of experiences, and of our organizing ourselves into groups" (1985, p. 32). For example, sociologist and physician Nicholas Christakis (1999) explains how self-fulfilling prophecy applies to a physician-patient relationship in the use of symbols. Physicians, in making predictions about a terminally ill patient's prognosis, can change the patient's perception in a way that promotes the predicted outcome. If the doctor were to give a favorable prognosis to a cancer patient, it would probably lead to a favorable outcome. If the patient's morale is good, she or he will probably have better nutrition habits and be stronger to fight off potentially fatal infections better than might otherwise be true. The patient will not be cured, but his or her life may be prolonged.

Another example of self-fulfilling prophecy applying to doctors working with patients relates to compliance with therapeutic regimens (Christakis, 1999). If positive feelings are inculcated in an adolescent patient with newly diagnosed diabetes, research data suggest that this is one of the best predictors of a good outcome. But if the person is not upbeat, then the physician's job becomes changing the patient's negative view of the illness. Likewise, if a physician assumes (perhaps a false assumption) that a particular patient does not wish to be told "the truth" regarding prognosis, and the physician then does not "tell the truth," the definition of the situation by the physician then becomes reality, and the patient is *not* told and remains officially unaware. Thus, the social construction of reality: We define situations as real, and they become real in their consequences.

The symbolic interaction perspective can be summarized in what some have called the **ISAS** statement—*individual-level* behavior is in response to *symbols*, relative to the *audience*, and relative to the *situation*. ISAS stands for the initial letter of each of the four basic components, as described some 30 years ago by Vernon and Cardwell (1981). Death-related behavior of the dying person, and of

---

LISTENING TO THE VOICES | SELF-FULFILLING PROPHECY

When I* was a graduate student, I went to the hospital to donate blood on behalf of a fellow graduate student. Prior to taking my blood, two nurses each checked my temperature (glass thermometers were used back then, and were washed and reused). On both occasions my temperature was 104°F, a rather high temperature for an adult. Before going to donate blood, I was feeling very good. Now that two separate nurses, two authoritarian figures with medical credentials, had determined that I had a rather high temperature and therefore would not take my blood, I began to *feel* badly. Thus, the self-fulfilling prophecy was at work—I was defined as being sickly, began to

feel sickly, and indeed was getting sick! Prior to going home and going to bed with my "illness," I decided to purchase a thermometer, because I really did not feel feverish. After holding the thermometer in my mouth for the recommended three minutes, my temperature read 98.6°F (Obviously, the two nurses had not shaken down the thermometers and had placed them in water for sterilization with a temperature of 104°F). I began to *feel* better immediately. What others say and how we perceive what they say affects our behavior.

*George E. Dickinson

those who care about that person, is in response to meaning, relative to the audience and to the situation. Death-related behavior is shared, symboled (given meaning), and situated. It is socially created and not biologically predetermined.

Symbols.   A symbol is anything to which socially created meaning is given. Indeed, language itself is a system of oral symbols. For example, a group of Southwest Native Americans did not have a future and past tense in their language, only the present tense. Indeed this would have an impact on meaning and perception! The Yanomamo Indians in Venezuela and Brazil have a numerical system in their language which consists of "one," "two," and "more than two." Thus, if you were asking adults their ages, they would all be "more than two," making it difficult from our perspective in the United States to perceive adult age differentials. Another group had only three colors in their color spectrum, thus limiting somewhat their perception of colors. If, for example, you go to buy a new car or a new pair of shoes, they probably have many colors and numerous variations on each color—a brown shoe is not just brown, it is something, something brown. Thus, meaning assigned to a situation or object will vary significantly according to the perception.

Meaning is socially created and socially perpetuated; it is preserved in symbols or words. However, preserved words have to be rediscovered and reinterpreted if they are to be continually used in human interaction. Interaction is a dynamic, flexible, and socially created phenomenon. What death means to each of us results

George Dickinson

For the patient entering Saint Christopher's Hospice in London, the average stay is three weeks. Knowing the meaning of entering a hospice program confirms that death is forthcoming.

from our socially shared ideas. Generation after generation repeats the process with a somewhat different content—no book means the same thing to every reader. Death-related behavior and meanings are dynamic phenomena. For example, death to Baby Boomers is perceived somewhat differently than by earlier generations in the United States, as discussed in Chapters 3 and 11.

As human beings, we are more than stimulus-response creatures. We have the ability to use symbols—we have the ability to *interpret* the stimulus, assign meaning to it, and then *act* upon our interpretation. What is the meaning of the telephone-ringing stimulus? The telephone ringing typically invokes a response of answering. However, one may not be in a frame of mind to talk on the phone, and thus responds by not answering. The interpretation of the stimulus, given the moment or situation, was to *not* respond. We do not always respond in the same way to the same stimulus. The physician may give the patient bad news (stimulus), yet the patient may respond by interpreting what was said (the stimulus) with denial—the doctor made a mistake. She or he did not really mean that which was said. In other words, the patient believes what she or he wants to believe, and most of us in fact do not want to hear the doctor say that we have a terminal condition. Thus, through the interpretation (denial), the individual concludes that everything is okay.

Audience-Related Behavior. As a teenager, you probably did not behave in the same way in the presence of your parents or grandparents as you did when

George Dickinson

The meaning of "exit" from Saint Christopher's Hospice for patients typically means death. Thus, the meaning of exit varies significantly for patients and visitors.

among peers. Among our peers, we would likely be cool, whereas the presence of older adults might have brought out more sophisticated behavior. Our audience may even affect certain meanings we attach to interactions and words. In the same way, our audience affects our death-related meanings. The audience involved may be family, physicians, clergy, nurses, peers, or even strangers walking down the hall of the hospital. A dying person observes the behavior of people and listens to what they do or do not say. Decisions about the meaning of the dying person's own death and behavior while dying are often heavily influenced by the words and behavior of this audience.

The Situation.   Like other meanings, the definition of the situation is an attempt by the individual to bring meaning to the world. Because the situational definition always involves selective perception, the terminally ill patient will assign meaning to the environment and will respond to this symbolic reality. The thing to which the patient responds does not have existence independent of his or her definition. Therefore, each terminally ill patient will interpret the dying environment differently. This accounts for the different experiences of dying patients. This point will be elaborated upon in Chapter 6.

WORDS OF WISDOM | FINDING THE MEANING IN WORDS

Blood,
    and Pus …
    Entrails,
    Vomit,
    Dung, Spit, and Afterbirth.
    Disgusting words.

Love,
    And Soft.
    Mother.
    Kiss, Mood, and Friendship.
    Tender words.

Spring,
    And Smile.
    Dance,
    Play,
    Fun, Sing, and Beachball.
    Happy Words.

Death,
    And Fire.
    Pain,
    War,
    Divorce, Poverty, Hospital.
    Sad Words.

No. That's not right at all. You cannot
    String words together
    And say They're bad
    Or good.

Where are the verbs?
    Who are we talking about?
    What are the circumstances?

Vomit is beautiful to a mother whose child
    Had just swallowed a pin.
    Love is pain if you are a third party,
    Outside, looking in.
    Death is very nice for someone very old,
    Very ill, and ready.

And surely you've danced with a clod.
    Or had a sad spring.

No, Words aren't sad
    Or glad.

You are.
    Or I am.
    Or he is.

From *A Bag of Noodles* (p. 8), by W. A. Armbruster, 1972, St. Louis: Concordia. Copyright 1973 by Concordia Publishing House. Reprinted by permission.

If symbolic interactionists were interested in funeral rites and rituals, they might investigate one or more of the following research questions:

1. What are the social meanings that give rise to the attendance of funerals?
2. How are death-related behaviors influenced by the social audiences of the American funeral?
3. In what ways do families use the funeral service to project an image of family value commitments and group cohesiveness?
4. How do families use the funeral to demonstrate status differentiation?
5. What is the influence of the location of the funeral upon the meaning of the funeralization process?
6. How does the size of the funeral audience convey meaning to the bereaved and attribute social importance to the life of the deceased?

CONCLUSION

Why study death from a sociological perspective? The answers are many.

First, although it is not the primary goal of this book to prepare you for your own death or the deaths of loved ones, you may still learn many things that will be personally beneficial. Given the fact that you will have death-related experiences, it might be useful for you to employ sociological and social-psychological perspectives to acquire valuable insights into your own life and the lives of people about whom you care.

Second, as you read about the history of death-related customs in the United States and elsewhere and as you comprehend social class and ethnic variations in bereavement customs, the sociological perspective may provide you with a new appreciation for your own life experiences and family traditions as they relate to issues of dying, death, and bereavement.

Third, as you consider the ways by which the family socializes its members to develop death conceptualizations, attitudes, and feelings, the sociological and social-psychological perspectives may help you to understand more fully how your own abilities to cope are a function of your upbringing. Furthermore, as you study the material in this book, you may attain a more objective perspective on some of the decisions that may confront you and your family as, one day, you attempt to deal with death-related situations. The sociological and social-psychological perspectives, then, can make you more aware of the many options available to you regarding death-related decisions.

Finally, the study of dying, death, and bereavement is an important and interesting area of study to which social and behavioral scientists have much to contribute. A good reason to study dying and death from a sociological perspective is to learn more about the discipline of sociology. Social thanatology provides the student with an opportunity to investigate the meaning and application of many concepts developed in sociology and the other behavioral sciences.

SUMMARY

1. The American way of dying is typically confined to institutional settings and removed from usual patterns of social interaction.

2. American society has done little to formally socialize its members to deal with dying and death on the personal and emotional levels.

3. The "thanatology movement" was a concerted effort in the 1970s to bring about an open discussion and awareness of behaviors and emotions related to dying, death, and bereavement. The movement continues into the 21st century.

4. An increased emphasis on dying and death is reflected in today's media.

5. The issue of when death occurs is a difficult one to resolve because consensus does not exist in America regarding the meaning of life and death.

6. The American way of dying has changed considerably since 1900. Relative to earlier times, Americans are less likely to die of acute diseases. Approximately half of the deaths taking place in the United States can be attributed to two chronic diseases: heart disease and cancer.

7. Life expectancy has increased significantly in the past 50 years. This increase has had many effects upon Americans' understanding and ability to cope with dying and death.

8. Death and dying can be approached from the perspectives of biology, psychology, anthropology, philosophy, and sociology.

9. Meaning consists of symbols that are socially created and socially used. Some symbols refer to something else or have an empirical referent. Some do not.

10. The major function of both types of symbols is to permit humans to relate to each other and thus create shared behavior and meaning.

11. From a sociological approach, four theories are used in the study of death and dying: structural-functional, conflict, social exchange, and symbolic interaction.

12. Structural functionalists view society as a social system of interacting parts. Structural-functional theory focuses upon the issue of societal maintenance and social equilibrium.

13. Conflict theory is primarily concerned with issues related to social change and disequilibrium. Conflict theorists focus upon competition, conflict, and dissension that result when individuals and groups compete for limited societal resources.

14. Social exchange theory contends that human behavior involves a subjective and interpretative interaction with others that attempts to exchange symbolic and nonsymbolic rewards. Social exchange will always involve reciprocity, so that each individual involved in the interaction receives something perceived as equivalent to what is given.

15. From the symbolic interaction perspective, human beings are autonomous agents whose actions are based on their subjective understanding of society as a socially constructed reality.

DISCUSSION QUESTIONS

1. Why did death "come out of the closet" in the 1970s? What events related to the "thanatology movement" helped change the American awareness of dying and death?

2. What effect have the media had on present interest in death?

3. Why is death education from a sociological perspective important?

4. Compare the leading causes of death in 1900 with those today.
5. What affects life expectancy?
6. Discuss why women outlive men in most countries of the world.
7. How do the biological, psychological, anthropological, philosophical, and sociological approaches to death differ?
8. Compare and contrast the structural-functional and conflict orientations in death-related research.
9. Compare and contrast the symbolic interaction and social exchange orientations in death-related research.
10. Explain what is meant by the ISAS statement and describe the four components.

GLOSSARY

Chronic Disease: A noncommunicable, self-limiting disease from which the individual rarely recovers, even though the symptoms of the disease can often be alleviated. Chronic illnesses usually result in deterioration of organs and tissues that make the individual vulnerable to other diseases, often leading to serious impairment and even death. Examples of chronic illnesses include cancer, heart disease, arthritis, emphysema, and asthma.

Dysfunction: An event that lessens the adjustment of a social system. Dysfunctional features of a society imply strain or stress or tension. A dysfunction of modern medicine might be that it prolongs life, contributing to an overcrowded population, and also contributes to a frail and disabling existence for many in their old age.

Etiology: The assignment of a cause, such as the major causes of death in a country.

Eufunction: Consequences of behavior that are positive. For example, a eufunction of attending a funeral is that it shows concern and care for the survivors.

Existentialist Philosophy: Based on the doctrine that existence takes precedence over essence and holds that human beings are totally free and responsible for their acts, and that this responsibility is the source of the dread and anguish that encompass them.

Extended Family: A kinship group comprising the conjugal (married) family plus any other relatives present in the household, such as a grandparent or uncle.

Fictive Kin: Individuals referred to using kinship terms but who are not related via kinship. A godparent, for example, has no kin tie, yet may be called Mother or Father or the best friend of one's father may be called Uncle.

Gerontophobia: The fear of growing old.

Infant Mortality Rate: The number of deaths of children under one year of age per 1,000 live births.

ISAS: A shorthand presentation of the symbolic interactionist's statement: Behavior of the individual is in response to symbols, relative to the audience, and relative to the situation.

Latent Function: Latent functions are consequences that contribute to adjustment but were not intended. For example, a latent function of attending a funeral is that it becomes a family reunion. Distant relatives from various geographical locations often show up for a funeral. Latent functions are the sociological explanations of a given action.

Life Expectancy: The number of years that the average newborn in a particular population can expect to live.

Longitudinal Study: A study that looks at two or more points in time (e.g., obtaining the attitudes of physicians toward end-of-life issues in 1980, then repeating the study 30 years later). This differs from a cross-sectional study, which is one point in time (e.g., ascertaining death attitudes of college students in 2010).

Manifest Function: Manifest functions are objective consequences that contribute to adjustment and were so intended. They are the official explanations of a given action.

Morbidity: The rate of occurrence of a disease.

Mortality Rate (Death Rate): The number of deaths per 1,000 population.

Phenomenology: The philosophical study of phenomena without any attempt at the nature of being; studying "the thing itself."

Road Scholar: An informal educational program for individuals 60 years of age and older. Courses cover a variety of topics often having to do with the geographical region and featuring instructors who are local experts. They typically last one or two weeks. Several thousand offerings are available in countries around the world and are often conducted at colleges and universities.

Self-Fulfilling Prophecy: A situation defined as real that becomes real in its consequences when individuals act to make it so.

Sex Ratio: The number of males per 100 females.

Socialization: The social process by which individuals are integrated into a social group by learning its values, goals, norms, and roles. This is a lifelong process that is never completed.

Symbol: Anything to which socially created meaning is given.

Thanatology: The study of death-related behavior including actions and emotions concerned with dying, death, and bereavement (literally, the study of death).

SUGGESTED READINGS

Bailey, S., & Flowers, C. (2009). *Grave expectations*. Kennebunkport, ME: Cider Mill Press. A funny and irreverent, yet helpful, book on various rites of passage that occur at the end of life.

Barry, J. M. (2004). *The great influenza: The epic story of the deadliest plague in history*. New York: Viking. An account of the spread of history's most lethal influenza virus, which killed as many as 100 million people in the United States and Europe, beginning in the winter of 1918.

Barry, V. (2007). *Philosophical thinking about death and dying*. Belmont, CA: Thomson/Wadsworth. A book that explores the nature of death, survival of death (afterlife beliefs and their bases), and voluntary death (suicide, euthanasia, and futile medical treatment).

Belshaw, C. (2008). *Annihilation: The sense and significance of death*. Durham, UK: Acumen Publishing. How our death relates to the death of the brain is explored in detail. Belshaw also considers the view that death is often bad for us as well as the grounds for thinking that one death can be worse than another.

Cann, C. (2014). *Virtual afterlives: Grieving the dead in the twenty-first century*. Lexington: University of Kentucky Press. Examines mourning practices in the United States and compares them to practices in Asia and Latin America.

Cohen, J. A., Mannarino, A. P., & Deblinger, E. (2006). *Treating trauma and traumatic grief in children and adolescents*. New York: The Guilford Press. The book encourages children to talk about thoughts and feelings about the intentionality of the death and addresses such issues as shifting from an active relationship to a relationship of memory. It uses learned optimism to focus on present and future, rather than the past.

Dickinson, G. E., & Leming, M. R. (Eds.). (2014). *Annual editions: Dying, death, and bereavement 13/14*. An anthology covering topics on issues in dying and death, dying and death across the life cycle, the dying process, ethical issues of dying, death and suicide, funerals, and bereavement.

Erasmo, M. (2012). *Death: Antiquity and its legacy*. New York: Oxford University Press. A classics professor explores classical and contemporary approaches to death and internment across the ages.

Erickson, K. (2013). *How we die now: Intimacy and the work of dying*. Philadelphia: Temple University Press. Talks about what we can learn from those who work with the dying, rather than just about the dying process itself.

Gilbert, S. M. (2006). *Death's door: Modern dying and ways we grieve*. New York: W.W. Norton & Company. Provides a comprehensive overview of literary prose on death and dying. This prominent critic, poet, and memoirist explores our relationship to death through literature, history, poetry, and societal practices.

Green, E. C., & Ruark, A. H. (2011). *AIDS, behavior, and culture*. Walnut Creek, CA: Left Coast Press. Anthropological methods are applied to AIDS prevention in sub-Saharan Africa. The authors describe successful community programs that have involved local traditional initial sexual experience and male circumcision, which are associated with HIV decline.

Green, J. W. (2008). *Beyond the good death: The anthropology of modern dying*. Philadelphia: University of Pennsylvania Press. Taking an anthropological approach, Green examines the changes in the concept of death over the last several decades and determines that the attitudes of today's Baby Boomers differ significantly from those of their parents and grandparents.

Kagan, S. (2012). *Death*. New Haven, CT: Yale University Press. Philosophy professor Kagan examines the myriad questions that arise when we confront the meaning of mortality: How should I live in the face of death? Will I exist after I die? Is suicide ever justified? Is fear of death appropriate?

Kellehear, A. (2007). *A social history of dying*. Cambridge: Cambridge University Press. The book is divided into four sections: the stone age, the pastoral age, the age of the city, and the cosmopolitan age. Kellehear argues that our ways of dying today are likely the interactive product of tradition and contemporary forces.

Luper, S. (2009). *The philosophy of death*. Cambridge: Cambridge University Press. This is a discussion of the basic philosophical issues concerning death and a critical introduction to the relevant contemporary philosophical literature.

Manheimer, E. (2012). *Twelve patients: Life and death at Bellevue Hospital*. New York: Grand Central Publishing. As former medical director at Bellevue, Dr. Manheimer uses stories taken from case histories to humanize issues like immigration, teen suicide, obesity, and the cost of health care.

Mollica, R. F. (2008). *Healing invisible wounds: Paths to hope and recovery in a violent world*. Nashville, TN: Vanderbilt University Press. This book reveals how trauma survivors, through the telling of their stories, teach us how to deal with the tragic events of everyday life. When violence occurs, there is damage not only to individuals but to entire societies, and to the world.

Morgan, J. D., & Laungani, P. (Eds.). (2005). *Death and bereavement around the world: Death and bereavement in Asia, Australia and New Zealand*. Amityville, NY: Baywood Publishing Company. This is volume four in the series. It concentrates on the consolidation of ancient cultures and modern methods of caring for the dying and grieving.

Oppenheimer, G. M., & Bayer, R. (2007). *Shattered dreams? An oral history of the South African AIDS epidemic*. New York: Oxford University Press. The book describes the AIDS epidemic in South Africa and shares hopes, dreams, disappointments, and insights of this situation.

Ray, R. (2008). *Endnotes: An intimate look at the end of life*. New York: Columbia University Press. This book explores how individuals construct meaning through their interactions with others, as Ruth Ray, a gerontologist, befriended an 82-year-old man suffering from Parkinson's and shared the account in *Endnotes*.

Rynearson, E. K. (Ed.). (2006). *Violent death: Resilience and intervention beyond the crisis*. New York: Routledge. This book focuses on aiding and supporting the vulnerable survivors of violent death regardless of how it occurred.

Sofka, C. J., Cupit, I. N., & Gilbert, K. R. (Eds.) (2012). *Dying, death, and grief in an online universe*. New York: Spring Publishing Company. Provides current information about "thanatechnology," the communication technology used in providing death education, grief counseling, and thanatology research.

Sumiala, J. (2012). *Media and ritual: Death, community and everyday life*. New York: Routledge. The author draws on both media studies and the sociology of religion to analyze recent events such as the death of Michael Jackson and school shootings.

Swedlund, A. C. (2010). *Shadows in the valley: A cultural history of illness, death and loss in New England, 1840–1916*. Amherst: University of Massachusetts Press. An examination of the history of mortality in several small communities in western Massachusetts.

Taylor, M. C. (2009). *Field notes from elsewhere: Reflections on dying and living*. New York: Columbia University Press. This book chronicles Mark Taylor's inverted journey through a critical illness as he went from death to life.

I mourn not those who lose their vital breath; But those who, living, live in fear of death.
—**The Ancient Greek Anthology**

Death is not the enemy; living in constant fear of it is.
—**Norman Cousins, Head First**

Death is nothing at all,
I have only slipped into the next room
I am I and you are you
Whatever we were to each other, that we are still.
—**Henry Scott Holland, taken from a sermon first preached in 1910**

THE AMERICAN EXPERIENCE OF DEATH

CHAPTER **2**

George Dickinson

How do you react when you first learn of a major tragedy such as the terrorist attack on the United States on September 11, 2001, or on Madrid on March 11, 2004, or on London on July 7, 2005; the shooting at Columbine High School; the crash of TWA Flight 800; the Indian Ocean earthquake causing the tsunami on December 26, 2004; the bombings at the Boston Marathon on April 15, 2013; the mudslide near Seattle, Washington on March 22, 2014; the explosions of space shuttles Challenger and Columbia; the Oklahoma City bombing; tornadoes and hurricanes destroying properties and lives; or the death of someone you did not know personally but felt that you did (e.g., Charles Schulz of *Peanuts* fame, Princess Diana of England, Michael Jackson, Ronald Reagan, Nelson Mandela, or Ted Kennedy)? Our "gut reaction" is often one of disbelief, mixed with feelings of sorrow for those affected. We get caught up in the event, especially when the media continue their focus. We may initially wish that we could in some way help them. But with the passing of time, alas, sometimes sooner than later, we are "over" the event and go on with life.

Journalist Harry Schatz (1976), writing a quarter of a century ago, recalled his mother weeping bitterly around 1940 over a news report on the radio about a tornado in Texas that killed several children. She kept saying over and over, "The poor children." Schatz wondered what his mother, an immigrant from Russia who came to Brooklyn in 1900, had in common with these children in Texas, because she was so far removed from them in culture, language, and thought. Yet her compassion reached out over thousands of miles to share personally in the heartfelt grief and anguish of a distant tragedy to unknown children. Schatz was moved to this recollection of 35 years earlier while reading a newspaper item about a particularly brutal murder. He found himself casually turning the page and doing a double take as he realized that the event had not triggered any emotional response in him. He was startled and overcome by a sense of guilt at his reaction. He asked himself if he were slowly and insidiously being programmed to accept as commonplace and normal such events and if his human qualities were disintegrating into moral decay. Certainly, he did not react as his mother had years before.

Imagine what Harry Schatz's mother's reaction would have been to that same event today with the visual images of television! She was grieving over the radio account of the children's deaths with only a picture in her head as to what was happening. Perhaps we would all like to think that we would react like Harry Schatz's mother and show concern and compassion in such a situation. After all, a tragedy has occurred. But in our fast-paced world of today, do we really "have time" for such reflection? It worried Harry Schatz that maybe his reaction is the direction toward which our society is headed. A distant death, be it of a single individual or a large number of individuals, can be brought home via visual images such as television. The hype of the media, particularly television, to sensationalize events such as death to improve their ratings and to be "number one," obviously plays into their coverage. Television reporters are notorious for holding the camera on a grief-stricken person when a much more humane approach might be to give the person privacy at that point. For the moment, we may be caught up in the event, as was Schatz's mother, but nowadays we soon move on.

Indeed, moving on in life is what might be necessary in our postindustrial society. We cannot take the time to get involved with the many distant deaths, some very tragic, that are reported to us by the media. Yes, we may occasionally send flowers or leave a note by the former location of the World Trade Center in New York, or in London or Madrid, by a fence in Oklahoma City, or in Los Angeles by Michael Jackson's Neverland Ranch. Probably we mean well with such an action, and the behavior helps us in a therapeutic way to move on through this event. Perhaps moving on does not mean that we do not care or are callous, but that responding and getting involved with these happenings simply is not fitting for our secularized, heterogeneous, specialized, fairly densely populated society of the 21st century.

Yet regarding emotional trends in the United States, it has been observed that the reactions to the portraits and descriptions in the *New York Times* of deceased persons following the terrorist attack on New York City in 2001 demonstrate our capacity to feel real sorrow for people we do not know (Democratization of Death, 2002). Because we feel that certain kinds of death are not appropriate to Americans, we need to express a kind of grief to which we do not ordinarily try to relate. To some extent, we are developing a better capacity to grieve for strangers, as long as they are Americans!

Anthropologist Colin Turnbull (1972) noted that in our society the individual is paramount. The family has lost much of its value as a social unit, and religious beliefs and practice no longer bind us into communities, says Turnbull. Kathy Charmaz (1980) argued that this individualistic perspective on death has led American society to abdicate most of its social responsibilities to the dying and their family members. She asserted that in America we are assured of an inequitable distribution of health care and a lack of social concern for the dying precisely because Americans value individualism and privacy.

Charmaz (1980) noted that beliefs in individualism, self-reliance, privatism, and stoicism are ideological and justify the ways in which dying is handled. The ideological view of dying as a private affair, something that should be the responsibility of the family, relieves other social institutions, notably health and welfare organizations, from the necessity of providing comprehensive services.

| WORDS OF WISDOM | I HEARD A FLY BUZZ |

I heard a Fly buzz—when I died—
The Stillness in the Room
Was like the Stillness in the Air—
Between the Heaves of Storm—

The Eyes around—had wrung them dry—
And Breaths were gathering firm
For that last Onset—when the King
Be witnessed—in the room—

I willed my Keepsakes—Signed away
What portion of me be

Assignable—and then it was
There interposed a Fly—

With Blue—uncertain stumbling Buzz—
Between the light—and me—
And then the Windows failed—and then
I could not see to see—

Despite the fact that the individual is responsible for taking care of dying, 75 percent of deaths in the United States today occur in institutional settings—primarily hospitals, yet also nursing homes, with more than a third spending a minimum of 10 days in an intensive care unit before they die (Brabant, 2003). The figure has gradually risen since 1949, when it was 50 percent, to 61 percent in 1958, and to 70 percent in 1977 (Nuland, 1994). Grandfather seldom dies at home today, where he has spent most of his life, but instead in a rather impersonal institutional setting. Yet, studies (Hays, Gold, Flint, & Winer, 1999; Thomas, 2009) in the United States and various other countries revealed that the majority of individuals consistently express a preference to die at home. The removal of death from the usual setting prompted Richard Dumont and Dennis Foss nearly four decades ago to raise the question: "How is the modern American able to cope with his own death when the deaths he experiences are infrequent, highly impersonal, and viewed as virtually abnormal?" (1972, p. 2). In some cases it makes the dying feel insignificant, like the person in Emily Dickinson's poem "I Heard A Fly Buzz."

DEFINING DEATH

"The Wizard of Id" defines *death* as a "once in a lifetime experience." Although one cannot refute this definition, it would not suffice in the medical or legal professions. In the first edition of *Encyclopaedia Britannica* death was defined accordingly: "Death is generally considered as the separation of the soul and body in which sense it stands opposed to life which consists in the union thereof" (1768, v. 2, p. 309). Such an 18th-century definition would obviously not suffice today. Until the 1960s there was little real controversy about what it means to be dead in the public policy sense (Veatch, 1995). Now, for the first time, however, it is a matter of real public policy significance to decide precisely what is meant when someone is labeled as dead. Today, matters have changed because technologies have greatly extended the capacity to prolong the dying process and to use human organs and tissues for transplantation or research.

A headline in a newspaper ("Doctors 'Kill' Patient," 1988) read "Doctors 'Kill' Patient to Save Her Life." The article went on to say that surgeons had put the patient into a coma, stopped her heart, chilled her by 40 degrees, and drained her body of blood for 40 minutes to save her from the aneurysm pressing on her brain. She was placed into a sort of suspended animation that allowed surgeons to cure a hard-to-reach, high-risk aneurysm once considered inoperable. As one surgeon noted, "It may be the surgery of the future in cases where bleeding poses the greatest risk to the operation." Thus, the headline said the physicians "killed" to "save" the patient. Defining death is indeed a confusing issue!

The diagnosis of death has vacillated throughout history between the centralist theory and the decentralist theory (Powner, Ackerman, & Grevnik, 1996). The centralist theory proposed that a single organ contains the vital life force and, if it fails, the person would die, a prominent view before the 18th century. Decentralism proposed that the entire body and every organ and cell possess the life force. This theory gained precedence when physicians discovered that a failed organ could be resuscitated. The modern theory of brain death, however, has resurrected the centralist theory.

The definition of death has changed significantly from the 1960s, from strictly physical criteria to a debate about the value of life and the essence of human qualities (Prior, 1989). That debate relocated the origin of death in the brain rather than in the respiratory system, and the first test of death became one that questioned the receptivity and awareness of the dying patient (Gervais, 1987). Total irreversible loss of all brain function is death. Part of the reason is that if that condition exists, the heart and lungs cannot work (Kaplan, 2008). To determine if a patient's brain has ceased to function, physicians generally look for three signs: the patient must be in a deep coma with a known cause; the patient must not be able to breathe independently; and the patient must have no reflexes associated with a part of the brain called the brain stem (Monaghan, 2002). Death, once so clearly defined as a physical fact, became entangled in a discussion about the value of life. The case of Terri Schiavo in Florida in 2005 certainly highlighted the whole dilemma of brain death and ethical questions surrounding such situations (see Words of Wisdom box).

INTERNATIONAL DEFINITIONS

Defining death is a difficult assignment. In the 1950s, the United Nations and the World Health Organization proposed the following definition of death: "Death is the permanent disappearance of all evidence of life at any time after birth has taken place" (United Nations, 1953). Thus, as death can take place only after a birth has occurred, any deaths before a (live) birth cannot be included in this definition. The latter is called a *fetal death* and is defined in the following way (Stockwell, 1976):

> Death (disappearance of life) prior to the complete expulsion or extraction from its mother or a product of conception irrespective of the duration of pregnancy; the death is indicated by the fact that after such separation the fetus does not breathe or show any other evidence of life, such as beating of the heart, pulsating of the umbilical cord, or definite movement of voluntary muscles.

Not all countries observe the definition of death recommended by the United Nations. In some countries, infants dying within 24 hours after birth are classified as stillbirths rather than as deaths or are disregarded altogether. In some other countries, infants born alive who die before the end of the registration period (which may last several months) are considered stillbirths or are excluded from all tabulations. As Robert Hertz (1960) observed, they are considered still a part of the spirit world from which they came; therefore, their death is often not accorded ritual recognition, and no funeral is held. In some areas, the baby is not given a name until he or she lives one year. The child does not "exist" until given a name. Thus, if the child dies before achieving one year of age and therefore was not named, then the child did not exist. If the child did not "live," he or she could not "die"! Thus, one is not "alive" until officially registered or named, and one cannot be legally "dead" if never alive!

A survey of 80 countries (Monaghan, 2002) showed that all 70 countries with guidelines require irreversible coma and lack of brain-stem reflexes for a diagnosis of brain death. Yet countries differ on how long a patient needs to be monitored and how many physicians must diagnose brain death. Some countries require

WORDS OF WISDOM | THE TERRI SCHIAVO CASE/HER CONDITION: DOCTOR EXPLAINS THE "PERSISTENT VEGETATIVE STATE"

Sabin Russell, Chronicle Medical Writer

Terri Schiavo, the brain-damaged Florida woman whose husband and parents are battling over whether to let her die or keep her alive, exists in the rare and paradoxical "persistent vegetative state," where neurologists can find a patient to be at once unconscious but alert.

With her cerebral cortex apparently destroyed, Schiavo is almost certainly unable to return to a state of awareness—that is, knowing who she is and where she exists in time and place.

These are traits, like the faculty of speech, governed by the cerebral cortex, which, doctors say, has died in Schiavo's case.

But she is decidedly not brain-dead.

Her brain stem—a more primitive and hardy core of nerve tissue that controls her breathing and cycles of wakefulness—is very much alive, as evidenced by her ability to breathe without a respirator and her open eyes, which can appear to track a moving object.

Defining the differences in these states of being is the work of neurologists such as Dr. Wade Smith, director of the Neurocritical Care Unit at the UCSF Medical Center. He regularly teaches a course on the subject to young medical students, many of whom are entering a lifetime of counseling families on difficult choices of life and death in the midst of trauma and tragedy.

Without her feeding tube or water, and barring legal intervention, Schiavo probably has less than two weeks to live. But in a vegetative state, does she feel hunger, thirst or pain? "As a neurologist, I would say no," said Smith.

A person who is unconscious will feel no more hunger or pain than a patient who has undergone general anesthesia. The awareness functions of the higher brain are no longer a factor. "It is why anesthesia works," Smith said.

But people in a persistent vegetative state will nonetheless react to a "painful" stimulus in a reflexive manner. Pinch, and they will flinch.

A patient in a persistent vegetative state would never be considered a candidate to donate an organ; hearts, lungs, livers and kidneys can be harvested only from the brain-dead.

Neurologists are armed with a long list of descriptors for the varying states of mental being. At one end is brain death, and at the other is consciousness. But in between, the medical meanings can sow confusion:

Because Schiavo's eyes can open, she is, by definition, alert.

With eyes that open and close in accord with cycles of sleep, it even could be argued that a person in a persistent vegetative state can be simultaneously unconscious and awake. A brain scan, however, would show no activity that we normally associate with wakefulness.

A person in a vegetative state is defined as one who is arousable—which means merely a reflexive response to a stimulus—but unaware. It is not necessary to wake up to be aroused. If a vegetative state exists for six weeks to three months, it is deemed a "persistent" vegetative state.

A person in a persistent vegetative state is not in a coma. A person in a coma is unconscious, and unable to be aroused.

There is a similar distinction defining stupor. A person in a stupor shows some response to stimulation but will easily slip back into a state where he or she cannot be aroused.

Sleep, strictly defined, is a state of unconsciousness from which a person can be aroused. But there is another halfway state—somnolence—where a person can wake up enough to carry on a brief conversation before falling back into sleep.

The strange limbo of people in vegetative states poses a dilemma to those who must make life-and-death decisions in the state of full awareness.

"If two physicians sit down and examine a person like Terri Schiavo, there can be a disagreement," said Smith.

In the medical literature, there are no cases of someone recovering from what is defined as clinical brain death, but there are five published cases of "recovery of some sentient function" among patients who were diagnosed as being in a persistent vegetative state.

Typically, the decisions that are made about the future of people in vegetative states are made by family members.

"Most of these discussions never leave the intensive care unit," Smith said. "That's the way it should be."

confirmation that the brain has stopped functioning such as an electroencephalo-gram or EEG, which should show no electrical activity in the brain. The EEG is the most validated test implemented in many countries' guidelines.

The Canadian guidelines closely follow those of the United States regarding the use of brain death. Finland was the first European country to accept brain-death criteria. All EU countries have now accepted the concept of brain death (Kellehear, 2009). Yet, while the required clinical signs are uniform, less than half of the European countries require technical confirmatory tests and approximately half require more than one physician to be involved. Confirmatory tests are not manda-tory in many developing-world countries because of their unavailability, notes Allan Kellehear. In Asia, neurological criteria for death has not been uniformly accepted. While China has no legal criteria for death, Japan officially recognizes brain death, though the public is reluctant to accept it. India and the United Kingdom require brain-stem death, while New Zealand and Australia have accepted whole-brain-death criteria.

For the Kaliai of Papua, New Guinea (Counts & Counts, 1991, p. 193), dying is almost complete "if the breath smells of death, if the person stares without blink-ing or shame at another person's face, is restless and must be moved frequently, or loses bladder or bowel control." Dying is complete "when breathing stops, when the heart ceases to beat, and when the eyes and mouth hang open." The Kaliai also believe that persons dying or dead may return to life at any time after the dying process begins. The Kaliai have no generic term for *life*. Thus, the whole question of life and death is more complicated than it might initially appear.

AMERICAN DEFINITIONS

By traditional definition, death is when heartbeat and breathing stop. But medical advances have made this definition of death obsolete. Under this definition, many people living today would have been considered dead because their hearts stopped beating and their breathing stopped as a result of heart attacks. With respirators to artificially breathe for a patient, along with other life-support equipment, many lives are saved. A diagnosis of **brain death** allows respirators to be turned off when the brain is totally and irreversibly dead. The issue of who is alive and who is dead is literally a life-and-death question because of advances in transplant sur-gery and increased medical-legal issues (Evans, 2011). How death is defined can determine whether the discussion of the termination of treatment is the issue, and which would lead to death, or the discussion of continuation of treatment that would prolong a life.

Leon Kass (1971, p. 699) defined *death* simply as "the transition from the state of being alive to the state of being dead." Simple enough, death can be regarded as a discontinuous event, yet the issue is clouded by many factors (Bruno, Ledoux, & Laureys, 2009). Traditionally, death has clinically meant the irreparable cessation of spontaneous cardiac activity and spontaneous respiratory activity. The functions of heart and lungs must cease before one is pronounced dead by this definition.

In 1968 a committee of the Harvard Medical School (Blank, 2001) claimed that the ultimate criterion for death is brain activity rather than the functioning of heart and lungs. The Harvard report notes that death should be understood in terms of a permanently nonfunctioning brain, for which there are many tests. With brain death the body would not be able to breathe on its own because breathing is controlled in the brain. If the brain is dead, any artificially induced heartbeat is merely pumping blood through a dead body. Brain death is classically caused by a lesion which increases intracranial pressure and hence causes intracranial circulation to cease and damages the brain stem due to herniation (Bruno, Ledoux, & Laureys, 2009).

In 1981 President Ronald Reagan created the President's Commission for the Study of Ethical Problems in Medicine and Biomedical and Behavioral Research to study the ethical and legal implications of the matter of defining death (Crandall, 1991). The commission examined three possible ways of determining death: death of part of the brain, death of the whole brain, or nonbrain ways to determine death, such as absence of heartbeat. The commission decided on a definition of death that includes the traditional nonbrain signs of death and a whole-brain definition. Its proposal to determine death, which was accepted by both the American Bar Association and the American Medical Association (Blank, 2001), notes that an individual who has sustained either irreversible cessation of circulatory and respiratory functions or irreversible cessation of all functions of the entire brain, including the brain stem, is dead. A determination of death must be made in accordance with accepted medical standards.

The majority of states have adopted the proposal or one similar in content. Had the commission decided on death of part of the brain, then neocortical death (the irreversible destruction of neural tissue in the cerebral cortex, critical for intellectual functioning) could have been operative. Likewise, death of the brain stem (which controls breathing and heartbeat) would not qualify as "death," as happened in the case below.

A case in Florida ("Baby Born," 1992) involved a baby girl, Theresa Ann Campo Pearson, born without a fully formed brain. She had only a partially formed brain stem and no cortex, the largest part of the brain. Her parents asked the courts to declare her brain dead, but to no avail. The Florida Supreme Court said it did not have the constitutional authority to hear the case. Theresa Ann did not have an entire brain and thus did not fit the legal definition for declaring one dead: "irreversible cessation of all functions of the entire brain, including the brain stem." The parents had wanted to donate their daughter's organs, but by the time the 10-day-old girl died, her organs had deteriorated so much that they could not be transplanted.

The whole-brain definition of death (or brain-stem death) has become the accepted standard of practice in most Western nations, yet it continues to be surrounded with controversy (Blank, 2001). Even though the supercellular brain function is irreversibly destroyed, isolated brain cells may continue to live and emit small electrical potentials measurable by an electroencephalogram (EEG). The concept of whole-brain death also ignores spinal cord reflexes, but requires cessation of lower brain-stem activity, thus contradicting the definition that "all functions of the entire brain" be dead, observed Robert Blank. Furthermore, whole-brain death assumes that the brain is the integrating organ of the body whose functions cannot be replaced, even though intensive care units increasingly have become surrogate brain stems with the use of respirators and other treatments to replace respiratory, hormonal, and other regulatory functions. Robert Truog (1997) proposes two

George Dickinson

Though the definition of death is obviously not a uniform one, this poster on a telephone pole in England certainly grabs an individual's attention. A picture is worth a thousand words.

alternative approaches to the whole-brain formulation: (1) higher brain criteria where individuals who have permanently suffered the loss of all consciousness are dead; and (2) return to the traditional tests for determining death—the permanent loss of circulation and respiration.

Various tests can be given to determine any signs of brain activity (Monaghan, 2002). The determination can be made reliably by competent physicians. They search for any hint of normal brain-controlled reflexes. They turn the head and even put cold water into the ear to look for any sign of movement. They search for any sign of the pupils of the eye responding to light. One test is to touch the cornea of the eye and see whether this triggers a blink. The respirator is also stopped briefly to look for any sign of spontaneous breathing. They can give the patient some carbon-dioxide gas to inhale. In a normal person the gas would trigger the brain stem to direct the body to begin breathing again. Absence of brain activity as signified by a flat EEG is another criterion for brain death. The tests can be repeated 24 hours later to make certain that the absence of these life signs is not temporary. The EEG is the most validated of the tests (Kellehear, 2009). The brain death prognosis is totally predictable and uniform: Brain-dead people never regain consciousness, much less any recovery, and suffer cardiovascular failure within a short time (Jasper, Lee, & Miller, 1991).

According to one neurologist (Cope, 1978), when death is declared on the basis of cessation of breathing and heartbeat, the brain is deprived of blood supply and dies within a matter of minutes. Death of the brain is what is final and absolute. On the other hand, one may lose all brain function, but it is not an irreversible loss. For example, a person having taken an overdose of barbiturates may have a temporary absence of any signs of brain activity, but this is not brain death because it is reversible. The importance of making an accurate diagnosis of brain death is obvious (Henneman & Karras, 2004). What is not obvious, however, is the potential harm to patients' families that could result when the process used to make the diagnosis is less than optimal. Well-intentioned, yet uninformed, clinicians can cause unnecessary stress for patients' family members by prolonging the diagnostic process or inaccurately interpreting findings.

Perhaps as a result of President Reagan's commission, there has come to be increasing acceptance of a brain-oriented criterion for determining death as the "irreversible loss of all brain function" (Youngner, Landefeld, Coulton, Juknialis, & Leary, 1989, p. 2205). The cerebral death signals the death of the person when specific higher brain functions cease, not all brain activity (Blank, 2001). This definition assumes that without consciousness human life no longer exists. Nonetheless, Raanan Gillon (1990) argued that the criteria for brain death are a compromise between extreme versions of two concerns. On the one hand, the compromise is with a concept of death as complete disintegration of the human organism's biological functioning, with all living components dead or at least totally disintegrated and

dissociated from each other. On the other hand, brain death criteria are also a compromise with the concept of death as cessation of personal existence, because brain death criteria will classify as alive some humans who are dead as persons. A clinical example is a vegetative state that involves permanent loss of consciousness and of the capacity and potential for consciousness. Because a capacity for consciousness is a necessary condition for being a person, an individual in a permanent vegetative state cannot be a person, yet according to brain death criteria (whether brain-stem death criteria or whole-brain criteria) that same human being is unequivocally alive.

THE MEANING OF DYING AND DEATH

The meaning of death varies from culture to culture. In some societies, an individual who dies is not recognized as "dead" until a year has passed. This "wait" gives the family time to prepare for the funeral. In the meantime the body is kept in the house (hut) and "given" food, drink, and cigarettes. Socially, the person is not dead, because he or she has yet to be defined as such. On the other hand, in a few cultures, such as the place of the dead in India, an individual may be alive and talking (biologically alive), but is socially dead. Indeed, an individual in a vegetative state is technically alive, yet socially dead. Are not some residents in nursing homes or individuals on death row in correctional institutions biologically alive yet socially dead?

The philosophical meaning of death suggests that the objective reality of death, the biological process that occurs when someone dies, is very different from the individual, subjective experience of death. For the one who is dying, death is momentous. It is the end of the world, the demise of one's very existence. Yet, objectively, the death of an individual is just a part of the cycle of living and dying. It is neither momentous nor particularly noteworthy. The philosopher Fingarette observes that "death itself is nothing, but the thought of it is like a mirror" (Fingarette, 1996, p. 5).

Let's take a closer look at the social meaning of death. Social meaning is the most important component of every aspect of dying and death considered in this textbook, as earlier noted by Glenn M. Vernon (1970) in the 1970s. Recall from Chapter 1 the discussion of symbols and meaning in our presentation of the symbolic interactionist perspective. Social meaning consists of symbols that are socially created and socially used. Some symbols refer to something else or have an empirical referent. Some do not. The major function of both types of symbols is to permit humans to relate to each other and thus create shared behavior and meaning. If dying is perceived to be primarily a biological process, then defining dying and death simply as biological or physical processes would seem to be appropriate. Most people would probably agree with the "biology is primary and social meaning is secondary" interpretation. We do not.

Dying, in fact, is much more than a biological process. It is truly one of the most individual things that can happen to the body, and what happens takes place exclusively within the skin of the one person. However, with reference to the meaning of dying, the dying process is one of the most social experiences that one can have—all human bodies exist within a social and cultural context. When a person dies, many things other than internal biological changes take place. For nearly all human beings, every act of dying influences others. At the same time, the dying are influenced by their environment and those around them. Consequently, the act of dying is a social or shared event.

DERIVING MEANING FROM THE AUDIENCE In Chapter 1 we discussed the importance of the audience in establishing meaning. It is essential for those who interact with the dying to understand the acute awareness of the dying person of those around him or her. This is especially true given the American tendency not to talk directly about death or dying. The dying person is therefore always trying to interpret the behavior and non–death-related words spoken in his or her presence. This awareness of the audience includes, but is not limited to, the following:

1. What people are willing to talk about with me—and what they avoid
2. Whether they are willing to touch me, and how they touch me when they do
3. Where I am, or maybe where others have located me—hospital, nursing home, intensive care unit, isolation unit, or my room at home
4. Tangible and verbal gifts that others give me
5. What people will let me do, or expect me to do, or will not let me do
6. The tone of voice that people use when they talk to me
7. The frequency and length of visits from others
8. Excuses that people make for not visiting
9. The reactions of others to my prognosis

The audience to which the dying person relates may also be supernatural. Symbol users are not restricted to the natural empirical world nor are they restricted to the world of the living. If believers realize that those who have died have an existence in another realm or that there is a life after death, this belief is real to them and has consequences for their behavior.

People who are approaching death may involve themselves in a gradual replacement of a living audience with a supernatural or other-world audience. As we have demonstrated, the dying relate to many audiences.

DERIVING MEANING FROM THE SITUATION Where a person dies is also given meaning. As patients come to grips with their terminal condition, the manner in which they define the situation (and respond to it) will have a tremendous impact on their dying. Dying in a nursing home or hospital is different from dying in the home, in bed, surrounded by a loving family and feelings of belonging (see Chapter 7). If patients view the institutional death setting as a supportive environment, their coping behavior may be helped. On the other hand, if they feel all alone and if they have defined the place of dying as a foreign environment, adjustment will not be facilitated (Leming, Vernon, & Gray, 1977).

DEATH AS A LOST RELATIONSHIP The sociological perspective emphasizes the social-symbolic nature of human interaction. The key factor that unites biological entities into a social group entity is shared meaning: Many of the goals that one person wants to achieve and many of the experiences that one person wants to have require shared and coordinated cultural meanings. Symbols are the means by which socially created meaning is shared in the process of human interaction. Symbols are words or gestures that stand for something else by reason of association.

A specific death has distinct meanings wherever the deceased had meaningful relationships, and the death of one person has extensive social consequences. With a

George Dickinson

This photograph is much more than a man digging a hole in the ground. Consider your emotional response to this photograph.

death in a husband-wife **dyad,** one half of that entity dies. If the couple has two children, one fourth of the family dies. One thirty-thousandth of the community dies, and one three-hundred-and-ten-millionth of the nation dies. When one person dies, his or her spouse loses a husband or wife, children lose a parent, and friends lose a friend. Yet, the person who dies loses all human relationships. Thus, one could argue in numerical terms that the dying person is losing more than anyone else.

If a person is identified with many social roles or positions and that person's death creates a situation where many roles or positions are vacated, then many persons or role occupants die in a single death. Furthermore, we may agree that, although one biological body dies, ownership of that body is difficult to determine. "Who owns my body?" is a question often posed by people who are dying. Related to the basic ownership question are "Who is qualified to make decisions about my body?" and "If I am the one who is dying, what right do any others have to tell me what to do with my body?" Because humans are social animals, ownership is a creation of symbol-using people. Joint ownership patterns are created and exist for most people. Therefore, body ownership might be considered shared, and the decisions related to it would be joint decisions.

The elimination or departure of a person from the ranks of "the living" leaves a hole in the midst of the living. Certain meaning is lost, while new meaning is added—behavior that previously involved the dead person is literally no longer possible because it also has ceased to exist. Established interaction or behavior patterns are disrupted, an event that demands attention. The funeral process involves activities, rites, and rituals associated with the final disposition of the dead person. This process usually reduces the social disruption caused by death insofar as the rituals are acceptable to those involved and are performed according to societal norms. The living are concerned not only with the death of a person but also with what happens to them as a result of that death. Although the biological person may be gone, the meaning remains just so long as the living grant the symboled immortality or meaning immortality to the deceased (Lifton & Olson, 1974).

In the social commentary immediately following the death of Diana, Princess of Wales, a British broadcaster was asked by an American interviewer if the funeral ceremony would be a "royal funeral." He replied:

> Diana is much more than a royal—royals are two a penny. Princess Di is a star the likes of which we have only seen in James Dean, Marilyn Monroe, Elvis Presley, and John Kennedy. Her immortality will transcend the House of Windsor, not because she was born to it, but because it has been bestowed upon her by the British people and the people of the world.

Granting or creating such immortality is one of the things symbol-using beings can do. A person may even be granted considerably greater significance than he or she had while alive, as in the case of Princess Diana. For the bereaved, changing the meaning of a lost relationship is essential to the process—and is accomplished with varying degrees of ease—depending on the nature of the relationship.

CREATING AND CHANGING DEATH-RELATED MEANING   Biological bodies are created; they live, and they die. Bodies of meaning are also created, live, and die. Biological continuity occurs through a process of biological transmission or transference. Meaning (culture) continuity occurs through a process of socially symboled transmission. The socialization process occurs as biological bodies are transformed into social beings and as we teach our children how to behave in what our society considers to be a human way.

It is important for caregivers and loved ones of dying persons to realize that their own death-related meanings will shape their behavior toward the dying. And, as we discussed in Chapter 1, these behaviors may affect the meanings that the dying ascribe to their own death. To avoid thoughtless actions that may affect an individual's ability to die with dignity and integrity, those administering to dying persons should always question the following commonly held assumptions:

1. Those with whom you work necessarily share your meaning of death.
2. Meanings that were helpful to earlier generations are equally functional today.
3. Meaning remains constant and does not change.
4. Dying biologically is all that is happening.
5. Knowing about the biological aspects of dying will in and of itself provide knowledge about how humans expect to behave in death-related situations.
6. The terminally ill person is the only one having death adjustment problems.
7. The dying person has somehow stopped meaningful living during the terminal period.
8. Talking is the only way for the caregiver to communicate that she or he cares.

WORDS OF WISDOM	DO NOT DISDAIN DEATH

Do not disdain death, but be content to accept it, since it too is one of the processes which nature ordains. For dissolution of life is part of nature, just as it is part of nature to be young and to be old; to grow up and to grow old; to grow first teeth, then a beard, then grey hairs; to beget, to be pregnant, and to give birth; and to move through the rest of the natural functions which the seasons of your life bring. A reasoning person, therefore, will not be careless nor over-eager nor disdainful toward death, but will await it as one of nature's inevitable processes.

Marcus Aurelius (121 A.D.–180 A.D.). Taken from *Death: Philosophical soundings* (p. 162), by H. Fingarette, 1996, Chicago, IL: Open Court.

Creation of new meaning is always possible. Death-related meaning is no different from any other type of meaning. It is important to remember that this meaning is created by humans, not discovered in the world. All meanings are subject to change. However, a well-established meaning is difficult to change. Meaning, for example, is frequently defined as sacred and is, therefore, more likely to be protected and perpetuated than changed. Crises or traumatic conditions may be necessary for a change in death-related meaning. As noted in Chapter 1, many of the contemporary death-related meanings, including the rituals involved in adjustment, were created by ancestors who experienced dying in quite different social circumstances than those found today. Furthermore, considering the dramatic changes in health, longevity, and health care, our ancestors experienced death in somewhat different biological bodies. Therefore, it is not surprising that discontinuities have developed in American death-related meanings and experiences.

Any aspect of death-related behavior can be changed if there is enough societal (or subsocietal) support. One person can change death-related meaning for himself or herself, but it is difficult to maintain and sustain the new meaning if a **significant other** does not support, legitimatize, or validate it.

Like the ripples caused by dropping a stone into a lake, changes in death-related meaning will inevitably have consequences that penetrate other areas of living. Change in the sacred components of dying and death may come in through the back door, so to speak. Cremation may gain increased acceptance, not as a direct result of changes in religion, but rather as a result of the unavailability of space for earth burials. Likewise, changes in life-prolonging, or death-prolonging, procedures may result more from the availability of technological devices than from changes in religion or **mores**—the "must" behaviors that a society believes are for the good of that society.

THE AMERICAN EXPERIENCE OF DEATH

The experience of death varies from individual to individual, from culture to culture, and through history. For example, the story of the American experience of death would likely not be the same, if told from an Afrocentric perspective with the public lynchings in the Deep South, as it would if the story were told from a different perspective. This diversity is due to both psychological and sociological factors. In this section, we will explore the sociological factors that have influenced the dramatic changes in attitudes from the Puritans to contemporary society. These changes reflect, in part, a changing way of life in America. Historians such as James Farrell (1980) write about the transformation of the American way of death. Three major shifts in perspective in the American experience of death are identified: "the living death" (1600–1830), "the dying of death" (1830–1945), and the "resurrection of death" (1945–the present).

LIVING DEATH (1600–1830)

Historically, changes in the American way of life contributed to how we viewed death and responded to dying and death. Between 1600 and 1830, for example, death was a living part of the American experience. Rural Americans during these earlier years were well acquainted with death. People died in their homes and would be reminded of death by the tolling of the funeral bell in the village church-yard or simply passing by the church, surrounded by the cemetery, as they walked to and from their shops and homes. Symbols on gravestones in the early 18th-century cemeteries included the crossbone and skull (stark reality of death), which was replaced by the soul face and later the soul face with wings attached (symbolic of an afterlife).

The New England Puritans believed that a sovereign god ruled over the earth and displayed sovereignty by intervening in the natural or social world—and one such providence was death. They knew that they deserved death and that the last enemy on earth was death: Death was painful, a punishment for sin, and brought separation from their loved ones. Unlike modern thanatologists, the Puritans encouraged each other to fear death. They increasingly used that strong human emotion to rouse people from their psychological and spiritual security. Puritans knew that they would die, but not *when* they would die, or if they were among the elect. Therefore, they admonished themselves and each other to be constantly prepared for death. Like modern thanatologists, they felt that an awareness of death could improve the quality of their lives and that they had a role to play in the work of God's redemption. Indeed, for them, dying, death, and bereavement were opportunities to glorify God by demonstrating human dependence on divine providence.

Most Puritan deaths occurred at home. The family sent out for midwife-nurses to care for the corpse, ordered a coffin, and notified friends and relatives not already present. Puritans considered the corpse a mere shell of the soul. They simply washed it, wrapped it in a shroud, and placed it into the coffin. Friends and relatives visited the home to console the bereaved. Prayers were offered, not for the soul of the deceased, but for the comfort and instruction of the living and to glorify God. The Puritan funeral acknowledged the absence of the deceased and drew the community members together for mutual comfort. After the funeral, the mourners were to return to their callings and resume their life's work.

In 1802 Nathaniel Emmons delivered a magnificent funeral sermon entitled "Death without Order." In it, he reviewed the Puritan orthodoxy of death, observing that "in relation to God, death is perfectly regular; but this regularity he has seen proper to conceal from the view of men." Emmons saw the uncertainty of death as a demonstration of God's sovereignty and human dependence and as a way of teaching people "the importance and propriety of being constantly prepared for it." However, he also saw that, despite the fact of death's disorder, multitudes of Americans had resolved "to observe order in preparing to meet it" (Emmons, 1842, pp. 3, 29). Between the 1730s and the 1830s, such orderly Americans were influenced by the Enlightenment, the American Revolution, Unitarianism, and Evangelism—and all of these were influenced by an underlying market revolution. These movements slowly but surely reformed the Reformed Tradition and gave Americans an eclectic tradition from which to fashion new beliefs and behavior about dying, death, and bereavement.

THE DYING OF DEATH (1830–1945)

Between 1830 and 1945, American thinking toward dying and death, rather than a "living death" as previously thought, was more of a "dying of death," as noted by an English author in 1899. This process brought "the practical disappearance of the thought of death as an influence bearing upon practical life" ("The Dying of Death," 1899) and the tactical appearance of funeral institutions designed to keep death out of sight and out of mind. This change was in part due to the rise of a new American middle class.

Both ideas and institutions were the product of a new American middle class—a group of people trying to distinguish themselves from the European aristocracy and from the American common people, somewhere "in the middle." The middle class wanted death with order. One strategy for achieving death with order was the ideology of separate spheres. In the course of the 19th century, middle-class people separated management from labor, men's work from the home, and women's work from men's. They also tried to separate death from life, both intellectually and institutionally. Increasingly, specialists segregated the funeral from the home and the cemetery from the city. The rural, landscaped cemetery movement began, for example, with Mount Auburn Cemetery in Cambridge, Massachusetts, in 1831.

Both separation and specialization were strategies of control, an increasingly important idea in Victorian society. Nineteenth-century Americans worked for self-control, social control, and control over nature. They saw self-control as the key to character and sexual control as the key to marriage. In separating their homes from their shops, they tried to control both spheres of their lives. The home would be a controlled environment for reproduction and socialization, while the workplace would be a controlled environment for increased production and time discipline. Schools served as a transition from one controlled environment to another, while asylums provided controlled environments for societal deviants. Science and technology attempted to make the whole continent a controlled environment. Therefore, it should not surprise us that the same class that practiced birth control should also devise forms of death control (Howe, 1970; Hale, 1971; Rosenberg, 1973).

Belief in the beauty of death and the funeral was new to the 19th century as the middle class used its aesthetic awareness to beautify corpses, door badges, caskets, casket backdrops, hearses, horses, funeral music, cemeteries, monuments, mourning customs, and death itself. Death was made so artistic that it almost became artificial and thus less fearful. By the end of the 19th century the obituary began to serve as a funeral notice, and later the telephone allowed people to notify others about funerals without leaving their house, a more impersonal means than face-to-face, as was the case in the mid-1800s. Consolation literature, which showed mourners that they were not alone in their sorrow and showed the emotional and moral benefits of their sorrow, allowed people to share their grief without sharing it with people they knew. Such literature included obituary poems and memoirs, prayer guidebooks, hymns, and books about heaven. These writings inflated the importance of dying and the dead by every possible means.

Because of scientific breakthroughs, an insistence that death should be painless developed with the belief that all people could eventually attain an "easy," natural death at an advanced age. With the discovery of ether in the 1840s and the introduction of the word *painkiller* in the 1850s, Americans applied physical and mental anesthetics to kill the pain of death. A major reason for people to think seriously about death was in part eliminated. Death began to be treated as an occurrence of old age, an idea that encouraged people to postpone or preempt preparation for death. Some medical researchers even asserted that old age is a curable congenital disease and posed the question, "Why not live forever?" Americans were advised not to take death personally, but rather to accept it as a part of human progress. Thus, the emphasis was turned from death and the deceased to survivors and posterity. The whole idea of life insurance developed out of these ideas (see Chapter 11). By defining death as part of evolutionary progress propelled by a merciful God, Americans were allowed to approach death optimistically. Thus, we have the "dying of death" in America.

THE RESURRECTION OF DEATH (1945–THE PRESENT)

On August 6, 1945, the United States dropped a single atomic bomb on the Japanese city of Hiroshima. The bomb destroyed everything within 8,000 feet, blowing bodies at 500 to 1,000 miles per hour through the air, and killing 70,000 people. The heat of the explosion ignited clothing within a half-mile radius and trees up to a mile and a half away. The subsequent shock wave ruptured internal organs. The effects of the radiation disfigured or killed thousands more, and the whole world bore the scars of the blast. This event ushered in the era of atomic weapons and our mode of thinking about death began to change, wrenched violently back to the basic fact of death. And so America experienced the "resurrection of death."

In a predictive 1947 article on the social effects of the bomb, Lewis Mumford had this to say about the Atomic Age:

> Life is now reduced to purely existentialist terms: existence towards death. The classic other worldly religions undergo a revival; but even more quack religions and astrology, with pretensions to scientific certainty flourish; so do new cults…. The belief in continuity, the sense of a future that holds promises, disappears; the certainty of sudden obliteration cuts across every long-term plan, and every activity is more or less reduced to the life-span of a single day, on the assumption that it may be the last day…. Not a single life will yet have been lost in atomic warfare; nevertheless death has spread everywhere in the cold violence of anticipation. (Mumford, 1947, pp. 9–20, 29–30)

"For the first time in six centuries," wrote Edwin Shneidman (1973, p. 189), "a generation has been born and raised in a thanatological context, concerned with the imminent possibility of the death of the person, the death of humanity, the death of the universe, and, by necessary extension, the death of God." "The bomb," said philosopher William Barrett (1958, p. 65), "reveals the dreadful and total contingency of human existence. Existentialism is the philosophy of the atomic age." Indeed, the bomb did lead many postwar adults to the philosophy of existentialism, which began with the reality of death, worked through anxiety and alienation, and culminated in the "God is dead" theology of the 1960s. The threat of "megadeath" was a constant reminder of the fragility of life and uncertainty of existence—ideas that the Puritans could surely appreciate. The sheer fact that atomic weapons are available to numerous countries places massive destruction of entire societies in the limelight.

The threat of the end of the world took away some of the traditional consolations of dying, including three conceptions of immortality: immortality through reproduction, immortality through the continuity of nature, and immortality through creative endeavors. As Albert Einstein, whose theories helped give birth to the bomb, said, "The atomic bomb has changed everything except the nature of man." By the 1960s and 1970s, a resigned fatalism had set in. Television ads showed children no longer asking what they wanted to do *when* they grew up, but *if they* grew up.

Thus began the effort to avoid a nuclear war, which, throughout the 1950s and 1960s, seemed almost inevitable. This fear of nuclear war culminated in the Cuban Missile Crisis in 1962, when John F. Kennedy, the president of the United States, faced down Nikita Khrushchev, the premier of the Soviet Union, in a battle of nerves that could easily have led to the annihilation of life as we know it.

Today, despite a slowdown in the arms race, the threat of a nuclear war is always present. Although the immediacy of this threat has waned, there are other, smaller wars that are a reality in many countries, claiming lives on a daily basis. The terrorist attack on the United States on September 11, 2001, with subsequent attacks on Madrid and London, certainly "resurrected" death and was a reminder of how quickly and unexpectedly death can come to many individuals. And there are many other reminders of our mortality, and the painful and brutal form that death can take. Beginning with the Vietnam War in the 1960s, we have seen the devastation of AIDS, ethnocide in Africa and the Balkans (a grisly reminder of the Holocaust), the interminable round of deaths and retaliations in the Middle East, and the deaths from the drug wars in Central America. The conflicts in Iraq and Afghanistan are reminders of death from fighting and from frequent car bombs.

The vivid reminder of death via television coverage of caskets coming home from war was declared a "breach of security," going back to the George H. W. Bush administration in 1991 when President Bush imposed it during the Persian Gulf War, since such pictures had earlier helped to create much sentiment against the Vietnam War. The Obama administration softened this policy and ended the ban on media coverage (with family approval) of the bodies of soldiers returned to the United States. Most families have allowed reporters and photographers to witness the solemn ceremonies that mark the arrival of flag-draped caskets (Chase, 2009). Such a vivid scene brings death closer to home. With the media reporting daily on the rising death toll from disease and warfare all over the world, we are only too aware that death surrounds and shapes our very existence.

DEATH ACROSS CULTURES | DEATH AND THE MUSICAL REQUIEM

Since the 15th century there have been more than 3,000 compositions related to the Latin mass of the dead. The purpose of requiem compositions has changed over time, as the original liturgical functions were intercession on behalf of the dead and an emphasis on the power, importance, and grandeur of the decedent, especially for kings and other rulers. Since World War I, however, many requiem compositions are not dedicated to individuals but to large groups of the deceased, often the victims of specific wars, genocides, or other human-made catastrophes.

Requiems are no longer about the afterlife but about the here and now. What matters is not what happens after death but what caused it. Such a shift in purpose may result from the requiem no longer having its original religious context but being thoroughly secularized today. It is now an intercultural symbol of death and mourning in music. Biological death cannot be prevented, yet untimely, human-made death could. If requiems are linked to the mass slaughter of individuals, the holocaust, or even victims of AIDS, they remind audiences of these events and their consequences, aiming to do more than merely entertain or console them.

Taken from W. Marx (2012). "Requiem sempiternam"? Death and the musical requiem in the twentieth century. *Mortality, 17*(2), 119–129.

Yet in this broad range of time from 1945 to the present, the emergence of the hospice movement and palliative care in the latter part of the 20th century somewhat challenges this "resurrection of death." As British sociologist Clive Seale (1998) notes, hospice care incorporates a critique of the modern way of death, which is perceived to involve a taboo. This revivalist alternative proposes an elevation of the private experiences of dying and bereavement so that these are brought into the field of public discussion. This revivalist psychological discourse enables individuals faced with death and bereavement to engage in practices such as psychotherapy that involve claims to membership in an imagined human community of resurrective practice.

CONTEMPORARY ATTITUDES TOWARD DEATH

Most individuals in our society do not need to fear the violent death that is so much a part of life in many parts of the world. Many Americans, however, are afraid of death, violent or otherwise, and seek to deny it. This overwhelming fear is palpable in the words of Ken Walsh (1974), in *Sometimes I Weep*, "Lord if I have to die, Let me die; But please, Take away this fear" (p. 18).

DENIAL OR ACCEPTANCE OF DEATH?

"The denial of death" is a phrase that Freud used in his 1915 essay "Our Attitude towards Death" (Dollimore, 2001). Freud argued that before World War I the attitude to mortality had amounted to a denial, which he said took the form of forgetting. Death was put aside, eliminated from life. A problem with death denial is that no matter how much we try to suppress death awareness, anxiety about death can still manifest itself in a variety of symptoms such as depression, worrying, stress and conflicts (Yalom, 2008).

But are Americans in the 21st century denying or accepting of death? One may look at the prevalence of life insurance policies and wills as evidence that Americans are quite accepting of death and are taking measures to deal with that eventuality. However, as discussed earlier, this may, in fact, be an attempt to deny death by avoiding the one aspect of death that can be controlled, social death. Through life insurance, we continue the accumulation of wealth, and through wills we control the distribution of property. The idea that modern society is death denying is widespread (Seale, 1998). For example, Illich (1976) emphasizes the medicalization of death, which, he feels, makes death an alien experience, cut off from the rest of life, a foreign agent. When an individual finally dies, it is the ultimate form of consumer resistance to medical domination of the experience, notes Illich. The denial of death thesis is highlighted by Baudrillard (1993), who argues that there is an irreversible evolution from savage societies to our own: The dead gradually cease to exist as they are thrown out of symbolic circulation into ghettos known as cemeteries.

Philippe Aries, in his classic book entitled *The Hour of Our Death* (1981), discusses the evolution through which death was viewed in the Western world and says that today there has been a gradual replacement of community-oriented personal identity with today's radical individualism. He argues that dying and death have become remote from ordinary experience and that we deny death's existence so that we no longer develop personal and communal resources to give it meaning. Culture's loss of spirituality enhances death's meaninglessness, Aries argues. Thus, Aries provides a slightly different analysis of the death-denying situation that varies somewhat from the "we live in a death-denying society" commentary. Despite the "resurrection of death" in the 1940s, there are some who still see evidence that we are a death-denying society (Dumont & Foss, 1972).

First, we prefer to obscure the dying process. Instead of talking about death directly, we employ **euphemisms** for death as buffers rather than the stark words *dying*, *dead*, and *death* (Harvey, 1996). Examples of these euphemisms are *succumbed, passed away, was taken, went to heaven, departed this life, bit the dust, kicked the bucket, was called home by God, croaked, was laid to rest, cashed in, expired, ended his or her days, checked out, ran out of time, brought down the curtain, went on to glory, is pushing up daisies, was taken by the Grim Reaper, heard the trumpet call,* and *is six feet under.*

The following joke captures this American approach to death:

> John came home from a trip and said, "How's the cat?" His brother Bob replied, "The cat's dead." John then said, "Don't be so blunt. You should ease into such information." "Okay," said Bob, "I opened the window, the cat got on the ledge, and the next thing I knew, it had fallen to its death." Then John said, "By the way, how is Mom?" Bob answered, "I opened the window...."

Second, we have a taboo on death conversation. A friend of George Dickinson recently wrote a two-page letter describing her family's summer activities. She noted that she and her husband were on vacation and that he became ill. He was taken to the hospital for surgery. She said, "He *had* the best of care. I *loved* him so." She never said that her husband had died, but it was obvious. It is difficult to say those words—*dead, dying,* and *death.*

To address the issue of limited death conversations, "Death Cafes" are cropping up. A long-standing tradition in Switzerland, these cafes migrated to Great Britain in 2011 and then to the United States in 2012 (Ward, 2013). Death cafes are held in 40 U.S. cities—free-wheeling and free of dogma and conclusions. Not meant to be a grief or therapy group, the idea is to increase awareness and make the most of our finite lives. Fear of dying and society's trepidation about discussing the topic of death are two recurring themes. With an emphasis on happiness, success, youth, health, and active lives, a tendency exists to marginalize those who are ill, aging, or dying, notes Bill Ward. Discussions at the death cafes tend to bring up more tough questions than easy answers, but the topic is being addressed. Additionally, the Dying Matters Coalition was set up in the United Kingdom in 2009 to create a wide range of resources to help people start conversations about dying, death, and bereavement. Dying Matters noted that 79 percent of individuals in the United Kingdom are uncomfortable discussing dying, even with those who are closest to them (Mills, 2013). Caitlin Doughty, in her book *Smoke Gets in Your Eyes: And Other Lessons from the Crematory* (2014), sees a cultural shift happening around what she calls "death awareness." More open and public discussions about mortality and loss are occurring. Certainly the Internet has made grief more public and casual (Seligson, 2014).

Third, **cryonics** (body freezing, derived from the Greek word meaning "cold") suggests a denial of death. In this procedure, the person's body is frozen in dry ice and liquid nitrogen. A very expensive process initially with additional high annual costs, cryonics has not caught on. Of the three main U.S. cryonics organizations (Alcor in Arizona, Cryonics Institute in Michigan, and the American Cryonics Society in California) and the Cryonics Society of Canada in Toronto, for example, Alcor had 88 frozen bodies and 903 members who had completed financial and legal arrangements by September 2009 (Alcor Extension Foundation, 2009).

Cryonics was recently highlighted by the media with the flap over whether former American baseball great Ted Williams's body should be cremated or placed in a cryonic state. Cryonics is based on the idea that someday a "cure" will be found for the "deceased" and that he or she can then be treated with the "cure" and thawed out. Thus, one does not die, but rather he or she is put into the cooler and brought back at a later date.

The first body frozen was that of James Bedford (Alcor Extension Foundation, 2009), a 73-year-old psychologist from Glendale, California, who was frozen in 1967. There are a dozen or so cryonics foundations today in North America, Europe, and South America. Most individuals pay for cryonics through a life insurance policy. Alcor suggested in 2009 that a life insurance policy of $150,000 would be needed for body preservation and $80,000 for neuropreservation (head only).

Fourth, we do not die in America; rather, we simply take a long nap. Caskets have built-in mattresses, some strawlike and others innerspring, and the head of the deceased person rests on a pillow in the casket. The room in the funeral home where the body is "laid out" is referred to by some as the "slumber room." Certainly, no one ever heard of having pets *killed* by the veterinarian; we have them "put to sleep." As is discussed in Chapter 11, the word *cemetery* derives from the Greek word *koimeterion*, which means "sleeping place" (Crissman, 1994). Besides "rest in

BEAUTIFUL MEMORIES
OF
A LOVING DAUGHTER
AND SISTER
OLIVIA CLARKE
BORN 10th AUGUST 1995
FELL ASLEEP 6th JULY 2005

Smile because she has lived,
Cherish her memory and let her live on
and be full of the love you shared.

"Until we are together again."

George Dickinson

Gravestones sometimes read "rest in peace" (RIP) or may, as in this photo, simply say "fell asleep." Putting a photo of the deceased person on the stone is becoming more popular in the Western world, as on this gravestone in England. In the 1730s and 1740s a portrait of a deceased individual was occasionally carved into the gravestone, especially for individuals such as clergy.

peace," cemetery markers sometimes also make other references to sleeping. Some markers for burials from the 19th and early 20th centuries have a crib or bed effect—headboard, footboard, and sideboards, sometimes with a "double bed" for a married couple. Dying does not occur in America; one just goes to sleep.

Fifth, unlike nonliterate societies, we call in a professional when someone dies. The funeral director takes the body away, and we do not see it again until it is ready for viewing (that is, if a viewing is to occur). When we see the body again, the cosmetic job is such that the person looks as "alive" as possible under the circumstances. When viewing a body at a funeral home, George Dickinson once heard a woman make the comment that the dead person "looks better than I have seen her look for years!" This is the "dying of death" idea discussed earlier.

Sixth, in many places in the United States, if a casketed burial is to occur, the casket is not lowered into the ground until after the family and friends have left the cemetery. It is not easy to watch a casket lowered into the ground because this reminds the viewers of the finality of death. There are several "bad trips" when a significant other dies—initially being told of the death, seeing the body for the first time, seeing the casket closed for the last time, and seeing the casket lowered into the ground. One can avoid the last "trip" by leaving the cemetery.

The death-denying society argument has its supporters today, yet sociologists such as American sociologist Talcott Parsons point out some limitations of this thesis (Parsons, 1978). He noted that in modern society there are efforts to construct most deaths as natural by relieving the physical suffering of dying and controlling premature death. Both of these would introduce uncertainty, if not prevented and controlled in these ways, and would result in a designation of death as unnatural. Parsons feels that science helps to construct death as a natural event. Embalming, for example, is often equated with a denial of the harsh physical reality of death, yet in fact is a symbolic affirmation of the value of a natural death, occurring at the end of a long life and akin to falling asleep.

Allan Kellehear (1984) provides a strong rebuttal to those who argue that we are a death-denying society. He argues that the phrase "death-denying society" is psychiatric in origin and implication, and that the term itself is ill-defined and too broad and various for specific application. Kellehear says that the fear of death is not universal because not everyone has this fear. There are so many fears of areas in ordinary life included under "fear of death" that such a fear can arguably be read as a fear of life. Fear of death belongs mainly to adults, as the fear of death in children is learned, like the fear of snakes or the dark or anything else that popular culture cares to impose on children. Fear of death is diminished and sometimes replaced in the elderly who fear dependency and disability more than death itself (see Chapter 3). The process of dying, rather than the event of death, tends to be feared more when individuals discuss their fear of death.

Since the **medicalization of death**, dying has become an embarrassing business, a situation of awkwardness, not because of denial, notes Kellehear, but because of the prevalent image of death widely accepted. The dying today are second-class citizens alongside drug addicts, disliked ethnic groups, and convicts. The only denial still involved is one that means depriving the dying of their former economic and social status. Medicalizing death means "the transformation of the dying role into a low status, technology intensive and potentially contaminating situation in need of sanitizing" (Kellehear, 1984, p. 717). Contrary to popular belief, corpse cosmetics and other funeral practices are not a denial of death but an affirmation of normal capitalist marketing strategy. This accepts death as part of its way of life, but also as a possible rich source of earthly revenue, states Kellehear. A reluctance to speak of death does not mean denial of death. If death talk is upsetting, then it makes sense to avoid the subject and emotion aroused by it. The avoidance of such talk does not stem from death denial but from aspirations for smooth relations and conduct.

Whichever side of the argument one takes, denial or acceptance, this paradox seems to exist. Americans have an obsessive fascination with death and death-related phenomena, as evidenced through our popular culture via television, cinema, music, print media, and jokes, all of which are fraught with thanatological content (Durkin, 2003). The paradox of denial, yet acceptance, is seen through our craving for some degree of information and insight concerning death. One could also argue that the treatment of death as entertainment and humor is simply

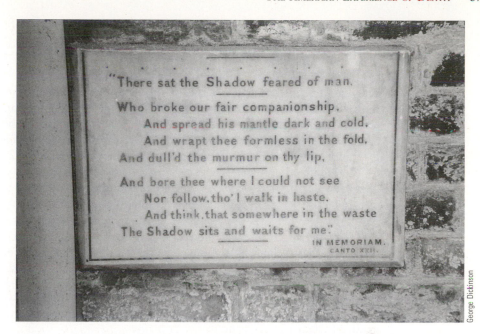

George Dickinson

This poem by Alfred Lord Tennyson appears near the entrance to Highgate Cemetery in London. Indeed, death (the Shadow) separates us from significant others and "waits for" us. This is a rather appropriate marker to remind visitors to the cemetery that we too will someday encounter the "Shadow."

an extension of, or another configuration of, death denial. As death is a disruptive event, both for the individual and for society, all societies must construct mechanisms to deal with death's problematic impacts (Blauner, 1966). It appears that the thanatological themes in U.S. popular culture function as a mechanism to help deal with death, notes Keith Durkin (2003). Finally, the tremendous amount of exposure to dying and death through our popular culture may make us more accepting of these phenomena. This saturated environment of thanatological concerns may function to make us accustomed to the difficulty of dying and death, therefore diluting or counteracting anxiety about these phenomena. Are we a death-accepting or a death-denying society? What do you think?

FEARING DEATH?

In this book the point is made often that death per se has no meaning other than that which people give it. And some Americans give it a positive meaning. For example, B. F. Skinner, the psychologist who popularized behavior modification, approached his death from leukemia with no apparent regrets. The 86-year-old Skinner, just a few weeks before he died, said with a laugh (1990), "I will be dead in a few months, but it hasn't given me the slightest anxiety or worry or anything. I always knew I was going to die." Likewise, 75-year-old psychiatrist Timothy Leary noted, a few months before his death from prostate cancer in 1996, that he was "waiting for his graduation." Leary (Mansnerus, 1995) stated that he was "looking forward to the most fascinating experience in life, which is dying." He said that dying must be approached the way that life is lived—with curiosity, hope, fascination, courage, and the help of your friends.

Anthropologists will tell you that there are many cultures that share the attitude of these prominent Americans. For example, the Chinese look upon death with pleasure rather than fear. This view is evident in the quotation from a fourth-century Chinese philosopher shown in the Words of Wisdom box. It is natural to think that the fact that we are going to die should indeed influence how we live, fearful or otherwise, notes Yale philosopher Shelly Kagan (2012).

Although there are certainly many cultures and many Americans who approach death without fear, a surprising number of cultures, including our own, attach fearful meanings to death and death-related situations. Why is it that so many people have less-than-positive views of death? Philip Slater (1974) suggested that in societies in which individuality is a high cultural priority, the fear of death logically follows. People in industrial and postindustrial societies lose the connectedness based on community provided in other societies. Thus, the fear of death may be the price paid for living in a society whose ideology rests on the type of individualism experienced in the United States. When in his late 80s, German sociologist Norbert Elias (1985) wrote *The Loneliness of Dying*, in which his gravest concern about dying was not physical suffering but the progressive withdrawal of those close to death to a setting behind the scenes of social life. His fear was that of dying alone in a hospital ward isolated under a barrage of medical equipment and surrounded by rigid hospital rules. His fear was loneliness.

Death in the United States is viewed as fearful because Americans have been systematically taught to fear it. Horror movies portray death, ghosts, skeletons, goblins, bogeymen, and ghoulish morticians as things to be feared. *Sesame Street* tries to create a more positive view of monsters by showing children being befriended by Grover, Oscar, and Cookie Monster. With the possible exception of Casper "the Friendly Ghost," death-related fantasy figures have not received much in the way of positive press. Instead of providing positive images, our culture has chosen to reinforce fearful meanings of death. Cemeteries are portrayed as eerie, funeral homes are to be avoided, and morgues are scary places where you "wouldn't be caught dead."

The fear of death may be incorporated in the minds of some individuals with the fear of being buried alive. To be taken into the blackness when life is extinct is a dreadful enough prospect, noted Robert Wilkins (1996), but for the Grim Reaper to come calling before the appointed hour is to condemn the yet-living to a seeming eternity of suffocative horror. Though scattered accounts of individuals being

WORDS OF WISDOM | NOTHING IS LACKING IN DEATH

Chuang Tzu was dying, and his disciples wanted to give him a grand burial. But Chuang said that all the ceremonial needs were already there: "Heaven and earth will be my outer and inner coffin, the sun and moon my pair of jade discs, the stars in their constellations will be my pearl jewels, the myriad things of the world will be my mourners. Nothing is lacking. What could you add?"

Taken from *Death: Philosophical soundings* (p. 123), by H. Fingarette, 1996, Chicago, IL: Open Court.

buried alive are noted over the past several centuries (Wilkins, 1996), most medical personnel remain skeptical. When the Paris cemetery Les Innocents was moved from the center of the city to the suburbs in the latter part of the 19th century, the number of skeletons found face down in their coffins convinced the lay public that premature burial was common. Skeptics explain such "movement" as being the result of externally imposed movements such as the coffin bumping into walls as it is being negotiated down narrow stairwells or of sudden movements produced by bolting funeral horses. Nonetheless, the fear of being buried alive has persisted over the centuries. Such fear indeed might contribute to the overall fear of death.

One of the reasons why classes in dying and death often incorporate field trips to funeral homes and cemeteries is to confront negative death meanings and fantasies with firsthand, objective observations. The preparation room at the mortuary is a good example. If you have never been to one, think about the mental image that comes to your mind. For many, the preparation room is a place that one approaches with fear and caution—something like Dr. Frankenstein's laboratory, complete with bats, strange lighting, body parts, and naked dead bodies. The great disappointment for most students as they walk through the door is that they find a room that looks somewhat like a physician's examination room. "Is that all there is?" is a comment often heard after visiting the preparation room.

Sociologist Erving Goffman (1959) observed that first impressions are unlikely to change and tend to dominate the meanings related to subsequent social interaction patterns and experiences. Some of us want to retain untrue fearful meanings of death, even when confronted by positive images. On a recent field trip to a funeral home, one student refused to enter the preparation room. While other students were hearing a description of embalming procedures, she discovered a bottle labeled "skin texturizer" and became nauseated. Other students had a difficult time understanding her problem because they had had positive experiences.

Fearful meanings can also be ascribed to death because of a traumatic death-related experience. Being a witness to a fatal auto accident, discovering someone who has committed suicide, or attending a funeral where emotional outbursts create an uncomfortable environment for mourners can increase death anxiety for individuals. However, such occurrences are rather uncommon for most people and do not account for the prevalence of America's preoccupation with death fears.

As will be discussed in Chapter 3, when the life cycle and death fear are discussed, the age group with the highest death anxiety tends to be middle-aged men, while those 65 years and older tend to have a relatively low death anxiety. The elderly seem somewhat accepting that they have lived their "life span" and that one does not live forever. Yet, middle-aged men (seemingly more than women) are at a point where they realize that they are *approaching* the end and may feel that time is running out before they are ready, thus may be anxious. For younger individuals, particularly college-age students, death is not something on the horizon for them personally, as their life is before them not behind them, thus their death anxiety tends to not be high. The occasional death of a peer is a reminder that death can occur within any age group, yet the frequency of such is overwhelmingly with the elderly.

CONTENT OF DEATH FEARS   When one speaks of death fear or death anxiety, it is assumed that the concept is unidimensional and that there is a consensus about its

meaning. Such is not the case, however, because two persons may say that they fear death, and yet the content of their fears may not be shared. Death anxiety is a multidimensional concept and is based upon the following four concerns: (1) the death of self, (2) the deaths of significant others, (3) the process of dying, and (4) the state of being dead. This model's more elaborate form (Leming, 1979–1980) shows eight types of death fears that can be applied to the death of self and the deaths of others:

1. Dependency
2. The pain in the dying process
3. The indignity in the dying process
4. The isolation, separation, and rejection that can be part of the dying process
5. The leaving of loved ones
6. Concerns with the afterlife
7. The finality of death
8. The fate of the body

From the eight dimensions of death anxiety (Table 2.1) we can see that the content of fear will be influenced by whose death the individual is considering. From a personal death perspective, one may have anxiety over the effect that one's dying (or being dead) will have on others. There might also be private worries about how one might be treated by others. From the perspective of the survivor, the individual may be concerned about the financial, emotional, and social problems related to the death of a significant other.

Because many factors related to the experience of death and death-related situations can engender fear, we would expect to find individual differences in the

TABLE 2.1 | THE EIGHT DIMENSIONS OF DEATH ANXIETY AS THEY RELATE TO THE DEATHS OF SELF AND OTHERS

Self	Others
Process of Dying	
Fear of dependency	Fear of financial burdens
Fear of pain in dying process	Fear of going through the painful experience of others dying
Fear of indignity in dying process	Fear of not being able to cope with physical problems of others
Fear of loneliness, rejection, and isolation	Fear of being unable to cope emotionally with problems of others
Fear of leaving loved ones	Fear of losing loved ones
State of Being Dead	
Afterlife concerns	Afterlife concerns
Fear of unknown	Fear of the judgment of others—"What are they thinking?"
Fear of divine judgment	Fear of ghosts, spirits, devils, etc.
Fear of the spirit world	Fear of never seeing the person again
Fear of nothingness	Fear of floating in space—being in a void all alone away from others
Fear of the finality of death	Fear of the end of a relationship
Fear of not being able to achieve one's goals	Guilt related to not having done enough for the deceased
Fear of the possible end of physical and symbolic identity	Fear of not seeing the person again
Fear of the end of all social relationships	Fear of losing the social relationship
Fear of the fate of the body	Fear of death objects
Fear of body decomposition	Fear of dead bodies
Fear of being buried	Fear of being in cemeteries
Fear of not being treated with respect	Fear of not knowing how to act in death-related situations

© Cengage Learning

type and intensity of death fear, including social circumstance and past experiences. However, with all of the potential sources for differences, using a death fear scale developed by Michael Leming, consistently high scores are found for the fears of dependency and pain related to the process of dying, and relatively low anxiety scores are found for the fears related to the afterlife and the fate of the body (see Practical Matters box).

Approximately 65 percent of the more than 1,000 individuals surveyed had high anxiety relative to dependency and pain, and only 15 percent had the same

PRACTICAL MATTERS | LEMING FEAR OF DEATH SCALE

Read the following 26 statements. Decide whether you strongly agree (SA), agree (A), tend to agree (TA), tend to disagree (TD), disagree (D), or strongly disagree (SD) with each statement. Give your first impression. There are no right or wrong answers. In any given subscale, a score of 3.5 or higher means slightly fearful of death.

I. Fear of Dependency
1. I expect other people to care for me while I die.
   SA A TA TD D SD
   1 2 3 4 5 6

2. I am fearful of becoming dependent on others for my physical needs.
   SA A TA TD D SD
   6 5 4 3 2 1

3. While dying, I dread the possibility of being a financial burden.
   SA A TA TD D SD
   6 5 4 3 2 1

4. Losing my independence due to a fatal illness makes me apprehensive.
   SA A TA TD D SD
   6 5 4 3 2 1
   Total of 4 scores _____ divided by
   4 _____

II. Fear of Pain
5. I fear dying a painful death.
   SA A TA TD D SD
   6 5 4 3 2 1

6. I am afraid of a long, slow death.
   SA A TA TD D SD
   6 5 4 3 2 1
   Total of 2 scores _____ divided by
   2 _____

III. Fear of Indignity
7. The loss of physical attractiveness that accompanies dying is distressing to me.
   SA A TA TD D SD
   6 5 4 3 2 1

8. I dread the helplessness of dying.
   SA A TA TD D SD
   6 5 4 3 2 1
   Total of 2 scores _____ divided by
   2 _____

IV. Fear of Isolation/Separation/Loneliness
9. The isolation of death does not concern me.
   SA A TA TD D SD
   1 2 3 4 5 6

10. I do not have any qualms about being alone after I die.
    SA A TA TD D SD
    1 2 3 4 5 6

11. Being separated from my loved ones at death makes me anxious.
    SA A TA TD D SD
    6 5 4 3 2 1
    Total of 3 scores _____ divided by
    3 _____

V. Fear of Afterlife Concerns
12. Not knowing what it feels like to be dead makes me uneasy.
    SA A TA TD D SD
    6 5 4 3 2 1

13. The subject of life after death troubles me.
    SA A TA TD D SD
    6 5 4 3 2 1

14. Thoughts of punishment after death are a source of apprehension for me.
    SA A TA TD D SD
    6 5 4 3 2 1
    Total of 3 scores _____ divided by
    3 _____

VI. Fear of the Finality of Death
15. The idea of never thinking after I die frightens me.
    SA A TA TD D SD
    6 5 4 3 2 1

16. I have misgivings about the fact that I might die before achieving my goals.
    SA A TA TD D SD
    6 5 4 3 2 1

17. I am often distressed by the way time flies so rapidly.
    SA A TA TD D SD
    6 5 4 3 2 1

18. The idea that I may die young does not bother me.
    SA A TA TD D SD
    1 2 3 4 5 6

continued

19. The loss of my identity at death alarms me.
SA A TA TD D SD
6 5 4 3 2 1
Total of 5 scores _____ divided by
5 _____

VII. Fear of Leaving Loved Ones

20. The effect of my death on others does not trouble me.
SA A TA TD D SD
1 2 3 4 5 6

21. I am afraid that my loved ones are emotionally unprepared to accept my death.
SA A TA TD D SD
6 5 4 3 2 1

22. It worries me to think of the financial situation of my survivors.
SA A TA TD D SD
6 5 4 3 2 1
Total of 3 scores _____ divided by
3 _____

VIII. Fear of the Fate of the Body

23. The thought of my own body decomposing does not bother me.
SA A TA TD D SD
1 2 3 4 5 6

24. The sight of a dead body makes me uneasy.
SA A TA TD D SD
6 5 4 3 2 1

25. I am not bothered by the idea that I may be placed in a casket when I die.
SA A TA TD D SD
1 2 3 4 5 6

26. The idea of being buried frightens me.
SA A TA TD D SD
6 5 4 3 2 1
Total of 4 scores _____ divided by
4 _____

Developed by Michael Leming (1979–1980). Religion and death: A test of Homans' thesis. *Omega, 10*(4), 347–364.

high level of anxiety relative to concerns about the afterlife and the fate of the body (Leming, 1979–1980). Thus, it is the process of dying—not the event of death—that causes the most concern.

DEATH FEARS, GENDER, AND AGE Two consistent findings associated with the literature on death fear or anxiety are the age and gender effects (Russac et al., 2007). The age effect refers to young adults often reporting higher levels of concern over mortality than older adults. The gender effect refers to women typically reporting higher levels of death anxiety than men. Yet, gender differences in death fear are one of the most frequently noted and poorly understood findings in thanatology (Dattel & Neimeyer, 1990). Judith Stillion (1985) suggested that the discrepancy in these findings simply reflects the greater tendency of women to admit troubling feelings.

Literature (Fortner & Neimeyer, 1999) on the relationship between death anxiety and age demonstrates an inverse relationship (i.e., the older the person, the less death anxiety). This is especially true when middle-aged persons are compared with the elderly—death anxiety declines from middle age to older age. Within the elderly cohort, however, this trend with age does not hold as death anxiety stabilizes during the final decades of life, according to B. V. Fortner and R. A. Neimeyer. The paradox presented here via the age effect is the question of why younger people, with their lives before them, should be overly concerned about death, yet older persons, much closer to the end of their lives, express less anxiety. Robert Kastenbaum (2000) suggests that people become less anxious with age because death does not threaten as many of their values and/or there is a continued developmental process through which individuals "come to terms" with mortality.

Other research (Rasmussen & Brems, 1996) confirmed that age and psychosocial maturity are significantly and inversely related to death anxiety. Psychosocial maturity was found to be a stronger predictor than age, however, perhaps suggesting why some studies have revealed only moderate correlations between age and death anxiety. Possibly age alone cannot account for the decrease in death anxiety among the elderly; the combination of aging and the achievement of greater psychosocial maturity may serve to decrease death anxiety.

In summary, death fears are not instinctive; they exist because our culture has created and perpetuated fearful meanings and ascribed them to death. They are also a function of the fact that death is not an ordinary experience challenging the order of everyday life in society. They are a function of occasional firsthand encounters with deaths so unusual that they become traumatic.

RELIEVING DEATH ANXIETY THROUGH RELIGION George Homans (1965) suggested that when individuals encounter death, the **death anxiety** they experience is basically socially ascribed (learned). Death fears can be likened to the fears of other things—snakes or electricity, for example. If we believe that we are in a dangerous setting,

© Andreas Gradin./Shutterstock.com

Horror movies gross large profits for the movie industry. Not unlike "rubber-neckers" on the highway, the sight of a bloody scene in a movie appears to be fascinating to many people. To be frightened by death and horror seemingly is thrilling to some individuals.

| WORDS OF WISDOM | OF DEATH |

Men fear Death as children fear to go in the dark. As that natural fear in children is increased with tales, so is the other.

A man would die, though he were neither valiant nor miserable, only upon a weariness to do the same thing so oft over and over.

From *Of death*, by Francis Bacon (1561–1626).

we react accordingly. Some religions, with an emphasis on immortality of the soul and a belief in a coming judgment, increase the level of death anxiety for individuals. However, after individuals have fulfilled the requisite religious or magical ceremonies, they experience only a moderate amount of anxiety. Homans brings the perspectives of anthropologists Bronislaw Malinowski and A. R. Radcliffe-Brown together by drawing four conclusions:

1. Religion functions to relieve anxiety associated with death-related situations.
2. Death anxiety calls forth religious activities and rituals.
3. To stabilize the group of individuals who perform these rituals, group activities and beliefs provide a potential threat of anxiety to unite group members through a "common concern."
4. This secondary anxiety may be effectively removed through the group rituals of purification and expiation.

Summarizing the relationship between **religiosity** and death anxiety, we can arrive at the following theoretical assumptions:

1. The meanings of death are socially ascribed—death per se is neither fearful nor nonfearful.
2. The meanings that are ascribed to death in a given culture are transmitted to individuals in the society through the socialization process.
3. Anxiety reduction may be accomplished through social cooperation and institutional participation.
4. Institutional cohesiveness in religious institutions is fostered by giving participants a sense of anxiety concerning death and uniting them through a common concern.
5. If the religious institutions are to remain viable, they must provide a means for anxiety reduction.
6. Through its promise of a reward in the afterlife and its redefinition of the negative effects of death upon the temporal life of the individual, religion diminishes the fear that it has ascribed to death and reduces anxieties that are ascribed to death by secular society.

To test the empirical validity of these assumptions, Michael Leming (1979–1980) surveyed 372 randomly selected residents in Northfield, Minnesota, concerning death anxiety and religious activities, beliefs, and experiences. Subjects were divided into four categories based on a religious commitment scale developed by

C. Glock and R. Stark (1966) and J. Faulkner and G. F. DeJong (1966). Approximately 25 percent of the respondents were placed into each category—the first consisted of those who were the least religious, and the fourth consisted of those persons who were the most religious.

The graph shown in Figure 2.1 gives the mean fear of death scores for each of the levels of religious commitment. The relationship between the variables of religiosity and death anxiety is **curvilinear**—persons with a moderate commitment to religion have added to the general anxiety that has been socially ascribed to death from secular sources. The persons with a moderate commitment receive only the negative consequences of religion—what Radcliffe-Brown called a "common concern." These persons acquire only the anxiety that religion is capable of producing and none of the consolation. On the other hand, the highly committed individual has the least anxiety concerning death. Religion, as Malinowski predicted, provides individuals with a solace when they attempt to cope with death attitudes.

Religiosity seems to serve the dual function of "afflicting the comforted" and "comforting the afflicted." We have discovered that religion, when accompanied with a high degree of commitment, not only relieves the dread that it engenders, but also dispels much of the anxiety caused by the social effects of death. Recent studies (Ens & Bond, 2007) found that as inner conviction and spirituality increased, death anxiety decreased; thus highly spiritual persons tend to have lower death anxiety. With the exception of the fear of isolation, Leming found that persons who had the strongest religious commitment were the least fearful with regard to the various types of death concern. Furthermore, in each of the eight death fear types, the strength of commitment was the most significant variable in explaining the relationship between religion and the fear of death. Robert

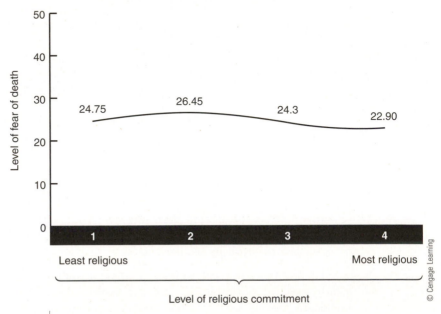

FIGURE 2.1 | Mean Fear of Death Scale by Level of Religious Commitment

Kavanaugh (1972, p. 14) seems to have empirical support for his statement, "The believer, not the belief, brings peace."

CONTEMPLATING ONE'S OWN DEATH

Psychiatrist Robert E. Neale, in his book *The Art of Dying* (1973), suggested that the reader look in the mirror and imagine herself or himself as a corpse. Such a suggestion should certainly catch the attention of the reader! Neale asks the reader to reflect on her or his own intimacy with death by answering a series of questions (see Practical Matters box).

A lack of intimacy with death is not irrevocably bad, noted Neale, but it can lead to making too much or too little of death. There are no right answers to whether one thinks too often of death or thinks of death too infrequently. Perhaps some individuals make too little of death, whereas others make too much of it. We are unable to distinguish between them in any concrete case, says Neale.

To contemplate the meaning of one's own death, philosopher Herbert Fingarette (1996) liked Tolstoy's story "The Death of Ivan Ilyich." Initially, Ivan Ilyich reacts to the possibility of impending death in a way that perhaps most of us do, with denial. He thinks that maybe this is not death after all, but the pain is merely the effect of some malfunctioning organ that the doctors can probably correct. Soon Ilyich realizes that these hopes are feeble straws, yet he is impelled to grasp at them to continue hoping. At the end of the story Ilyich is forced to confront his terror, and a voice within him asks what he wants. He answers that he wants to live, pleasantly, as before. This dialogue with himself leads Ivan Ilyich to a reexamination of his life, a review of the "pleasures" for which he has lived. He realizes that he has deceived himself and now sees that his life has been a commitment to a selfish inhumanity. He had enjoyed exercising power without compassion. Facing death, he comes to understand the truth of his life. Fingarette notes that Ivan's "confrontation with death" turns out to be a retrospective exploration and revelation to him of the meaning of his life. Seeing the truth of his life, he is able at last to face death and see its truth. At the end, the dying Ivan Ilyich asks himself, "Where is it? What death?" Tolstoy answers, "There was no fear because there was no death."

Philosopher Martin Heidegger (1927/1962) acknowledged that we know we are going to die, yet we accept this inauthentically. Our social existence represents

---

| PRACTICAL MATTERS | QUESTIONS TO REFLECT ON ONE'S OWN INTIMACY WITH DEATH |

1. Have you ever seen an embalmed body? Yes __ No__
2. Have you ever seen an unembalmed body? Yes __ No __
3. Have you ever seen a person die? Yes __ No __
4. Have you ever been in a situation in which you thought you would die? Yes__ No__

5. Have you had a member of your family or close friend die? Yes __ No __
6. Have you ever attended a funeral? Yes __ No __
7. When you were a child, was death talked about in your family? Yes __ No __

an evasive attitude to death. If we are to live authentically, therefore, we must realize that death is not the end of life, but rather the inner possibility of being. Accordingly, death, despite its universality, has an individualizing effect: We each have to die (and therefore live) in our own way.

Malcolm Boyd (1997) stated that the worst way to live is to fear living and the worst way to die is to fear dying. He suggested confronting fears and staying open to others, opting for service instead of selfishness, learning to forgive, relinquishing hatred, and practicing loving.

You might consider writing your own **obituary** to allow yourself to look at your own values and priorities in life. What is it you consider to be important? This assignment will literally force you to consider that someday you indeed will die. What is it you would like to have accomplished during your lifetime? Though your life is ahead of you, if a typical college-aged student, project ahead. In your obituary consider how you would like to be remembered by others. Were your goals in life to control others, to be rich, to give orders rather than take them, to be a status-seeker, and so forth? Or would you prefer to be remembered in other ways, such as a compassionate individual who went about doing good, trying to help those in life who had less than you did, devaluing material aspects of life and seeking something more meaningful than tangible products?

After writing your obituary, stop and reflect on what you have said about yourself. Do you feel good about how you have said you wish to be remembered? If yes, then perhaps you are content with your life and can die feeling good about self, having reached Erik Erikson's last stage of the life cycle of integrity, as discussed in Chapter 3. If no, then maybe you would wish to reevaluate where you are going with your life. An exercise such as writing one's own obituary is not an easy assignment and one that you are not often asked to do, but it can be most helpful in deciding if you are heading in the direction in which you wish to go. As with Ivan Ilyich, you may see the "truth" of your life and thus be able to face death and indeed have "no fear."

CONCLUSION

Although the American way of dying is being discussed and researched more today than in previous decades, this discussion often poses as many questions as answers. Aristotle (trans. 1941, p. 975) observed long ago that "death is the most terrible of all things"; yet it will not go away and, thus, we must learn to cope with it. The following questions do not have simple and straightforward answers: When does death take place? Who should determine the timing of a particular death? Who in society should be responsible for defining the meaning of life and death? When is a death an accident, and when is it a suicide?

Americans have developed a paradoxical relationship with death: We know more about the causes and conditions surrounding death, but we have not equipped ourselves emotionally to cope with dying and death. The American way of dying is such that avoiding direct confrontation with dying and death is a real possibility for many persons. What we need is the ability to both understand and cope with dying and death. The purpose of this book is to provide an understanding of dying, death, and bereavement that will assist individuals to better cope with their own deaths and with the deaths of others. As Rabbi Earl Grollman noted (Foderaro, 1994): "The important thing about death is the importance of life. Do what you have to do now. Live today meaningfully."

SUMMARY

1. The definition of death varies from culture to culture and throughout history. It has vacillated between the centralist theory that places the life force in a single organ and the decentralist theory that places the life force in every cell of the body. The most recent definition uses the centralist theory of brain death. However, establishing the criteria for brain death has not been a simple process. The definition of death is further complicated by the fact that there is not a consensus on the meaning of life and death.

2. Death-related meaning permits sharing death-related behavior. The death-related behaviors of the dying individual and of those who are significant to that individual are in response to meaning, relative to the audience and to the situation. It is a phenomenon that is socially created, not biologically predetermined.

3. Because most people participate in various social groups, they are involved in many different interaction patterns. Consequently, even though it is but one biological body that dies, many "role holes" or vacancies are left by the death of a single individual. Bodies die and so do social relationships and social networks.

4. Death meaning includes evaluations of whatever those involved decide to evaluate. Evaluators may include values believed to be absolute, abstract, and situational. Defining values such as "living is always preferable to dying" as abstract rather than as absolute helps explain the relativity of situational values and is likely to lead to fewer adjustment problems when an individual confronts dying.

5. Dying is a social process. The person who is dying is living and is involved in living experiences with others.

6. Evolving death meanings are part of the general cultural changes taking place in contemporary society. Much of this change is centered around a discounting of biological influences upon social behavior.

7. The American experience of death has taken us from "living death" (1600–1830), characterized by a constant preparation for and fear of death, to the "dying of death" (1830–1945), an era in which death was denied, removed from daily activities, and relegated to the funeral homes, and the emphasis was on ways to postpone or preempt death. The next stage, the "resurrection of death" (1945–the present), was prompted by the dropping of the atomic bomb, when death was once again on everyone's mind and nuclear disaster was imminent. Today, we see evidence of the denial of death from the Victorian era, as well as the fear of death from the Puritans.

8. Evidence supporting the view that the United States is basically a death-denying society includes the widespread use of euphemisms for death, taboos on death conversations, the fascination with cryonics, caskets built for comfort, cosmetically enhanced corpses, and internment after guests leave the gravesite. Yet there is not agreement among thanatologists today on the issue of death denial or death acceptance as a society.

9. Death in the United States is feared because we have been taught to fear it.

10. Death anxiety is a multidimensional concept. There are eight types of death fears: dependency; pain of dying; indignity of dying; isolation, separation, and rejection; leaving loved ones; concerns with the afterlife; finality of death; and fate of the body.

11. Religion can relieve death anxiety if an individual is highly committed to it. A moderate commitment will increase anxiety about death.

12. One way to prepare for death and reduce anxiety is by contemplating one's own death.

DISCUSSION QUESTIONS

1. How has the definition of death changed over the years? What complications has this created for the American way of dying?

2. Discuss the differences between biological and symbolic death.

3. What arguments are offered in the rejection of this premise: "In death, biology is primary; meaning is peripheral"? Evaluate and discuss.

4. Each act of dying has three interconnected characteristics: shared, symboled, and situated. How does this relate to the statement that more than a biological body dies?

5. Discuss the implications of the following quote: "Even though it is but one biological body that dies, many 'role holes' or vacancies are left by the death of that one person."

6. In making decisions about the death meaning, how does the treatment of the dying patient affect that patient's understanding of death and his or her role in the dying process?

7. Discuss the evolution of death experiences in American life from 1600 to 2015.

8. What effect has the dropping of the atom bomb had on American death conceptions?

9. What factors have contributed to the American avoidance of dying and death?

10. Discuss whether you think the United States is basically a death-denying or a death-accepting society.

11. What is meant by this statement: "Death anxiety is a multidimensional concept"?

12. As one faces imminent death, one becomes increasingly aware of the social nature of life. This increase in awareness can lead to a life review during which one realizes how extensively one lives with, through, and for others. Speculate as to why this change in perspective takes place.

GLOSSARY

Brain Death: The brain is totally and irreversibly dead. This is sometimes referred to as the Harvard definition of death.

Cryonics: A method of subjecting a corpse to extremely low temperatures through the use of dry ice and liquid nitrogen.

Curvilinear: Referring to a type of nonlinear relationship between two variables where at a certain point, associated with the increasing values in the independent variable, the relationship with the dependent variable changes. A scattergram graph of this relationship will look like either a letter U or an inverted U.

Death Anxiety: A learned emotional response to death-related phenomena characterized by extreme apprehension; used synonymously with *death fear*.

Dyad: Two units regarded as a pair (e.g., a husband and a wife).

Euphemism: A word or phrase that is considered less distasteful than other words or phrases.

Medicalization of Death: Modern medicine defines illness and health procedures needed (for example, tonsil removal was fairly routine with children in the 1940s and 1950s because medical science deemed it so) and determines procedures for the dying which may prolong or shorten life. Not unlike childbirth via convenient-for-the-doctor Caesarean sections, modern medicine somewhat controls when we die and can control premature deaths.

Mores: Ways of society that are felt to be for the good of society. These are "must" behaviors that have stronger sanctions than a folkway (e.g., eating three meals per day) but that are not as severe as laws.

Obituary: Notice of a death, usually with a brief biography.

Religiosity: The extent of interest, commitment, or participation in religious values, beliefs, and activities.

Significant Other: A person to whom special significance is given in the process of reaching decisions.

SUGGESTED READINGS

Dickinson, G. E., & Leming, M. R. (Eds.). (2014). *Annual editions: Dying, death, and bereavement 13/14* (14th ed.). New York: McGraw-Hill. An anthology covering topics on issues in dying and death.

Faust, D. G. (2008). *This republic of suffering: Death and the American Civil War*. New York: Alfred A. Knopf. During the Civil War, approximately 620,000 soldiers died. Faust details the logistical challenges involved when thousands were left dead and chronicles the efforts to identify, reclaim, preserve, and bury battlefield dead. She shows how the war victimized civilians through violence that extended beyond battlefields.

Gillick, M. R. (2000). *Lifelines: Living longer, growing frail, taking heart*. New York: W. W. Norton & Company. Takes the reader on a journey into the end-of-life stories of four composite people. Public policy discussions of frailty and disability in an aging world are also included.

Hayslip, B., & Peveto, C. A. (2005). *Cultural changes in attitudes toward death, dying, and bereavement*. New York: Springer Publishing Company. The authors discuss the factors that influence death attitudes and the impact of sociocultural change on death attitudes.

Merridale, C. (2001). *Night of stone: Death and memory in twentieth-century Russia*. New York: Viking. During the 20th century, Russia, Ukraine, and other territories of the former Soviet Union experienced more bloodshed and violent death than anywhere else on earth—50 million dead in an epic of destruction that encompassed war, revolution, famine, epidemic, and political purges. This book is an emotionally wrenching picture of untold millions who were forbidden to mourn their loved ones.

Tolstoy, L. (1960). *The death of Ivan Ilyich and other stories*. New York: New American Library (Original work published 1886). Provides an insightful view of the difficulties and confusions that may have an impact on a survivor and/or the dying person.

Volk, T. (2002). *What is death? A scientist looks at the cycle of life*. New York: Wiley. An organic and holistic look at the meaning of death and its relationship to life.

Of death I try to think like this—
The Well in which they lay us
Is but the Likeness of the Brook
That menaced not to slay us,
But to invite by that Dismay
Which is the Zest of sweetness
To the same Flower Hesperian,
Decoying but to greet us—

I do remember when a Child
With bolder Playmates straying
To where a Brook that seemed a Sea
Withheld us by its roaring
From just a Purple Flower beyond
Until constrained to clutch it
If Doom itself were the result,
The boldest leaped, and clutched it—

GROWING UP WITH DEATH/ GROWING OLD WITH DEATH

CHAPTER **3**

© Kim Reinick/Shutterstock.com

In the Emily Dickinson poem above, death is seen not as the "menace," but rather as that which "invites" us (Cooney, 1998). Invites us where and to do what?: to the "Flower Hesperian" (the evening flower); to the "Zest of sweetness" (is this not life itself?). The "boldest" among us leap to clutch at the flower of life. Death disguised ("decoying") as a "roaring" "Sea" is there "to greet us" and to provide that obstacle to overcome. In this chapter we will talk about death with the hope that we can help each other better understand death conceptualizations at different stages of the life cycle—from childhood through old age.

Sigmund Freud traced conceptions of death to our earliest feelings concerning sexuality and to our fears of being punished for them. Alfred Adler had several brushes with death himself and, as a child, suffered from a debilitating disease. When Adler formulated his theories concerning the human psyche and its development, he attributed our need to strive and overcome to our early sensitivity to weakness and death.

The **ego psychologists** later departed somewhat from Freud and credited the individual with a greater ability to manage the stresses and problems of life. Yet, they recognized that humans raise a whole set of defenses against the idea of death. Both children and adults have the power to distort their perceptions according to their needs. For example, individuals have ways of denying harsh or painful thoughts, and people often believe and see what they want to believe and see, thus transforming images of their imaginations into reality.

In his Pulitzer Prize-winning book entitled *The Denial of Death,* Ernest Becker (1973) argued that fear and denial of death are basic dynamics for everyone. He asserted that we struggle to find meaning in life through heroic efforts. If we discover that heroic efforts are not possible, the dilemma is avoided by building elaborate systems to explain the dilemma away. Some even flee into neurosis or a psychotic break. Becker felt that the fear of death is a basic problem of meaning with which we all struggle.

Though the subject of death was not a major concern in his writings, Swiss developmental psychologist Jean Piaget was probably instrumental in nudging psychologists to employ better methods of research in the developmental approach toward understanding concepts of death (Ginsberg & Opper, 1988). Through keen observations of his children and others, Piaget postulated that it is not until the early teen years that one is capable of genuinely abstract thought processes.

Since the publication of Herman Feifel's *The Meaning of Death* in 1959, interest and research on the conceptualization of death have grown. Representing **psychoanalysis, behaviorism, humanism,** and other points of view, much has now been written on a **developmental approach** to death attitudes and awareness.

It is not our intent in this chapter to suggest that age should be seen as the sole determiner of one's death concept. Many other factors influence cognitive development, such as level of intelligence, physical and mental well-being, previous emotional reactions to various life experiences, religious background, other social and cultural forces, personal identity and self-worth appraisals, and exposure to death or threats of death. Though an age-based approach is followed in this chapter, other important factors should not be ignored.

| LISTENING TO THE VOICES | NOT DEAD, JUST "HIBERNATING" |

A college student reflecting on a childhood experience as a six-year-old gives this account of her pet's death and burial; the story suggests that she was not aware of the permanency of death as a child.

> I had a cute little hamster. I think we had been talking about hibernation in school, because when I came home and found the hamster lying there, I thought he was hibernating! My next door neighbor, Sharon, came over and said that if we warmed him up then he would come back to life. So I put him in my electric blanket to see if he would wake up. He never did and by that time my mom realized what had happened, so we had to bury him. My brother and I put him in the shoe box, put a blanket in there to keep him warm and a picture of us to keep him company. Then we buried him in the backyard.

From *First childhood death experiences* (pp. 176–177), by G. Dickinson, 1992, *Omega, 25.*

CHILDHOOD

Despite a new openness to death in the United States, children are generally not encouraged to express themselves on the topic of dying and death. Children approach the unknown in a rather naïve way. Their very naiveté makes their questions very difficult to answer. Children usually have no prior experience on which to base their reactions; thus, their earliest experiences are unique to each of them. They typically watch to see how others react following a death, then react accordingly. With no norms to follow, their behavior is often simply a mimic of others' behavior.

To even be talking about children and death within the same sentence seems inappropriate, because death among children is much less frequent than death among older adults, yet death can occur with children and adolescents, as noted in Table 3.1. There are about twice as many deaths in the first year of life as there are in the next 13 years total. Then the death rate rises rapidly following puberty because of the large number of deadly accidents, homicides, and suicides in the 15–25-year age group (UMM, 2011). One learns to crawl before walking; typically one is supposed to grow up (and indeed grow old) before dying. Nonetheless, children are confronted with death, probably not their own but that of others, and they need to understand and cope. Dying and death have an impact on our lives throughout the life cycle. Death is in the news, in the movies, in popular songs, and on television. The subject of death at all ages is difficult to avoid. It is ubiquitous.

HOW DO CHILDREN LEARN ABOUT DEATH?

Children learn about death in a variety of ways. The first death experience is often that of a dead bird or any road kill or a dead insect (ant, roach, or beetle) or spider or a pet. In a study of college students' recollections of their first death experiences, George Dickinson (1992) found that 28 percent of his subjects' first death experiences involved a pet.

The death of an animal is a good opportunity for parents to help children learn their first lessons of death by answering questions. The children may have had the

TABLE 3.1	TOP THREE CAUSES OF DEATH FOR CHILDREN AND ADOLESCENTS BY AGE GROUP*

0–1 Year of Age

1. Developmental and genetic conditions present at birth

2. Sudden infant death syndrome

3. All conditions associated with prematurity and low birth rate

1–4 Years of Age

1. Accidents

2. Developmental and genetic conditions present at birth

3. Cancer

5–14 Years of Age

1. Accidents

2. Cancer

3. Homicide

15–24 Years of Age

1. Accidents

2. Homicide

3. Suicide

*Taken from *Death among children and adolescents*. (2011). University of Maryland Medical Center web site at www.umm.edu. Accessed on October 5, 2012.

pleasure of watching the family pet grow from a small puppy, kitten, or guinea pig into a large dog, cat, or guinea pig. Such an experience is typical of the maturation process: After birth, one grows, finally reaches full size, and then ages and dies. With pets having a shorter lifespan than humans, children often observe a pet going through the life cycle. Thus, a pet's death is a realistic orientation to death within the animal kingdom.

Of course, the death of a family member, such as a grandparent (the greatest number of earliest death recollections of these college students involve grandparents, according to Dickinson, 1992), can bring death too close to home. The child then learns quickly by observing others and experiences the various feelings she or he has. Or the first death experience can occasionally be that of a classmate or a teacher, thus validating the importance of school systems socializing teachers to death issues and encouraging them to serve as good role models for the students by being communicative and open to their questions.

More impersonal death experiences would include those in the mass media: movies, television, and the news. Though exposure to death by these means may occur often and sometimes continuously, the experience is usually less personal. Nonetheless, deaths publicized by the media can be bothersome to children, especially if open communication is not available to vent feelings. The Oklahoma City bombing; the Columbine High School shootings; terrorist attacks on the United States,

Great Britain, and Spain; Hurricane Katrina on the Gulf Coast; the tsunami horror from the Indian Ocean earthquake; and the Boston Marathon bombings are disasters for which there are direct victims, family members, first responders, and indirect victims (all the rest of us). Media coverage of disasters is always intense. In crisis work with children, pictures or print media were found to be the most powerful stimulant because the children could study the material for any length of time (Raphael, 2003).

Children are also introduced to death through religion. No matter what religious orientation one has, death plays a role. For Jews or Christians, for example, the Bible is filled with death accounts and talk of death: the celebration of Easter for the Christian community is about death.

Children's storybooks sometimes contain accounts of death, some of which are rather gruesome, for example, Mother Goose nursery rhymes. Other more innocent stories may include death themes and may be written specifically for children to teach them about death. Children's death concepts change with age—older children are considered to have more accurate death concepts, are affected by cognitive level (those more cognitively mature have relatively more accurate death concepts), and are affected by experience (those having more experience with death are more likely to believe in personal mortality).

PERSONAL EXPERIENCES Children most often confront death at an early age. Their vivid recollections of these first death experiences testify to the impact that such experiences had on these individuals (Kastenbaum, 2009). George Dickinson (1992) found, in analyzing 440 essays of recollections of college students (average age of 24) about their first childhood death experiences, that the average age at the first experience was eight years and that the recollections were quite clear. The college students recalled a variety of emotions upon learning of the death of a significant other: relief, confusion, disbelief, fear, happiness, anger, sadness, emptiness, shock, and guilt. One student, writing about an experience at age 10, "had extreme fear" until someone explained to her about what would happen to the body. Some children were upset because they were unable to go to the hospital and say goodbye to the deceased.

Guilt experienced at an early age can stay with a person into adulthood. For example, the responsibility of a four-year-old girl was to feed the dog, Charlie, every night. One night she forgot to feed Charlie, and the next day Charlie did not come home. She said:

> I just knew he had run away because I had not fed him. I felt terribly guilty, but I kept my mouth shut and did not tell my parents of my grave mistake. A few days passed, and Charlie still did not come back, even though every night I tried to feed him. I can literally remember calling his name and crying and yelling that I was sorry. My parents finally told me that Charlie would never be coming back because he had been run over. Well, this information devastated me. I put all the blame on myself. I just knew that Charlie had been run over because I had failed to feed him that night. I kept all this to myself until recently when my family was discussing the history of our family pets. Even though now I know that I am not responsible for Charlie's death, I still wish that I would have fed him that night. (Dickinson, 1992, p. 172)

Within the family setting, one's primary socialization and perhaps the most intense experiences of a lifetime occur. Death is one of these experiences. When a family member dies, the family rallies around, and extended family members appear who have not been seen since the last death in the family. Indeed, a latent function of a death in the family is that of a family reunion.

The family is the setting where we often first experience a death, most typically that of a grandparent. With the death of a grandfather, for example, a small granddaughter or grandson soon becomes aware that Grandpa is immobile, noncommunicative, cold, and lying in a box called a casket. The child observes the reactions of others, then usually responds accordingly. Sometimes adults may be laughing in the kitchen during the "mourning period," something that often is repulsive to the small child. "How could they be so irreverent?" the child may ask. "Why aren't they sad all the time, over Grandpa's death?"

It is very helpful when parents pay special attention to small children at the time of a death in the family, as indicated by this account of a nine-year-old boy on the death of his grandfather:

> The day of my grandfather's death, my dad came over to my aunt and uncle's house where my brother and I were staying. He took us into one of the bedrooms and sat us down. He told us Granddaddy Doc had died. He explained to us that it was okay if we needed to cry. He told us he had cried, and that if we did cry we wouldn't be babies, but would just be men showing our emotions. (Dickinson, 1992, pp. 175–176)

Thus, this parent made the little boys feel comfortable about crying, if they so felt the need (a healthy orientation from a parent). Without parents being especially attentive to small children, a child may needlessly suffer. For example, one child was afraid that her dead uncle would come back and bother her, so she slept with the covers over her head at night.

A seven-year-old girl felt that her parents had prepared her for her grandfather's funeral when they told her that he had gone to heaven and that she would meet him there someday. However, when they went to the funeral home and saw his body, she said, "You told me that Grandaddy was in heaven. No he is not, he is right here." She was upset that her grandfather was lying there not talking to anyone (Dickinson, 1992, p. 173). Thus, parental explanations do not always suffice in the concrete thinking of small children.

MASS MEDIA Television and movies expose children to countless deaths each week, whether of the cartoon characters Cow and Chicken or Wile E. Coyote, or in other TV shows or feature films. Movies and television programs have been criticized for "provoking heightened fear reactions in children" (Cantor et al., 2001, p. 218), yet the orientation is often one of death being temporary. How many times does Wile E. Coyote die within a 30-minute period? Road Runner continues to escape death, but barely (the bird with nine lives). One does not die off on a soap opera but simply reappears in a few weeks on another network playing another role. Or someone killed in a movie appears again in a few months in another movie. Death is not permanent, but only a temporary situation, as a child might interpret it from television and the movies.

Children's films do not always acknowledge deaths, especially those with missing parents. Sometimes there is an acknowledgment of death, but it is not grieved, as in *Bambi*. By contrast, in *The Lion King*, death is acknowledged, and the young character grieves and displays a gamut of typical grieving emotions ranging from self-blame and anger to profound sadness (Cox, Garrett, & Graham, 2005). In their **content analysis** of 10 Disney classic animated full-length feature films (*Snow White and the Seven Dwarfs, Bambi, Sleeping Beauty, The Little Mermaid, Beauty and the Beast, The Lion King, The Hunchback of Notre Dame, Hercules, Mulan,* and *Tarzan*), Cox, Garrett, and Graham conclude that there are 23 death scenes. Some portrayals of death in these films send ambiguous messages about death and may be confusing to young children; results indicate that the antagonists (bad guys) indeed deserve to die. Yet, watching films in which characters die may help children understand real death in a way that is less traumatic and threatening.

Hollywood is also seemingly killing off numerous animals in movies (Death in Children's Movies, 2007), perhaps as a lesson to children about how sad things can happen in life. Movies and children's books where horses have died include *Phar Lap, Black Beauty, My Friend Flicka, My Pal Trigger,* and *The Red Pony.* Ever since Bambi's mother was killed by a hunter, innocence has been a target. In *Old Yeller,* a young boy has to shoot his ailing pet dog as a way to show he is "becoming a man." Nemo's mother and several hundred brothers and sisters are eaten in the first 10 minutes of *Nemo.* The Lion in *Narnia* sacrifices his life. Three of the huskies in *Snow Dogs* are killed. Children's movies often also have a "let's kill the villains" theme (e.g., *Little Mermaid, Snow White,* and *Sleeping Beauty*).

Television highlights sensational national and world events such as school shootings, major natural disasters, terrorist attacks, and war. These realistic, though often gruesome, scenes bring death live into the home; this death is not reversible. The tragic event of September 11, 2001 was broadcast repeatedly into many children's homes. Such repeated exposure has been reported to have had a significant negative effect on children's mental functioning (numerous studies cited in Hunter & Smith, 2008). With events like the wars in Iraq and Afghanistan and the constant threat of terrorism, it is unlikely that present-day American children can escape hearing about or seeing images of death at a young age. However, in some fictional television programs, death has a fairy-tale quality (Silverman, 2000). The protagonists are not really hurt by the blows inflicted on them and frequently come back alive, like Wile E. Coyote. Media deaths do not allow children to learn about the true consequences of someone dying or to learn that death is part of the real world. They also do not learn how to cope with death when such a reality does occur.

Newspapers, like television, display death across the front pages, in addition to other parts of the paper. Pictures of death often accompany the write-ups in the newspaper. Certainly the image of violent death—the picture of the fireman carrying the body of the small child out of the federal building in the Oklahoma City bomb disaster or the collapse of the World Trade Center after the terrorist attack—is ingrained in the memory of U.S. residents. Indeed, the numerous incidents of elementary and high school children being killed by classmates in school are all too common stories in the media today.

As noted earlier, books often have death scenes in them. Books can be sources of comfort and can provide valuable information to children. Books are written both for parents to read to their children and for the children to read themselves. Books can serve as good sources for broaching the subject of death. *Charlotte's Web,* for example, is a classic tale of survival, hope, life, and death (Cairney, 2011). Author

E. B. White shows through Wilbur (the pig), Fern (the little girl), and Charlotte (the spider) how death is part of life, yet how death is not the end. Life goes on.

Nursery rhymes and fairy tales expose children to the topic of dying and death. At the time when these children's classics came into existence, the death of family members, often siblings, was a much more immediate reality than it is today, and the stories reflect this grim reality. Mother Goose nursery rhymes not only describe peaceful deaths but include horrible deaths. Such gruesome deaths should be a cause for concern among parents, yet we might read one of these stories to our children at bedtime, only to find later that they have difficulty going to sleep. No wonder! For example, the following prayer, with a suggestion of fatalism, has kept many a child awake at night:

Now I lay me down to sleep,
I pray the Lord my soul to keep;
And if I die before I wake,
I pray the Lord my soul to take.

Nature's destruction of the body is pointed out in this nursery rhyme:

She saw a dead man on the ground;
And from his nose unto his chin,
The worms crawled out, the worms crawled in.

The tale of Solomon Grundy points out the brevity of life:

Solomon Grundy,
Born on a Monday,
Christened on Tuesday,
Married on Wednesday,
Took ill on Thursday,
Worse on Friday,
Died on Saturday,
Buried on Sunday.
This is the end of Solomon Grundy.

The deterioration of the body after death comes across grotesquely in this verse:

On looking up, on looking down
She saw a dead man on the ground;
And from his nose into his chin,
The worms crawled out, the worms crawled in.
Then she unto the parson said,
Shall I be so when I am dead?
O yes, O yes, the parson said,
You will be so when you are dead.

Then there is the familiar fairy tale about the giant in "Jack and the Beanstalk," as he repeated with lust:

Fee-fi-fo-fum,
I smell the blood of an Englishman.
Be he alive, or be he dead,
I'll grind his bones to make my bread.

And Hansel and Gretel roast the witch in her own oven. Dorothy is trapped in Oz until the Wicked Witch of the West could be liquidated. Traditional versions of many fairy tales from many different cultural traditions have not been afraid of death as a theme in narrative and have dealt with death in graphic detail (e.g., *Little Red Riding Hood, The Three Little Pigs, Jack and the Beanstalk, The Gingerbread Man*, and the *Little Match Girl*), though today it is common for such tales to be sanitized and death pushed into the background of the narrative (Cairney, 2011). For example, before Walt Disney highlighted *The Three Little Pigs*, the wolf had consumed the first two for lunch, but the third boiled the beast and ate him for supper—and lived happily ever after.

Mother Goose nursery rhymes have been a source of death education for many children, as gruesome as they may sometimes be. The average collection of 200 traditional nursery rhymes contains approximately 100 rhymes which "personify all that is glorious and ideal for the child" (Baring-Gould & Baring-Gould, 1967, p. 20), yet the remaining 100 rhymes "harbor unsavory elements." Following is a partial list of such from Mother Goose:

- Two cases of choking to death
- One case of death by devouring
- One case of cutting a human being in half
- One case of decapitation
- One case of death by squeezing
- One case of boiling to death
- One case of death by hanging
- Four cases of killing domestic animals
- Seven cases related to the severing of limbs
- One case of devouring human flesh

Thus, the media are filled with orientations toward death and dying, a primary source of death information for children.

RELIGION It appears that children's religious beliefs are determined by their cognitive developmental status (Wass, 1995). What children mean by concepts such as "heaven" and "God" reflects their thinking and reasoning modes at the particular developmental level at which they are functioning. Hannelore Wass notes that some young children view God very concretely as a father figure, except that he is

---

LISTENING TO THE VOICES | GOD WANTS A DEAD KITTEN?

The father of one four-year-old girl told her that her kitten "went to heaven to be with God." She responded, "Why does God want a dead kitten?"

A three-year-old became angry when told that her grandmother had gone to heaven, and responded, "I don't want her in heaven, I want her here!"

From "First Childhood Death Experiences" (p. 173) by G. Dickinson, 1992, *Omega, 25.*

much taller, stronger, and older and has larger hands than Dad. God gives out food and lets dead people be angels and fly around. Other children view God as imposing restrictions. Some children do not believe in an afterlife.

Children, depending on their particular religious upbringing, if any, may be exposed to ghosts, the devil, hell, and concepts that may be scary for them. One six-year-old girl (Dickinson, 1992) was afraid of her great-grandmother who had "big brown moles all over her face and sat in a wheelchair." When told that she had died, the little girl became "infatuated with her ghost." When the television shows went off the air at night and the "black and white fuzzy stuff" came on, she used to imagine that she would hear her great-grandmother's voice or see her face in the television screen.

To tell a child of an ill-defined place where everyone goes after death could be frightening. However, the majority of individuals in a study (Dickinson, 1992) of 440 college students who were asked to recall their first childhood death experience seemed content with the explanation that the deceased had gone to heaven, which was described as a "happy place." Given a child's world of fantasy—the "magic world of Disney" mentality—the response "gone to heaven" seemed to suffice for the majority of these respondents. Thus, religion can play a part in children's understanding of death.

CHILDREN'S UNDERSTANDING OF DEATH

As will be noted in Chapter 9, children growing up in an inner city experience violence and death all too often. Other children in the United States, however, have less frequent exposure to death as a natural part of the life cycle; nonetheless, death is an aspect of the life cycle from early childhood to old age.

Largely gone are the days when most children lived with or near their grandparents, experiencing their death in the home. Urbanization and mobility have contributed to a separation from grandparents and, consequently, from the process of their dying. Death has been removed from the home to more institutionalized settings—the hospital or nursing home. This is summed up by the statement of the little boy who said, "I don't want to go to the hospital because that is where you go to die." That is where Grandfather was taken the last time that the little boy saw him alive. After going to the hospital, Grandfather turned into a cold, stiff corpse. The process of the grandfather's dying was not part of the daily routine in the family's home.

As was pointed out in Chapter 1, gone are the days when the majority of children were reared on farms and experienced the life cycle daily—animals were born, and animals died. Birth and death were commonplace in the socialization of children reared on farms. Gone are the days when siblings and relatives were born at home and died at home.

Not being exposed to dying and death makes today's children view this process and event as very removed and "foreign." This underscores the natural tendency to deny the facts of death. Just how do children conceptualize death? In the following discussion children are divided into age groups, but one must remember that ages and stages are in no way absolute. Each child's understanding of death is affected by his or her experiences and the family's cultural heritage.

| WORDS OF WISDOM | JEAN PIAGET'S COGNITIVE STAGES OF DEVELOPMENT |

Developmental psychologists' stages of development by years differ slightly, yet Jean Piaget's is certainly among the best known. The child's understanding of death changes as she or he develops, as explained by Piaget's cognitive stages of development. Piaget focuses primarily on the child as an active constructionist of her or his own knowledge. The stages focus on the ways in which children's thought processes change over time, regardless of the environment in which they grow and develop. At each stage, mental operations evolve from learning based on simple sensory and motor activity to logical and abstract thought. The first stage, sensorimotor, occurs from birth until approximately two years of age. At this stage, children come to understand the world through their senses and motor manipulations. During the second stage—the preoperational stage, usually ages two to seven—magical thinking, symbolic reasoning, egocentricity, reversibility, and causality characterize children's thinking. According to Piaget's theory, children in the preoperational period do not yet have the mental capacity to fully comprehend the concept of death because they have not yet developed the ability to conserve, a prerequisite to the mature conception of death.

Piaget's third stage of development, concrete operational, usually includes ages seven to 12 years. The child in this stage in relation to death is very curious and realistic and seeks information. At this stage children demonstrate logical thought through the use of mental operations to solve concrete problems. In the fourth stage, called formal operations, adolescents (age 13 and up) see death as a natural process that is very remote from their everyday life and something they cannot control. By this stage, the young person has developed the capacity for abstract thought and scientific reasoning. Teenagers are often preoccupied with shaping their own life and deny the possibility of their own death. Such a rationalization to explain why death does not apply to her or his situation can be interpreted as denial, but it emotionally supports the child as she or he faces the challenges of illness.

Sources: S. B. Hunter and D. E. Smith. (2008). Predictors of children's understandings of death: Age, cognitive ability, death experience and maternal communicative competence. *Omega*, 57, 143–162; H. Ginsberg and S. Opper. (1988). *Piaget's theory of intellectual development* (3rd ed.). Englewood Cliffs, NJ: Prentice-Hall; D. Martini. (2009). *Helping children cope with chronic illness*. Chicago, IL: American Academy of Child & Adolescent Psychiatry. Retrieved August 7, 2009 from www.aacap.org.

BIRTH TO AGE THREE   By age six months an infant perceives differences in caregivers and the degree to which physical and emotional needs are being met. One cannot have a concept of death until the beginning of thought, as evidenced by the emergence of symbolic function between 18 and 24 months. Though sounding very Freudian, perhaps the origin of death anxiety is in the traumatic separation from the mother at the time of birth. The delight of a child's playing peek-a-boo lies in the relief of the intermittent terror of separation. Under age five death is perceived as separation, but separation from one's caregiver is a terrifying thought.

Erik Erikson (1963) observed that an infant "decides" early in life whether the universe is a warm and loving place to be. Upon this primitive, yet momentous, subconscious decision is based to a large degree on the ability to deal with the threats and difficulties of later life. We must not assume that the small child has no concept or grasp at all of death, and we must be concerned about the effects of a given death upon his or her life. Children's capacity to grieve is related to their level of **cognitive development**. Until a child has developed a sense of what Piaget called object constancy, she or he cannot feel loss. Object constancy refers to the notion that objects continue to exist even when they are no longer visible. Very young children have no concern for objects, or people, that are not visible to them.

In their minds, these objects or people do not exist, so they cannot experience any sense of loss. It is only at about 18 months that children have reached a stage where they do understand the significance of objects and can grieve over something that has been "lost"—whether it is a toy or a loved one.

A toddler recognizes that a pet is alive and a table is not. Although the two-year-old child dying in a hospital has no real concept of his or her death, the child has a real appreciation of the altered patterns of care and the separation from usual caregivers. The toddler may notice the absence of the individual without truly understanding what has happened. An infant/toddler can sense when there is excitement or sadness or when a significant person is missing, but has no understanding of what he/she cannot see.

AGES THREE THROUGH FIVE   The clearest and most consistent observation from the literature on death concepts of children is that children as young as three have very definite ideas of what death means, though there is considerable disagreement as to the nature of those ideas. Most children three to five years old lack an appreciation that death is a universal phenomenon and is a final and complete cessation of bodily functions. Though death is seen as temporary and/or partial and even reversible, the child is very curious to know and understand the practical and concrete aspects of all that surrounds death. For some children, death is believed to be life under changed circumstances. They cannot grasp, for example, that the body in the casket is not aware of them.

Children in this age group may think of death as punishment for bad behavior or that they are to blame for the death, thus a feeling of guilt. It is important, therefore, to talk with the children in concrete, simple terms and reassure them that they did not cause the death.

They will likely have difficulty verbalizing, and thus may act out their feelings. They will typically only show sadness for short periods of time and display "magical thinking" regarding where their deceased grandfather may be at the moment, for example.

The permanency of death is not clear in early childhood, as evidenced by this story about a three-year-old child whose father had been killed six months earlier in an auto accident. One day the little boy's mother came in and said, "I have a surprise for you." The little boy anxiously replied, "Is Daddy coming home?" On another occasion this same little boy was playing with his LEGOs and built a house. He placed a person inside the house and was asked to identify the individual. He replied, "Oh, that is Daddy. He is asleep for 100 years." Daddy is not dead; he is simply away on a trip or engaged in a deep sleep.

When George Dickinson's daughter was five years old, she attended the funeral of an elderly friend of the family. After the service and the final viewing, the casket was closed and sealed. At this point she said, "But, Daddy, how is Mrs. Kirby going to breathe in there?" Again, it is difficult for a five-year-old to grasp a concept of death. Death is not permanent, since one still needs to breathe!

The practicality of young children is apparent in this story of a five-year-old whose mother had died. While waiting in the airport to fly from Kentucky to Texas to bury her mother, she looked up at her father and said, "Daddy, can you cook?" This was a traditional family in which the father brought home the bacon,

| WORDS OF WISDOM | ERIKSON'S DEVELOPMENTAL STAGES |

Erik Erikson organized life into eight stages, extending from birth to death.* The age categories vary, yet the ages seem appropriate. His psychosocial theory considers the impact of external factors on personality development, as an individual passes through this series of interrelated stages over the entire life cycle. The eight stages are:

1. Infancy/oral sensory (birth–18 months). The infant forms a loving and trusting relationship with the caregiver, or a sense of mistrust might develop. A sense of trust gives confidence for the future, yet a failure to experience trust may contribute to a feeling of worthlessness.

2. Early childhood/toddler/muscular-anal (18 months–3 years). The child learns to master skills for her- or himself, such as grasping and walking, and has the opportunity to build self-esteem and autonomy. Yet, the child in this stage can be vulnerable. If shamed in toilet training, for instance, he or she may feel shame and doubt.

3. Play age/preschooler/locomotor (3–5 years). Child becomes more assertive, takes more initiative, and mimics adults. If frustrated over goals and natural desires, however, guilt may be experienced.

4. School age/latency (6–12 years). The child must cope with the demands to learn new skills or risk a feeling of failure and incompetence.

5. Adolescence (12–18 years). Development begins to depend upon what one does, rather than what is done to an individual, as was previously the situation. It is important to begin establishing a philosophy of life.

6. Young adulthood (18–35 years). Development of intimate relationships is important or feelings of isolation may occur.

7. Middle adulthood (35–65). Creative and meaningful work and issues of family are important. It is significant to transmit values of culture through children. Life changes occur, thus causing some to become self-absorbed and somewhat stagnant.

8. Older adult (65–death). If an individual can look back on her or his life with content and happiness, a feeling of integrity results and death is accepted as the completion of life. Yet for some adults, they may fear death as they struggle to find a purpose of their lives.

*A ninth stage of development was added by Erik Erikson's wife, Joan Erikson, when she was in her 90s. This stage refers to "Old older adults" in their 80s and 90s. In this age category an individual may no longer have a retrospective accounting of life to date, as loss of capacities and disintegration may demand almost all of one's attention. One's focus may solely be concerned with daily functioning, however positive or negative an individual may feel about her or his previous life history. There is also much sorrow with which to cope, as one in the 80s and 90s is apt to have experienced many losses of close relationships. In addition, it is obvious that death's door is nearby and beginning to open.

Sources: E. Erikson. (1968). *Identity: Youth and crisis*. New York: W. W. Norton & Company; E. Erikson. (1950). *Childhood and society*. New York: W. W. Norton & Company; E. Erikson. (1997). *The life cycle completed*. New York: W. W. Norton & Company.

| LISTENING TO THE VOICES | A FIVE-YEAR-OLD CHILD'S MEMORY OF A DEAD BODY |

A college student in George Dickinson's death and dying class shared her recollection, as a five-year-old girl, of the death of a great aunt. She said that her dad lifted her up at the funeral home to see and touch the body. She said, "I can still feel it … cold and sleek. Her face was peaceful with lots of bright makeup. I had discovered what death was, or so I thought…. It was bright pink lipstick, blue satin bedding, and cold hands that smelled like powder."

Calvin and Hobbes   by Bill Watterson

Calvin and Hobbes cartoon

___

and the mother cooked it. The child's concern was whether or not the father could fill the vacant role.

AGES SIX TO TWELVE From research over a quarter of a century ago, Israel Orbach and associates' study (1985) of six- to 11-year-old children's concepts of death in humans and animals showed that death in humans is easier to comprehend than death in animals. They noted that seemingly the comprehension of animal death is acquired chronologically later than that of death in humans. Apparently animal death is a more difficult concept than human death. Because animals are known to be a favorite identification object of young children, the death of a pet may be more meaningful and more personal to a child than that of a generalized other human being. Thus, children may erect more defenses against death of animals. Another explanation noted by Orbach and associates is that human and animal deaths can be regarded as two specific examples of the complicated concept of life and death. Although a child may comprehend the meaning of death to a specific example of life, such as among humans, he or she may not comprehend it in another sphere of life, such as plants and animals. This reflects a lack of understanding of the broader concepts of life, birth, death, and animate and inanimate objects.

In this age range the child begins to view the world from an external point of view (schoolmates, teachers, other adults, the media, and readings), and language skills are becoming communicative and less egocentric. This is the first major separation from home, and the child is entering the world of school where teachers and adults, other than parents, become the models for identification. Magical thinking persists, yet the child's ability to test reality increases. A sense of moral judgment continues to develop. The child is mastering social skills.

Death is personified as a live person or some variation such as an angel, a skeleton, or a circus clown. Death is connected with violence. Besides television, their interest in violent death could be related to killings of worms and insects by young children or by their playing cops and robbers or other gun-related games involving killing.

Death is viewed as a taker or spirit that comes and gets you. There is fear that death is contagious. George Dickinson remembers in the early 1950s, when he was

George Dickinson

Children in a small rural elementary school in southeastern United States were instructed by their teacher to draw a picture of death. Above is a sample of their depictions. Death may be viewed as the taker of life, like the Grim Reaper.

a child of 11 or 12, that polio was rampant. When polio was diagnosed in two of the children living next door to him, he waited for the "thing" (the Grim Reaper?) to visit his house next. Polio usually did not kill, nonetheless, this experience was frightening for a child, as "it" weaved throughout the neighborhood.

EXPLAINING DEATH AND DYING TO CHILDREN

Psychologist and former priest Robert Kavanaugh (1972) referred to children as "little people." He sees them as compact cars instead of Cadillacs, traveling the same roads of life and going the same places as big cars. Although they are more vulnerable and fragile, they have all the parts and purposes of big people. They are ready and capable to talk about anything within the framework of their own experience. Little people can handle any situation adults can handle comfortably and should do anything big people should do, as long as they are physically able, noted Kavanaugh. Though very small children will not understand terms like "dying" and "death," they can be helped to cope with the absence of the individual by our simply being there for them (Ireland, 2010).

The fact is, children want to know about death. In speaking to a group of fourth graders on the topic of dying and death, George Dickinson's frustration was trying to decide whom to call on when a dozen or more hands at a time were up from these 40 children wanting to ask questions. Each had his or her own concerns, and they were sincere, legitimate interests. He remembers one little boy asking if it were true that one would die from getting embalming fluid on the skin. Dickinson tried to assure him that this would probably not be a cause of death, but it was a genuine concern for this child.

Adults need to support, comfort, and help the child to express grief, as children have no previous experience on how to react. An attempt by adults to protect children from the reality of death reinforces the perception that death is either not real or too frightening to examine or that the ending of life is not worth noting with reverence and respect. Some parents' main objective seems to shift from explaining and teaching to protecting (Cox, Garrett, & Graham, 2004–2005). For instance, rather than telling children why and how people die, they may focus on downplaying the emotionality, seriousness, and reality of death. Children also have vivid imaginations, which may lead to more fear and confusion if death is not discussed. Accordingly, children who are excluded from discussions concerning the death of someone close to them and who are not allowed to participate in funeral or memorial services are likely to feel isolated and may try to fill in missing information regarding death with their imaginations or by media messages (Yang & Chen, 2006). Such imaginations-run-wild reactions may lead to negative evaluations and reactions toward death. Encouraging children and adolescents to discuss death-related situations as they occur and to express reactions is important.

When we big people fail to give little people credit and try to shield them from information because they are unable to take it, could it not be that we big people feel uncomfortable talking to children about dying and death, and thus avoid the topic altogether? Avoidance is not the best solution, as noted in the example of the needless worry and frustration of a four-year-old girl who was not told about her puppy's traumatic death by a mowing machine until two or three weeks afterwards (Dickinson, 1992). She was allowed to search for him frantically every day. She even put out food and would worry at night that he was cold or hungry.

Adults should allow children to talk freely and ask their own questions, without any adult speeches; let them ramble, talk crudely if they wish, change the subject, or present unanswerable questions without being squelched. The child should always be supported in a "comforting way" with an assurance of a continued relationship. It is important to create an atmosphere of openness and give the impression that there is no right or wrong way to feel (Lyness, 2012).

BE HONEST AND OPEN Being honest and straightforward with children in talking about dying and death is a good rule of thumb to follow. Respond to the child's questions immediately with concrete, rather than abstract, answers. If your answer is unsatisfactory, the child will probably ask a follow-up question in a few seconds or a few hours. If the child asks something you cannot answer, be honest and say you do not know. If you cannot explain something, find someone who can.

Answer directly, but do not be too detailed with your responses. When a small child asks where he or she came from, you do not go into minute detail about the sperm and the ovum forming a zygote. You state that you came from your mother. If the child is not satisfied, another question will follow and can be answered accordingly.

If parents are open to discussing death, opportunities will present themselves. As noted earlier, dead flies, mosquitoes, birds, and animals beside the road can lead into a general conversation about death, which will not be fraught with emotional turmoil. When the explanation is postponed until the death of someone or something deeply loved by the child, the acceptance of the reality of death will be more difficult. In explaining the death of an individual such as a grandfather, however, use basic and concrete terms (Lyness, 2012). If not a sudden death but a lingering one, the child might be told that her/his body was no longer working and that doctors could not fix it. If a sudden death, then the explanation might be simply that the body stopped working. "Dying" or "dead" means the body stopped working.

The burial of a deceased pet by parents and children together can be a positive learning exercise in relating to death. The animal is cold, still, and not alive—it is dead. That is reality. When George Dickinson's children were young, they had guinea pigs. These animals have a short life span; thus, many guinea pig funerals were held in the backyard with the entire family participating. For other family members to be present at a time of grief helps the child to better bear this burden. Grief shared is grief relieved.

AVOID EUPHEMISMS Try to avoid euphemisms when talking to a child about death. Use words like *dead*, *stopped working*, and *wore out*—simple words which establish that the body is biologically dead. Though they have good intentions, many adults often hinder children's understanding of death by using confusing terms and abstract language to explain the concept to them. They may say that someone has "passed away," which does not convey a realistic portrayal of death to children (Willis, 2002). One child was told that Grandfather's heart was bad and stopped working. This seemed to satisfy the child. Couched in a different way, however, another child was told, "Grandfather can breathe easier now," implying that Grandfather is still breathing. This child wanted to join his grandfather because he had asthma and would welcome the chance to breathe easier. He was told, however, that he was too young to join Grandfather (Dickinson, 1992)!

Dead is a difficult word to say, but to use a euphemism such as *went to sleep* may make it difficult for the child to go to sleep at night. If Grandad went to sleep and is then buried in a box underground or destroyed through cremation, that situation is probably not what the small child's goals are in life! Thus, the objective is to stay awake and *not go to sleep!*

To use euphemisms such as *went away* or *departed this life* or *passed away* may cause the child to expect the deceased to return. After all, he or she is simply on a trip! Even to suggest that "he's gone to heaven and will live forever" is confusing to a child when this is said through heavy tears and with upset, emotional feelings. Be cautious, however, regarding a child's questions about death, as they may sound much deeper than they actually are (Lyness, 2012). For example, a five-year-old who asked where someone who died is now, probably is not asking whether there is an afterlife but might be satisfied hearing that the decedent is now in the cemetery. Yet, this could be a time to share your beliefs about an afterlife or heaven if indeed that is part of your belief system.

A four-year-old girl was told that her younger sister was "too sick to live with us so she went to visit Grandmother in heaven and would never come home again."

WORDS OF WISDOM	LAMENT
	Edna St. Vincent Millay

Listen children:
Your father is dead.
From his old coats
I'll make you little jackets;
I'll make you little trousers
From his old pants.
There'll be in his pockets
Things he used to put there,
Keys and pennies
Covered with tobacco;
Dan shall have the pennies
To save in his bank;
Anne shall have the keys
To make a pretty noise with.
Life must go on,
And the dead be forgotten;
Life must go on,
Though good men die;
Anne, eat your breakfast;
Dan, take your medicine;
Life must go on;
I forget just why.

She was frightened for months afterward because "whenever anyone did not feel well, I thought they would go away forever too" (Dickinson, 1992, p. 173). Rabbi Harold Kushner (1981) cautioned, however, that to try to make a child feel better by stating how beautiful heaven is and how happy the deceased is to be with God may deprive the child of a chance to grieve. By doing such, we ask a child to deny and mistrust his or her own feelings, to be happy when sadness is desired. Kushner noted that the child's right to feel upset and angry should be recognized. One should also be cautious that the child not feel that the deceased *chose* to leave and go to heaven. It should be made clear that the person did not wish to leave or desert, as the child may feel.

Anthropologist Colin Turnbull (1983) described death as being like it was before birth—a state of nothingness. Ask someone to describe what it was like before birth, and you will probably not receive much of an answer. To suggest that after death there is no place to describe—a nothing—would deny children a defined place to imagine. But then with the vast imagination of children, perhaps this might be more creative for them than a vague description of a "big house in the sky."

After you have made an effort to avoid euphemisms and gotten across the fact that the person is dead, the next step is to explain what is now going to happen. Tell the child about the body being moved from the hospital to the funeral home. Alert the child that funeral arrangements will be made (if that is the case), and a funeral will follow. Outline the format of a funeral itself, then talk about going to the cemetery (if that is the case) and burying the body in the ground.

SHOW EMOTION It is important that the child knows it is okay to show emotion when someone dies. Because it is a very sad time, the child should be told that everyone is upset and that many individuals may be crying. If the child feels like crying, he or she should be assured that crying is okay. Crying is normal. It should also be explained that because some individuals do not show emotion does not mean that they did not love the dead person.

You probably cannot overhug a child during these times. It is important to always reassure the child that you care for him or her. Hugs and tears are very compatible expressions. Do not apologize for crying. Your crying in front of the child gives assurance that crying is okay. Children often cry when hurt, thus crying when someone dies seems natural. Crying can help the child feel relieved. Displaying emotions may make the dead person or pet seem more worthy. Tears are a natural tribute paid to the deceased. The child misses the one who is gone and wishes the person were still around.

Unfortunately, some parents do not allow their children to cry when death occurs (Dickinson, 1992). A 15-year-old recalls being spanked with a hairbrush for crying over the death of her puppy. A 10-year-old was smacked by her uncle to make her stop crying at the funeral. Some children were told that it was not grown-up to cry. For many, their first childhood memories of death included recalling that it was the first time they had seen their fathers cry. No one commented that it was the first time they remembered seeing their mothers cry.

Several college students recalled when they were children (Dickinson, 1992) that they watched others' reactions to a death, then responded accordingly. Because children are watching adults, a parental role model of expressing one's self in front of children might be very beneficial to their socialization. Children then would not feel a need to hide in the closet to cry or cry in their pillows at night, as some reported. One child noted the positive experience of his parents and sibling sitting down together to have a good cry at the death of his pet—this pointed out to him the warmth of the family in sharing this event. It was not a burden he carried alone; his family grieved with him. It is comforting to know that others care.

According to former professional football player Rosie Greer, it is okay to cry. He sings a song in which he notes it is all right for women, men, girls, *and* boys to cry. For little boys to cry is *not* "sissy." If Rosie Greer says it is all right, it must be okay! Anthropologist Ashley Montagu (1968) agreed with Rosie Greer when he stated that women have a superior use of their emotions because of their ability to express themselves through crying.

Some men feel that "real men" do not cry because it is not macho to cry. Such men will wear dark shades on the cloudiest of days, if they fear they may cry at a funeral. Several years ago, George Dickinson's teenage daughter attended a funeral of a friend with him. After leaving the cemetery on a very cloudy day, she noted that nearly all the men at the graveside service were wearing dark glasses. Dark glasses help to hide tears.

What favors are we doing our children by teaching them that it is not okay to cry? It is okay to laugh, why should it not also be okay to express feelings through crying?

ADOLESCENCE

Adolescence is the training period between childhood and adulthood in the life experience of humans. This period reflects a time of changes and challenges as one goes through a time of transition into adulthood. The social concerns of teenagers mirror universal concerns including "staying alive" and "living well" (Bibby, 2001).

Adolescence—from age 12 to age 19—encompasses the acquiring of formal logical thought, the onset of biological sexuality, the growth of physical structure, continued acquisition of adult social skills, clarification of ethics and values, and

Adolescence is a time of transition from childhood to adulthood. Death is not something that adolescents often think about. Their lives are at the alpha stage, filled with dreams and the future, whereas the elderly are at the omega stage of their lives, filled with memories and the past.

ability to make long-term commitments to persons and goals. Adolescence is often a time fraught with anxiety, rebellion, and indecision. As adolescents' understandings of death develop, many variables shape and determine their death anxiety. The understanding of death emerges when individuals are able to conceptualize the meaning of life (Ens & Bond, 2007).

Adolescents generally do not think in terms of their distant future. Their struggle with present life experiences, especially the concern with their own identity, and anxiety about successes during the immediate future, evidently occupies so much mental energy that thinking about what life will be like at age 45 or 70 is nearly impossible for them. Standing on the threshold to adulthood, adolescents are more focused on the years of living that lie ahead. Before them are dreams and aspirations.

The death of a peer for adolescents may be very difficult as it emphasizes the adolescent's own vulnerability and mortality. Such a death is a grave injustice as the individual has died before reaching the fullest potential and opportunity to experience life. Because adolescents' identities are shifting toward members of the peer group, away from family, the loss of a peer can upset identity formation that may already be unstable.

Schools today are aware of the importance of having an opportunity to express one's self when a classmate dies and are doing a commendable job of dealing with the death of a student. They bring in counselors and have them available for anyone desiring to talk. They provide numerous opportunities for classmates to mourn and deal with the death. Mike Leming and George Dickinson, though not grief counselors, have been invited into high schools after the death of a student, simply to serve as a catalyst in the classroom to encourage the students to "talk about" the deceased and express their feelings.

By 11 or 12 years of age, individuals are able to move from the use of language and ideation that is concretely oriented to an abstract level of thought. No longer wedded to the concrete, the adolescent can now use conditional statements such as "if-then." Ideas can be taken apart and put back together in new ways. They formulate their own theologies about life after death.

Our society's discomfort with the process of aging, illness, and death does not contribute in a positive way toward the adolescent's image of a future. Because being young is envied and growing old is feared in America, to "grow old along with me" may not mean that "the best is yet to be" in the minds of adolescents. Consequently, an immunity toward physical deterioration and death may develop in adolescence. When the death of a significant other occurs, previous experience has likely not prepared the adolescent for the feelings of rage, confusion, loneliness, guilt, and disbelief that accompany a personal loss. If peers have not experienced a similar loss, they may have difficulty being supportive of others in such a situation.

IDENTITY CRISIS AND DEATH ANXIETY

Being gripped by questions such as "Who am I?" and "How do I fit into the scheme of things?" the adolescent struggles with good and evil, love and hate, belonging and loneliness, and thoughts of life and death. These can be disturbing issues to the adolescent who is not sure whether he or she is a child or an adult, because identity signals come from all directions, adding to the confusion of one's self-concept. The perception of a self under reconstruction is granted most keenly as one experiences the possibility of failure, loss, catastrophe, and death.

Death anxieties and fears, like masturbation and other sexual issues, are universally experienced and discussed with equally uninformed peers. One of the tasks of adolescence is to begin to grapple with the meaning of life and death and to emerge with a philosophical stance that promotes optimism for the future. For an adolescent in this metamorphosis of life, this is not an easy assignment, especially because one's attitude toward the future may involve more pessimism than optimism.

MEDIA INFLUENCES

Movies provide a medium for facing death as the viewer identifies with characters. Movies often feed unnecessary fears and support unhealthy death attitudes, yet can also provide a challenge to personal attitudes toward death and give the viewer a healthy perspective (Niemiec & Schulenberg, 2011). Adolescents often see dying and death depicted in the media as violent, "cool," distant, or unnaturally beautiful. According to the National Center for Children Exposed to Violence (2005), by age 18 the average American child will have viewed about 200,000 acts of violence on television alone. The National Television Violence Study (2003) found that nearly two out of three television programs contained some violence, averaging about six violent acts per hour. There is concern by some individuals that children will become desensitized to such violent acts and learn to see them as valid responses. Indeed, the "lesson" often is that aggression is used by the "good guys" to gain rewards that are unavailable to others. Some research (National Center for Children Exposed to Violence, 2005) supports the idea that violent thoughts and behavior increase after exposure to violent films, music, television or video games. The argument supporting this assertion is observational learning—learn by imitating what one sees.

Death abounds in horror and action films. Adolescents seem especially fascinated by movies about violent death. The *Scream* and *I Know What You Did Last Summer* series of movies, popular among adolescents in the late 1990s and into the 21st century, are examples of movies with violent deaths. Also, George A. Romero's *Land of the Dead*, *The In Crowd*, *Dark Water*, *Wild Things*, *Halloween Resurrection*, and *Urban Legend* were horror movies popular with adolescents. With DVDs, downstreaming from the Internet, and other technological capabilities, adolescents' opportunities to view violence and uncensored death, in the privacy of their own homes, increase significantly.

A 2012 animated horror movie entitled *Frankenweenie* deals with dead animals yet "gives hope" by bringing them back to life, even if in the form of a monster!. Though not a movie for young children, this movie had appeal for adolescents. The boy in the movie uses science to resurrect his dead dog Sparky. The boy's reanimation formula is used by classmates to reanimate their own animal, resulting in a mummy hamster, vampire cat, and mutant sea monkeys.

Other films, however, in which death plays a major role, are less violent. *A Walk to Remember*, for example, was popular with adolescents. A favorite movie-industry theme now is a person dying with some part of his or her life unfinished, thus posing a problem (Wells, 2000). *Flatliner*s, a film about medical students experimenting with near-death experiences, is really concerned with their own deeply repressed and unresolved problems as they die. Other near-death experience films include *Saved by the Light*, *Resurrection*, *Ghost Dad*, *Dead Again*, *Made in Heaven*, and the classic, *It's a Wonderful Life*. Lingering and romantic deaths reminiscent of 19th-century literature, though now caused by cancer rather than tuberculosis, have been popular in films such as *Dying Young* and *Beaches*. A blockbuster 1999 movie entitled *The Sixth Sense* is about a little boy who "sees" and "relates to" ghosts. Our society seems to have a renewed fascination with ghosts.

Destructive themes in rock music are not unlike those of many movies viewed by adolescents. Death themes in rock music might be therapeutic (Wass, Miller, & Redditt, 1991). The themes should be taken metaphorically rather than literally. Rock lyrics may sometimes provide the means for dealing with issues of death and for managing the anxieties created by these themes. Wilson Frank and the Cavaliers' "Last Kiss," for example, is about a teenager's girlfriend dying in an auto accident in which he was driving: "Oh where oh where can my baby be? The Lord took her away from me. She's gone to heaven so I've got to be good, so I can see my baby when I leave this world." He talks about giving her a last kiss, as he held her close. He closes with, "I lost my love, my life that night."

Pornographic rock, in which fantasy has taken over and death is dehumanized and distorted, is evident in some lyrics (Nordheim, 1993). Groups that produce hateful and destructive lyrics have become popular. For example, Guns N' Roses sold 15 million copies of their album *Appetite for Destruction*. Everlast's "What

It's Like" is about violence and death via drugs and guns. Mentioned in the lyrics is a "kid named Max" who pulled out his gun and wound up dead, leaving his wife and kids caught in the midst of pain. "What It's Like" is about loss and emptiness being the result of death. "Ends" (both of these songs are on their *Whitey Ford Sings the Blues* album) basically says that the end justifies the means—to take a life to achieve one's goal is okay. It is the end (result) that counts, no matter how one gets there ("Sometimes kids get murdered for the ends").

According to the police in Colorado Springs, Colorado, a linkage exists between violence and hip-hop (Frosch, 2007). After a series of shootings and with a rising homicide rate, the police are saying that gangsta rap is contributing to the violence by luring gang members and criminal activity to nightclubs. The actions of the police to shut down hip-hop angered the hip-hop community, primarily blacks and Latinos.

In Korn's "Dead Bodies Everywhere" a son talks about his parent who seems to view him as a nothing, a "cipher." "You make me feel like no one," he says, "Dead bodies everywhere." The teenager seems to be saying that life is depressing and oppressive. Verve Pipe's "The Freshmen" is about death via an accident: "We fell through the ice." Marcy Playground's "One More Suicide" depicts Christopher O'Malley's jumping off the bridge into the river. "One more suicide," the lyrics say over and over. Other popular song titles with teenagers include Onyx's "Betta Off Dead," Paula Cole's "Road to Dead," Scarface's "Hand of the Dead Body," Stone Temple Pilots' "Dead & Bloated," Street Military's "Dead in a Year" (from their *Don't Give a Damn* album), House of Pain's "Back from the Dead" ("I'm the resurrected, skip the autopsy 'cause I never O.D.'d, I'm back from the dead"), and Insane Clown Posse's "Dead Body Man."

LEARNING ADULT RITUALS

Society is created and held together by **rituals**—morning wake-up rituals, going to school/work rituals, eating rituals, religious rituals, political rituals, and others. Dying, death, and bereavement all are ritualized by society as a way of containing and giving meaning to feelings of loss. The granting of adult responsibility and privilege to adolescents varies from culture to culture and group to group. Because older people all too often do not teach children the adult rituals for handling dying, death, and bereavement, it becomes one of the personal tasks of adolescence to acquire this knowledge. Being unfamiliar with the proper rituals to follow at the time of a death, an adolescent may feel especially anxious: "What do I do? What do I say?" George Dickinson recalls his own son, as a high school student, getting ready to go to the funeral home for visitation after the death of a classmate. He asked what he should say to the parents. Having not previously experienced the death of a peer, he was not sure what the "correct" behavior was. We have to be socialized as to the sanctioned rituals in our society.

Just as for younger children, adolescents learn the "proper" way to respond in dying and death situations by watching others. For example, Dickinson has had numerous students tell him that they did not know how to react to a death, but if others were crying, then they would usually mimic such behavior. On the other hand, if others were crying and the individual could not, she or he often felt guilty.

"Why am I not crying, since that seems to be the thing to do?" they may ask themselves.

As Karla Holloway (2002) noted in *Passed On: African American Mourning Stories*, dying and death historically have very persistently figured into the experiences of black Americans. Being around death and dying has not been unusual, and the experience of death has been worked into the culture's iconography and included as an aspect of black cultural sensibility. Holloway observed that the formative years, the waning years, and each day between have traditionally been haunted by one spiritual's refrain: "soon one morning, death will come a-calling" (2002, p. 6). Thus, dying and death remain in the cultural memory of blacks in America today and has cut across and through decades and centuries in the collective minds of black Americans.

COMMUNICATING ABOUT DEATH

Whenever an individual is going through an unstable or stressful period—especially when the stress has to do with basic feelings about self-worth, identity, and capability—the thought of death is particularly difficult to manage. Crucial to positive outcomes is the manner in which parents, friends, peers, teachers, and others enable the adolescent to deal with the anxiety caused by the thought of death and the actual death of friends, acquaintances, and loved ones. Particularly in the adolescent years, parent–child relationships are often fraught with conflict and alienation, and adolescents often prefer to discuss important ideas and emotional issues with their peers, not their parents (Wass, 1995).

Though parents may feel uncomfortable talking about death with their children, it is important that they try, because open and honest communication is far more helpful than silence or evasion and is best achieved in an atmosphere of mutual trust and respect. With proper preparation and support, younger adolescents can be helped to become aware of death in manageable ways, and older adolescents can be helped to impart a meaning to death and life that transcends everyday events and infuses the future with hope.

PRACTICAL MATTERS | IMPROVING FAMILY COMMUNICATION ABOUT DEATH

1. Be aware of the adolescent's concerns about death and be open to discuss anything that he or she feels like exploring.
2. Listen actively and perceptively, keeping your attention on the adolescent and the apparent feelings underlying his or her words.
3. Accept the adolescent's feelings as real, important, and normal.
4. Use supportive responses that reflect your acceptance and understanding of what the adolescent is trying to say.
5. Project a belief in the adolescent's worth by indicating that you are not attempting to solve his or her problems, but are instead trying to help the adolescent find his or her own solutions.
6. Be willing to take time to enjoy each other's company and to provide frequent opportunities for talking together.

Source: Adapted from J. N. McNeil. (1986). In talking about death: Adolescents, parents, and peers (pp. 197–198). In C. A. Corr & J. N. McNeil (Eds.), *Adolescence and Death*, New York: Springer Publishing Company.

ADULTHOOD

We have already observed that one of the main problems with following a developmental scheme in the explanation of how people think, feel, and integrate life's experiences is that there are so many possible combinations of factors in any given life. This is particularly true in studying the adult stage of life. The longer one lives, the more complex the picture becomes as the probability of additional factors influencing a particular person increases. It is important to keep in mind that research findings describing a particular population, age group, or cross section of individuals should not be seen as more than a description of that particular population sample. Individuals may vary widely within a single sample.

Also, different studies may portray different profiles of a single population. This reflects the theoretical approach of the researcher and limitations of present knowledge, methodologies, and conclusions.

Another limitation to studying adults' concepts of death includes the lack of a satisfactory definition of "maturity" in relation to the concepts of death. Yet another limitation of the developmental perspective is that researchers have not focused on adult conceptualizations of death. Although Erik Erikson and others have studied development over the life span, few have studied the relationship between the stages depicted in these studies and adult ways of thinking about death. The major exceptions to this trend are found in studies of mentally and terminally ill populations.

Finally, a limitation is faced in determining what *adult* actually means. Biologically, though we may stop growing taller at a certain age, great changes continue in the human body. Psychologically, it is even more difficult to arrive at any definition of adulthood that will be generalizable to a significant percentage of the population. One must conclude that the word *developmental* takes on a more individualistic character. Thus, one should expect that age will be less influential in explaining death conceptualizations for adult populations than for other age groups.

WORDS OF WISDOM | A SEASON FOR EVERYTHING

For everything there is a season,
and a time for every matter under the heaven,
a time to be born, and a time to die;
a time to plant, and a time to pluck up what I planted;
a time to kill, and a time to heal;
a time to break down, and a time to build up;
a time to weep, and a time to laugh;
a time to mourn, and a time to dance;
a time to cast away stones, and a time to gather stones together;

a time to embrace, and a time to refrain from embracing;
a time to seek, and a time to lose;
a time to keep, and a time to cast away;
a time to rend, and a time to sew;
a time to keep silence, and a time to speak;
a time to love, and a time to hate;
a time for war, and a time for peace.

Source: Ecclesiastes 3:1. The Holy Bible (Revised Standard Version). (1962). New York: Oxford University Press.

YOUNG ADULTHOOD

From the discussion of the development of the adolescent's intellectual understanding of death, one could expect the young adult, roughly between the ages of 20 and 44, to have a good grasp of the universality, inevitability, and finality of death. The young adult should know, at least intellectually, that death is an entirely possible event for anyone at any moment, yet death probably seems far away to most young adults, especially during the early years of young adulthood. Certainly, the death of a celebrity at a young age should be a wake-up call that one can die as a young adult. Yet, many will observe that such an event is indeed unique and unlikely to happen, thus not a big worry. At this stage of life, one is just entering the arena of a somewhat independent life where capabilities and skills can be tested and pride can be taken in positive results. Hopes, aspirations, challenges, and preparation for success in life are the focus at this age.

As noted earlier in this chapter, males in the age category of late adolescence and young adulthood are dying by violence, a situation that seems unnecessary and wasteful. Violent crimes in this age group have increased dramatically, especially beginning in the 1980s. Homicide is the second-leading cause of death after accidents for adolescents and young adults, and homicide is the leading cause of death for adolescent and young adult black males (NAHIC, 2007).

Again, it is evident that though age may have some influence over the way that one thinks, individual circumstances and external forces have a more powerful influence upon one's thoughts about death. The older that one grows, the more apparent it becomes that one needs to reflect on life as much as one needs to engage in it. Such a practice can provide the individual with greater life satisfaction. As one moves into middle adulthood, however, it can have both positive and negative results.

MIDDLE-AGED ADULTHOOD

Although there appears to be no agreement among social scientists as to the exact beginning of middle age, most seem to suggest that this period of life starts between ages 40 and 45. The U.S. Census Bureau defines middle age as being ages 45 to 64, whereas Vera in the Broadway play *Mame* describes middle age as being "somewhere between 40 and death." Middle-aged individuals are sometimes referred to as the **sandwich generation** because they may be supporting and/or caring for their own children as well as for their parents. They are caught between these two groups.

Carl Jung (1923) suggested that the primary goal of the second half of life is to confront death. Jung (1971) believed that middle adulthood is the time of greatest growth for most people—a time of integration of undeveloped dimensions of personality. He described the major task of the middle years to be reassessing and giving up the fantasy of immortality and omnipotence that carries us through earlier years when our own death is incomprehensible. Through the increasing awareness of body changes, loss of significant others, and children moving toward independence, we shift from a future-oriented perspective to an inescapable confrontation—a conscious awareness of death and mortality (Douglas, 1991). Erik Erikson's theory of ego development (1963) in these years suggests a concern for the next generation and an acknowledgment of mortality.

For the person who has lived 40 or 50 years, life brings with it the advantages of experience. Promotion to supervisor, foreman, or analogous status rankings in one's work or social milieu demonstrates that one gains greater political and social power during middle age. Not everyone is promoted, however, and even those who are, as well as their less fortunate colleagues, become gradually aware that physical vitality has now begun to wane.

Robert Fulton and Greg Owen (1988) described this group born after World War II (**Baby Boomers**) as primarily experiencing death at a distance. Unlike earlier generations, they were probably born in hospitals and are no longer likely to die from infectious diseases. Unlike their parents and grandparents, this Baby Boom generation experienced the maximum benefits of an urbanized and technologically advanced society. Fulton and Owen further note that as the commercial meat processing industry removed the slaughtering of animals from the home, so did modern health care institutions shield this group from general exposure to illness and death. Death became invisible and abstract. This generation was the first in which a person could reach adulthood with only a 5 percent chance that an immediate family member would die.

With Baby Boomers now reaching retirement age, death is becoming a major topic, perhaps an obsession, with many people of this generation. The Baby Boomers are demanding more meaningful deaths and burials in the 21st century, just as they took control of the childbirth experience in the 1970s (Dickinson, 2012c). With cremation rates rising, a gradual rejection of traditional burial practices is on the horizon (see Practical Matters box in Chapter 11 entitled "Baby Boomers and Personalized Death Trends").

Thanks to improved diets, healthier lifestyles, and unprecedented advances in medical care, the Baby Boomers are living longer than their parents did. The images of being struck down in one's prime and of the untimely accident or death are pervasive for this age group. Denial, hostility, and depression are factors that often accompany an illness of the middle aged.

PANIC AND DENIAL The "panic" begins when one realizes that the idealized self, fulfilled and successful, may never be attained. Those whose job or self-concept depends upon youthful physical vigor especially suffer from this recognition of possible unfulfilled dreams. No amount of jogging reverses the effects of time. No patent medicine can undo the damage from wear and tear. The only hope is to make the best of what energies and experience remain and to focus on what one does best.

Death is a salient issue for midlife adults. One of the major midlife tasks to be completed is to accept death as a reality. Failing health, deaths of parents and close friends, and changes in physical appearance contribute to a heightened awareness of death. Midlifers watch their peers die quick deaths from cardiovascular disease and prolonged deaths from cancer. Fitness becomes a necessity to help them cope with the increasingly familiar threat of death in their own lives. Individuals in the middle years may begin an exercise program in the hope of contributing to their longevity. With heredity being a given, an individual can at least try to increase longevity by behavior and eating habits. Exercise, though it takes time, may contribute to the prolongation of life and may indeed help to make the "last of life the best to be."

Another developmental influence on death concerns among adults is the longevity of their parents. As long as one's parents are living, there is a buffer between the person and death—one's parents are supposed to die first. After one's parents die, however, this buffer is gone, and one's own generation becomes the genealogical line of descent to die. What effect does that have? Does it affect the level of death anxiety?

REFLECTION AND ACCEPTANCE   Awareness of death changes people's lives in middle age, causing them to become more philosophical about their lives and to reevaluate values and priorities. In the process of confronting their own mortality, people deepen their capacities for love and enjoyment and ultimately acquire more meaning in their lives. Carl Jung (1933) believed that adults become more introspective and concerned with meaning during midlife and concluded that they experience an inner transformation after recognizing previously suppressed aspects of their personality. He maintained that those who successfully come to grips with life become more individuated.

Middle-aged adults are aware of getting closer to the "day of reckoning" and will need to evaluate values, meanings, and sense of self-worth in the face of finitude. The individual becomes increasingly conscious of thoughts of painful death, of the dying process, and of ceasing to be as a person. An awareness develops of the meaning of absence to significant others—spouse, children, other relatives, and friends.

Thus, death at this point in life often carries with it some of the same sense of injustice and anger that the younger adult experiences with thoughts of death. Now, however, these thoughts are tempered with the recognition that many more forces exist that could bring death home to the individual.

PERSONAL GROWTH   Although the mature adult needs to give up fantasies of immortality, omnipotence, and grandiosity, there still needs to be a sense of accomplishment—a fulfilling of oneself and one's plans for family and personal enterprises. Consciousness of time and death, therefore, makes little difference for the middle-aged adult. One strives to put all of one's skills and experiences to the best possible use.

After one is into the decade of the 50s, a turning point is reached at which one's finitude becomes even more evident. One develops an increasing consciousness that one no longer measures time from birth as much as one measures time until death or until the end of one's most productive years. Focusing upon what one wants most to do before retirement or death causes one to avoid those things considered extraneous and/or uninteresting. All of this is not to say that increasing age means cessation of personal growth. On the contrary, as long as there is life, growth can occur for the individual who has the will to live and the will to give of self to others.

OLDER ADULTHOOD

Death is expected for the elderly person. Indeed, in contemporary American society, death is primarily something that happens in old age. Almost two-thirds of the individuals who die in the United States each year are 65 or older. Death is seen as appropriate for very elderly persons who have lived their allotted span of life. Though growing old is a natural part of being human, the circumstances in which most Americans age and die are increasingly "unnatural" and certainly unprecedented (Cohen & Kass, 2006). Death comes on the doctor's watch and in high-tech surroundings, almost always following years of chronic illness, typically following decisions about additional medical intervention, often made on behalf of patients incapable of making decisions for themselves by caregivers who do not always know how to honor those who have lost their most human qualities.

Although some have observed that "growing old is hell," others look forward to the autumn of their lives. As Goethe stated, "To grow old is in itself to enter upon a new venture." Aging professional athlete Satchel Paige noted many years ago that growing old is really mind over matter—as long as you do not mind, it does not matter. In the movie *Citizen Kane,* Mr. Bernstein said to Mr. Thompson, a reporter, "Old age. It's the only disease, Mr. Thompson, that you don't look forward to being cured of." Francis Bacon Sr. (1561–1626) summed it up this way: "I will never be an old man. To me, old age is always 15 years older than I am." Thus, individuals tend to have different reactions to growing old.

Demographers divide older adults into subgroups: 65–74, 75–84, and 85 and over (Novak, 2009). The oldest age group will be among the fastest growing in the future. This group will grow more than four times in size from 1990 to 2030 and more than six times in size from 1990 to 2050. By 2050, nearly one-fourth of the older population (aged 65 and older) will be 85 years old or over. Whereas 65 has been the marker of old age since the beginning of the Social Security system, age 55 is becoming a meaningful lower age limit for the older adults because of the lowering age of retirement.

For the elderly approaching the "end of the tunnel," a reflection on the life behind them, remembering the good times and the journey they have traveled, perhaps together as a couple, can certainly present a feeling of integrity. As poet Robert Browning said, "Grow old along with me! The best is yet to be, the last of life, for which the first was made."

LISTENING TO THE VOICES | ACCEPTING DEATH

My Dear Children,

This seems a strange gift for this time of the year. This lovely ball was given me by the friend who made it, and I want it to hang for many years on your happy tree. If I live until another Christmas, I would be 97 which is too long to stay in this devastated world which my generation has made. I am ready to depart any time. God has been wonderfully good to me. I have had all any one could ask for—love and care and now every comfort in this shadowing time. I say with Cardinal Newman, "So long thy hand hath led me, sure it will lead me on." I know that you are leading lives full of meaning, and my blessings go

with this bright ball. I believe a circle has no beginning and no end.

Sincerely your friend,
Lucie Reid

After living next door to Lucie Reid for two years in western Pennsylvania, George Dickinson's family moved. For several years after that, they received a beautiful hand-decorated Christmas ball from Mrs. Reid. The ball usually arrived in mid-December. One year, however, the ball arrived in mid-July. This letter was enclosed in the package. Lucie Reid did indeed die before "another Christmas."

Life can be compared with a train traveling through a tunnel. There is a point at which the train is leaving the tunnel rather than entering it. As older relatives and friends die, one cannot help but become aware that he or she is not immune to death. As noted earlier, when no older generation exists and the members of one's own generation die with increasing frequency, the scarcity of time becomes a reality. Therefore, for the person older than age 60, the end of the tunnel is in sight. The letter from Lucie Reid (see Listening to the Voices box), sent to George Dickinson and his wife, suggested that the "end of the tunnel is in sight" for her. She seems accepting of the fact and ready to die.

ACHIEVING INTEGRITY   As noted earlier in this chapter, Erik Erikson's theory of human development divided the life cycle into eight stages from infancy to old age. Each stage of development presents a crisis in understanding of one's self, of one's purposes, and of relationships with others. The developmental task at each stage is to resolve the crisis successfully; then the person can progress to the next stage of maturity. The eighth stage is called maturity or older adulthood. Erikson argues that a person must have been successful to some degree in resolving the seven crises that come before to resolve the eighth stage. The task of the final stage of life is to achieve integrity—a conviction that one's life has meaning and purpose and that having lived has made a difference. Having reached the eighth stage, an individual can look back on her or his life and indeed have a feeling, not unlike that of the Apostle Paul in the Bible, that he or she has "fought a good fight" and completed the tasks, and feel a sense of completeness or wholeness. Having the task of life now behind one (completed) helps to reduce anxiety that there is yet more to accomplish and therefore puts one in a "ready" position to die.

C. D. Ryff's findings (1991) in studying the elderly fit well with Erikson's idea of integrity: As people achieve integrity, they view their past less critically and become content with how they have lived their lives. Ryff found that older adults seem to see themselves as closer to really being the persons they wanted to become than does any other age group. With integrity, an individual should have reached the acceptance stage of the dying process, as will be noted by Kübler-Ross in Chapter 5, and feel pretty secure about self. A good self-concept, which completion of Erikson's eight stages should bring, allows the dying individual to have self-confidence, whether facing death or some other challenge in life. If the argument is correct that one dies as one lives, then the dying process for the individual who has reached the stage of integrity should be rather positive, as one's demeanor carries over to the dying process. Such an attitude and behavior would make the dying patient much easier to relate to for caregivers.

A death notice in the Philadelphia Gazette in 1752 (Archives) of Mr. William Bradford suggests that he probably reached the stage of integrity, as he neared the end of his 94 years of life. He came to the United States around 1680 and settled where Philadelphia now stands. He was "a Man of great Sobriety and Industry, a real Friend to the Poor and needy, and kind and affable to all…. Being quite worn out with old Age and Labour, his Lamp of Life went out for want of Oil" (a beautiful euphemism for death!). It is very likely that Mr. Bradford felt good about his self and the life that he had lived.

A connection exists between engaging in a **life review**, a concept advanced by Robert Butler, and achieving integrity. The life review is triggered by the realization that one has reached the end of life and that death is near. The life review serves to prepare a person for dying—preparation that may decrease the fear of death. In one study (Haight, 1992), homebound older adults who were part of a program that assisted people in remembering and reviewing their lives showed significant improvements in life satisfaction and positive feelings and a decrease in depressive symptoms, compared with homebound older adults who did not participate. These changes were still evident two months after the program ended.

A life review sometimes presents an opportunity to return to places one had frequented earlier in life and of which the individual has fond recollections. In the public television production from the 1970s entitled *Dying*, mentioned in Chapter 1, the Reverend Bryant, dying of liver cancer, takes his two adult sons and his wife back to the places he lived as a child: the old homeplace, the cemetery where his parents were buried, and the place where he worked as a young boy. He was re-living his earlier life with his children. "This is where your dad spent his formative years of life, and I want to introduce them to you, as I am about to exit this life," the Reverend Bryant seemed to be saying. He was having a life review ("This is your life, Reverend Bryant," just like the old television show entitled *This Is Your Life*). The life review gives an individual an opportunity to achieve "closure" on her or his life (a way of "putting the ribbon" on the package of life, drawing down the final curtain of the play). Life is over, and, with integrity, the actor feels that she or he has performed well.

DIMINISHING DEATH FEARS   The topics of death and sex tend to cause anxiety to many throughout the life cycle. Sex is typically about the beginning of life, and death is about the end. Both death and sex are part of the life cycle, yet sex is easier for many individuals to discuss than death (Bursack, 2012). Indeed, like their younger contemporaries, older adults have some anxieties and concerns as they approach death. They fear a long, painful, and disfiguring death or death in a vegetative state, hooked to sophisticated machines while hospital bills devour their insurance and savings. They fear that their families will be overburdened by their prolonged care and the expense that it involves. And they dread losing control of their lives by consignment to nursing homes (McCarthy, 1991). Older persons tend to think of death more often than do younger adults, but older persons appear to have less fear and anxiety concerning death. Elderly individuals' death-related experiences and attitudes and reactions toward death are likely to influence their adjustment toward their own aging process (Missler et al., 2012). Accordingly, older persons discuss death more openly and accept it more peacefully, traits characterized by Elizabeth Kübler-Ross in her stages of the dying process. Older individuals have a more positive approach to death and are more aware of their own finitude (Roy & Russell, 2006).

Several factors account for the fact that death anxiety in adults decreases as age increases (Breytspraak, 2008). Factors that may contribute to lower anxiety are a sense that goals have been fulfilled, coming to terms with finitude, living longer than expected, and dealing with the deaths of friends. In an extensive review of research on this topic, Fortner and Neimeyer (1999) concluded that the most powerful predictors regarding older adults' fears of personal death turned out to be their level of "ego integrity" or life satisfaction, the feeling that they had lived long and well.

Because older people are more likely to have fulfilled their goals in life, they are perhaps less fearful of death. For others, death might threaten personal achievement. According to Kastenbaum (2000), coming to terms with death may represent a "final task" of psychological development. The decrease in death anxiety with age may result from a diminished quality of life, thus making death more appealing, or from greater religiosity. Those older people who have not fulfilled their

WORDS OF WISDOM	WARNING

Jenny Joseph

When I am an old woman I shall wear purple
With a red hat which doesn't go, and doesn't suit me.
And I shall spend my pension on brandy and summer gloves
And satin sandals, and say we've no money for butter.
I shall sit down on the pavement when I'm tired
And gobble up samples in shops and press alarm bells
And run my stick along the public railings
And make up for the sobriety of my youth.
I shall go out in my slippers in the rain
And pick the flowers in other people's gardens
And learn to spit.

You can wear terrible shirts and grow more fat
And eat three pounds of sausages at a go

Or only bread and pickle for a week
And hoard pens and pencils and beermats and things in boxes.

But now we must have clothes that keep us dry
And pay our rent and not swear in the street
And set a good example for the children.
We must have friends to dinner and read the papers.

But maybe I ought to practice a little now?
So people who know me are not too shocked and surprised
When suddenly I am old, and start to wear purple.

J. Joseph. (1987). Warning. In S. Martz (Ed.), *When I am an old woman I shall wear purple: An anthology of short stories and poetry*. Watsonville, CA: Papier-Mache Press.

goals are more likely to have either made their goals more modest or somehow to have rationalized their lack of achievement. It is also possible that older people come from an age cohort whose members were better socialized as children to deal with death. They may have become somewhat more able to imagine a world without them in it as they experience the death of others and consider what they will leave behind. Many older persons have, therefore, in some ways come to terms with their own finitude.

Differences in Prevalence of Death Fears.   Heightened death concerns among women have been reported by the majority of researchers who have examined gender differences (Russac et al., 2007), though others have detected no gender differences. By age 60, however, death anxiety and fear for both men and women appears to have stabilized at a low level.

Some evidence (Wass, 1979) shows that widowed individuals tend to be more fearful of death than elderly persons who are married or remarried. Persons living alone have a greater fear of death than those living with a family. Death anxiety tends to be higher among older individuals in institutions, according to a study in the Netherlands (Missler et al., 2012). Elderly persons with only a grade school education have a greater fear of death than do those with a college education. Elderly persons with low incomes exhibit a greater fear of death than do those with higher incomes. High levels of death anxiety for the elderly seem to be associated with poor physical and mental health.

Differences in death anxiety and fear appear to be more a function of religiosity than of age. Because older persons are more likely to be religious, a sense of comfort should be provided as they approach death. The older person is more likely to believe in an afterlife and to rely on a faith in God as a coping strategy in dealing with death. The ceremonial act of regular attendance at religious services contributes to lower death anxiety, perhaps due to group support, whereas private "worship" such as prayer or scripture reading have a lesser impact on death anxiety. Intrinsically religious individuals (those for whom religion serves as a framework for life by providing both meaning and value) are known to report lower levels of death anxiety (Jonas & Fischer, 2006). Years ago French sociologist Emile Durkheim in *The Elementary Forms of Religious Life* (1915) argued that it is the ceremonial acts that are mainly responsible for the life-enriching effects of religion. Thus, shared meanings and values through group participation could certainly give support to an individual regarding the fear of death.

CHOOSING A PLACE TO DIE   Death for the older person becomes a normal and acceptable event. The crisis for the elderly is not so much death itself, but how and where the death will take place. The prospect of dying in a foreign place in a dependent and undignified state is a very distressing thought for older adults. They do not wish to be a financial or physical burden on anyone, yet the options of care may be limited.

The question of physically relocating the elderly to another place is a sensitive issue among **gerontologists**. Those favoring relocation believe that relocating is often less detrimental than leaving the elderly where they currently live. They believe that moving older persons from substandard facilities, for example, will be

DEATH ACROSS CULTURES | ABKHASIANS, THE LONG-LIVING PEOPLE

While the debate may continue as to why some individuals live longer than others, it is well known that the Abkhasians on the coast of the Black Sea in Abkhasia live to ripe old ages. Some 125,000 in numbers, these people live in a country about half the size of New Jersey. They do not have a phrase for "old people." Those older than 100 years of age are called "long-living people"—and most work regularly, performing light household tasks, working in the orchards and gardens, and caring for the animals.

Not only do a large percent of the Abkhasians live a very long time, they also seem to be in fairly good health in their "old age." Close to 40 percent of aged men (older than 90) and 30 percent of aged women have good vision and do not need glasses for any sort of work. Nearly half have "reasonably good hearing." Most have their own teeth. Their posture is unusually erect.

Why do these people live so long? First, they have no retirement status but simply decrease their expected workload as they grow older. Second, they do not set deadlines for themselves; thus there is no sense of urgency except in emergencies. Third, they consider that overeating is dangerous; fat people are regarded as sick. Milk and vegetables make up 74 percent of their diet. The aged average an intake of 1,900 calories per day—500 less than the U.S.

National Academy of Science recommends for those older than 55. They do not use refined sugars; they drink water and honey before retiring in the evening. They eat a lot of fruit, and meat is eaten only once or twice per week. They usually cook without salt or spices. The wine they drink is a dry red wine not fortified with sugar and has a low alcohol content. Few of them smoke, and they do not drink coffee or tea. Their main meal is consumed at lunch (between 2 and 3 p.m.), and their supper is light. Between meals they eat fruit or drink a glass of fermented milk (which is like buttermilk) with a high food value and useful for intestinal disorders. They have a relaxed mood at mealtimes and, when eating, take small bites and chew slowly, thus ensuring proper digestion.

Fourth, they avoid stress by reducing competition. Fifth, they exercise daily. Sixth, their behavior is fairly uniform and predictable. Seventh, they practice moderation in everything they do.

Much of the above is advice we have often heard but failed to put into practice. Perhaps we should adhere to some of the Abkhasians' practices. Who knows, we might even live longer and be more healthy—if that is a goal to be sought.

Source: Adapted from Benet, Sula. (1974). *Abkhasians: The long-living people of the Caucasus*. New York: Holt, Rinehart, & Winston.

beneficial. Those opposed to relocation of elderly persons argue that relocating is traumatic and will generally increase their risk of death.

Although most elderly individuals would prefer to die in their homes in familiar surroundings, the majority continue to die in hospitals and nursing homes, as previously noted. Many people simply cannot afford the luxury of dying at home. At the same time, the institutionalized settings are better equipped to handle dying individuals, and the family is removed from the considerable strain of caring for a dying family member in the home. The elderly are more likely to be separated from family and friends as they die. For them, therefore, the dying process may involve a fear of isolation and loneliness.

CONCLUSION

Rather than trying to protect the young from death talk, we need more open communication channels with children on this topic. Children from an early age have a concept of death. We need to stop pretending that children cannot handle this topic.

It is okay for children (and adults!) to show emotions such as crying when they feel so inclined. It is okay to feel the way one feels—moral judgments should not be attached to feelings. We should try not to be judgmental of others' feelings.

Whoever or whatever (in the case of a pet) has died, children especially need support at this time. One college student recalled the occasion of her mother's unexpected death in an accident. She was five years old at the time and said she felt so insecure when she heard the news. She said, "It was like my whole security blanket crumbled." Indeed, her world had collapsed.

Psychosocial studies of developmental concepts of death are themselves in a stage of infancy. Significant progress is being made, however, in understanding how children and adolescents experience loss at various stages of their development. Adolescence is a particularly vulnerable period with regard to facing death.

People can and do manage much of what happens within their minds. An individual gathers information and insights, helps himself or herself to cope, and gives aid to others in need. At whatever stage of the life cycle, one can be helped to face both life and death more positively.

More research is needed to explore the relationships between death conceptualizations, gender differentiation, and place within the adult life cycle. Some stereotypes concerning older men and women are being discredited, but our conclusions are tentative, and more empirical research is needed.

Growing older pushes one to depend more upon educational, intellectual, and social skills than upon physical prowess. Feeling that one is useful and contributes to the well-being of others, as well as having a healthy understanding of death, contributes significantly to meaningful living and dying.

Some pessimism can be found in older people, but disengagement from life is not necessarily a universal trait. The fear of death lessens with age, but the thought of death increases. The way that time is used changes, as does the meaning that an individual finds in life's experiences. Yet, living a happy and meaningful life is one of the ways to develop a positive and healthy view of death. Perhaps Alex Keaton of the 1990s television show *Family Ties* best summed up the developmental approach to death when he said, "Children die with opportunities and dreams; old people die with achievements and memories."

SUMMARY

1. Due in part to an urbanized and industrialized society, children are removed from death, other than through the rather artificial means of the media.

2. The permanency of death is unclear to young children, who tend to see death as reversible.

3. Children can take about anything adults can dish out to them, including the topics of dying and death.

4. Honesty and openness in relating to children about death is very important.

5. Avoid euphemisms when talking to children about death.

6. It is okay to express emotions through crying. Adults can sanction this behavior by not hiding their own tears when they are sorrowful. By seeing big people cry, children can know that crying is normal behavior.

7. During adolescence the sense of personal identity is most vulnerable, and concepts and feelings of death are powerfully influenced by that vulnerability.

8. The young-adult stage has been labeled the "novice phase" because a strong sense of need to train one's self in the art of reaching one's fullest potential is contradicted by any thought of death.

9. The "panic phase" begins during the middle years when one realizes that the idealized self that one longed to develop may not actually happen.

10. According to Erik Erikson, the task of the final stage of life is to achieve integrity.

11. The crisis for elderly persons is not so much death, but rather how and where the death will take place.

12. Jean Piaget and Erik Erikson present stages of development that help to explain perceptions on dying and death at different age levels.

DISCUSSION QUESTIONS

1. Discuss your first childhood memory of death. How old were you? Who or what died? What do you remember about this event?

2. As a child growing up, how was death talked about in your family?

3. Discuss the various perceptions of death as one goes from birth through age 12. What shortcomings do you find with a life-cycle approach?

4. List as many euphemisms as you can to identify dying and death. Why do you think euphemisms are used with death?

5. Discuss why adults tend to avoid talking about death with children.

6. What are some of the factors, other than age, that influence death conceptualizations?

Why are these factors important in understanding the ways that people conceptualize death?

7. In your opinion, have death conceptualizations changed much in the past few decades? Do you see death conceptualizations changing much in the next few decades?

8. What are some death themes in contemporary adolescent music? How do you explain death themes in music?

9. How do you explain the popularity of some movies with brutal death scenes?

10. Why would education tend to reduce one's anxiety about death?

11. How do Piaget and Erikson's developmental stages differ? How are they similar?

GLOSSARY

Adolescence: Stage of life commonly defined as the onset of puberty when sexual maturity or the ability to reproduce is attained.

Baby Boomers: Individuals born between 1946 and 1964.

Behaviorism: A school of psychology that focuses chiefly on overt behavior rather than on inner psychological dynamics that cannot be clearly identified or measured.

Cognitive Development: Development of processes of knowing, including imagining, perceiving, reasoning, and problem solving.

Content Analysis: A methodology used in the social sciences to systematically study the makeup of something such as the theme of dying and death in

movies or the role of women as caregivers in early 20th-century novels.

Developmental Approach: Branch of social sciences concerned with interaction between physical, psychological, and social processes and with stages of growth from birth to old age.

Ego Psychologists: Theorists and therapists who moved away from Freud toward putting more emphasis in their therapy upon the coping strategies and strengths of the person than upon the more elusive dynamics of the libido and the unconscious.

Gerontology: The study of the biological, psychological, and social aspects of aging.

Humanism: Psychological model that emphasizes an individual's phenomenal world and inherent capacity for making rational choices and developing to maximum potential.

Life Review: Robert Butler's term suggesting a reverence for what once was and for a time for judgment. It includes looking back over an individual's life and perhaps tracing back her or his steps in earlier years, a review of one's life, as death draws near, and is a therapeutic technique to help elderly persons.

Psychoanalysis: A school of theory and therapy that concentrates upon the unconscious forces behind overt behavior, dealing principally with instinctual drives and their dynamics in the individual's inner psyche. Its principal focus is the interplay of transference and resistance between psychoanalyst and client.

Ritual: The symbolic affirmation of values by means of culturally standardized utterances and actions.

Sandwich Generation: Refers to middle-aged individuals who are caught up in the middle. They are caught between their young adult children who still may need support and their elderly parents who also may need support, both physical and financial.

Suggested Readings

Abela, J. R. Z., & Hankin, B. L. (Eds.). (2008). *Handbook of depression in children and adolescents.* New York: The Guilford Press. Offers a comprehensive, scientific overview of contemporary models of the etiology, treatment, and prevention of depression in children and adolescents.

Bearison, D. J. (2006). *When treatment fails: How America cares for dying children.* Oxford: Oxford University Press. The author uses narrative theory to analyze the care of 20 children at a pediatric teaching hospital. Narrative theory is interested in how individuals construct interpretations to give meaning to experience. In addition, narratives are told from the perspective of many medical professionals.

Cruikshank, M. (2013). *Learning to Be Old.* Lanham, MD: Rowman & Littlefield. An examination of what it means to grow old today. The book questions social myths and fears about aging, sickness, the social roles of the elderly, and ageism.

Dickinson, G. E., & Leming, M. R. (2014). *Annual editions: Dying, death and bereavement 13/14* (14th ed.). New York: McGraw-Hill. An anthology with articles related to dying, death, and bereavement.

Jackson, M., & Colwell, J. (2001). *A teacher's handbook of death.* New York: Brunner-Routledge. Provides teachers with methods to facilitate open discussions of death and ways of talking with children about what happens when someone they know dies. Also offers strategies for talking about death in the context of different school lessons to prepare students of all ages.

Kerr, M. M. (2009). *School crisis prevention and intervention.* Upper Saddle River, NJ: Merrill. The author has gathered very good resources for school crisis and has interwoven guidelines for crisis planning with vignettes.

Loe, M. (2011). *Aging our way: Lessons for living from 85 and beyond.* New York: Oxford. Follows the everyday lives of 30 elders living at home, most alone, to understand how they create and maintain meaningful lives for themselves.

McCullough, D. (2008). *My mother, your mother: Embracing "slow medicine," the compassionate approach to caring for your aging loved ones.* New York: Harper/Collins. Geriatrician McCullough advocates for careful anticipatory "attending" to an elder's changing needs rather than waiting for crises that force acute medical interventions.

Stevenson, R. G., & Cox, G. R. (Eds.). (2007). *Perspectives on violence and violent death.* Amityville, NY: Baywood Publishing Company. The sections of this edited volume by two veteran thanatologists include the nature of violence, encounters with violence, empirical studies, and alternatives to violence or coping with violence. The book takes an international perspective on the many manifestations of violence.

Valent, P. (2002). *Child survivors of the Holocaust.* New York: Brunner-Routledge. Valent, himself a child survivor and psychiatrist, explores the memories of those survivors he interviewed, and addresses the trauma facing contemporary child victims of abuse worldwide.

Religion provided me with answers to problems I didn't even know I had.
—**St. Olaf College student, 1977**

Death radically challenges all socially objectivated definitions of reality—of the world, of others, and of self.... Death radically puts in question the taken-for-granted, "business-as-usual" attitude with which one exists in everyday life.... Religion maintains the socially defined reality by legitimating marginal situations in terms of an all-encompassing sacred reality.
—**Peter Berger,** *The Sacred Canopy*

PERSPECTIVES ON DEATH AND LIFE AFTER DEATH

CHAPTER **4**

Pool/Getty Images News/Getty Images

All living creatures are part of a cycle of life and death. This is a natural process that occurs with regularity and is fairly easy to explain in biological terms. However, humans have never been satisfied with just a biological explanation for this cycle of life and death. Across cultures and throughout history, humans have constructed belief systems that explain death in cosmological, spiritual, and/or religious terms. These explanatory systems, even though widely diverse, usually deal with death in some way because, like no other social event or situation, death inherently challenges the "taken-for-granted" meanings of all societies.

Some belief systems view death as a continuation of life, with spirits of the dead inhabiting the world alongside the living. For example, the traditional African attitude toward death is "positive and accepting and comprehensively integrated into the totality of life" (Barrett, 1992). Other belief systems regard death as a transition to another level of existence—an afterlife—which has very little to do with the living. The Christian concepts of heaven and hell represent one example of this type of belief. For the Navajos, heaven is earth, where all goodness resides. Death leads to an underworld that is similar to earth and considered the place of all evil. Precautions must be taken to ensure that the dead are buried properly so they do not come back to haunt the living.

There are also those who believe that, after death, there is nothing, and death is meaningless. Although this view is held by many people, there is no culture or dominant belief system in which this nihilistic perspective on death is normative. One of the most feared, distressing, and anxiety-provoking deaths is a death that is perceived as meaningless. The search to find meaning in life and death is a task that all people share.

THE NEED TO LOOK BEYOND DEATH

From a symbolic interactionist perspective (see Chapter 1), meanings are created and reproduced by humans. These meanings supply a base for activities and actions (behavior is in response to meanings) and provide order for the people who share a given culture. Peter Berger (1969) suggested that the human world has no order other than that created by humans. To live in a world without the order contributed by one's culture would force an individual to experience a meaningless existence. Sociologists refer to this condition as **anomie**—"without order."

Many situations in life challenge the order on which social life is based. Most of these situations are related to what Thomas O'Dea (1966) referred to as the three fundamental characteristics of human existence: uncertainty, powerlessness, and scarcity.

Uncertainty refers to the fact that human activity does not always lead to predictable outcomes. Even after careful planning, most people recognize that they will not be able to achieve all of their goals. Less optimistically, the 21st century's Murphy's Law states "Anything that can go wrong will go wrong." The human condition is also characterized by *powerlessness*. We recognize that there are many situations in life and events in the universe over which humans have no control—among these situations are death, suffering, coercion, and natural disasters. Finally, in *scarcity*, humans experience inequality with regard to the distribution of wealth, power, prestige, and other things that make a satisfying life. This inequality is the basis for the human experience of relative deprivation and frustration. The three experiences of uncertainty, powerlessness, and scarcity challenge the order of everyday life and are, therefore, marginal to ordinary experiences. According to O'Dea (1966, p. 27), such experiences "raise questions which can find an answer only in some kind of 'beyond' itself." Therefore, **marginal situations**, which are characteristic of the human condition, force individuals to the realm of the transcendent in their search for meaningful answers.

Berger (1969, pp. 23, 43–44) claimed that death is the marginal situation par excellence:

> Witnessing the death of others and anticipating his own death, the individual is strongly propelled to question the ad hoc cognitive and normative operating procedures of "normal" life in society. Death presents society with a formidable problem not only because of its obvious threat to the continuity of human relationships, but also because it threatens the basic assumptions of order on which society rests. Death radically puts in question the taken-for-granted, "business-as-usual" attitude in which one exists in everyday life. Insofar as knowledge of death cannot be avoided in any society, legitimations of the reality of the social world in the face of death are decisive requirements in any society. The importance of religion in such legitimations is obvious.

It is **religion**, or a transcendent reference, which helps individuals remain reality-oriented when the order of everyday life is challenged. Contemplating death, we are faced with the fact that we will not be able to accomplish all of our goals in life. We also realize that we are unable to extend the length of our lives and/or control the circumstances surrounding the experience and cause of our deaths. We are troubled by the fact that some must endure painful, degrading, and meaningless deaths, whereas others find more meaning and purpose in the last days of their lives than they experienced in the years preceding "the terminal period." Finally, the relative deprivation created by differential life spans raises questions that are unanswerable from a "this world" perspective.

Religious-meaning systems provide answers to these problems of uncertainty, powerlessness, and scarcity created by death. O'Dea (1966, pp. 6–7) illustrates this function of religion:

> Religion, by its reference to a beyond and its beliefs concerning man's relationship to that beyond, provides a supraempirical view of a larger total reality. In the context of this reality, the disappointments and frustrations inflicted on mankind by uncertainty and impossibility, and by the institutionalized order of human society, may be seen as meaningful in some ultimate sense, and this makes acceptance of and adjustment to them possible. Moreover, by showing the norms and rules of society to be part of a larger supraempirical ethical order, ordained and sanctified by religious belief and practice, religion contributes to their enforcement when adherence to them contradicts the wishes or interests of those affected. Religion answers the problem of meaning. It sanctifies the norms of the established social order at what we have called the "breaking points," by providing a grounding for the beliefs and orientations of men in a view of reality that transcends the empirical here-and-now of daily experience. Thus, not only is cognitive frustration overcome, which is involved in the problem of meaning, but also the emotional adjustments to frustrations and deprivations inherent in human life and human society are facilitated.

Religious systems provide a means to reestablish the social order challenged by death. Our society has institutionalized the continued importance of religion by creating funeral **rituals** that have a religious quality about them. To demonstrate the widespread tendency toward religious perspectives, a Gallup Poll declared in 2013 that 14 percent of Americans claimed no religious identity, and another 2 percent did not answer the question of religious preference. Furthermore, 76 percent of Americans in 2012 told Gallup interviewers that they identified with a Christian religion. That percentage includes 41 percent who said they were Protestant, 10 percent who were "other Christian," 23 percent Roman Catholic, and 2 percent who named another Christian faith (Mormon). These results suggest that more than nine out of 10 Americans who identify with a religion (83 percent) are Christian in one way or another. It appears that in the United States most people rely upon religious and spiritual paths in coping with life and its challenges.

Although many people do not follow a formal religion, there are very few who do not have some kind of spiritual concern with transcendent meanings that may be shared by a number of religious groups. These nonaffiliated individuals, or what Glenn Vernon called the "religious 'nones'" (people who responded "none" when asked the question "To which religion do you belong?"), often borrow from many religious traditions and spiritual perspectives as they formulate supernatural and transcendent interpretations for death and after-death possibilities. According to the Gallup Poll in 2013, 14 percent of Americans fall into this category. There are even some religious groups that have institutionalized this broader, more inclusive spiritual orientation to religion and death. Included among these groups are Unitarian-Universalists, Baha'is, Unity Church members, and even some Quakers. From a functional point of view, people with a spiritual point of view are similar to the religious, but they are more individualistic and private or subjective in the way in which they apply their transcendent views.

DIVERSITY IN PERSPECTIVES

William Cowper once said, "Variety's the very spice of life, that gives it all its flavour." If Cowper is correct, a look at various cultures and how they view dying and death indeed will produce a lot of "flavours." If we did not have cultural diversity, however, would not life be rather dull?

CROSS-CULTURAL VIEWS

Cross-cultural studies of death reveal that most societies seem to have a concept of soul and immortality. For some, a belief in a soul concept explains what happens in sleep and after death. For others, a belief in souls explains how the supernatural world becomes populated (Tylor, 1873). Much of the death ritual performed is related to the soul and is often an appeasement of the beings in the spiritual world. This spiritual dimension of death plays a significant role in the social structure of societies that hold such beliefs.

In studying death cross-culturally we are reminded of other similarities, such as beliefs in ghosts. For many groups, souls of the dead become ghosts, who may take up residence indefinitely, or after a short period of wandering may cease to exist. Some people converse with ghosts and make offerings to them. This is sometimes referred to as the "cult of the dead" (Taylor, 1988). These ghosts are not worshipped but simply maintain a relationship with a person. They act as guides and protectors and may also confer power. In some societies, to have a guardian spirit is a positive experience. For others, like Navajo and Apache Native Americans, ghosts are responsible for sickness and death and are to be avoided (Cox & Fundis, 1992). Ghosts are not unique to nonliterate cultures. Many sane, sober individuals in the United States have had "encounters" with ghosts.

The continuity between the living and the dead is elaborated in ideas of reincarnation and in other ways in non-Western as well as Western societies. For example, many Japanese think that a person's spirit belongs to the same family and the same local community before and after death (Nagamine, 1988). A person's spirit gradually fades away from its family as time goes on. The Japanese do not clearly distinguish the dead from the living and seem to recognize the continuity between life and death. Especially in rural Japan, the dead remain an integral part of life and offer constant solace to the living (Kristof, 1996).

The Japanese are not alone in believing that the dead remain a vital part of their lives. For instance, in some parts of southern India, where many Hindus mourn the dead for 13 days, each person serves part of a meal on a banana leaf to provide for the needs of the dead person's soul (Tully, 1994). The meal is laid out in the open, and if the crows eat it, that means that the offering has been accepted.

The Hopi Indians viewed life and death as phases of a cycle (Oswalt, 1986). Death was an important change in individual status because it represented an altered state for the person involved. The Hopi believed that a duality of being existed in each person—a soul and a body expressed as a "breath-body." A person dying on earth literally was reborn in the afterworld. Corpses were washed and given new names before burial. The breath-body's pattern of existence would supposedly be the same in the domain of the dead as it had been on earth—when it died in the afterworld, it would be reborn on earth.

The Igbo in Nigeria (Uchendu, 1965) believe that death is important for joining their ancestors. Without death, there would be no population increase in the ancestral households and thus no change in social status for the living Igbo. The lineage system is continued among the dead. Thus, the world of the "dead" is a world full of activities.

The Ulithi of Micronesia are not morbid or defeated by death, according to William A. Lessa (1966). Their rituals afford them some victories, and their mythology provides a hope for a happy life in another realm. Though their gods are somewhat distant, these gods assure that the world has an enduring structure, and their ancestral ghosts stand by to give more immediate aid when merited. Thus, after the Ulithi express their bereavement, rather than retreat, they spring back into their normal work and enjoyment of life.

For the Dunsun of northern Borneo, few events focus more on the beliefs and acts concerned with the nonnatural world than does the death of a family member (Williams, 1965). Death is considered a difficult topic that everyone fears talking about, yet it must be prepared for because it causes great changes.

The Abkhasians on the coast of the Black Sea (Benet, 1974) view death as irrational and unjust. The one occasion when outbursts of feeling are permitted is at a funeral—wailing and scratching at one's flesh are permitted (see Box on page 116).

Bernard Pierre Wolff/Science Source

This Japanese funeral emphasizes the social status of the man who has died. The living honor the dead and thereby create a symbolic community consisting of their ancestors and living members of their communities.

Case Study: The Sacred World of Native Americans

While *religion* is a word that has no equivalent in the languages spoken by Native Americans, they are intensely spiritual people whose spirituality is exhibited in their art and everyday living (Mander, 1991). In the daily life of the Native American, spirituality is an inevitable part of every facet of life. Spirituality and prayer are inseparable to the Native American. To separate spirituality and prayer would mean cultural death for the Native American. Every day is a holy day because all that Native Americans undertake begins and ends with the influence of a spiritual kinship with nature.

The Native American looks to this spirituality to explain the successes and failures of daily life. The guiding spirit or force, which rules nature, can help him or her through daily life and its struggles. As a consequence, Native Americans do not view death as something to be ignored, but rather as a natural part of life.

For the Native American, death is the natural end of life—it needs to be natural and the dying person should be allowed to die without restrictions. Therefore, the confrontation with one's destiny is the confrontation with death and with one's spirituality.

Attitudes toward life and death are learned and spiritual teachings are one source of learning. The concepts of mortality, soul, and afterlife are important for Native Americans and, as a consequence, they place a great emphasis upon prayer and spiritual rituals. While the individual is essentially spiritual, the Native

American will turn to a priest or a shaman or other qualified person to deal with the sacred world because they have the knowledge of how to deal with the sacred. The ritualistic prayers and chants of the Navajo have their parallels in other religions, as does the centrality of belief and faith for religious practice. However, a major difference for Native American spirituality is that the sacred words of the Navajo are passed down in an oral tradition, even though in recent years, these sacred words have been collected in a written form (Wyman, 1970; Gill, 1981).

While generalizations are dangerous, the variety of cultural expressions of dying and death do have some commonalties for the various Native American tribes. Most tribes express a willingness to surrender to death at any time without fear. The Lakota Chief Crazy Horse was noted for his chant before going into battle, "Today is a good day to die." Every day is to be lived as if it were one's last day. One must enjoy life and live fully. Just as one cannot buy land, one cannot buy life. Death is waiting. One cannot escape. One does not seek death before its time, nor does one avoid death or try to delay its occurrence. No one is ever truly alone. The dead are not altogether powerless. There is no death, but rather simply a change of worlds (Steiger, 1974).

Once one dies, the ghost becomes of great concern. One does not use the name of the dead. Ghosts may be too busy, and to interrupt them may cause anger (Stoeckel, 1993). Not to honor the dead also invites their wrath. For the Navajo, all sickness and disease, whether mental or physical, is believed to be due to supernatural causes. To appease their ancestors, the Apache set aside part of their fields as the fields of the dead. They would not cultivate these fields for three years, even if it meant starvation for the family (Mails, 1991).

To the Lakota everything was filled with spirits and powers that controlled or otherwise affected the lives of the people. Humans are no better or worse than other living things. All things have a spirit, and illness and death are a part of life, which must be accepted. If one is to be in balance or harmony with nature, one must die when it is one's time. A person should not have to sacrifice dignity by crying out for interventions that will only prolong the natural process of dying because it is better to die today than to be shamed by living tomorrow.

The lessons of the Native American are clear. A natural death is better than an artificial one involving futile interventions. For the Native American, medicine heals the spirit as well as the body. When the Native American dies, it was his or her time to die. It was not the fault of the shaman or the family. In the past, white physicians did not have the techniques and medicines that they now have to prolong the dying process. Then, they too relied on supernatural interventions. Today, when a person dies, the physician loses a patient, and death is portrayed as the enemy of modern medicine. For the Native American, death is natural and not an enemy, just as suffering is a part of life that cannot be avoided. Modern medicine's emphasis upon winning has led to guilt, experienced by the patient, family, and physician. From a Native American perspective, neither the physician, the family, nor the patient has lost. Rather, the patient died, as will we all.

Thus, attitudes toward death vary significantly among societies around the world—from seeing death as a continuation of life to the end of everything. We will now turn to the role of death in creating religious beliefs and rituals.

(*Adapted from an article written by Gerry Cox and used with his permission.*)

RELIGIOUS INTERPRETATIONS OF DEATH

Until this point, we have provided a functional perspective on religion—we have defined religion in terms of what religion does for the individual and society. This perspective focuses on the consequences of religion rather than on the content of religious belief and practice.

We will now attempt to provide substantive perspectives on religion and death. The substantive perspective to religion attempts to establish what religion is. Substantive definitions of religion endeavor to distinguish religious behaviors from nonreligious behaviors by providing necessary criteria for inclusion as religious phenomena. The substantive definition of religion most frequently used by sociologists is the one formulated by Emile Durkheim (1915) in *The Elementary Forms of Religious Life*:

> A religion is a unified system of beliefs and practices relative to sacred things; that is to say, things set apart and forbidden—beliefs and practices that unite into one single moral community called a church, all of those people who adhere to them.

In this definition, Durkheim designates four essential ingredients of religion—a system of beliefs, a set of religious practices or rituals, the sacred or supernatural as the object of worship, and a community or social base. Most substantive definitions of religion used by contemporary sociologists will incorporate these four essential ingredients. However, some sociologists have argued that to be inclusive of phenomena that most people consider religious, it might be proper to exclude the necessity of having a sacred point of reference. Buddhism, for example, does not have a supernatural being to which beliefs and rituals are oriented. We will now consider five religious traditions and explain how each tradition attempts to interpret the meaning of death-related experiences and provides funeral rites and rituals for the bereaved. The five religious traditions considered are Judaism, Christianity, Islam, Hinduism, and Buddhism.

JUDAISM

Death in the Jewish tradition came into being as a result of Adam's and Eve's sin, which caused them to be expelled from the Garden of Eden. When Adam and Eve ate the fruit from the "Tree of Conscience," they received the curse of pain in childbirth, the burden of work, and the loss of physical immortality. According to the biblical account in Genesis (2:4–3:24), although death was a punishment, it also brought the ability to distinguish between good and evil as well as the power and responsibility to make decisions that have a future consequence. According to J. Carse (1981):

> Adam and Eve lost their immortality, but acquired consciousness instead. God drove them out of Paradise into death, but also into history. God's design for the people of Israel is to save them from their enemies in order that their history might continue.

In this description we gain an understanding of the importance of God's covenant with Abraham—that Abraham would become the father of many nations and that God would have a special relationship with his descendants forever (Gen. 17). Consequently, immortality was to be found in one's identity within the group.

| **PRACTICAL MATTERS** | USING RELIGIOUS SYMBOLS IN A FUNERAL HOME |

Forty years ago while on a class field trip to a funeral home, Michael Leming complimented the funeral director on the beautifully painted pastoral scene hanging on a wall. The funeral director said it was a very unusual wall hanging and took it down to show the framed velvet reverse side, which could also be displayed. He then took out a box containing a cross, a crucifix, and a star of David that could be hung on the velvet backing. The funeral director told the class that he changed the hanging whenever the religious affiliation of the deceased varied. Since the time of that field trip, Leming has become very conscious of the way that funeral homes extensively use religion to provide comfort for the grieving. Consider the following:

1. Within the funeral home, *chapel* is the name given to the room where the funeral is held.
2. Most memorial cards have the 23rd Psalm on them.
3. The music that one hears on the sound systems within most funeral homes is religious in nature.
4. Wall hangings found in most funeral homes usually have religious content.
5. Funeral homes often provide Christmas calendars, complete with Bible verses and religious scenes, for certain religious groups and other interested members of the community.

Among contemporary Jews, including those who consider themselves religious, opinion differs regarding personal immortality. Some contend that there is no after-*life*, only an after-*death*—the dead go to Sheol, where nothing happens, and the soul eventually slides into oblivion. Other Jews believe in a resurrection of the soul, when individuals are brought to a final judgment (Carse, 1981). Still, for others there is a real ambivalence regarding the immortality of the soul. Carse cites the writings of Rabbi Leona Modena (1571–1648), who stated:

> It is frightening that we fail to find in all the words of Moses a single indication pointing to man's spiritual immortality after his physical death. Nonetheless, reason compels us to believe that the soul continues.

Regardless of the content of Jewish beliefs regarding the immortality of the soul, Jewish funeral customs and rituals emphasize that God does not save us, as individuals, from death, but saves Israel for history, regardless of death (Carse, 1981).

In 2012 Rabbi Shmuley Boteach (cited by Christian Century, 2012) provides a very contemporary interpretation, "Heaven misses the point of religion, while I do not deny its existence, I believe that for a Jew his job is to think about this world rather than the next. We need to make the earth itself more heavenly without any thought of reward for having done so."

CHRISTIANITY

Although Christianity shares much of the historical and mythical foundations of Judaism, there are many distinct differences in the Christian approach to death, afterlife, and funeral rituals. For the Christian, death is viewed as the entrance to eternal life and, therefore, is preferable to physical life. There is a strong belief in the immortality of the soul, the resurrection of the body, and a divine judgment of

© iStockphoto.com/Boryak

A religious orthodox Jew prays at the western wall in Jerusalem. Jewish funeral customs and rituals emphasize that God does not save us, as individuals, from death, but saves Israel for history, regardless of death.

one's earthly life after death, resulting in the eternal rewards of heaven or the punishments of hell. For the Roman Catholic there are four potential dispositions of the soul after death—heaven, hell, limbo, and purgatory. According to McBrien, "some will join God forever in heaven; some may be separated eternally from God in hell; others may find themselves in a state of merely natural happiness in limbo; and others will suffer in purgatory some temporary 'punishment' still required of sins that have already been forgiven" (1987, p. 443).

For the Christian, the teachings of Jesus and the Apostle Paul are the most important sources in arriving at a theology of life after death. Jesus declares to his followers:

> I am the Resurrection and the Life, he who believes in Me, though he die, yet shall he live, and whoever lives and believes in Me shall never die. (John 11:25, RSV)

In the 15th chapter of the first letter to the Corinthians, the Apostle Paul discusses the significance of Jesus' resurrection for the Christian believer. In this chapter Jesus is portrayed as the "first born from the dead" and as "the one who has destroyed death" (1 Cor. 15:26, RSV). Paul declares (1 Cor. 15:52–58, RSV) that at the end of history:

> The dead will be raised imperishable, and we shall be changed. For this perishable nature must put on the imperishable, and this mortal nature must put on immortality.

When the perishable puts on the imperishable, and the mortal puts on immortality, then shall come to pass the saying that is written:

> Death is swallowed up in victory.
> O death, where is thy victory?
> O death, where is thy sting?

The sting of death is sin, and the power of sin is the law. But thanks be to God, who gives us the victory through our Lord Jesus Christ.

Christians employ two basic, and somewhat paradoxical, perspectives when facing death. The first has just been described—that through faith in Jesus Christ, the Christian has victory over death and gains eternal life with God. The following passage (Rom. 8:31–39, RSV) provides us with an example of this orientation:

> What then shall we say to this? If God is for us, who is against us? He who did not spare his own Son but gave him up for us all, will he not also give us all things with him? Who shall bring any charge against God's elect? It is God who justifies; who is to condemn? Is it Christ Jesus, who died, yes, who was raised from the dead, who is at the right hand of God, who indeed intercedes for us?

> Who shall separate us from the love of Christ? Shall tribulation, or distress, or persecution, or famine, or nakedness, or peril, or sword? As it is written:

>> "For thy sake we are being killed all the day long; we are regarded as sheep to be slaughtered."

> No, in all these things we are more than conquerors through him who loved us. For I am sure that neither death, nor life, nor angels, nor principalities, nor things present, nor things to come, nor powers, nor height, nor depth, nor anything else in all creation, will be able to separate us from the love of God in Christ Jesus our Lord.

The second perspective on death employed by Christians emphasizes the experience of true human loss. This approach is exemplified by Jesus (John 11:32–36, RSV) as he responds to the death of his friend Lazarus:

> Then Mary, when she came where Jesus was and saw Him, fell at his feet, saying to Him, "Lord, if you had been here, my brother would not have died." When Jesus saw her weeping, and the Jews who came with her also weeping, He was deeply moved in spirit and troubled; and He said, "Where have you laid him?" They said to Him, "Lord, come and see." Jesus wept. So the Jews said, "See how He loved him!"

C. S. Lewis, in his book *A Grief Observed* (1961, p. 24), illustrated how Christians use this paradoxical double perspective:

> What St. Paul says can comfort only those who love God better than the dead, and the dead better than themselves. If a mother is mourning not for what she has lost but for what her dead child has lost, it is a comfort to believe that the child has not lost the end for which it was created. And it is a comfort to believe that she herself, in losing her chief or only natural happiness, has not lost a greater thing, that she may still hope to "glorify God and enjoy Him forever." A comfort to the God-aimed, eternal spirit within her. But not to her motherhood. The specifically maternal happiness must be written off. Never, in any place or time, will she have her son on her knees, or bathe him, or tell him a story, or plan for his future, or see her grandchild.

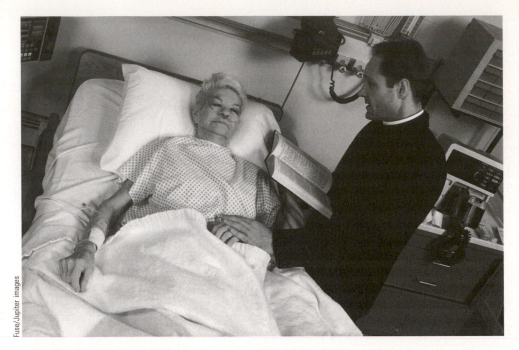

Fuse/Jupiter images

Religious-meaning systems help provide individuals with a transcendent point of reference whereby a loss created by death is compensated for by a system of other-worldly gains. In this photo, the woman receives last rites from her priest.

ISLAM

As in Christianity, life after death is also an important focus within the Islamic tradition. Earthly life and the realm of the dead are separated by a bridge that souls must cross on the Day of Judgment. After death, all people face a divine judgment. Then they are assigned eternal dwelling places where they will receive either eternal rewards or punishments, determined by the strengths of their faith in God and the moral quality of their earthly lives. According to the Qur'an, there are seven layers of heaven and seven layers of *alnar* ("Fire of Hell"), and each layer is separated from the layer above by receiving fewer rewards or greater punishments. The fundamental reason that individuals might be condemned to a life of torment in the Fire of Hell is a lack of belief in God and in the message of his prophet Muhammad. Other reasons include lying, being corrupt, committing blasphemy, denying the advent of the Judgment Day and the reality of the Fire of Hell, lacking charity, and leading a life of luxury (Long, 1987).

Like Jews and Christians, followers of Islam believe that God is fundamentally compassionate and place a similar emphasis on God as just. Therefore, individuals are held accountable for moral integrity at the time of their death. The primary expression of the Islamic concern for justice and accountability is found in the belief in assignment to paradise or damnation. Accordingly, the Qur'an provides very vivid sketches of both paradise and hell. However, many Islamic theologians also stress that God's judgment is tempered with mercy, that the angel Gabriel will

Muslim women praying in the cemetery.

intercede on behalf of those condemned to punishment, and that they will eventually be pardoned. The following prayer (the opening of the Surah of the Qur'an) illustrates the Islamic perspective on divine justice and mercy:

> In the name of God, the Merciful, the Compassionate. Praise belongs to God, the Lord of all Being, the All-merciful, the All-compassionate, the Master of the Day of Doom.
>
> Thee only we serve; to Thee alone we pray for succour, Guide us in the straight path, the path of those whom Thou has blessed, not of those against whom Thou art wrathful, nor of those who are astray.

HINDUISM

Unlike the religious traditions that we have previously discussed, Hinduism does not have a single religious founder nor a single sacred text. Although the Vedas are recognized by almost all Hindus as an authoritative source of spiritual knowledge, Hinduism is not dogmatic. Many theologies and religious approaches exist within the Hindu tradition. Many gods also exist, each viewed as an aspect or manifestation of a single ultimate reality, but it is not essential to believe in the existence of God to be a Hindu (Srinivas & Shah, 1968). Essentially, Hinduism is a system of social customs imbued with religious significance.

Three concepts are central to an understanding of Hinduism—*karma*, *dharma*, and *moksha*. *Karma* refers to a moral law of causation; it suggests that an individual's actions produce results for which the individual is responsible. *Karma* also refers to the balance of good and bad deeds performed in previous existences. *Dharma* are the religious duties, requirements, and/or prescriptions. The extent to which one fulfills one's *dharma* determines one's karma. In turn, *moksha* is the reward for living a saintly life. The main ways of achieving *moksha* are overcoming spiritual ignorance by acquiring true knowledge, performing good deeds, and living a life of love and devotion to God (Srinivas & Shah, 1968).

The central doctrine affecting death-related attitudes and behavior in the Hindu religion is reincarnation and the transmigration of souls (*samsara*). For the Hindu, one's present life is determined by one's actions in a previous life. Furthermore, one's present behavior will shape the future. According to R. W. Habenstein and W. M. Lamers (1974, p. 116), "The ultimate goal of the soul is liberation from the wheel of rebirth, through reabsorption into or identity with the Oversoul (*Brahma*)—the essence of the universe, immaterial, uncreated, limitless, and timeless." The way in which one would experience reabsorption is to overcome spiritual ignorance (*avadya*) by ultimately coming to know the truth of oneself and the truth of God. It is *moksha* that brings reabsorption and ends *samsara*, the cycle of reincarnation and transmigration of souls.

By way of contrast, whereas Jews, Christians, and Muslims believe in the immortality of the soul and hope for an afterlife, Hindus hope that their soul will be absorbed at death. For the Hindu, the goal is not to experience life after death, but rather to have one's soul united with the Oversoul. Punishment for the devout Hindu might be to *have* "everlasting spiritual rebirth."

Within the Hindu tradition, death brings two possibilities—liberation or transmigration of the soul. Neither is inherently fearful, even though separation from one's friends and loved ones may cause sadness and personal loss. Somewhat analogous to the dual perspectives of Christianity, death for the Hindu individual brings hope of something better but also the human loss of being separated from a dead loved one.

BUDDHISM

There are many types of Buddhism, but the most popular are *Theravada*, *Mahayana*, and *Tantrayana*. *Theravada* Buddhism is the dominant religion in Southeast Asia (Myanmar, Thailand, Laos, Vietnam, Cambodia, and Sri Lanka). *Mahayana* Buddhism is primarily practiced in Korea, China, Japan, and Nepal. *Tantrayana* has traditionally been the dominant form of Buddhism in Tibet, Mongolia, and parts of Siberia. Although there are differences among each of these sects and many others found throughout the world, Buddhists in general find a common heritage in the life of Siddhartha Gautama ("The Enlightened One" or "Buddha").

Buddha was born in 563 B.C. as a prince in northern India. When he was a child, his thoughts were preoccupied with the finitude of human existence. Unsuccessfully, his family tried to shelter him from human suffering and death. At the age of 29 he left his life of privilege and began to search for personal salvation. After rejecting physical asceticism and abstract philosophy, he attained a state of enlightenment through the process of intense meditation. For the remaining 50 years of his life, he served as a missionary and preached his message of salvation to all people regardless of social position and gender (Pardue, 1968).

The Buddhist message of salvation, taught by Buddha in his first sermon, is the "Four Noble Truths." The first of these truths is that all human existence is characterized by pain and suffering in an endless cycle of death and rebirth. The second truth is that the cause of the agony of the human condition is desire for personal satisfaction, which is impossible to obtain. The third truth is that salvation comes by destroying these desires. By completely destroying ignorance, one experiences

Michael R. Leming

Devoted Hindus in Bali make pilgrimage to temple, the home of the Hindu gods. One third of a woman's life in Bali will be spent either preparing for or performing Hindu religious rituals.

| DEATH ACROSS CULTURES | TEACHING REINCARNATION IN THE BHAGAVADGITA |

Death and birth are seen by us Hindus as gateways of exit and entry to the stage of this world. This process is described in one of the most striking and famous analogies in the entire *Bhagavadgita*. It is a verse which we movingly recite when the body is being cremated, and it has given comfort and solace to the grieving Hindu heart for centuries.

> Just as a person casts off worn-out garments and puts on others that are new, even so does the embodied-soul cast off worn-out bodies and take on others that are new.

This analogy is full of suggestions, but I can only briefly draw your attention to a few of these. First, a suit of clothing is not identical with the wearer. Similarly, the body, which is likened here to worn-out garments, is not the true being or identity of the human person. Second, there is the similarity of a continuity of being. When a worn-out suit of clothing is cast off, the wearer continues to be. Similarly, with the disintegration of the physical body, the indweller, that is, the Self, continues to be. Finally, there is the parallel with the timing of the change. A suit of clothing is changed when it no longer serves the purpose for which it was intended. Similarly, the physical body is cast off when it no longer serves the purpose for which it came into existence.

This does not mean that the Hindu is not saddened by death or does not lament the loss it brings. There is a mysterious element in death which will always sadden and hurt. While we ache and long for the beauty, tenderness, and companionship of a dear one, we are deeply comforted in the knowledge that all that is good and true and real in him or her continues to be. What has passed away is what has always been limited by time.

Source: Annant Rambachan, Department of Religion, St. Olaf College. By permission of the author.

enlightenment, and the cycle of transmigration of the soul is broken. Finally, in the fourth truth, one can experience perfect peace through the eightfold path to enlightenment. In the words of Pardue:

> For this purpose the proper [meditation] is the "eightfold path," an integral combination of ethics (*sila*) and meditation (*samadhi*), which jointly purify the motivations and mind. This leads to the attainment of wisdom (*prajna*), to enlightenment (*bodhi*), and to the ineffable *Nirvana* ("blowing out"), the final release from the incarnational cycle and mystical transcendence beyond all conceptualization. (1968, p. 168)

Buddhism assists its practitioners by encouraging them to detach from striving for living, holding onto relationships, and attempting to avoid pain and sadness. From a Buddhist perspective it is by accepting the impermanence of life, relationships, and desirable things and then letting go of them that one finds enlightenment.

In the Buddhist meditation entitled "The Way of the Mystic," Sister Medhanandi (1998) claimed:

> There are two kinds of death for the living, one that leads to death and one that leads to peace, to enlightenment. When we carry around a lot of wreckage in the mind, we are not putting down the burden. We are identifying with and caught in self-view, "I am an abused person" or "a grief-stricken person," or "Five of my friends have died from AIDS and I just can't face life." That's a death that leads to death.
>
> But if we can meet the present moment with mindfulness and wise reflection, we can begin to put down that burden, surrender it, and allow ourselves to receive the next moment with purity of mind, letting the conditions that arise and our attachment to them die. That kind of death leads to enlightenment.

| DEATH ACROSS CULTURES | THE STORY OF KISAGOTAMI |

In the story of Kisagotami, this teaching is beautifully brought to life. She had married into a wealthy family, in spite of her poverty and unattractive appearance, and finally won the acceptance of her in-laws when she bore a little son. Suddenly, the child died. Nothing could be more tragic for Kisagotami. She refused to accept that her little son was dead. In her desperation, she came to see the Buddha, cradling the infant in her arms, believing that the Blessed One could revive him.

The Buddha asked her to procure a small quantity of mustard seeds from a house where no one had ever died. When she could find no household that had been spared death's unremitting hand, the insight into the impermanence of all conditioned phenomena arose in her mind. And so, Kisagotami was able to go beyond "the death of sons," beyond sorrow. (S.N. [*Samyutta Nikaya*] 5)

Source: From The way of the mystic, by Sister Medhanandi, 1998. Available online at http://www.buddhanet.net/mystic.htm

She also said in an article entitled "The Joy Hidden in Sorrow":

In pain we burn but, with mindfulness, we use that pain to burn through to the ending of pain. It's not something negative. It is sublime. It is complete freedom from every kind of suffering that arises; because of a realization—because of wisdom—not because we have rid ourselves of unpleasant experience, only holding on to the pleasant, the joyful. We still feel pain, we still get sick and we die, but we are no longer afraid, we no longer get shaken.

When we are able to come face to face with our own worst fears and vulnerability, when we can step into the unknown with courage and openness, we touch near to the mysteries of this traverse through the human realm to an authentic self-fulfillment. We touch what we fear the most, we transform it, we see the emptiness of it. In that emptiness, all things can abide, all things come to fruition. In this very moment, we can free ourselves. We are not diminished by tragedy, by our suffering. If we surrender, if we can be with it, transparent and unwavering—making peace with the fiercest emotion, the most unspeakable loss, with death—we can free ourselves. And in that release, there is radiance. Jelaluddin Rumi wrote: "The most secure place to hide a treasure of gold is some desolate, unnoticed place. Why would anyone hide treasure in plain sight? And so it is said: Joy is hidden in sorrow" (Medhanandi, 1996).

Like the Hindu, for the Buddhist the goal is not to experience life after death but rather to experience nirvana—which has the property of neither existence nor nonexistence. According to Margaret Ayer (1964, p. 52), nirvana is the "state of peace and freedom from the miseries of the constantly changing illusion which is existence."

The location of nirvana is to be found in the image of the flame when a candle is "blown out"—it is in a place beyond human understanding. According to Ayer (1964, p. 53), whenever people achieve nirvana "they 'will be seen no more.' It is through loss of desire, selfishness, evil, and illusion that this state of wisdom, holiness, and peace is reached." Buddhism contends that whereas physical death causes one to experience life again in a transmigrated form, "death to this world" (via nirvana) provides the gateway for ultimate happiness, peace, and fulfillment. Unlike Hinduism and the other religious traditions that we have considered, the ultimate goal of Buddhism is a state of consciousness and not a symbolic location for the disembodied soul.

Michael R. Leming

The central object of worship in Buddhism is the Buddhist "trinity."—the Buddha, the Dharma (teachings of the Buddha), and the Sangha (the disciples of the Buddha).

TEMPORAL INTERPRETATIONS OF DEATH

Even though the funeral industry and most people in the United States tend to merge religious and death meanings, temporal and existential interpretations of death also provide a means for protecting social order in the face of death. Such interpretations tend to emphasize the empirical, natural, and "this world" view of death.

According to Glenn Vernon (1970, p. 33), "when death is given a temporal interpretation and is seen as the loss of consciousness, self-control, and identity, the individual may conclude that he or she can avoid social isolation in eternity by identifying him or herself with specific values, including religious ones." If we define religion as a system of beliefs and practices related to high-intensity value meanings and/or meanings of the supernatural (Vernon, 1970), then it is possible for individuals with temporal orientations to be "religious" in their outlook without affirming an afterlife. Furthermore, because any death has many consequences for the persons on whom it impinges, we would expect that even religious persons would assign some temporal meanings to death.

Vernon pointed out that individuals whose interpretations are primarily temporal share the following beliefs and attitudes:

1. They tend to reject or de-emphasize a belief in the afterlife.
2. They tend to believe that death is the end of the individual.
3. They tend to focus on the needs and concerns of the survivors.
4. They tend to be present-oriented for themselves, but present- and future-oriented for those who will continue after them.
5. Any belief in immortality is related to the activities and accomplishments of the individual during his or her lifetime—including biological offspring and social relationships that the individual has created.

There is a strong temptation to view the person who has a temporal orientation as being very different from the person who finds comfort in a religious interpretation of death. In fact, both will attempt to restore the order in their personal lives and that found in society, by placing death into the context of a "higher" order.

For the individual with religious commitments, protection from anomie and comfort for anxiety are to be found by being in a relationship with the supernatural. For the person with a secular or temporal orientation, these same benefits are found in becoming involved with other people, projects, and causes. These involvements, although not pertaining to the supernatural, still provide a frame of reference that transcends the finite individual—a person may die, but his or her concerns will continue after death.

An existentialist perspective on death, as discussed in Chapter 1, suggests that one creates meaning for one's own life and death. Because one lives in a hostile world where one will eventually die, the primary human task is to find meaning in an otherwise meaningless world. This perspective is well illustrated by the classic article "On Death as a Constant Companion," which appeared in *Time Magazine* in 1965:

> Alone with his elemental fear of death, modern man is especially troubled by the prospect of a meaningless death and a meaningless life—the bleak offering of existentialism.
> "There is but one truly serious philosophical problem," wrote Albert Camus, "and that is suicide." In other words, why stay alive in a meaningless universe? The existentialist replies that man must live for the sake of living, for the things he is free to accomplish. But despite volumes of argumentation, existentialism never seems quite able to justify this conviction on the brink of a death that is only a trap door to nothingness.

SYMBOLIC IMMORTALITY

From an existentialist perspective, the creative efforts that comprise one's life can provide individuals with personal meaning for both life and death. One way this is accomplished is through the creation of a symbolic immortality, which actually includes surrogate forms of immortality—such as the continuity of history, the permanence of art, or even the biological force of sex. **Symbolic immortality** (Lifton & Olson, 1974) refers to the belief that the meaning of the person can continue after he or she has died. For the religious, symbolic immortality is often related to the concept of soul, which either returns to its pre-existent state, goes to an afterlife, is reincarnated in another body, or is united with the cosmos. For the person whose primary orientation is temporal, symbolic immortality is achieved by being remembered by others, by creating something that remains useful or interesting to others, or by being part of a cause or social movement that continues after the individual's death.

One of the reasons that many parents give for deciding to have children is the need for an heir—someone to carry on the family name. Research has demonstrated that in the United States, families with only female children are more likely to continue having children (in hopes of producing a male offspring) than are families with only male children. For the ancient Hebrews, the cultural institution of the **levirate marriage** required that a relative of the deceased husband have sexual intercourse with his dead relative's widow to provide a male heir. If it is possible to pass on something of oneself to one's children, then children are one method of providing symbolic immortality.

Within a given society there are high-status and low-status types of death. Giving one's life in defense of family or country is generally considered high status. In previous times, dying in childbirth was considered to be a high-status death for females. As the dying of martyrs dramatically illustrates, a death may be willingly entered into if it is meaningful. Given the right configuration of meaningful components (see Chapter 1), not to die would be more difficult. For example, dying may be preferable to defining oneself and being defined by others as a coward or a traitor. Dying is acceptable if it furthers "the cause." People may, in fact, literally work themselves to death to obtain a promotion, an artistic achievement, or public recognition. Whatever the specific content, the key factor of concern is the meaning involved. If the meaning is right, dying may be evaluated as a worthwhile thing.

The death of an individual brings about a change in influence over the members of his or her group or society. Heroes are more likely to come from the ranks of the dead than of the living. Hero meaning is a symboled-meaning component. One cannot attain the status of hero by oneself. Society bestows this rank only upon certain of its members—one type of symboled immortality. Death is often a part of the process by which heroes are made. When something dies, something else is born or created. The death of Jesus Christ has expanded in significance over time. It is likely that Christ's death has had a greater impact upon humanity than did his life. The Roman Catholic Church grants sainthood only to persons who have been dead for many years.

WORDS OF WISDOM

You can keep your things of bronze and stone, just give me one person who will remember me once a year.

Source: Damon Runyon

Following her death Mother Teresa was beatified by Pope John Paul II and given the title "Blessed Teresa of Calcutta." **Beatification** is a recognition accorded by the Catholic Church of a dead person's accession to heaven and capacity to intercede on behalf of individuals who pray in his or her name. Beatification is the third of the four steps in the **canonization** process, which declares a deceased person to be a saint. The process leading up to her beatification was the shortest in modern history. In early 1999—less than two years after Mother Teresa's death—Pope John Paul II waived the normal five-year waiting period and allowed the immediate opening of her canonization cause. Typically, The Roman Catholic Church grants sainthood only to persons who have been dead for many years. On April 27, 2014, in an unprecedented double ceremony, Pope Francis declared two of his 20th century predecessors, Popes John Paul II and John XXIII, saints in the eyes of the Roman Catholic church.

> "We declare and define Blessed John XXIII and John Paul II to be saints and we enroll them among the saints, decreeing that they are to be venerated as such by the whole church,"

A parallel perspective can be found in non–church-based **civil religion**. Civil religion is concerned with the transcendent meanings that support the state and provide it with a super-empirical or supernatural identity. Governments may grant pardons to convicted persons years after their deaths. Meaning is a flexible, yet powerful, thing. Furthermore, from the perspective of American civil religion, only the dead can be honored by their images appearing on U.S. postage stamps. These stamps can now even be declared by the U.S. Postal Service to be "Forever Stamps."

Investing oneself in relationships with others also ensures that one will be remembered after death. Some will argue that if we have influenced the lives of others, something of us will continue in their lives after we die. Organ donations supply a tangible method for providing this type of symbolic immortality. In this way, one can even ensure that a part of his or her physical self can continue in another person. Currently there is an increasing tendency for individuals to donate their organs and tissues upon death to the living. In many urban areas, kidney foundations, eye banks, and transplant centers will supply donor cards and, when death occurs, will arrange for transplants.

One of the reasons why some people write books (even books on death and dying) is to promote their own symbolic immortality. As long as their books can be read, their influence will outlive their biological body. The same is true for television and motion picture stars. Cable TV is keeping Lucile Ball alive for as long as they continue to rerun *I Love Lucy*—which is aired more than once each day, 365 days per year. Also, each year the youthful Judy Garland is resurrected from the dead as *The Wizard of Oz* is shown on television and through Netflix.

Some deceased celebrities are making considerably more money after their deaths than they did when they were alive. According to Dorothy Pomerantz, in 2013 the top-earning dead celebrities were: 1. Michael Jackson ($160 million), 2. Elvis Presley ($55 million), 3. Charles Schultz ($37 million), 4. Elizabeth Taylor ($25 million), 5. Bob Marley ($18 million), 6. Marilyn Monroe ($15 million), 7. John Lennon ($12 million), 8. Bettie Page (1950's pin-up model and manufacturer of vintage and retro clothing for modern women) and Albert Einstein ($10 million) 10. Theodor Geisel (Dr. Seuss) and Steve McQueen ($9 million). All I can say is, "That's living!"

The body of Mother Teresa, "the once and future saint," lying in state before her funeral.

Great inventors, political leaders, and athletic "hall of famers" are also given immortality when we use their products, remember their accomplishments, and celebrate their achievements. In the case of medical practitioners and bionic inventors, not only do the living remember their accomplishments, but also these accomplishments extend the lives of those who provide the dead with immortality.

The person's finite identity is protected by the group's permanence. Just as Standard Oil is the legacy of John D. Rockefeller, John H. Leming and Son's Insurance Agency will remain even after John H. Leming and his sons are dead (providing that the new owners feel that it is in their business interest not to change the name of the company). [As a side note, Michael Leming's father died in 1983 and his oldest brother finally sold their father's business in 2009 and died in 2014, but John H. Leming and Son's Insurance Agency continues to this day. The new owner is Lino Cambaliza.]

The same can be said for people who give themselves to political movements and causes. Marx, Lenin, Stalin, and Mao—as leaders of communism—will be remembered by future communists, despite the efforts by present officials to accomplish the contrary. We even provide infamous immortality to villains, murderers, and traitors. It seems ironic that most of us spend a lifetime working for immortality, when the guns of John Wilkes Booth, Lee Harvey Oswald, James Earl Ray, Osama bin Laden, Adam Lanza, and Aaron Alexis supplied their owners with our everlasting remembrance.

For the existentialist, a lifetime is spent searching for and creating meaning. The search for meaning is a task that all people share. Furthermore, significant others are involved, as individuals try to create meaning for themselves. In many respects, the meaning created turns out to be meaning for the group or society. Symbolic immortality is something that only the living can give the dead. Yet people live with the faith that their survivors will remember them and perpetuate the meaning of their lives after they die. Like religious interpretations of death, temporal meanings enable individuals to protect themselves and their social order from death.

For the existentialist, the construction of symbolic immortality can give life a purpose and a sense of fulfillment, but for the critics of existentialism, even symbolic immortality cannot outwit death. From their point of view, there is also a

---

PRACTICAL MATTERS | HOW CELEBS MAKE A LIVING AFTER DEATH

(CBS) When Michael Jackson died in 2009, he had nearly a half a billion dollars in debts. Since then, it's been a great year for his career: lawyers for his estate say they have lined up merchandising deals worth $100 million, and surging record sales and other income will produce another $500 million.

And this is not unusual. Decades after their demise, some departed stars continue to work on new projects and draw more income than they ever made while drawing breath. And there is a growing legion of agents and managers willing to represent them. Dead celebrities can be just as lucrative as many live ones, and in some cases, a lot less trouble.

No other agent in the world represents more famous people than Mark Roesler: stroll down Hollywood Boulevard with him and he'll point out 62 of his clients who are immortalized with their own stars on the "Walk of Fame," stars such as Errol Flynn, Gloria Swanson, and Ginger Rogers.

His client list includes some of the biggest names of the 20th century: actresses like Ingrid Bergman, Bette Davis, Natalie Wood, and Marilyn Monroe, baseball legends Babe Ruth and Lou Gehrig, and singers Ella Fitzgerald and Billie Holiday. According to Roesler, "We're a business agent for about 250 entertainment, sports, music and historical clients. But most of those are deceased." You could call Roesler's business a William Morris agency for the departed. The business is called "CMG" (CMG Worldwide) and it is headquartered far from the glitter of Hollywood in an office park on the fringes of Indianapolis.

His real clients are the heirs and estates of the dearly departed, who ultimately approve or reject the merchandising deals that CMG puts together. The product endorsements run the gamut from paraphernalia to the pinnacle of postmortem prestige. And Roesler has licensed more than 200 deals with the U.S. Postal Service.

The agency has created web sites for all its deceased clients and maintains and revives their fan clubs. According to Roesler, "We get at least 15 million hits a day that come through this building, for the different clients that we represent." It is all part of a legal and entertainment niche that Roesler pioneered more than 25 years ago, after graduating from law school. Until Roesler came along in the early 1980s, a celebrity's right to control or profit from their good name was buried along with them. Their heirs had virtually no say in how their loved one's image or persona was used and no claim to any of the monies they generated.

So Roesler set about to trying to change that in courts and in state legislatures around the country, helping to establish what is now recognized as the postmortem right to publicity. "We have the right to prevent our name, our likeness, our image, our signature, our voice, from being used in some commercial fashion," Roesler explained.

Elvis, many think, is the perfect business model for the Michael Jackson estate. Elvis is the all-time king of afterlife income and still pulls in $55 million a year. But then Elvis is more than a dead celebrity. He is also a destination, at $33–$70 a head to visit his Tennessee home, "Graceland." Graceland and the rest of the Elvis realm are now controlled by billionaire entertainment entrepreneur Robert Sillerman. Sillerman doesn't just represent Elvis, he owns Elvis: four years ago he spent $100 million to buy 85 percent of the rights to the Presley estate. According to Sillerman, "It turned out to be a wonderful deal for us and for the Presley family." Sillerman doesn't think there will ever be another phenomenon quite like Elvis, who has turned out to be relatively recession proof. Some parts of his business are actually up.

Source: 60 Minutes, September 27, 2009 (CBS News). Available online at http://www.cbsnews.com/video/watch/?id=5345034n&tag=contentMain;contentBody

| WORDS OF WISDOM | OZYMANDIAS |

I met a traveler from an antique land
   Who said: "Two vast and trunkless legs of stone
   Stand in the desert.... Near them, on the sand,
   Half sunk, a shattered visage lies, whose frown,
   And wrinkled lip, and sneer of cold command,
   Tell that its sculptor well those passions read
   Which yet survive, stamped on these lifeless
   things,
   The hand that mocked them, and the heart that fed;

And on the pedestal these words appear:

'My name is Ozymandias, king of kings:
   Look on my works, ye Mighty, and despair!'

Nothing beside remains. Round the decay
   Of that colossal wreck, boundless and bare
   The lone and level sands stretch far away."

Source: Percy Bysshe Shelley

more looming problem for those whose immortality depends upon others: What would happen if a nuclear holocaust were to occur, and there were no survivors?

NEAR-DEATH EXPERIENCES

In the past, one of the assumptions that most people made about the field of thanatology was that the real "experts" were not among us—they were dead. With the publication of Raymond Moody's *Life After Life* (1975), many people have stepped forward to challenge this assumption. Having been near death or having been declared clinically dead by medical authorities, a number of survivors of these experiences have "returned from the dead" to tell us that they now know what it is like to be dead and that they possess empirical evidence to support a rational belief in the afterlife.

DEFINING A NEAR-DEATH EXPERIENCE

The near-death experience (NDE) is defined as an experience of a person near death (e.g., a person in a coma or a person who has experienced a heart attack), in an unconscious state, reporting that "something happened." In his now famous book, Moody brought national attention to NDEs. Throughout the book, Moody creates a composite of the many accounts by individuals who have survived the experience of being near death or being declared clinically dead. This ideal model was constructed by interviewing more than 150 people who had had NDEs. Moody reports, and other scientists support (Ring, 1980; Johnson, 1990), the similarity of most of these "life after death" accounts. Although no two NDEs are identical, there appears to be a pattern of experiences that include many, if not all, of the following nine traits (Perry, 1988):

1. *A sense of being dead.* At first many people don't realize that the experience that they are having has anything to do with being near death. They find themselves floating above their body and feeling confused. They wonder, "How can I be up here, looking at myself down there?"
2. *Peace and painlessness.* An illness or accident is frequently accompanied by intense pain, but suddenly during a near-death experience the pain vanishes.

According to research by psychologist Kenneth Ring (1980), 60 percent of people who have had a near-death experience report peace and painlessness.

3. *Out-of-body experience.* Frequently people feel themselves rising up and viewing their own bodies below. Most say that they are not simply at a point of consciousness but seem to be in some kind of body. Ring says that 37 percent of the research subjects have out-of-body experiences.

4. *The tunnel experience.* This generally occurs after an out-of-body experience. For many, a portal or tunnel opens, and they are propelled into darkness. Some hear a "whoosh" as they go into the tunnel, or they hear an electric vibration or humming sound. The descriptions are many, but the sense of heading toward an intense light is common to almost all tunnel experiences. Twenty-three percent of Ring's subjects reported entering darkness, which some described as entering a tunnel.

5. *People of light.* After people pass through the tunnel, they usually meet beings of intense light that permeate everything, as these beings fill the people with feelings of love. As one person said, "I could describe this as 'light' or 'love,' and it would mean the same thing." They frequently meet with friends and relatives who have died, though the glowing beings can't always be identified. In Ring's research, 16 percent saw the light.

6. *Being of Light.* After meeting several beings of light, there is usually a meeting with a supreme Being of Light. To some this is God or Allah, to others simply a holy presence. Most want to stay with him or her forever.

7. *The life review.* The Being of Light frequently takes the person on a life review, during which his or her life is viewed from a third-person perspective, almost as though watching a movie. Unlike watching a movie, however, the person not only sees every action, but also its effect on people in his or her life. The Being of Light helps put the events of life into perspective.

8. *Rising rapidly into the heavens.* Some people report a "floating experience," in which they rise rapidly into the heavens, seeing the universe from a perspective normally reserved for satellites and astronauts.

9. *Reluctance to return.* Many find their unearthly surroundings so pleasant that they don't want to return. Some even express anger at their doctors for bringing them back.

One study by Borjigin, Lee, Liu, Pal, Huff, and Klar (2013) performed continuous electroencephalography in rats undergoing experimental cardiac arrest and analyzed changes in power density, coherence, directed connectivity, and cross-frequency coupling. They identified a transient surge of synchronous gamma oscillations that occurred within the first 30 seconds after cardiac arrest and preceded isoelectric electroencephalogram. Gamma oscillations during cardiac arrest were global and highly coherent. Moreover, this frequency band exhibited a striking increase in anterior–posterior-directed connectivity and tight phase-coupling to both theta and alpha waves. High-frequency neurophysiological activity in the near-death state exceeded levels found during the conscious waking state. They concluded that the mammalian brain can, albeit paradoxically, generate neural correlates of heightened conscious processing at near-death (Borjigin et al., 2013).

In a 2012 study by Pranab Bhattacharya (2013, p. 151), the author claims that there is cross-cultural evidence for near-death experiences:

> Near death experiences (NDEs) have been reported throughout [the] world in essentially all cultures, including amongst the believers of the Hindu religion. The contents of NDEs are independent of the gender, age, profession, religion, belief of soul, belief in angels of death or ghosts and belief in death kingdom and heaven, of people who experienced it. The frequency of occurrence is estimated to be between 5% to 48% in adults, and around 85% in children who experienced near-death situations. This frequency may be higher still, perhaps even 100 percent, were it not for the dreamlike and dissociative character of these experiences, and the amnesia-prone participation of the temporal lobe cortex of brain, causing a clear tendency to forget the NDE.
>
> A number of experiences can be very similar to NDEs, such as review of one's life in this planet, or an out-of-body experience (OBE) in which the physical body and its surroundings are observed from various external vantage points, often from above, such that the body is passing through a deep dark tunnel, or seeing flash of light equal to thousands of sun for pure souls. The experience of seeing God and conversing with him, seeing alien lands, seeing dead relatives or someone's future, can all be regarded as similar in nature. Many individuals have reported horror experiences as well. Numerous cases are existing in which the reality of the OBE-observation can be independently "verified," by external conditions, situations, people, objects, etc. Even people who are non-religious, subsequent to NDE experiences have displayed a markedly decreased fear of death, and a corresponding increase in the belief in "life after death" and re-incarnation.
>
> Certain elements of NDE experiences can be induced by drugs, such as hallucinogenic substances and anesthetic drugs like ketamine, and electrical stimulation of the right temporal lobe or the limbic system has also produced such effects. The possibility that the hallucinogenic transmitters (and endorphins) of the brain themselves play a role in the NDE has been postulated. Nevertheless, there are NDE-elements, such as the frequently reported quick life-reviews, and the acquisition of external, verifiable information about the physical surroundings, that cannot be explained. Wish-fulfillment, death-denial or fighting against death, and other defense mechanisms of the brain, are also not adequate explanations. The large body of NDE data now points to genuine evidence for a non-physical reality.

According to Rodabough and Cole (2003), approximately 23 million Americans have, by prevailing medical definition, died briefly or come close to death, and 8 million of these have had some sort of near-death experience (NDE). The phenomenon is so widespread that in 1978 the International Association for Near-Death Studies (IANDS) was founded by researchers and those who have had NDEs. Today, the organization is found on every continent but Antarctica and has many local chapters in 38 states in the United States. IANDS publishes a quarterly newsletter, *Vital Signs*, maintains a speaker's bureau, sponsors an annual North American conference, and produces a wide variety of educational materials. IANDS also encourages interest in research and professional applications through its publication of the quarterly *Journal of Near-Death Studies* and a program of small grants to encourage scholarly research on NDEs.

According to Holden (2009), by 2005 there had been over 800 refereed journal articles on NDEs on more than 65 research studies involving over 2,500 near-death experiences of people claiming to have had an NDE. These empirical studies addressed the following NDE-related topics: content and aftereffects of NDEs, the circumstances and incidences surrounding the experiences, and the characteristics of near-death experiments. The more than 30 years of scientific research on the topic of NDEs also attempts to provide a more comprehensive understanding of death and the nature of consciousness.

In a more recent review of research related to near-death experiences of resuscitated patients by Cant, Cooper, Chung and O'Connor (2012), it was reported that near-death experiences are reported in only 4–9 percent of general community members and up to 23 percent of critical-illness patients. According to Cant and colleagues, one explanation is that paranormal visions that include seeing bright lights, a tunnel, and having feelings of peace may be a stage of enlightenment as death approaches. However, a more objective explanation points to neurochemical changes in a stressed or dying brain as explanation for nearly all the elements of near-death experiences. However, if this is so, NDE should occur in all patients who are critically ill and near death, which is not the case.

Explaining Near-Death Experiences

The following three questions become relevant as we contemplate the relationship between NDEs and other material presented in this chapter:

1. Are NDEs real, and what do they tell us about the dying process and/or being dead?
2. Can afterlife beliefs be empirically supported by these NDE accounts?
3. How do religious beliefs affect the content of near-death or clinical death experiences?

In reviewing the scientific research from more than 80 empirical studies of NDEs conducted between 2003 and 2013, we have discovered that scientific consensus relative to the answers to the questions above remains problematic. To all of our questions, we must temper our responses by saying that, if not real, these experiences are very real to those who have had them. Not unlike being in love, NDE phenomena are extremely subjective and do not lend themselves to verification by others. Typical of this type of conclusion is the research by Stevenson, Cook, and McClean-Rice (1989), which concluded that, of those claiming the NDE, only 45 percent were judged by medical experts to have had serious, life-threatening illnesses or injuries; the rest were judged to have had no life-threatening condition. Our conclusion is that it is difficult to doubt that the individual has experienced something. However, even if we are unable to prove, or disprove, that the individual has died, all that we can say scientifically is that the individual claims to have had the experience of "being dead."

Neurosurgeon Alexander Eben (2012) believed that there is a scientific explanation for near-death experiences—until he had one of his own. What was unusual about his near-death experience was that his cortex, the part of the brain that makes us human, was inactivated during a seven-day coma. He has no doubt that his inner self was alive and well during that time. In his bestselling autobiography, *Proof of Heaven: A Neurosurgeon's Journey into the Afterlife,* Dr. Alexander claims that through most of his near-death journey, he was accompanied by a young woman. Without using words, she conveyed a three-part message to him: "You are loved and cherished, dearly, forever." "You have nothing to fear." "There is nothing you can do wrong." As a result, Dr. Alexander now claims he wants to spend the rest of his life studying consciousness and show that humans are much more than their physical brains (2012).

Returning to our analogy of being in love, all that can be known is that a person says that he or she is in love. Whether or not the person really is cannot be determined empirically. From the perspective of the individual, it does not really matter, because a situation that has been defined as real will have very real behavioral consequences.

Barbara Walker (1989) claimed that many having an NDE have post-NDE depression due to the fact that health care providers and family members often deny the validity of the NDE experience. It is not uncommon for persons having an NDE to tell a nurse or psychiatrist about their near-death event, only to be told that they were hallucinating (Walker, 1989). Kenneth Ring (1980) observed that a considerable number of divorces resulted from a spouse's inability to relate to the NDE.

If we try to determine the validity of a belief in an afterlife, we are confronted with problems similar to those encountered in the NDE. What can science say about beliefs in the afterlife? Scientifically, afterlife beliefs cannot be proved or disproved with NDE evidence. **Science** is based upon the principle of **intersubjectivity**. This means that independent observers, with different subjective orientations, must agree that something is "true" based upon their separate investigations. Unfortunately, the opportunity to experience the afterlife (and to return) is not uniformly available to all observers. Research by Tillman Rodabough (1985) gives many metaphysical, physiological, and social-psychological alternative explanations to account for the NDE. Therefore, although those who have had these experiences may feel rationally justified in their beliefs, the evidence that they use is not scientifically based. Science, at this point, can neither verify nor refute afterlife beliefs.

In responding to our last question concerning the effects of religious beliefs on the content of near-death or clinical death experiences, there is some scientific evidence that can provide limited answers. According to Raymond Moody (1975) and R. R. Canning (1965), the content of afterlife experiences is largely a function of the religious background, training, and beliefs of the individuals involved. Only Roman Catholics see the Virgin Mary and the host of Catholic saints, whereas encounters with Joseph Smith are reserved for Mormon believers. Like dreams, continuity exists between experiences in "this world" and experiences in the "afterlife." For example, personages in the afterlife are dressed in attire that would conform to the individual's cultural customs and beliefs. Also, relatives appear to be at the same age that they were when they were last seen. In fact, there is so much continuity between this world and the "other world" that persons with afterlife experiences report few, if any, surprises. In response to this evidence, Kathy Charmaz (1980) raised the following question: Is the consciousness reported

"... And you say you first noticed them just
after you had a 'near death experience'?"

in these NDEs a reflection of a shared myth or evidence for it? In a more recent Australian study, Allen Kellehear (1996) reported that NDEs are confined largely to societies in which historic religions are dominant. His study argued that social and historical accounts are the best explanations for similarities of NDEs.

Even with the many disagreements regarding the NDE and its etiology, there is consensus that NDEs have profound effects upon the lives of most of those having the experience. According to P. M. Atwater (Johnson, 1990), 65 percent of individuals who have had an NDE make significant changes in their lives, and 10 percent make radical changes. According to Skip Johnson (1990), people with an NDE have the following aftereffects as a result of the phenomenon:

1. *Loss of fear of death.* People report that they no longer fear the obliteration of consciousness of self.
2. *Sense of the importance of love.* An NDE can radically change people's value structure. They see the importance of brotherly love.
3. *Sense of cosmic connection.* People feel that everything in the universe is connected. Many have a newfound respect for nature and the world around them.
4. *An appreciation of learning.* People gain a newfound respect for knowledge, but not self-gain. Many often will embark on new careers or take up serious courses of study.
5. *A new feeling of control.* People feel that they have more responsibility for the course of their lives.

6. *A sense of urgency*. Some people realize the shortness and fragility of their lives.
7. *A better-developed spiritual side*. This leads to spiritual curiosity and abandoning of religious doctrine purely for the sake of doctrine.
8. *Reduction in worries*. Some people feel more in control of life's stresses and are able to be more forgiving and patient.
9. *Reentry syndrome*. Some people have difficulty adjusting to normal life, especially those who undergo intensive changes in values that disrupt their former lifestyle. Some people report developing psychic abilities that can be scary to them and/or family and friends.

Finally, there is evidence that individuals (even if they have attempted suicide) who have had NDEs do not try to bring about an end to their lives to return to the "life beyond." In fact, most individuals find new reasons for living as a result of these experiences (Flynn, 1986; Greyson, 1997). Cant et al. (2012) claim that nearly all patients report positive psychological outcomes after a near-death experience. As a result of this research, the authors encourage nurses to support patients during a time of crisis by assisting them and their families to comprehend the experiential event using effective communication and listening skills (Cant et al., 2012).

CONCLUSION

Religion is a system of beliefs and practices related to the sacred—to what is considered to be of ultimate significance. The cultural practice of religion continues because it meets basic social needs of individuals within a given society. A major function of religion, in this regard, is to explain the unexplainable.

For most "primitive," or less-complex, societies, events are explained by supernatural rather than by rational or empirical means. Thus, an eclipse of the sun or moon was said to be a sign that the gods had a message for humankind. These supernatural explanations were needed, in part, because there were no competing rational or scientific explanations. Technologically and scientifically advanced societies tend to be less dependent upon supernatural explanations. Yet such explanations are still important, especially when scientific explanations are incomplete. It is not uncommon to hear a physician say, "Medical science is unable to cure this patient; it's now in God's hands." Or, a physician might say, "It's a miracle that the patient survived this illness; I can't explain the recovery." Thus, one might suggest that religion takes over where science and rationality leave off. We depend less on religious or supernatural explanations than do nonliterate societies, but nonetheless we rely on them when knowledge is incomplete.

Yet even in secular societies, religion and spirituality often play significant roles by helping individuals cope with extraordinary events—especially illness, dying, and death. Not only does religion help restore the normative order challenged by death, but also strong religious and spiritual orientations can enable individuals to cope better with their own dying and the deaths of their loved ones. For others, strong commitments to a temporal orientation may fulfill many of the functions provided by a religious or spiritual worldview.

SUMMARY

1. Religion helps individuals, when the order of everyday life is challenged, by providing answers to problems of uncertainty, powerlessness, and scarcity created by death.
2. Religious systems provide a means to reestablish the social order challenged by death.
3. When one encounters death, the anxiety experienced is basically socially ascribed.
4. Religion provides individuals with solace when they attempt to cope with death.
5. Regardless of the content of Jewish beliefs regarding the immortality of the soul, Jewish funeral customs and rituals emphasize that God does not save us, as individuals, from death, but saves Israel for history, regardless of death.
6. Christians employ two basic, and somewhat paradoxical, perspectives when facing death. The first is that through faith in Jesus Christ, the Christian has victory over death and gains eternal life with God. The second perspective on death employed by Christians emphasizes the experience of true human loss.
7. The primary expression of the Islamic concern for justice and accountability is found in the belief that assignment to paradise or damnation is based upon the strengths of the individual's faith in God and the moral quality of their earthly lives.
8. Hindus believe in the immortality of the soul and hope for an afterlife, but their hope is that their soul will be absorbed at death. For the Hindu, the goal is not to experience life after death, but rather to have one's soul united with the Oversoul. Punishment for the devout Hindu might be to *have* "everlasting spiritual rebirth."
9. Buddhism assists its practitioners by encouraging them to detach from striving for living, holding onto relationships, and attempting to avoid pain and sadness. From a Buddhist perspective, it is by accepting the impermanence of life, relationships, and desirable things, and then letting go of them, that one finds enlightenment.
10. Temporal interpretations of death provide a means for protecting the social order by emphasizing the empirical, natural, and "this-worldly" view of death.
11. Symbolic immortality is evidenced by offspring carrying on the family name, by donating body organs, and by having accomplishments or achievements (positive and negative) remembered by others.
12. Near-death or afterlife experiences are influenced by the individual's religious background, cultural beliefs, and prior social experiences.

DISCUSSION QUESTIONS

1. How does religion function to provide a restoration of the order challenged by the event of death?
2. Explain the following statement: "Religion afflicts the comforted and comforts the afflicted."
3. How can symbolic immortality and temporal interpretations of death provide a source of anxiety reduction for those who face death?
4. How can organ donations provide symbolic immortality for donors and their loved ones?
5. Do accounts of near-death experiences provide empirical evidence for afterlife beliefs? Why or why not?
6. What are the similarities and differences in Jewish, Christian, Islamic, Hindu, and Buddhist beliefs about death and funeral practices?

GLOSSARY

Anomie: A condition characterized by the relative absence or confusion of values within a group or society.

Beatification: A recognition accorded by the Catholic Church of a dead person's accession to heaven and capacity to intercede on behalf of individuals who pray in his or her name.

Canonization: The process within the Catholic Church that declares a deceased person to be a saint.

Civil Religion: The form of non-church secular religion that is concerned with the transcendent meanings that support the state and provide it with a superempirical or supernatural identity.

Intersubjectivity: A property of science whereby two or more scientists, studying the same phenomenon, can reach the same conclusion.

Levirate Marriage: An institution typified by the Hebrew requirement that a relative of the deceased husband must have sexual intercourse with the deceased's widow to provide a male heir.

Marginal Situations: Unusual events or social circumstances that do not occur in normal patterns of social interaction.

Religion: A system of beliefs and practices related to the sacred, the supernatural, and/or a set of values to which the individual is very committed.

Rituals: A set of culturally prescribed actions or behaviors.

Science: A body of knowledge based upon sensory evidence or empirical observations.

Symbolic Immortality: The ascription of immortality to the individual by perpetuating the meaning of the person (the self).

SUGGESTED READINGS

Berger, P. L. (1969). *Sacred canopy: Elements of a sociological theory of religion*. New York: Doubleday. Excellent treatment of the role of religious worldviews as they relate to life crises. Death is discussed as the ultimate marginal situation to normal social functioning that calls forth religious meaning systems.

Bhattacharya, P. (2013). Is there science behind the near-death experience: Does human consciousness survive after death? *Annals of Tropical Medicine and Public Health*, 6(2), 15.

Borjigin, J., Lee, U., Liu, T., Pal, D., Huff, S., & Klarr, D. (2013). Surge of neurophysiological coherence and connectivity in the dying brain. *Proceedings of the National Academy of Sciences of the United States*, 110(35), 14432ff.

Bryant, C. D. (2003). *Funeralization in cross-cultural perspective*. In C. D. Bryant (Ed.), *Handbook of death & dying* (pp. 611–693). Thousand Oaks,CA: Sage Publications. This is a series of essays dealing with funeralization, including articles on the Native American, Hindu, Muslim, Japanese, Chinese, European, and Jewish ways of death.

Cooper, R., Chung, C., & O'Connor, M. (2012). The divided self: Near death experiences of resuscitated patients—A review of literature. *International Emergency Nursing*, 20(2), 88.

Eben, A. (2012). *Proof of heaven: A neurosurgeon's journey into the afterlife*. New York: Simon and Schuster. An excellent discussion of the near-death experience by a neurosurgeon who has personally experienced this phenomenon.

Eliade, M. (Ed.). (1987). *The encyclopedia of religion*. New York: Macmillan. Academic resource on religious rites, beliefs, and traditions. This reference tool is written from an interdisciplinary perspective by international scholars of religion.

Garces-Foley, K. (Ed.). (2006). *Death and religion in a changing world*. Armonk, NY: M. E. Sharpe. This book is an excellent edited collection of articles on death from the perspective of many religious traditions.

Goss, R. (2005). *Dead but not lost: Grief narratives in religious traditions*. Walnut Creek, CA: AltaMira Press. This book provides many perspectives on grieving from the experiences of many religious traditions.

Gouin, M. (2010). *Tibetan rituals of death: Buddhist funerary practices*. Abingdon, UK: Routledge. A comprehensive survey of the available literature on funerary practices in Tibetan societies.

Gowan, D. E. (2003). Christian beliefs concerning death and life after death. In C. D. Bryant (Ed.), *Handbook of death & dying* (pp. 126–136). Thousand Oaks, CA: Sage Publications. This article discusses the emergence of life-after-death beliefs in Judaism and Christianity.

Green, J. W. (2008). *Beyond the good death: The anthropology of modern dying*. Philadelphia: University of Pennsylvania Press. An anthropologist provides an excellent interpretation of the ways in which Americans react when death is at hand for themselves or for those they care about.

Heaven above or below? (2012). *The Christian Century, 129*(14), 9.

Holden, J. M., Greyson, B., & James, D. (Eds.). (2009). *The handbook of near-death experiences*. Santa Barbara, CA: Praeger. A comprehensive anthology of near-death experiences.

Holden, J. E. (2009). *Near-death experiences*. In C. D. Bryant and D. Peck (Eds.), *Encyclopedia of death and the human experience* (pp. 773–776). Thousand Oaks, CA: Sage Publications. This article provides a great summary of the current state of social scientific knowledge regarding near-death experiences.

Morgan, J. D., & Laungani, P. (Eds.). (2002). *Death and bereavement around the world*. Amityville, NY: Baywood Publishers. This edited work provides analysis of death-related behavior from the perspective of the major religious traditions. It also provides valuable bibliographic references and information.

Moody, R. A., Jr. (1975). *Life after life: The investigation of a phenomenon—Survival of bodily death*. Boston: G. K. Hall. Patient accounts of near-death experiences shared with Moody, a physician and philosopher. Moody objectively presents the cases.

Pomerantz, D. (2013). Michael Jackson leads our list of the top-earning dead celebrities. *Forbes Magazine* (on line at http://www.forbes.com/sites/dorothypomerantz/2013/10/23/michael-jackson-leads-our-list-of-the-top-earning-dead-celebrities/). Accessed October 23, 2013.

Ring, K. (1980). *Life at death: A scientific investigation of the near-death experience*. New York: Coward, McCann and Geoghegan. Provides an understanding of the near-death experience.

Rodabough, T., & Cole, K. (2003). Near-death experiences as secular eschatology. In C. D. Bryant (Ed.), *Handbook of death & dying* (pp. 611–693). Thousand Oaks, CA: Sage Publications. This is an extensive review of current research on near-death experiences especially as it relates to temporal interpretations of death.

Stanworth, R. (2004). *Recognizing spiritual needs in people who are dying*. New York: Oxford University Press. Based on stories shared by 25 dying individuals, this book teaches those who work with the dying how to listen attentively to their stories.

After the death of his wife, Jenny, Forrest Gump stated, "Mamma always said that dying is a part of life. I sure wish it wasn't."
—**Forrest Gump**

I'm not afraid of dying. I just don't want to be there when it happens.
—**Woody Allen**

THE DYING PROCESS

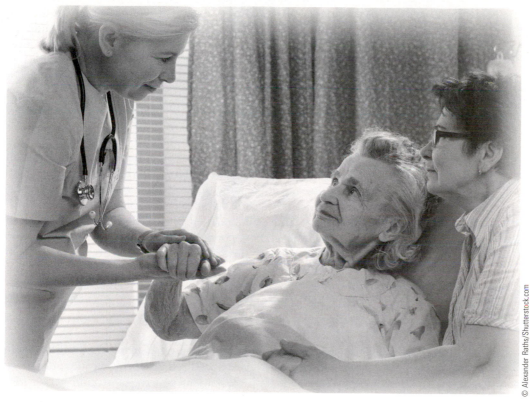

We tend to fear the dying process more than the event of death: the pain, the isolation, the loss of physical strength and nervous system activity, the loss of favorite activities, the possibility of institutional care, the loss of independence, and the regression to a state of extreme dependency. As Morrie Schwartz, who was dying of amyotrophic lateral sclerosis (ALS), or Lou Gehrig's disease, said to his former student Mitch Albom in the bestselling book entitled *Tuesdays with Morrie* (Albom, 1997, p. 49), "The ultimate sign of dependency is having someone wipe your bottom." The loss of one's physical capabilities can be demeaning. One of the themes that Professor Schwartz was trying to teach to his former student is that this is the time to reflect on life's lessons and share them with others. Then those left behind, when Schwartz was gone, would ponder these lessons and grasp what is really important in life.

Yet, being immobile and limited to one's bed leaves a lot of time for thinking. On a positive note, such time presents the opportunity for the patient to get her or his house in order and complete any unfinished business, such as saying goodbyes, preparing a will (if he or she did not already have one), and making amends for any wrongdoings committed earlier in life. The time of idleness allows for reflection and recalling memories of days gone by. The patient may have some idea of the doctor's prognosis as to the amount of time left to live and thus may plan accordingly.

Because the process of dying is typically not something that we experience often in life, the dying person may have few clues as to how she or he should act. We want to behave as expected and follow the norms of our society, but not knowing the "rules for dying" may be frustrating for the individual. Indeed, we are typically not socialized to the process of dying, so the dying patient may simply learn while in the process. Yet, as individuals in the United States are living longer today, the movement is rapidly toward a society in which old age and death are more common than birth and childhood, thus we may have experienced more deaths of others in a lifetime than previously and therefore become more sophisticated about the dying process (Erickson, 2013). Additionally, as dying today tends to be more gradual, with more chronic deaths than acute deaths, an individual can see her/his own death coming, thus perhaps make some preparation for the inevitable. For example, sudden deaths leave little time for goodbyes, a realization that the end is near, or opportunities to intentionally achieve closure on life (Fishman, 2010).

On the other hand, medical personnel may be limited in their ability to relate to the dying person, because they have been socialized to "prolong life and relieve suffering." Thus, we may find that both the patient and those around her or him may be uncomfortable with their roles as caregivers to the dying. This awkwardness is often compounded by the patient's lack of awareness of the prognosis. The situation is made even uncomfortable by the fact that perhaps the medical personnel involved are not aware of what the patient and family know, and thus mum is the word: "I'm afraid to say anything for fear I might disclose information that was not previously known to others."

Friends and family may also be uncomfortable because, as discussed in Chapter 1, they have had little exposure to death. In fact, they may be so uncomfortable that they do not assist the dying person at all.

| TEACHING WHILE DYING

Mitch Albom visited his former sociology professor, Morrie Schwartz, on Tuesdays for the last 14 weeks of Professor Schwartz's life and shared the account in *Tuesdays with Morrie.* In dying a lingering death from ALS, Morrie noted that you should "be grateful that you have been given the time to learn how to die." He showed a complete lack of self-pity.

Professor Schwartz said that the way you get meaning into your life is by loving others, devoting yourself to your community around you, and devoting yourself to creating something that gives you purpose and meaning. (Money is *not* the most meaningful thing!) Without love, "we are birds with broken wings."

Professor Schwartz reminded us of the Buddhist saying that each of us has a little bird on our shoulder every day that asks, "Is today the day? Am I ready? Am I doing all I need to do? Am I being the person I want to be?"

Morrie said, "Aging is not just decay..., it's growth. It's more than the negative that you're going to die, it's also the positive that you understand you're going to die, and that you live a better life because of it."

One achieves respect by offering something that one has, observed Schwartz. You do not have to have a big talent. There are lonely people in hospitals and nursing homes who only want companionship. Play cards with a shut-in and find new respect for yourself, *because you are needed.*

"Death is as natural as life," said Morrie. "It's part of the deal we made." Everything that gets born dies. Death ends a life, not a relationship.

Adapted from *Tuesdays with Morrie,* by M. Albom, 1997, New York: Doubleday.

DEATH MEANINGS

The meaning of our dying will depend to a great extent upon the social context in which the dying occurs. Meanings are the basic component of human behavior because individuals respond to the meanings of phenomena rather than to the phenomena themselves. Meanings are both socially created and socially perpetuated.

The major types of meanings to which individuals respond in death-related situations are the following: time meanings, space meanings, norm and role meanings, value meanings, object and self-meanings, and social situation meanings (Vernon, 1972). We will now look at each of these types of meanings and their effects upon the dying process.

TIME MEANINGS: DEALING WITH THE PROGNOSIS

As anthropologist Colin Turnbull once told George Dickinson, "Americans want to live forever. If you ask individuals how long they wish to live, they will probably say somewhere in the upper 90s because they do not suspect they will reach 100. But if they reach 100, then they will aim for 101, 102, etc." Turnbull said that African groups know when it is time to die, and they accept this. Perhaps Richard Pryor summed it up well when he said, "Even if I live for 100 years, I'll be dead a lot longer."

In thinking about the dying process, the first thing that comes to our awareness is the concept of time. We are confronted with the fact that time, for the terminally ill patient, is running out. Yet, when does the dying process begin? Are we all not dying, with some reaching the state of being dead before others? From the moment of our births, we are approaching the end of our lives. We assume that terminally

| LISTENING TO THE VOICES | About Living |

I invited to class a woman whose diagnosis was terminal cancer. Deeply impressed by the visitor's positive outlook on the time left to her, one student noted on the following day, "When I came into the session I expected to meet someone who was dying. Instead, I realized that woman is more alive than I am. The discussion wasn't about dying at all, it was about living."

Dialectic on dying (pp. 182–189), by J. M. Boyle, 1981. In M. M. Newell et al. (Eds.), *The role of the volunteer in the care of the terminal patient and the family*. New York: The Foundation of Thanatology/Arno Press.

ill patients will experience death before other individuals, but this is not always the case. Because it is possible to diagnose diseases from which most people die, we can assume that patients with these diseases are "more terminal" than individuals without them. The terminally ill patient is very much concerned with the time dimension of his or her physical existence. "How long do I have to live?" is more than a question found on soap operas; it is a query that general practitioners hear, on average, six times a year and oncologists face as often as 100 times a year (Sharlet, 2000). In general, patients want to know if their illness is terminal. However, physicians traditionally may have felt unprepared to tell their patients of a terminal diagnosis. According to studies by sociologist and physician Nicholas Christakis (1999), doctors usually ignore prognosticating, and when they do, they overestimate their patient's remaining days by a factor of two to five. Christakis noted that prognosis is akin to prophecy. Prophets seek to shape the future they foresee by foretelling it, revealing good fortune or warnings of doom. In so doing, prophets can alter the actions of those to whom they prophesy. Christakis found that only 14 percent of doctors in his study included substantive discussion of the future in their meetings with patients. Most physicians stick to an unspoken code: Don't foresee. If you do, do not foretell.

With more emphasis on consumer rights today, physicians are more likely to tell patients of their diagnosis than was the case in the early 1960s. In a 1961 survey of physicians (Oken, 1961), only 10 percent favored telling a patient with cancer his or her diagnosis, whereas in 2008, 98 percent of physicians surveyed said that their usual practice is to disclose the diagnosis to terminally ill patients that they will die (Daugherty & Hlubocky, 2008). This change could reflect the American Medical Association's shift in policy in 1980 of encouraging physicians to tell the patient his or her prognosis:

> The physician must properly inform the patient of the diagnosis and of the nature and purpose of the treatment undertaken or prescribed. The physician must not refuse to so inform the patient. Previously, the AMA had left the decision to the discretion of the physician. Additionally, under the Federal Patient Self-Determination Act of 1991, hospitals must inform patients of their rights to make their own decisions about their medical care (Blackhall et al., 1995); physicians today commonly tell their patients their diagnoses, no matter how dire. (Dickinson & Tournier, 1994)

A study (Dickinson & Tournier, 1994) of physicians' attitudes toward terminally ill patients revealed more of an openness toward informing the patient of his

| PRACTICAL MATTERS | BREAKING BAD NEWS |

Suggestions for breaking bad news to patients by Sean Morrison and Jane Morris (1995) follow a six-step protocol:

1. Arrange to meet in a private setting where you will not be interrupted.
2. Establish what the patient (and/or family) already knows.
3. Identify how much the patient (and/or family) wants to know.
4. Share the diagnosis and prognosis with the patient (and/or family). Present the various treatment options available and provide a realistic appraisal of the benefits and burdens of each.
5. Respond to the patient's (and/or family's) feelings, and identify and acknowledge their reactions.
6. Formulate a plan of care and establish a contract for the future.

Adapted from *I don't know what to say: How to help and support someone who is dying*, by R. Buckman, 1992, New York: Viking Press; R. S. Morrison & J. Morris (1995, July), When there is no cure: Palliative care for the dying patient. *Geriatrics, 50*, 45–50.

or her prognosis after 10 years of practicing medicine than soon after graduation from medical school. A follow-up study (Dickinson, Tournier, & Still, 1999) 10 years later showed a continued openness to communicating with terminally ill patients and their families. A study (Seale, 1991) in England suggested a general preference for more openness between physicians and dying patients and their families about illness and death. This openness should be tempered by the consideration that bad news needs to be broken slowly in a context of support while recognizing that not everyone wishes to know all.

Of the medical specialties, oncologists have the highest percentage of patients with terminal illnesses. Oncologists give bad news to patients some 35 times per month on average, telling a patient that she or he has cancer, that the tumor has returned, or that no further treatment would be helpful (Groopman, 2002). There seemingly is no agreement among specialists about how to deliver such news. More than 40 percent of oncologists withhold a prognosis from a patient if he or she does not ask for it or if the family requests that the patient not be told. A similar number speak in euphemisms, skirting the truth, noted Jerome Groopman (2002).

Kerry Gasperson (1996), in her study of first-year medical students' attitudes toward delivering bad news in a clinical context, observed that a prominent concern of these students is how to deliver news about the terminal nature of a condition. The professionals seem to differ regarding the use of euphemisms when delivering bad news. Maguire and Faulkner (1992) endorsed the use of euphemisms to ease the patient into the truth, whereas Timothy Quill (1991) strongly advised against using euphemisms when communicating a diagnosis to the patient. When the message is not clear, some clinicians think, communication difficulties may result. For example, stories are abundant about patients, who when told that they have a malignancy, are relieved to know that it is not cancer (Sell et al., 1993). Ignorance is not bliss regarding such communication, as a malignancy *implies* cancer. Or perhaps we have selective hearing and hear what we want to hear.

Some ethnic groups in the United States prefer that the patient not be told the prognosis. A study of 800 elderly patients in California (Blackhall et al., 1995) found that immigrant families from South Korea and Mexico were far less willing

LISTENING TO THE VOICES | SHARING THE EXPERIENCE OF DYING

Some of the most beautiful human interactions I have witnessed have occurred between dying patients and supportive families. Sometimes the quality of human interactions in the terminal phase far exceeds anything the patient or family experienced prior to diagnosis.

I strongly feel the dying patient should be told as much as he or she wants to know. The family should also be encouraged to share feelings with the patient in an open manner. Nothing is worse than dying alone. The terminal patient whose family won't broach the subject, or who is afraid to upset his or her family or doctors with fears and feelings, does die alone.

A physician's comment from a survey of 1,093 physicians, "Death education and physicians' attitudes toward dying patients" (pp. 167–174), by G. E. Dickinson and A. A. Pearson, 1980–1981, *Omega*, *11*(2).

to let their terminally ill family members make decisions about medical care than were either black or white Americans. Black and white respondents were about twice as likely as Korean Americans to be honest with the patient about his or her terminal status and about one and a half times as likely as Mexican Americans. For immigrants from South Korea and Mexico, the role of the family was larger and that of the individual was smaller than is generally the case in the United States. In Japan a similar situation exists as doctors rarely tell dying patients that they are terminally ill (Kristof, 1996). As a sort of buffering mechanism, many Asian Americans have been taught to mask behavior and avoid eye contact when self-disclosure is occurring.

Though many physicians in the past have not favored telling their patients that they are dying, Elisabeth Kübler-Ross (1969) suggested that the patient should be told, but that the prognosis should not include the amount of time the patient has left. As Janice Rosenberg (2000) observed in her article entitled "Art of Prognosis Becoming an Increasingly Valued Skill," experts agree that currently there is no accurate method for predicting when a patient will die. Prognosticating is especially difficult for noncancer diagnoses, though prognostication improves for any illness when physicians know their patients as people. Specifying an amount of time dispels hope and serves no function other than telling the patient that the condition is very serious and life threatening, advised Kübler-Ross. If the patient presses for time specificity, a range of time should be given, such as: "Sixty percent of the patients with your disease live as long as 3 to 5 years." This statement provides hope without deluding the patient. Also, it is honest. Medical science cannot say with any certainty that a person will live one year—some die sooner, and others live longer (some even outlive the physician!). Physicians should tell patients they have limited time left when they still have enough energy to think about projects and goals, noted Rosenberg (2000).

Thus, after the physician has given the news to a patient, there is no wonder that when the physician asks the patient if there are any questions, the patient often has none. Many times the patient does not clearly understand what the doctor said. Indeed, as patients we need to be more assertive and to inquire because it is our lives that are at stake and our right to know. However, it is not easy for a physician to give a patient a "bad prognosis." At a palliative care conference in

Sheffield, England, on June 6, 1999, Dame Cicely Saunders shared that a patient said to her, after she told him he was going to die: "It's hard to be told. It's hard to tell, too, isn't it?"

Perceptions about the course that dying will take are referred to as **dying trajectories**. Staff interaction with patients is closely related to the expectations formed about the patients' dying. Trajectories may range from lingering to sudden. Quick trajectories usually involve acute crises (Fulton & Metress, 1995). In real life, patients perceived to be in a lingering death trajectory may suffer a loss in social worth, and medical personnel may give up on them, contributing to a downward slide of the patient's health. On the other hand, even when nurses know that a patient is terminally ill, their behavior is determined by the physicians' official message. The patient may then be treated as if there were little chance for recovery. This alteration in treatment is predicated upon the physician's decision but is not carried out until he or she communicates it. If the patient does not die on schedule, the family and staff may experience stress (Shneidman, 1980).

SPACE MEANINGS: ISOLATION AND CONFINEMENT

Even when the patient has not been told of a terminal condition, he or she will eventually become aware of it. Many times factors related to social space will give the patient clues that the condition is terminal. The dying seem to know when the end is near by observing themselves and the people around them. Within the hospital, there are areas where the very ill are treated. When one is moved into the intensive care unit or onto an oncology ward, it becomes obvious that all is not well and that death is a real possibility. When being moved from a double room to a single-patient room, 17-year-old Tom Nelson, who was dying of cancer (as seen on CBS's *Living With Death*), said, "Isn't this what they do to you when you are about to die?"

Confinement to a health care institution conveys a tremendous amount of meaning to the patient. For the most part, the patient is alone. Through spatial meanings the individual is informed that he or she is removed from those things that give life meaning and purpose—family, friends, and job. For the terminally ill patient this is the first stage of **societal disengagement**—the process by which society withdraws from, or no longer seeks, the individual's efforts.

Within any health care setting, the patient's confinement serves to diminish his or her social and personal power. According to Rodney Coe (1970), three processes occur within the institution to accomplish this—stripping, controlling resources, and restricting mobility. The process of *stripping* takes place when the patient is issued a hospital gown and stripped of any valuables for safe keeping. When this is done, the patient's identity is also stripped. Most factors that differentiate patients with regard to status in the larger society are taken from them to create the primary status of patients—all patients look alike. According to Coe (1970, p. 300), "Every distinctly personalizing symbol, material or otherwise, is taken away, thus reducing the patient to the status of just one of many." Perhaps this helps to explain why physicians are known to be such bad patients—they are forced to relinquish their physician status when they are admitted to the hospital.

The second process is *controlling resources*. When the patient is denied access to medical records and important information about the events of the hospital, personal power is greatly diminished. Personal power is the ability to make decisions that determine the direction of one's own life. Without all of the information concerning one's situation and the place of one's confinement, it may not be possible to make important decisions. One method used by hospitals in controlling resources is to deny all patients and their families access to medical charts and records—an "ignorance is bliss" mentality.

The third process is *restricting mobility*. Not being able to leave one's room or bed further reduces the patient's personal power. The patient is put into a position of dependency upon others. Confinement of this type greatly affects the patient's autonomy. It also makes it possible for others to withdraw from the patient.

Social space is very important in the process of patient disengagement. This disengagement can be accomplished by two methods—the patient can withdraw from others, and others can withdraw from the patient. If the patient is debilitated by illness, energy may not be abundant enough to continue normal patterns of social interaction. The loss of physical attractiveness can also cause the patient to withdraw. Some patients, knowing that their condition is terminal, may disengage themselves as a coping strategy to avoid having to see all that their death will take from them. They may disengage as a sign of their acceptance of social death—"I'm as good as dead" (anticipatory death).

When significant others withdraw from the patient, the second method of patient disengagement will take place. In this situation the process of disengagement is something beyond the patient's control. Family members and friends can refrain from visiting the patient as a sign of their acceptance of social death. The "terminal" label can stigmatize the patient, and others may disagree or treat the individual differently.

Orville Kelly, founder of Make Each Day Count (a support group for persons with cancer), told the story of being invited by a friend to dinner. The table was set with the finest china and silverware—with one exception. Orville's place setting consisted of a paper plate and plastic fork, spoon, and knife. Mr. Kelly was the guest, but he had cancer, so he was supposed to use disposables so no one would be contaminated!

Randall Wagner, an active volunteer with the American Cancer Society, recalls a situation at a high school football game when his leukemia was in a state of remission. A friend had a Thermos of hot chocolate but had only two cups. The first cup was given to Randall. After he had finished drinking, he returned the cup to be filled for someone else. Nobody would drink out of Randall's cup, however. He was told to "just keep it."

Many individuals are afraid of catching cancer. One does not catch cancer. Likewise, many individuals are afraid to be physically near someone with AIDS for fear of acquiring the disease. Because AIDS is primarily passed through sexual intercourse, intravenous drug needles, blood transfusions, or from a mother to her embryo or fetus, one is not going to acquire AIDS by sitting next to or by physically being in the room with someone with AIDS. Because many do not know how to relate to persons with cancer or AIDS or to the terminally ill in general, they withdraw as a method of coping with their inadequacy. In doing this, patient disengagement is accentuated.

NORM AND ROLE MEANINGS: EXPECTED DYING BEHAVIOR

Norms are plans of action or expected behavior patterns felt to be appropriate for a particular situation. **Roles** are plans of action or expected behavior patterns specifying what should be done by persons who occupy particular social positions. Applied to the death-related behavior of the dying patient, norm definitions would involve the general expectation that the dying patient should be brave and accept the fact that life will soon end. The patient is not supposed to cry or become verbal in regard to feelings about his or her death. Nurses often sanction such behavior by giving less attention to patients who deviate from this norm. Elisabeth Kübler-Ross (1969, pp. 56–57) gave the following example of one such deviant:

> The patient would stand in front of the nurses' desk and demand attention for herself and other terminally ill patients, which the nurses resented as interference and inappropriate behavior. Since she was quite sick, they did not confront her with her unacceptable behavior, but expressed their resentment by making shorter visits to her room, by avoiding contact, and by the briefness of their encounters.

Role meanings differ from norm meanings in that they specify, in a detailed fashion, what behavior is expected of persons who occupy specific social positions. For example, if a wife and mother who is a principal provider in the family is dying, it is expected that she do all that she can before death to provide for the financial needs of her family. It would probably be expected that she make arrangements for her funeral, ensure that her bills are paid, finalize her will, and establish a trust fund for her children.

As noted earlier, one of the important aspects of norm and role meanings is their relationship to the process of societal disengagement by which society withdraws from the individual (Atchley, 2004). The individual can also withdraw from societal participation and choose not to perform the roles that he or she performed before the terminal diagnosis. This type of disengagement goes beyond withdrawing from interaction patterns with others and refers to a withdrawal from the social structure (such as quitting one's job and taking a trip around the world).

In addition to the disengagement process, patients are expected to acquire the **sick role** (Parsons, 1951). They are expected to want to get better and to want to seek more treatment even though everyone realizes that such treatment only prolongs death and not life. This role disengagement can create conflict within the family, especially when the patient has accepted his or her death (and even longs for it), whereas family members are unwilling to let the patient go. The case of Terri Schiavo in Florida in 2005 illustrates this point. Her biological family wanted to keep her alive, yet her husband's determination to remove the feeding tube and let her die made him a criminal suspect in the eyes of many individuals. As Kübler-Ross (1969) documented, families many times cannot comprehend that a patient reaches a point when death comes as a great relief and that patients die easier if they are allowed and helped to detach themselves slowly from all of the responsibilities and meaningful relationships in their lives. John Harvey stated:

> There is a kind of freedom that comes with a recognition that your time is limited. Long-term consequences seem less important. A certain kind of fearless authenticity often comes from a confrontation with one's mortality. (1996, p. 173)

Often, we, the survivors, are the stumbling blocks for dying persons. We do not want the dying person to die; thus, we will not let go and allow her or him to say goodbye and break away.

Value Meanings: Reassessing the Value of Life and Death

Values, like all other meaning systems, are socially created. They are not inherent in the phenomena, but rather they are applied by humans to the phenomena. Death per se is neither good nor evil. Humans do ascribe value, however, to the different types of deaths. Each of these value meanings has important behavioral consequences.

Most people in our society view death as being intrinsically evil and, therefore, something to fear. We tend to see death as an intruder—the spoiler of our best plans. Thus, in the past the medical profession has attempted to delay death in favor of life. People are kept on machines to prolong life, even if death is inevitable. We assume that people wishing to die are mentally ill or irrational because we see death as something to be avoided. Yet, most terminally ill individuals eventually come to view death as a great blessing, when they have finally accepted the fact that they are going to die. This is possible because humans are able to create hierarchies of values in which value meanings take on relative meanings. To the terminally ill patient, dignity is more highly valued than is life with pain, indignity, and suffering. Consequently, death may be ascribed positive value for the dying patient who has accepted the inevitability of his or her death.

Daniel Goleman (1989) cited research showing that confronting individuals with the fact that they will die makes them cling tenaciously to their deepest moral values. They tend to become more moralistic and judgmental. They are harsher toward those who violate their moral standards and kinder toward those who uphold them. Open-minded persons, however, become even more tolerant of those whose values differ from theirs. Researchers (Goleman, 1989) note that these findings give the fear of death a central role in psychological life and that a culture's model of "the good life" and its moral codes all are intended to protect people

---

LISTENING TO THE VOICES | Don't Abandon the Patient

When I say I feel as comfortable with a dying patient as with any other, and that I do not find treating a dying patient unpleasant, I do not mean that I am anaesthetized to the fact that they are dying, and do not have feelings about the patient which are different from my feelings about a patient whom I know will get well. Anesthetizing of feeling is the method which we physicians employ initially in dealing with the pain—ours and theirs—involved in treating a dying patient. But this passes, and when one accepts the patient as part of life, and

not someone who is no longer a real part of the world (or a frightening part of the world), then caring for the dying patient becomes (though often sad) neither unpleasant nor something one wishes to avoid. To abandon the dying patient is the worst thing that can be done—both for the patient and the doctor.

A physician's comment from a survey of 1,093 physicians, "Death education and physicians' attitudes toward dying patients" (pp. 167–174), by G. E. Dickinson and A. A. Pearson, 1980–1981, *Omega, 11.*

from the terror of death. Cultures prescribe what one must do to lead a good life, and if one leads a good life, he or she will be protected from a tragic fate at death.

OBJECT AND SELF-MEANINGS: ACCEPTING THE SELF AS TERMINAL

The previous discussion focused on meanings that have been applied to an object—the dying patient. From a biological perspective, the person is a living organism—a physical object. From a social-psychological perspective, the patient is a social object—a self.

An important part of accepting one's self as terminal is exploring the meaning of that condition with others. With death close, dying patients may still want to express what they experience and to convey their fantasies about death. The presence of the physician in this last phase of life may be crucial for a peaceful death. Such a presence may allow the patient not only to die with self-respect, but also to feel less lonely. Sociologist Arthur Frank (1991), who himself had two serious illnesses by the age of 40, believes that too many ill individuals are deprived of conversation. He noted that too many believe that they cannot talk about their illness: They simply repeat what they know from the medical staff. When ill persons try to talk in medical language, they deny themselves the drama of their personal experience. Frank noted that sick persons need to talk about their hopes and fears and the prospect of death. But because such talk embarrasses us, we do not have practice with it and, lacking practice, we find such talk difficult and avoid it altogether.

As death draws near, individuals often recognize one who has already died coming for them. Occasionally, angels or religious figures are mentioned as beckoning to them. A friend, who works with chronically and terminally ill children in a hospital setting, told George Dickinson about a terminally ill child who said the "pretty woman with flowers standing in the corner of the room" was calling the child to come with her to play with the children. Though no one was standing in the corner of the room, the friend acknowledged the child's "observation." "I am not ready to go yet," the little girl replied. A couple of days later, the child again said that the woman "in the corner of the room" was signaling for her to come and play with the children, who were laughing and having a good time. The child said, "I am now ready to go with them." She died a few hours later. Thus, there is a connection between the living and the dead as one prepares for the journey.

DETACHMENT FROM THE LIVING   Many patients who are defined as having a terminal condition begin to view themselves as being "as good as dead." Ideally, the dying person has reached Erik Erikson's final stage of the life cycle of integrity believing that his or her life has had meaning and purpose, and the person is ready to break away. Life is over, and the dying patient is accepting of that. The patient has accepted the terminal label, has applied it to a personal understanding of who she or he is, and has experienced anticipatory death. Families tend to see their loved ones as being in bereavement. This symbolic definition of the patient is reinforced by the role disengagement process, the spatial isolation of the dying patient, and the terminal label placed upon the patient by the physician and other medical personnel. The patient seems to take on a status somewhere between the living and the dead.

Elisabeth Kübler-Ross stated that in the end terminally ill persons have a need to detach themselves from living persons to make dying easier. The following example illustrates this point:

> She asked to be allowed to die in peace, wished to be left alone—even asked for less involvement on the part of her husband. She said that the only reason that kept her still alive was her husband's inability to accept the fact that she had to die.
>
> She was angry at him for not facing it and for so desperately clinging on to something that she was willing and ready to give up. I translated to her that she wished to detach herself from this world, and she nodded gratefully as I left her alone. (1969, p. 116)

When an individual's condition has been defined by self and others as terminal, all other self-meanings take on less importance. Although a given patient may be an attorney, a Democrat, a mother, a wife, or a Presbyterian, she tends to think of herself primarily as a terminally ill patient. The terminal label becomes her **master status** because it dominates all other status indicators. Consequently, most of the symbolic meanings previously discussed become incorporated into the individual's self-meaning.

KÜBLER-ROSS'S FIVE STAGES OF DYING   Acquiring the terminal label as part of the self-definition is not an easy task for the individual. In her bestselling book *On Death and Dying* (1969), Elisabeth Kübler-Ross delineated the following five stages that patients go through in accepting their terminal self-meaning: denial, anger, bargaining, depression, and acceptance. The stages are based on interviews with approximately 200 adult patients during a period of less than three years.

In the first stage, *denial*, the patient attempts to deny that the condition is fatal. This is a period of shock and disbelief (e.g., the patient wants to believe that the physician must have read the wrong x-ray). The patient wants to believe that a mistake has been made. One may seek additional medical advice, hoping that the terminal diagnosis will be proven false. When the diagnosis is verified, the patient may often retreat into self-imposed isolation.

The second stage, *anger*, is a natural reaction for most patients. The patient may vent anger at a number of individuals—at the physician for not doing enough,

---

| **WORDS OF WISDOM** | ELISABETH KÜBLER-ROSS'S STAGES OF THE DYING PROCESS |

Psychiatrist Elisabeth Kübler-Ross outlined the stages of the dying process, as noted below:

1. *Denial (shock and disbelief):* An individual is unable to admit that her or his medical condition is terminal and that death is forthcoming.
2. *Anger:* The pain of loss is projected onto others.
3. *Bargaining:* One makes a last-ditch effort to hold on to life and pleas with whomever or whatever for survival.

4. *Depression:* The reality of imminent death hits home.
5. *Acceptance:* An individual comes to the reality of forthcoming death and makes preparation accordingly.

From E. Kübler-Ross. (1969). *On death and dying.* New York: Macmillan Publishing Company.

at relatives for outliving the patient, at other patients for not having a terminal condition, and at God for allowing the patient to die. In other words, a **scapegoat** is sought—the patient seeks someone or something on which the blame can be placed. If an individual becomes stuck in this stage, it encroaches upon the rights of others to move on the process.

When one has incorporated the terminal label into self-meaning, an attempt to bargain for a little more time may result. This third stage, *bargaining*, may include promises to God in exchange for an extension of life, followed by the wish for a few days without pain or physical discomfort. For example, in the movie *The End*, Burt Reynolds decides to commit suicide by drowning. He swims far out into the ocean but begins to change his mind about wanting to die. He bargains with God and prays to God to make him the best swimmer in the world so that he can swim back to shore. In return for God's granting his wish, Reynolds promises to give most of his income for the rest of his life to the church. As he draws nearer to shore and realizes that he is going to survive, his "promises" to God are reduced to 50 percent of his income, then 40, and downward, until he reaches the shore and reduces his promise to almost nothing. Bargaining generally includes an implicit promise that the patient will not ask for more if the request to postpone death is granted. The promise is rarely kept, however, noted Kübler-Ross (1969).

In the fourth stage, *depression*, the patient begins to realize that a mistake was not made, the x-rays were correctly read, and the prognosis is not good. The patient realizes that meaningful things of life—family, personal accomplishments, and often a sense of dignity—will be lost as death approaches. This is a time to "get one's house in order" and to begin breaking away. Contrary to perhaps other situations of depression when one might be thought to need counseling, this is a rather positive stage in that it opens the door to face the reality that death is forthcoming very soon. Is not the truth here depressing, yet reality can help an individual to move on?

In the fifth stage, *acceptance*, the patient accepts death as a sure outcome. Although not happy, this acceptance is not terribly sad, either. The patient is able to say, "I have said all the words I have to say and am ready to go."

Kübler-Ross's stage theory of dying can also be applied to divorce and grief. It can address any loss, whether of the individual directly involved or peripheral individuals. The theory is not without its critics, however (e.g., Charmaz, 1980; Kellehear, 1990; Copp, 1998; Greenberg, 2003; Kastenbaum, 2009), yet others give support (e.g., Maciejewski et al., 2007; Marcu, 2007). Some reject the developmental nature of the sequential stage approach because it lacks universality—not all patients manifest all five stages of behaviors. Others have noted that the stages are not mutually exclusive—some patients may bargain, be depressed, and be angry at the same time. Research indicates that men and women grieve differently, a factor not accounted for in the theory (Greenberg, 2003). Others have observed that the order of the stages is more arbitrary than Kübler-Ross would have us believe—dying patients may go from denial to acceptance, followed by depression and anger (i.e., an individual may not follow the stages in one-two-three order). It should also be remembered that individuals are unique and no two individuals necessarily go through crises in life the exact same way.

Dying with dignity or self-respect does not always happen. Positive self-meanings can be created and sustained with the help of others. These others, who become the social audience for dying-related behaviors of the patient, include family, physicians, clergy, nurses, peers, or even strangers walking down the hall of the hospital.

Kathy Charmaz (1980) argued that the stages emanate from preconceived psychiatric categories imposed upon the experiences rather than from the data. She noted that what originated as a description of reality often becomes a prescription for reality. Respected thanatologist Robert Kastenbaum (2009) has pointed out that no evidence has been put forward to support the stages and that individuals do not actually move from stage one through to stage five.

The stages also do not adequately take into consideration the perspective of the patient. Anger, for example, may be vented at others because they have withdrawn from the patient in an attempt to cope with the loss of someone for whom they care. On the other hand, the anger may be directed at someone who happens to be at the wrong place at the wrong time. For example, George Dickinson recalls a colleague who verbally assaulted him following the death of the colleague's father. As Dickinson was teaching a course entitled Death and Dying, the colleague lashed out that Dickinson knew nothing about death and dying. There was no withdrawal from the colleague, but he needed someone to release his frustrations on—wrong place, wrong time for Dickinson!

The bargaining behavior of the patient may be motivated by a need for support from caregivers, rather than by a hope for an extension of time. This third stage may be a plea for help to assist the individual in moving through this phase. Depression may be a function of the severity of the physical condition of the patient rather than an emotional response to the terminal condition (Charmaz, 1980). As the disease progresses, the strength of the patient will diminish and be evaluated by others as psychological depression. In actuality, the patient may be depressed not by dying, but rather by the physical effects of the illness.

An individual may start out in one stage and never move out of it. Denial may be evidenced until the very end of the illness. Yet some individuals may not move from the anger stage: "I am only 25 years of age and it is not fair that I am dying." Such an attitude may not change. Others may, however, begin with acceptance and stay there: "I told you I was sick, and now there is evidence from the physician to support my assertion."

With regard to the stages of dying, we must conclude that dying behavior is more complex than five universal, mutually exclusive, and linear stages. Kübler-Ross has helped us, however, in understanding that each of the five behaviors is a "normal" coping strategy employed by dying patients. It may be that it is the social situation that accounts for the similar coping strategies of dying patients and that in other cultural settings different patterns of behavior will be found. A stage theory like this can help patients, caregivers, and professionals to cope with dying and death. Indeed, a health care professional in a hospital or hospice room with a terminally ill patient in the anger stage venting verbal assaults toward her or him can know that this is normal. The health care worker can hopefully then not take the verbal jabs personally and therefore not vent back at the patient with even stronger verbal language. The individual should simply walk away, if the assaults become unbearable. In other words, get out rather than lose your cool and unplug the life-support equipment of the angry patient!

Others, such as Charles Corr, have proposed a theory in which the various tasks that dying people may face or need to work on are outlined. Corr's four broad areas are physical, psychological, social, and spiritual. Likewise, Ken Doka has proposed a theory that emphasizes how the tasks or concerns of individuals with terminal illnesses change from one phase of the illness to another—the pre-diagnostic to acute to chronic to terminal phases (Corr & Corr, 2002). Thus, different theories and lessons learned have emerged regarding an individual with a terminal illness, with the application of the theories perhaps varying with the type of illness.

SOCIAL SITUATION MEANINGS: DEFINITION OF THE ENVIRONMENT   As the dying person comes to grips with a terminal condition, the way in which the social situation is defined will have a tremendous impact upon the process of dying. If the hospital is viewed as a supportive environment, patient coping may be facilitated. However, if the patient feels all alone in the place of confinement and if the hospital is defined as a foreign place, personal adjustment to dying and death will be hindered. The hospital is not home by any sense of the imagination. Compared with a hospital, there is nothing like home sweet home when one is confined to a bed. In the environment of the home, the patient has familiar smells, sights, and perhaps sounds. In the privacy of the home the patient can probably be on her or his own schedule, rather than that of a hospital. In a hospital, for example, the patient may be awakened simply because it is the time that the hospital serves breakfast (with no regard for the possibility that the patient perhaps finally fell asleep only an hour or two before). At home, the schedule can be somewhat more flexible. Also at home, unlike in a hospital, the patient can have pets in the room—even in the bed, if she or he wishes.

The bureaucracy, with all its rules and regulations, may seem inappropriate and inconvenient to a patient, yet the patient must abide by the rules—dying or not. In the hospital, the patient is basically told what to do. For example, George

| The Care of the Dying Patient

No matter how we measure his worth, a dying human being deserves more than efficient care from strangers, more than machines and septic hands, more than a mouth full of pills, arms full of tubes and a rump full of needles. His simple dignity as man should merit more than furtive eyes, reluctant hugs, medical jargon, ritual sacraments or tired Bible quotes, more than all the phony promises for a tomorrow that will never come. Man has become lost in the jungle of ritual surrounding death.

From *Facing death* (p. 6), by R. E. Kavanaugh, 1972, Baltimore: Penguin Books.

Dickinson was in the hospital for surgery. Feeling an urge to visit the toilet for the purpose of urinating, he got up out of the bed and attempted to walk to the bathroom (though in for surgery he was feeling very healthy). A nurse walked by his room about the time his feet hit the floor and literally screamed at him, "Get back in that bed! What do you think you are doing?" He immediately dived back into the bed to avoid further wrath from this sergeant nurse. A couple of hours later, the same nurse, in a calmer voice this time, said, "Mr. Dickinson, you can now 'empty your bladder'." Little did she know that her previous verbal attack had canceled his desire to "empty his bladder." However, her commanding voice had quickly socialized him to the hospital rule of bladder emptying; thus, since the rule enforcer now said that he could go, he went (right on command!).

Like all other meaning systems, the definition of the social situation is an attempt by the individual to bring order to his or her world. Order contributes to efficiency. Thus, order is important to the bureaucratic setting of a hospital. Because situational meaning always involves selective perception, the terminally ill patient will create the meaning for the social environment and will respond to this meaning and not to the environment itself. Hospital rooms are generally bland and uniform. The bland environment of the room may carry over to the patient's perception of the entire hospital setting—including the staff's demeanor. They may be perceived as cold and unfeeling. Each terminally ill patient will not only experience death in a different environment, but also will have a unique interpretation of the social situation. This accounts for the different experiences of dying patients. The hospice movement (see Chapter 6) is an attempt to create a more positive and supportive social situation in which dying can take place.

RELATING TO THE DYING PERSON

Relating to a dying individual is really not easy for any of us. For medical personnel there is the failure of not being able to make the person well again. As surgeon Pauline Chen (2007) notes, few premedical students choose the field of medicine to care for the dying; rather, they believe they will save others from the inevitability of death. For family members and friends there is the sorrow in anticipating a world without that special person. We turn now to a discussion of being in the presence of someone who is dying.

Reynolds, Dan/CSL, CartoonStock Ltd

MEDICAL PERSONNEL

Death anxiety is significantly linked with both age and experience for physicians as a whole. The longer a physician has practiced medicine, the less death anxiety he or she has. In a 10-year follow-up study (Dickinson & Tournier, 1994) and again in a 20-year follow-up (Dickinson, Tournier, & Still, 1999) of physicians soon after they graduated from medical school, the younger, less experienced physicians reported higher levels of death anxiety and felt less comfortable with dying patients.

Physicians may have a greater need than others to reject death, as they take an oath to prolong life. Some physicians may compensate for their unconscious personal fear of illness and death by distancing themselves from dying patients. Such "avoidance mechanisms" are certainly a way of relating to a confrontation with suffering and death. These situations can be addressed, however, through training sections. For example, a study (Melo & Oliver, 2011) of 150 health care workers (including physicians) in Portugal, following a six-day course on communication, emotional and spiritual support for patients, and personal introspection on death anxiety, revealed that such education helps reduce the health care workers' need to protect themselves by avoiding meaningful helping relationships with dying patients.

American medicine is often indicted because physicians are educated to treat diseases rather than people and to deal with patients impersonally rather than holistically. However, in a study some 35years ago of 1,012 physicians, George Dickinson and Algene Pearson (1979a) concluded that those physicians with a high probability of dealing with dying patients (such as oncologists) were more open with their patients than were physicians who practiced medicine in areas in which there was a lower probability of dealing with death (e.g., obstetricians and gynecologists). Because this study suggested that differences exist among medical specialties in relating to dying patients, it is possible that one factor influencing the selection of a medical specialty is the medical student's personal understanding and feelings concerning dying and death issues. Yet, Michael Bell (1996), in the latter stages of a fight against terminal cancer, observed that this priority of technical skills over bedside mannerisms may be exactly what we prefer that our doctors develop, because one's very life may depend on such skills and knowledge. The death of a patient may be viewed by the physician as a professional failure and thus generate guilt. Therefore, it is important for medical students to develop technical skills necessary to deal efficiently with terminally ill patients and their families.

THE SOCIALIZATION OF PHYSICIANS Surgeon Sherwin Nuland, in his book *How We Die* (1994), shared the story (early in his practice) of one of his patients, a 92-year-old woman with heart disease who was being treated for an acute digestive tract disorder. Dr. Nuland persuaded the patient to have an operation for the disorder, though she argued that she had lived long enough and did not want any further intrusions into her body. The young doctor was following the medical code to "prolong life and relieve suffering." Yet, he was not listening to the patient's system of logic to use this sudden illness as a gracious way to die. The patient survived the operation, yet died two weeks later of a stroke. She told Dr. Nuland that he had let her down by not allowing her to die in due course without the pain and complications experienced from the operation. Dr. Nuland said he learned from this experience that for dying patients "the hope of cure will always be shown to be ultimately false." He learned that he needed to listen to the logic of his patients and not rely solely on his own medical code of logic. Dr. Nuland himself died of prostate cancer on March 3, 2014 (Gellene, 2014). His daughter recalled how he told her he was not ready for death because he loved life. Not afraid of dying, he said, yet not ready to leave a beautiful life.

In addition to listening to one's patients, physicians need to be cognizant of the complexity of culture inherent within any society (Yapp, 2012). An individual's cultural beliefs and values are main determinants of decision making. Without the proper context in which to base end-of-life discussions, the intersect of diverse cultures and health care policy may result in misunderstanding, with regard to treatment decisions, and in tense interactions. Deeply embedded cultural beliefs are prevalent throughout the health care system. Though a physician cannot be knowledgeable of all cultural beliefs and behavior, simply being aware of differences within a health care setting can go a long way toward understanding and agreement between medical personnel, the patient, and the patient's family.

Americans are poorly socialized with regard to issues of dying and death, as noted in Chapter 1. One might expect that the early socialization experiences of physicians would be similar to those of other members of society. This is exemplified by the following story told to George Dickinson by the late Dr. Charles B. Huggins, Nobel recipient for cancer research and professor at the University of Chicago School of Medicine, about his first day as a student in medical school:

> After the gross anatomy professor finished the initial lecture, the class went to the laboratory to begin work on the cadavers. My **cadaver** was a female. After taking one look at the body (having never seen either a dead woman or a naked woman), I said to myself, "I should have gone to law school after all."

Even though this Nobel recipient entered Harvard Medical School in 1920 (at the age of 18), many first-year medical students today have had similar experiences and feelings and find that their first exposure to death is an impersonal experience in an anatomy laboratory. Human dissection in a gross anatomy laboratory, typically occurring during the first semester, is a rite of passage for all future medical doctors (Dickinson, Lancaster, Winfield, Reece, & Colthorpe, 1997). Students must deal with the psychological and social aspects of cutting and touching a dead body and must struggle with questions of their own and others' mortality. In her excellent book *Body of Work: Meditations on Mortality from the Human Anatomy Lab*, Christine Montross (2007) said that she cried when they sawed into the skull of her cadaver and removed the brain. In addition to their own anxieties about death, students in gross anatomy classes are subjected to horror stories from upperclassmen and classmates. Cadaver stories portray, and thus help create, a world of outsiders and insiders and emotionally weak or strong medical students (Hafferty, 1991). A norm within medicine is that one must have emotional strength to survive the mental stress brought on by practicing medicine; strength begins and must be displayed in the gross anatomy laboratory. Students have a range of different coping mechanisms to deal with the stress related to the experience of dissection of humans, the most commonly reported being the discussion with family and friends about the emotions and experiences that the students are encountering (Robbins et al., 2008).

Entry into medical school involves movement from a largely lay to a largely medical culture, noted Frederic Hafferty (1991) in his study of the socialization of first-year medical students. Experiences such as the dissection of human cadavers in the anatomy laboratory provide students with opportunities to internalize a variety of attitudes, values, motives, and rationales with respect to both their current role as student and their future role as physician. These experiences give students

---

WORDS OF WISDOM | DENIAL OF DYING

Before 1959 when Herman Feifel wanted to interview the dying about themselves, no doubt for the first time, hospital authorities were indignant. They found the project "cruel, sadistic, traumatic." In 1965 when Elisabeth Kübler-Ross was looking for dying persons to interview, the heads of the hospitals and clinics to whom she addressed herself protested, "Dying? But there are no dying here!" There could be no dying in a well-organized and respectable institution. They were mortally offended.

From *The hour of our death*, by P. Aires, 1981, New York: Knopf.

an opportunity to assess their progress in identity transformation and skill building (Hafferty, 1991).

Situational Adjustment.   Students entering the medical profession come in with certain attitudes and feelings toward patients that will be shaped and continually processed until their attitudes comply with those of the medical profession itself. Howard Becker (1964) referred to this "molding" as **situational adjustment.** As one moves in and out of social situations, learning the requirements for success in each situation, with a strong desire to deliver the required performance, the individual becomes the kind of person that the situation demands. Thus, the medical student is molded into the medical profession by learning what is expected and then doing it. Becker notes that much of the change in an individual is a function of the interpretive response made by the entire group—the consensus that the group reaches with respect to its problems.

The process of situational adjustment accounts for changes that people undergo, but people also exhibit some consistency as they move from situation to situation. Becker (1964) referred to this consistent line of activity in a sequence of varied situations as *commitment.* A variety of commitments constrain one to follow a consistent pattern of behavior in many areas of life.

Whether one's medical socialization took place many decades ago, as with Dr. Huggins, or whether it took place in the 21st century, as with Dr. Montross, medical training transforms laypersons into something else—physicians (Hafferty, 1991). When we speak of socialization, we are talking about the structure, the method, and the route by which initiates move from one status to another and acquire the technical skills, knowledge, values, and attitudes associated with the new position or group. Thus, one must attain a new cultural base but must also facilitate movement away from the old status.

Gender or Occupational Related Socialization? In a study comparing the attitudes of physicians and nurses toward death, Campbell, Abernethy, and Waterhouse (1984) found differences that were tied to professional roles, not to gender. Nurses tended to see a more positive meaning in death than did physicians—death as rebirth rather than as abandonment and as tranquil rather than as frightening. Because physicians make the crucial decisions, whereas nurses carry out the orders, the implication of patient death as professional failure is reduced for nurses.

Becker's (1964) emphasis on situational adjustment, followed by a commitment to consistency, seemed to suggest that socialization to occupational role overcomes earlier gender role socialization. This is the conclusion of Sylvia Ashley-Cameron and George Dickinson (1979) in explaining the different attitudes of female nurses and physicians regarding dying patients. Even if women have traditionally been socialized to be the more nurturing gender and to be more sensitive and responsive to the needs of others, female physicians tend to be less sensitive and responsive than female nurses. Thus, the differences between physicians and nurses regarding death and terminally ill patients tend to be more a function of the role expectations of the particular medical occupation than of gender.

Jack Kamerman (1988) cautioned, however, that as nurses are given greater responsibility in diagnosis and treatment, their attitudes toward death may move closer to those of physicians. In addition, as nurses become more susceptible to the strains that physicians experience at a patient's death, it is possible that they will retreat further behind the shield of professional detachment, particularly if nursing follows the medical model of professional status.

By contrast, a study by S. C. Martin, R. M. Arnold, and R. M. Parker (1988) showed that gender differences do play a role in how male and female doctors communicate with patients. Perhaps this is because medical schools have not traditionally placed heavy emphasis on patient interviewing, and female doctors therefore default to the less-directed communication style that is used predominately by females in American society. Women's styles of speech are typically less obtrusive, men are more likely to interrupt women than vice versa, and women are more likely to allow such interruptions. The conflicting paradigm to this less-directed communication style of women is that physicians are socialized to dominate the physician–patient interaction. The physician is to control the flow and topics of conversation. Female physicians tend to integrate these two models of communication by being more egalitarian in their relationships with patients, more respectful, and more responsive to patients' psychosocial issues (Martin, Arnold, & Parker, 1988).

The successful communication skills of female physicians are supported by George Dickinson and Algene Pearson (1979b). In their study of more than 1,000 physicians, they found that female physicians related better to dying patients and their families than did male physicians. A 1986 follow-up to this study (Dickinson & Tournier, 1993) showed even more striking differences between males and females after a decade of practicing medicine. Perhaps the traditional "feminine characteristics" of gentleness, expressiveness, responsiveness, and kindness help to explain these differences.

Martin, Arnold, and Parker (1988), in studying gender and medical socialization, reported that medical students regard female faculty physicians with clinical responsibility for patients to be "more sensitive, more altruistic, and less egoistic" than males; nurses believe that female physicians are more humanistic and have greater technical skill in communicating with patients; and patients also perceive female physicians as more humanistic, more empathic, and better listeners. If more physicians and other medical personnel were to display these human qualities with a greater frequency, patients might receive more help and support in their dying.

AWARENESS CONTEXTS   Whether medical personnel are male or female or nurse or physician, communication with terminally ill patients is of utmost importance. In a classic work by sociologists Barney Glaser and Anselm Strauss (1965), four awareness contexts in interacting with a dying patient are presented. These researchers defined the **awareness context** as what an interacting person knows of the patient's defined status, and his or her recognition of the patient's awareness of a personal definition. The awareness contexts are closed, suspicion, mutual pretense, and open.

Closed awareness is usually the first context. To maintain the patient's trust and yet to keep him or her unaware of the terminal condition, the staff may construct a fictional future biography. If the patient has not been told the truth, or has been told but did not want to hear, this often places an additional burden on the hospital nursing staff in working with the patient for an eight-hour shift. Because the physician often directs the medical team, nurses are sometimes forced to work in a closed awareness context regardless of their own views. Because most patients are able to recognize death-related situational and spatial clues, this context tends to be unstable, and the patient usually moves to the suspicion or full awareness contexts.

Suspicion awareness is a context for control between the patient and the medical staff. The patient suspects that he or she is dying but receives no verification from the staff. Nurses must use teamwork to refute this challenge.

Mutual pretense often follows and requires subtle interaction with both patient and staff "acting correctly" to maintain the pretense that the patient is not approaching death. A game continues to be played whereby all concerned act as if they know nothing about the terminal condition of the patient.

If mutual pretense is not sustained, *open awareness* follows in which the patient and everyone else knows that the patient is dying. This is the context found in hospice programs (see Chapter 6). Ambiguities may develop in this context, however. The patient is obligated not to commit suicide and to die properly. Dying properly is difficult, however. Nurses and other medical staff expect proper dying, but the patient usually knows no model to follow, noted Robert Blauner (1966). If the patient is dying in an unacceptable manner, then difficulties are faced when he or she is trying to negotiate for things from the staff.

Ashley-Cameron and Dickinson (1979) found that nurses working with dying patients seem to be comfortable in a closed awareness context. Because nurses spend more time with patients in the hospital than do physicians, a closed awareness context may produce a more comfortable setting for the nurse. Closed awareness may not be easy to pull off, however; thus mutual pretense often is the awareness context found in medical settings. A study in England, however, produced a very different finding regarding awareness contexts. Open awareness was preferred in a study of 548 physicians and nurses (Seale, 1991). Eighty-one percent of these British practitioners said that they found it easier to work with dying patients if the patients were aware of their terminal condition. Perhaps the openness of the 1990s and/or the difference in cultures is reflected in these seemingly contradictory findings over different time periods.

Whether a person is a nurse or a physician, the ability to cope with terminally ill patients does not come easily. Society has not prepared one well for such interaction. Some obviously react better than others. Working with patients with AIDS may be especially difficult due to the fear of acquiring AIDS. The key to good relations with dying patients is personally coming to grips with death. One should also have good communication skills. If the dying patient could be viewed as a person living *with* cancer, not dying *from* cancer, and treated as a human being, certainly the trauma of the dying process could be eased. In the end, medical personnel, the patient, and the family would benefit.

FAMILY AND FRIENDS

It is most important that the family member or friend make contact with the dying individual, with the frequency of contact being somewhat dependent on the closeness of the relationship. Just being with the dying individual or making contact in other ways reminds the patient that you care and that the relationship is important to you. When you make contact, it is not as crucial what you say as it is that you touch base.

Not only do family members and friends need to be cognizant of the ongoing human needs of the terminally ill patient, they need to continue to maintain relationships and continue to incorporate the dying person within the network of family and friends (Doka, 1993). The patient's family often plays a pivotal role in end-of-life decision making, relying on their understanding of the patient's diagnosis, prognosis, and alternate treatment options provided by the primary care physician (Yapp, 2012).

It is important for family and friends to openly discuss their relationships with the dying person. If family members or friends have never verbalized to the dying person what she or he really means to them, this would be a good time to do so. If difficult to say in words, then write it down and share these feelings with the individual. Also, recall the good times you have experienced together. Reminiscing may provide meaningful interaction for the present and may mitigate subsequent grief.

Family and friends may participate with the dying individual in other ways in which they feel comfortable. For example, some may choose to be involved in personal care, monitoring medication, providing massages, or helping with grooming. Others may feel comfortable in social conversation, reading aloud, or watching television or listening to music together. Yet others may find meaning for themselves and the ill person by participating in religious rituals together. Family and friends should recognize the need to allow the dying individual the freedom to decide whether or not she or he wishes to pursue various activities (Doka, 1993).

As the family member or friend moves through the process of dying, assuming the death is not sudden and unexpected, the caregivers become aware of physical changes and sometimes increased pain. Today, the typical dying trajectory is lengthening as the proportion of deaths from chronic illnesses is increasing. The degree of awareness of various family members and friends concerning the seriousness of the illness is important in relating to the sick person. That is to say, does everyone have the same degree of knowledge concerning the prognosis of the illness? The meaning(s) associated with knowledge of impending death make a significant contribution to the type of awareness displayed, and meanings are likely to be contingent on social context (Young, Bury, & Elston, 1999).

Ira Byock (1997), former president of the American Academy of Hospice and Palliative Medicine, stated that people who are dying also have a responsibility to their family and friends. It is very important for them to make an effort to reconcile strained relations. A dying person may wish to be left alone, but that individual should recognize that family and friends may suffer because of isolation from the dying person. It is the dying patient's right to be left alone, since death is primarily a personal experience, but Byock notes that the friends and relatives may have their own personal experiences with regard to the patient's dying. Within the social context of dying, both the patient and family and friends should keep each other's feelings in mind. Byock (1997) thinks of family as a process: Family is marked by feelings of mutual connection, appreciation, and caring. He noted that individuals also have both relatives and friends who qualify as family in this sense of the word. He suggested that patients notice who shows up and who patients miss when they do not. Even for individuals who have been utterly alone for years, it may be nurses or aides or volunteers who become family to the dying patient, observed Byock.

| PRACTICAL MATTERS | SIGNS OF APPROACHING DEATH AND WHAT TO DO TO ADD COMFORT |

Hospice exists to support the family's desire to aid a dying loved one in familiar surroundings. This time period is a very difficult one for families. The following was devised to help alleviate some of the fears of the unknown. The information may help caregivers prepare for, anticipate, and understand symptoms as patients approach the final stages of life. It is important to note that some symptoms may appear at the same time and some may never appear.

Symptom: The hospice patient will tend to sleep more and more and may be difficult to awaken.

Action: Plan activities and communication at times when he or she seems more alert.

Symptom: You may notice your loved one experiencing confusion about time, place, and identity of people.

Action: Remind your family member of the time, day, and who is with him or her.

Symptom: Loss of control of bowel and bladder may occur as death approaches, as the nervous system changes.

Action: Ask the hospice nurse for pads to place under the patient and for information on skin hygiene. Explore the possibility of a catheter for urine drainage.

Symptom: Arms and legs may become cool to the touch, and the underside of the body may become darker as circulation slows down.

Action: Use warm blankets to protect the patient from feeling cold. Do not use electric blankets since tissue integrity is changing and there is danger of burns.

Symptom: Due to a decrease in oral intake, your loved one may not be able to cough up secretions. These secretions may collect in the back of the throat causing noisy breathing. This has been referred to as the "death rattle."

Action: Elevate the head of the bed (if using a hospital bed) or add extra pillows. Ice chips (if the patient can swallow) or a cool, moist washcloth to the mouth can relieve a feeling of thirst. Positioning patient on his or her side may help.

Symptom: Hearing and vision responses may lessen as the nervous system slows.

Action: Never assume the patient cannot hear you. Always talk to the patient as if he or she can hear you.

Symptom: There may be restlessness, pulling at bed linens, having visions you cannot see.

Action: Stay calm, speak slowly and assuredly. Do not agree with inaccuracy to reality, but comfort with gentle reminders to time, place, and person.

Symptom: Your loved one will not take foods or fluids as the need for these decreases.

Action: Moisten mouth with a moist cloth. Clean oral cavity frequently. Keep lips wet with a lip moisturizer.

Symptom: You may notice irregular breathing patterns, and there may be spaces of time when no breathing occurs.

Action: Elevate the head by raising the bed or using pillows.

Symptom: If your loved one has a bladder catheter in place, you may notice a decreased amount of urine as kidney function slows.

Action: You may need to irrigate the tube to prevent blockage. If you have not been taught to do this, contact the hospice nurse.

Source: Hospice in the Home Program, Visiting Nurse Association of Los Angeles.

THE STRESS OF DEALING WITH A DYING FAMILY MEMBER   As the family member or friend's health deteriorates and the individual moves into the terminal phase, family members and friends begin to cope emotionally and in other ways with the now-expected death and the ever-growing burdens of care (Doka, 1993). Although denial may exist with some friends and family members, most will probably recognize the possibility of death. These individuals will rally around the dying person, sometimes causing resentment between family members and/or friends.

A terminally ill person in the household places strain on family interaction. Primary caregivers are often fatigued, both mentally and physically, from lack of sleep and the strains of relating to a bedridden person—for example, lifting, turning, and changing diapers. Financial stress is also placed on the shoulders of some family members. After a while, family members and friends may wish (though they probably won't voice the thought) that the person would just go ahead and die. A videotape entitled *Dying*, shown on national public television, presents Bill, a man in his early 40s who is dying from cancer, and his wife Harriet. Harriet actually says to Bill at one point, "Why can't you just go ahead and get this over with?" She shares a similar thought with her therapist. Harriet is frustrated over Bill's dying and feels that the man who said "I do" is leaving her "with two young boys to raise alone with drugs and everything." She notes that if he could go ahead and die now, she could then have time to find the boys another father, and therefore have help in raising them through the troublesome teen years! Besides anger and resentment, family members and friends may feel guilt about their own inability to be perfect all the time for the ill individual, noted Doka (1993).

How can family and friends find ways to reduce the stress of relating to a terminally ill person? Melodie Olson (1997, p. 212), nurse practitioner and professor, suggested that coping skills include "anything one can use to alter the relationship between the person and the environment to change the negative results of psychophysiologic consequences of stress." Family members and friends may find counseling and self-help groups useful, suggested gerontologist Doka (1993). They may also seek effective stress-reduction strategies: improving problem solving; increasing planning and communication; withdrawing from stressful situations or delegating; modifying unrealistic expectations of self; and examining lifestyle management such as good diet, regular exercise, social support, relaxation, and meditation. Certainly, the last suggestion of maintaining one's own health is perhaps the primary coping technique one can use to combat the stresses of caring for a dying person.

It is important for family and friends to allow themselves respite (relief). An individual should not try to do it all, but should call on other family members or friends to help. With a bountiful number of family members and friends, a division of labor can be established and a share-the-care plan can be devised. However, if family members and friends are few, having the ill person in hospice will help provide volunteers to give respite. Whatever the case, an individual should not try to care for a dying person 24 hours per day, seven days per week.

HELPING CHILDREN COPE WITH A DYING PARENT   When a parent is dying, families need to mobilize help from any source and make this help work for them, suggested thanatologist Phyllis Silverman (2000). Family and friends can assist with the details of daily living such as being there when children come home from school and helping with daily routines of cooking, doing laundry, and other chores. Hospitals with policies that allow healthy children to visit and encourage the children to decorate the room provide greater opportunities for interaction and participation in the life of the dying parent. With these more liberal hospital policies, children "feel more connected to what is happening around them," according to Silverman (2000, p. 194).

Older school-aged children need more time to prepare for a parent's death than do younger school-aged children, observed social worker Grace Christ (2000) from her study of 88 families and their 157 children (ages 3 to 17) who experienced the terminal illness and death of one of their parents from cancer. Regularly updated information about the parent's condition seemed to provide children with a helpful context. Visits with the dying parent can be meaningful, even when little verbal communication with the parent is possible. Visits provide concrete evidence for the reality that the parent is dying and give the children an opportunity to say goodbye, noted Christ. Being told about the parent's probable death in a timely way allows for **anticipatory grief** and helps the child to cope with the period of mourning after the death.

Children like to draw; thus, drawing pictures is an excellent way for them to express themselves. Elisabeth Kübler-Ross tells the story of a family in which the mother was terminally ill with cancer. Dr. Kübler-Ross was invited to their house to talk to the children. The father and mother were afraid to tell them and felt that the children knew nothing of the approaching death. Kübler-Ross sat down with the little girl, gave her some crayons and paper, and asked her to draw a picture. The little girl drew a dining table with four chairs, one of which was leaning against the table, rather than upright like the other three. Kübler-Ross asked her to explain the drawing.

The little girl said that the three upright chairs were for her father, her brother, and herself. "The turned-down chair is for Mommy. She is dying in the hospital and won't be coming back home." This was a child who supposedly was in the dark about the situation of her mother's dying. By asking her to draw a picture, she was able to explain freely her perception of what was happening. The little girl was not in a closed awareness context but was functioning in a mutual pretense environment.

A child whose parent is dying often finds it useful to talk with other children who are having the same experience. Support groups such as I Count Too can provide an opportunity for children to express themselves in the presence of other children who are experiencing a similar situation in their own families. I Count Too provides counselors who sit in on the sessions to give guidance and advice to the children. Children experiencing the dying of a parent need intervention, because they typically have no history of such a process. Counselors can provide tools through which children can work toward coping with the dying of a parent. Without such help, these children are like a child sitting in a sandbox with no toys with which to play. We must provide the toys!

DYING WITH DIGNITY

Just what is meant by *dying with dignity*? For the cowboy in the days of the "old West," dying with dignity might have meant dying "with one's boots on" while riding the range. For a soldier in war, to die "in the trenches" fighting for a cause or one's country would probably have dignity. For some early-day Eskimos, if an elderly member of the hunting party whose health was failing could not make it back to the camp, then he would be abandoned to die with dignity. After all, for him to hold the hunting party back would not be appropriate. Thus, he would probably be remembered as having died with dignity. An individual who goes into a burning building for the fifth time to rescue children and does not return would have died with dignity. Most of us today, however, are not cowgirls or cowboys, soldiers, hunters gathering food for the family, or individuals rescuing children from burning buildings; thus, such deaths would not be fitting for us, either with or without dignity.

Cross-culturally, what might be considered dying with dignity may vary. The following story was told to George Dickinson by a nurse in England. An elderly Chinese patient with cancer appeared to the hospital staff to be in pain. Because she spoke no English, the staff, via communication with the woman's daughter, confirmed their suspicion. Thus, they placed a shunt in her arm and administered pain medication. Soon after this action by the staff, however, the daughter removed her mother from the hospital, took her home, and ripped out the shunt. Though her mother was in pain, the daughter had incorrectly communicated with the hospital staff. The daughter did not intend for the staff to control the pain, because her mother's religion would have her die in pain as part of the appropriate dying process for her culture. Pain was necessary for her mother to die with dignity and, therefore, have a good afterlife. Thus, cultural variations produce different appropriate ways of dying, and to die in pain is not a prerequisite for dying with dignity in most parts of the world.

Dying is a human activity that is carried out in a normative manner. Each individual learns from society the meaning of death and the proper ways to die. One hopes to die what one's culture considers to be a good death. For example, anthropologist Margaret Mead seemed to close out her life in the way that she had lived it. A recorded interview with her on public television a few months before her death revealed a woman who gave the last moments of her life as energetically, as intelligently, and as openly as she had given the early years of her life (Cuzzort & King, 2002). Her conversation was animated—no brooding soul turned inward by the thoughts of impending death. At the end of the program she briefly revealed her attitude toward life and work and death. Then she arose and began to walk backward slowly, into the shadowy and dark recesses of the stage setting, still facing the camera. She smiled a charming smile, waved her cane, and said, "Goodbye, goodbye, goodbye"—knowing that it was the final farewell as she stepped into the darkness. As Cuzzort and King (2002, p. 390) noted, it was an act, yet a wonderful act. It summed up the way that she lived, wrote, and felt about people.

Though there is no good role model to follow in dying, perhaps the late Senator Hubert H. Humphrey of Minnesota came close by maintaining some control over final events in his life. Even in his dying, he kept a positive attitude and frequently made telephone calls to individuals to wish them well. Calling former President Richard M. Nixon, for whom many in Humphrey's position would have had limited affection, to wish him a happy birthday, epitomizes Humphrey's behavior during the final days of his illness. Because an **appropriate death** is generally consistent with past personality patterns, Senator Humphrey died as he had lived. Likewise, Jacqueline Kennedy Onassis spent her last few days of life as she had lived, with poise and dignity.

Avery Weisman (1993) outlined four characteristics of an appropriate death: (1) *Awareness* of impending death comes at about the same time as learning that nothing else can be done to "save" the patient; (2) *acceptance* of death may depend on the person speaking with a patient and with whom a patient is ready to speak; (3) *propriety* refers to the nonmedical features of illness that distinguish a good death from something more objectionable and refers to what a patient decides is right and proper based on community expectations and standards; (4) *timeliness* is the question of when is the best time to die.

The concept of an appropriate death has been suggested as an alternative goal for those working with the dying (Hooyman & Kiyak, 1988). An appropriate death means that the individual dies as he or she wishes to die. The individual maintains as much control over dying as possible to make it meaningful. Although an appropriate death is usually only partially achieved, those working with the dying can assist them in exerting such control.

Anne Hawkins (1990), in an article on pathographies about dying, noted various constructions of death in the literature, including those of de Beauvoir, Conley, and Wertenbaker. Simone de Beauvoir's model of death (1965) described her mother's death from cancer as an "easy death," one in which suffering was avoided. Herbert Conley (1979) described death as one's finest hour, when death is accepted as a part of life with avoidance of self-pity, concentrating on the needs of others and maintaining dignity and self-composure. Conley's description of death fits Elisabeth Kübler-Ross's concept in *Death: The Final Stage of Growth* (1975). Lael Wertenbaker (1957) offered the model of a heroic death, a manly death in which pain is endured as a test of bravery, courage, and heroism. In an Australian study of 100 dying individuals, Allen Kellehear (1990) found that most of the terminally ill people engaged in some form of personal preparation for death (funeral arrangements, for example) and seemed to have an implicit conception of

DEATH ACROSS CULTURES | DYING IN SCOTLAND

Rory Williams examines the issue of a good death with elderly residents of Aberdeen, Scotland. He asks, "How is one defined as dying?" There is dying as defined by the doctor and confirmed by the event of death. Secondly, there is the situation where someone died before they were recognized to be dying, so that the recognition had to be reconstructed retrospectively. Then there is the situation where somebody recognized as effectively dying failed in fact to die. These three sorts of dying have a definite relation to ideals about good and bad deaths. To die "the proper way" is on the one hand to go quickly, easily, quietly, and unconsciously. Yet, this idea clashes with the idea that the dying should be cared for. Too quick a death might

be "good for the one that's gone but bad for those left behind." Thus, the "properness" of the death depends on whether it is in the interest of the living or that of the dying. Thus, two broad ideals of dying well exist: going as quickly and unconsciously as possible and going only after an affectionate reunion with kin. However, both ideals of death are threatened from the third sort of dying when people lived on who were considered as good as dead and who were thought to be better off dead. Thus, there are two kinds of "good dying" and one kind of "bad dying."

From *A Protestant legacy: Attitudes to death and illness among older Aberdonians* (pp. 98–100) by R. Williams, 1990, Oxford: Clarendon Press.

the **good death** that emphasized looking after the needs of the survivors in a practical way. A good death appears to be one that is appropriate at a particular time and place.

According to the Institute of Medicine (1997), a good death is one that is free from avoidable distress and suffering for patients, families, and caregivers, is in general accord with patients' and families' wishes, and is reasonably consistent with clinical, cultural, and ethical standards. Hospice coordinators, studied by D. C. Smith and M. F. Maher (1993), reported that dying individuals who experienced a good death tended to have the following qualities: having a sense of control, discussing the practical implications of dying, exploring an afterlife, talking about religious or spiritual issues, reviewing the past, having a sense of humor, not avoiding painful truths, taking an interest in personal appearance, benefiting from the presence of significant others, and participating in physical expressions of caring. In a study of adults with lung cancer, Travonia Hughes and colleagues (2008) discovered that the respondents reported that the good death had four themes: dying while asleep, death as pain-free, peaceful passing, and dying quickly (see Table 5.1). As Beverly McNamara and colleagues (1994) noted, the good death may be viewed as a complex set of relations and preparations, rather than as relating to a fixed moment in time. Thus, from this perspective, the good death is not a single event, but a series of social events.

A review of the literature on good death identified common attributes of a good death from the point of view of patients, health care providers, and families: relief from pain and suffering, being aware of dying, accepting the timing of one's death, acceptance and autonomy, preparing for departure, keeping hope alive, and making the decisions about when to die (Granda-Cameron & Houldin, 2012). For physicians, a good death for patients encompassed expected, peaceful, and timely death; rational/coherent, appropriate comfort care; and effective communication with family, patient, and providers. Overall, researchers Granda-Cameron and Houldin concluded that a summary of a good death includes the following: pain and symptom management, patient's dignity, family presence, family support,

| TABLE 5.1 | FREQUENCY OF LUNG CANCER PATIENTS' THEMES REGARDING ATTRIBUTES OF A GOOD DEATH |

Theme	Number of Patients Mentioning the Theme ($N = 100$)
Asleep	84
Pain-free	74
Peaceful	27
Quick	16
With family around or cared for	12
With God	6
Free of regret	4
At home	3

Source: T. Hughes, M. Schumacher, J. M. Jacobs-Lawson, and S. Arnold. (2008). Confronting death: Perceptions of a good death in adults with lung cancer. *The American Journal of Hospice and Palliative Medicine, 25,* 39–44.

THE DYING PERSON'S BILL OF RIGHTS

- I have the right to be treated as a living human being until I die.
- I have the right to maintain a sense of hopefulness, however changing its focus may be.
- I have the right to be cared for by those who can maintain a sense of hopefulness, however challenging this might be.
- I have the right to express my feelings and emotions about approaching death in my own way.
- I have the right to participate in decisions concerning my care.
- I have the right to expect continuing medical and nursing attention even though "cure" goals must be changed to "comfort" goals.
- I have the right not to die alone.
- I have the right to be free from pain.
- I have the right to have my questions answered honestly.

- I have the right not to be deceived.
- I have the right to have help from and for my family in accepting my death.
- I have the right to die in peace and dignity.
- I have the right to retain my individuality and not be judged for my decisions which may be contrary to the beliefs of others.
- I have the right to expect that the sanctity of the human body will be respected after my death.
- I have the right to be cared for by caring, sensitive, and knowledgeable people who will attempt to understand my needs and will be able to gain some satisfaction in helping me face my death.

From *Cancer care nursing* (p. 33), by M. Donovan and S. Pierce, 1976, New York: Appleton-Century-Crofts.

awareness of death, and good communication among patient, family, and health care team. As these are attributes of the ideal good death, the definition of a good death is subjective and therefore may vary from individual to individual. For some individuals, for example, fighting to stay alive until the very end may be the mark of a good death.

In their article "The Promise of a Good Death," Ezekiel and Linda Emanuel (1998) concluded that developed countries have the capacity to make a good death the standard of care. They argued that (1) society is focused on ensuring a good death, (2) physicians have more powerful medications and other interventions to alleviate pain than ever before, (3) clinicians are recognizing the multidimensional aspects of dying and the importance of attending to concerns other than pain, (4) hospice is widely available and increasingly used, (5) advance-care planning is strongly endorsed, and (6) medical schools, hospitals, and professional organizations are committing themselves to training physicians to improve the care of the dying. If indeed societies place more emphasis on a good death, hopefully death fears can be reduced significantly.

THE DYING CHILD

According to Robert Kavanaugh (1972), "little people" enjoy the same human rights as their bigger counterparts. They have a right to know if they have a fatal condition. When kept in ignorance, children, like adults, will rarely grow beyond the initial stage of denial and isolation. When this occurs, children are robbed of the peace and dignity that can be theirs in the final stage of acceptance and resignation.

WORDS OF WISDOM   "LISTENING" IS THE KEY TO "TELLING" ABOUT DYING

How do we tell a child about his or her dying condition? Kavanaugh stated that to adults brave enough to listen, this is not a valid question. The child will do the telling, if we create an atmosphere in which the child can make all appropriate deductions. The child's talk will flit in and out of the awful revelation.

Who should do the telling or serve as a catalyst for it? Kavanaugh said that anyone strong enough to take the consequences by being a regular visitor, a trusted confidant, and a patient listener can be the catalyst. Many adults cannot qualify.

From *Facing death* (pp. 142–143), by R. E. Kavanaugh, 1972, Baltimore, MD: Penguin Press.

RELATING TO THE DYING CHILD

The prognosis of death should be made known to children as soon as it is clear and final (Kavanaugh, 1972). Physicians, nurses, and families obviously need time to bring their own emotions under control. We now know how to treat the dying child kindly. Knowledge is kindness; ignorance is cruelty. The child is the patient whose life is being lost and whose concerns are preeminent. Kavanaugh noted that when children have known the truth about their condition and were allowed to talk about it openly, they have been as brave as any adult.

Reasons for the preceding advice given by Kavanaugh (1972) are based on the following observations.

First, the dying child is no ordinary child. The ordinary process of maturing quickens through lengthy illness with confinement, suffering, and deprivation. Children who have been ill for a long time usually exhibit a maturity beyond their calendar years.

Second, children's consciences are more tender and concerned than most adults are in a position to know. Children on their deathbeds who are uninformed of their fate will often have guilt for the sadness and poorly veiled tears that they witness around their bed. They can sense the phoniness around them, and they may believe that they are being punished for something evil that they have done. Their isolation is heightened in their heavy concerns. Nothing is sadder than a dying child learning of his or her fate from playmates.

Finally, moderately aware and normally alert children know what is predicted for them in the signs that they see—recognizing their plight in memories of dying scenes on television. They may ponder why the physician comes so often, why everyone is so nice to them, and why they are receiving all of the gifts. A review (O'Halloran & Altmaier, 1996) of studies on death awareness among children who are healthy, chronically ill, and terminally ill reveals that children with life-threatening diseases demonstrate increased understanding of death. In contrast, healthy and chronically ill children appear to require a certain age, cognitive development level, or intelligence threshold to understand these concepts. As anthropologist Myra Bluebond-Langner (1989) observed from her work with dying children, all terminally ill children become aware of the fact that they are dying before death is imminent. Yet, Bluebond-Langner noted that for these children the acquisition and assimilation of information are a prolonged process. Either a child shares what he or she knows about dying, or the final weeks and months become a lonely vigil, a sentence to fear and guilt, confinement, and confusion.

The impact of health care professionals working with dying children tends to produce burnout, compassion fatigue, secondary traumatic stress, and vicarious traumatization, according to Ungureanu and Sandberg (2008). Pediatric therapists in their study found some solace in the work as summed up by this nurse: "After the experience with dying children, life in general became more valuable for them" (p. 82). Another nurse said, "I learned to enjoy small things, so I am very happy when I get up and the sun is shining" (p. 82). Yet another nurse commented about the shift she experienced about death as a result of working with dying children: "My role has helped me to see that death is not always the worst thing to happen. Death can be a blessing versus living with a bad injury or illness" (p. 84).

By observing and talking to children with leukemia aged three to nine in a hospital, Bluebond-Langner (1978) concluded that most of them knew not only that they were dying, but also that this was a final and irreversible process. Children may wish and need to be open about their condition, but adults are often traumatized by children's openness and honesty. Bluebond-Langner noted that the age of the children is not as important in their self-awareness of their dying as is their experience with the disease and its treatment.

Though one may not know what to say to dying children or may not feel skilled in this area, it is imperative that we show support to them and let them know that we care. Support of others is important throughout life, whether relating to a terminally ill person or otherwise. This point is illustrated in the following story told by Rabbi Harold Kushner (1985):

> A little boy had gone to the store, but was late in returning home. His mother asked, "Where were you?" He said, "I found a little boy whose bicycle was broken, and I stopped to help him." "But what do you know about fixing bicycles?" his mother asked. "Nothing," the little boy replied, "I sat down and cried with him."

Many times, we may not know how to "fix the bicycle," but like the little boy in the preceding story, we can give support to the individual in other ways.

HELPING THE CHILD COPE WITH DYING

Children sometimes consciously keep their feelings and responses secret in an attempt to protect their parents (Robinson & Mahon, 1997). Such secrecy could make it difficult to help a child cope with her or his dying. Gerry Cox (1998) advocates the use of humor, art, and music to help children express their feelings related to dying.

Adults who keep family secrets are less inclined to use humor. Humor and laughter are the opposite of secrets, noted Cox. To laugh together is a positive form of sharing and social support. Laughing is typically done in the company of others, not when alone. Laughing is contagious, as is crying. A more relaxed person, not one always taking self too seriously, is better able to laugh at one's self. Laughing together is viewed as caring and being supportive, observed Cox.

How do you make a dying child laugh? To make the child laugh it is useful to know what she or he particularly enjoys. Watching various cartoons on television might provoke laughter in some children. For others, play therapy, using a favorite toy, might produce an enjoyable, even humorous, situation, as the child's mind forgets the illness for a short while and goes into the never-never world of play and creativity.

For some children, a visit from a clown to the hospital or home can produce a funny setting. Other children might enjoy relating to a puppy or kitten and find joy in the playfulness of the small animal.

Music is another form of expression that does not require oral fluency, said Cox (1998). Music is a therapeutic tool. Joyful music can encourage laughter and perhaps move one's emotions from despair to happiness. Though words are not essential for music to be helpful, the use of words can be powerful. Catchy tunes that repeat themselves such as a round (e.g., "Row, Row, Row Your Boat" or "Three Blind Mice") can create a "happy" environment for children. Small children simply enjoy having someone sing to them, as they like having someone read to them. Anything to divert their thinking from the unpleasantness of their disease can only enhance the situation for the dying child.

The use of humor, art, and music allows children to remove the sense of distance between themselves and others and to enhance their self-esteem and lower death anxiety, noted Cox (1998). These therapies are effective ways to provide social support for dying children. Various therapeutic techniques, from storytelling to humor, exist for aiding these children. Dying children need to say goodbye by writing a letter, making a picture, or sending up a balloon with a message, suggested Cox. And, perhaps most important, adults need to support these efforts.

WORDS OF WISDOM | CHILD LIFE SPECIALISTS

Child life specialists are child development experts who work to ensure that life remains as normal as possible for children in health care settings and other challenging environments. They promote effective coping through play, self-expression activities, and age-appropriate medical preparation and education. As advocates of family-centered care, child life specialists work in partnership with physicians, nurses, social workers, and others to meet the unique emotional, developmental, and cultural needs of each child and family. Child life specialists generally work in pediatric inpatient units, and often in specialty areas like the emergency department, surgical and intensive care units, and outpatient areas. Child life specialists exist in countries around the world, with some 4,000 found in the United States.

It is the job of the child life specialist to make it easier for the child to be in the hospital setting and to make it more fun. "Fun" in a hospital setting does not seem to fit, yet child life specialist Kelly Schraf tells about an eight-year-old girl being readied for surgery on her cleft palate. One of the scariest parts of surgery for a child is the anesthesia mask. So, Ms. Schraf brings a mask and tells the little girl that the doctor is going to put "sleepy air" through a hole in the mask and informs her that it is sometimes stinky. She then offers the little girl special smells to put inside to make it smell good—bubblegum, strawberry, and cotton candy. She chose bubblegum and applied a generous layer of it to her mask. By the end of the visit, the little girl was grinning!

Taken from J. Gold (2012, July 23). Child life specialists help sick kids be kids. *Kaiser Health News*; Child life: Empowering children and families (2012). Rockville, MD: Child Life Council. www.childlife.org.

Parents of the Dying Child

Parents of the dying child will likely have to cope with considerable anxiety and uncertainty. In the diagnostic period, parents learn what they and the child must face—the possible threat of a sustained illness and even death. The family's financial state is likely to be adversely affected by the costs of the illness. During the chronic phase of the illness, continued stress with occasional points of crisis occur. Family life tends to revolve around the disease, sometimes producing neglect for other children in the family.

Therapists working with children with life-threatening diseases are exposed to posttraumatic stress disorder symptoms of the parents and siblings (Ungureanu & Sandberg, 2008). Professionals and parents caring for dying children should be aware of the emotional impact such caregiving can have on their personal well-being and how it can also impact relationships outside of the home, in the workplace.

A study of 133 families of children who died (Weidner et al., 2011) identified seven dimensions of end-of-life care that were important to parents: respect for the family's role in caring for their child, comfort, spiritual care, access to care and resources, communication, support for parental decision making, and caring/humanism. Achieving a balance between alleviating the child's pain and discomfort and maintaining his or her ability to interact with family members was also important to the parents in the study. Additionally, the parents wanted honest and factual information and wanted the right information at the right time.

Though parents can never be really prepared for the end, as the dying child continues to lose ground, they reach a point at which resignation begins to replace hope for survival, and the inevitable end slowly comes into focus (Knapp, 1986). The character of hope changes from hoping for survival to hoping for a full range of living in the time available and for a comfortable, pain-free death. When parents consciously accept the fact that death is imminent, they become totally absorbed in the life of that child and try to make each day a memorable occasion. At this point, parents have to make decisions about where the child will die: at home, in the hospital, or in hospice (Silverman, 2000). They are often faced with decisions about continuing life support and using antibiotics.

It is difficult for the parents to finally accept that their child's death is forthcoming. This out-of-order aspect of the child dying before the parents is difficult to comprehend. The unnaturalness of the child dying before the parents may also be complicated by death resulting from the child dying suddenly and unexpectedly in an accident. Parents' dreams of the child growing into an adult and relating to them as big people are shattered as the child lies dying. Parents often wonder what their child would have been like as an adult. What would he or she have done for a living? Would he or she have married? Would their child have had children? Where would he or she have lived?

Parents of a dying child often take a protective approach toward communication in which they attempt to shield the child from the implications of the illness, noted Kenneth Doka (1993). Such a strategy may not work because children have access to a wide range of information about their condition: internal health cues, external treatment cues, information from books and television, and input from ill peers.

Doka noted that the question "Am I going to die?" from a dying child may not be a request for information but a call for reassurance; thus, it is important to understand what the child is really asking. Visual art can be helpful in a child's communicating with an adult. Very young children can use different crayon colors to illustrate what they may be thinking. They can draw sad or happy faces to express themselves. Play therapy may also be appropriate to use in working with dying children. By giving the child a doll, she or he can play doctor and release frustrations of being stuck with a needle by doing the same to the doll.

SIBLINGS OF THE DYING CHILD

Everyone within the nuclear family will be affected by the dying child's illness (Doka, 1993). Avoidance of the child is one way that some family members might react. Siblings might be prohibited by parents from playing with the ill child. Siblings of the dying child may definitely feel neglected by parents and other family support members, because their energies are being directed toward the ill child, not the siblings. Parents are not as free to play with the other children and go places with them. Even if free, the parents probably have little energy left to spend with the other children, because they may have been up most of the night with the ill child.

Individuals outside the immediate family or extended family members can play a significant role in giving attention to the siblings of the dying child by taking them to a movie or to the park, anything to show attention and divert the mind for a few hours from the situation at home or in the hospital. Teachers of siblings of a dying child should be notified of the situation at home, so that they can give special attention to the sibling and better understand why he or she may not be acting normally. The sibling at school, for example, might throw a fit or hit another child, when normally he or she would not act in this way. "Look at me," the sibling is probably saying. "I exist too and would like some attention."

Depending on their ages, however, siblings of a dying child can play a key role in helping within the family setting. As Phyllis Silverman (2000) noted, children need to be involved and seen as active members of the family, as helpers, and as grievers. They can assist with chores and relate to the dying sibling by playing with her or him or reading to her or him. Parents should encourage siblings of the ill child to participate in and contribute to the situation. Martinson and Campos (1991) found that siblings who were allowed to participate in caring for a sibling who was dying of cancer felt pride and pleasure about their ability to help.

Other children in the family need to know what is going on so that they can cope with the changes as the family adjusts to these new circumstances (Silverman, 2000). Physical, as well as behavioral, changes need to be anticipated. For example, siblings need to be forewarned if their dying sibling is receiving chemotherapy that will cause hair loss. Skin color may vary as the disease progresses. The dying child may have swelling of various joints or places in the body, depending on the nature of the illness.

The terminal illness of a child in the family often changes the course of the sibling relationship, bringing the siblings closer together or reducing tensions between them (Silverman, 2000). Children may mature rapidly and develop an understanding beyond their years as they learn to ask questions and take in what is going on.

PRACTICAL MATTERS | EIGHT LESSONS FROM THE END OF LIFE

Sociologist Karla Erickson, from her ethnographic work with nursing assistants who work with individuals with terminal illnesses, outlines eight lessons learned about end of life.

1. Avoid overtreatment. Medical technology can extend life beyond all common sense. Medical interventions indeed may give one more days/ months to live, but these interventions do not ensure quality of life.

2. Choices are there to be made. Sometimes individuals have the false conviction that death is forever avoidable because of heroic medical interventions. Often it is the family members and friends, not the dying person, who seek more time at any cost. Shift the focus away from heroic measures and toward an acceptance of a death that is planned, peaceful, and deliberate.

3. Fear a bad death, not death itself. Pain can be managed via hospice, allowing for both comfort and control in one's final days.

4. Death can be seen coming. The dying process often has identifiable steps which may include a diminished interest in details about the wider world, decline of eating and drinking, talk of an upcoming journey, spotting of knees and lower legs, coolness of extremities with skin turning grayish or bluish, "death rattle" developing, and a slowness of breathing.

5. Make plans for death. Making plans, often written down, gives a greater chance that wishes will be honored and the dying more peaceful.

6. Small acts matter. Simple behaviors such as pulling up the covers or turning on a fan or feeding ice chips produce final opportunities to express regard and love. Do not hesitate to touch the individual.

7. Never too late to say what you mean. Frank conversations such as saying good-bye, offering apologies and forgiveness when needed, and communicating one's deepest feelings should not be avoided. Talk *to* the dying, rather than *about* them.

8. Accompanying others through death is a privilege. Such interaction with end-of-life issues makes one grateful for the opportunity and helps overcome one's fears of death, live more fully, and do meaningful work. This experience is rewarding and reminds an individual of the preciousness of life.

Taken from K. A. Erickson (2013). *How we die now: Intimacy and the work of the dying*. Philadelphia, PA: Temple University Press.

They are pushed to confront issues and feelings that would not ordinarily be expected of children their age. Indeed, with a dying child in the household, siblings may grow up quickly in numerous ways.

CONCLUSION

Because the thought of dying is stressful, it seems appropriate that we be aware of dying and death. With more awareness, we hope that a greater acceptance of dying and death will result. Through a better understanding of death meanings, our coping with dying and death should be enhanced, whether we are a consumer of the medical system or a health care professional working within the system. It is difficult to relate to the dying if we ourselves have not been sensitized to our own death.

Kübler-Ross's stages of the dying process are not meant to be a formula for relating to an individual with a terminal illness, rather a model by which it becomes obvious that a patient's behavior is normal and fits into a pattern. The stages may not be followed as described, yet they can be beneficial to family and health care workers in establishing where a patient is at a point in her or his illness.

Likewise, Glaser and Strauss's awareness contexts of dying can give others a clue as to where the perceptions of death are at a point in an illness. Such awareness directly impacts how individuals relate to a patient and how the patient might relate to others. We often play games in communicating with a dying patient. We know, the physician and nurse know, and the patient knows (probably whether told or not) when the condition is terminal, but we often exist in anything but an open awareness context. No one lets the others know that he or she knows. Death talk remains taboo.

For children to die is completely out of sequence in the dying process. Grandparents die, then parents, then children, *not* children before the others. Children are often braver than we give them credit, and it is often the parents and other family members who are not so brave. Siblings of dying children must not be forgotten, as they may literally feel left out when a sibling is dying. With so much attention and energy going to the dying sibling, other siblings may be having a "pity party" for themselves and seemingly saying, "Look at me, I need your attention as well."

What is an appropriate death or good death for one may not be for another individual. To die on the battlefield might define a good death for some, yet for another to die in one's own bed while sleeping at night, having had a good day, might constitute a good death. What would be your idea of a good death?

SUMMARY

1. The meaning of dying depends upon the social context in which it takes place.
2. Time, space, norm, role, value, self, and situation are important components of the meaning of our dying.
3. One may go through stages in accepting his or her terminal self-meaning.
4. Dying a good death makes it easier on everyone.
5. According to Glaser and Strauss, four different awareness contexts exist between medical personnel, patients, and the patients' families: open, mutual pretense, suspicion, and closed.
6. The dying child needs support, as do the siblings of the dying child.
7. Societal disengagement refers to society withdrawing from an individual.

8. Kübler-Ross divides the dying process into five stages: denial, anger, bargaining, depression, and acceptance.
9. Socialization of physicians regarding communicating with dying patients and their families has been very limited historically.
10. For family and friends relating to a dying person, it is imperative that they first of all try to maintain their own health to better cope with the stress of taking care of the dying individual.
11. Primary caregivers to a dying individual must share the responsibility of care with others to reduce stress.
12. Music, humor, and art can be positively used as therapy with terminally ill children.

DISCUSSION QUESTIONS

1. What is meant by this statement: The meaning of dying will depend upon the social context in which the dying takes place?

2. Would you prefer to live with a person with a terminal disease or a person who is chemically dependent? Discuss the advantages and disadvantages of each.

3. How could steps be taken to overcome the diminished social and personal power of the hospital patient? Are such limitations on patients necessary for an orderly hospital?

4. Discuss this statement: The terminally ill eventually come to view death as a blessing.

5. You have just been told that you have inoperable cancer. Discuss how you think you would react. In what ways would you change your life?

6. How do you think you would relate to a dying child, if you were a parent?

7. Discuss Glaser and Strauss's four awareness contexts. Which do you think most often exists in a medical setting with a dying patient?

8. How do you think a physician should go about telling someone that she or he has a terminal illness? How would you prefer to be told this information?

9. Discuss how societal disengagement applies to a dying person.

10. Discuss Kübler-Ross's five stages of the dying process. If you know of someone who is dying, try to "fit" her or him into the appropriate current stage.

11. How does research cited in this chapter show similarities or differences in how male and female physicians relate to dying patients and their families?

12. If you have experienced the death of someone close to you, what was the awareness context during most of that individual's illness? Would you have preferred another context?

13. Suggest ways in which art, humor, and music can be therapeutically used with dying children.

GLOSSARY

Anticipatory Grief: Experiencing grief before a death actually occurs; grief work aimed at loosening attachment to the dying, making loss less painful when it occurs.

Appropriate Death: A person dying as he or she wished to die. The death is generally consistent with past personality patterns.

Awareness Context: What each interacting person knows of the patient's defined status and his or her recognition of the others' awareness of a personal definition.

Cadaver: A dead body. Human cadavers are used in medical schools for the purpose of dissection to learn the parts of the body.

Dying Trajectory: Perception about the course that dying will take.

Good Death: An "appropriate" death at a particular time and place.

Master Status: The status (position) most important in establishing an individual's social identity.

Norm: A plan of action or expected behavior pattern thought to be appropriate for a particular situation.

Role: Specified behavior expectations for persons occupying specific social positions.

Scapegoat: A person, group, or object upon whom blame is placed for the mistakes of others.

Sick Role: A set of characteristic behaviors that a sick person adopts in accordance with the normative demands of the situation.

Situational Adjustment: The process by which an individual is molded by the group into which he or she is seeking acceptance. The person learns from the group how to continue successfully in a situation.

Societal Disengagement: A process whereby society withdraws from or no longer seeks the individual's efforts, as distinguished from social disengagement in which the individual withdraws from society.

SUGGESTED READINGS

Chen, P. W. (2007). *Final exam: A surgeon's reflections on mortality*. New York: Alfred A. Knopf. *Final Exam* follows liver transplant surgeon Pauline Chen over the course of her education, training, and practice as she grapples with the problem of mortality.

Erickson, K. A. (2013). *How we die now: Intimacy and the work of dying*. Philadelphia, PA: Temple

University Press. Written from a sociological perspective, this ethnography gives excellent insight into the complexity of dying from the point of view of nursing assistants who work with the dying.

Evans, A. R. (2011). *Is God still at the bedside? The medical, ethical, and pastoral issues of death and dying*. Grand Rapids, MI: William B. Eerdmans Publishing Company. An ethicist-theologian offers a perspective on the web of issues surrounding dying and death.

Frawley, M. H. (2003). *Life in the sick room: Harriet Martineau*. Peterborough, Canada: Broadview Press. This edited work from the writings of Harriet Martineau (1802–1876), a prolific writer of sociological works from England, chronicles some of her writings. Martineau lost her hearing as a teenager and had numerous health issues in her life. She wrote *Autobiography* some 25 years before she died, thinking that she was dying at the time.

Geist, M. E. (2008). *Measure of the heart: A father's Alzheimer's, a daughter's return*. New York: Springboard Press. A caregiver gives up her career to move back home to help care for her father deteriorating from Alzheimer's disease. This book should be a good read for caregivers as well as health care professionals to help them better understand what family caregivers confront.

Groopman, J. (2004). *The anatomy of hope: How people prevail in the face of illness*. New York: Random House. Takes the reader into the lives of people at pivotal moments when they reach for and find hope.

Gutkind, L. (2011). *Twelve breaths a minute: End-of-life essays*. Dallas, TX: Southern Methodist University Press. These 23 essays are testimonies from skilled and experienced professionals who encounter death regularly and from devoted caregivers who know it intimately.

Kellehear, A. (2009). *The study of dying*. Cambridge: Cambridge University Press. A multidisciplinary approach to dying, giving insights from medicine, the social sciences, and the humanities. Studies how individuals behave just prior to their death and how that conduct is influenced by physical, psychological, cultural, and spiritual factors.

Lewis, M. J. (2006). *Medicine and care of the dying: A modern history*. New York: Oxford University Press. An examination of the evolution of the care of the dying in the United Kingdom, the United States, Australia, Canada, and New Zealand. An overview of the practice of palliative care is described in the wider context of the history of medicine.

Montross, C. (2007). *Body of work: Meditations on mortality from the human anatomy lab*. New York: Penguin Press. Christine Montross shares her experiences as a first-year medical student in gross anatomy and in relating to patients, some of whom were dying.

O'Rourke, M., & Dufour, E. (2012). *Embracing the end of life: Help for those who accompany the dying*. Toronto: Novalis. Practical suggestions for caregivers of the dying.

Ratner, T. (2008). *Reflections on doctors: Nurses' stories about physicians and surgeons*. Kaplan Publishing: New York. This edited volume tells it like it is from nurses' points of view regarding their relationship with physicians at the bedside of patients and in the operating room.

Schuurman, D. (2003). *Never the same: Coming to terms with the death of a parent*. New York: St. Martin's Press. This self-help book aims to encourage adults who have experienced the death of a parent in childhood or adolescence to reflect on and come to some understanding of this event.

Stanworth, R. (2004). *Recognizing spiritual needs in people who are dying*. Oxford: Oxford University Press. The book points out that spiritual care is not about something abstract but is about helping another human being to understand that they are loveable.

"I'm afraid of the pain."
"I don't want to be alone when I'm dying."
"I'm afraid of a long, protracted period of suffering."
"I don't want to die in a hospital. Let me die at home."
"I'm not afraid for myself, but I am worried about the effect of my death on those I love."
—**Frequent responses to the question "Does dying frighten you?"**

Hospice doesn't help people die; hospice helps dying people.

LIVING WITH DYING

CHAPTER **6**

Many individuals are living for years with a serious illness. The chances of an illness going into remission or even going away are more likely this year than last, and the same will be true of next year because of scientific and medical advancements each year. In the early stages or remission phase of a chronic illness, when one is asymptomatic or experiences minor symptoms and little pain, it may be possible to continue to maintain a good quality of life, even one that approaches the standard of life before the diagnosis (Doka, 1993).

With a population better educated on health issues, the stigma of cancer (the "big C" disease) and other illnesses has diminished considerably in recent years. This allows people with life-threatening illnesses to live openly with their illnesses. Without the stigma, individuals can acknowledge their illness to others, making it easier to deal with their impending death. In this phase of relative health, an individual can begin to plan for the inevitable decline, researching long-term care options, making financial arrangements, and completing an advance directive and a will.

As discussed in Chapter 5, Arthur Frank (1991) observed that to seize the opportunities offered by illness we must live actively and must *talk* about the illness. In that way, as individuals and as a society, we can begin to accept illness fully. Only then can we learn that it is nothing special. Do not curse your fate, but rather count your possibilities, advises Frank (1991). Being ill is just another way of living, but by the time we have lived through illness, we are living differently. As a friend who was dying of lung cancer told George Dickinson, "I look at a sunrise and sunset and watch the tide come in and go out in a different way than before my illness." Indeed, in dying, one is living differently; dying changes the perspective on events in life.

The "changed perspective" for 11 women with breast cancer was one of a positive rather than a negative orientation (Vargens & Bertero, 2012). The four main themes from interviews with these 11 women were: death as a main concern, a reevaluation of life, living a normal life with support, and living until death. They were aware that they were living with death by their side all the time. Their priorities in life shifted. The reevaluation provided opportunities to strengthen their own integrity as well as self. The key to it all was that they were living until they died—still alive and living until death.

It is amazing how positive an outlook many individuals living with life-threatening diseases have. Many of them continue to hold jobs, go to and from treatments, and basically live life to the fullest. In some cases, the person who is ill needs to help friends deal with the illness. A former colleague of George Dickinson's, who was dying of cancer, decided to continue teaching. Rather than continuing as if nothing was happening, he decided to be open about his condition and active in encouraging his friends and colleagues to be part of the process of dying and to *talk* about it. This communication helped both the patient and friends cope with the dying. He sent a letter to friends to come see him ("keep me company," as he put it), and noted that if "you don't know what to say," come anyway. He further said, "You don't have to avoid talking about health matters: My situation is real, and I'm not ignoring it. But I *am* determined to be upbeat and positive." Many individuals avoid visiting someone who is seriously ill for fear that they will not know what to say or, worse yet, will say the wrong thing. However, it is very important that you make contact with a significant other who is seriously ill. Even if you do not know what to say, go anyway to let the person

WORDS OF WISDOM	THE CHALLENGE OF LIVING (WHILE DYING)

The challenge of the dying process is the challenge of *living while dying* rather than *dying while living*.

From *The grace in dying,* by K. D. Singh, 1999, Dublin, Ireland: Newleaf.

know that you care. Dying of ALS, Professor Morrie Schwartz said to his former student (Albom, 1997, p. 36), "I may be dying, but I am surrounded by loving, caring souls."

Recently George Dickinson was asked to visit a 76-year-old woman who was in the latter stages of cancer. Her breast cancer had spread throughout her body, and the prognosis was not good. She wanted him to assist in planning her funeral and writing her obituary so that her adult children would not have to worry themselves with these matters. The day he met this delightful, cheerful individual, Dorothy Sutch, her mobility had decreased to the point that she was confined to a wheelchair. They talked for about an hour. When the assignments were completed, she threw her hands into the air and, with a smile on her face, said, "Now, I'm ready!" Her preparation for death was complete. She then talked about how dying was "like a journey." "It's happening so fast," she said. "It's exciting, unfolding before my eyes." He visited her two more times that week, and by the end of the week, she was dead. Indeed, it happened "so fast." Though he had not had the pleasure of knowing Mrs. Sutch, Dickinson later learned from her many friends that she had lived as one who cared and tried to make life enjoyable for those around her. In her dying, she again had been cheerful and caring. She died as she lived, as we all basically will.

We can learn much from persons with life-threatening diseases. As noted in Chapter 1, physician Elisabeth Kübler-Ross says that working with the terminally ill has made her appreciative of life. Each day that she awakens she is thankful for the potential of another day of life. George Dickinson learned much about living and dying from his one week's acquaintance with Dorothy Sutch.

In dying, one often lives life to the fullest. It has been said that some individuals live more in their final months of life, knowing that the end is near, than they had lived in their previous years. Living with dying can be a most meaningful experience, both for the patient and for those with whom she or he comes into contact.

UNDERSTANDING AND COPING WITH THE ILLNESS

Though we may have limited experience with and may be initially uncomfortable in relating to someone who is dying, simply being aware of what that person may be going through should enhance our understanding of that individual's situation. Through better awareness, we hope we can better cope with the illness.

THE LOSS OF PHYSICAL FUNCTIONS

Depending on the particular illness, an individual with a terminal diagnosis will begin to lose bodily functions. Morrie Schwartz, dying of ALS in Albom's *Tuesdays with Morrie* (1997), well describes his gradual inability to dance, a favorite leisurely activity for him. Then he has difficulty simply walking and soon he is bedridden. Then comes the loss of the use of his arms and his ability to swallow and the capability to urinate and defecate without help—all such everyday events for us, *when healthy*. His speech became impaired. Simply turning over in bed became impossible without help. As small children, we are socialized to be independent; then in dying, we again become dependent. The sheer physical and psychological strain on caregivers of the dying can indeed be stressful.

If the patient has lost physical functions and is at home, various modifications within the home environs need to occur. A hospital bed, which can be adjusted up and down, will help the bedridden individual get into different positions. A wheelchair and/or walker should be available, if the patient has some mobility. Oxygen may be required. A bedpan will be necessary, if the patient does not have tubes to eliminate waste. As noted elsewhere in this book, a financial drain on the family may occur. In addition, the physical aspects of turning and lifting the patient will take their toll on family members after a while. The psychological strain involved in relating to a dying individual will be stressful over time for caregivers. The patient will have health hygiene needs with which others will need to assist. The list of the needs of one who has lost physical functions is almost endless.

THE LOSS OF MENTAL CAPACITY

With some illnesses, such as a stroke, an individual may lose the ability to think. Or medications and treatments may cause a very sick person to be confused and unable to think. If the patient has brain damage, as is sometimes the situation with a stroke, the brain function may be reduced to that of a small child. The ability to speak, to write, or to respond in any meaningful way may be lost. In extreme cases, the individual is alive but may have virtually no mental capacity. It is, therefore, almost impossible for the individual to express specific wants and needs. Health care providers must guess what the patient needs, and guesswork is not an exact science.

A disease affecting an estimated 5.3 million Americans today is **Alzheimer's disease** (Alzheimer's Association, 2009); every 70 seconds someone develops Alzheimer's in the United States. The chance of developing Alzheimer's disease doubles every year after the age of 65; by the age of 85, one out of two people suffer from the disease (Fishman, 2010). Alzheimer's is a chronic, degenerative, dementing illness of the central nervous system and involves personality change, emotional instability, disorientation, memory loss, loss of verbal abilities, and an inability to care for oneself (Weiss & Lonnquist, 2009). Alzheimer's is difficult to determine and is often not diagnosed until late in the illness. Some patients with Alzheimer's are violent and this makes caring for them difficult. Awareness of Alzheimer's has increased significantly in recent years, and Alzheimer's research is well under way today.

DYING OF CANCER AND HEART DISEASE

CANCER   The term **cancer** refers to a group of diseases that are characterized by an uncontrolled growth and spread of abnormal cells (Weiss & Lonnquist, 2009). Some cells are noncancerous (benign); others are cancerous (malignant). About one person in three now living in the United States will be diagnosed with cancer at one point (this does not include skin cancer). As noted in Chapter 1, cancer is the second-leading cause of death in the United States.

Causes of cancer are largely chemicals in the air, the water we drink, and the food we eat. We basically develop cancer because of bad habits, bad working conditions, and bad luck—of the genetic draw—noted Robert Proctor in *Cancer Wars* (1995). The use of tobacco is a bad habit that contributes to cancer. G. L. Weiss and L. L. Lonnquist (2009) noted other causes of cancer. Too much sun contributes to most of the 1 million cases of nonmelanoma skin cancer diagnosed each year. Excessive alcohol consumption is the main cause of cirrhosis of the liver, which puts one at high risk for developing liver cancer. Excessive exposure to radiation can contribute to cancer. Exposure via occupational hazards and environmental pollution correlates with cancer. More recently, bad diets, lack of physical activity and obesity together are negatives to good health (Neergaard, 2013). Besides the above-mentioned cancer etiologies, heredity does play a role in some cancers, though most cancers are not inherited.

Cancer rates and causes vary (Proctor, 1995). The leading cancer killer in the United States is lung cancer, directly related to smoking; in Japan it is stomach cancer (perhaps caused by use of charcoal in preparing food); in India, cancer of the mouth is common, perhaps from chewing betel nuts and tobacco leaves; and in China, moldy bread contributes to esophageal cancer. Religion may play a part in cancer rates as Mormons have lower death rates from cancer than non-Mormons (Mormons do not smoke). Cervical cancer is rare in nuns, but common in prostitutes.

The rate of cancer cases and deaths for all cancers combined have declined in the United States since 1990, according to the American Cancer Society (2008). The greatest decline in cancer incidence rates has been among men, who generally have higher rates of cancer than women. Among men, cancer death rates dropped by 1.8 percent a year between 2000 and 2009 and by 1.4 percent a year among women (Neergaard, 2013). The declines are thanks mostly to treatment advances and better screening. Another trend is the increasing rate of survival for persons diagnosed with cancer. Individuals are living longer *with cancer*.

As seen in Table 6.1, the top four types of cancer—lung, prostate, breast, and colorectal—account for approximately half of all cancer deaths. Cancer incidence and mortality rates are higher for blacks than for whites. For males, the lungs, prostate, and colon are the most common sites for cancer, but lung cancer is by far the most lethal. For females, the lungs, breast, and colon are the most common cancer sites, but lung cancer is the most fatal.

Rates of cancer are probably going down because of better education and thus more frequent screening and earlier detection. Recent court decisions regarding smoking, followed by a media emphasis on antismoking education, have no doubt contributed to lower cancer rates, particularly those of lung cancer. Smoking is also associated with cancer of the mouth, pharynx, larynx, esophagus, pancreas, uterus, cervix, kidney, and bladder.

Patients with cancer may experience fever, followed by chills and sweats as the body attempts to regulate internal temperature (CancerNet, 2000). The major causes of fevers in patients with cancer are infection, tumor-associated factors

TABLE 6.1 | ESTIMATED U.S. CANCER DEATHS, 2008

Type of Cancer	Men % of Deaths from Cancer	Women % of Deaths from Cancer
Lung & bronchus	31	26
Prostate	10	—
Breast	—	15
Colon & rectum	8	9
Pancreas	6	6
Liver & intrahepatic bile duct	4	2
Ovary	—	6
Leukemia	4	3
Non-Hodgkin lymphoma	3	3
Esophagus	4	—
Urinary bladder	3	—
Uterine corpus	—	3
Kidney & renal pelvis	3	—
Brain/ONS	—	2
All other sites	24	25
	Number of deaths for men = 294,120	Number of deaths for women = 271,530

Note: ONS other nervous system.

Source: American Cancer Society. (2008). www.cancer.org

(e.g., acute leukemia, Hodgkin's disease, lymphomas, bone sarcomas, and hypotha-lamic tumors), allergic or hypersensitive reactions to drugs, and allergic or hyper-sensitive reactions to blood component therapies.

Anorexia, the loss of appetite or desire to eat, is the most common symptom in patients with cancer and may occur early in the disease or later as the cancer grows and spreads. The patient may also suffer from cachexia anorexia, a wasting condi-tion in which the patient has weakness and a marked and progressive loss of body weight, fat, and muscle. Maintenance of body weight and adequate nutritional status can help patients feel and look better and maintain or improve their performance status. It may also help them to better tolerate cancer therapy (CancerNet, 2000).

WORDS OF WISDOM | DEATH HUMOR—MINNESOTA STYLE: THE COOKIES

Ole was at home, dying in bed. He smelled the aroma of his favorite chocolate chip cookies baking. He wanted one last cookie before he died.

He fell out of bed, crawled to the landing, rolled down the stairs, and crawled into the kitchen where his wife, Lena, was busily baking cookies. With wan-ing strength he crawled to the table and was just barely able to lift his withered arm to the cookie sheet.

As he grasped a warm, moist, chocolate chip cookie, Lena suddenly whacked his hand with a spatula.

"Why?" Ole whispered. "Why did you do that?"
"They're for the funeral."

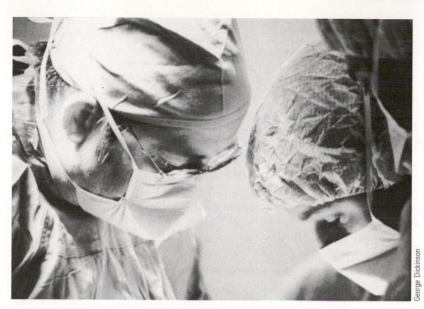

George Dickinson

Surgery is one of several options for cancer treatment today.

Treatment options for patients with cancer generally include surgery, chemotherapy, and radiation. Sometimes all three options are chosen. Side effects of treatment can be nausea, vomiting, oral complications (mouth sores, dry mouth, and bleeding gums), delirium (confused thinking and hallucinations), fatigue, fever, constipation (impaction and bowel obstruction), swelling caused by blocked lymph nodes, diarrhea, sleep disorders (insomnia and disturbed sleep cycle), itching, and hair loss.

HEART DISEASE The leading risk factors for cardiovascular disease (CVD) are cigarette smoking, high cholesterol, high blood pressure, Type A behavior pattern (competitive, aggressive, and tense), family history, diabetes, obesity, physical inactivity and lack of exercise, advancing age, being male, and stress. Heart disease now occurs much later in life (usually past the age of 70). About twice as many males die from heart disease as females, yet men are more likely than women to have a favorable prognosis if they survive the first serious heart attack (Cockerham, 2012).

Heart disease remains the leading cause of death in the United States, as noted in Chapter 1, yet death rates from heart disease have fallen 31 percent since 1999 (Dunham, 2008). In addition, death rates from stroke, the third-leading cause of death in the United States, have fallen by 29 percent. Better control of cholesterol (likely contributed to statin drugs) and blood pressure (more individuals are getting treatment today), declining smoking rates, and better medical treatments all contributed to the dropping death rates. Though this is good news, there is concern that the accelerating rates of obesity could reverse this progress, as government figures show that 26 percent of Americans are obese (Dunham, 2008).

TREATMENT OPTIONS

EVALUATING TREATMENT OPTIONS AND SYMPTOMS

George Dickinson recalls a story told to him by a sociologist a few years ago. After examination and testing, his neurologist informed the sociologist that he had a malignant tumor in his brain. The patient immediately said to his doctor, "That means I will die, doesn't it, Doctor?" The physician replied, "Not necessarily. We have surgery, **chemotherapy**, and **radiation therapy**. We will try all three, if needed." These treatment options suggested to the patient that indeed there were alternatives from which to choose to fight this cancer. The doctor's encouragement of treatment options gave hope to the patient and a willingness to fight the disease.

After a diagnosis of a life-threatening disease, individuals will need to make decisions about treatment options (Doka, 1993). Treatment plans differ according to the illness. Some treatment plans will be relatively unobtrusive and may require little monitoring, yet others may cause lifestyle modifications. Heart disease, for example, may cause an individual to take daily medication, go on a special diet, and basically simplify his or her lifestyle and try to reduce stress. With other conditions such as cancer, options, as noted above, may be surgery, chemotherapy, and/or radiation. The order in which these treatments are given may vary, depending on the particular health status of the patient. Certainly any one of these treatments or a combination of them can contribute to lifestyle modifications and limitations. Perhaps the good news about treatment options for basically all illnesses is that medical advancements are occurring annually; thus, one's chances of survival are extended as the onset of a disease is delayed.

Whatever therapies are available, feelings and fears about them should be explored, noted Kenneth Doka (1993). Surgery may be feared because of anesthesia, loss of consciousness, postoperative pain, visible damage (scars, for instance), loss of functional abilities, or financial considerations. The procedures of chemotherapy and radiation may cause anxiety and concern over side effects.

Many women with a tumor in the breast choose to have the entire breast removed, even though study after study has shown that women with small to moderate-sized tumors fare just as well with a lumpectomy—removal of just the tumor, leaving the breast intact (Proctor, 1995). Fear of breast cancer has led some women to have their breasts removed, thinking this will increase their

---

PRACTICAL MATTERS | STEPS TOWARD CANCER PREVENTION

1. Ban smoking.
2. Impose stiffer taxes on tobacco sales.
3. Initiate stiffer supervision of pesticides and alternatives to petrochemical agriculture.
4. Better regulate loopholes that allow nonfood pesticides to be used on crops we eat.
5. Empower OSHA to deal with indoor pollutants such as radon and second-hand smoke.

6. Allow the FDA to limit the substances added to tobacco in manufacturing cigarettes.
7. Modify trade practices at home that encourage cancer abroad.
8. Curtail the export of asbestos.

Adapted from *Cancer wars*, by R. T. Proctor, 1995, New York: Basic Books.

chances of survival. Men have not yet begun to fear their prostate glands, though early signs of such fears are on the horizon, according to Proctor in his book on politics and knowledge about cancer.

Another treatment option for various illnesses is **immunotherapy** (treatment by stimulation of the body's own immune system). The immune system includes a complicated system of organs, tissues, blood cells, and substances to fight off infections, cancers, and other illnesses. Alternative forms of health care are controversial, as noted by the Words of Wisdom box entitled "Alternatives Gain Ground." Alternative forms of health care include **chiropractic medicine**, which relates to health problems based on blockages of the flow of vital energy caused by malalignments of the vertebrae. **Homeopathic medicine** (the use of natural drugs to treat patients) is built around the belief that if a small amount of a natural substance will precipitate symptoms in a healthy person, a similar amount in a sick person will aid in recovery. **Acupuncture** (using needles along meridians in the body to redirect energy flow) is a traditional Chinese medical treatment going back nearly 5,000 years and is a holistic system that understands health in the context of the relationship between the human body and nature. **Faith healing** is the use of the power of suggestion, prayer, and faith in God to promote healing. Such healing involves mind over matter, whether the healing is from snake-handling groups in Appalachia or members of the Christian Science community. **Folk healing** (the use of folk remedies to heal the sick) exists in part because of a dissatisfaction with professional medicine and may include treatments such as ginger tea, honey, sassafras, salt, and butter. Folk healers include those referred to as root doctors, spiritualists, or voodoo practitioners. Some Native Americans practice folk healing by believing that diseases are caused by soul loss or spirit possession and use herbalists and bone setters to treat the sick.

TREATING DRUG SIDE EFFECTS

Some anticancer drugs cause nausea and vomiting because they affect parts of the brain that control vomiting and/or irritate the stomach lining. The management of nausea and vomiting caused by chemotherapy is an important part of care for patients with cancer. There is no single best approach to alleviating these symptoms in all patients. Therapy must be tailored to each individual's needs, including factors such as the type of anticancer drug administered, the general condition of the patient, and age.

For treating loss of appetite and weight loss for patients with cancer, changes in diet and consumption of foods, taken through a tube or orally, can help. Researchers can evaluate the effect of the drugs alone or in combination on nausea and vomiting and assess differences in the quality of life among patients. Toxic effects related to the use of drugs can also be assessed.

Though controversial, the use of marijuana for controlling chemotherapy-induced nausea and vomiting in patients with cancer is generating much interest. The U.S. Food and Drug Administration (FDA) approved the use of marijuana for treatment of nausea and vomiting associated with cancer chemotherapy in patients not responding to standard drugs. The U.S. Supreme Court ruled, however, on June 6, 2005, in a 6-to-3 decision, that federal authorities may prosecute patients whose doctors prescribe medical marijuana despite the state laws that allow its use. The ruling

WORDS OF WISDOM | ALTERNATIVES GAIN GROUND

Future doctors and nurses are learning about acupuncture and herbs along with anatomy and physiology at a growing number of medical schools. It's another example of how alternative medicine has become mainstream.

And it's often done with Uncle Sam's help.

The government has spent more than $22 million to help medical and nursing schools start teaching about alternative medicine, lesson plans that some critics say are biased toward unproven remedies.

Additional tax money has been spent to recruit and train young doctors to do research in this field, launching some into careers as alternative medicine providers.

Doctors need to know about popular remedies so they can discuss them nonjudgmentally and give competent advice, the government says, and many universities and medical groups agree.

"Patients are using these things" whether doctors think they should or should not, and safety is a big concern, said Dr. Victor Sierpina, an acupuncturist at the University of Texas Medical Branch of Galveston who heads a group of academics who favor such training.

But to critics, it's like teaching Harry Potter medicine. Students are being asked to close their eyes to science principles that guide the rest of their training in order to keep an open mind about pseudoscience, they say.

"I'm concerned about the teaching of illogical thinking to medical students" and lending credence to biologically implausible theories like distance healing and energy fields, said Dr. Stephen Barrett, a retired physician who runs *Quackwatch*, a Web site on medical scams.

Teaching about alternative medicine implies acceptance of it and "potentially creates more gullibility and less critical, objective thinking," said Dr. Wallace Sampson, editor of the journal *Scientific Review of Alternative Medicine*.

The real issue is not whether alternative medicine should be taught, but how, said Dr. Joseph Jacobs, former head of the federal Office of Alternative Medicine.

"The parallel here is creationism versus science," Jacobs said. "If the topic is taught objectively, to help students communicate with patients, it's a good idea. If it's being taught as part of an advocacy, for acceptance among physicians, I think that's a little bit bogus."

Sometimes the line is blurry.

Some schools have close ties to alternative medicine providers or advocates who shape information on the schools' Web sites or classes for students and the public. Two examples:

The University of Arizona's Center for Integrative Medicine has medical residency programs in hospitals around the country, partly sponsored by well-known advocate Dr. Andrew Weil, the center's founder. A private group that promotes such care, the Bravewell Collaborative, gives scholarships for dozens of the Arizona schools' students to get hands-on training in integrative care clinics.

The University of Minnesota offers medical students an elective course in alternative healing methods at a Hawaiian medical center founded by a philanthropist-advocate of such care, although students pay their own transportation and living expenses.

M. Marchione (2009, November 2). Alternatives gain ground. Associated Press.

does not invalidate state law, however, in the states that have removed criminal penalties. As of November 2012, these 18 states, plus the District of Columbia, have enacted laws that legalize medical marijuana: Alaska, Arizona, California, Colorado, Connecticut, Delaware, Hawaii, Maine, Massachusetts, Michigan, Montana, Nevada, New Jersey, New Mexico, Oregon, Rhode Island, Vermont, and Washington. Use of marijuana for medicinal purposes, therefore, can continue without much fear of prosecution by local authorities, and federal law enforcement does not have the resources to bring more than a few token cases to court.

Smoked marijuana for medical purposes remains controversial. Proponents for the legal use of medical marijuana argue that marijuana is a safe and effective

treatment for dozens of conditions such as cancer, AIDS, multiple sclerosis, pain, migraines, glaucoma, and epilepsy. Proponents say that thousands of deaths annually from legal prescription drugs could be prevented if medical marijuana were legal; no one has ever died from an overdose of marijuana (Carter et al., 2011). Marijuana does not have to be smoked to be effectively used as medicine but can be ingested orally or applied topically in a liniment. If smoked, it does not cause lung cancer, though it can potentially irritate bronchial mucosal membranes. Opponents of medical marijuana argue, however, that marijuana is too dangerous to use and that various FDA-approved drugs make the use of marijuana unnecessary. They claim that marijuana is addictive, leads to harder drug use, injures the lungs, harms the immune system, damages the brain, interferes with fertility, impairs driving ability, and sends the wrong message to kids. They also say that people who argue for medical use are actually using it for recreational pleasure (Medical Marijuana, 2009).

A nonsmoked medication has a better chance of being accepted by the medical establishment and approved by the government, observed Guterman (2000). The Food and Drug Administration has approved the drug Marinol, taken as a pill, for appetite stimulation and relief of vomiting in patients with AIDS and cancer. The pill can be difficult for a person with severe nausea to keep down, and Marinol takes hours to take effect, whereas smoked marijuana's instant hit of cannabinoids in the bloodstream allows patients to monitor their own dose, say the supporters of legalized medical marijuana. And the politics of medicine continues!

PAIN AND SYMPTOM MANAGEMENT

The care of the dying should have its fullest expression in helping patients cope with the technologically complicated medical environment surrounding them at the end of life. The physician must seek a level of care that optimizes comfort and dignity, as not all dying patients succumb peacefully. Many patients experience that which is called "terminal restlessness," an anguished death which may be characterized by the symptoms of increased physical, emotional, or spiritual pain (Mazzarino-Willett, 2010). Terminal restlessness may be observed during the last days and/or hours before death. Patients may exhibit symptoms of agitation, restlessness, involuntary twitching or muscle spasms, and physical irritability in addition to impairment to their level of consciousness. Hallucinations, paranoia and/or nightmares may occur. Controlling pain, whether physical, emotional or spiritual, is certainly an important aspect of this art of medicine. Pain is always subjective. Especially with physical pain, there is no way to distinguish pain occurring in the absence of tissue damage from pain resulting from tissue damage (Connor, 2009). Pain is what the patient says it is. Caregivers must always be alert to the reality that pain is multidimensional.

Phenomenology may help illuminate experiences of pain and suffering in end-of-life illness. Certainly it is characterized by recognition of the sociality in the structure and meaning of psychic life, an account of motivation and desire, and it focuses on subjectivity (Morrissey, 2011). Phenomenology expands the consciousness of pain by helping to understand the complexity of the emotional and sensory processes involved in pain experiences and others' apprehensions of the responses of the experiences and responses thereto.

In today's medical environment, more and more doctors are integrating pain management into their treatment of dying patients. This pain and symptom management approach first took hold within hospice settings, which developed in response to the medical model approach to dying. As sickness progresses to death, dying patients may require palliative care of an intensity that rivals even that of curative efforts. Death education in medical schools must strive to help the student know how to make the dying patients' last few days more comfortable. As in natural childbirth—the other occasion on which the role of the physician and nurse is not to cure but to make the patient comfortable—natural death involves demedicalization to the extent that death is defined as a natural rather than a medical event (Walter, 1995).

Traditional medical care is often based upon a **PRN** (for the Latin *pro re nata*) approach, which means that medication is given "as the situation demands." In practice, this means that often a person must first hurt and ask for relief before pain management can be administered. This approach is responsible for much of the suffering endured by the terminally ill. British physician Phil Hammond observed (Hammond & Mosley, 1999) that orientation to pain management in medical school is "often pitiful" and that many students become doctors with "quaintly stoical views about pain" and a hesitation to use pain management.

In 1958 Cicely Saunders (1992, p. 20) developed an alternative method of pain control that is still being used today, especially in hospice. She wrote:

> Here at St. Joseph's Hospice, as elsewhere at that time, one saw people "earning their morphine," and it was wonderfully rewarding to introduce the simple and really obvious system of giving drugs to prevent pain happening—rather than to wait and give them once it had occurred. Here too there was the potential for developing ideas about the control of other symptoms, and also for looking at the other components of pain. But first of all I must salute the Sisters of St. Joseph's and the compassionate matter-of-factness of their dedicated care. Together we began to develop the

DEATH ACROSS CULTURES PAIN IN DIFFERENT ETHNIC GROUPS

The subjective experience of pain is powerfully contoured by culture. In the early 1950s, a study of responses to pain among men in a veteran's hospital in New York revealed that these individuals accepted pain without complaint. These Americans had lived in the New World for many generations so that they no longer had ties with their countries of origin and thus were "Old" Americans. Irish Americans in this study also accepted pain without complaint, but the two groups differed greatly in how they experienced the meaning of pain.

There is a difference between private and public pain. Left alone, the Old Americans might collapse into tears, but never in public. In the presence of nurses and doctors, to admit to pain was permitted because the professional situation transformed complaints into purposeful discussion. They had faith in their doctors and were fairly optimistic about their outcomes.

On the other hand, the Irish American patients differed. They lacked the optimism of Old Americans. In public, an Irish patient masked pain, but this was also done when talking to medical practitioners. The Irish American felt helpless, guilty about becoming ill, and very pessimistic about the future.

Italian American and Jewish American veterans tended to display highly emotional responses and

groaned and cried and complained and talked in detail about how they suffered. They shared a cultural trait that permitted vivid expressions of pain. The Jewish American patient tended to experience a future-oriented anxiety, and pain was taken as a frightening warning of ultimate possible doom. This patient needed reassurance from the physician, yet this same patient was skeptical toward the doctor. Jewish pain tended to be associated with powerful existential concerns and ultimate eschatological issues.

The Italian American patient was equally vocal, yet the patient showed great trust in doctors and hospitals. Instead of a future orientation, the Italian American patient experienced a present-oriented apprehension and begged for strong analgesics to quiet the pain.

Though this research by anthropologist Mark Zborowski on pain was conducted a half century ago and methodologically has some flaws, the findings at least hint at the impact of culture on behavior. Learned behavior indeed varies culturally, thus producing different outcomes.

From *Magic, science, and health: The aims and achievements of medical anthropology*, by R. Anderson, 1996, Fort Worth, TX: Harcourt Brace College Publishers.

appropriate way of caring, showing that there could be a place for scientific medicine and nursing. We could illustrate an alternative approach to the contrast between active treatment for an illness (as if to cure it were still possible) or some form of legalized euthanasia.

PAIN MANAGEMENT A pain management approach is based on the belief that a patient should not hurt at all. Regular medication is, therefore, given before the pain begins. The aim is to erase the memory of the pain that has been experienced and to deal with the fear of pain in the future. Pain medications are standardized to the needs of the patient. The aim is to control the pain and other symptoms without sedating the patient. Every symptom is treated as a separate illness because only when each symptom is under control can a patient begin to find fullness and quality of life.

Pain among Americans seems to be accepted, whether the individual is terminally ill or not. For example, a Gallup poll (Goldenberg, 2000) of 2,000 people found that regular pain was a pervasive part of American life and that few people sought medical attention for it. Sixty percent of those surveyed agreed that pain was "just something you have to live with." Of those adults surveyed, 42 percent reported experiencing pain every day. Only half said they had seen a doctor in the past three years for help with their pain.

On the other hand, not everyone agrees that pain should be accepted. For example, George Dickinson was talking with a British palliative medicine doctor in South Yorkshire, England. In the course of the conversation, the physician asked Dickinson's opinion of physician-assisted suicide. Dickinson said he felt that in certain circumstances, such as a patient in the latter stages of AIDS, such action was justified. He said, "especially if the patient's pain is uncontrollable." At this point the palliative medicine doctor snapped back with, "But why should the patient be in pain?" To this British palliative medicine physician, there is no excuse for anyone to have pain. So much for Dickinson's argument of occasionally favoring physician-assisted deaths! The physician is correct. Some would argue that there is no excuse for pain in the 21st century with opiates and appropriate adjuvant medications to control pain. Morphine, for example, is an excellent painkiller.

Anxiety and depression are part of the chronic nature of pain. The patient is anxious about the pain returning—anxious because of the meaning of pain. When awakening in the morning with a stiff neck, the patient with cancer assumes **metastases** (spread of the disease), whereas the patient without cancer assumes that she or he slept incorrectly. The degree of perceived pain is totally different in these two situations. One must take into account the anxiety these patients suffer, along with the chronic depression caused by chronic pain. Additionally, many patients have more than one type of pain; thus it is critical for staff to assess each pain source carefully (Connor, 2009).

Most cancer pain responds to pharmacological measures. Oral administration (by mouth) of analgesic drugs is preferred, and they are given regularly to prevent recurrence of pain. A step-by-step approach is used, beginning with a nonopioid analgesic such as aspirin. At the second step a weak opioid such as codeine is added, and when this proves inadequate a strong opioid is substituted for the weak opioid. Morphine is often the recommended strong opioid analgesic. If pain returns consistently before the next regular dose is due, the regular dose is increased. If patients are unable to take drugs orally, alternative routes are rectal and subcutaneous (Hanks et al., 1996).

| LISTENING TO THE VOICES | ENDURING PAIN |

An elderly male patient with cancer would not take his pain medication. The nurse asked to talk to him. She learned that he was in the landing at Normandy in World War II and endured much pain and hardship. Many of his comrades died in the invasion. Out of his feelings for those comrades and those memories of the landing, he told the nurse that he could and would endure the pain. The pain brought back memories, and he must "bear it" out of respect for those fallen fellow soldiers.

Told to George Dickinson by a nurse in the spring of 1999 in Sheffield, England.

Bone metastases are the most common cause of cancer-related pain (Bomanji, Britton, & Clarke, 1995) and have been found in up to 83 percent of patients with carcinomas of the prostate, breast, or lung at the time of death. People with painful bone metastases may survive a number of years and can experience chronic severe pain and impaired mobility; thus, pain control has a significant effect on their quality of life (Day, 1999). Strontium therapy, administered by intravenous injection, is used in pain relief for patients with bone metastases from tumors in the prostate and breast. Carried by the bloodstream, a single injection of strontium treats all painful sites simultaneously and delays the development of new pain.

In addition to traditional medical treatment of pain, art and music therapies can work to help control pain. In the medical model, it is common to address physical problems first. In a pain management approach, pain is the first thing to be treated. Art and music therapists offer interventions to assist the patient in balancing the focus of attention away from the physical sensation of pain. The therapist engages the patient using art or music, refocuses energy from the pain, and ultimately makes the person more balanced by the decrease of the perception of pain.

MANAGING DEHYDRATION Some symptoms are associated with dehydration, but each is easily treated. Thirst, experienced in a small number of dehydrated patients at the end of life, is easily palliated with lip moisturizer, sips of water, ice chips, and hard candies for the patient to suck (Hoefler, 2000). In a small percentage of patients, electrolyte imbalance may lead to neuromuscular irritability and twitching, both of which are easily treated with sedation. Nausea, sometimes a symptom of dehydration, is treated with antiemetics. Normally, however, dehydration actually reduces nausea, vomiting, and abdominal pain. In addition, decreased urine output from dehydration may decrease the incidence of urinary tract infection.

DEATH ACROSS CULTURES | RESISTANCE TO PAIN MEDICATION

The Ugandan government makes and distributes its own morphine for use in hospitals, but poor management means the supply is erratic. Morphine is cheap, effective and simple, and easy to manage. Yet, according to the World Health Organization, every year more than 5 million people with cancer die in pain without access to morphine. It costs only $2 per week.

In well-off countries like the United States there is enough morphine to treat 100 percent of the individuals in pain, yet in low-income countries it is only 8 percent. In many countries (up to 150) morphine is almost impossible to get. Some governments do not provide it, or they strictly limit it, because of concerns it will be diverted to produce heroin. And many physicians are reluctant to prescribe morphine, fearing addiction.

In India, whether or not one can get morphine depends on where the individual is treated. In some parts of the country, only 1–2 percent with cancer pain are treated with morphine. Patients often think that morphine equals death, and they recoil at the suggestion of using morphine to control their pain. Many oncologists will not send patients to a clinic where morphine is available, for that reason. These medical doctors do not want to give up on their patients when it comes to giving them hope. Saying that the doctor is referring a patient to a palliative specialist is indirectly saying that there is medically nothing more to do for the patient.

On a brighter note, however, there are improvements on the way. Overall in low-income countries, morphine consumption is up tenfold since 1995.

Taken from J. Silberner (2012, December 7). Public Radio International, Cancer Series, Part V: Dispensing Comfort.

Pulmonary secretions also decrease with dehydration; thus, the patient experiences less coughing, congestion, choking, and shortness of breath. Dehydrated terminally ill patients often report less pain and discomfort than patients receiving medical hydration; thus, their need for pain medication is less (Hoefler, 2000). Dehydration also reduces swelling in the body, thus improving the person's sense of well-being and reduces pressure on tumors, if they exist.

ORGAN TRANSPLANTATIONS

Many individuals today are "living with dying" while waiting for an organ transplant, yet others are living because of organ transplants already received. Some 95,000 U.S. patients are currently waiting for an organ transplant; every day, 17 individuals die in the United States while waiting for a transplant of a vital organ (Twenty-five Facts about Organ Donation and Transplantation, 2009). Thus, these individuals likely live rather anxious lives. Organ and tissue donations have prolonged many individuals' lives, while improving significantly the quality of their lives. Particular tissues and/or organs in one's body may simply cease functioning and would cause the death of the individual, if not for new organs or tissues taken from someone else or some other animal.

Some organs and tissues come from individuals who have been legally pronounced dead, whereas others come from living persons. Well over two dozen different kinds of tissues and organs are used for transplantation, including eye corneas, skin, bones, tendons, bone marrow, kidneys, livers, pancreases, blood vessels, intestines, lungs, and hearts. The first face transplant in the United States occurred in December of 2008, as 80 percent of a woman's horribly disfigured face was replaced with that of a female cadaver (Marchione, 2008). With advances in medical technology and preservation techniques, vital organs may be procured and transported hundreds of miles to a recipient center for transplantation. Hearts and lungs can be preserved for up to six hours before transplantation, and kidneys can be preserved up to 72 hours (United Network for Organ Sharing, 2009).

Major organ transplants such as kidneys, livers, pancreases, and hearts began in the 1950s and 1960s. Today, there is an increasing tendency for individuals to donate their organs and tissues. As is shown in Table 6.2, for example, in 2008 a total of 27, 958 organ transplant procedures were performed in the United States, the overwhelming majority coming from deceased organ donors (78 percent), with the others coming from living donors (New York Organ Donor Network, 2009). Compared to deceased donor transplantation, at least for kidneys, living donation is less expensive and results in better graft survival, lower rates of acute rejection, and improved recipient survival and long-term functioning (McGrath, Pun, & Holewa, 2012).

The National Conference on Uniform Laws developed the Uniform Anatomical Gift Act to help answer questions concerning transplantations, and by 1971 all 50 states had adopted the act with only minor variations (Sade, 1999). The act permits a person older than 18 years of age to donate part or all of his or her body upon death. If the body or body part has not been donated before death, a family member may agree to do so, with permission based on the priority of the relationship. The priority is as follows: spouse, adult son or daughter, parent, adult sibling, legal guardian, or other party responsible for disposition of the body.

| TABLE 6.2 | NUMBER OF TRANSPLANTS PERFORMED IN THE UNITED STATES IN 2008 FROM DECEASED AND LIVING DONORS |

Type of Transplant	Number of Transplants
Kidney (no pancreas)	16,514
Liver	6,318
Kidney-pancreas	836
Intestine	185
Heart	2,163
Heart-lung	27
Lung	1,478
Total Transplants in the U.S.	27,958

Source: New York Organ Donor Network, 2009.

Various problems affect human transplantations. Whether or not the donor and donee tissues are compatible can be a problem. There is often a shortage of donors, thus producing a difficult decision in selecting among potential recipients. The lack of a legal definition of death in some states prevents surgeons from removing healthy organs when brain activity has stopped but the heart and lungs are still functioning. The lack of a nationwide communications network to coordinate information presents another problem for human transplantations.

Another issue with organ transplants involves the definition of death, a topic discussed in Chapter 2. When is the donor dead? For decades, organs typically have been removed only after doctors determine that a donor's brain has stopped

Vittoriano Rastelli/Historical/Corbis

Since their beginnings in the mid-20th century, organ transplants have allowed thousands of individuals to live normal lives. Rapid means of transportation have enhanced the process of getting an organ to the recipient.

working completely. In brain-death donations, the donor is kept on a ventilator to keep oxygen-rich blood flowing to the organs until they are removed. Yet, a relatively new procedure involves removing all life support, leaving little brain activity, a procedure not meeting the criteria for brain death. This procedure is called *donation after cardiac death* and is being encouraged by the federal government, organ banks, and others as a way to make more organs available and give more families the option to donate (Nano, 2008). The controversy with this procedure, however, is that state laws stipulate that donors must be declared dead before donation, based on either total loss of brain function or heart function that is irreversible. Others argue that the definition of death is flawed and that more emphasis should be on informed consent and the chances of survival in cases of severe brain damage.

Other problems include the question of who will pay for expensive transplantation procedures. A few public insurers have decided in favor of coverage for heart and liver transplants. Kidney transplants have been covered since 1973 under Medicare's end-stage renal disease program. Though the costs of solid organ transplants are high, the benefits are substantial when one considers that they may be lifesaving.

As medical technology has progressed and transplantations have become more successful, the demand for transplantation has greatly increased. In 2008 for instance, 6,229 Americans died because of the shortage of organ donors (New York Organ Donor Network, 2009). If we add to this problem the fact that, increasingly, health maintenance organizations (HMOs), insurance companies, and other third-party providers will not pay for transplantation procedures, we face a situation in which those who receive organ and tissue transplantations are primarily those who can afford such surgery. Such an unequal distribution of resources is a good example of the conflict perspective, discussed in Chapter 1.

A solution to the organ donor shortage suggests following the example in Spain where, in 1979, they made the switch to the "opt-out" method of donation (Ayres & Nalebuff, 2007). In such a plan, everyone is a donor, unless an individual

PRACTICAL MATTERS | Uniform Donor Card

Uniform Donor Card of (name of donor). In the hope that I may help others, I hereby make this anatomical gift, if medically acceptable, to take effect upon my death. The words and marks below indicate my desires. I give:

a. _____any needed organs or parts
b. _____only the following organs or parts—specify the organs or parts—for the purposes of transplantation, therapy, medical research, or education:
_____

c. _____my body for anatomical study if needed

Signed by the donor and the following two witnesses in the presence of each other:

Signature of donor	Date of birth of donor
Date signed	City and State
Witness	Witness

opts to not be. Currently in the United States, there is an "opt-in" system, usually from filling out a form for your driver's license. Now in Spain 80 percent of Spaniards are potential organ donors, with virtually no waiting list. Opt-out also is the norm in other European countries.

PALLIATIVE CARE

The phrase **palliative care** (controlling pain) was first coined by Canadian surgeon Balfour Mount in 1974 (Clark, 2007). Palliative care seeks to satisfy the needs of patients and their families in several domains including the physical, psychological, social, and spiritual. Palliative care helps the person live out his or her final days in as comfortable and dignified a manner as possible. Palliative care is both person-centered and family-centered and interdisciplinary (Morrissey, 2011). However, palliative care is not restricted to individuals in a hospice program and indeed is much broader than that. Though much of this chapter entitled "Living with Dying" concentrates on hospice, it is because the hospice way of palliative care is an excellent model to follow throughout the health care system in the 21st century. Such *caring* treatment needs to be diffused throughout medical care. For example, recent studies of 3,702 nursing home residents nationwide on kidney dialysis and 323 individuals with advanced dementia (Chang, 2009) conclude that many frail, elderly Americans in nursing homes are suffering from futile care at the end of their lives, whereas palliative-care options should probably be made instead.

Pioneers of the modern hospice movement have always maintained that the principles of palliative care, so successfully employed in the management of

"He's our new Palliative Specialist!"

There are many physicians, like me, who specialize in continuity of care. A single board does not certify us. Our unique expertise is the accumulated understanding of our patients; not merely of their diseases, but of their lives. With that understanding comes trust, slowly developed over the years. We treasure the special moments when a trusting relationship is what patients and their families need most to reach an important decision. These moments sustain me and keep my passion for medicine alive.

From *An old man's friend*, by W. A. Hensel, 2000, *Journal of the American Medical Association, 283*(14), 1793–1794.

patients with cancer, can and indeed should "not only be facets of oncology but of geriatric medicine, neurology, general practice and throughout medicine" (Derry, 1997). However, Sally Derry noted that there is still a long way to go before the principles of palliative care become widely used in the management of patients with diseases that are acknowledged to be chronic and progressive.

Evidence-based practice (defined as the use of innovative and methodologically robust research in combination with both clinical expertise and patient and family input to best inform the treatment decisions made by health care professionals and interdisciplinary teams) is an important aspect of palliative care and is growing in the United States and other industrialized countries, yet hospice programs have yet to implement widespread initiatives, despite professional hospice organizations endorsing and encouraging such efforts (Sanders et al., 2010). Evidence-based practice does not mean ignoring clinical experience or patient and family wishes, which are common components of palliative care. Rather, such practice means integrating clinical expertise and findings with the best available external evidence from systematic research. The evidence about effective palliative care services is rather like a sieve—there are gaps and holes in our knowledge, said I. J. Higginson (1999). Patients with advanced illness are among the most difficult to study, and there is no second chance to get care right. It is imperative, therefore, to use existing research through appropriate systematic reviews to maximize the value of these data for the care of patients and their families. Phil Hammond (Hammond & Mosley, 1999) in his advice-giving, somewhat tongue-in-cheek book entitled *Trust Me (I'm a Doctor)* stated that evidence-based medicine should be an international concept, with hard science "trumping" irrational cultural beliefs every time. But it is not, noted Hammond. He stated, for example, that if you have a stomach ache, in France you get a suppository, in Germany you visit a health spa, in the United States they cut your stomach open, and in Britain they put you on a waiting list.

Palliative daycare is a rapidly growing aspect of palliative care since the first palliative daycare unit was established in 1975 in Sheffield, England. Such daycare is not unlike child daycare or geriatric daycare except that these individuals need special care due to the various manifestations of their illnesses. Positive aspects of a palliative daycare facility include the patient's getting out, engaging in various activities, and associating with others. Services received by the patients include the presence of a physician, nursing care, bathing, occupational therapy,

George Dickinson

Saint Christopher's Hospice in London, England is the model hospice for hospice programs around the world. Besides having 66 beds for hospice patients, Saint Christopher's is an educational center for teaching health care professionals about palliative care.

arts and crafts, aromatherapy, massages, hairdressing, and relaxation. Clearly, the services provided by daycare centers for patients and their families are of considerable value and have been acknowledged as an integral and vital aspect of palliative care.

Palliative medicine is now recognized in several countries. The leader in palliative medicine is the United Kingdom. Palliative medicine in the United Kingdom has been recognized as making a significant contribution, not only to the practice of cancer medicine, but also to the care of those with terminal disease. The United Kingdom was recently rated as leading the world in a "quality of death" ranking across 40 countries, possibly because of its early start in the field (Seymour, 2012). Patients treated by a team with skill and expertise in the care of the dying will value such care whatever the primary specialty of the medical team. These patients can gain from pain relief, symptom control, communication, and family support— the skills of palliative medicine practitioners. The health care system in the United States is rapidly beginning to see the benefits of palliative medicine and palliative care and is moving in this direction.

THE HOSPICE MOVEMENT

The **hospice movement** shares all the goals of palliative care. **Hospice** has developed as a response to fears related to the dying process and the ways in which death is typically handled in institutional settings. Primary patient concerns that hospice care directly addresses are the problems related to symptom and pain control, the apprehension caused by having others in control of one's life, and the anxiety about being alone at the time of death. The primary goals of hospice care are to promote patient–family autonomy, to assist patients in obtaining pain control and real quality of life before they die, and to enable families of patients to receive supportive help during the dying process and the bereavement period. The National Hospice Organization, which recently changed its name to the National Hospice and Palliative Care Organization (NHPCO), has grown from serving 158,000 patients in the United States in 1985 to 1,580,000 in 2010. The percentage of hospice patients covered by the Medicare hospice benefit versus other payments was 84 percent in 2010, and the median length of service was 19.7 days (NHPCO, 2012). The hospice movement proclaims that every human being has an inherent right to live as fully and completely as possible up to the moment of death. Some traditional health care, emphasizing the *curing* but not so much the *caring* of the patient at any cost, has ignored that right.

U.S. hospice programs, as of 2010, were 36 percent not-for-profit, 58 percent for-profit, and 6 percent government-run (e.g., U.S. Department of Veterans Affairs medical centers). The majority (67 percent) of patient care in hospice is provided in one's "home" (private residence, nursing home, residential facility), with the remaining patients receiving care in a hospice inpatient facility (22 percent) or an acute care hospital (11 percent) (NHPCO, 2012). Types of inpatient hospice agencies include free standing/independent (58 percent), part of a hospital system (21.3 percent), part of a home health agency (19.2 percent), and part of a nursing home (1.4 percent). The demographic composition of most hospice patients in the United States is white, female, and over the age of 75 years, as noted in Table 6.3. Cancer had the highest percent of hospice admissions in 2010 with 36 percent (NHPCO, 2012). As noted in Table 6.4, other top diseases include heart disease, dementia, and lung disease.

TABLE 6.3 | PERCENTAGE OF HOSPICE PATIENTS BY RACE, GENDER, AND AGE IN 2010

Patient Race	Percentage
White/Caucasian	77.3
Multiracial or Other Race	11.0
Black/African American	8.9
Asian, Hawaiian, Other Pacific Islander	2.5
American Indian or Alaskan Native	0.3
Patient Gender	
Female	56.1
Male	43.9
Patient Age Category	
Less than 24 years	0.4
25–34 years	0.9
35–64 years	16.1
65–74 years	15.9
75–84 years	27.9
85+ years	38.9

Source: NHPCO Facts and Figures: Hospice Care in America, 2011 Edition (2012). National Hospice and Palliative Care Organization.

TABLE 6.4 | PERCENTAGE OF U.S. HOSPICE ADMISSIONS BY PRIMARY DIAGNOSIS IN 2010

Primary Diagnosis	Percentage
Cancer (malignancies)	35.6
Noncancer diagnoses	64.4
Heart disease	14.3
Debility, unspecified (includes frail elders with multiple illnesses and steady deterioration)	13.0
Dementia, including Alzheimer's disease	13.0
Lung disease, including chronic obstructive pulmonary disease	8.3
Stroke or coma	4.2
Kidney disease, including end-stage renal disease	2.4
Non-ALS Motor neuron	1.2
Liver disease	1.9
Amyotrophic Lateral Sclerosis (ALS)	0.4
HIV/AIDS	0.3
Other diagnoses	5.4

Source: NHPCO Facts and Figures: Hospice Care in America, 2011 Edition (2012). National Hospice and Palliative Care Organization.

Looking at the global picture as of 2006, 115 of the world's 234 countries have established one or more hospice-palliative care services, yet only 35 (15 percent) have achieved a measure of integration with other mainstream service providers (Clark, 2007). Though only about half of the world's countries have hospice-palliative care services, these countries encompass 88 percent of the global population. Where the need is greatest, however, the fewest hospice-palliative care services exist, notes British medical sociologist David Clark.

Many physicians have been trained to focus on restoring the patient to health. Accordingly, many patients are subjected to a series of operations designed to prolong life, even though a cure is sometimes impossible, as in the case of a rapidly progressing cancer. Most hospice patients have previously had some surgery, chemotherapy, or radiation treatments. In hospice, however, cure goals for patients are changed to comfort goals, and every patient has a significant role in all health care decisions.

Because of the emphasis on quality of life, hospices pay attention to many facets of pain reduction, discussed earlier, including but not limited to physical pain. Hospice medical professionals have spent considerable time in developing a variety of methods of pain control that subdue not only what the patient describes as pain, but also the symptoms related to the illness.

Much of this emphasis on pain control has developed despite the practice by some professionals of sedating patients in pain. Quality of life cannot be achieved if the patient is knocked out or has become a zombie. Physicians try to find the point at which the pain is managed, but sedation has not occurred. Such pain management has necessitated considerable retraining of health care leaders.

Hospice personnel also deal with social, psychological, financial, and spiritual pain. Terminally ill patients may experience social abandonment or personal isolation that comes when friends and acquaintances stop visiting them because of an inability to cope with issues of death, a lack of knowledge about what to say or do, or simply a lack of awareness of what the patients are experiencing. It is ironic that the dying patient's need for social support and companionship comes at a time when he or she may often be more alienated than at any other time in life.

Financial problems are also experienced by patients and their families who face large hospital and medical bills at a time when family income may be diminished. Finally, there is a spiritual pain that people experience when they seek answers to existential questions and ultimate meaning and purpose in the face of suffering. "Why did God allow this to happen to me?" and "Why do bad things happen to good people?" are questions frequently asked by patients and their families.

Critical questions are: What constitutes quality of life? What do people most want to accomplish before they die? What do they most want to do? When one of Michael Leming's hospice patients in Northfield, Minnesota, was asked that last question, he said that he had always wanted to take a helicopter ride. With the help of the local NBC television station, they made this dream a reality. Like this patient, almost everyone has unfinished business in life. Some may wish to renew relationships with friends or family members. Others may desire to put their own affairs in order, to write their memoirs, to plant a garden, to watch the sunset, or to plan their own funeral service. Then there is the 37-year-old woman who, in her last months of life, studied for her real estate license examination, passed the test, and with the help of her husband sold two houses. Thus, at age 37, she found her first job, *while dying*. On the other hand, the concern of the dying individual might be for others, as was true of a 40-something-year-old man in

DEATH ACROSS CULTURES | UK AND US PRISON HOSPICE PROGRAMS

The intent of prison hospices is to afford terminally ill inmates the right to approach death with dignity, unshackled and supported in the most appropriate way possible. Prison hospices were introduced in the United States in the 1980s, with an estimated 69 prison hospice programs currently in existence. The United Kingdom has only one prison hospice which opened in 2004. Overall in Europe, there are relatively few prison hospice systems operating today. The fact that there is limited literature on prison hospices might suggest an attitude of less value being placed on end-of-life care for incarcerated individuals.

As with hospice programs outside of prison settings, volunteers plan a major role within prison hospice programs. Many of the volunteers are drawn from inmate populations rather than lay people. In the United Kingdom with its one prison hospice program, the most commonly used method for end-of-life care at present appears to be the transfer of prisoners to a hospital or hospice within the community or compassionate release. Studies report that

the British system is lagging behind the United States in the delivery of end-of-life care in the form of on-site facilities. With prison hospices, the concern is not only for comfort but containment and support in a safe environment is essential.

The United States has shown that within-prison hospices can be cost efficient. These programs can also be emotionally rewarding for both the patient and the inmates supporting them. Hospice programs for the prison population are still in their infancy, thus have their problems in particular with the use of narcotics for pain release and issues of trust both on the part of the inmates and the healthcare professionals. Yet the provision of end-of-life care is increasingly being made available for the prison population, especially in the United States.

Source: K. Stone, I. Papadopoulos, and D. Kelly (2012), Establishing hospice care for prison populations: An integrative review assessing the UK and USA perspective, *Palliative Medicine*, 26(8), 969–978.

Kentucky whose wife had not leaned to drive, and his worry was that she could not be independent or able to drive the kids to school and do other tasks. He asked the social worker to give his wife driving lessons so that she should acquire her driver's license. The social worker obliged, and the dying man's wife passed the driving exam a few days before he died. Thus, his dying wish was granted.

THE HOSPICE TEAM

Hospice care is provided by an interdisciplinary **hospice team,** with each discipline having something to contribute to the whole. All disciplines work together, each in its own area of expertise.

Each interdisciplinary hospice team includes several layers or levels of care. The center of the team is composed of the *patient and his or her family.* The hospice movement emphasizes the need for people to make their own decisions with the supportive help of health care professionals and other trained persons. A vital part of the process is the *patient's own physician*—the professional who will continue to be in charge of the care of the patient and write medical orders when necessary.

The next layer of the team includes the hospice's professional caregiving staff. This consists first of *physicians*, required to direct medical care. *Nurses* comprise the next layer. Registered nurses are responsible for coordinating the patient's care. Licensed practical nurses and nurse's aides are also included—especially in inpatient settings.

The *hospice social worker* constitutes a pivotal part of the team. The social worker spends considerable time working with families, thus enabling family members to communicate with each other. Although family members may be aware that the patient is dying, they may never have discussed it with each other or with the patient. The social worker may spend time dealing with social problems, such as alcoholism and marriage problems, and working with the children or grandchildren of patients.

Pastoral care is a basic aspect provided by the team. A larger hospice may employ a *chaplain*, who will direct pastoral care to patients and their families, counsel other members of the caregiving team on spiritual issues, and try to involve clergy of the community in the care of their own people. In smaller hospice programs, all of the care may be provided by local clergy who work closely with the hospice staff.

Financial counseling is a service provided by the hospice team. Because patients and families have often exhausted their financial resources at the time of care, attention is given to seeking other forms of third-party reimbursement, such as those provided by Medicare or Medicaid or private insurance companies, and to seeking other programs for which the patient may be eligible.

Another layer of the hospice team includes a variety of health care professionals or other key leaders in the community whose help may be called upon during the illness. A *psychiatrist* or *psychologist* may be needed to provide expert counseling help. *Nurses, home health aides*, and *homemakers* employed by public health nursing agencies—such as visiting nurse associations—may be sought to provide special continuing health care or to share in the provision of patient care.

Physical, *speech*, and/or *occupational therapists* may be helpful in working with the patient to ensure maximum daily functioning. Finally, the services of a *lawyer* and/or *funeral director* may be required to help the patient settle personal affairs and provide for the needs of survivors after his or her death.

Artists are increasingly recognized as important members of the hospice team. The Connecticut Hospice pioneered the development of an arts program that considers the arts as a means to help patients find meaningful fulfillment during their last days. In many programs artists in areas such as metalwork, photography, pottery, drama, dance, and music work with patients interested in such self-expression. For instance, a music therapist would strive to comprehend the perceived meaning of illness and dying and the specific implication of music. During illness, the music therapist would access the music that in some way holds significance or value for the patient and family, as therapists encourage the use of preferred music in strategies as a means for coping (Magill, 2009).

Trained *volunteers* are "an absolutely essential part of the team-based approach to caring for the dying" (Claxton-Oldfield, Gosselin, & Claxton-Oldfield, 2009, p. 47). Many volunteers have experienced the death of someone close and find that this provides them with an opportunity to serve others. Some volunteers are retired health care professionals such as physicians or nurses. Others are nonprofessionals who are deeply interested in the needs of dying patients and their families. Specific reasons cited in a recent study for volunteering include: to help ease the pain of those living with a life-threatening illness, to help others cope with death and dying, to support the philosophy of hospice

George Dickinson

Dame Cicely Saunders, founder of the modern hospice movement, in her office at Saint Christopher's Hospice in London, England (with George Dickinson). With degrees in nursing, social work, and medicine, indeed Dr. Saunders was well qualified for her work! After retiring as director of St. Christopher's Hospice, she continued to play an active role with speaking engagements and fund-raisings until her death in 2005 at the age of 83.

care, and to gain personal experience with the death of a loved one (Planalp & Trost, 2009).

Before volunteers begin a hospice program's extensive training program, they are interviewed by the volunteer coordinator and may be asked to complete specially designed questionnaires that assess their feelings and sensitivity toward dying persons. In addition to the initial volunteer training program, every hospice program has regular in-service training to maintain and update the volunteers' skills. Some volunteers work in patient-care tasks such as providing transportation, sitting with a patient to free family members to get out of the house for a while, carrying equipment, or providing bereavement counseling to family members after the death of the patient.

The unpaid volunteer has the double benefit of being identified by the patient and family as being knowledgeable but not having the professional status that can create a social distance. By being a stranger who provides a listening ear, without emotional involvements or professional entanglements, the volunteer can support the patient and family members as can no other participant in the social network of dying—"stranger and friend" at the same time.

The volunteer also functions in an advocate role by acting on behalf of the dying patient and family. Sometimes patients and their family and friends are afraid to challenge or ask questions of physicians and other medical personnel. The volunteer, who has become a trusted friend and confidant, can often speak up for patients and their families and make their needs known to those responsible for their care. Michael Leming once served as the primary volunteer for a male patient who was undermedicated. When the patient complained to his nurse regarding his pain, he was told to "brave it out." Knowing the medical system, Leming was able to contact the appropriate individuals, who were able to have his pain medications reevaluated.

The final role served by volunteers is that of educator. Most individuals in our society have not had many personal experiences with death. The hospice volunteer can learn from each experience in working with dying patients and pass on insights that may be helpful to patients and families. The volunteer can help dying persons and their families and friends understand that the dying process is usually complex, stressful, and disordered. In addition, most patients and families have a strong need to have their feelings and experiences validated. Patients and families who have a difficult time understanding their feelings, emotions, and experiences during the dying process should be assured that they are quite normal.

Within the community at large, a number of influences either assist with patient care or help to make it possible. Family members or friends are urged to participate in the patient's care as much as possible. When family members cannot provide as much care as may be needed at certain times, hospice personnel will try to meet the patient's needs by exploring all possible options to do so. The patient–family support system is the most significant factor in the dying process for many patients, but elements of the system also include numerous close or distant relatives, friends, neighbors, members of local churches, and/or other civic groups.

Hospice programs depend on a high degree of community interest and support. Bringing this about requires a planned program of public information. The concept of hospice must be sold to the medical community and to members of the larger community. Specific activities require not only financial support (especially while the hospice program is developing), but also a willingness to testify before regulatory agencies about the granting of hospice accreditation, Medicare certification, and/or approval to begin offering services to the people of the area.

THE PATIENT–FAMILY AS THE UNIT OF CARE

One of the distinguishing features of the hospice concept of care is that, whenever possible, hospice enables patients to make decisions about how and where they want to live their lives. **Patient-centered care** is nonjudgmental, unconditional, and empowering.

One of the patients of the hospice program where Michael Leming volunteers provides an example of this philosophy of care. This patient had adult children in town but lived as a single person with his dog. He had lung cancer and desired to die at home alone. He was also a smoker and heavy alcohol drinker. The hospice program agreed to honor the patient's desires whenever it was possible. Therefore, a hospice nurse visited his house every four hours, and members of the police department checked in on him every hour from 10 p.m. to 7 a.m. The patient's pain was kept under control without sedation. A hospice volunteer (who also happened to be a licensed vocational nurse) visited the patient two to three times each day, and she, along with friends and family members, met the patient's requests for liquor and cigarettes. Visits from all members of the hospice team never lasted longer than 10 minutes. The patient died as he wanted—in his home, free from pain, and in control of his own care.

Although not every member of the hospice team, or the patient's family, would have chosen to die as this patient did, everyone respected his right to make decisions regarding his care. The hospice philosophy states that patients and their families have the right to participate in decisions concerning their care and that they should not be judged because their decisions are contrary to the beliefs of their caregivers.

Traditional health care has concentrated on the patient and ignored the family. Perhaps many health care workers would say, if given an opportunity to state their opinions confidentially, that they would prefer family members to stay away. Traditional ratios of physicians, nurses, social workers, or chaplains to those needing care have been based on an assumption that only the patients need attention. Although hospice staff, to be sure, are not given the responsibility to meet physical needs of family members, they do have tremendous concern for the social, psychological, and spiritual needs of the family.

Hospices challenge the health care system to provide an adequate ratio of professional staff members to patients. For example, in the state of Connecticut the public health code, in its regulations for hospice licensure, stipulates that at all hours of the day or night there must be at least one registered nurse for every six patients and at least one nursing staff member (licensed practical nurse or nurse's aide and a registered nurse) for every three patients.

Family care, however, involves much more than numbers of staff. It requires that health care workers know how to cope with the fears, worries, tears, and turmoil of family members and when to speak, when not to speak, and what to say.

Hospice helps dying individuals rather than helping individuals die. Though dying, the individual with a terminal illness is still living. We need to remember that the person is living and treat her/him accordingly.

It requires that they take time to listen to determine how they may be most helpful.

Hospice care is costly care due to the number of staff people involved. It challenges society as a whole to give priority to such care because of the right of the dying to quality of life. A harried nurse in a traditional hospital setting, trying to meet the needs of perhaps a floor of patients at night, is not being granted the time required to sit with a dying patient for whom night is especially fearful. Neither does this nurse have the time to be of assistance to husbands, wives, or children struggling with grief.

The interdisciplinary team supports the staff person within each discipline by enabling the resources of the entire team to come into play in meeting family needs. For example, a night-shift nurse who is asked questions relating to spiritual care might wish to give an answer at the time that the question is asked. This nurse will, however, also have the resources of the chaplain to determine the best methods to meet patient needs. In hospice care the patient–family unit is involved in decision making. This poses crucial questions to caregivers who may be accustomed to making decisions and having everyone go along with what they have decided.

The Cost of Hospice Care

Even though hospice care is personalized to meet the needs of each of its patients (involving an entire team of professional and volunteer caregivers), it is also very cost effective because more than 90 percent of hospice care hours are provided in patients' homes, thus substituting for more expensive multiple hospitalizations. Medicare rates for fiscal year 2011 were $147 per day for routine home care and

| PRACTICAL MATTERS | HOW TO LIVE WITH A LIFE-THREATENING ILLNESS |

With the help of hospice you can:

1. Talk about the illness. If it is cancer, call it cancer. You can't make life normal again by trying to hide what is wrong.
2. Accept death as a part of life. It is.
3. Consider each day as another day of life, a gift from God to be enjoyed as fully as possible.
4. Realize that life never is going to be perfect. It wasn't before, and it won't be now.
5. Pray, if you wish. It isn't a sign of weakness; it is your strength.
6. Learn to live with your illness instead of considering yourself dying from it. We are all dying in some manner.

7. Put your friends and relatives at ease. If you don't want pity, don't ask for it.
8. Make all practical arrangements for a funeral, will, etc., and make certain your family understands them.
9. Set new goals; realize your limitations. Sometimes the simple things of life become the most enjoyable.
10. Discuss your problems with your family, including your children if possible. After all, your problems are not individual ones.

From O. Kelly, *Make Each Day Count* newsletter.

$652 per day for general inpatient care to treat symptoms that are not manageable outside the hospital (Medicare Payment Policy, 2011). The primary source of payment for hospice services is Medicare, as Medicare certifies 93 percent of hospices in the United States (NHPCO, 2012). Medicare in 2010 served 84 percent of hospice patients, whereas private insurance covered 8 percent, the Medicaid benefit served 5 percent, and other payment sources (donations, grants, and private pay) covered 3 percent (NHPCO, 2012). As noted in Table 6.5, the overwhelming majority of hospice patients die in their place of residence.

Hospice care is insured by the Medicare Hospice Benefit, enacted in 1982, provided that the hospice is Medicare certified. A hospice program must undergo a rigorous evaluation of the services that it provides to become Medicare-certified and must agree to directly provide the following services: nursing care, medical social services, physician services, counseling, and volunteer services. In a recent study of Medicare patients in matched hospice and nonhospice cohorts, for all diseases except prostate cancer and stroke, mean cost was lower for patients who chose

TABLE 6.5   LOCATION OF HOSPICE PATIENTS AT DEATH (IN 2010)

Location of Deaths	Percentage
Patient's place of residence	66.7
Private residence	41.1
Nursing home	18.0
Residential facility	7.3
Hospice inpatient facility	21.9
Acute care hospital	11.4

Source: NHPCO Facts and Figures: Hospice Care in America, 2011 Edition (2012). National Hospice and Palliative Care Organization.

hospice over nonhospice programs (Pyenson, Conner, Fitch, & Kinzbrunner, 2004).

Hospices have always had the ability (and under Medicare rules, the obligation) to discharge some patients. Data from 2009 and 2010, as reported by the Medicare Payment Advisory Commission, reported that 20 percent of hospice patients are discharged alive each year (Span, 2014), with one-third of discharges initiated by patients themselves, and two-thirds by hospices.

The General Electric Company was the first major employer in the United States to provide a hospice benefit for its employees. Coverage for hospice is provided to more than 80 percent of employees in medium and large businesses. Furthermore, the majority of private insurance companies offer a comprehensive hospice care benefit plan, and major medical insurance policies, provided through insurance companies and offered to employees as part of a benefit package, also underwrite hospice coverage in most instances. However, many hospice programs still rely on grants, donations, and memorials to meet the needs of their patients and families who are not covered by Medicare, Medicaid, and insurance reimbursements (NHPCO, 2012).

Hospice leaders hope to make it possible for any person of any age suffering from a terminal illness to be eligible for the coverage of costs related to hospice care. They are also firm in their conviction that such care saves considerable money in the long run. Many patients currently hospitalized would not need hospitalization if hospice services were available for patients and families. A basic societal question is whether as Americans we believe enough in quality of life for the dying to be willing to make it possible.

Though hospice care requires a higher ratio of staff to patients than that usually provided in health care programs, it should be remembered that the cost is nonetheless lower than that for other forms of care. Because the majority of

WORDS OF WISDOM | ANIMAL HOSPICES

The emergence of a pet hospice movement is an assimilation of human hospice. It also signifies more careful attention to the manner of pets' deaths and treatment of their pain. Like human hospices, animal hospice care aims to provide a terminally ill animal with comfort and palliative care so that the animal can live out its final days with as much quality of life as is possible. The focus of attention shifts from cure to care, with death being accepted as the inevitable outcome. Animal hospice can be viewed as an alternative to premature euthanasia. Animal hospice is largely in the home, with occasional visits to vets if necessary. There are a few free-standing hospices, however, which provide care for elderly, ill, and injured animals. Hospice care

is not free and is typically more expensive than immediate euthanasia. Thus, money may be a barrier to the widespread use of hospice for animals. Unless the animal is dropped off at a sanctuary like BrightHaven in Santa Rosa, California, which relies entirely on donations and is for senior and disabled animals, or access is available to Colorado State University's hospice center, pay for hospice care is inevitable. Certainly hospice care for animals offers a gentler way to approach death and offers the possibility of a good death for animal companions.

Taken from J. Pierce's *The last walk: Reflections on our pets at the end of their lives* (2012). Chicago, IL: University of Chicago Press.

hospice patients are able to remain at home for much, if not all, of their illness, the costs of patient home care, when compared with any forms of inpatient care, are proportionately low. Because of the level of services provided, hospice **inpatient care,** however, will normally be higher than that provided in a nursing home, but lower than inpatient care in a general hospital setting.

Public Attitudes

The hospice movement began at a time when public consciousness of dying and death issues had reached an all-time high. It afforded an opportunity to do something tangible for other people, and many took advantage of the chance to volunteer for an active role. At the same time, increasing public awareness of dying and death gave rise to considerable publicity in the media. This helped provide public support when regulatory agencies held hearings on granting approval for hospice services.

Public attitudes toward care of terminally ill persons and their families will play an increasingly important role in the future. These attitudes will help to determine whether health care professionals will, in fact, broaden the scope of care to encompass the family and strengthen their skills in dealing with dying patients. Patients and families are, after all, consumers. In this age of consumer awareness it is becoming increasingly evident that those who purchase services can control to some extent the types of services available. Health care professionals are becoming increasingly responsive to the desires of their clients. The most important factors causing caregivers to seek improvement of skills will be the desires of those they serve. At the same time, especially in areas of competition among hospitals, consumer awareness will play an important part in encouraging such institutions to humanize the care that they give.

Many physicians, nurses, social workers, clergy, and other personnel at hospitals and nursing homes have heard about hospice care and have taken the initiative to secure specialized training and to incorporate the hospice philosophy into routine treatment of their patients. When any popular movement arises, an immediate question is whether it will become institutionalized to such an extent that the original spirit will be lost as it adjusts to the reality of regulation, control, and payments of costs. The hospice movement is currently at that juncture. There is every cause for hope that one of two things will happen: Either hospices will continue to provide the specialized care for dying patients, or the health care system itself will change to incorporate many of the improvements represented by the hospice movement. Certainly palliative care throughout the medical system would be a win–win for everyone.

Hospice programs in the United States seem to be giving that which the majority of Americans desire—an opportunity to die at home under the care of family. Yet, sometimes people are reluctant to use hospice programs. Many may not be familiar with the hospice philosophy or even be aware of the existence of hospice in their community. Caregivers might be reluctant to discuss hospice with the family for fear the topic might upset them. Recent studies on caregiver–physician communication revealed infrequent discussions of hospice by physicians

(Yapp, 2012). Yet, physicians in South Carolina (Sanders, Burkett, Dickinson, & Tournier, 2004) concluded that medical doctors overwhelmingly initiate referrals to hospice, rather than the patient or family. Entering a hospice program may be viewed as giving up. Such misinformation might contribute to one's failing to enter a hospice program when indeed the individual could benefit from hospice. Entering a hospice program very late in the illness denies the patient counseling services and social support that hospice can provide to increase the quality of the patient's remaining time.

As noted earlier in this chapter, hospice participants are not equally distributed by race, as is noted by the underuse by African Americans and Latinos. One suggested reason for this underuse by African Americans includes more of a sanctity-of-life philosophy. Going into hospice would then be viewed as a preference for nonaggressive treatment. The hospice focus on quality rather than quantity of life goes against the common African American belief in longevity and the redemptive nature of suffering. Other factors include lack of knowledge and trust (given the Tuskegee experiment, for example), spiritual beliefs, and lack of ethnic minority employees in hospice (Conner, 2012). Cecile Yancu and colleagues (2010) suggest that the hospice philosophy represents core values of the dominant white culture such as independence, yet the African American culture embraces caring for and being cared for by a family member. Among factors that might influence an underuse of hospice care by Latinos are their beliefs about death and end-of-life care, language difficulties, lower referral rates by physicians, the hospice caregiver requirements, and experiences in the health care system prior to hospice admission (Colon, 2012). While a lack of knowledge regarding hospice is common to the general public, it may be particularly so to African Americans and Latinos, note Yancu and colleagues.

EVALUATION OF HOSPICE PROGRAMS

There is an increasing emphasis on measuring the outcomes of hospice care. An end-result outcome is something experienced by the patient, the most important of outcomes. A health care provider wants to find out if the outcome was related to the interventions delivered. Stephen Connor (2009) points out the efforts and shortfalls of trying to evaluate hospice programs. He states that one way of measuring outcomes is to ask the patient to rate how she or he is feeling. With a median length of service of hospice patients of only 19.7 days in 2010, obtaining ratings is challenging, because communication is generally diminished as death approaches (NHPCO, 2012). Efforts to measure psychosocial and spiritual care outcomes have been difficult to do, as there are too many variables to understand the information. Also, population data vary widely, and some family members may hesitate to open up about their feelings and anxieties. Connor notes that there are currently no empirical measures of spiritual growth in use. Greater use of qualitative indices rather than quantitative measures, which are often used, might prove to be more useful. Data on symptom control would help to show how good hospices are. Such data could then be compared with data on care provided outside the hospice venue.

LISTENING TO THE VOICES | GRIT AND GRACE: MILITARY MAN BATTLES HEALTH ISSUES

"A case worker once asked, 'To what do you attribute your ability to recover?' That's easy: military mentality," Cliff Davis said.

Davis' pancreatic cancer has spread through his blood to his lungs. His hospice nurse admires his fortitude but wishes he could admit that the intense pain is unnecessary.

His first stroke in 1996 was small, but it was quickly followed by a massive one that knocked out his legs, voice, arms and most of one side of his body. He fought back, though, only to suffer two more strokes.

Today, he is in a wheelchair relying on extra puffs of oxygen provided through plastic tubes. His new pain medication is working, and it doesn't leave him groggy.

Davis, 62, is not an ordinary man. Before these most recent health problems, he spent more than 23 years in the military: four tours in Vietnam, seven years in Special Ops, "visiting countries."

He's been shot twice, once in the leg, once in the arm, and stabbed in the back with a bayonet. He went down in a chopper that lost its tail rotor to small-arms fire in Vietnam, crushing vertebrae. He was the only survivor of a crew of five. He's received four Purple Hearts. Four.

As a combat medic, he dodged bullets to get to wounded soldiers. When he was transferred from the Army to the Air Force, joining an air rescue and recovery team, his missions were to fly deep into North Vietnam, Cambodia and Laos to retrieve downed pilots.

In Charleston, Davis would become a physician's assistant. He would become a master parachute rigger, a dive master, an underwater photographer, a small-arms expert, a burn specialist.

He would join nine Masonic orders. He would clean up the mess at Jonestown, Guyana, where Jim Jones and more than 900 others committed "revolutionary suicide."

He would have two daughters, divorce and eventually remarry, in 1986, after meeting Sherri, after "walking on air" during the first Thanksgiving holiday they spent together. He would go fishing three or four times a week. He loves fishing.

Later, Davis would volunteer at the VA hospital and help design a special line of utilitarian clothing for disabled people.

"It's amazing how, in his health predicament, he's thinking about others," Sherri Davis said.

Today, he tires easily, but he is clear-eyed, dogged and among people who love him.

"It's that military mentality that keeps him going, going, going," family friend Theresa Winger said.

A. Parker. (2009, October 11). Grit and grace: Military man battles health issues. *The Post and Courier*, Charleston, SC.

Cliff Davis, diagnosed with pancreatic cancer, has faced his illness, and the others that preceded it, with determination. A decorated military man, he attributes his persistence, regular volunteerism, and selfless dedication to others to his "military mentality." Mr. Davis died on December 11, 2009.

In conclusion, despite various concerns cited above, hospice today appears to have established itself fairly well within the overall health care establishment, stated famed American thanatologist Robert Kastenbaum (2009). Many health care professionals and administrators have become persuaded that hospice does what it claims to do and does so in a rational, accountable, and cost-effective manner, noted Kastenbaum.

CONCLUSION

A diagnosis of heart disease and cancer today does not necessarily mean a death sentence. With medical advancements in recent years, many individuals in our midst are "living with dying." Though they may have a terminal illness, they may live many years with this disease. We have made progress in being more open about the "C" word (cancer). Such illnesses are being viewed more as attacking individuals in all walks of life. Medical treatment options are more numerous, perhaps offering more encouragement to patients. The ability to transplant numerous tissues and organs also contributes to the options for one with a terminal diagnosis.

As the American way of life has changed from a primary group orientation to a more secondary, impersonalized orientation, so has dying shifted from the home to the hospital or nursing home setting—away from kin and friends to a bureaucratized setting. The birth of the hospice movement in the United States might be considered a countermovement to this shift. As we seek out primary group relations in our secondary-oriented society, we seek to die in the setting of a familiar home rather than in the sterile environment of a hospital. Perhaps we are evidencing a return to a concern for each other—a dignity to dying may be on the horizon.

Hospice is a return to showing care and compassion. It is a revival of neighbors helping neighbors—a concept so often lost in our urbanized society. Hospice consists of professionals literally going the extra mile and coming to one's home when needed—medical personnel actually making house calls. Hospice, for example, encourages children younger than age 14 to be present with the terminally ill person rather than making them wait in the hospital lobby. Hospice is a grassroots movement springing up in small communities, as well as in larger urban settings, to provide better health care. To paraphrase the words of Robert Kavanaugh (1972), the hospice concept of care helps us to unearth, face, understand, and accept our true feelings about death and provides us with the opportunity to live joyfully and die as we choose. In short, hospice is a movement that transforms our awkwardness in death situations into a celebration of life.

With federal money now covering most hospice expenses and with rigid government requirements for approval of hospice programs, it is important that every effort be made to prevent hospice programs from being strangled by the bureaucracy from which they receive financial assistance. Hospice programs must also continue to make the patient–family unit the focus of their care and treat these clients in a nonjudgmental and unconditional manner, thus empowering them as autonomous human beings.

SUMMARY

1. Hospice is a specialized health care program that serves patients with illnesses such as cancer during the last days of their lives.
2. Palliative care concerns pain control but also is sensitive to the social, psychological, spiritual, and cultural aspects of the dying person's needs.
3. Pain perception varies with cultural background.
4. The hospice team is made up of various health care professionals, the primary care-giver, and volunteers.
5. The hospice movement supports the inherent right of every human being to live as fully and completely as possible up to the moment of death.
6. Though the majority of Americans today die in hospitals and nursing homes, the majority of individuals enrolled in hospice programs die at home.
7. Much of the cost of hospice care is covered by Medicare.
8. Treatment options today are more numerous than in the past, giving patients hope that if the first one does not work, there is now something else to try.
9. Persons with terminal illnesses have treatment options other than surgery, chemotherapy, and radiation. These alternatives include immunotherapy, homeopathic medicine, chiropractic medicine, acupuncture, faith healing, and folk healing.
10. Pain *can* be controlled, but political issues get in the way, preventing widespread use of morphine in the United States. Legalizing marijuana for medical purposes is also controversial, yet marijuana can be useful in preventing nausea caused by certain cancer treatments.

Discussion Questions

1. What is hospice care? How does it differ from the treatment given by most acute-care hospitals? Identify the major functions of a hospice program.
2. Discuss issues related to the family as the unit of care in hospice programs. How do hospices try to achieve quality of life for each of the patients they serve?
3. Discuss the pros and cons of legalizing marijuana for medical purposes.
4. Pain control is not a goal of all dying patients. Discuss the cultural ramifications of controlling pain.
5. Do you feel that bereavement care should be offered to the families of the terminally ill even after the individual has died? Justify your answer in terms of medical, emotional, and financial considerations.
6. If you were terminally ill, would you consider entering a hospice program? Explain your answer and refer to specific reasons such as cost, family burden, and imminent death.
7. Discuss the special opportunities and challenges in providing hospice services to patients with AIDS.
8. Discuss, from both a societal and a personal point of view, some of the problems affecting human organ transplantations.
9. Should society make eligible for a transplant only those persons who can afford it, or should the potential societal contribution of the individual be taken into account?
10. Discuss some of the problems involved with dying individuals such as the loss of physical and mental functions.
11. Have you ever been around someone who had Alzheimer's disease? If so, discuss your reactions. How can you relate to a patient with Alzheimer's disease, if he or she is in the latter stages of the illness?

GLOSSARY

Acupuncture: Involves the use of needles to redirect the flow of energy within the body to treat illness.

Alzheimer's Disease: A chronic, degenerative, dementing illness of the central nervous system.

Cancer: Refers to a group of diseases that are characterized by an uncontrolled growth and spread of abnormal cells.

Chemotherapy: Treatment of cancer with certain chemicals that attack and destroy certain types of cancer cells.

Chiropractic Medicine: A therapeutic approach to healing that involves a hands-on manipulation of bones in the spinal column to relieve pressure on nerves.

Evidence-Based Practice: The conscious, explicit, and judicious use of current evidence in making decisions about the care of individual patients.

Faith Healing: Uses the power of suggestion, prayer, and faith in God to promote healing.

Folk Healing: Primarily uses "folk remedies" passed down orally from generation to generation and common ingredients found with a particular group of people to treat illness.

Homeopathic Medicine: Involves the use of natural drugs to treat patients.

Hospice: A specialized health care program that serves patients with life-threatening illnesses, such as cancer, during the last days of their lives.

Hospice Movement: A response to fears related to the dying process and the institutionalized ways in which death is typically handled in institutional settings.

Hospice Team: An interdisciplinary team of professionals and volunteers who work together to contribute their expertise to the quality of patient care.

Immunotherapy: Treatment by stimulation of the body's own immune system.

Inpatient Care: The type of institutionalized care that is required, for example, as an illness progresses and that may be provided in a hospice facility.

Metastases: The spread and invasion of cancer cells to other organs or tissues. When this happens, the disease is said to have metastasized.

Palliative Care: Care designed to give the patient as pain-free a condition as possible. In addition to physical needs, the patient's social, psychological, cultural, and spiritual needs are considered.

Palliative Daycare: A facility, not unlike child daycare or geriatric daycare, for patients with special needs—sort of like hospice on a day-by-day basis.

Patient-Centered Care: A distinguishing feature of the hospice approach, which enables patients to make decisions about how and where they want to live their lives. Patient-centered care is nonjudgmental, unconditional, and empowering.

PRN, or *pro re nata*: A traditional medical approach that means that medication is to be given "as the situation demands." In practice, it means that patients must first hurt and ask for relief before pain management can be administered.

Radiation Therapy: Treatment using x-rays to destroy cancerous tissues.

SUGGESTED READINGS

Armstrong-Dailey, A., & Zarbock, S. (2008). *Hospice care for children.* New York: Oxford University Press. This resource emphasizes practical topics and covers the entire range of issues related to hospice care from psychological stress to pain and symptom management.

Byock, I. (2012). *The best care possible: A physician's quest to transform care through the end of life.* New York: Penguin Group. A well-known palliative care physician shows how doctors and nurses can shape the way families experience loss.

Buchwald, A. (2006). *Too soon to say goodbye.* Random House: New York. Pulitzer Prize recipient Art Buchwald shares his life's story while a patient in hospice. He writes with frankness, dignity, and humor, as he lives life to the fullest, though dying.

Callone, P. R., Kudlacek, C., Vasiloff, B. C., Manternach, J., & Burnback, R. A. (2006). *A caregiver's guide to Alzheimer's disease: 300 tips for making life easier.* New York: CaringConcepts. The book gives answers to caregivers' and family members' frequently asked questions, from settling legal issues to appropriate ways to tell children about what is happening to grandma or grandpa.

Chapkis, W., & Webb, R. (2008). *Dying to get high: marijuana as medicine*. New York: New York University Press. Two sociologists trace the history of the use of marijuana as medicine in the United States.

Clark, D., Small, N., Wright, M., Winslow, M., & Hughes, N. (2005). *A bit of heaven for the few? An oral history of the modern hospice movement in the United Kingdom*. Lancaster, UK: Observatory Publications. Explores how hospice ideas were quickly taken up in many places, how provision came to be expanded to numerous areas, and the challenges that remain if all in need are to receive the care they require.

Connor, S. R. (2009). *Hospice and palliative care: The essential guide*. New York: Routledge. This book addresses the fundamentals of hospice and palliative care, including the goals of community involvement, symptom management, the business side of hospice, and the manner in which death and bereavement are addressed.

Fine, P. G. (2008). *The diagnosis and treatment of breakthrough pain*. New York: Oxford University Press. This book is ideal for palliative care physicians, pain management specialists, and oncologists, as well as for primary care physicians and internists on the frontlines of care.

Finlay, G. B. (Ed.) (2007). *Dying, bereavement and the healing arts*. Philadelphia, PA: Jessica Kingsley Publishers. This edited volume describes several successful programs pioneered by artists, writers, nurses, musicians, therapists, social workers, and chaplains in palliative care settings.

Fuss, D. (2013). *Dying modern: a meditation on elegy*. Durham, NC: Duke University Press. A Princeton professor of English, Diana Fuss focuses mainly on American and British poetry of the past two centuries and explores modern poetry's fascination with pre- and postmortem speech, pondering the literary desire to make death speak in the face of its cultural silencing.

Goldman, L. (2009). *Great answers to difficult questions about death: What children need to know*. Philadelphia, PA: Jessica Kingsley Publishers. This book explores children's thoughts and feelings on the subject of death and provides parents and others with guidance on how to respond to difficult questions.

Hartley, N., & Payne, M. (Eds). (2008). *The creative arts in palliative care*. Philadelphia, PA: Jessica Kingsley Publishers. Focuses on designing objectives for the creative arts in palliative care and demonstrates the theory and principles in practice, with detailed case studies.

Heyse-Moore, L. (2008). *Speaking of dying: A practical guide to using counseling skills in palliative care*. Philadelphia, PA: Jessica Kingsley Publishers. A book written to give guidance to caregivers with an emphasis on communication skills.

McMullin, J., & Weiner, D. (2009). *Confronting cancer: Metaphors, advocacy, and anthropology*. Santa Fe, NM: SAR Press. Anthropologists examine the experiences of individuals confronting cancer and reveal the social context in which prevention and treatment may succeed or fail.

Watson, M., Lucas, C., Hoy, A., & Wells, J. (2009). *Oxford handbook of palliative care*. New York: Oxford University Press. Covers all aspects of palliative care in a concise and succinct format.

I left medical school ... with no idea how to break bad news or manage the sick patients I was about to be confronted with.
—**Phil Hammond, MD**

It is a great thing to die in your own bed, though it is better still to die in your boots.
—**George Orwell**

DYING IN THE AMERICAN HEALTH CARE SYSTEM

CHAPTER 7

George Dickinson

233

Within the American health care system, most individuals believe that the sole responsibility for their terminally ill family members lies with the physician. Yet, Chinese individuals believe that the family *and* the physician are responsible for the patient's treatment, though the physician is viewed as an authority figure (Tanner, 1995). The American way of life emphasizes individual-centeredness of autonomy, assertiveness, and independence within a youth-oriented perspective, whereas Chinese values display generational continuity, family solidarity, respect for elders, and situation-centeredness—individual impulses are subordinate to the will of the family as a group, observed Jane Tanner (1995). In China, when the patient, medical staff, and family members know that the patient is dying, they tend to practice mutual pretense (discussed in Chapter 5) not unlike in U.S. hospitals. In the Chinese culture, mutual pretense is the predominant context in some hospitals, noted D. Lin (1992).

Often traditional Chinese medicine (acupuncture and herbal treatments) is used in conjunction with Western medicine, and all Chinese physicians today receive some training in both. Because of the expense of Chinese hospitals, family members may do considerable labor such as washing floors, bringing in food, and taking the patient to laboratories or for x-rays. Such assistance may be supported by giving the family member time off from work (with pay), according to Schneider (1993).

Whereas dying in the United States primarily occurs in an institutional setting, in general, dying at home has been the traditional arrangement for most Chinese families. Dying at home allows more latitude for the Chinese to handle dying in culturally appropriate ways (Hsu, O'Connor, & Lee, 2009). Also, there is a Chinese belief that people who die away from home will become lonely spirits, ghosts in the wild. In addition, in dying at home, the dying individual is often integrated into the round of daily life and is involved in living while dying. The Chinese feel that a symbolic completion of their shared relationship and solidarity results when the person dies at home. Thus, dying varies significantly in different cultures, yet there are similarities. Let's now take a more detailed look into dying within the American health care system.

THE MEDICAL MODEL APPROACH TO DYING

The **medical model** in the United States is basically the idea that, when sick, we go to a physician to be made well. If we are terminally ill, however, the doctor cannot make us well. Thus, dying does not fit the medical model of being made well—the patient living with a terminal illness is often avoided or negatively sanctioned. Western medicine focuses on living and keeping individuals alive, or assisting them to die. This model ignores the fact that illness may be part of the dying process. When one treats the illness, one may be trying to cure the patient who is living with a disease for which there is no cure. With people living longer today, one is much more likely to live with a chronic disease that eventually leads to death. Modern medicine is more oriented toward cure than prolonged illness.

By ignoring the terminal nature of an illness within the medical subculture and failing to accept the fact that dying is normal within the life cycle, the medical model views the dying patient as a deviant. When one is labeled "deviant," an entire interactional framework is created within which dying takes place where "normals" respond to the deviant, dying patient. Regardless of whether or not the individual is responsible for the deviant label, the stigmatized individual is still discredited and treated with less respect than normal persons.

Talcott Parsons, an American sociologist and one of the best-known structural functionalists and one of the early medical sociologists (1951), observed that the physician does not view the patient in a holistic sense, but rather views her or him as someone with a particular disease or ailment that needs to be treated. Thus, the physician's goal is instrumental—use medical expertise to cure the patient's disease and return her or him to health and normal social functioning. Illness is a rejection of the societal value of health, noted Parsons (1958). A structural-functional perspective (as discussed in Chapter 1) assumes that most individuals are healthy most of the time, and those who are ill must be "made well" by the medical system, so that society can function smoothly.

DYING AS DEVIANCE IN THE MEDICAL SETTING

As we have discussed, the medical profession's attitudes toward patients, and in particular dying patients, are functional to the reinforcement of the view of the "physician as healer" and functional with respect to maintenance of order in the medical subculture. The dying patient is a deviant in the medical subculture because death poses a threat to the image of the "physician as healer." George Dickinson recalls submitting a manuscript on physicians' attitudes toward terminally ill patients (and referring to them as deviants within the medical subculture) to a medical journal a few years ago and the editor writing back, stating that dying patients cannot be called "deviants" (not a good word to use within a medical setting!).

In observing patients within a hospital setting, Daniel Chambliss (1996) argued that the patient is first separated from home, work, and even from biography. For most, illness (terminal or otherwise) is abnormal in that it is an unusual breakdown in a normally healthy daily routine. The hospital patient is dressed in a gown. He or she becomes an object of looking and talking. Hospital life for patients is an endless round of being looked at, listened to, touched, and poked and prodded. This is especially true in a medical school setting with medical students and residents shadowing a physician.

The patient's rights to privacy are forfeited. Chambliss (1996) suggested that medicine needs the object it can treat—defined problem, curable disease, and the grateful patient who believes in the doctor, the nurses, and the hospital. Patients not fitting this sort of object (who have a chronic or incurable illness or who are noncompliant) challenge more than medicine's effectiveness. They challenge medicine's entire worldview. They are indeed deviant to the medical system.

Death also creates embarrassing and emotionally upsetting disruptions in the scientific objectivity of the medical social system. Thus, the disruption caused by death in the medical social system, if not controlled, could lead to a great deal of conflict.

LABELING THEORY   A major school of thought explaining deviance is **labeling theory**. The perspective of this school of thought does not focus on the act or the actor, but rather on the audience observing. Erving Goffman (1963) described the stigmatized person as one who is reduced in the minds of others from a "whole and usual" person to a "tainted and discounted" one. Therefore, the key to the identification of deviance is found in the audiences' labeling the individual as deviant. Thus, in analyzing the dying patient as deviant in the medical subculture, the medical audiences who interact with and participate in the labeling of the patient must be examined.

With such labeling, an entire interactional framework is created within which the normals relate to the deviant. Given the idea that all dying patients are deviant, imagine the stigma attached to persons living with HIV and AIDS. Individuals living with AIDS have a double deviant label—*that of dying and having the HIV virus*. Many HIV/AIDS-related issues are cast as moral ones (Smith, 1996). Such characterizations occur because HIV infection is spread through unpopular and/or illegal behaviors, such as homosexual relations or intravenous drug use. Because HIV infection is spread through sex and can be fatal, it evokes certain base-level fears in people. Those affected by the disease are subjected to moral pronouncements and may be discriminated against, noted J. M. Smith (1996). Because discrimination disenfranchises people from meaningful participation in society, groups typically discriminated against may not have any incentive to comply with social norms (Becker, 1963).

A former student of George Dickinson's, who had AIDS, desired volunteer work at a funeral home because of his fascination with death. His request was denied. He later, rather jokingly, said that he did not know if the refusal came because he was gay or because he had AIDS. How individuals become infected with HIV is frequently of more concern to some individuals in the evaluation of positive adaptation than is the stage or progress of the ailment. Thus, whether the patient acquired AIDS through sexual activity, intravenous drug use, or through a blood transfusion, the stigma of deviance tends to apply.

According to pioneer medical sociologist Elliot Freidson (1972), when a person is labeled deviant, the stigma interferes with normal interaction. Although other individuals may not hold the deviant person responsible for her or his stigma, they are nonetheless usually bothered or repulsed by it. Therefore, the assumption can be made that the deviant person elicits certain aversive attitudes from the audience with whom interaction occurs. These aversive attitudes may be of sufficient strength to elicit attempts to manage them and to decrease aversion through avoidance behavior.

DEVIANCE RESULTS IN PUNISHMENT   The primary reaction to deviance of any type is punishment of some sort. French sociologist Emile Durkheim (1961) observed that the primary purpose of punishment is not to punish the deviant himself or herself, but rather to affirm in the offense the rule that the offense would deny. Thus, one could suggest that illness is a rejection of the societal value of health, or that the terminally ill patient to the physician is the antithesis of healing and getting well as taught in medical school.

For the patient with AIDS, **homophobia** may be present in the community. The patient may feel extreme guilt or shame for being ill or may blame himself or herself for "getting what was deserved." On the other hand, numerous individuals are working today to change the stigma of having AIDS. AIDS stigma and discrimination exist worldwide, although they manifest themselves differently across countries, communities, religious groups, and individuals. They occur alongside other forms of stigma and discrimination (AVERT, 2014). As of 2012, UNAIDS reported that 61 percent of countries now have some form of legislation in place to protect individuals living with HIV from discrimination (UNAIDS, 2013). However, no policy or law alone can combat HIV/AIDS-related discrimination. Stigma and discrimination will continue as long as societies as a whole have a poor understanding of HIV and AIDS and of the pain and suffering caused by negative attitudes and discriminatory practices. The prejudice that lies at the core of the HIV/AIDS-related discrimination needs to be addressed at the community and national levels, with AIDS education playing a crucial role (AVERT, 2014). Much headway has been made in the treatment of HIV, the virus that causes AIDS, with far fewer individuals dying of AIDS today than 20 years ago, when the film *Philadelphia* was in movie theaters and elevating the national conversation about the disease and those living with it. Yet, HIV is still spreading, and prevention and treatment efforts continue (Gordon, 2013). Often the worst part of living with HIV, however, is the stigma and discrimination experienced. Hopefully, through better education, the negative stigma will diminish and individuals can be seen as individuals, not lumped together in a group and not discriminated against because they may be "different" from the majority.

The dying person can seldom assume normal role functioning, although he or she views the illness as undesirable and has tried to cooperate with medical personnel to get well. The dying person, therefore, is permanently cast into a deviant role due to the inability to respond to treatment and get well.

NORMALIZATION OF DYING IN THE MEDICAL SETTING

The process of dying is no less normal than the process of living; they coexist in the same world. Dying must be considered a normal state in and of itself—*Death: The Final Stage of Growth*, as Elisabeth Kübler-Ross entitled one of her books. It is the world of patienthood that is artificial. Modern medicine, with its technology, has created industrial complexes called hospitals, where the environment is so artificial that both patient and staff must work to normalize the dying process. **Normalization of dying** refers to maintaining roles, relationships, and identity, though dying. In dying, both for the patient and significant others, living a life as normal as possible is a real challenge. Roles and relationships may have to be modified. Bodily changes in the patient are in order and must be accepted. Self-image of the patient may be affected. Others may tend to be overly protective of one with a terminal illness, noted Kenneth Doka (1993).

Normalization of the life of the dying person is not the same as normalization of the life of a patient, observed van Eys (1988). To be dying is normal, and normality is determined by the patient. In attempting normalization for the patient, the staff tries to keep the patient from slipping into an exaggerated state of patienthood. In normalization for the dying, the dying person must teach the staff that there is a normal process going on.

Hospitals are places in which dying could be allowed, noted van Eys. There are many things patients need at the time of their dying to make their time left more tolerable. There is room in hospitals to allow for the dying. Normalization is not really based on ethnic preferences, social patterns, and familial behavior. Normalization targets self-esteem, goal orientation, and abatement of loneliness. It generates a community that allows patients to continue to have full participation in human commerce.

The dying person redefines her or his community. The dying person is the high priest of his or her own temple, said van Eys. It is as if life has a holy of holies where only those initiated into that inner sanctum are allowed. It transcends the human construct of the hospital and the secular state of patienthood. Hospitals will be able to accommodate dying persons only if the staff recognizes this transcendent element of the dying process. The current view of the human body as a machine that can be fixed without the spirit leaving the driver's seat makes it difficult to accommodate dying.

While trying to normalize dying, as living is normalized, hospitalized patients may find themselves at a loss as to how they can handle events surrounding their illness (Schroepfer, 1999). In the depersonalized setting of a hospital, control over tasks that are normally performed by the patients themselves is relinquished to health care professionals. A power differential is also created between the patient and physician because doctors possess the information, knowledge, and skills required for dealing with patients, thus again affecting the patient's sense of control, stated Schroepfer.

DYING IN A TECHNOLOGICAL SOCIETY

A cultural value that makes it difficult to decide to let nature take its course and allow death to happen is the **technological imperative** (Freund & McGuire, 1999), a prevalent idea in most Western societies, especially the United States. Such a concept urges that, if we have the technological capability to do something, we should do it. This idea implies that action in the form of the use of an available technology is always preferable to inaction. However, medical ethicist Daniel Callahan (2000) uses the notion of technological brinkmanship to capture the series of practical and moral dilemmas that have developed around the contemporary dying process; because medical technology can prolong lives beyond the point of all sense, what, then, is the option to avoid such brinkmanship? Callahan wonders.

Routine social processes of new technologies or experiments soon are considered standard in clinical practice. For example, by the early 1970s, the mechanical ventilator or breathing machine was standard equipment in medical centers and community hospitals in the United States. The technology was available and thus should be used. The breathing machine contributed to the creation of the intensive care unit and immediately was considered essential technology there (Kaufman, 2000).

Progress, as enabled by science and manifested in technology, is an enduring feature and primary value in modern medicine, observes Sharon Kaufman (2000). The most powerful form of progress in contemporary clinical practice is the techno-logical imperative, criticized for decades for being a means without an end, an activity carried out in the absence of reflexive guidance. The technological imperative shapes the goals-on-the-ground of medical practice and frames its gaze, notes Kaufman. Yet clinical medicine has ends—to save lives and manage the course of disease. That moral imperative also guides and rationalizes medicine's specific practices, especially reliance on and commitment to uses of technology. These means and ends exist as basic assumptions and give meaning both to medical practitioners' actions and public perceptions of progress and "the good" in medical practice.

The technological imperative is embedded in institutional responses to health crises, observed P. E. S. Freund and M. B. McGuire (1999). Institutional decisions to invest in technology are often driven by financial incentive—government subsidies, for example. Most hospitals have created several high-technology wards (e.g., coronary and neonatal intensive care units) equipped with high-tech equipment. Such expensive high-tech equipment requires costly specialized support workers and thus has caused the costs of health care to escalate significantly.

Sociologist and physician Nicholas Christakis (1999) stated that the explosive growth in both the amount and sophistication of technology deployed by physicians to combat disease has given them unrealistic expectations about their own abilities. Thus, physicians tend to regard death as a personal failure. Christakis said that it is not surprising that the technological forces arrayed to treat serious illness, in an effort to control death by postponing it, have in recent years come to be focused on controlling death by managing and predicting it. Such management is expressed in the increasing technicalization of euthanasia, for example. With various

Medicine in the 21st century does indeed involve advanced technology. A surgical facility today resembles a *Star Wars* movie set, unlike an operating room in the early 20th century.

medical technologies emerging today, the relevance of prognosis for physicians has increased. For example, an obstetrical ultrasound may reveal information about the internal anatomy of a baby that would not otherwise be known until it is born. Also, the advent of genetic testing technology provides yet another important new arena for prognostication for physicians. Analysis of a person's genes may reveal relevant medical outcomes years or decades in advance of manifestation.

In addition to technological advances in medicine, there has been a change in the place where death occurs. Early in the 20th century, death took place in the home, where the dying person was surrounded by family and friends. In the 21st century, death more likely takes place "offstage" in institutions—hospitals, nursing homes, and other extended-care facilities. There is also a widespread belief that modern medical technology and know-how can save us from physical ills. The belief seems to be that death need not occur in the foreseeable future and death is a reversible event, such as when actors on the stage are "killed" but reappear in subsequent productions. The technology of cryonics, discussed earlier in Chapter 2, illustrates this point well.

When people began to question the price society seemed to be paying for medical progress (hopelessly ill or brain-damaged patients being kept alive by respirators, feeding tubes, and pharmacologic maneuvers that could not be halted without court orders), living wills were created in 1967, as discussed in Chapter 12 (Grady, 2000b). The documents were an effort (the success of which might be questioned) to wrench dying from technology's grip. Nonetheless, the 21st century is a time of high-tech "everything"; thus, modern medicine is in vogue with its high-tech emphasis, whatever the consequences might be.

THE ENVIRONMENT OF THE DYING PERSON

Choosing a place to die, if one has such a luxury, is not a difficult choice for most Americans. The overwhelming majority of U.S. citizens have traditionally preferred to die at home, as noted in Chapter 1. Let's take a brief look at the environs in which one will probably die: hospital, home, nursing home, and hospice inpatient care.

HOSPITAL   More than 75 percent of adults are hospitalized at some point during the year before they die, and almost 60 percent of adults see a physician at least five times during the last year of their life (Christakis, 1999). In the United States, one-third of Medicare spending goes to the final year of life, and one-third of that goes to the final month of life (Jacoby, 2012). The amount spent in the intensive care unit is climbing. Between 2007 and 2010, Medicare spending on patients in the last two years of life jumped 13 percent, to nearly $70,000 per patient (Gorenstein, 2013). Thus, hospitals and doctors become frequent visiting places for persons toward the end of life.

From a societal perspective, as Talcott Parsons and Renee Fox (1952) suggested, hospitalization both protects the family from many of the disruptive effects of caring for the sick in the home and operates as a means of guiding the sick and injured into medically supervised institutions where their problems are less disruptive for society as a whole. A hospital is a bureaucracy characterized by specialization, rationalization, development of power through expert and specialized knowledge (knowledge secretly protected), and depersonalization. As German sociologist Max Weber (1968) observed, the more the nature of bureaucracy is developed, the more the bureaucracy is dehumanized. The more the hospital succeeds in eliminating purely personal and emotional elements from its daily operation, the closer the bureaucracy comes to perfection (Moller, 1996).

Modern societies are inherently bureaucratic societies, argued Weber. As dying becomes bureaucratized, it takes place in specialized institutions and the social role of "formal caretakers of the dying" emerges, noted D. W. Moller (1996, p. 25). With such bureaucratization, the responsibility of the community and family to care for the dying person is basically a thing of the past. As dying is prolonged by an active process of medical treatment, a hostility is established between medical technology and death. In such a medicalized situation, death becomes transformed into an enemy to be defeated. The depersonalization is also found in physicians' being socialized to remain emotionally neutral and undisturbed in the presence of dying and death.

David Sudnow, in his classic **ethnography** *Passing On: The Social Organization of Dying* (1967), compares a public to a private hospital. Sudnow revealed how medical staff respond to death in a standardized, routinized manner. Thus, the bureaucratization of death within an institutional setting contributes significantly to an impersonal way of dying. The normality of dying within an institutional setting, therefore, is basically redefined to fit the model of bureaucracy.

Dying in a hospital traditionally gives patients very little control over the circumstances of their daily lives and the course of their dying (Kamerman, 1988). The alienation of dying patients in the United States today is a "self-exacerbating blend of naturally and societally occurring powerlessness" (Moller, 2000, p. 126). We all know that we must someday die, because that is the nature of life, yet death becomes perceived through societal orientation as a thief that robs us of our most precious possession. The awareness contexts of suspicion, closed awareness, and mutual pretense (especially mutual pretense) of Barney Glaser and Anselm Strauss, discussed in Chapter 5, often exist in a hospital setting.

HOME   In both community and clinical samples, a majority of individuals consistently express a preference to die at home, according to studies cited by Judith Hays

---

LISTENING TO THE VOICES | LOW STATUS FOR HOSPITALS AND PHYSICIANS

Writing in 1950 from his own experiences in European hospitals, George Orwell observed that the previous 50 years or so had brought a great change in the relationship between doctor and patient. Prior to the 20th century, a hospital was popularly regarded as much the same thing as a prison, and an old-fashioned, dungeon-like prison at that. A hospital was a place of filth, torture, and death, a sort of antechamber to the tomb. Only the destitute would have thought of going into such a place for treatment. The whole business of doctoring was looked on with horror and dread by ordinary people. From the 19th century one could collect a large horror-literature connected with doctors and hospitals. The dread of hospitals probably still survives among the very poor.

From How the poor die, by G. Orwell, 1950. In *Shooting an elephant and other essays*, New York: Harcourt Brace.

and colleagues (1999). Adults sampled in Australia and Italy favored home death over hospital death by three to one, and in a multiethnic Los Angeles sample, Anglo American and Japanese American adults preferred home death to hospital death also by three to one, African Americans by two to one, and Mexican Americans by five to three. Clinical studies of preference for place of death have been conducted among terminally ill patients in the United Kingdom, the United States, Canada, and Japan, and a preference for home death was reported in 54 to 74 percent of dying patients when the only alternative was a hospital setting. When inpatient hospice care was also an alternative, 53 to 58 percent of those dying of cancer and 32 percent of those dying of AIDS expressed a preference for a home death (between 15 and 29 percent preferred a hospice death).

Though home is the preferred place to die, as noted earlier, the majority of Americans die in an institutional setting, such as a hospital or nursing home. The majority of individuals dying in a hospice setting in the United States do die at home, under the care of hospice. Yet, not everyone who dies at home is enrolled in a hospice program. Dying at home with hospice as a choice has only been available since the early 1970s.

The first hospital in the United States was established in Philadelphia in 1713. By 1873 there were only 178 hospitals in the United States, but by the year 2007, the number had increased to 5,708 (Cockerham, 2012). Thus, for most of the history of the United States, dying in a hospital was not an option, because they barely existed. Even with hospitals proliferating in the 20th century, many rural areas still did not have adequate access to a hospital; thus, dying at home was about the only option. Also, without life-support equipment, which only came into existence in the latter half of the 20th century, in many cases hospitals could do nothing more for the dying person than could be done at home. Thus, dying at home was the way individuals died in early America. Nursing homes were not available on every corner until late in the 20th century; thus, again, the home was the place where individuals died.

George Dickinson can remember his great-grandmother dying at home back in the 1950s. There were no nursing homes in the community, hospice did not exist, and she did not need hospital care. She was dying of old age in her 90s. Thus, the option was to die at home. Different relatives took turns sitting up with her and caring for her various needs. The doctor made house calls. She died in familiar environs and surrounded by her family. In a way, this was the good old days.

NURSING HOME An estimated one in four Americans who reach the age of 65 dies in a nursing home (Waldrop & Kusmaul, 2011), and over 40 percent of individuals aged 65 years and over are expected to spend some time in a nursing home (Olson, 2003). Chronic care nursing home residents have entered the living-dying interval with its three distinct phases: an acute crisis (generally prior to entering), a chronic living-dying phase (influenced by the diagnosis and coexisting conditions), and a terminal stage (Waldrop & Kusmaul, 2011).

A nursing home provides a type of help to residents who require care with activities of daily living. Nursing homes offer the services of physicians, nurses, nurses' aids, speech and occupational therapists, social workers, and social activity directors. A nursing home is a place for individuals who do not need to be in a hospital, yet cannot be cared for at home. Regulatory barriers, however, make it difficult for nursing homes to provide specialized end-of-life care as they are mandated to maintain or promote residents' physical and psychological functioning. Such a mandate overlooks the reality that all permanently placed residents will die and that physical and functional decline, weight loss, anorexia, and symptom exacerbations are part of the dying process (Waldrop & Kusmaul, 2011).

Indeed, death is an everyday occurrence in nursing homes and retirement communities. Sociologist Jaber Gubrium (1975) used **participant observation** in studying residents in a nursing home. His ethnography was entitled *Living and Dying at Murray Manor*. Gubrium noted that nursing home residents view the nursing home as their *final* place of residence. When one is sent from home to the hospital, there is at least the hope of returning home—not the case when one is sent to a nursing home. The situation in a nursing home is not unlike that in a hospital—very regimented and bureaucratized. The daily routine is set by the administration, and it is followed. Residents tend to define their futures in terms of death. Vivid signs of dying in Murray Manor include physical crises such as heart attacks, being transferred to another floor to die, and daily reports from those with dying spouses to other residents in the home. After death, the bodies are taken out through the front lobby at night—by day, they are taken to the basement via the elevator to the service ramp.

Research has documented poor care at the end of life in U.S. nursing homes and underscored the need for creative and innovative solutions (Waldrop & Kusmaul, 2011). A personnel shortage exists and is in part due to low pay, meager fringe benefits, and difficult working conditions at many nursing homes. Nursing homes are licensed by states and must meet federal standards in order to participate in **Medicaid** and **Medicare**. Though hospice benefits are available in nursing homes, utilization is often short and referrals are typically made late in the dying process.

www.hospice.com

The Connecticut Hospice, Inc. was the first hospice in the United States and the first palliative teaching hospital-hospice to receive palliative certification (without recommendations for improvement) from The Joint Commission.

HOSPICE INPATIENT CARE Hospice inpatient care is in a facility that typically is either a freestanding hospice, such as St. Christopher's or the Connecticut Hospice, or is within a hospital or a nursing home. Such inpatient hospice care usually becomes necessary for several reasons. To bring a patient's pain and symptoms under control, a stay of a few days in an inpatient facility may be necessary or helpful. The family taking care of the patient at home may become exhausted and need a few days' rest while the patient is cared for elsewhere. Home care may be inappropriate at a given stage of the illness because of the patient's condition or home situation. It is hoped that upon admission to an inpatient hospice facility, patients will be able to move back and forth from home care to inpatient care at various stages of the illness.

When admitted to hospice inpatient care, special efforts are made to make the patient feel as much at home as possible. When taking a tour of St. Luke's Hospice in Sheffield, England, the medical director was showing George Dickinson the admittance area in which the patients were brought into hospice. The gurney was awaiting the arrival of a new patient. Sticking up under the top sheet was a bed warmer! The patient coming into the hospice within the next few minutes would indeed receive a warm welcome. Psychologically, how fitting for a hospice: It is all about *care*.

Whereas traditional medicine in recent years has tended to concentrate care in specialized hospitals or in nursing homes, hospice care returns the focus to the family. Because the family is the unit of care within the inpatient hospice facility, sufficient space must exist for a large number of family members to congregate. In addition, such care requires a homelike environment—the idea is to make the facility as much like a home away from home as is possible. Patients are encouraged to bring favorite possessions with them such as pictures, a favorite chair, or plants.

No arbitrary visiting restrictions are placed on those wishing to see hospice patients. One may visit at any time of day or night. Visitors of any age, including young children, are not restricted in their visitation. Furthermore, family pets, such as dogs or cats, may come as well. The goal of an inpatient hospice facility is to provide a homelike environment where the patient and his or her family can appreciate the joys of social relationships.

The inpatient facility of the Connecticut Hospice in Branford, Connecticut, which inaugurated hospice care in the United States in 1974, illustrates the preceding principles. As the structure was built then, the family room was off limits to staff and was provided solely for the comfort of family members. Hospice care places considerable emphasis on the tastiness, attractiveness, and nutritional value of food prepared for patients. The Connecticut Hospice employed a gourmet chef trained in Paris to supervise its food preparation. Kitchens containing a refrigerator, a microwave oven, a stove, and a sink were available for use by families. Washing machines were likewise maintained for their use. Large living rooms with fireplaces were available. Ten rooms with four beds in each helped patients develop social support systems among family groups. There were four single bedrooms, as well. Spacious corridors next to patient rooms contained plants and areas for family gatherings. A common room and chapel were used not only for religious services but also for presentations by various kinds of artists. Operated by volunteers, a beauty parlor was available to help patients feel better about themselves. When the

LISTENING TO THE VOICES — DYING IN A HOSPICE ENVIRONMENT

After a week in the hospital and a few days at a regional medical center, we found out that my father had a rare type of cancer. It only took 10 weeks to take his life, but he did put up a fight. He tried chemo twice, but it was not working. Toward the end of my father's illness, we checked him into hospice. My mother and I could not take care of him the way he needed anymore. It was just becoming too much. The people at the hospice program could not have cared for my father any better. They were like angels sent down from heaven to make his last days comfortable.

They would read to him, talk to him, etc. They let my whole family stay with him night and day.

The hospice program is wonderful in the help that they provide for the patient and the family. They helped us cope with the loss, and still call to see how we are doing. Actually, a few of the employees came to my father's service.

Source: From a letter received on November 12, 1997, from Courtney Schomp, a former student in George Dickinson's class entitled "Death and Dying."

patient died, he or she was taken to a viewing room for the family members. As of 2001, the Connecticut Hospice enlarged its facility, yet continues with its care of the physical, psychological, social, and spiritual needs of its patients. The Connecticut Hospice has expanded its mission to include research and has added a training affiliate called the John D. Thompson Hospice Institute for Education, Training and Research, Inc. This new facility was the first palliative teaching hospital to receive the national joint commission's "Advanced Palliative Certification." The Connecticut Hospice remains a nonprofit home care agency but now is a hospital with 52 beds, still located in Branford, Connecticut.

Inpatient hospice facilities seemingly serve as a last resort for patients with terminal illnesses. The stay in inpatient hospice facilities is rather brief, with many individuals dying within a day or two of arriving, and the average stay before death is about three weeks.

END-OF-LIFE EDUCATION IN MEDICAL AND NURSING SCHOOLS

As noted in Chapter 1, medical education has historically offered only limited assistance to the medical student encountering death for the first time (Dickinson, 2011). In an analysis of major medical textbooks (Carron, Lynn, & Keaney, 1999), the conclusion was that little was said about what to expect or how to care for people near death. Out of 122 medical schools, the number of full-term death and dying courses in the United States increased from seven in 1975 to 21 in 2010 (Dickinson, 2011). Eighty percent of medical schools in 1975 offered death education in the form of an occasional lecture or minicourse; however, this number had increased to 100 percent by 2010 (Dickinson, 2011). In a study of over 1,000 oncologists, however, 73 percent said prognosis communication education regarding patients with terminal illnesses was either absent or inadequate in their medical school experiences, and 96 percent said such should be part of cancer care training (Daugherty & Hlubocky, 2008). Thus, an emphasis on dying and death in medical schools is still somewhat limited today, yet the exposure for students is increasing.

In addition to death and dying offerings, 99 percent of U.S. medical schools in 2010 addressed the issue of palliative care (Dickinson, 2011b).

With more complex medicine today, medical professionals have been able to keep patients alive much longer than before (Parker et al., 2012). With the expansion of technology and services for the dying patient, however, comes more responsibility for the primary caregivers. Additional training may be needed to lower the apprehension of physicians in understanding and/or administering end-of-life care.

The nursing care of seriously ill patients and their families is a vital part of interdisciplinary, holistic end-of-life care (Kirchoff, 2003). Of all health professionals, nurses are in the most immediate position to provide care, comfort, and counsel at the end of life for patients and families. Nurses spend the most time with patients and their families at the end of life, compared with other health care providers. Nurses serve as advocates for patients with end-stage illnesses and their families, collaborate with multidisciplinary team members regarding patient outcomes, and provide nursing care based on patient care goals developed by working with patients and their families (Dickinson, 2007b). Yet studies of nurses' attitudes toward end-of-life care reveal that nurses often feel apprehension in dealing with terminally ill patients (Parker et al., 2012).

In comparing palliative and end-of-life care emphasis in U.K. and U.S. undergraduate nursing programs and medical schools, the average number of teaching hours in both nursing and medical schools in the United States is about the same: 14 and 17, respectively (Dickinson, 2007a; Dickinson, 2012a). On the other hand, in the United Kingdom, the average number of hours in nursing schools is 45 and in medical schools 20 (Dickinson & Field, 2002; Dickinson, Clark, & Sque, 2008) (see Table 7.1 for topics covered). Child life specialty programs in the United States average 33 hours on end-of-life issues, thus more than nursing and medical school curricula in the United States, but not more than U.K. nursing programs (Parvin & Dickinson, 2010).

Gross Anatomy Lab in Medical Schools   The University of Massachusetts Medical School integrates sessions on dying and death into the gross anatomy courses. These are typically taught during the first few weeks of the course. During this course, students are encouraged to identify and articulate feelings about death and about the experience of dissecting a cadaver. Without this discussion, the students tend to become desensitized to the human dimension of death. After all, the initial viewing of a human cadaver may be quite unnerving for some medical students. For example, a study (Hancock et al., 2004) found that 10 percent of students exhibited a stress reaction a week following their first dissection, with percentages dropping over the next several months, down to zero at two years.

In a study (Dickinson, Lancaster, Winfield, Reece, & Colthorpe, 1997) of first-year medical students in the gross anatomy laboratory, pre- and post-surveys of the 84 students revealed that 54 percent showed less death anxiety after completing gross anatomy, 29 percent had increased fear of death, and 17 percent experienced no change. Thus, the gross anatomy course tends to desensitize students to death, at least according to this study. A more recent pre/post gross anatomy study by Brent Robbins and colleagues (2008) concluded that, over time, the students became increasingly habituated to the cadaver's presence and were thus more

TABLE 7.1 | END-OF-LIFE TOPICS COVERED IN THE CURRICULA OF U.S. AND U.K. MEDICAL AND NURSING SCHOOLS, IN PERCENTAGES

Topics Covered	U.S. Medical N = 99	U.S. Nursing N = 407	U.K. Medical N = 24	U.K. Nursing N = 52
Attitudes toward death and dying	95	97	100	100
Communication with dying patients	99	92	89	98
Communication with family members	95	92	96	96
Grief and bereavement	90	98	92	92
Spiritual/ cultural aspects of dying	82	91	67	86
Psychological aspects of dying	91	90	92	84
Experience of dying (e.g., pain, anxiety)	80	86	79	94
Social contexts of dying (e.g., family care)	79	80	89	73
End-of-life hydration	63	74	67	59
End-of-life nutrition	67	73	58	61
Relating to patients with AIDS	44	71	37	49
Neonatal issues	29	70	33	27
Euthanasia	48	66	89	71
Advance directives	93	98	75	61

Sources: G. E. Dickinson and D. Field. (2002). Teaching end-of-life issues: Current status in United Kingdom and United States medical schools. *American Journal of Hospice and Palliative Care, 19*, 181–186; G. E. Dickinson. (2007). End-of-life and palliative care issues in medical and nursing schools in the United States. *Death Studies, 31*, 713–726; G. E. Dickinson. (2011). Thirty-five years of end-of-life issues in US medical schools. *American Journal of Hospice and Palliative Medicine, 28*(6), 412–417; G. E. Dickinson, D. Clark, and M. Sque. (2008). Palliative care and end of life issues in UK pre-registration, undergraduate nursing programmes. *Nurse Education Today, 28*, 163–170.

capable of feelings such as a sense of awe and amazement about the process of dissecting a human being.

Although the reduction of death anxiety is good, this is often part of a general numbing that occurs during medical school that may affect doctors' ability to relate to dying patients as people. To address this problem, some medical schools are trying to treat cadavers as people rather than objects. Christine Montross (2007) observes that in medical schools in Thailand the cadavers are treated with much respect. The gross anatomy students learn the cadavers' names and something about them and refer to the cadavers as Great Teacher. In the United States, several medical schools routinely hold a memorial service at the end of gross anatomy lab for the cadavers that are dissected. Through such a dedication service the students give thanks to those who, in death, taught them about life.

In recent years, there has been controversy over the ethics and effectiveness of using dead human beings as a learning tool. There has been some concern that work on cadavers may have negative consequences for students that may outweigh the benefits of using a human body (Robbins et al., 2008). A few medical schools, the University of California at San Francisco, for example, spare the students all hands-on contact with a cadaver by having them inspect cadaver structures that have already been dissected by an instructor (Zuger, 2004). Some medical schools have switched over to virtual cadavers.

Yet, some research suggests that learning anatomy with actual cadavers generates more knowledge, even if the students' scores on the written portion of the exam are no better than students receiving only lecture and readings (Jones et al., 2001). In addition, it has been speculated that learning from a dead human's body could instill a sense of compassion in the student which may generalize to work with living, breathing people (Robbins et al., 2008). In the study by Robbins and colleagues, they observed that the students became increasingly curious about the unique structure of the body and how the cadaver's body reflects but also diverges from the textbook models of the body, perhaps another argument for using dead human beings over virtual cadavers.

DEVELOPING A SENSITIVITY TO SOCIAL AND PSYCHOLOGICAL NEEDS

The medical training of most physicians historically seems to be primarily concerned with the patient's physical state rather than with the patient's social and psychological needs. Granted, the patient wants (or demands) a physician knowledgeable of her or his physical state, yet such knowledge is only part of the picture. Some changes are occurring, however, as evidenced by Dartmouth Medical School's interest in facilitating the medical students' understanding of the patient as a bio-socio-psychological being. In Dartmouth's program, medical students are required to learn from the terminally ill patient what it is like to be dying. Thus, the patient is used as teacher. The inclusion of a humanistic emphasis in death education should help both the dying patients and the medical students.

At the Yale School of Medicine, a seminar is offered for first- and second-year medical students on problems in communicating with dying and seriously ill patients. Very sick patients are used as teachers. Approximately one-third of the first-year students take this elective course. Goals for students in the course are to

(1) learn to talk with, and listen to, sick persons; (2) learn to establish a professional relationship without the intrusion of friendship; (3) ascertain the meaning of compassion without sentimentality and the need for humility in the context of the physician's ignorance; (4) learn of our common frailty as human beings, the finality of death, and the need that we all have for companionship when death is near; and (5) enrich the students' understanding of those in their care.

DEVELOPING COMMUNICATION SKILLS

Many practicing physicians feel their education and training have not adequately prepared them to value palliative care, nor have they been helped to learn how to cope with their own feelings about death and possible grief over a patient's loss (Cowell, Farrell, Campbell, & Canady, 2002). A survey of more than 600 physicians (Dickinson, 1988) in 1986 revealed that the majority (78 percent) agreed that more emphasis in medical school should be placed on communication skills with terminally ill patients and their families. Results from another study of 350 family physicians in South Carolina (Dickinson, 1988) found that the majority felt that their medical education was inadequate in helping them relate to terminally ill patients and their families. More recently, a study of medical students (Hesselink et al., 2010) suggested that more attention can and should be paid to education on end-of-life care in the medical curriculum so that students are well prepared to provide adequate end-of-life care. Additionally, a national study of fourth-year medical students at 62 medical schools that reported responses (Billings et al., 2010), found that students reporting more exposure to curricula on end-of-life care felt more prepared and rated their end-of-life care education higher than those not so exposed.

In a Harris survey (Richardson, 1992), 91 percent of 1,501 practicing dentists, nurses, pharmacists, physicians, and veterinarians said that teaching students how to communicate effectively with patients and their families is very important, yet 36 percent gave their schools a poor rating in this area. More recently, Sirmons, Dickinson, and Burkett (2010), in a survey of 319 South Carolina dentists found that only 5 percent agreed that dental school well prepared them for "relating to patients with end-of-life issues," yet 50 percent of the 52 U.S. dental schools responding, out of 58 U.S. dental schools, said that "dying, death and bereavement is an important topic that should be included in dental school." Similarly, a recent survey of 347 veterinarians in South Carolina (Dickinson, Roof, & Roof, 2009) revealed that 35 percent agreed that their "veterinary training well prepared them for relating to owners of terminally ill animals." Seventy-five percent of these veterinarians said that veterinary schools should "place more emphasis on communication skills with owners of terminally ill animals."

Seeing the need for better communication skills with patients, the Association of American Medical Colleges in 1991 added two 30-minute essays to the admission test to help medical schools evaluate communication skills (Altman, 1989). As of the fall of 2004, all medical students in the United States have to pass a new exam that measures clinical skills, but the exam also attempts to gauge how well doctors communicate with patients. Similarly, the American Medical Association (Montgomery, 1996) announced in 1996 a major effort, beginning in 1997, to teach doctors how to aid the dying by helping patients and their families plan for dying, by providing effective ways to reduce suffering, and by treating psychiatric complications. In response to a mandate by the Liaison Committee on Medical Education (LCME, 2010) that the curriculum of a medical education program "should cover all organ systems, and must include the important aspects of prevention, acute, chronic, continuing, rehabilitative, and end-of-life care," progress is being made in the United States toward inclusion of end-of-life issues in medical schools.

In 2000, the American Medical Association concluded that treating terminally ill patients with kindness, tact, and good medical judgment may come more naturally to some doctors than to others, but the AMA decided that the skills can be learned (Grady, 2000a). A two-and-a-half-day course entitled "Education for Physicians on End-of-Life Care" has been developed. The doctors learn details of pain control that were not taught in medical school, as well as how to use drugs to ease shortness of breath, nausea, seizures, agitation, and other problems. They are urged to encourage family members to lie in bed with a dying patient if they wish. They are taught how to turn off a respirator in a way that does not leave the patient gasping for air. They learn how to help prepare a patient's family for the physical changes that take place as a person dies.

In a study of 441 family medicine practitioners in South Carolina (Durand, Dickinson, Sumner, & Lancaster, 1990), those who reported having been taught concepts concerning terminally ill patients and their families while in medical school had a more "positive" attitude toward death than did those who received no instruction. Likewise, surveys of medical students graduating from the University of Pittsburgh with personal or professional experience with death had more positive attitudes and higher knowledge scores regarding helping dying patients and their families than those who did not have such experiences (Anderson et al., 2008). Herbert Fraser and colleagues (2001) similarly found, from surveys completed by 262 senior-year medical students, support for the development of formal curriculum on end-of-life issues and emphasized the importance of clinical exposure to terminally ill patients. Likewise, in a study of 166 first-year medical students before and after exposure to hospice (Wechter et al., 2013), a significant change in attitude was noted after the observational experience. Thus, exposure to individuals with terminal illnesses seems to have merit for medical students.

| WORDS OF WISDOM | MORE EMPHASIS ON DEATH EDUCATION NEEDED IN MEDICAL SCHOOL |

Physicians are not being taught to recognize when patients are reaching the final stages of life, to manage symptoms associated with dying, or to attend to the special needs of dying patients. Medical students learn to fight disease and are taught to view death as a failure. They are trained to respond to illness with aggressive care, to subject patients to invasive tests, procedures, and machines, until death is recognized as imminent.

The dying process and the dying person are largely ignored. Such practices will persist if the system for educating doctors and other health care workers continues to emphasize the basic sciences and does not teach students how to care for the dying.

Adapted from Medical education must deal with end-of-life care (pp. A56–A57), by F. Cohn, J. Harrold, and J. Lynn, May 30, 1997, *The Chronicle of Higher Education.*

Physicians themselves may be of therapeutic value to their patients, simply by talking regularly to them and assuring them that they will not be abandoned as death approaches (Cohn, Harrold, & Lynn, 1997). Medical students must be trained to understand that this role is important and involves skills that can be taught, including the management of care as a patient moves among different settings (e.g., hospital, home, nursing home, and hospice) and how to offer spiritual and emotional support.

Treating a dying person may involve a period of mixed management in which a patient continues to need some aggressive care, even while the doctor places increased emphasis on palliation, noted F. Cohn, J. Harrold, and J. Lynn (1997). Physicians can help patients by raising and discussing issues and working out ways of incorporating visits from family members, members of the clergy, or counselors into the plan of care.

Ethicists Cohn, Harrold, and Lynn (1997) suggested that medical education requires the development of specific competencies to bring about better end-of-life care. For example, when students learn to conduct patient interviews, they must learn to address emotional and spiritual issues as well as medical concerns. Students should learn not only about the existence of living wills and durable power-of-attorney forms (see Chapter 12), but also how to discuss them with patients. Students should be taught to ask and appreciate what is important to the patient and must learn how to care for themselves as they face the task of caring for dying people.

A good start in reorienting medical education should be to increase students' contact with dying patients, recommend Cohn, Harrold, and Lynn (1997). As discussed earlier in this chapter, at Yale Medical School students in the seminar on

© Alexander Raths/Shutterstock.com

Good communication skills for health care professionals should be emphasized in their medical education, just as technical skills are stressed. Rapport with patients is very important in helping them cope with illness.

dealing with very sick patients are assigned to a dying person. In suggesting more contact with dying patients, Mermann (1997) noted that what may be missing in a medical school lecture is the unique individuality of each of us, especially when confronted by the prospect of disaster, with its accompanying anxiety and fear, and the possibilities of pain and suffering. Students could also accompany nurses and social workers on home visits to dying patients. Students could take the hospice training and volunteer to work on a weekly basis with a dying person—such an assignment is common in England but less so in the United States.

Regarding the "telling" of a patient of her/his life-ending disease, a study of 1,137 physicians (Daugherty & Hlubocky, 2008, p. 5989) revealed that those "telling" were relatively evenly divided between saying they "always discuss their patients' prognosis because they need to know it" (42 percent) and either saying they "ask their patients if they want to know their prognosis and discuss it if they say yes" or "only discuss it if the patients ask about it" (48 percent for the latter two responses). Regarding whether or not a specific time frame is given, 43 percent reported "always" or "usually," and 57 percent reported "sometimes," "rarely," or "never" giving a time frame.

Good rapport with dying patients might be an intrinsic quality of certain medical students that is reinforced by experience, or it might be something in which medical students can be given instruction. Though values, ethics, and communication skills may be presented and learned in medical school, there is no guarantee that they will be carried over into clinical practice.

Ira Byock (1997), former president of the American Academy of Hospice and Palliative Medicine, said that care for dying patients is still inadequately taught. He notes that the culture of curative medicine is entrenched within medical universities, with attention directed toward preserving life at all costs. The values of caring are subordinate. Considerations of money, physical comfort, human dignity, and the quality-of-life experiences are, at best, secondary, according to Dr. Byock. The problem is not that medical professionals are heartless but rather that the goals of end-of-life care are often not well considered, if at all.

A survey of medical education deans by Amy Sullivan and colleagues (2004) found that 84 percent described end-of-life care education as "very important" and supported incorporating more end-of-life care teaching into the curriculum. The deans supported integrating teaching end-of-life care into existing courses or clerkships rather than required end-of-life courses. Such a recommendation also was confirmed by Horowitz and colleagues regarding integration of end-of-life issues (2013), advice that is apparently being followed both in medical schools in the United States and in the United Kingdom (Dickinson & Paul, 2014).

THE COST OF DYING

It is not too unusual to hear of a medical situation where a patient in the latter stages of life has numerous procedures performed in an effort to prolong life. The final medical bill may be in six figures, only to sometimes give the individual a few more days or weeks of life, time that may not produce much in the quality of life. Was it worth it? Most Americans still feel every person, regardless of age, income, or level of insurance, deserves life-prolonging care. But one of every seven health care dollars spent each year is used during the last six months of life. Is this really the most effective way to allocate health care resources?

The cost of health care in the United States in 2010 was $2.19 trillion, more than three times the $714 billion spent in 1990, and over eight times the $253 billion spent in 1980 (Centers for Medicare and Medicaid Services, 2009; MEC/PAC, 2012); thus approximately 16.2 percent of the U.S. gross domestic product (GDP) is spent on health care. This is among the highest of all industrialized countries. Without reform, there is general agreement that health costs are likely to continue to rise in the foreseeable future. Since 1999, employer-sponsored health coverage premiums have increased by 119 percent, placing increasing cost burdens on employers and workers (Kaiser Family Foundation, 2008).

Government programs such as Medicare and Medicaid account for a significant share of health care spending. Public health expenditures made up about 40 percent of the health care dollars in 2010 (Medicare and Medicaid), with the remainder split between private and out-of-pocket spending (46 percent and 14 percent, respectively) (MEC/PAC, 2012). Hospital care accounts for the largest share of health expenditures (31 percent), whereas physician and clinical services account for 21 percent. Prescription drugs account for only 10 percent of total expenditures but are one of the fastest-growing segments of health expenditures. Factors that are driving health care costs up include prescription drugs and technology, chronic disease, an aging population, and administrative costs (U.S. Health Care Costs, 2009).

Costs associated with illnesses are varied but can be characterized as direct and indirect (Smith, 1996). **Direct costs** typically include those for a wide variety of medical care from hospital to home-based care (e.g., services of the medical staff, medication, and use of the hospital facility). **Indirect costs** may include overhead costs of running the hospital (e.g., postage, utilities, and upkeep of the building) or costs associated with lost productivity (e.g., money lost from one's not having a job).

The United States' method of health care financing can only be described as a shell game, according to former Secretary of Health, Education and Welfare Joseph Califano (1986). The pea represents the cost of health care and the shell represents the vendors (e.g., federal government and insurance companies) who provide funding for health care. Ultimately, the American people pay all health care costs, whether for a terminal or a chronic illness, but health care dollars are directed from the people to the providers in three main patterns: (1) direct payment by the consumer, (2) private insurance, and (3) government taxation at various levels (Freund & McGuire, 1999).

Proposals by Altman and Levitt (2003) to contain costs include the following: (1) investment in information technology such as electronic medical records; (2) improvement in quality and efficiency such as decreasing unwarranted variation in medical practice and unnecessary care; (3) adjusting provider compensation to physicians to ensure that fees paid reward value and health outcomes; (4) rather than volume of care, government regulation, as with the Medicare program of controlling per capita spending; (5) prevention by providing financial incentives to workers to engage in wellness and prevention to decrease long-term costs of treatment; (6) increasing consumer involvement in purchasing with greater price transparency to make consumers more price sensitive and more prudent purchasers; and (7) altering the tax preference for employer-sponsored insurance by eliminating or changing the tax exclusion for employer-sponsored health care to help finance the costs of expanding coverage, as well as reducing incentives for the most generous and therefore expensive health plans. Certainly not everyone will agree with these proposals, but they give food for thought and thus debate.

As noted above, 46 percent of health care costs come from private insurance. The first private insurance, Blue Cross, was established in 1930. With a company like Blue Cross (later Blue Shield was added for nonhospital services), an individual (or her or his employer) pays a premium, usually monthly. When illness occurs, the individual receives medical treatment. The bill is then sent to the insurance company which pays most (or all, in some cases) of the medical bill. This **third-party payment** is very popular in the United States today. Thus, the three parties are the patient, the physician or hospital, and the insurance company.

In the latter part of the 20th century, **managed care organizations** entered the scene. This form of private insurance controls spending on health care by screening prospective utilization of insured medical care. Expenditures are also managed by limiting coverage to care provided by doctors and health facilities. Demands of patients and their advocates are balanced against the price purchasers are willing to pay (Wholey & Burns, 2000).

The public welfare system, funded via government taxation, for health purposes includes Medicare and Medicaid. In 2010, 40 percent of all health dollars spent in the United States were paid for by Medicare and Medicaid (MEC/PAC, 2012). Because 70 percent of all people who die in the United States each year are 65 or older, they are probably covered under Medicare. Medicare was created by legislation enacted in 1965 to finance acute medical care mainly for elderly Americans (Freund & McGuire, 1999). Part A of Medicare is hospital insurance; Part B is supplementary medical insurance for physicians' services, outpatient care, laboratory fees, and home health care. As of 2006, Medicare also provides prescription drug coverage. Chronic illness and long-term care are not part of the

DEATH ACROSS CULTURES | HEALTH CARE IN SWEDEN

Sweden has demonstrated that a socialized system of health care delivery can be effective in a capitalist country through the formation of a national health service. The Swedish National Health Service (NHS) is financed through taxation. Taxes in Sweden have been the highest in the world, but Sweden remains one of the world's most egalitarian countries. Universal health insurance would eliminate the worries of finances for one dying in Sweden. George Dickinson had a friend visiting in Sweden who had to have major surgery. Her bill for the entire hospital stay and surgery was less than $5. She had to pay for a long-distance telephone call from her hospital room!

Physicians in Sweden are paid according to the number of hours worked rather than the number of patients treated. Sweden is one of the leaders in the world in palliative medicine; thus one's pain in dying would largely be eliminated. General hospitals are owned by county and municipal governments. Drugs are either free or inexpensive. Sweden remains committed to universal and equal access to health services paid by public funding.

Taken in part from *Medical sociology,* by W. C. Cockerham, 2012, Upper Saddle River, NJ: Prentice Hall.

Medicare plan; thus, many individuals do not have adequate medical insurance to cover their expenses. Such a void in medical coverage in our society has been a major political issue for several years. The Affordable Care Act (ACA) is addressing this situation, however. As of March 31, 2014, 7.1 million Americans signed up for health coverage, another 4.5 million signed up for Medicaid under the ACA's expansion of eligibility, and 3 million young adults ages 18 to 25 gained insurance under their parents' coverage (*The Week,* 2014). Though still receiving continued opposition from some political groups, the ACA is chipping away at reducing the number of Americans who have not previously had health care coverage.

For individuals with low incomes in the United States, Medicaid provides financial assistance with health care costs regardless of age. Medicaid was created at the same time as Medicare, but functions differently. Medicaid is perceived as "charity" rather than "deserved aid," noted P. E. S. Freund and M. B. McGuire (1999). Each state operates its own Medicaid program but receives federal contributions ranging from 50 to 80 percent based on per capita income of the states. Despite Medicare and Medicaid, an estimated 48 million Americans have not had insurance to cover medical expenses, yet the federal health care plan (the ACA) passed by Congress and signed into law in March of 2010 is working toward reducing this concern. Many individuals have not had health insurance coverage simply because they could not afford it, yet some individuals simply do not believe in health insurance (pay as it comes, with no regard for what will happen if a *major* illness occurs, or "God will take care of you").

In the United States health care is closely linked to employment status—medical insurance often comes with a job as a fringe benefit. Blacks and Hispanics are particularly likely to be uninsured because they are disproportionately represented among the working poor, having jobs that pay no insurance benefits and living in regions of the country where labor is less commonly unionized and Medicaid insurance is unavailable (Ginsberg, 1991). Looking at young adults, irrespective of ethnic origin, those without private health insurance are 35 percent more likely to die than those with insurance (Rogers, Hummer, & Nam, 2000). Thus, having private health insurance coverage, as opposed to none, enhances one's chances of survival. Mortality risks are reduced when income is used to purchase health insurance and to buy goods and services that will promote health and prevent disease. There is most definitely a correlation between income and mortality.

Because one's major health care costs normally occur during the last few months of life, the process of dying can be expensive. The current health care cost crisis has caused many people to become paupers simply because they are incurably ill and are not dying quickly enough (Byock, 1997). Driven by Medicare and the insurance industry to reduce expenditures associated with the last year of life, hospitals, clinics, and health maintenance organizations (HMOs) have instituted an array of cost-containment measures reflecting a de facto strategy of "doing more with less," noted Ira Byock. Costs of care are therefore being shifted onto the backs of the patients and families in need.

Ira Byock further stated, "To be terminally ill or elderly in America today is to be reminded frequently that you are a drain on the nation's resources"

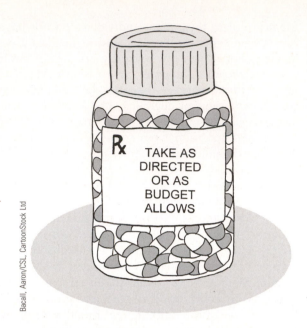

Rx TAKE AS DIRECTED OR AS BUDGET ALLOWS

(1997, p. 242). Byock also said that the message to the elderly and the incurably ill is: "Limit your use of resources and get out of the way to make room for those who are younger, vigorous, and still able to contribute to society" (p. 242). Our society places value on producers and not so much on those who are strictly consumers. Terminally ill and elderly persons are high consumers but are often unable to be productive, and thus they are considered less worthy recipients of medical dollars.

Though an accurate crystal ball for predicting the future is not available, according to geneticist Ricki Lewis (2013), by 2060 the ability to use gene testing to predict a patient's future health, coupled with genetic interventions, may have reached an unprecedented level of precision, with profound repercussions. With diseases stopped immediately before they take off, health care costs could plunge as a longer-lived, physically fit population emerges. Stay tuned!

In the meantime, common challenges to health care systems around the world, as pointed out by sociologists Greg Weiss and Lynne Lonnquist (2009, p. 390) include the following:

1. What is the optimal level of involvement of the national government in the health care system and in what ways should the government be involved?
2. Should there be both a public and private health care sector?
3. What is the optimal number of physicians within the system, and what should be the distribution between primary care physicians and specialists?
4. Given considerations of cost and equitable distribution, what is the optimal commitment that should be made to the incorporation of health care technologies?
5. How can health care cost increases most reasonably be controlled?

CONCLUSION

While one may think that the cost of living is expensive, the cost of dying is *not* inexpensive. Should our society have more Jack Kevorkians to help reduce the high costs of dying by shortening the period of the dying process? Should the government simply stop funding medical care costs for very sick elderly persons? The answer to both of these questions for most Americans would probably be an emphatic "no." With escalating health care costs in the United States, we have to take a careful look at expenditures at the end of life so that we can better manage these expenses.

It is encouraging that health care professional schools are leaning more toward a humanitarian approach to health care in the socialization of their students. They are placing more emphasis on palliative care, and thus more treatment of the whole person—physical, social, psychological, and spiritual. Options for dying today include hospitals, nursing homes, inpatient hospices, and at home. Certainly an individual may not have a choice as to where she or he dies, but if the option presents itself, the choices today are greater than in the distant past. We must work toward a point in medical treatment where the dying patient is a normal aspect of the medical subculture and is treated warmly as a human being, not just the "abdominal cancer case in Room 614."

Suggested Research Topics Related to End-of-Life Issues

1. The Current Status of Kidney Transplants
2. Technology to Extend Life Expectancy to 100 by 2075?
3. Grief Therapies: Art, Play, Theater, Music, Narratives
4. College Students' Knowledge/Opinions Regarding "Do Not Resuscitate"
5. Post-Cancer Reengaging in Society: Views of Young Adults
6. Roadside Memorials
7. Removal of Driver's License from the Elderly: Death Contributor?
8. Cadavers or Electronic Gross Anatomy Labs?
9. Police Officers' Perspectives on Death
10. College Students' Death Anxiety
11. Meta-analysis of Literature on Pain Management
12. Death Certificates: Need for Reform
13. College Students' Definitions of Palliative Care and Hospice
14. Attitudes toward Euthanasia
15. Suicide Information via the Internet
16. Cross-Cultural Differences in Final Body Disposition
17. Ethnic Differences Regarding Dying and Death
18. Children's Literature and Death
19. Death Certificates: Need for Reform?
20. Disenfranchised Grief: Changing Attitudes Regarding Pet Deaths
21. Death and Humor
22. Pros and Cons of the Affordable Care Act

SUMMARY

1. The process of dying is not extraordinarily different from the process of healthy normal living.
2. Dying patients are considered deviants in the medical subculture because they do not fit the norm of sickness—they cannot be made well.
3. The focus of labeling theory is on the audience doing the labeling, not the person being labeled.
4. Medical and nursing schools have traditionally offered very little death education, yet the 21st century shows promise.
5. In a few medical school offerings on relating to seriously ill patients, a dying patient is used as teacher.
6. Medicare and Medicaid are programs to assist the elderly and poor, respectively, with health care needs.
7. The cost of health care in the United States is approximately 2.3 trillion dollars annually.

Discussion Questions

1. Discuss why the process of dying is not really different from the process of healthy normal living.
2. How does labeling theory tend to make one a deviant?
3. Why is a person with HIV/AIDS a "double deviant?"
4. Why is a dying patient considered to be a deviant in the medical subculture?
5. Why do you think that medical and nursing schools have traditionally offered very little death education?
6. Discuss why a terminally ill patient might be very effective in socializing medical students on how to relate to seriously ill patients.
7. Discuss whether or not elderly patients should be prevented from having costly treatments paid for by the federal government, when that money could apply toward the health care needs of children in poverty pockets in the United States.

Glossary

Direct Costs: Usually include costs for services of care such as the medical staff, the use of the hospital room or surgery room, medicine, and use of equipment.

Ethnography: The systematic description of a culture based on firsthand observation; literally, *writing* about a *culture*.

Homophobia: An intense dislike of or prejudice against homosexuals.

Indirect Costs: Typically include overhead expenses such as utility bills for the hospital or costs of billing and may also include lost wages for the time the patient cannot work.

Labeling Theory: The theory based on the idea that whether other people define or label a person as deviant is a critical determinant in the development of a pattern of deviant behavior.

Managed Care Organizations: Groups that use administrative processes or techniques to influence the quality, accessibility, utilization, costs, and prices or outcomes of health services provided to a defined population by a defined set of providers.

Medicaid: A federal and state program that uses general revenues to fund health care for the poor. Eligibility is tied closely to economic status.

Medical Model: When sick, an individual goes to a physician to be made well.

Medicare: A federal program of health insurance for persons 65 and older.

Normalization of Dying: Maintaining roles, relationships, and identity, though dying.

Participant Observation: A method of data gathering, particularly popular in cultural anthropology, in which the researcher "lives with" and observes the group that she or he is studying.

Technological Imperative: A concept urging that if we have the technological capability to do something, we should do it. Action in the form of the use of an available technology is always preferable to inaction.

Third-Party Payment: A health care payment scheme in which the patient pays a premium into a fund and the doctor or hospital is paid from this fund for each treatment provided to the patient.

SUGGESTED READINGS

Bertman, S. L. (2009). *One breath apart: Facing dissection.* Amityville, NY: Baywood. The book was put together to help medical students confront their anxieties and ethical concerns of taking gross anatomy.

Boulis, A. K., & Jacobs, J. A. (2008). *The changing face of medicine: Women doctors and the evolution of health care in America.* Ithaca, NY: Cornell University Press. Using both quantitative data and in-depth interviews, the authors try to understand how women have changed the way medicine is practiced by providing insight into the historical and social trends that will ultimately lead to the way medical care is delivered in the United States.

Byock, I. (2012). *The best care possible: A physician's quest to transform care through the end of life.* New York: Avery Books. A noted U.S. palliative care doctor describes the stories of patients of various ages and outlines provisions of effective, good end-of-life care.

Camosy, C. C. (2010). *Too expensive to treat? Finitude, tragedy, and the neonatal ICU.* Grand Rapids, MI: William B. Eerdmans Publishing Company. The author goes into the emotionally charged and expensive world of the neonatal intensive care unit to examine health care rationing in the United States.

Chapple, H. S. (2010). *No place for dying: hospitals and the ideology of rescue.* Walnut Creek, CA: Left Coast Press. The author, an anthropologist and bioethicist with two decades of professional nursing experience, shows how dying is a management problem for hospitals, occupying space but with few billable encounters and of little interest to medical practice or quality control.

Chen, P. W. (2007). *Final exam: A surgeon's reflections on mortality.* New York: Alfred A. Knopf. Pauline Chen, a liver transplant surgeon, shares some of her early experiences, both good and bad, as a doctor.

Connidis, I. A. (2010). *Family ties & aging.* Los Angeles, CA: Pine Forge Press. An up-to-date account of relating to the elderly and their needs in the 21st century.

Dickinson, D. (2012). *Changing the course of AIDS: Peer education in South Africa and its lessons for the global crisis.* Ithaca, NY: Cornell University Press. An account of individuals who are doing something to combat the spread of HIV in their communities.

Ferrell, B. R., & Coyle, N. (2008). *The nature of suffering and the goals of nursing.* New York: Oxford University Press. This volume gives voice to the suffering that nurses witness in patients, families, colleagues, and themselves.

Groopman, J., & Hartzband, P. (2012). *Your medical mind: How to decide what is right for you.* New York: Penguin Group. An oncologist and endocrinologist aim to empower patients to become active participants in decisions about their health care.

Kaufman, S. R. (2005). *…And a time to die: How American hospitals shape the end of life.* Chicago, IL: University of Chicago Press. Drawing on over two years of ethnographic research, Sharon Kaufman explores the ways American hospitals contribute to "unwanted" types of dying.

Kosoko-Lasaki, S., Cook, C. T., & O'Brien, R. L. (2009). *Cultural proficiency in addressing health disparities.* Boston, MA: Jones and Bartlett Publishers. Identifies our most vulnerable populations and offers guidelines on how to avoid cultural incompetence and promote cultural proficiency.

Manheimer, E. (2012). *Twelve patients: Life and death at Bellevue hospital.* New York: Grand Central Publishing. Former medical director at Bellevue Hospital, The author uses stories taken from case histories to humanize issues like obesity, immigration, teen suicide, and the cost of health care. Dr. Manheimer was the medical director at Bellevue from 1997 to 2012.

Masten, J., & Schmidtberger, J. (2011). *Aging with HIV*. New York: Oxford University Press. This book, written by a psychotherapist and a physician, serves as a guide for gay men in middle age and beyond with HIV.

Montross, C. (2007). *Body of work: Meditations on mortality from the human anatomy lab*. New York: The Penguin Press. A well-written treatise on Dr. Montross's experiences in medical school.

Pronovost, P., & Vohr, E. (2010). *Safe patients, smart hospitals*. New York: Hudson Street Press. Peter Pronovost, a professor at Johns Hopkins School of Medicine, created a five-step checklist in the hopes of slowing the rate of infections patients often die from in hospitals.

Reid, T. R. (2010). *The healing of America: A global quest for better, cheaper, and fairer health care*. New York: Penguin Books. The author visits countries around the world (e.g., France, Britain, Germany, and Japan) to provide examples of affordable health care systems.

Vukmir, R. B. (2008). *The ER: One good thing a day*. Lanham, MD: University Press of America. This is the story in the emergency room behind the scenes with the ancillary staff (registration, technicians, aids, and housekeepers), the cohesive group that gets the hard jobs done.

It is silliness to live when to live is torment; And then have we a prescription to die when death is our physician?
—**William Shakespeare,** ***Othello***

All substances are poisons; there is none which is not a poison. The right dose differentiates a poison and a remedy.
—**Paracelsus**

BIOMEDICAL ISSUES AND EUTHANASIA

CHAPTER **8**

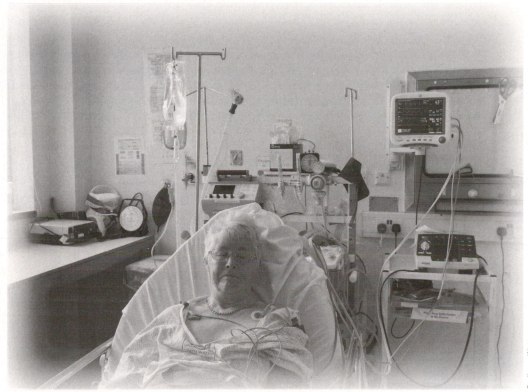

Realimage/Alamy

A 90-year-old woman came to the emergency room of a New York City hospital in critical condition with severe chest pain due to a serious heart attack confirmed with an EKG (Jauhar, 2000). The children of the woman were told that if nothing were done to help their mother she might die. Angioplasty (in which a catheter is threaded through an artery in the groin to open the blockage cutting off blood flow to the heart) probably offered her the best chance of surviving this crisis, but it was risky owing to her age, especially in an emergency. Their mother could die on the table or suffer brain damage or end up on a ventilator. The outcome was uncertain. She could be treated with a thrombolytic drug to dissolve blood clots, but this simple injection carried a risk of bleeding into her head. "So what should we do?" they asked. The children could not decide. Perhaps, they suggested, they should ask their mother, since she was awake and seemingly competent. The physicians could keep her alive, but only she should decide whether the effort would be worth it, the doctors and family thought. When the options were explained to the patient, she wanted assurances they could not give. In the end, she looked out at the room of white coats, slowly shook her head, and said she did not know. Because neither the family nor the patient could make a decision, the patient was taken to the cardiac care unit for observation. Who decides and what is decided for such life and death decisions? Such questions are not easy for anyone to answer.

Physicians, depending on their specialty, are confronted daily with these kinds of situations regarding life and death matters. Perhaps it is no wonder that physicians generally have shorter life spans than do other professionals, given the life and death nature of the decisions they must make. How long should specialists work to save the life of a premature baby who weighs slightly more than one pound at birth? Should a teenager be kept alive with life-support equipment after a skull-shattering auto accident? What is the "right" decision regarding a 90-year-old adult receiving life support? Even the most medically savvy people find dilemmas such as these wrenching and bewildering (Guyer, 1998).

With the technical transformation of medicine, it is now possible to extend the life of critically ill patients, restoring some to their former health, allowing others to live in a severely disabled state, and prolonging the dying process for still others. Such an ability to extend life raises ethical issues about quality of life and resource allocation, issues complicated by a lack of consensus in our diverse society. If the use

---

PRACTICAL MATTERS | MAKING LIFE AND DEATH DECISIONS

1. A newborn is brain-dead. Should the baby receive life support, as the religious mother wishes?
2. A baby has massive internal defects. Doctors recommend treatment after treatment, but the parents say, "Enough is enough."
3. A teen with a painful, fatal disease is ready for death; her parents aren't.

4. An old woman makes an adult child her power of attorney for health care. When that child says to let Mom die, another child says no.

Adapted from When decisions are life-and-death, by R. L. Guyer (1998, February 6–8), *USA Weekend*, p. 26.

of technology merely prolongs the life of a severely ill patient with no hope for cure, the costs to society are enormous, with very little or even negative return to the individual patient, noted E. W. Mebane and colleagues (1999). Yet, for religious or other reasons, some individuals may feel compelled to keep the individual alive at all costs. By keeping the person alive, a "cure" may be developed, and the patient may then have a good quality of life. Some may also ask, "Who am I to say that the person should be allowed to die?"

Many factors are included in decision making regarding **biomedical issues**. The mere financial costs of medical procedures developed through scientific research often present a dilemma. According to Donald Joralemon (1999), biomedicine has been associated with some negative issues such as little impact on some health problems afflicting the majority of the world's population, limited access to biomedical care for some due to a scarcity of medical personnel, and severe constraints on public health budgets. On the other hand, however, biomedicine has had positive outcomes (e.g., increased life span, eradication of smallpox, organ transplants, and effective treatments for many other diseases). In this chapter, we will address biomedical issues that we face in the 21st century. The issues are usually not clear-cut but fall in the grey zone; so who decides, and upon what grounds?

ETHICAL BEHAVIOR

Regarding ethical behavior, we must realize that medical ethics cannot really do more than lay out the issues clearly for all to see and argue about. If we expect medical ethics to give answers, this would be cause for disappointment, observed Arthur Zucker (2000), law and ethics editor of *Death Studies*.

WHAT IS ETHICAL BEHAVIOR?

Ethical behavior refers to a conscious reflection on moral beliefs and seems typically to be applied to specific cases in a setting as opposed to the nature of the setting itself (Chambliss, 1996). In professional settings such as hospitals, ethical behavior is usually based on the codification of moral principles by an occupational group and may reflect the group's long-range self-interest in its image as a servant of the community. In medicine, the use of bioethical language has made moral debates more abstract, rights-driven, individualistic, and centered on discrete cases. Ethical behavior is acting after trying to answer the question, "What should be done?" More practically speaking, the question is, "What can be done?" Ethics often assumes that people are autonomous decision makers sitting in a fairly comfortable room trying logically to fit problems to given solution-making patterns, notes D. F. Chambliss. But that is not the real world where decisions have to be made; that is the hypothetical world.

Inside hospitals, decisions are driven not by academic problem-solving techniques but by the routines of life in a professional bureaucracy. Too often the response to the patient is strictly based on the rules of the hospital without taking into consideration the individual situation. For example, the elderly mother of a friend of George Dickinson's had been comatose for several months when an emergency health situation occurred. This 95-year-old woman was rushed from the nursing home to the emergency room in the hospital and had major surgery performed on her. Her "living will" was ignored, and the hospital staff did what they had *learned* to do, according to the "medical rules." The surgery "successfully" kept her alive. She lived another three weeks and died in the intensive care unit. If students were exposed to "real settings" ("live scenarios") they would find the "exercise" less academic and more of a real-life situation and would respond to "What *can* be done?" and not "What *should* be done?" In the case of the friend's mother, *no action*, other than keeping her comfortable, would have been appropriate.

Yet, in the popular media and in political debates (from 2009 and beyond) over health care reform related to the Affordable Care Act (ACA), or so-called "Obamacare," one of the provisions for seniors was to be counseling for end-of-life provisions and for the creation of advance directives. Those opposing the bill (mostly members of the Republican Party, particularly "Tea Party" members) tried to raise fears that "death panels" would be established that would determine the care of "Grandma" based upon utilitarian concerns—the government might want to kill the elderly in order to save on medical costs. When an impersonal "someone else's mother" is concerned, most people would have no problem with "no action," but if it is the issue of "*my* mother," the issue is not so easy. As a result, many seniors are suspicious of advance directives because they fear that their future options for health care may be limited.

Chambliss (1996) observed that ethical behavior in health care is inseparable from the organizational and social settings in which ethical problems arise. He noted that ethical problems are not random or isolated failures of the system; they are often in fact fundamental, if unintended, products of that system. People work in organizational and professional roles and settings, and these shape their behavior; ethical decisions are not made in some hypothetical "free choice zone." For example, nurses and physicians have very different professional roles, though both treat patients. Doctors have vastly more power, give the orders, are rewarded for scientific expertise, and hold the legal responsibility for the patient's well-being. Nurses, on the other hand, carry out the doctors' orders, are largely rewarded for organizational skills, and handle much of the day-to-day work of patient care. Because of training, physicians tend to lean toward aggressive treatment and are often reluctant to stop treatment, yet the nurses may be getting frustrated dealing with a patient whose condition is not improving and who is begging to be left alone. Thus, there is a moral dilemma between the more powerful and the "ones carrying out the orders," noted Chambliss (1996).

George Dickinson recalls back in the 1960s when kidney dialysis machines were scarce, not abundant as they are today. A medical intern lamented to him that dialysis treatment for a young woman in his hospital had been stopped, which meant that she would die. Her dialysis was stopped because they had only three machines but had four persons in need of dialysis; thus, **triage** was practiced because of limited sources. A "bureaucratic decision," based on the policy of the hospital, had deemed that she must "come off" and thus would die. Stopping kidney dialysis leads to uremic poisoning and death. It was "someone else's turn" to have the machine. As Chambliss said, an organization made the decision—very impersonal in what is a very personal situation.

BIOETHICISTS

Patients today believe that they have a *right* to take part in their own care—they expect their wishes to be a determining factor in doctors' decisions. The right of patients to refuse or discontinue treatment is now an acknowledged aspect of American medicine. The appropriate allocation of scarce resources (deciding who will get treatments or organs when there are not enough to meet existing needs) is a current issue in this time of budgetary constraints.

How does one make these difficult decisions regarding various moral problems posed by modern medical science and technology? In our very specialized society today, we are not alone in decision making regarding biomedical issues. There is a specialist to assist with these difficult medical decisions—a **bioethicist**. Such a specialist is a recent addition to the world of medicine (DeVries & Subedi, 1998). Hired by hospitals and academic medical centers, bioethicists are called on for their philosophical and legal expertise to help with difficult decisions about medical treatments and end-of-life care. This medical consultant meets privately with the patient, family, and medical staff to collect facts and beliefs about the patient's illness and wishes, and to discuss how each individual sees the situation (Guyer, 1998). Often this fact-gathering search ends in the "most justifiable decision," and the case is ended. However, it is not a perfect world, and in some situations the bioethicist gathers all parties together to talk. If this fails, the hospital ethics committee will conduct an official review and make formal recommendations. Bioethicists have no medical or legal authority, notes Ruth Guyer.

Guyer (1998) described a bioethicist as someone with a bioethics degree. According to the Hastings Center (2011) 26 universities in the United States offer master's degrees in bioethics and 19 offer Ph.D.s in bioethics or a doctorate in philosophy, law, medicine, or religion with an emphasis on bioethics. The bioethicist is trained in philosophy, law, psychology, religion or humanities, and social science disciplines and learns to "talk the talk" of medicine when working in hospitals. A bioethicist can also be a doctor, nurse, or other clinician who has studied moral philosophy. George Dickinson recalls meeting an individual working in a hospital in Boston in the early 1980s who introduced himself as a "moral philosopher." After talking with him, it was obvious that he was what today is called a bioethicist.

In 1992 the Joint Commission (formerly the Joint Commission on Accreditation of Healthcare Organizations) mandated that health care organizations come up with some way of addressing bioethical concerns. By the year 2000, 95 percent of general hospitals offered ethics consultation or were starting up a consult service (O'Reilly, 2008).

Many individuals hope this new bioethical presence in health care settings will make medicine more humane and more ethical. However, a sociological view suggests that bioethicists might begin with good motives, but that their position in the structure of health care may come to color their decisions. If we consider the interests of those who are represented by bioethicists, most bioethicists claim that they represent the patient, protecting her or his autonomy against the power of medicine. However, journalist Ruth Shalit (1997) came to a very different conclusion. She noted that a swelling number of bioethicists who are cashing in on their ethical

Harris, S/CSL, CartoonStock Ltd

expertise and marketing their services to managed care executives are eager to dress up cost-cutting decisions in labels and lofty principles.

As R. DeVries and J. Subedi (1998) observed, when looking at the organizational location of bioethics, that the presence of bioethicists in medical institutions leads to an affinity between bioethicists and other professionals there. They liken the role of the bioethicist to that of a public defender in the American legal system. The formal role of each is to represent the interests of a client in a large and confusing bureaucracy, but, like public defenders, bioethicists must maintain good relationships with other members of that bureaucracy, many of whom are working against the interest of their clients. Given this organizational situation, bioethicists will be inclined to represent the interests of medical professionals and medical institutions over those who are merely passing through—the patients and their families—said DeVries and Subedi.

USE OF THE BODY IN MEDICAL RESEARCH AND TRAINING

The debate in the United States about the marketing of fetal tissue for medical research and drug production is an example of how culturally specific conceptions of the person influence where the boundary is drawn between alienable and inalienable body parts (Joralemon, 1995). Perhaps transplanting organs, together with genetic engineering, artificial reproduction, therapeutic cell lines, and mechanical implants, will be enough to overwhelm the intuitively unambiguous connection we feel to our bodies.

Embryonic stem cell research is providing an exciting potential treatment for a host of diseases, but the controversial research has left many conservative lawmakers hesitant about its use (Nelson, 2000). During an eight-year ban on public funding of such research during the Bush administration, funding for research efforts was effectively transferred to the private sector. Since 2001, Congress had numerous bills pending to address this issue. The more conservative argument questions the federal government's sponsoring the "killing of human embryos." This approach concludes that all humans should be treated as persons and that research must not deliberately harm any humans, including embryos—at the moment of conception a new and unique individual comes into being and needs protection and nourishment to become an adult person (O'Mathuna, 1996). A more liberal argument, such as that of Joseph Fletcher, in his well-known book *Situational Ethics* (1966), views every situation as ethically unique so that we can never say that a certain act is always right or always wrong. Fletcher argued that it is ethical to treat different humans in different ways.

On March 9, 2009, President Obama lifted the eight-year-old ban on federal funding for embryonic stem cell research, putting the weight of his office, he said, on the side of scientists who believe "these tiny cells may have the potential to help us understand, and possibly cure, some of our most devastating diseases." The executive order marked the third time in his young administration that Obama had reversed Bush-era policies at the intersection of public-health goals and ethical questions about the nature of human life (Wilson, 2009).

Embryo stem cells are desirable because of their ability to be transformed into any cell type, which establishes a blank slate waiting to be filled in by the appropriate instructions. Theoretically, the cells could serve as a ready supply of replacement tissue for damaged or missing cells in patients with diseases such as diabetes, Huntington's chorea, and Alzheimer's disease, or in those who have suffered spinal cord injuries; thus, they have a tremendous potential for the medical field. National Institutes of Health researchers have discovered a method of generating virtually unlimited supplies of specialized neurons from mouse embryonic stem cells. The most promising source of these human stem cells is leftover embryos discarded from *in vitro* fertilization clinics.

Cells have also been derived from aborted fetuses, raising a number of objections, noted B. Nelson (2000). Fetal stem cells share many properties with their embryonic counterparts, though technically there are some differences. Opponents of using discarded embryos to harvest stem cells have based their arguments on their belief that life begins at the moment of conception. Thus, researchers who harvest cells from frozen four- or five-day-old embryos, therefore destroying them, are then committing the act of abortion.

The individuals who perform the tasks of removing body parts are called technicians and are employed by companies that retrieve body parts, also known as harvesters, such as the Anatomic Gift Foundation in Illinois (O'Meara, 1999). Because of a federal law prohibiting the sale of human tissue or body parts, the traffickers have worked out an arrangement to expedite the process and remain legal. The harvesters receive the fetal material as a donation from the abortion clinic. In return, the clinic is paid a site fee for rental of laboratory space where technicians perform the dissections. The harvesters then donate the body parts to the researchers and, rather than pay for them, the researchers donate the cost of the service with a formal price list. No laws are broken, because no body parts from aborted fetuses are sold. Normally, the fetus is dissected, and the body parts are shipped to their destination.

Nearly 75 percent of women choosing abortion agree to donate the fetal tissue (O'Meara, 1999). Anti-abortionists agree that if women think that something good is coming out of the abortion, it is easier for them to make the decision to abort. Thus, the ethical issue is difficult to separate from the scientific issue.

Human cadavers have been used for training medical students in the United States since the 19th century. Initially, such usage of dead bodies was illegal, but the necessity of using cadavers for gross anatomy laboratories in medical schools eventually won approval. Although using cadavers for learning purposes may not be agreed to by all individuals, dead bodies are not supposed to be commodities. Donors usually will their bodies with the thought of advancing science.

Willed-body programs with surplus cadavers are not allowed to sell them for a profit, only at cost. However, with universities drawing up price lists—$350 for a full leg, $500 for a torso—and private companies vying for the specimens, the competition for cadavers is heating up (Kowalczyk & Heisel, 1999). The family giving the body to science trusted the medical school to act ethically and use the body for students to learn; yet donors sign contracts allowing medical schools to use their bodies for any legal purpose. Normally, medical schools have enough cadavers and sometimes have "leftovers." Computer software for learning gross anatomy is becoming more popular in medical schools; thus, the day may come when medical schools need fewer human cadavers.

Cadavers are also used by the U.S. Department of Transportation to measure the effectiveness of various crash protective devices in automobile collisions (Schiedermayer, 1994). The use of cadavers in crash testing has been significantly curtailed in recent years, however, and they are now used primarily to calibrate the dummies. Only a few dozen cadavers are used annually for testing crash protection devices in car collisions.

Though not familiar to the public, doctors in training in emergency rooms sometimes perform an unnecessary invasive procedure, just for practice and without informed consent, on patients for whom resuscitation efforts appear to be failing, according to an article in *The New England Journal of Medicine* (Kaldjian, Wu, Jekel, Kaldjian, & Duffy, 1999). Though the practice does no physical harm and one-third of the doctors in training say they approve of it, the procedures are indeed questionable to some. Physicians opposed to the practice argue that it is medically unnecessary and violates patient's rights or dignity, and the procedure could be learned in other settings. According to L. C. Kaldjian and colleagues, the practice reduces patients to "mere objects of use in education" and may reinforce unethical attitudes among interns and residents, especially because it is done without the patient's consent. Such "practice procedures" involve threading a tube into the femoral vein in the groin, a procedure best learned on a live patient because pulsating vessels are used as landmarks to guide the insertion of the tube. Though using the human body and its parts in medical research and training may be beneficial to society, such usage lacks the approval of all segments of society.

ORGAN TRANSPLANTATION

Attitudes toward organ transplantation illustrate different people's cultural constructions of the sacred, noted British sociologist Clive Seale (1998). In the United States, rationality about the body under the tutelage of medicine has gone far, and

| WORDS OF WISDOM | TEACHING SURGERY WITHOUT A PATIENT |

To avoid some of the ethical dilemmas of using live patients or dead ones, surgical residents can perform "operations" on machines rather than patients in a growing number of teaching hospitals. The cost is more than a million dollars just in simulation equipment. The system is designed to train residents in laparoscopy or minimally invasive surgery performed through small incisions. The simulator makes the surgeon feel like she or he is actually cutting into tissue.

The students observe what they are doing on a television screen. Residents can also practice life-support techniques on a lifelike simulated patient who has a chest that rises and falls with each "breath." Some physicians are leery of these simulation techniques and are not convinced that a computerized system, no matter how realistic, can teach residents the skills they will need in a real operating room. Yet, such simulated procedures do not carry the anxiety of working against the clock in a real setting on a live patient. With our highly computerized society, who knows what the 21st century will bring to medicine in the operating room. Stay tuned!

From Teaching surgery without a patient, by K. S. Mangan (2000, February 25), *The Chronicle of Higher Education, 45*, pp. A49–A50.

the transplantation of body parts is accepted as the rational solution to otherwise certain death for some individuals. Likewise, the use of animal organs is acceptable, though in some countries, for example Israel and Denmark, this is resisted.

In Japan, there is great resistance to human organ transplantation, deriving from Japan's equivocal position as a society that on the one hand seeks to break free from religious and traditional forms, but at the same time is suspicious of the wholesale importing of Western approaches to matters of life and death, observed Seale. In Japan, only one heart transplant had ever been performed before 1995, and the doctor responsible was prosecuted for murder (Lock, 1995). The Japanese have religiously inspired concerns about the wrath of ancestors whose bodies have been violated. The Japanese resistance to organ transplantation can also be equated with earlier opposition to life insurance in the United States and European countries and involves similar issues about what is to be considered as sacred and beyond human calculation.

Some of the ethical issues involved in organ transplantations are entitlement criteria for acceptance to waiting lists such as age, ownership of cadaver organs, criteria for allocation of scarce organs, surgery on healthy people as live donors of organs, purchase of organs from the poor to serve the rich, and abortion for viable replacement tissue. If solid organs from different animals were available for transplantation purposes, called **xenotransplantation**, many of these ethical hazards could be diminished (Koshal, 1994). Besides the rejection problem with many animal tissues, the use of animals for transplant purposes has its ethical problems also. Do animals have rights? If so, are these rights overridden by the rights of humans? Should xenotransplantation, involving the intentional death of an animal, be considered before exhausting all the possibilities of human organ replacement? Are there ethical issues in using healthy animals for their organs and killing and consuming dead animals for food?

In the past 10 years there has been a significant surge of interest in xenotransplantation. Attention has been focused on the pig, which is already sacrificed in large numbers for food consumption. Pig heart valves have been used for the last

30 years. Current efforts are under way to breed pigs specifically for organ donation purposes. A survey in the United States suggests that 51 percent of those polled would be willing to receive an animal organ if needed (Koshal, 1994).

The initial donor policy in the United States was one of pure voluntarism— donation was made legal and it was hoped that volunteers would come forward (Weiss & Lonnquist, 2003). Courts ruled that competent adults could voluntarily donate organs to relatives, and minors could donate only with parental and judicial consent. Yet, the demand for organ donations soon exceeded the supply, and a more assertive organ donation policy of encouraged voluntarism evolved.

In the United States individuals have to give permission for organ donation. They can do this when they apply for a driver's license or create a living will, or they can have a family member do it for them (with a durable power of attorney) at the time of death. Canada is now considering an "Opt Out" system, where it is assumed you will be a universal donor unless you request not to be (Eggertson, 2012).

Organ donation has led to some serious ethical issues. With the current demand for human organs for transplantation purposes much greater than the supply, a tendency exists in some parts of the world to sell certain organs. Though such selling is unlawful in the United States, a company in Germany routinely sends a form letter to all persons listed in the newspaper as having declared bankruptcy, offering $45,000 plus expenses for a kidney that is then sold for $85,000 (Weiss & Lonnquist, 2003).

Many ethical, moral, and social questions are raised by issues related to organ donation: Is it okay for one to sell body parts for money? Is it appropriate for a small number of people to benefit from public financing of an expensive technology, when a larger number of people could benefit from expenditures on a greater diversity of less expensive problems? What criteria should be used to select organ transplant recipients, and are such criteria consistent with the values of a democratic society? What are the medical and quality-of-life outcomes associated with organ transplantations? There are probably more questions than answers on the ethics of organ donations, because numerous opinions exist regarding these questions.

A readers' panel of *Nursing Standard* raised the question (2011) regarding the appropriateness for families of donors to receive compensation to help pay for the donor's funeral expenses. Further clouding of this issue is a recent court decision in the United States that blood stem cell donors are entitled to be compensated for their donation. Judge Andrew Kleinfeld (cited by Fallis, 2012) ruled:

> When the "peripheral blood stem cell apheresis" method of "bone marrow transplantation" is used, it is not a transfer of a "human organ" or a "subpart thereof" as defined by the statute and regulation, so the statute does not criminalize compensating the donor.

As a result, according to Fallis (2012) a coalition has formed, including patients needing bone marrow transplants, a doctor from Minnesota who specializes in bone marrow treatments, and MoreMarrowDonors.org, a California-based non-profit organization that wants to give donors $3,000 in scholarships, housing allowances, or gifts to charities in exchange for their donations.

Laura Spinney (2011) discusses new trends in organ transplantations and a new "Red Market" for the "Black Market." Her book describes the global trade in body parts for "health care professionals" in surgery, fertility, and adoption industries. According to Spinney (2011), the World Health Organization estimates that 10 percent of world organ transplants are obtained illegitimately. She cites and reviews research by investigative journalist Scott Carney on emergent world-wide trends in the selling of organs, blood, eggs, and surrogates. Spinney discusses many ethical, social, and medical issues caused by the ability of medicine practitioners to do transplants in a world where there is inequality (both for individuals and nations) of financial resources for would-be donors and recipients.

There is also the question of ownership as it applies to the corporeal part of the person, notes anthropologist Donald Joralemon (1995). Legal scholars have observed that the courts and public opinion have come a long way toward recognizing property rights and commercial value in some of what the body produces during life—for instance, blood, hair, semen, and ova. Exactly which body parts are considered to be integral to personhood varies by culture. For example, selling hair may be a legitimate economic venture in America, but in some societies (e.g., where voodoo is practiced) it would subject the individual to the risk of sorcery, according to a more inclusive notion of bodily integrity, stated Joralemon.

To take organs from a dead body, when is the person actually dead so that the organs can be removed? As discussed in Chapter 2, the definition of *dead* is not so easy to determine, yet the "brain dead" definition seems acceptable in many states in the United States and in many European countries today, for organ donation purposes. Research (Callender, 1987) in minority communities on attitudes toward organ donation, however, reveals widespread suspicion that brain death might be declared prematurely to provide access to organs for transplantation.

In 1992, the University of Pittsburgh Medical Center adopted a controversial plan that added a new category of potential donors—"non-heart-beating cadavers" (Joralemon, 1999). This category refers to persons whose lives will probably end when mechanical support is withdrawn but will not meet the criterion of brain death as long as life support is maintained. If the patient or next of kin requests that life support be withdrawn and if there is a request that the organs be donated for transplantation, these procedures would be followed: (1) Take the person to an operating room where life support is withdrawn; (2) administer medication to minimize any pain; (3) declare the person dead; and (4) procure organs immediately after death.

The National Organ Transplantation Act of 1984 authorized the creation of a national organ and transplantation network. United Network for Organ Sharing (UNOS), a nonprofit organization, won the federal contract to operate this network and a system of regulations governing the distribution of organs from donors. Prior to the development of formal federal regulations, organs were generally distributed in local areas first, even if there were sicker patients elsewhere. Later federal regulation mandated a set of performance goals for organ allocation, generally emphasizing patients' medical urgency over local matching. Transplant surgeons and UNOS objected to government agencies setting the rules, saying allocation of organs is a medical decision that should be made by medical experts. UNOS lobbied Congress to modify these rules on distribution. In response, Congress imposed a one-year moratorium on adherence to the regulations. However, that moratorium expired in October of 1999, and the regulations were

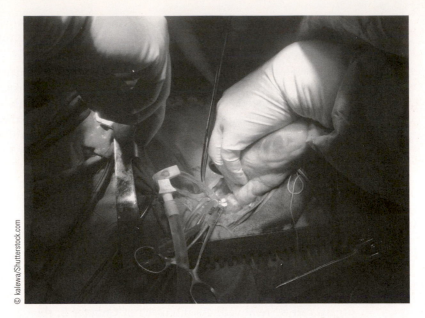

© kalewa/Shutterstock.com

Organ transplantations have become common in American medical practice.

implemented. For a current review of research in this area see F. G. Miller's (2011) article "Death and Organ Donation: Back to the Future."

According to United Network for Organ Sharing (2009), as of November of 2009, there were approximately 105,000 candidates waiting for organs. To give a perspective on the issue of donor shortages, in the entire year of 2008 there were 27,965 transplants (21,747 from diseased donors and 6,218 from living donors). This would mean that if there were no new candidates for organs, it would take approximately four years to eliminate the backlog of candidates waiting for organs. As of this writing, the most current data available show that, from January 1 through July 23 of 2009, there were 16,679 transplanted organs from 8,541 donors—most donors providing more than one organ.

In addressing the issue of donor shortages of organs, surgeon Robert Sade (1999) suggested that the most common reason for nondonation is denial of consent by the donor's family. Sade recommended that patients, before their deaths, or families, after the deaths, could select a category of recipients or a specific recipient for donated organs. This would be similar to the process involved in estate planning, where assets are distributed to designated groups or causes. By designating organ recipients, the donor could feel comfortable that the donation was consistent with her or his personal beliefs and values.

Going a step further, British ethicist John Harris (Jenkins, 1999) suggested that bodies should become public property on death, allowing surgeons to harvest usable organs without going through the traumatic process of asking grieving relatives for consent. Harris's idea fits the Western cultural construction of donation of blood and body parts as a gift of life. The idea is that individuals live within a community of faceless, statistically defined, risk-bearing individuals whose obligations to each other are based on an abstract sense of common humanity, rather than particular ties of blood or kinship, noted Clive Seale (1998).

Harris also recommended that a commercial market in live organs could be set up and that people should be able to trade their body parts to cut down on waiting lists. Perhaps husbands and wives who divorce in the near future will make organ donations a part of their divorce settlement (Rumbelow, 1999). The monetary value of a previously donated organ to a spouse would figure into the settlement.

Government officials argue that if organ transfers were left to the free market, evil results would ensue (Brown, 1996). The worst-case scenario is that people would begin murdering others and stealing their body parts to profit from the sale of their organs. Though the idea of people buying and selling human organs in a free market is repulsive to many individuals, the government has no right to interfere, suggests a conservative argument on this issue.

Catherine Hollander (2013) asks the question, "What's the Fairest Way to Dispense Donated Organs?" Samia Madwar (2011) addresses this issue by claiming that UNOS is now considering a number of factors in deciding who gets an available organ, other than the egalitarian criteria of the recipient's place in line without regard to wealth, class, or race. Among those criteria are the likelihood of a successful transplant, age of the recipient, and the age of the donor. These criteria can sometimes be less than objective, even if they produce more successful outcomes.

Regarding social class variations in the receiving of organs, the whole idea of considering a future where the rich have the ability to secure organs from the poor is indeed problematic. Paying $15,000 may be considered a reasonable price to a wealthy individual, and it would be almost irresistible to a potential donor whose annual salary equals that amount. Thus, the whole question of who would receive an organ poses ethical questions. Is it a mere coincidence that the former New York Yankees baseball great Mickey Mantle or *I Dream of Jeannie* and *Dallas* television star Larry Hagman received liver transplants "almost immediately" when needed, whereas individuals who were not celebrities died while on waiting lists for this same organ? (Both Mantle and Hagman died within a short period of time following their transplants.) In our very political world, who one is can indeed often make a difference. Money and influence can definitely be factors in the distribution of organs today. The question of ethics remains, and it is one that will have to be addressed in the years to come.

EUTHANASIA

The meaning of **euthanasia** today is not the same as the original Greek meaning of "a good death" or "a gentle death," because the current meaning is the administration of death to the dying—a hastening or advancing of death. Today, patients and their families are demanding increasing control over their health care and finalization of life. A desire for more personal control reflects the trend in Western society. People in Australia, Canada, the United States, the United Kingdom, and other parts of Europe and the world fear that they will not be able to die in this gentle, easy way, nor "in control."

Your insurance only covered the removal of the damaged organ...you'll have to put the transplant in yourself!

In 1998, Michael Manning wrote a book entitled *Euthanasia and Physician-Assisted Suicide: Killing or Caring?* in which he traced the history of the word euthanasia. He argued that "The term euthanasia originally meant only 'good death,' but in modern society it has come to mean a death free of any anxiety and pain, often brought about through the use of medication. Most recently, it has come to mean 'mercy killing'—deliberately putting an end to someone's life in order to spare the individual's suffering."

According to Euthanasia.ProCon.org (2013), the Pro-Life Alliance defines euthanasia as "Any action or omission intended to end the life of a patient on the grounds that his or her life is not worth living," while the Voluntary Euthanasia Society defines it as "A good death brought about by a doctor providing drugs or an injection to bring a peaceful end to the dying process."

A recent Gallup opinion poll (Saad, 2013) discovered that 70 percent of Americans are in favor of allowing doctors to hasten a terminally ill patient's death when the matter is described as allowing doctors to "end the patient's life by some painless means." However, only 51 percent support it when the process is described as doctors helping a patient "commit suicide." In 2005 (Gallup) and in 2000 (Clark et al.) found that 75 percent in the United States, 76 percent in Australia, 77 percent in Canada, and 82 percent in the United Kingdom favor some form of euthanasia.

Yet, the rapid development of technology has threatened individuals' sense of control (Kelner & Bourgeault, 1993). Whereas new technologies have made it possible to sustain life longer, patients often have limited say in initiating these complex treatments. If an individual, for example, is in very bad health, has a poor medical prognosis, and desires to die, should the medical profession have the option of assisting that individual with death?

Distinguishing between killing and letting a person die seems easy, at least on the surface, noted Margaret Battin (1994). *Killing* involves intervening in an ongoing physiological process that would otherwise have been adequate to support life, whereas *letting die* involves not intervening to aid physiological processes that have become inadequate to support life. Yet, there are ambiguous cases. For example, removing a respirator may seem to be letting the patient die, or it might be viewed as killing the patient. Despite grey areas, to grant that there is a difference between killing and letting die that is adequate enough to support assertions against killing, is not to grant that there is or must be a moral difference.

A now rather famous article by James Rachels (1975) in *The New England Journal of Medicine* challenged the conventional conception of the moral difference between killing and letting die with this case: Smith and Jones will each inherit a considerable fortune from their respective six-year-old cousins, should the cousins die. One evening, while his cousin is taking a bath, Smith sneaks into the bathroom and drowns him. Meanwhile, Jones is also planning to drown his own cousin, who is also taking a bath, but as Jones sneaks into the bathroom the child hits his head and slips under the water, and Jones does nothing to save him. Now both children are dead. Smith has killed his cousin; Jones has merely allowed his cousin to die. Clearly a conceptual and causal difference in what the two men did exists, yet is there a moral difference? Not really, because they both acted despicably, and it is no excuse for Jones to say that he did not kill his cousin, he "merely" let his cousin die.

Richard Trammell (1975), however, argued that there is a difference between Jones's killing and Smith's letting die. He said that the difference is like trying to taste the difference between two fine wines when they are mixed with green persimmon juice. There is a difference between the wines, but in those overpowering circumstances one cannot see what it is. Just so with the drowned cousins—because the behavior of both men is so repugnant, it has a masking effect, and we cannot see the difference in what they did.

Battin (1994) said that in some cases killing is right—the *coup de grace* granted a mortally wounded, dying soldier who cannot be saved on the battlefield, or abortion to save the life of the mother. The **double effect** is another example when killing may be right. According to Euthanasia.ProCon.org (2013), double effect is an ethical principle that claims it is acceptable if a morally good action has a morally bad side effect, provided that the side effect wasn't intended (even if it was foreseen). The principle is used to justify situations in which a doctor gives a patient drugs to relieve painful symptoms, when doing so may shorten the patient's life. Michael Leming was a hospice volunteer for a patient who was administered a strong dose of liquid morphine by her physician with the intended goal of reducing her severe pain, while the physician was clearly aware that such a dose would hasten his patient's death. She died within 48 hours of the injection.

It must be recognized that the distinction between killing and letting die does not succeed in carrying the moral weight often placed on it. Even if one could always draw clear conceptual lines between the two, the conceptual distinction almost always brings along with it an unjustified moral distinction. To describe a procedure as "mercy killing" could infer that because it involves killing it is wrong, thus taking the moral baggage along with the conceptual distinction, though the inference does not follow. Each case needs to be argued on its own merits, suggested Battin (1994).

SANCTITY-OF-LIFE VERSUS QUALITY-OF-LIFE DEBATE

Occasionally, a newspaper article will say that the survivor of a serious automobile accident is in a "vegetative state." Has the patient died? Is the patient no longer human? Now consider the situation in which an individual is assaulted and put into an irreversible coma. Has the assailant committed murder or assault and battery? If this same patient is disconnected from life-support equipment or **intubation** (tube feeding) and dies, who caused the death—the person who committed the crime or the medical personnel who withdrew the life support or feeding device? In general, people respond to such questions about medical conditions from one or two orientations concerning the meaning of life. The first orientation emphasizes the **sanctity of life**, whereas the second emphasizes the **quality of life**.

SANCTITY-OF-LIFE VIEW Euthanasia from a sanctity-of-life orientation would contend that all "natural" life has intrinsic meaning and should be appreciated as a divine gift. As a consequence, human beings have the obligation to prolong life. Hessel Bouma and colleagues (1989) noted that the Hippocratic tradition in the medical profession recognizes that physicians' responsibility to terminally ill patients is to simply "mitigate their suffering while allowing them to die." Furthermore, the Hippocratic Oath, reacting against an earlier practice of actively and intentionally hastening the deaths of terminally ill patients, forbids the giving of "a deadly drug" to dying patients. According to the Hippocratic Oath, part of the essential qualification for being a doctor is an absolute rejection of killing the patient (Jeffrey, 1993). The following two quotations illustrate the sanctity-of-life orientation:

> The quality of all our lives suffers, it insists, unless every human life is considered inviolable because of the very fact of its existence. A dying patient's relationship to those about him or her symbolizes the relationship of all individuals to one another. To practice direct euthanasia, even at the request of the patient, is to weaken the claim of each one of us to the right to have others respect and not violate us. (Weber, 1981, p. 49)
>
> No horror against life is impossible once we have allowed anyone but the Creator to usurp sovereignty over life. Whom the gods would destroy they first make mad. Legalized euthanasia is such madness. (Morriss, 1987, p. 149)

Mebane et al. (1999) examined racial differences in patient preferences for end-of-life treatments and physician-assisted deaths and reported that black patients tend to request more life-sustaining treatments, view physician-assisted deaths more negatively, and place a higher value on longevity than white patients.

Likewise, physicians share similar preferences for end-of-life treatments with their patients: Black physicians are more likely than white physicians to request aggressive medical treatments for themselves in the persistent vegetative stage and organic brain disease situations. The African Americans in the Mebane and colleagues study tended to be taking a sanctity-of-life perspective. They preferred living "at all costs."

Religion may be a factor in the sanctity-of-life view. R. Kalish and D. Reynolds (1976) concluded that African Americans are reported to be more religious than whites; thus, black patients and physicians might feel guilty if treatment were stopped, based on religious reasons. Mebane and colleagues (1999) argued that longevity is valued as an intrinsic good more for blacks than for whites, based partly on the history of oppression experienced by older blacks and the idea that suffering may be an expected part of life rather than a reason to terminate life.

QUALITY-OF-LIFE VIEW   From a quality-of-life perspective, the concept of quality is even more difficult to define. Who should decide when life has quality? *Quality* is a relative term, and its meaning often changes as one progresses through her or his life cycle and as a result of social circumstance. Would not your quality of life be diminished if you were paralyzed, blind, and/or deaf? Yet, many people with these disabilities experience a life of quality filled with purpose and meaning.

The quality-of-life orientation holds that when life no longer has quality or meaning, death is preferable to life. What may be quality to one may not be to another. Some individuals might prefer death over life, for example, if they were in a permanently paralyzed state, yet others may find quality in such a situation and consider themselves to have a high quality of life. Others might consider the inability to walk to have no quality and would choose death to life. To accept any frailty in life might be more than some individuals can bear, and they might consider such a life with no quality unbearable and would therefore prefer death.

The research on perceptions of end-of-life issues by race, reported by Mebane and colleagues (1999), suggests that whites in their study follow more the quality-of-life view. They found that white physicians are more likely to want physician-assisted death for themselves in the persistent vegetative state and organic brain disease scenarios. Black doctors are less likely to view tube feeding as "heroic" and less likely to report positive attitudes toward advanced care planning than white physicians. (Advanced care planning would limit treatments.)

The hospice movement takes a quality-of-life view, in that the goal is for the patient to live with dignity even while dying. Hospice philosophy suggests that the individual have a good death, free of pain, with a sense of control over her or his life.

Though advanced medical technology aids in prolonging life, the quality of life may be limited; thus, modern technology is a curse in the eyes of some quality-of-life supporters. Individuals supporting the sanctity-of-life view should see advanced medical technology as a positive development, in that life can be extended, no matter the quality. However, Mebane and colleagues (1999) found that African Americans, who are sanctity-of-life advocates, according to their findings, are more likely to have negative perceptions about hospital care and have some anxiety about life-support equipment being stopped prematurely *because of* their race. Thus, some aspects of life-support technology are *not* viewed positively

DEATH ACROSS CULTURES | BRITISH COURT PERMITS MAN TO HELP WIFE DIE

LONDON, Nov. 30 (Reuters)—The husband of a critically ill British woman won a legal fight on Tuesday to take his wife to Switzerland to end her life, a ruling that throws Britain's tough laws against assisted suicide into confusion.

A High Court judge said it was not for him to decide whether the husband of a woman identified only as "Mrs. Z" should be allowed to take her abroad to end her life, and over-turned a temporary ban.

"I have decided to discharge the injunction," Justice Mark Hedley said in his ruling.

While effectively lifting the ban on anyone helping Mrs. Z travel abroad, the judge said it was for the police to decide whether to charge anyone who assisted her to die.

"The court should not frustrate indirectly the rights of Mrs. Z," he added,

"The role of Mr. Z is now a matter for the criminal justice agencies."

Belgium, the Netherlands and Switzerland have passed laws allowing assisted euthanasia, but Britons risk up to 14 years in jail for helping a patient die, a penalty that is among the harshest in Europe.

Mrs. Z, who in 1977 received a diagnosis of a degenerative brain disease, could not make the trip to Switzerland without help.

The case was brought to court by a local council that provides home care for Mrs. Z over fears that it would be liable if it allowed her to leave the country and end her life. Last week the council was granted a temporary ban it had requested.

British court permits man to help wife die. November 30, 2004. Copyright © 2004 Reuters Limited. All rights reserved.

among the African Americans in the study. Looking at the overall impact of technology on death, Ivan Illich (1976, pp. 207–208) illustrates the dilemma:

> Today, the man best protected against setting the stage for his own dying is the sick person in critical condition. Society, acting through the medical system, decides when and after what indignities and mutilations he shall die. The medicalization of society has brought the epoch of natural death to an end. Western man has lost the right to preside at his act of dying. Health, or the autonomous power to cope, has been expropriated down to the last breath. Technical death has won its victory over dying. Mechanical death has conquered and destroyed all other deaths.

THE RIGHT TO DIE If we conceive of the right to die as being like most other rights (e.g., the right to property, the right to an education, the right to travel, the right to go about our daily business in peace, and so on), then we should, as in all these other cases, have a right to die and be able to call upon the resources of the community to help us to fulfill that right, noted Scottish psychologist John Beloff (1989).

Although the overwhelming majority of those who are terminally ill fight for life to the end (Hendin, 1997), some individuals, whether terminally ill or otherwise, may wish to die and feel that it is their right to die. H. M. Chochinov, K. G. Wilson, and M. Ennis (1995) reported that 44 percent of 200 terminally ill patients reported occasional wishes that death would come soon. Indeed, does an individual have a right to die? The media have highlighted the actions of Jack Kevorkian, the retired pathologist, who has assisted individuals who wished to die. In fact, Kevorkian was a champion of the right-to-die issue throughout the 1990s. Between 1999 and 2007, Kevorkian served eight years of a 10-to-25-year prison sentence for second-degree murder. He was released on June 1, 2007, on parole due to good behavior. For a critical review of the actions of Jack Kevorkian,

Jack Kevorkian has advocated for severely ill people's right to commit suicide and claims to have assisted over 100 individuals with their deaths. Kevorkian served eight years of a 10-to-25-year prison sentence for practicing AVE (second-degree murder). He was released on June 1, 2007, on parole due to good behavior, and he died on June 3, 2011 at the age of 83.

Derek Humphry, Philip Nitschke, and other early leaders in the "Right to Die" movement to assist dying patients who wished to end their lives, see the article entitled "Who Is Leading Us There?" (Fenigsen & Fenigsen, 2012).

In 1996 the U.S. Ninth Circuit Court of Appeals in the state of Washington ruled that, based on the Fourteenth Amendment, an individual has a right to determine when and how she or he will die, as this court determined that it was unconstitutional for the state to prohibit physicians from prescribing life-ending medication for use by terminally ill patients.

The whole question of the right to die was in the news in the United States and throughout the world during the 1990s, and the issue is continuing into the 21st century. The Terri Schiavo case in Florida in 2004 and 2005 focused on this topic in the media and all three branches of the federal government. The difficult decision to forgo (withhold or withdraw) life-sustaining treatment, however, has been lacking in the medical education of physicians (Faber-Langendoen & Bartels, 1992). A major emphasis in medical education is placed on the diagnosis and treatment of diseases, yet traditionally there has been limited guidance given for physicians regarding which treatments to discontinue once a decision to forgo treatment has been made. Medical education should teach students and residents to grapple with the complexity of forgoing treatment, noted K. Faber-Langendoen and D. M. Bartels (1992).

Federal legislation now allows one to rightfully refuse medical treatment. The **Patient Self-Determination Act (PSDA)** is a federal law mandating that health care providers inform patients about their right to refuse and accept treatment, even when they have lost decision-making capacity (Timmermans, 1999). The purpose of the Patient Self-Determination Act is to inform patients of their rights regarding decisions toward their own medical care, and ensure that these rights are communicated by the health care provider. Specifically, the rights ensured are those of the patient to dictate their future care (by means such as a **living will** or **power of** attorney), should they become incapacitated. The PSDA has been criticized because the timing of informing people—as they enter the hospital for treatment—is bad. Thus, the right to die is being wrestled with in the courts, in legislative bodies, in ballot boxes, and in the media. This is a topic that probably will not go away anytime soon.

As will be discussed in Chapter 12, a living will provides a vehicle by which individuals make their intentions known concerning the withholding of medical treatment. In allowing a patient to die, palliative care becomes of utmost importance. Medical personnel must try to keep the patient comfortable and attempt to respond to psychological, social, and spiritual needs. Palliative care personnel should frame chronic pain in strictly behavioral terms by developing an appreciation for the cognitive and evaluative aspects of the experience, suggested Joseph Kotarba (1983) in researching chronic pain management. The physician should inquire into the patient's cognitive resources for normalizing suffering. If the patient is a member of a religious group, for example, the usefulness of that membership should be stressed. If such membership is lacking, the individual should be encouraged to seek a compatible source of meaning, said Kotarba.

Many palliative care professionals cite weakness and fatigue as some of the most disabling symptoms of terminal illness because there is little that can be done to alleviate this state of powerlessness, observed anthropologist Beverly McNamara (1997). These manifestations of terminal illness are just a few of many symptoms that will influence the dying person's conception of self. Nausea, vomiting, breathlessness, constipation, diarrhea, edema (swelling), smell from infections and wounds, confusion, and pain are also associated variously with disease and with intervention procedures. Knowledge of approaching death further complicates self-knowledge and the capacity to act to change circumstances. Some individuals view euthanasia as a dignified death, whereas others believe it is unlawful killing. "Does dying 'suddenly' mean that people would rather not have control over the circumstances of their death?" asked McNamara.

The terminally ill person's access to power should be seen in contrast to that of the health professionals who care for him or her. A doctor, for example, who is dying of cancer can rather easily negotiate her or his own medical management with the aid of colleagues, but an Australian Aboriginal person is unlikely to benefit from very basic forms of support, noted McNamara. The degree to which terminally ill persons engage in decisions about their dying and forthcoming death will be mediated by their access to power and to the place and time of their death. In the last days of life, individuals will be better able to maintain some degree of authority if resources are channeled into facilities that support their needs and wishes.

Many dying people fear pain, but also feel dissatisfied with the debilitating side effects of pain-controlling medications such as morphine. A dilemma is posed for the health professional: How can terminally ill persons articulate their needs and desires if their cognitive functions are altered by medications, and yet, how can they focus on autonomous decision making if they are incapacitated by pain? asked McNamara. A cautionary note about the reliance on technology's capacity to provide answers to suffering is that overuse of medical technology will not only anesthetize people to pain, it may also anesthetize them to the act of dying.

If an individual wishes to die "naturally" and not be kept alive by extraordinary means, then *allowing* that person to die constitutes passive euthanasia. As noted earlier, the hospital today is not a "place where you go to die" but rather is a "place where you go and they will *not let you die.*" Passive euthanasia is *letting* the patient die. On the other hand, if the patient wishes to be *helped* to die through physician-assisted death or active voluntary euthanasia, this is active euthanasia and is illegal in the United States, except for Oregon, Washington, Montana, and Vermont, where physician-assisted death is legal. In Montana, unlike the other three states, physician-assisted suicide is legal through a court ruling, not by a vote of the people nor act of the state legislature.

Despite strong early support for the Death with Dignity Act in Massachusetts, in the final vote in the 2012 statewide election (Question 2 on the ballot) the law was defeated by the voters—49 percent "Yes" and 51 percent "No." But the foundation for support in Massachusetts has been built, and the Death with Dignity Act will undoubtedly come back to the voters for approval in the near future (Sandeen, 2012).

Worldwide, Physician Aid in Dying (PAD) is legal in the following five countries (Euthanasia.ProCon.org, 2013): Germany (legal since 1751 but active voluntary euthanasia is illegal), Switzerland (legal since 1943 but active voluntary euthanasia is illegal), and active voluntary euthanasia and PAD are both legal in the Netherlands (2001), Belgium (2002), and Luxembourg (2008). For an excellent review and evaluation of current practices of PAD in the countries listed above, see Robin Gibson's "The case for euthanasia and physician-assisted suicide" (2012). Now let's look more closely at these two types of euthanasia.

PASSIVE EUTHANASIA

The action of allowing a patient to die is often referred to as **passive euthanasia**. Passive euthanasia may involve withholding medical treatment or removing life-support equipment. Such action is supported by people who affirm both sanctity-of-life and quality-of-life perspectives and is indeed legal in the United States and in other countries around the world. The patient is being allowed to die naturally. She or he was only being kept alive because of the life-support equipment and/or medication. Though action was involved to remove the support, the patient still died on her or his own accord.

Allowing a patient to die has not historically been easy. There have been several situations in which court action was needed to allow the individual to die. In the 1970s one such well-publicized case was that of Karen Ann Quinlan, a college student who mixed alcohol with other drugs and went into a comatose state. After several months of a negative medical prognosis, the courts ruled that she could be allowed to die by removal of life-sustaining equipment.

DISTINGUISHING BETWEEN ORDINARY AND EXTRAORDINARY MEASURES If one accepts the legitimacy of passive euthanasia, as set forth in a living will, it is clear that the use of extraordinary measures is not required of attending physicians, for terminally ill patients and members of their family. However, differentiating between ordinary and extraordinary measures is not easy. **Ordinary measures** of preserving life are all those medicines, treatments, and operations that offer a reasonable hope of benefit for the patient and that can be obtained and used without excessive expense, pain, and other inconvenience (Bouma et al., 1989). **Extraordinary measures** of preserving life are all those medicines, treatments, and operations that cannot be obtained without excessive expense, pain, or other inconvenience or that, if used, would not offer a reasonable hope of benefit.

These measures introduce other ambiguous terms—*reasonable hope*, *excessive expense*, *excessive pain*, and *excessive inconvenience*. What is reasonable and what is excessive? Who among us can make decisions using these highly subjective concepts? Which of the following groups should society allow to determine these matters: politicians, physicians, lawyers, judges, members of the community at large, or patients and their families?

Many within the medical community have defined *ordinary measures* as any procedures that are "usual and customary" medical practice and extraordinary measures as any procedures that are exotic, innovative, and medically experimental. However, it is clear that this distinction provides us with little help because technology has a way of imposing itself upon modern society in an amoral fashion, and what has been considered innovative and experimental becomes usual and customary with time. Consider the situation in which an infant with Down's syndrome starves to death because parents will not permit a physician to perform a simple operation that would allow food to be digested by the child. The fate of this infant is a classic example of what sociologists call **cultural lag** (not all aspects of a culture change at the same pace; one aspect "lags behind" another)—a condition created by rapid medical technological development and relatively static definitions of biological life and death.

Medical practice today can prolong life (or postpone death) without having precise definitions of a point at which meaningful living has stopped and death has occurred. This condition would be of little significance if medical science were unable to maintain some biological functioning of the individual by artificial means, but such is not the case. Compared with the horrors of senility and extended terminal treatment because of modern medicine's ability to prolong life, the terminally ill patient may indeed welcome death.

Furthermore, even if one were able to distinguish between ordinary and extraordinary measures, would it be acceptable for patients (or someone acting on their behalf) to request the withdrawal of ordinary life support such as food and water? This is an issue that tends to divide persons who affirm a sanctity-of-life perspective and favor passive euthanasia.

NUTRITION AND HYDRATION Withholding and withdrawing artificial nutrition and hydration is a perplexing and emotional issue in end-of-life care, observed Z. Huang and J. C. Ahronheim (2000) in their update on these issues. Since the Karen Ann Quinlan case in 1976, as mentioned earlier, in which the court authorized termination of life support for a patient in a persistent vegetative stage, courts have realized an important distinction between killing and letting die. This has enabled the patient's wishes to be upheld without the patient or care provider being held criminally liable.

Commentators have questioned the practice of giving intravenous fluids to terminally ill patients and have explored whether dehydration causes distressing symptoms in dying persons (Huang & Ahronheim, 2000). Yet, studies on artificial nutrition and hydration at the end of life suggest that tube feeding is the rule rather than the exception (Hoefler, 2000). Many elderly and infirm patients spend their last days in long-term and tertiary-care facilities, and many of these institutions find tube feeding to be more cost efficient than hand feeding patients who have trouble feeding themselves. In addition, reimbursement rates are higher for tube-fed patients (Hoefler, 2000). The decision to tube feed is often influenced by regional and cultural preferences (Amella, Lawrence, & Gresle, 2005). Caregiver anxiety about withdrawing food and fluids also is a factor, at least partly because of the important role that food plays in our culture. Physicians also get caught up in the cross-cutting currents about whether to feed hopelessly ill and demented patients in their care or not, especially when family members are clamoring for the physicians to "Do something!" In addition, physicians' concern with legal liability may cause anxiety. Many physicians also seem to be confused about medical-ethical issues associated with hydration and artificial nutrition—many agree that tube feeding can be withdrawn from patients in a vegetative state at the request of a surrogate, but are reluctant to do so, thinking that they would either be violating professional norms or their action would lead to increased patient discomfort (Hoefler, 2000).

Nutrition and hydration problems call for expertise in dealing with terminally ill patients. From an ethical perspective, at what point is the patient not given food and water? Nutritional deficiencies usually result from nausea, mouth soreness, or anorexia (Gentile & Fello, 1990). However, a person can live quite a while without food. Hydration is more critical. A human being cannot live for any length of time without fluids. Intravenous (IV) therapy questions are often raised by families. If a patient has been ambulatory but is weakening and is dehydrated, one or two liters of fluid can enhance the quality of life temporarily, if that is what the family and patient desire. If, however, the patient is in an extremely debilitated state, is bedridden, and is barely conscious, starting IV fluid replacement could be considered invasive and will probably not add to the quality of life, noted M. Gentile and M. Fello.

When food and fluids are withheld or withdrawn, patients die of dehydration, not starvation (Hoefler, 2000). Death by starvation is a long, arduous process involving major weight loss and body wasting. Bodily functions begin to shut down, compromising the production of white blood cells, thus weakening the immune system. Toward the end of the starvation process, the intestines begin to fail, leading to uncontrollable diarrhea. Eventually, the heart muscle gives out, and the person dies of cardiac arrest. The process of starvation can drag out for weeks.

Unlike starvation, death via dehydration is relatively quick and painless—dehydration tends to be the direct cause of all "natural" deaths (not associated with violent trauma or acute infection). According to Gentile and Fello (1990), slow dehydration caused by ingestion of less and less fluids, and finally no fluid, does not usually create much discomfort. Research on dehydration suggests that patients who stop taking in food and fluids slowly sink into a state of unconsciousness over the course of several days and die peacefully shortly after that (Hoefler, 2000). Usual symptoms of the slow dehydration process are slight elevation of body temperature and drying of the mucous membranes such as the nose and mouth. Rapid dehydration, however, which is usually caused by prolonged vomiting, severe diarrhea, or mechanical evacuation of stomach contents, can be significantly more uncomfortable. Without IV fluid replacement, a person will weaken and die quickly. Some patients choose to forgo IV therapy knowing full well the eventual outcome, observe Gentile and Fello.

To let a person die from dehydration or lack of nutrition is not agreeable to all groups. For example, some Orthodox interpretations of Jewish law conclude that death cannot be hastened through withdrawing or withholding artificial nutrition and hydration or other treatments that may prolong life (Rhymes, 1996). Some patients and family members may have religious concerns based on a misunderstanding of the requirements of their religion.

In the end, however, according to James Hoefler's thorough study of the topic (2000), there is little doubt that the prevailing sentiment among those who have had experience with death and dehydration is that the benefits of dehydration outweigh the burdens for certain patients, namely, those who are terminally ill, irreversibly unconscious, and severely demented.

CPR VERSUS DNR   **CPR** (cardiopulmonary resuscitation) follows a sanctity-of-life view in that the idea is to prolong life—or restore the person to life. On the other hand, **DNR** (do not resuscitate) would follow a quality-of-life view for an individual in a terminal condition.

In the early 1960s, a leading group of resuscitation researchers agreed that CPR—the combination of securing an open airway, mouth-to mouth ventilation, and chest compressions—was the most promising resuscitation technique. CPR addressed the often fatal problem of an obstructed airway (Timmermans, 1999). Before CPR was introduced in 1960, health care providers alternated between resuscitating aggressively and preparing for impending death on the basis of the patient's social viability (e.g., age and "moral character"). In contrast, legislators and biomedical researchers have now designed an emergency system in which care providers are required to save lives whenever medically indicated (Timmermans, 1999). Now, only people who choose not to be resuscitated should be exempt from a reviving attempt. No longer should anyone systematically withhold potentially beneficial care from groups with a low social value. To do so would mean passive euthanasia for some and aggressive but futile lifesaving efforts with needless suffering for others, whose presumed social value is higher, noted S. Timmermans (1999).

Timmermans (1999), in his book on the myth of CPR, asks whether resuscitation protocols and legal protections, instituted since the 1960s, have in fact eliminated the social inequality of emergency room care. As a sociologist using participant observation of CPR, Timmermans noted that the initial impression of the emergency room staff is based on the circumstances of the CPR, the social characteristics of the patient, the biomedical indicators, and the treatment. The team members then ask if the patient has a reasonable chance of survival and how long they will work on the patient. When the patient arrives in the emergency room (ER), the physician may declare her or him **DOA** (dead on arrival). Legal changes since the 1960s have mandated that when someone calls 911, a resuscitative effort begin and be virtually unstoppable until a physician sees the patient in the ER.

Unless the patient shows obvious signs of death, the **EMTs** (emergency medical technicians) or paramedics need to begin CPR as prescribed by their protocol. Upon arrival at the ER, the physician cannot legally stop the lifesaving attempt until all protocols are exhausted. Stopping sooner would qualify as negligence and grounds for malpractice. Timmermans (1999) observed, however, that the ER staff do not always aggressively revive all clinically viable patients, nor do they give up on all biologically dead patients. Thus, there is a lack of consistency, depending on the circumstances.

We are in danger of abusing the procedure of CPR today (Rahman, 1989). At some point, we must stop adding to the suffering of dying patients by pounding on their chests and possibly breaking bones, just to extend their lives by a few days or weeks. With the routine use of CPR today and the estimated survival rate of 1 to 3 percent of victims who have cardiac arrests out of hospitals, the efforts appear to be rather futile (Cummins, Oranto, & Thies, 1991). Why batter the sternum of a drastically debilitated, 85-year-old woman who is suffering from pneumonia or try to revive with CPR a patient who has been sick for years with diabetes, hypertension, heart disease, or kidney failure and who has just had an operation for a perforated intestine? The idea of beating on the chests of elderly sick persons such as these is morally and physically repugnant, notes Rahman. Yet, the procedure of CPR continues to be used indiscriminately. When the question of using CPR, in the event it is needed, is not raised beforehand, under the duress of the moment, it is difficult for families to make a quick decision; thus, the physician simply follows the routine of performing CPR.

Whereas CPR is a method for sustaining life, DNR is a formal order, written on the patient's chart, which indicates that, if the patient's heart stops, no attempt will be made to resuscitate the patient. Approved nationwide guidelines explicitly allow doctors to write DNR orders. Officially approved procedures can be adhered to, and all parties involved are legally protected (Chambliss, 1996). Such orders are quite common, and on any given day in a large hospital a number of patients will have "DNR" or "No Code" orders on their charts. Approximately 70 percent of patients who die in the hospital have a DNR order in place (Council on Ethical and Judicial Affairs, 1991).

Officially, physicians are responsible for signing DNR orders. A DNR order states that a particular patient whose breathing and circulation fails should not be revived. By making such a call and then following it through, the medical personnel are allowing the patient to die. Once the DNR order is signed by the doctor, the nurse is organizationally responsible for monitoring the death. He or she must sit and watch the patient turn blue and listen to the final gurgling sound of the patient "drowning in his or her own juices" (Chambliss, 1996).

Physicians, following the Hippocratic Oath to prolong life, sometimes have difficulty with DNR orders. Such resistance is especially true for young doctors, interns, and residents, observed D. F. Chambliss. In general, it seems that everyone involved with a DNR order has some desire to sidestep responsibility for this awesome decision. Families may prefer to have the doctors decide, because they are the experts, yet staff will argue that families should make the decisions themselves. A good example of the difficulty with DNR orders is found in journalist T. M. Shine's *Fathers Aren't Supposed to Die: Five Brothers Reunite to Say Good-Bye* (2000).

Shine and his four brothers were having problems with the fact that their dad was dying and were debating the issue of the DNR decision. One of Shine's brothers suggested that, instead of signing a statement "Do not restart this man's heart under any circumstances," the statement could read "At least try once." Indeed, both family and medical staff are reluctant about the DNR order.

CPR, on the other hand, fits the medical model in U.S. hospitals to prolong life and make the patient well. Yet, CPR can be controversial if the patient prefers death, and death is denied to her or him. Outside of a hospital setting and in an emergency situation, however, EMT personnel will routinely apply CPR if required by the individual's condition.

ACTIVE EUTHANASIA

The previous section discussed the issue of allowing the patient to die. Such a procedure involves passive euthanasia and is indeed legal in the United States and most countries of the world. We now turn our attention to **active euthanasia,** whereby the individual is helped to die. Such help can come from a physician giving advice about how the patient can take her or his life and/or giving a prescription for a lethal dosage of a medication to end the person's life. This action is typically referred to as **physician-assisted death (PAD)**, sometimes called physician-assisted suicide or **PAS**, because in essence the individual takes her or his own life. A more aggressive form of helping an individual to die is called **active voluntary euthanasia (AVE)** whereby, upon request by the patient, the physician ends the patient's life through direct action (e.g., injecting a lethal dose of a drug into the patient).

There is an obvious analytical distinction between active and passive euthanasia. However, after one accepts the legitimacy of letting people die by terminating life-sustaining treatment, it is an easy step to accept the legitimacy of assisting terminally ill patients to commit suicide or to perform active euthanasia. A telephone survey (Huber, Meade-Cos, & Edelen, 1992) of 200 Midwestern respondents determined that 90 percent favored some kind of personal control over death circumstances (e.g., passive and active euthanasia).

| PRACTICAL MATTERS | WAYS TO AVOID EXCESSIVE CPR |

Suggestions to avoid excessive use of CPR include the following:

1. The public and professionals must be educated to honor a peaceful death, if so desired.
2. Patients should sign living wills.
3. Patients with incurable diseases should be tactfully approached by their families and physicians about the use of aggressive measures.

4. If a patient is opposed to resuscitation, the primary physician must write clearly on a patient's chart, "DNR."

Adapted from Routine CPR can abuse the old and sick by F. Rahman (1989, February 27), *Minneapolis Star and Tribune,* p. 9A.

DEATH ACROSS CULTURES | EUTHANASIA IN THE NETHERLANDS

In November 2000, the lower chamber of Parliament in the Netherlands voted 104 to 40 to allow doctors to help patients die, under certain restrictions. The bill became law in April 2001. Before this vote, AVE was not legal, but it was decriminalized in the Netherlands, as long as the physician followed certain guidelines. Because this meant that an act of PAD or AVE would not be prosecuted, the country was identified as one in which the patient had the opportunity to have a physician help end his or her life. The new law sets out strict guidelines, not significantly different from the decriminalized ones, which are as follows:

1. The request must be voluntary and well considered.
2. The patient must be facing unremitting and unbearable suffering, but does not have to be terminally ill.
3. The patient must have a clear and correct understanding of the situation and prognosis.
4. The physician and patient must reach the conclusion that there is no reasonable alternative acceptable to the patient. The decision to die must be the patient's own.
5. The physician involved must consult at least one other independent doctor who has examined the patient.
6. The physician must end the patient's life in a medically appropriate manner.
7. The physician must report the cause of death as euthanasia or suicide.
8. The patient must be an adult. Young people from 12 to 16 must have parental consent. Children under 12 are not eligible under the present guidelines.

Following are some of the facets of Dutch society that have influenced attitudes toward end-of-life care. Ninety-nine percent of the population has medical coverage, and all have coverage for catastrophic illness. Malpractice suits are almost unheard of in *continued* the Netherlands because of high trust of and esteem for physicians. Forty percent of patients in the Netherlands die at home, compared with 15 percent in the United States. Dutch churches have less political power than those in the United States. The Dutch often act together for the collective good. Almost all Dutch citizens have a general practitioner, many of whom make house calls.

Studies show that about half of the physicians in the Netherlands have performed euthanasia. Nonetheless, they seem to have approached it circumspectly, almost reluctantly. Out of an estimated 25,000 requests for information, 9,000 patients actually make a formal application to their physician, and 3,000 have it carried out. Many patients are reassured to find their doctors receptive to ending their suffering, but in the end decide to die a natural death. The fact is that despite the social acceptance of euthanasia, these deaths only comprise 2 percent of the 135,000 annual deaths in Holland.

Part of the reason for this relatively small percentage of physician-assisted and AVE deaths may be a reluctance to file reports of the cases. Between 1990 and 1995, more than half of the occurrences of AVE and PAD went unreported. Reasons given for not reporting by the doctors were "too much paperwork," objections to the "criminal" overtones of the investigation that follow the reporting, and reluctance to put the grieving family "through the additional burden of interrogation" that was a part of the investigation. Thus, the reporting structure for PAD and AVE was changed in November of 1998: A three-person panel composed of a physician, a lawyer, and an ethicist reviewed all cases and made recommendations regarding investigation. Under the 2000 legislation, physicians will continue to be accountable to a panel of peers, including legal, medical, and ethical experts—no change from 1998. The hope has been that this change will result in greater compliance with reporting.

The Netherlands has also been criticized for its presumed lack of palliative care and the existence of very few hospices in the country. From a sociological point of view, one may be tempted to interpret the shift toward autonomy-based requests for euthanasia as a byproduct of a liberal society, with its emphasis on self-government, control, and rational choice. Others will say that more emphasis on patient autonomy fits perfectly into the process of emancipation of the patient than has been going on since the beginning of the 1970s. J. B. Vander Veer, after completing

continued

DEATH ACROSS CULTURES | EUTHANASIA IN THE NETHERLANDS *continued*

his study of euthanasia in the Netherlands, concluded that the Dutch system would not work in the United States because the societies are too different. The United States lacks universal health insurance and coverage for catastrophic illness, and the "slope seems too slippery" in the United States.

With euthanasia now being legal in the Netherlands, not "just *decriminalized*," it will be of interest to observe if the rate of physician involvement with patient deaths changes.

Adapted from Euthanasia in the Netherlands, by J. B. Vander Veer (1999, May), *Journal of the American College of Surgeons, 188*(5), 532–537; Slippery slopes in flat countries—A response, by J. J. M. van Dalden (1999, February), *Journal of Medical Ethics, 25,* 22–24; Dutch becoming first nation to legalize assisted suicide, by M. Simons (2000, November 28), *New York Times*, A3.

Clive Seale (1997) summarized the current causes for the high support today of AVE and PAD: (1) As medical technology has increased, so have fears of inappropriate life-sustaining treatment evolved; (2) the rise in support is an aspect of the decline in religious faith and the emergence of rationalistic approaches to matters of life and death; (3) as social bonds weaken, people opt for more individualistic solutions to problems; and (4) the population has aged, resulting in more lingering suffering than in the past. Seale noted that increasing secularization and rising levels of education are likely to lead to even more increased levels of public support in the future.

ORGANIZATIONS SUPPORTING ACTIVE EUTHANASIA   In the United States, the **Hemlock Society** was founded in 1980 to assist terminally ill individuals in the act of **self-deliverance**. In 1981 the Hemlock Society published a book written by its co-founder and director, Derek Humphry, entitled *Let Me Die Before I Wake: Hemlock's Book of Self-Deliverance for the Dying.* The first 10 chapters of this book provide detailed case studies of individuals who chose self-deliverance as an alternative to passive euthanasia (often with the help of family members and friends). The remainder of the book provides a bibliography and an extended discussion of personal and legal issues related to the decision to take one's life. Needless to say, this book was controversial in some segments of American society.

A year earlier, the English counterpart of the Hemlock Society—**EXIT** (founded in 1935 as the British Voluntary Euthanasia Society)—published *A Guide to Self-Deliverance* (1980). This book was unique because it provided the reader with detailed instructions to successfully and painlessly commit suicide. In the United States this book would have been construed as an attempt to intentionally aid and/or solicit another to commit suicide—a felony in some states. In 1991, Derek Humphry published *Final Exit: The Practicalities of Self-Deliverance and Assisted Suicide for the Dying.* The most controversial chapter was entitled "Self-Deliverance via the Plastic Bag." This chapter provides detailed instructions for self-deliverance including inserting one's head into a plastic bag "firmly tied around the neck with either a large rubber band or a ribbon," after one takes a sufficient number of pills to induce unconsciousness. The instructions came without a guarantee, but Humphry could predict satisfaction by stating that "in the vast majority of cases, the drugs work." After *Final Exit* gained wide public

"After reviewing how much your treatment is costing this hospital, I'm going to recommend assisted suicide."

attention through *The Wall Street Journal*, it began an 18-week journey on the *New York Times* bestseller list, and eventually more than a million copies were sold (Nuland, 1998).

Until the publishing of *Final Exit*, the Hemlock Society had not been willing to be explicit in providing suicide information to its members. However, with public sentiment turning more favorably toward active euthanasia, the Hemlock Society began to take a more active role in promoting self-deliverance. In 1991 the Hemlock Society became a strong advocate of Washington state's Initiative 119, which would have legalized PAD had the bill passed, and contributed more than $300,000 in support of the bill (Gabriel, 1991). This bill was subsequently passed in 2008, making at that time Oregon and Washington the only two states that allowed terminally ill patients to end their lives by taking medications prescribed by a physician.

After Humphry retired as executive director of the Hemlock Society in 1992, he created the Euthanasia Research and Guidance Organization (known as **ERGO!**). This nonprofit educational corporation sought to improve the quality of background research in physician-assisted deliverance for persons who are terminally or hopelessly ill and who wish to end their suffering. As well as conducting opinion polls, ERGO! develops and publishes guidelines—ethical, psychological, and legal—for patients and physicians to better prepare them to make life-ending decisions. The organization supplies literature to and does research for other right-to-die groups worldwide and also briefs journalists, authors, and graduate students who are interested in right-to-die issues. In addition, ERGO! is willing to counsel dying patients (and their families), provided that they are competent adults and are at the end stage of a terminal illness. Persons suffering from unbearable mental illness are referred by ERGO! to alternative forms of skilled help.

The Hemlock Society and ERGO! have been careful to differentiate suicide from self-deliverance and self-deliverance from mercy killing. Suicide is often condemned socially and religiously as a selfish act or as an "overreaction of a disturbed mind." Suicide is considered by many to be an irrational act that is a permanent solution to what is often a temporary problem. In contrast, self-deliverance should be considered a positive action taken to provide a permanent solution to long-term pain and suffering for the individual and her or his significant others faced with a terminal condition.

Self-deliverance is a completely voluntary act on the part of the patient, whereas mercy killing involves other people's behavior that may or may not be sanctioned by the patient. Laws in the United States do not distinguish between murder and mercy killing, and therefore mercy killing is predominantly treated as a criminal offense regardless of the motivations of individuals involved. The defense (other than for a physician) sometimes used in criminal court for a mercy killing is "temporary insanity." The Hemlock Society, ERGO!, and EXIT wish to eliminate the stigma attached to self-killing and to extend to all individuals the

"Would you also like to purchase the extended warranty?"

Andy White/CSL, CartoonStock Ltd

right of active, rational, and voluntary euthanasia whenever the dying process promises only unrelieved pain and a life devoid of dignity, meaning, and purpose.

In 2003 Hemlock changed its controversial name to "End of Life Choices," and a year later merged with the organization Compassion in Dying to form the national organization **Compassion & Choices**. The present surviving organization, Compassion & Choices, is a nonprofit organization located in Denver, Colorado, working to improve patients' rights and choices at the end of life. Primarily, Compassion & Choices litigates patient cases related to ensuring adequate end-of-life care and choice. Through litigation, Compassion & Choices works for terminally ill patients' rights to receive pain and symptom management, to voluntarily stop life-sustaining treatments, to request and receive palliative sedation, and to choose aid in dying under state and federal constitutional protections.

THE SLIPPERY-SLOPE ARGUMENT   Persons who oppose active euthanasia often do so on the grounds of what they call the **slippery-slope-to-Auschwitz argument** (Potter, 1993; Wilkinson, 1995). They allege that the justification for withholding some treatments also applies to withholding any and all treatments from those whom society considers unworthy. It is further argued that if a society wishes to extend to any of its members the right to active euthanasia, none of its members is protected from being killed. Such was the case under Hitler when German society's "unproductive," "defective," and "morally and mentally unfit" persons—Jews, homosexuals, and the handicapped—were sent to Auschwitz and other places of confinement to be "euthanized" (murdered).

The slippery-slope argument alleges that society moves toward a disregard for human life in predictable steps. The first step is the acceptance of passive euthanasia when medical technology and surgical procedures are withheld from (or rejected by) chronically and/or terminally ill patients. The second step occurs when ordinary or customary procedures are withdrawn—such procedures would include intravenous delivery of food and water. We arrive at the third step when society accepts the legitimacy of suicide (self-deliverance). The fourth step is when these "rights" are extended to patients with nonterminal illnesses. In the fifth step, the medical system will assist patients in self-deliverance or active euthanasia (as is presently the situation in Holland), like a beloved animal that is "put to sleep." In the sixth step all of these provisions are extended to nonterminally ill persons and/or persons acting on their behalf. Finally, we arrive at the seventh step—"Auschwitz"—when society is willing to accord these "privileges" to the unwilling, incompetent, and/or undesirable.

To prevent society from reaching such a level of desensitization, people who embrace the slippery-slope argument argue that no type of euthanasia should be considered socially acceptable—in other words, to be safe, society should "stay off the slope entirely!" However, other biomedical scholars who favor AVE and PAD (Hill, 1992; Logue, 1994; Ogden, 1995; Weijer, 1995) argued that if patients voluntarily permit and/or cooperate with their physicians to cause a premature death, their rights and dignity as persons will be enhanced, and a Nazi-like program of extermination will be unlikely. As evidence to this argument, in retrospective interviews with 87 relatives to describe the experiences of patients who died by euthanasia or physician-assisted suicide in the Netherlands, Georges and colleagues (2007) discovered that relatives believed, in 92 percent of the cases, that the euthanasia contributed favorably to the quality of the end of life, mainly by preventing or ending suffering. However, Robert Twycross (1996) counters that once the "slipping" begins, it simply continues. He observed that the Dutch have already slipped a long way down the slope, in that both voluntary and involuntary euthanasia exist in the Netherlands and that both physical and psychiatric disorders are grounds for euthanasia.

In a recent report of research in the Netherlands, Fenigsen and Fenigsen (2012) on the Government-Ordered Surveys of Euthanasia (1995–1996 and 2000–2002) have made the two following conclusions based on empirical evidence: (1) the practice of euthanasia in the Netherlands has more or less stabilized, and (2) the number of cases of voluntary euthanasia gradually increased. One might conclude from this empirical research that, if there is a "slippery slope to Auschwitz," the speed is not increasing as much as some critics predicted.

PHYSICIAN-ASSISTED DEATH As previously noted, the issue of active euthanasia and physician-assisted death was highlighted in the media during the 1990s. Dr. Jack Kevorkian's "death machine" in Michigan was the subject of numerous headlines.

In 1991, as noted earlier, Derek Humphry's *Final Exit*, providing detailed information on how to "self-deliver," went on the market and became a bestseller. *The New England Journal of Medicine* published Dr. Timothy Quill's experience of assisting his patient "Diane" by prescribing barbiturates after she refused treatment options for leukemia in 1991 (see abridged version in the "Listening to the Voices" box). Voters went to the polls and courts made decisions regarding PAD throughout the 1990s. In the 21st century, this issue is not going away.

Short History Legal Measures Related to PAD. In 1991 the voters of Washington defeated Initiative 119 (54 percent to 46 percent), which would have legalized PAD. Likewise, in 1992 California voters failed to pass Proposition 161 (52 percent to 48 percent), an initiative similar to Washington's Initiative 119. Active euthanasia was legalized in the Netherlands in April of 2001. Soon thereafter,

---

LISTENING TO THE VOICES | MY PATIENT'S SUICIDE

I have long been an advocate of the idea that an informed patient should have the right to choose or refuse treatment, and to die with as much control and dignity as possible. Yet there was something that disturbed me about Diane's decision to give up a 25 percent chance of long-term survival in favor of almost certain death.

Diane and I met several times that week to discuss her situation, and I gradually came to understand the decision from her perspective. We arranged for home hospice care, and left the door open for her to change

her mind. Just as I was adjusting to her decision, she opened up another area that further complicated my feelings. It was extraordinarily important to Diane to maintain her dignity during the time remaining to her. When this was no longer possible, she clearly wanted to die. She had known of people lingering in what was called "relative comfort," and she wanted no part of it. We spoke at length about her wish. Though I felt it was perfectly legitimate, I also knew that it was outside of the realm of currently accepted medical practice and that it was more than I could offer or promise. I told

continued

LISTENING TO THE VOICES | MY PATIENT'S SUICIDE continued

Diane that information that might be helpful was available from the Hemlock Society.

A week later she phoned me with a request for barbiturates for sleep. Since I knew that this was an essential ingredient in a Hemlock Society suicide, I asked her to come to the office to talk things over. She was more than willing to protect me by participating in a superficial conversation about her insomnia, but it was also evident that the security of having enough barbiturates available to commit suicide, if and when the time came, would give her the peace of mind she needed to live fully in the present. She was not despondent and, in fact, was making deep, personal connections with her family and close friends. I made sure that she knew how to use the barbiturates for sleep, and how to use them to commit suicide. We agreed to meet regularly, and she promised to meet with me before taking her life. I wrote the prescriptions with an uneasy feeling about the boundaries I was exploring—spiritual, legal, professional, and personal. Yet I also felt strongly that I was making it possible for her to get the most out of the time she had left.

The next several months were very intense and important for Diane. Her son did not return to college, and the two were able to say much that had not been said earlier. Her husband worked at home so that he and Diane could spend more time together. Unfortunately, bone weakness, fatigue, and fevers began to dominate Diane's life. Although the hospice workers, family members, and I tried our best to minimize her suffering and promote comfort, it was clear that the end was approaching. Diane's immediate future held what she feared the most: increasing discomfort, dependence, and hard choices between pain and sedation. She called her closest friends and asked them to visit her to say good-bye, telling them that she was leaving soon. As we had agreed, she let me know as well. When we met, it was clear that she knew what she was doing, that she was sad and frightened to be leaving but that she would be even more terrified to stay and suffer.

Two days later her husband called to say that Diane had died. She had said her final good-byes to her husband and her son that morning, and had asked them to leave her alone for an hour. After an hour, which must have seemed like an eternity, they found her on the couch, very still and covered by her favorite shawl. They called me for advice about how to proceed. When I arrived at their house we talked about what a remarkable person she had been. They seemed to have no doubts about the course she had chosen, or about their cooperation, although the unfairness of her illness and the death were overwhelming to us all.

Diane taught me about the range of help I can provide people if I know them well and if I allow them to express what they really want. She taught me about taking charge and facing tragedy squarely when it strikes. She taught me about life, death, and honesty, and that I can take small risks for people I really know and care about.

Adapted from Death and dignity: A case of individualized decision making, by T. Quill (1991, May), *Harper's Magazine,* pp. 32–34; originally published in *The New England Journal of Medicine,* March 7, 1991, 324.

Belgium legalized such action (Switzerland allows PAD in carefully controlled situations). The Supreme Court of Canada in 1993 refused (by a vote of five to four) to allow a woman with amyotrophic lateral sclerosis (ALS) to have her physician legally assist in her death. In 1994 Measure 16 passed (51 percent to 49 percent) in Oregon, legally allowing physicians to hasten death for the terminally ill. The results, however, were tied up in court, and Oregon voted again on this issue in November of 1997, and it passed again (by a vote of 60 percent to 40 percent). In 2008, voters in Washington state passed the Washington Death with Dignity Act. This act allows terminally ill adults seeking to end their life to request lethal doses of medication from medical and osteopathic physicians. These terminally ill patients must be Washington residents who have less than six months to live. Thus,

| A GOOD DEATH

A good death does honor to a whole life.

Petrarch (1304–1374)

TABLE 8.1 | FOUR STATES WITH LEGAL PHYSICIAN-ASSISTED SUICIDE

State	Date Passed	How Passed (Yes Vote)	Residency Required?	Minimum Age	# of Months Until Expected Death	# of Requests to Physician
1. Montana	Dec. 31, 2009	Montana Supreme Court in *Baxter v. Montana* (5-4)	Yes	*	*	*
2. Oregon	Nov. 8, 1994	Ballot Measure 16 (51%)	Yes	18	Six or less	Two oral (at least 15 days apart) and one written
3. Vermont	May 20, 2013	Act 39 (Bill S.77 "End of Life Choices"	Yes	18	Six or less	Two oral (at least 15 days apart) and one written
4. Washington	Nov. 4, 2008	Initiative 1000 (58%)	Yes	18	Six or less	Two oral (at least 15 days apart) and one written

*No legal protocol in place.

Euthanasia.ProCon.org. (2013). http://euthanasia.procon.org/view.resource.php?resourceID=000132.

PAD is now legal (as of 2014) in four states: Oregon, Washington, Montana, and Vermont (see Table 8.1). The number of assisted suicide prescriptions and deaths has increased in the state of Oregon since 2009. The number of prescriptions for PAD in 2009 was 95, whereas in 2012 it was 115, a 21 percent increase. Accordingly, the number of deaths went from 59 in 2009 to 77 in 2012, a 30 percent increase (Schadenberg, 2013). Washington state numbers are somewhat similar, with 121 individuals in 2012 receiving the prescription for the lethal dosage, and 83 of these individuals dying from ingestion of the lethal dose (Dore, 2013).

The history of creating a legal right to PAD has only recently been resolved through political action of voters in the individual states. In 1996 the Ninth Circuit Court of Appeals in Washington ruled that the Washington statute prohibiting physicians from prescribing life-ending medication for use by terminally ill, competent adults who wished to hasten their own deaths was unconstitutional, because it violated the due process clause of the Fourteenth Amendment of the U.S. Constitution. Also in 1996, the Second Circuit Court of Appeals in New York reached a similar ruling that expanded on the Ninth Circuit Court's ruling, in finding that allowing terminally ill patients to die through physician removal of life-sustaining treatments, such as mechanical ventilators or artificial hydration and nutrition, while not allowing such patients to die by taking physician-prescribed medication, is a violation of the equal protection guarantee of the Fourteenth Amendment. The U.S. Supreme Court ruling in 1997 declared that terminally ill patients do not have a fundamental constitutional right to assisted death and upheld the New York and Washington laws that make it a crime for physicians to give life-ending drugs, even to mentally competent terminally ill adults who have made such a request. However, states retain the right to enact laws that legalize assisted death; thus, the matter was thrown back by the U.S. Supreme Court to the states. Physicians still maintain the right to provide terminally ill people with adequate pain relief medication, even if this shortens life.

In the Northern Territory of Australia in 1996 the world's first law permitting voluntary euthanasia was tested when the first person died after it was enacted. Six months later in 1997, however, after four acts of voluntary euthanasia were reported, the law was revoked. Currently, as noted earlier, the states of Oregon, Washington, Montana, and Vermont are the only places in the United States where PAD is legal, but other states have had similar initiatives on their ballots for voters to decide—most recently in the Commonweath of Massachusetts, where the bill was rejected in a very close vote (49 percent "Yes" and 51 percent "No."

Public Opinion and Physicians' Attitudes.   Both U.S. medical ethics literature and U.S. medical associations have traditionally condemned physician-assisted death (Jecker, 1994). The American Medical Association's Council on Ethical and Judicial Affairs stated that, although life-prolonging medical treatment may be withheld, "the physician should not intentionally cause death" (1992). In addition, the American Geriatrics Society (1994) has stated that physicians should not provide interventions that will intentionally cause the death of patients.

Though there is opposition to physician-assisted death, acceptance among health professionals is beginning to occur (Dickinson, Lancaster, Summer, & Cohen, 1997). For example, few people publicly criticized the case in which Dr. Timothy Quill assisted "Diane" in ending her life in 1991. In 1996 the 30,000-member American Medical Student Association filed a brief before the Supreme Court supporting physician-assisted suicide death.

Public opinion of physician-assisted death is also becoming more favorable in the United States and in other places, as noted earlier. In 1950 only 34 percent of people in the United States supported active euthanasia for incurably ill patients if they and their families requested it (Blendon, Szalay, & Knox, 1992). By 1973, however, a Gallup poll in the United States found that 53 percent of those interviewed said that physicians should have the legal right to painlessly end the life of a person with a terminal illness if the patient and family request it (Nagi, Puch, & Lazerine, 1978). In 1977, as many as 60 percent supported active euthanasia. By 1991, 63 percent of the U.S. population supported active euthanasia, and a majority (64 percent) favored allowing PAD (Blendon et al., 1992). By 1997, 67 percent in the United States supported PAD (Angell, 1997). In Australia, in 1996 (Hassan, 1996) 76 percent agreed that a doctor should comply with a terminally ill patient's

PRACTICAL MATTERS

JUNE 19, 2006
PUBLIC CONTINUES TO SUPPORT RIGHT-TO-DIE
FOR TERMINALLY ILL PATIENTS

At least six in 10 Americans support euthanasia, doctor-assisted suicide

by Joseph Carroll

Gallup's annual survey on Values and Beliefs, conducted May 8 to 11, 2006, finds that the vast majority of Americans continue to support "right-to-die" laws for terminally ill patients, whether that involves a doctor ending a patient's life by some painless means, or a doctor assisting a terminally ill patient to commit suicide. An analysis of Gallup data collected since 2003 shows that senior citizens, Americans who frequently attend religious services, those with lower levels of education, blacks, conservatives, and Republicans are most likely to object to euthanasia and doctor-assisted suicide.

Overall Results

The May poll asked two different questions to gauge opinions for the issue of doctor-assisted suicide.

Half of the respondents in the survey were asked this long-term Gallup trend question on euthanasia:

- "When a person has a disease that cannot be cured, do you think doctors should be allowed by law to end the patient's life by some painless means if the patient and his family request it?"

The other half of the respondents were asked this question that specifically mentions "suicide":

- "When a person has a disease that cannot be cured and is living in severe pain, do you think doctors should or should not be allowed by law to assist the patient to commit suicide if the patient requests it?"

Both questions find that more than six in 10 Americans support the notion of euthanasia or doctor-assisted suicide.

The longer-term trend question, asked since the 1940s, specifies that the doctor would be ending the terminally ill patient's life by some painless means. In 1947 and again in 1950, one-third of Americans said they supported euthanasia. This increased to 53 percent by 1973, and then from 1990 through the most recent survey, support has been much higher, fluctuating between 65 and 75 percent. The latest update shows a slight drop in support for euthanasia—from 75 to 69 percent—following the death of Terri Schiavo in 2005.

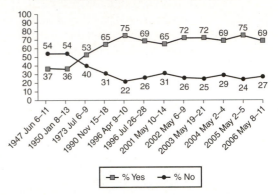

When a person has a disease that cannot be cured, do you think doctors should be allowed by law to end the patient's life by some painless means if the patient and his family request it?

Gallup has only asked the second question—which focuses on a doctor assisting the terminally ill patient to commit suicide at the patient's request—since 1996. At that time, roughly half of Americans (52 percent) supported the notion of doctor-assisted suicide. From 1997 through 1999, during the time when Doctor Jack Kevorkian was making headlines for assisting over 130 patients with committing suicide, roughly six in 10 Americans showed support for this. Support then reached its highpoint at 68 percent in 2001 before declining to the low- to mid-60 percent range from 2002 through 2004. In 2005, support declined to 58 percent before rebounding this year to 64 percent.

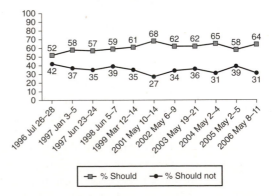

When a person has a disease that cannot be cured and is living in severe pain, do you think doctors should or should not be allowed by law to assist the patient to commit suicide if the patient requests it?

continued

PRACTICAL MATTERS	JUNE 19, 2006

PUBLIC CONTINUES TO SUPPORT RIGHT-TO-DIE FOR TERMINALLY ILL PATIENTS *continued*

Support by Subgroups

Since a majority of Americans have consistently shown support on these two measures, Gallup combined the results of its 2003 to 2006 surveys to get a better understanding of which groups Americans are most likely to support doctor-assisted suicide and euthanasia.

Group Support for Euthanasia/Doctor-Assisted Suicide
2003–2006 Aggregate

	Support doctor ending patient's life by painless means	Support doctor assisting patient to commit suicide
	%	%
Total sample	69	58
Gender		
Men	73	57
Women	65	58
Age		
18- to 29-year-olds	69	56
30- to 49-year-olds	72	63
50- to 64-year-olds	69	60
65 years and older	62	47
Race		
Whites	70	60
Blacks	56	38
Education		
High school or less	65	48
Some college	69	60
College graduates	76	70
Post-graduate education	73	69
Religion		
Protestants	61	50
Catholics	71	62
No preference	84	81
Church Attendance		
Weekly/almost weekly	54	39
Monthly	68	59
Seldom/never	80	72
Party Affiliation		
Republicans	63	50
Independents	71	61
Democrats	72	61
Political Ideology		
Conservatives	57	44
Moderates	74	65
Liberals	82	70

The table illustrates several key points about support for end-of-life issues:

- Catholics are more likely than Protestants to support both euthanasia and doctor-assisted suicide; support is even higher on both measures for those with no religious affiliation.
- Church attendance also plays a role in shaping views on euthanasia and doctor-assisted suicide, with frequent church-goers much less likely than those who attend services less frequently to support both methods.
- Republicans show less support on these two questions than do independents or Democrats.
- The results to the two questions also differ by self-described political ideology, with conservatives much less likely to support euthanasia and doctor-assisted suicide than moderates or liberals.
- Blacks are less likely than whites to support both methods to end a terminally ill patient's life.
- Americans with a college degree or post-graduate education are more inclined than those with less education to show support on both measures.
- Americans aged 65 and older are less likely than those who are younger to support either euthanasia or doctor-assisted suicide.
- Men are more likely than women to support a doctor ending a patient's life by some painless means; there is essentially no difference in support between men and women for a doctor assisting a patient to commit suicide.

request for a lethal dose; the Netherlands have shown a progressive increase in public support, with 92 percent of Dutch people supporting mercy killing in 1998 (Onstad, 2000); a Canadian survey (Edwards & Mazzuca, 1999) found that 77 percent support euthanasia in patients with terminal illness and severe pain; and in the United Kingdom 82 percent agreed in 1994 (Donnison & Bryson, 1995) that doctors should be allowed by law to end the life of a patient with a painful, incurable disease if such a request is made.

Why is there increased support for PAD and AVE? Possible answers to this question include (1) better educated populations, (2) more autonomy for individuals (an additional choice), (3) more technology contributing to more fear of misuse, (4) more secularized populations (a decline in religion) and thus a more rational approach, (5) an aging population contributing to more lingering suffering, and (6) the emergence of AIDS having an impact on younger, urban dwellers.

Because physicians would be involved with active euthanasia, we must wonder about their views on this topic. A comprehensive survey in 1993 of 938 physicians in Washington indicated that they are polarized: A slight majority favor legalizing physician-assisted death and euthanasia in at least some situations, but most would be unwilling to participate in these practices themselves (Cohen, Fihn, Boyko, Jonsen, & Wood, 1994). In a **replication** study of the Washington state survey, George Dickinson and colleagues (1997) found that 587 South Carolina physicians had attitudes remarkably similar to those of the physicians in Washington: Their attitudes were sharply polarized, with fewer than 15 percent giving a neutral response to the questions of euthanasia and assisted suicide. Another survey (Watts, Howell, & Priefer, 1992) of 727 geriatricians on their attitudes toward assisted suicide among patients with dementia found that 21 percent would consider assisting in the suicides of competent, nondepressed patients. H. G. Koenig (1993) stated that increasing support within the medical community for physician-assisted death comes from a recent decision by Michigan physicians to reverse their stand against the practice, preferring that it not be considered a felony. A study (Meier et al., 1998) of 1,902 physicians in the United States revealed that 24 percent would perform AVE and 36 percent PAD *if* it were legal. Regarding attitudes of both black and white physicians in the United States (Mebane et al., 1999), of the 502 physicians responding, 37 percent of white physicians and 27 percent of black physicians found PAD "an acceptable treatment alternative."

A systematic literature review of 39 articles related to U.S. physicians' attitudes concerning AVE and PAD (Dickinson, Clark, Winslow, & Marples, 2005) found that physicians' attitudes favoring legalization of PAD and AVE ranged from 31 to 71 percent. Responses regarding the acceptance of PAD ranged from 14 to 66 percent, whereas AVE acceptance responses varied from 23 to 63 percent. Eleven to 63 percent of physicians had received requests for PAD or AVE. If made legal, about one-third agreed to participate in PAD, with a lesser amount agreeing to AVE. Overall, less than 5 percent have performed AVE, with slightly higher percentages for PAD.

Physicians in other countries have also voiced their opinions about AVE and PAD in recent years. A study in Italy (Grassi, Magnani, & Ercolani, 1999) of 396 general practitioners found that only 15 percent gave responses somewhat or strongly favoring AVE and PAD. A Canadian study (Verhoef & Kinsella, 1996)

Fischer, Ed/CSL, CartoonStock Ltd

showed that physicians' agreement on legalizing AVE and PAD decreased over a three-year period (29 percent versus 15 percent and 50 percent versus 37 percent, respectively). In the United Kingdom (Clark et al., 2000), responses from 333 geriatric medicine doctors revealed strong opposition to legalization of AVE and PAD (80 percent agreed that AVE is never ethically justified and 68 percent agreed that PAD is never ethically justified).

Why is there such strong opposition to PAD and AVE on the part of physicians? Some answers given in the literature include the following: (1) Legalization would have a negative effect on palliative care and on physician–patient relationships; (2) legalization of AVE would follow legalization of PAD; (3) involuntary euthanasia would follow AVE, thus invoking the slippery-slope argument; (4) sanctity of life is emphasized in the Hippocratic Oath; and (5) a patient might be deterred from seeking medical advice, if a part of the physician's role was to "kill" patients.

A problem noted with PAD, when the patient is acting without the physician present, is the likelihood of serious accidents inherent in the patient's giving himself or herself an overdose. Vomiting may occur as the patient slips into a coma, with aspiration of the vomitus. If isolated patients change their minds, they may nevertheless choke to death in a panic, or, if rescued, die of pneumonia. With states of confusion resulting from morphine, barbiturates, or other compounds, some patients may experience terror or panic (Report of the Committee on Physician-Assisted Suicide and Euthanasia, 1996).

As noted earlier, public opinion polls regarding PAD and AVE and the opinions of physicians on these issues generally do not agree. Whereas the public overall seems to have the attitude that patients should have a right to have assistance from their doctors in ending their lives, physicians generally are not in favor of such action and in some cases are strongly against this practice.

Guidelines for Assisted Suicide. With Oregon, Washington, Montana, and Vermont as the only states in the United States where physician-assisted death is currently legal, the Oregon guidelines in the list below could be the model for similar states in the future, pending voter approval:

1. The patient must request a physician's assistance in suicide three times, the last in writing, with the statement dated and signed by the patient in the presence of two witnesses.

2. Two physicians must determine that the patient has a life expectancy of six months or less.
3. The physician must wait at least 15 days after the initial request, and at least two days after the final request, before writing the prescription for the lethal drugs.
4. The physician must determine that the patient is not suffering from a psychiatric or psychological disorder or depression causing impaired judgment.
5. At least one of the two witnesses to the patient's written request for the lethal prescription must be a person who is not a relative of the patient, does not stand to benefit from the estate of the patient, and is not an employee of the institution where the patient is being treated.

With widespread public interest in assisted suicide and euthanasia, whether legalized or not, the medical profession should work to improve the care of terminally ill patients through more effective control of pain and other symptoms. However, whether patients are terminally ill, pain-free or not, rational suicide may be the answer for some individuals.

Finally, we might ask the question as to why there seems to be less public controversy and opposition to the practice of physician aid in dying than for clinical abortion? Cathleen Kaveny (2011, p. 6) suggests the following answer:

> The anxiety created by the prospect of one's own very bad death is far more immediate and visceral than any concern that the United States might become a Nazi dystopia. People just do not want to die abandoned, alone, and in pain. They do not want to bankrupt themselves arranging for around-the-clock nursing care, or to become a physical and financial burden to their already overstretched children. In the context of these fears, the legal availability of PAS can become a security blanket, an emergency exit "just in case" their worst fears are realized.

However, rather than merely accepting the proliferation of laws to expand the access to physician aid in dying, Kaveny (2011, p. 6) proposes the following alternative:

> We must actively build a culture of life. Universal health care, including the latest techniques for pain control, must be widely available. In-home elder care and respite care need to be easily accessible. And churches, synagogues, and mosques ought to ramp up their ministries to the sick and the homebound. As Pope John Paul II recognizes in *Evangelium vitae,* protecting human dignity at the end of life is a matter of both justice and the works of mercy.

CONCLUSION

Physicians in the United States often press for more treatment than either a family or a terminally ill patient desires. On the other hand, some families insist, even in hopeless situations, that everything possible be done for the patient. Though these motives are kind, the results are often not, noted ethicist and physician Charles Culver (1990). Because of our ever-increasing technological sophistication, dilemmas and disagreements about allowing patients to die are frequent and will probably increase in the future. Bodies can be kept alive for a long time. There are cases in which further treatment would not be totally pointless, but the patient, were she or he competent to decide, would not desire further treatment. Or if the patient were truly altruistic, she or he might opt to die and donate organs so that others could live a quality life. But then deciding when one is dead, and to whom to give the organs once the individual is declared dead, pose other questions. Thus, the saga continues.

SUMMARY

1. Euthanasia, or "good death," must be defined within the context of one's understanding of the meaning of life.

2. The sanctity-of-life perspective of euthanasia contends that all natural life has intrinsic meaning and should be appreciated as a divine gift. As a consequence, human beings have the obligation to enhance the quality of life as it may exist.

3. The quality-of-life perspective of euthanasia contends that when life no longer has quality, death is preferable to living a life devoid of meaning.

4. Individuals try to facilitate a "good death" by two methods—passive and active euthanasia. Passive euthanasia involves a protocol whereby no action or medical intervention hastens death. Passive euthanasia usually involves the removal of medical technology or the refusal of medical intervention. Passive euthanasia is supported by people who affirm both sanctity-of-life and quality-of-life perspectives.

5. Active euthanasia requires a direct action to bring about death. Those who favor active euthanasia prefer to use the term self-deliverance when the life of a terminally ill patient is ended. Self-deliverance should be considered a positive action taken to provide a permanent solution to the long-term pain and suffering of the individual and his or her significant other who are faced with a terminal condition.

6. Persons opposing active euthanasia often do so with the "slippery-slope-to-Auschwitz" argument. They allege that the justification for withholding some treatments also applies to withholding any and all treatments from those whom society considers unworthy. The slippery-slope argument asserts that society moves toward a disregard for human life in predictable steps.

7. Because of the sensitive nature of the issue, human organ transplantation has not been widely discussed in society. Difficult decisions need to be made about what qualifies a person to be eligible to receive a transplant from a donor.

8. Physician-assisted death has been a major issue in the United States and throughout the rest of the world and will continue to be so throughout the 21st century.

DISCUSSION QUESTIONS

1. Compare and contrast passive and active euthanasia. What difficulties does implementing each of these approaches have for current social policy dealing with health care for the terminally ill?

2. If your parent were dying from an extremely painful and incurable form of cancer and decided that life was not worth living, what role would you be willing to play in assisting him or her in self-deliverance? Would you be unwilling to assist your parent? If not, what actions would you take in discouraging your parent's action?

3. Provide arguments for and against the removal of food and water from a terminally ill patient in a coma.

4. Discuss some of the problems affecting human organ transplantations from both a societal and a personal point of view.

5. Should society make eligible for a transplant only those persons who can afford it, or should the potential societal contribution of the individual be taken into account? Are there some medical procedures that should never be financed by insurance benefits or government funds because they are too expensive?

6. Discuss your reaction to Dr. Timothy Quill's assisting his patient, "Diane," who had leukemia, with her death.

7. Physicians are sworn to "prolong life and relieve suffering," yet physician-assisted death runs counter to prolonging life. The majority of people in the United States favor physician-assisted deaths. Discuss the pros and cons of physician-assisted deaths.

GLOSSARY

Active Euthanasia: A direct action that causes death in accordance with the stated or implied wishes of the terminally ill patient. Such action can be voluntary (patient requested) or involuntary (not requested by the patient).

Active Voluntary Euthanasia or AVE: Upon the request of the patient, a physician directly contributes to his or her death (e.g., by injecting a lethal dosage into the patient).

Bioethicist: A professional person who assists in decision making between biomedical practitioners and their patients regarding values and standards within a medical setting.

Biomedical Issues: Issues concerned with medical biology to the extent that social, psychological, and behavioral dimensions of illness are disregarded.

Compassion & Choices: A nonprofit organization working to improve patients' rights and choices at the end of life (formerly Hemlock Society).

CPR: Cardiopulmonary resuscitation—the act of using mouth-to-mouth resuscitation and chest compressions to try to restore one's breathing.

Cultural Lag: A situation in which some parts of culture change more slowly than others. A typical situation occurs when technology changes faster than the values of the culture.

DNR or Do Not Resuscitate: An order required by some hospitals for heroic care or other resuscitative measures to be withheld.

DOA: "Dead on arrival": A classification given to a patient upon arrival at the emergency room if the physician declares the patient dead. Thus, no resuscitation efforts would follow.

Double Effect: A physician giving a patient pain control medicine and increasing the dosage as the pain increases. Too much medication such as morphine may indeed kill the patient, and this is the "double effect" (i.e., the pain is being controlled, but a side effect of too much medication might cause the death of the patient).

EMTs: Emergency medical technicians.

ERGO!: Euthanasia Research and Guidance Organization.

Euthanasia: Literally, a "good death." Presently, those who favor active euthanasia, view it as a "good death brought about by a doctor providing drugs or an injection to bring a peaceful end to the dying process."

EXIT: A British society that promotes active euthanasia.

Extraordinary Measures: All of those medicines, treatments, and operations that cannot be obtained without excessive expense, pain, or other inconvenience or that, if used, would not offer a reasonable hope of benefit.

Hemlock Society: An organization in the United States that supported active euthanasia and is presently the organization Compassion & Choices.

Intubation: A medical procedure whereby the patient is fed through a tube placed in the stomach.

Living Will: A document stating that one does not want medical intervention if the technology or treatment that keeps one alive cannot offer a reasonable quality of life or hope for recovery.

Ordinary Measures: All of those medicines, treatments, and operations that offer a reasonable hope of benefit to the patient and that can be obtained and used without excessive expense, pain, or other inconvenience.

Quality of Life: The perspective that when life no longer has quality, death is preferable to living a life devoid of meaning.

Passive Euthanasia: The withholding of treatment, which, in essence, hastens death and allows the individual to die naturally.

Patient Self-Determination Act (PSDA): Went into effect on December 1, 1991, and requires all health care providers who receive federal funding to inform incoming patients of their rights under state laws to refuse medical treatment and to prepare an advance directive.

Physician-Assisted Death or PAD: A physician giving advice to a patient about taking her or his life and/or prescribing a lethal dosage of medication whereby the individual can take her or his own life (sometimes called physician-assisted suicide or PAS).

Power of Attorney: A written document that allows another to make decisions on behalf of the patient when the patient is unable to do so.

Replication: The repetition or duplication of the procedures followed in a particular study in order to determine whether or not the same findings are obtained at another time and/or place.

Sanctity of Life: The perspective that all natural life has intrinsic meaning and should be appreciated as a divine gift. As a consequence, human beings have the obligation to enhance the quality of life as it may exist.

Self-Deliverance: A rational and voluntary act of taking one's life; an alternative to the terms suicide and mercy killing.

Slippery-Slope-to-Auschwitz Argument: The belief that if a society extends to any of its members the right to active euthanasia, none of its members is protected from being killed. Auschwitz was a German concentration camp where people were killed during World War II.

Triage: A system of assigning priorities of medical treatment. Organ transplants are sometimes an example of triage: Who receives the organ and on what basis is the decision made?

Xenotransplantation: The transplantation of organs between different species.

SUGGESTED READINGS

DeVries, B. (Ed.). (1999). *End of life issues: Interdisciplinary and multidimensional perspectives.* New York: Springer Publishing Company. Examines the ways in which individuals at the end of life are influenced by aspects of their interpersonal and social environments.

Dickinson, G. E. (2005, February). Special issue: Ethical concerns involving end-of-life issues in the United States. *Mortality, 10*(1). An excellent group of social-science articles dealing with end-of-life issues. Topics include physician-assisted suicide, active euthanasia, Oregon's physician-assisted suicide law, suicide, prolonging life in vulnerable populations, and grieving related to end-of-life decisions.

Eggertson, L. (2012, November 6). Organ donation's "silver bullet"? *Canadian Medical Association Journal, 184*(16), E835ff.

Euthanasia.ProCon.org. (2013). State-by-state guide to physician-assisted suicide. http://euthanasia .procon.org/view.resource.php?resourceID=000132 (Last updated on: 5/28/2013 3:49:15 PM PST).

Accessed on November 6, 2013. An incredible resource to understand the complexities of the pros and cons of euthanasia and Physician Aid in Dying (PAD).

Fallis, J. (2012, January 10). Stem cell donations. *CMAJ: Canadian Medical Association Journal, 10*, E13ff.

Fenigsen, R., & Fenigsen, R. (2012). Dutch government-ordered surveys of euthanasia. *Issues in Law & Medicine, 28*(2), 237ff.

Fenigsen, R., & Fenigsen, R. (2012). Who is leading us there? *Issues in Law & Medicine, 28*(2), 333f.

Gibson, R. (2012, April). The case for euthanasia and physician-assisted suicide. *ISAA Review: Journal of the Independent Scholars Association of Australia, 11*(1), 55ff.

Hastings Center, 2011. Bioethics Graduate Programs. *TheHastingsCenter.org.* http://www .thehastingscenter.org/BioethicsWire/ BioethicsGraduatePrograms/Default.aspx. Accessed on November 8, 2013.

Hollander, C. (2013). What's the fairest way to dispense donated organs? *National Journal*

(July 25). http://www.nationaljournal.com/ njonline/ Accessed on November 8, 2013.

Kaveny, C. (2011, July 15). Dignity & the end of life: How not to talk about assisted suicide. *Commonweal*, 138(13), 6.

Madwar, S. (2011). United States officials propose further retreat from first-come, first-served organ donation. *CMAJ: Canadian Medical Association Journal*, 12(July), 639ff.

Prado, C. G. (2008). *Choosing to die*. Cambridge, UK: Cambridge University Press. This book addresses the difficult question of when and whether it is rational to end one's life in order to escape devastating terminal illness. It provides a multicultural perspective on end-of-life issues as to what is permissible and impermissible morally and legally.

Quill, T. E., & Battin, M. P. (Eds.). (2004). *Physician-assisted dying: The case for palliative care and patient choice*. Baltimore, MD: The Johns Hopkins University Press. This anthology contains 21 readings that argue for physician-assisted death. The arguments include patient autonomy, relief of pain and suffering, and nonabandonment.

Randall, F., & Downie, R. (2009). *End of life choices*. Oxford, UK: Oxford University Press. This book provides guidance on the ethical minefield that has developed for those who care for patients toward the end of life. Some of the issues discussed include current issues and controversies related to end-of-life care, advanced care planning, preferred place of care, ideas of "best interests," and physician-assisted suicide.

Saad, L. (2013). U.S. support for euthanasia hinges on how it's described: Support is at low ebb on the basis of wording that mentions "suicide." *Gallup Politics*. May 29, 2013. http://www.gallup.com/ poll/162815/support-euthanasia-hinges-described .aspx. Accessed on November 6, 2013.

Sandeen, P. (2012). Massachusetts voters deny rights to terminally-ill people. Death with Dignity National Center. http://www.deathwithdignity .org/2012/11/07/massachusetts-voters-deny-rights- terminally-ill-people. Accessed on November 7, 2013.

Smith, W. (2014). At the bottom of the slippery slope. In G. E. Dickinson and M. R. Leming (Eds.), *Annual Editions: Dying, death, and bereavement*. New York: McGraw-Hill Companies.

Spinney, L. (2011, June 9). Battling the body brokers: A hard-hitting book calls for greater transparency to deter the illegal trade in human blood, organs and eggs. *Nature*, 474(7350), 156ff.

Suicide is a permanent answer to a temporary set of problems.
—NBC Evening News

There is but one truly serious philosophical problem, and that is suicide. Judging whether life is or is not worth living amounts to answering the fundamental question of philosophy.
—Albert Camus

There are times in life when we would like to die temporarily.
—Mark Twain

Suicide should be seen not as a sudden isolated disaster but as a major event in an unhappy series, bringing in its wake grief certainly, but the possibility also of relief.
—Shepherd and Barraclough

SUICIDE AND OTHER SUDDEN, UNNATURAL TRAUMATIC DEATHS | CHAPTER 9

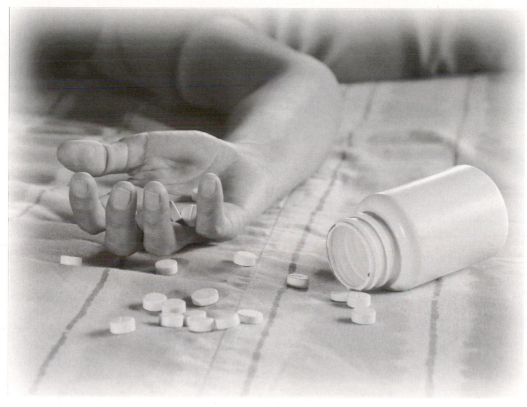

Any death is a shock, whether following a chronic condition or a sudden death. Even though someone may have been ill for a long time, the pronouncement of death remains nonetheless difficult to absorb. How much more difficult it is, then, to learn of an unnatural traumatic death through suicide, homicide, or an accident. As we have previously noted, accidents, suicide, and homicide rank at the top of the list of death etiology for adolescents and young adults, though this age group indeed has a very low death rate. The trauma of these kinds of deaths is certainly a shock to the surviving family and friends. Let's first take a look at suicide.

Physician-assisted suicide is a medical response to the wishes of terminally ill patients. This action involves a discussion and consideration between a doctor and the patient. Unassisted suicide is a very different matter. The decision is usually made in isolation by an individual who is often suffering from depression. Whereas physician-assisted suicide is a relatively recent phenomenon, unassisted suicide has been around for as long as recorded history. Yet, the word suicide is not in the Bible or in the pamphlet by John Donne (1644/1930) on self-homicide. The *Oxford English Dictionary* states that the word was first used in English in 1651 and is derived from the modern Latin *suicidium*, which stems from the Latin pronoun for *self* and the verb *to kill*. Though the word *suicide* was not used, the earliest recorded suicides known to Western culture were the deaths of Samson and Saul around 1000 B.C., and in the New Testament Judas Iscariot went out and hanged himself following the Last Supper after betraying Christ. Up to nine references to suicides can be found in the Old Testament, but these Biblical reports are done in a nonjudgmental, factual way and not reported as wrong or shameful and were not condemned (Colucci & Martin, 2008). The conditions that seemed to justify suicide include revenge (Samson's suicide), justice, shame (Judas's), and political or military defeat (Saul's). It was only in the fourth century, with St. Augustine's "City of God," that the official Catholic Church disapproval of suicide began to spread.

CHANGING ATTITUDES TOWARD SUICIDE

Historically, attitudes toward suicide have been somewhat mixed. In 1284, for example, the denial of Christian burial to those who had committed suicide was introduced, and indignities were often committed upon suicide corpses by public exposure (Colucci & Martin, 2008). Often those committing suicide were buried at crossroads because the constant traffic was thought to keep the ghosts down. Yet, early societies sometimes forced certain members to commit suicide for ritual purposes. For example, suicide was occasionally expected of the wives and slaves of deceased husbands or masters as an expression of fidelity and duty. In the ancient Roman Empire, suicide was socially acceptable until slaves started to practice it. Because the suicide of slaves caused a severe financial loss, suicide was declared a crime against the state (McGuire & Ely, 1984).

Judaism traditionally described suicide as a sin (Colucci & Martin, 2008). Modern Judaism, however, rests upon a long tradition of honoring heroic suicide to avoid rape, forced slavery, or idol worship. Until the middle of the 20th century, the Roman Catholic Church denied Catholic ritual and burial to those who took their own lives (*Codex Juris Canonici*, 1918). For Japanese kamikaze pilots in

World War II, suicide represented the great death. The Chinese regarded suicide as acceptable and honorable, particularly for defeated generals or deposed rulers. Suicide today is a crime under Islamic law and suicide attempts may lead to prosecution (Pridmore & Pasha, 2004).

In the 21st century, suicide bombings through the use of a vehicle or bombs attached to an individual have been all too common in Afghanistan and Iraq. The persons instigating these bombings feel that they will be rewarded in the afterlife for their deeds and will be heroes or heroines in the eyes of their own people. The 21st-century terrorist attacks on London, Madrid, and the United States were such suicide acts. Ellen Townsend (2007) suggests, however, that terrorist acts are not truly suicidal, though these acts are similar to altruistic suicides and share similar characteristics to others who die by suicide. These suicide terrorists are murderers, yet there is an absence of murderous intent in most suicides. Also, Townsend notes that depression tends to often be present in completed and attempted suicides, seemingly not the case in suicide terrorism. The suicide terrorist typically makes his or her decision to act as a suicide attacker at the level of the group, whereas an individual committing suicide tends to decide in isolation from others. Townsend concludes that the suicide terrorist could be considered an atypical variant of the category of altruistic suicide or as a new type of fatalistic-altruistic suicide. Whatever the label given, suicide terrorism tends to be different from other suicidal behavior and suicide terrorists are not suicidal in the same way as other committed suicides. Nonetheless, whatever label one wishes to use, these suicide terrorists *are* committing suicide.

Attitudes toward suicide changed radically when St. Augustine, drawing heavily from the philosophy of Plato and Aristotle, laid down rules against suicide that became the basis for Christian doctrine throughout the succeeding centuries. Societal opposition to suicide in Christian communities continued throughout the Middle Ages, until the Renaissance broadened thinking on the subject. Well into the 19th century, taking one's own life was still considered a grievous sin, but by the mid-20th century the taboo had lost some of its religious undertone (Goode, 2003). Yet today, suicides are often hidden, and in their aftermath individuals' behavior becomes awkward; people are uncertain what to ask and what to say. The ambiguous nature of suicide is evidenced by the fact that attempted suicide was considered a felony in several states, but since the 1990s this is no longer the case.

THE STIGMA OF SUICIDE

Historical records show that, during the Middle Ages, suicide stigmatization was fully institutionalized. As noted earlier, suicide corpses were regularly mutilated to prevent the unleashing of evil spirits. Suicides were denied burials in church cemeteries. The property of the suicide victims' families was confiscated and put into the control of local agents, and these families were excommunicated from the community. After a suicide loss, families often lost their land because they were unable to pay the heavy tithes imposed by the church (Feigelman, Gorman, & Jordan, 2009).

We have come a long way regarding the stigmatization of suicide since the Middle Ages, yet suicide is somewhat still morally proscribed by Western culture, so that families in which a suicide occurs may feel shame and individuals who

AP Images/STR

Soccer player killed in Iraq from a suicide bombing. Such bombings have become all too common in the 21st century in various places in the world.

committed suicide may be viewed as crazy, mentally ill, and psychologically stressed. In addition to the negative feelings, surviving relatives of suicide victims may be more grief-stricken because their loved ones willfully took their own lives. Society tends to reject suicide less under some specific conditions, such as a terminal illness, yet suicide continues to be viewed by many as an undesirable and abnormal act. The traditional idea in the United States is that the only legally, ethically, and morally correct response to suicide is intervention, with such ethical codes being found in many helping professions and their organizations (e.g., the American Psychological Association). Action to prevent suicide is expected and required, based on the assumption that anyone who considers suicide is suffering from some sort of mental problem.

Because suicide is still socially unacceptable, the family may lack the support systems normally available to those grieving other kinds of death—a traumatic loss is suffered, and the taboo act generates feelings of disapproval and shame. In fact, the stigma surrounding suicide has been implicated as a contributor to the amount of social support that suicide survivors receive, as well as survivors' willingness to talk about the suicide (Hung & Rabin, 2009).

The general attitude of the public toward suicide remains confused and sometimes contradictory. Extreme views on suicide range from total acceptance to total rejection of the right of an individual to commit suicide. At the heart of the debate is the question of whether people should be allowed to die without interference.

DEFINING SUICIDAL ACTS

The meaning of *suicide* continues to be problematic. One definition of suicide is a death arising from an act inflicted upon oneself with the intent to kill oneself (Andriessen, 2006). Most definitions share self-initiation, behavior with a fatal outcome, and intention or expectation to die. The *intention* aspect of the definition is more difficult to determine, as intention enables one to distinguish between accidental and suicidal behavior. Did Marilyn Monroe, Elvis Presley, and Michael Jackson *intend* to die or were their deaths *accidental*? We could, therefore, possibly classify the following persons as engaging in suicidal behavior:

1. A person who smokes cigarettes, knowing that the Surgeon General has determined that smoking is a major cause of lung cancer
2. A race car driver who races even though he or she knows that in any given race there is a good chance that someone will be killed
3. A person who takes a bottle of sleeping pills, hoping to call attention to self as one who has personal needs that are not being met
4. A person who refuses to wear a seat belt when studies show that wearing a seat belt reduces the death rate in automobile accidents
5. A person who continues to eat fatty foods after having suffered a heart attack
6. A person with advanced stage cancer who refuses chemotherapy or surgery
7. A person who overeats (or undereats) to the extreme that his or her health is directly affected
8. A person with high blood pressure who refuses to take medication to control it or fails to control food intake or to exercise
9. A person who ingests a mixture of alcohol and drugs

Probably most of us would not consider the preceding list of persons as suicidal. Rather, we would be concerned about whether the person in question was intentionally trying to kill himself or herself. Two measures of intent include (Linehan, 2000) implicit intent (intent inferred from the behavior itself) and explicit intent (intent directly communicated by the individual), yet such intent is often difficult to determine. *Suicide attempt* has different meanings to different individuals, yet such generally refers to self-injurious behaviors, whether suicide intent is present or not (Silverman, 2006).

There is a qualitative difference between a **suicide gesture** and a **completed suicide**. Suicide gestures are motivated by a need for aid and support from others, whereas completed suicides are acts of resignation. At this point, we have a problem of tautology—suicide gestures that mistakenly end in death are classified as intentional suicides, and unsuccessful suicide attempts are considered suicide gestures. In addition, suicides are sometimes classified as accidents or natural deaths as a favor to family members or as a method of providing a more positive view of the deceased person. Thus, there are deaths resulting from intentional acts of self-destruction that are recorded as natural deaths, and there are suicidal gestures accidentally ending in death that are recorded as suicides. Whether or not the act was intentional has to do with **suicidal ideation**, which refers to considering suicide, contemplating suicide, being prone to suicide, or having suicidal thoughts (Silverman, 2006). Obviously, there are problems in compiling suicide statistics!

SOCIAL FACTORS INVOLVED IN SUICIDE

The World Health Organization (WHO) (Kelland, 2012) estimates approximately 1 million individuals around the world commit suicide every year, a mortality rate of 16 per 100,000 or one death every 40 seconds. The WHO estimates that for

Doonesbury Cartoon.

every suicide around the world there are at least 20 suicide attempts. Industrialized countries tend to have a higher suicide rate than poor, developing countries, and the United States has a moderate suicide rate (11.2 per 100,000 in 2006) compared to other industrialized countries (e.g., Japan was 23.7 per 100,000 in 2006). Overall suicide rates range from 0.5 per 100,000 in the Philippines to 40.2 per 100,000 in Lithuania (Liu, 2009). In the United States in 2006, one person every 16 minutes committed suicide (American Foundation for Suicide Prevention, 2009). While suicide ranks globally as the 21st-leading cause of death, it is as high as the ninth top cause of death in women across Asia's "suicide belt," from India to China. Suicide ranks 14th in North America and 15th in Western Europe for causes of death (Cheng, 2012).

Suicide is rare in childhood, rises sharply during adolescence and early adulthood, and continues to rise with middle age and among elderly persons, most significantly for men after age 65. In fact, the suicide rate in men 65 or older is seven times that of females who are 65 or older. In the United States, women attempt suicide three times as much as men, yet men complete suicide at a rate four times that of women (American Foundation for Suicide Prevention, 2009). It is estimated that one woman in the United States attempts suicide every 78 seconds, yet a woman actually takes her own life every 90 minutes.

AGE, GENDER, REGION, AND MARITAL STATUS FACTORS

Suicide rates worldwide show a greater range for males, from 0.5 per 100,000 in the Philippines to 70.1 per 100,000 in Lithuania, in contrast to 0.3 per 100,000 in Peru to 24.7 per 100,000 in rural China for females (Liu, 2009). In most countries male suicide rates are higher than female suicide rates, with China being the only country where women have a higher suicide rate than men. This higher rate is attributed to the female suicide rate being higher in rural areas, yet slightly lower than the male rate in urban areas. The large rural population in China drives the overall male/female difference, notes Liu. In the United States in 2006, there were 26,308 suicides of males (17.9 per 100,000) and 6,992 suicides for females (4.6 per 100,000) (American Foundation for Suicide Prevention, 2009). Women are more likely than men to have strong social supports, to feel that their relationships are deterrents to suicide, and to seek psychiatric and medical intervention, perhaps contributing to their lower rate of completed suicide.

Between 1950 and 2004 worldwide, there appears to have been an increased trend in suicide rates among young men (ages 15–34) in general, yet suicide rates among middle-aged (35–60) and older women (60+) have significantly decreased (Liu, 2009). Greater gender equality and a greater acceptance toward working women have been suggested to have contributed to such decreases. Decreased toxicity of over-the-counter drugs has also been mentioned as a reason for decreases in female suicide rates around the world.

Regionally, in general, Europe has higher rates of suicide than other regions, while countries such as Latin America and the Caribbean tend to have lower rates (Liu, 2009). Among the subregions, Western Asia has the lowest overall male and female suicide rates, whereas Eastern Europe has the highest suicide rates.

In relation to marital status, the suicide rate is lower for individuals who are married than for those who never married and is highest for those who are widowed, divorced, or separated. Individuals with children commit suicide less than childless individuals. These data on marital and parental status fit Emile Durkheim's theory (discussed later in the chapter) that suicide is a function of social integration.

There were 633,000 emergency room visits from self-inflicted injury in the United States in 2009, though the actual number of deaths reported by suicide was 36,909. The means of committing suicide, listed in decreasing order of frequency, were shooting with firearms (especially handguns), which accounted for 6.1 deaths per 100,000 population (18,735 deaths), suicides by suffocation (2.9 deaths per 100,000 population; 9,000 deaths), and deaths by poisoning (2.1 per 100,000 population; 6,398 deaths), with other suicides coming from cutting and stabbing, jumping from high places, and drowning (Centers for Disease Control and Prevention [CDC], 2012). It has been well documented that attempts involving firearms and jumping from high places are more dangerous than drug ingestion, cutting, or stabbing (Zarkowski & Avery, 2006). For example, more than 90 percent of suicide attempts using guns are successful, whereas the success rate for jumping from high places is 34 percent and for drug overdose 2 percent (Stobbe, 2008). For elderly persons, not taking medication or taking medication irregularly or simply not eating can be added to the list of ways of committing suicide. Sometimes elderly persons might give up psychologically, thus triggering biological changes that increase potential for disease. Because men are more likely to use firearms in suicide attempts (80 percent of all firearm suicide deaths are white males) and women are more likely to use drugs, this obviously contributes to the higher completed suicide rate for males. When using firearms, women are more apt than men to shoot themselves in the body than in the head, possibly due to fears of disfigurement, note Stack and Wasserman (2009).

SUICIDE RATES BY ETHNICITY IN THE UNITED STATES (PER 100,000)

Native American	15.1
Caucasian	13.9
African American	5.0
Hispanic	4.9

A look at statistics (2012). Online Schools.

In looking at cross-national comparisons of suicide among adolescents and young adults (aged 15 through 24) in 34 of the world's wealthiest nations, 34 percent of these suicides were firearm-related (Johnson, Krug, & Potter, 2000). Methods of committing suicide, however, vary with cultures. For example, the most common suicide method among the older-than-65 age group of males in Italy is hanging and strangulation (De Leo, Conforti, & Carollo, 1997), followed by jumping from high places and submersion. For females in Italy, the most common method of suicide is jumping from high places. For young people in Italy, the most common method of suicide is hanging and strangulation.

In Hong Kong, the most common method of suicide is jumping from high places, followed by charcoal burning, since 2002 (Wong et al., 2009). Charcoal is burned on a barbecue grill inside a small room and inhaled by the individual. Death is from carbon monoxide poisoning from the burning charcoal. This method of suicide in Hong Kong has attracted a group of suicidal women who perhaps would not have died by suicide if this less painful and less violent novel suicide were not available, note Wong and colleagues.

SOCIOECONOMIC AND CULTURAL FACTORS

Suicide rates tend to be highest at both extremes of the socioeconomic ladder. Individuals in certain professions such as physicians, dentists, and lawyers tend to be particularly suicide-prone, whereas the suicide rates of teachers and members of the clergy are rather low. Physicians have the highest rate of suicide of any profession, with between 300 and 400 doctors taking their own lives each year in the United States, roughly one per day (Noonan, 2008). Unlike the general population, where male suicides overwhelming outnumber female suicides, the suicide rate among male and female doctors is the same. The high rates may be due to stress and the fact that physicians also have access to drugs with which to take their lives. Medical doctors also worry that they might lose respect, referrals, income, and their license if they admit to a mental health problem. Physicians are supposed to be the strong ones. To address this issue with physicians, medical schools all over the United States have launched programs that guarantee students who seek help that it will not appear on their records. The idea is to focus on medical students and residents, since depression often starts in young adulthood.

Additionally, dentists have high suicide rates. They also have access to drugs and do not always have a good image in the public's eyes and thus may suffer from low self-esteem. Just ask an adult to recall childhood images of trips to the dentist! The recollections are often not pleasant; thus, many adults have a negative association with dentists.

From an economic point of view, it is not that suicide creates only cost, and no benefits, to society (Ying & Chang, 2009). There are potential savings from suicides, including not having to treat the depressive and other psychiatric disorders of those who kill themselves, and avoidance of long-term pension, social security, and nursing home care costs. On the other hand, suicides cause losses in human capital and decrease the labor supply. Overall, suicide causes public unrest from the social point of view and creates a waste of human capital from the economic point of view. Suicide should be prevented based on humane considerations, not on the economic cost alone. Therefore, it is important for government authorities to take notice of the causes of suicide for future welfare and security policies.

Economic strain and suicide research has focused largely on unemployment, yet other strains such as loss of a home place, loss of a car, noxious social relationships, medical problems, death of a significant other, and involvement with the criminal justice system exist, according to Steven Stack and Ira Wasserman (2007). They applied **general strain theory** to a qualitative analysis of case files. A central finding of their study is that economic strain was typically combined with other strains in the genesis of suicidality. Certainly, the loss of one's job and home together could put tremendous pressure on an individual, as such losses impact the livelihood of the individual and her or his family.

The suicide rate for Native Americans is greater than the national average (15.1 per 100,000 for Native Americans in 2006) (CDC, 2009a), yet rates tend to vary from reservation to reservation. Rates also vary by age. Young Native Americans, for example, have a very high suicide rate. Suicide among Native Americans tends more often to be alcohol-related, and violent methods (firearms and hanging) are used more commonly than in the mainstream U.S. population.

There is also a difference in suicide rates between blacks and whites in the United States. Suicide rates among blacks are substantially lower (5.0 per 100,000 population in 2006) than among whites (CDC, 2007b). When broken down by gender, African American women have significantly lower suicide rates compared with the overall rate of men and other ethnic groups of women in the United States. This lower rate for African American women is consistent across age groups. One study (Gibbs, 1997) has identified the following as protective factors that reduce suicide risks among African Americans:

1. Church attendance or affiliation is associated with lower rates of suicide. The African American church has traditionally served as a source of social cohesion, social support, and stress reduction. Religious beliefs have probably influenced the cultural beliefs of many blacks that suicide is extraordinarily sinful and that suicide is a "white thing."
2. The central role of women in the African American community may be a significant factor in their low suicide rates. Black women have learned to cope with high levels of stress, poverty, and discrimination by forming strong social networks, sharing resources, assuming flexible family roles, and turning to religion as a source of comfort for their anger and depression.
3. Suicide rates tend to decline with age among African Americans. Elders are often treated with respect and dignity. They are more likely to live independently or in their extended families rather than in nursing homes or retirement communities. In addition, many blacks look forward to retirement from low-status, low-wage jobs.
4. Suicide rates for blacks tend to be lower in the South and in areas of less racial integration. The cohesive social environment of ethnic neighborhoods has reinforced the role of the extended family. In communities where African Americans are less integrated, they might experience greater social cohesion.

PRACTICAL MATTERS | SUICIDE RISK ASSESSMENT

A suicide risk assessment documents the clinician's thinking in evaluating a patient's risk for suicide by searching for factors which increase suicide risk. Suicidal intent and a suicidal plan are weighed. A history and examination of the patient's mental state is taken to determine suicidal risk and the treatment needed. A suicidal risk assessment integrates three elements—the patient's subjective history, objective clinical observations including any testing, and collateral history from family, friends, and the outpatient treatment team. The assessment is ongoing and weighs objective and subjective data. It considers changes in symptoms and integrates new information as hospitalization progresses. Suicide risk assessment begins at admission and continues throughout hospitalization.

Taken from M. J. Goldblatt and J. T. Maltsberger. (2011). Self-harming behavior and suicidality: Suicide risk assessment. *Suicide and Life-Threatening Behavior, 41*(2), 227–233.

The social and psychological benefits of a cohesive black community have been considerably weakened by major changes such as integration, urbanization, deindustrialization, and secularization of the broader society. These changes may have contributed to a declining sense of community and fewer social supports for vulnerable African Americans.

5. Suicide rates for blacks tend to be higher among those who live alone and are not involved in family units. The extended family and kin networks have played a major role in the survival of African Americans in a frequently hostile society by fulfilling important functions such as economic support, emotional support, and social support. The extended family serves as a buffer between its members and the stresses of the external environment, giving them a sense of identity, security, and support, thus reducing the risk of suicide among rural and low-income blacks.

More suicides occur in heterogeneous urban areas than in more homogeneous rural areas. Though nostalgic times such as the latter part of December are thought to drive higher suicide rates, the "renewal of life" times such as spring have consistently had the highest rates for suicide. The second-highest suicide rate is found in autumn, followed by winter, and then summer. The day of the week consistently found to be the day with most completed suicides is Monday. Why Monday as the suicide day is not known for sure. Durkheim (1897/1951) suggested that the rhythm of social life seemed to reproduce the calendar's division with a renewed activity when a new period is entered. The risk of suicide is lowest on Saturdays.

THEORETICAL PERSPECTIVES ON SUICIDE

Why do human beings kill themselves? The personal and sociocultural factors of suicide, taken individually and in combination, are so complex that it would be difficult to arrive at a definitive explanation. From a medical model perspective, an attempt or an act of suicide is considered abnormal behavior requiring clinical intervention. When suicide becomes a source of duty or even honor, however, as for a kamikaze pilot, a terrorist bomber, or for anyone sacrificing her or his life for the group or cause, motivations become more social and cultural. The most

WORDS OF WISDOM | EMILE DURKHEIM'S FOUR TYPES OF SUICIDE

1. **Egoistic suicide.** Occurs when persons are inadequately integrated into the society, such as intellectuals or persons whose talents or stations in life place them into a special category (celebrities or stars) and who therefore are less likely to be linked to society in conventional ways.

2. **Altruistic suicide.** Happens when individuals are overly integrated into the society (an exaggerated concern for the community) and are willing to die for the group, such as the Japanese kamikaze pilots of World War II; terrorists in the New York City, London, and Madrid attacks; or suicide bombings on the streets of various cities.

3. **Anomic suicide.** Results from the lack of regulation of individuals when the norms governing existence no longer control those individuals (who feel let down by the failure of social institutions), such as people in business who commit suicide after a stock market crash or middle-aged individuals who kill themselves after suddenly losing their jobs because of downsizing.

4. **Fatalistic suicide.** Later introduced by Durkheim, and it has recently received more attention. This type suggests that a person may receive too much control by society and feel oppressed by its extremely strict rules. Fatalistic suicides are a response to the despair engendered by a society allowing little opportunity for satisfaction or individual fulfillment.

significant contribution of sociologists to the study of suicide has been the sociological perspective itself—the insistence on seeing suicidal actions as in some way the result of social factors. According to sociologists, suicidal behavior can only be understood within the context of shared social values, beliefs, and patterns of social interaction.

DURKHEIM'S THEORY The noted French structural-functional sociologist, Emile Durkheim (1951), proposed a comprehensive theory of suicide that is a classic sociological study. Durkheim's purpose in *Suicide: A Study in Sociology* was not to explain why individuals commit suicide but to explain why suicide rates exhibit such stability—the proportion of individuals committing suicide within a given population remains relatively stable over long periods of time, and a similarity of patterns exists across different societies (Hughes, Martin, & Sharrock, 1995). The four types of suicide—egoistic, altruistic, anomic, and fatalistic—are related to Durkheim's conception of social integration.

As Durkheim noted, although we think of suicide as a supremely individual and personal act, it occurs within a social context that influences individual acts. He argued that religion protects against suicide by providing a community bound together by a common belief system of meaningfulness (Simonson, 2008). Durkheim stated that strong social networks formed in religious communities may decrease a person's risk of suicide by increasing the individual's social support (Robins & Fiske, 2009). He theorized that social integration offered protection against suicidal tendencies and conceptualized religious involvement as a form of social integration. He argued that individuals are less likely to commit suicide when they are bound to social groups and thus to group values, goals, norms, and traditions (Perry et al., 2012). Social regulation provides a controlling influence on individuals, therefore guiding their behavior through obligations to others and to the group and fostering a sense of comfort and security that minimizes threats to one's well-being. Those who are generally happy with their lives may be less vulnerable to any individual stressor because of their feeling that life is good, note Brea Perry and colleagues. Individuals with a sense of well-being may be engaged in a variety of roles, giving their lives meaning and providing outlets for tension release. Relationships formed with other individuals through contact with religious institutions may provide some protection from suicide by promoting a sense of belonging, said Durkheim. Yet, different types of society and social integration, via religion or otherwise, produce different rates of suicide.

A rising concern recently developed in Japan over a phenomenon known as **karoshi** (death from overwork), a sort of altruistic suicide. This obsession with overwork is found primarily in middle-aged men who work exceedingly long hours and become so exhausted that they literally die, usually from cardiovascular complications. Through their hard work, these men are demonstrating their loyalty to the company, being willing to stay and work after hours—overly integrated into the group. Many feel guilty when they take holidays and are away from their jobs. They die from stress and overwork for the good of the company. The Japanese government is addressing this issue by encouraging corporations to offer exercise programs in order to relieve stress.

Durkheim's work on suicide has been visited and revisited about as much as any 19th-century publication. Both empirical support and nonsupport for his thesis have been found. Criticisms have largely been directed at his theory and method of study, though some negatives have been directed to the statistical material that Durkheim used to support his argument. Some studies confirm Durkheim's family integration hypothesis, finding that marital and family stability are associated with lower suicide rates (Maimon, Browning, & Brooks-Gunn, 2010). Research outcomes from the early to the mid-20th century tend to support Durkheim's thesis, but more recent studies have challenged his theory (Colucci & Martin, 2008). Many current studies found that belonging to a certain religion showed a protective effect against suicide, as did the level of religious commitment and church attendance. Other studies, reported by Colucci and Martin, have shown that the association between religion and suicide is influenced by other variables, thus weakening the overall association between religion and suicide. Contrary to Durkheim's argument that Catholics commit suicide less often than Protestants, due to their better social integration and less individualism, elderly individuals in Catholic countries die more often from suicide compared to those from non-Catholic backgrounds. Additionally, a large proportion of deaths of Catholics, as reported by Durkheim, would have been categorized as suicides had they occurred to Protestants, but instead were reported as "sudden death" or "death from ill-defined or unspecified cause," due to Catholics' opposition to suicide at that time.

It is certainly easy in the 21st century to find criticism with a work completed at the end of the 19th century. Hindsight is always a good teacher and being a "Monday morning quarterback" is common today. Nonetheless, Durkheim's study of suicide not only developed a theory to explain such behavior from a societal perspective, but he also gathered data to try to support his theoretical framework. Such a combination was not happening too often in the 1890s.

本日の平均株価

Reuters/Corbis

A problem with many workers (primarily men) in Japan is that they are married to their jobs so that they literally kill themselves from overwork. They overdo it for the good of the company. Several Japanese businesses are currently addressing the problem, called karoshi (death from overwork).

CONFLICT THEORY From a conflict perspective, not everyone within a society has an equal access to power and material objects. Thus, some individuals are in a more favored position than others. One who feels left out, mistreated, or not in control of a situation may feel somewhat alienated. According to Karl Marx, **alienation** is a common state of consciousness in a capitalist society in which humans are more concerned about the products they create than the people who produce them.

Capitalism, Marx noted, separates laborers from nature, from their work, and from each other. In addition, workers are not allowed any creativity and can never realize their full potential. Thus, alienation results in a loss of self or a lack of self-realization. A worker on an assembly line, for example, performing the same task over and over again for eight hours, is likely to feel alienated, whereas a craftsperson, who takes a raw material in hand and creates a finished product, probably would not be inclined to feel alienation.

These constant feelings of alienation may lead over time to a feeling of power-lessness, which in turn may lead to helplessness and hopelessness. Suicide may be the individual's response to these intolerable feelings caused by alienation in a capitalist society.

DRAMATURGICAL PERSPECTIVE Derived from the general approach of symbolic interaction, the **dramaturgical perspective** uses the metaphor of drama to explain behavior. Rather than asking questions to find out about the motivations of the individual who committed suicide, someone using the dramaturgical approach would observe events related to the suicide. For example, after a suicide, family members and friends react in different ways. A dramaturgical analysis might examine those actions and interactions, paying close attention to nonverbal behavior. They would attempt to find the meaning of the suicide in these actions, not in interviews with the survivors.

Behavior is analyzed from the standpoint of the observer (**etic approach**) rather than from the subjective meanings of participants (**emic approach**). Whereas the dramaturgical analyst emphasizes action, the symbolic interaction emphasizes intention. Both perspectives lead to an examination of meanings derived from interaction.

An example of the dramaturgical approach is the film entitled *But Jack Was a Good Driver*. In this film, Jack, a high school student, has died in a single-car accident. As his two friends walk away from the graveside service at the cemetery, they begin to recall recent events involving Jack. They describe the actions of Jack in recent weeks. He had flunked a chemistry exam, broken up with his girlfriend, and given away his CD collection, all actions that were out of character for Jack. They conclude that Jack's "accident" was probably suicide (Jack was a good driver and should have been able to negotiate that curve) because his actions of recent weeks showed some evidence that he might have been planning for his death. The viewer learns about Jack's suicide through the conversation of his friends. The movie exemplified a dramaturgical approach.

EXISTENTIALIST PERSPECTIVE Existentialist philosophy, as discussed in Chapter 1, acknowledges that human existence is finite and that we all must face death. One must assume responsibility for his or her own actions and moral choices. This anticipation of death affects daily actions, bringing about an understanding of how death and life are linked. From an existentialist point of view, it is conceivable that the confrontation with death during a suicidal crisis may cause suicidal individuals to sense their aloneness and personal responsibility for their experience. The choice of death is considered as valid as the choice of life—both the result of reflection (Charmaz, 1980).

When one commits suicide, however, it is difficult to accept existentially because of feelings about the meaning of love, trust, and free will, observed Vaughn-Cole (2000). The thought that the person who committed suicide did not care about those left behind is hard to comprehend. Suicide shatters basic assumptions individuals have that the world is meaningful and benevolent and that the self is worthy. The survivors of suicide may feel social condemnation.

Existentialism is linked to the phenomenological method of inquiry. Using an emic approach, death is studied by asking a dying person about the meaning of death. In studying attempted suicide, one would conduct interviews with persons having attempted suicide to determine what they were thinking during their crisis (Charmaz, 1980).

PSYCHOLOGICAL PERSPECTIVE   Personality traits are linked with suicidal behaviors. The relationships among suicide attempts, depression, and impulsivity are well established in research on mental health, showing that both depression and impulsivity are positively associated with suicide and suicide attempts (Maimon, Browning, & Brooks-Gunn, 2010). Additionally, emotionality has been found to be associated with suicide attempts, as has substance use. Post-traumatic stress (PTSD), as described later in this chapter in the context of military personnel coming out of battle zones, is sometimes correlated with suicides. Another high-stress profession is that of firefighters. A recent study (Peluso, 2012) showed that firefighters' suicide rates were higher than the national average, based on data from suicide deaths among Chicago firefighters over a comparable time period. As Cornies (2012) observes, PTSD, depression, and suicide are real and very much underappreciated threats to the careers of first responders.

SUICIDE THROUGH THE LIFE CYCLE

Our perception of life and death and the way we respond to stress changes as we go from childhood to adolescence to adulthood to old age. These changes also affect the way we view suicide. Likewise, suicide rates vary. As noted earlier, suicide rates are high among adolescents and young adults and highest among elderly persons.

CHILDHOOD SUICIDE

Suicidal deaths are virtually nonexistent before age five and are rare between five and 14 years of age, as noted in Table 9.1. In fact, data for 1999 to 2005 reveal no suicides among children ages four and younger (CDC, 2008). As adults, we prefer to believe that children do not commit suicide, likely thinking that childhood is a carefree, happy time.

As with many other suicides, child suicides are easily mistaken for accidents. The causes of death themselves involve an element of chance: traffic accidents (sometimes running into traffic), falls (jumping) from high places, fatalities in handling guns, or drowning. Young children do not write suicide notes, thus compounding this difficulty because notes are a chief category of evidence that coroners use to verify suicide. Thus, the state of reporting suicides and suicide attempts of children and adolescents is filled with inconsistencies, confusion, and concealment.

Depression has long been recognized as a consistent correlate of suicidal ideation and attempts and thus is usually considered a strong risk factor for suicidal behavior. Suicidal behaviors are also related to impulsive thoughts and aggressive or violent behavior. Leilani Greening and colleagues (2008) examined impulsivity as a predictor of aggression in school-aged boys, possibly contributing to suicide. Their results revealed relationships between impulsivity, aggression, and depressive symptoms, and between suicidal ideation and aggression and depressive symptoms.

| TABLE 9.1 | SUICIDES IN THE UNITED STATES BY AGE FOR THE YEAR 2006 |

Age Group	Number of Suicides	Population	Rate*
5–14	219	40,260,779	0.5
15–24	4,189	42,268,411	9.9
25–34	4,985	40,182,221	12.4
35–44	6,591	43,554,885	15.1
45–54	7,426	43,226,850	17.2
55–64	4,583	31,556,836	14.5
65–74	2,384	18,909,923	12.6
75–84	2,075	13,057,166	15.9
85+	840	5,285,976	15.9
Total	33,300	298,754,819	11.2

*All rates are per 100,000 population.

Source: Figures from the National Center for Health Statistics for the year 2006. Taken from the American Foundation for Suicide Prevention, 2009.

It might be that poor impulse control leads to interpersonal problems, including poor peer relationships. Enduring social and personal adjustment problems can ultimately lead to depressive symptoms. Depressed mood may then increase one's risk for self-destructive behaviors including suicidal ideation. Self-destructive thoughts can then, in turn, increase one's risk for suicide attempts, note Greening and colleagues.

Suicidal behavior in childhood appears to be more closely associated with family turmoil and stress than is suicidal behavior in later life. Once a suicide attempt has been made, future attempts may be expected, if the first attempt occurs at a young age. Early pubertal development, not living with both parents, and poor self-worth may also contribute to a second suicide attempt.

ADOLESCENT SUICIDE

Developmentally, adolescents differ from younger youth and from adults in ways that may increase their risk for suicidal behaviors. They may be more impulsive and may have a different time perspective than adults and may focus more on shorter- rather than longer-term goals when making decisions (Daniel & Goldston, 2009). Adolescent suicidal behavior often occurs in the context of a family conflict, sometimes striving for autonomy, or in the context of academic and disciplinary problems, or as a result of disruptions in peer relationships, which increase in significance as youth get older.

A noticeable shift in suicide rate occurs at age 15 when it increases dramatically, as noted in Table 9.1. Suicide is a leading cause of death among adolescents and young adults from age 15 through 24. Adolescent and young adult suicide is not unique to contemporary times, but the rate more than tripled between the mid-1950s and late 1970s (from 6.3 per 100,000 in 1955 to 21.3 in 1977) (American Foundation for Suicide Prevention, 2009). The youth suicide rate generally leveled off during the 1980s and 1990s, and since the mid-1990s has been

WORDS OF WISDOM | BULLYING AND SUICIDE

There is a strong link between bullying and suicide, both in the United States and other countries. Bully victims are between two to nine times more likely to consider suicide than nonvictims. A study in the United Kingdom found that at least half of suicides among young people are related to bullying. Nearly 30 percent of students are either bullies or victims of bullying, and 160,000 children/adolescents stay home from school every day because of fear of bullying. Bully-related suicide can be connected to any type of bullying, including physical, emotional, cyber, sexting, or circulating suggestive or nude photos or messages about a person.

Source: Bullying and suicide. (2012). OLWEUS Bullying Prevention Program. http://www.bullyingstatistics.org/content/bullying-and-suicide.html. Accessed on October 16, 2012.

steadily decreasing. In recent years, copycat suicides (discussed later in this chapter) have called attention of the media to suicide and have caused parents and educational personnel to be even more aware of suicide.

Accidents and homicides are viewed by many individuals as disguised suicides. Many drug overdoses, fatal automobile accidents, and related self-destructive food and alcohol disorders are probably uncounted teen suicides. Thus, the total number of adolescent suicides may be greater than actually reported. Friedman and Kohn (2008) actually note that it is generally accepted that suicide data are underreported to avoid sociocultural stigma, to escape police enquiries and legal harassment, and to benefit from the insurance sector. In talking with an insurance adjuster, George Dickinson was told that single-car accident fatalities with adolescents are often difficult to determine as being an accident or a suicide (since most life insurance policies do not pay off if the death was a suicide, such a determination is important). The adjuster stated that adjusters always lean on the side of an accident, rather than suicide, as the death is difficult enough for the parents and other relatives, yet to think it was a suicide is even more difficult to swallow. He noted that a thorough investigation to try to prove suicide is simply not worth it. If in doubt as to whether or not it was an accident or a suicide, therefore, adjusters typically report the tragedy as an accident and pay off.

David Moller (1996) believed that attempted adolescent suicide must be interpreted as more than a cry for help. It is a response to the awareness of painful life circumstances and a fatalistic view of them. It is an absence of hope, with hope being the ability to imagine a different future. Therefore, a suicide attempt is an effort to restore a sense of personal autonomy and power in the face of perceived powerlessness. Hopelessness involves negative expectations for and feelings about the future (Zeyrek et al., 2009). Moller noted that a completed suicide for the adolescent makes a final and dramatic assertion of personal strength. This final act of life is ironically life-affirming—motivated by the adolescent's thirst for a life that is meaningful, stable, and autonomous. But the suicide must be successful. A failed attempt is considered much less favorably.

In a study of multiple suicide attempts among suicidal black adolescents (Merchant et al., 2009), the authors concluded that interpersonal orientation, specifically seeking out others for the purpose of getting feedback about one's self and needing the positive stimulation that others provide, differentiated adolescents with a history of multiple suicide attempts from those without such a history. Therefore, black adolescents who attend to the internal interpersonal workings and external social environment of adolescents may prove most effective at reducing subsequent suicide attempts for this population, note the authors.

WARNING SIGNS OF SUICIDE   Any unusual behavior on the part of an adolescent may be an indication that his or her life is not right and could be an early warning of more bizarre behavior in the future, such as attempted or completed suicide. For example, if an individual begins taking long walks at 2 a.m., when she or he had previously not done so, this is probably a sign that something is bothering him or her. Suddenly acting in a strange way, such as offering to give away a DVD collection without an obvious reason, is not normal.

Any stressful situation, especially those involving a loss, can trigger suicidal behavior. Breaking up with a boyfriend or girlfriend, moving to a new community and being the new kid in town, the death of a significant other, a divorce in the family or any other family dysfunction, or lack of academic achievement are all situations that could lead to a suicide attempt. Friends and family should be alert to the warning signs of suicide. Findings of a study based on a large national sample of American adolescents (Kidd et al., 2006) indicate that supportive social relations with peers, parents, and school have an interactive relationship in mitigating risk of suicide attempts for one specific high-risk group: boys who had a history of attempting suicide as well as poor peer relations. Protective effects of parental support against subsequent suicide attempts seem to be bolstered by positive school relations.

Sometimes an adolescent will commit suicide without having left any visible signs. Even if one is signaling warnings, they may not be obvious to those around. Because most of us are not trained in clinical psychology or psychiatry, we need to be alert to unusual behavior around us, so that we might be in a position to thwart an adolescent's suicidal efforts and help him or her seek professional counseling.

EXPLAINING ADOLESCENT SUICIDE   Sociological and psychological studies have indicated that both contextual factors (e.g., social disorganization, lack of family support, and poor peer relationships) and individual-level predictors (e.g., depression, emotionality, and impulsivity) are associated with suicide, suicide attempts, and suicide ideation (Maimon, Browning, & Brooks-Gunn, 2010). Adolescence is a time of frustration and confusion, a period of transition socially. Adolescents are learning to negotiate conflicts with their peers with less supervision from adults around them (Daniel & Goldston, 2009). Peer relationships are becoming more important, yet many of the peers in a chosen peer group may themselves be troubled, thus increasing the chances of behavioral and emotional difficulties for others. Adolescents are often treated like children and feel scared and uncertain; thus it is no wonder that the frustration in this age group can be so high. Suicide may become an available solution to an individual's problems. When stress is experienced, a constricted view of future possibilities and a momentary fix on present escapes may prevail. Impulsive action and limited alternative solutions to problems may turn suicidal fantasies into suicidal behaviors. Suicide is an answer (though permanent) to a set of temporary problems. The problem is gone, but so is life.

Because adolescents are caught between childhood and adulthood, the often intense and conflictual task of separating from the world of family can be more

LISTENING TO THE VOICES	THE "SOLUTION" IS SUICIDE

The following letter, written by a 17-year-old who took his life, apparently indicates that death was "the solution" to this adolescent's dilemmas of life. He lived with various problems but tended to keep his pain mostly to himself.

Dear Mom, Dad, and everyone else, I'm
sorry for what I've done, but I love you all
and I always will, for eternity. Please, please, please
don't blame it on yourselves. It was all my
fault and not yours or anyone else's. If I didn't
do this now, I would have done it later anyway.
We all die some day, I just died sooner.

Love,
John

After John's body was found, people began to piece together his life and were filled with the inevitable "if onlys." This reconstruction of an interpretation after the event of suicide follows the dramaturgical approach, as discussed earlier.

From *Helping suicidal adolescents: Needs and responses*, by A. L. Berman, 1986. In C. A. Corr & J. N. McNeil (Eds.), *Adolescence and death* (pp. 151–166) [Microfilm]. New York: Springer Publishing Company.

difficult when family dynamics interfere with a child's move toward self-sufficiency. Various factors may contribute to suicidality among adolescents. These factors include poor child–parental relationships; socioeconomic adversity; exposure to sexual abuse; high rates of neuroticism; alcohol, tobacco, and other drug use; delinquency and other antisocial behavior; bullying others and being bullied oneself; and, to a lesser extent, physical health (Epstein & Spirito, 2009). In listing factors associated with adolescent suicide attempts, however, Christopher Merchant and colleagues (2009) note that the strongest predictor of an adolescent suicide attempt is a previous attempt. In addition, psychiatric conditions that may contribute to suicidal behavior include depression and anxiety disorders.

Recent research on peer victimization of adolescents and suicide (Klomek et al., 2008), particularly bullying, concluded that the most common experience of being bullied involved having one's looks or speech belittled. Males were more likely than females to be physically bullied and to be belittled because of religion or race. Females were more likely than males to be the subject of rumors, sexual gestures, and meanness by use of the Internet. Generally, the more types of peer victimization the teen has been exposed to, the higher the risk for depression and suicidality. The most damaging impact of peer victimization is on self-concept. Therefore, for clinicians and others working with adolescents, interventions and prevention strategies that focus on enhancing self-concept may reduce peer victimization as well as depression and suicidality among adolescents.

How effective are telephone crisis service hotlines for helping prevent suicides among adolescents? Service utilization surveys of adolescents indicate that despite high awareness of hotlines and high satisfaction ratings among those who do contact hotlines, adolescents access hotlines infrequently (Gould et al., 2006). Commonly cited reasons for not using hotlines were thinking that the problem was not serious and that the problem could be solved by oneself. Females were more likely than males to call a hotline. Ironically, Madelyn Gould and colleagues found that objections to hotlines are strongest among the individuals most in need of help, thus this potential source of help appears to be inadequately tapped by adolescents. If hotline services hope to increase utilization among adolescents, they must work to promote a specific function that can be provided to adolescents in a way that fits with an adolescent's sense of her or his needs and is compatible with a teenager's lifestyle.

The Internet is a potential avenue for enhancing access to crisis services by youth. The Internet can be a source of education about warning signs (Mandrusiak et al., 2006). Hotline advocates should take advantage of the Internet's growing accessibility and teenagers' propensity to use it as a means to obtain help, suggest Gould and colleagues (2006). In recent years, e-therapy, as it has come to be known, has broadened to include instant messaging (IM). Web companies have introduced services that link clients with therapists (e.g., MyTherapyNet.com and HelpHorizons.com) where counseling can be conducted using IM (Lester, 2009). The Internet may be attractive as a medium for individuals with particular lifestyles and personality types. With the pros and cons of suicide information on the Internet, it at least can provide another option for prevention.

WORDS OF WISDOM   SUICIDE RATES IN THE MILITARY: RELATED TO POST-TRAUMATIC STRESS

Suicide rates of NATO combat operations soared in Afghanistan, especially in 2012, and are at an "epidemic stage," according to Defense Secretary Leon Panetta. Suicide rates outweighed combat deaths by a two-to-one ratio. Soldiers with multiple combat tours are more likely to commit suicide. Additionally, combat exposure, post-traumatic stress (PTSD), misuse of prescription medications and personal problems can all contribute to military suicides.

An example of PTSD is that of Army Private Daniel Rodriguez. Private Rodriguez and 50 other U.S. troops on the morning of October 3, 2009 in a U.S. Army northernmost base in Afghanistan were surrounded and outnumbered. The fighting was savage with point-blank killing as enemy fighters breached the walls. Rodriguez killed men and saw friends killed, including his best friend, shot in the head just an arm's length away. Rodriguez took a bullet through his shoulder and had shrapnel wounds to his neck and legs. After two days of fighting, the soldiers were evacuated. Eight Americans were killed and 22 wounded in one of the bloodiest battles of the Afghan war for U.S. troops.

Rodriguez was diagnosed with PTSD disorder. PTSD is the "new cancer," he said, and "everyone wants to know the cure for an incurable disease." In high school Rodriguez was considered the life of the party but came back from Afghanistan a subdued individual. He suffers night terrors and panic attacks and can't seem to get away from the battle. Many veterans, including Rodriguez, reject help from those who have not experienced combat. He had made a promise to his best friend that if he made it out of Afghanistan, he would pursue his dream of playing college football. He began working out and stopped drinking and made a video displaying his football potential and put it on YouTube. It went viral. Clemson coach Debo Swinney responded and offered him a position on the team. The college football experience is therapeutic for him and separates him from what he has been through. Being part of the team has given him hope and is helping him to get that sense of love back that he has wanted for so long. The social support he is receiving is most beneficial, as he missed the "team feel" he got from his army unit. He now has a new "team" and seemingly is adjusting.

Sources: E. Clifton (2012, June 8). Suicide rate in military at highest level in ten years. *ThinkProgress*; T. Sawchik (2012, October 20). Starting to heal with new team. *Post and Courier*, Charleston, SC, pp. A1, A6.

| PRACTICAL MATTERS | SUICIDE WARNING SIGNS AMONG YOUNG ADULTS |

Behaviors Requiring Immediate Response

1. Indicating intent to harm themselves (talking, threatening)
2. Seeking availability of or obtaining ropes, weapons, pills, or other ways to kill themselves
3. Talking or writing about death, dying, or suicide

Associated Behaviors Requiring Evaluation

1. Feeling hopeless
2. Expressing rage or anger; seeking revenge
3. Acting recklessly or impulsively or engaging in risky activities, seemingly without thinking

4. Feeling trapped, like there is no way out or nothing else will help
5. Increasing alcohol or drug use or abuse
6. Withdrawing from friends, school activities, community, and family
7. Expressing anxiety, agitation, an inability to sleep, or sleeping all the time
8. Exhibiting dramatic mood changes
9. Expressing loss of interest or reason for living; no sense of purpose or meaning in life
10. Acting "immaturely" and/or displaying disregard for others' safety, feelings, or property

From M. Silverman (2004, March). College student suicide prevention. *Student Health Spectrum*.

COLLEGE-AGED STUDENTS AND SUICIDE

After decades of debate over whether the rate of suicide was higher or lower among college-attending adults, a number of studies convincingly established the rate to be half the rate for young adults in the general population. Approximately 1,100 young adults kill themselves in U.S. colleges and universities each year (Joffe, 2008). The topic of suicide prevention is coming out of the closet on college campuses today through talks in auditoriums and growing conversations around campuses (Irvine, 2009).

With all of the pressures of college, what percentage of students contemplate suicide? In a study of 26,000 students at 70 colleges and universities in the United States, 15 percent reported having seriously considered attempting suicide and more than 5 percent reported making a suicide attempt at least once in their lifetime (APA, 2008). The majority of students described their typical episode of suicidal thinking as intense and brief, with over half of the episodes lasting one day or less. Reasons given for suicidal thinking were: (1) desiring relief from emotional or physical pain; (2) problems with romantic relationships; (3) desire to end their life; and (4) problems with school or academics.

As noted in the Practical Matters box, some behaviors require immediate responses whereas others need evaluation. The key for college and university faculty and staff is to be on the lookout for these signs, take them seriously, and seek help when warning signs are noted. With more than one thousand students in American college and universities committing suicide each year, this is a situation for which we need to be alert. The public health approach to suicide prevention on college campus is to increase mental health literacy specifically associated with suicide and attendant risk factors (Walker & Flowers, 2011).

When suicide attempts are a call for help and when a sharp object such as a knife is used, an individual will likely make the first attempt by slashing the wrists, far

away from major organs. As suicide attempts become more serious, the individual will typically stab closer to major organs. George Dickinson, when a freshman in college in Texas, recalls another freshman student down the hall in his dormitory who cut his wrists. Blood was everywhere. Earlier behavior by the student was unusual, yet [we] fellow freshmen were not attuned to signs of suicide. The individual certainly attracted attention with his call for help, with cuts far away from major organs, indicating perhaps that his intention was not for a completed suicide. Years later, when serving as a college dormitory head resident in Minnesota, Dickinson experienced a different type of suicide attempt by a student. This student had disappeared, and a search for him began. We found him lying in a ditch, semiconscious, having stabbed himself numerous times in the chest. He had stabbed himself very near major organs—heart and lungs—however his knife blade was short, thus did not penetrate any major organs. It seemed that this student's intent to terminate his life was definitely more serious than the freshman student in Texas.

ADULT SUICIDE

Though suicide rates are high among elderly persons, middle-aged adults are not immune to death by suicide. Contributing factors to suicide found in adults include feelings of hopelessness, personal losses such as health problems or bereavement, alcohol or drug abuse, and financial problems. In addition, unemployment, not unrelated to financial problems, is a major factor contributing to suicide among adults (Pirkis, Burgess, & Dunt, 2000). Additionally, for middle-aged men from disadvantaged backgrounds, they have lost a sense of identity and masculine pride and thus are much more likely to commit suicide than men from more advantaged backgrounds, according to a study in the United Kingdom (Kelland, 2012). The U.K. findings suggest that suicide is not simply a mental health problem but one of society's inability to adapt to men's needs when they are trying to deal with depression, anxiety, and other problems.

As noted in Chapter 3, middle-aged adults are in a stage of life during which the years remaining to live are shorter than those already lived. Thus, individuals in this age category may begin to panic over the fact that time is running out, and that they will not accomplish all they had hoped. This age group begins to see deaths of peers from natural causes (heart disease and cancer) and is therefore reminded that life is not forever. Also, being the sandwich generation, often having to deal with hanging-on young-adult children and aging parents at the same time, may prove too stressful for some. Suicide again may become a permanent solution for a temporary set of problems.

Some attempted suicides in adults may belong to a culture of risk in which the individual has not matured appropriately and may carry the adolescent developmental processes in an unresolved state. Overall, one would expect that the older people get, the less influence the risk-taking culture of adolescence would have on them. Attempted suicide among these adults would probably be failed suicide.

Alcoholism is a factor in about 30 percent of all completed suicides; approximately 7 percent of those with alcohol dependence will die from suicide (American Foundation for Suicide Prevention, 2009). Per capita alcohol consumption and suicide rates in the United States from 1950 to 2002 were analyzed by Jonas Landberg (2009). The results did not support the hypothesis of an alcohol–suicide relationship among U.S. males but did suggest a positive relationship between changes in per capita consumption and female suicide mortality rates. Given that females drink less and are less prone to binge drinking than men in the United States, the finding that changes in per capita consumption only affect female suicide rates is rather unexpected, notes Langberg. These findings imply that a change in U.S. per capita consumption would result in a change in female suicide rates, whereas the male rates would not be affected.

Most adult individuals who commit suicide tend to talk about their attempt before the act. Other than an outright statement such as "I am going to kill myself," more subtle verbal clues might include statements such as "I'm not the person I used to be," "You would be better off without me," "I can't stand it anymore," "Life has lost its meaning for me," and "Nobody needs me anymore." Behavioral signs of suicide might include getting one's house in order as if preparing to depart, frequent crying without explanation, sudden irregular sleeping habits, loss of appetite, inability to concentrate, sudden changes in appearance, or sudden withdrawal from society. Though the signs may appear obvious after a completed suicide, at the time we might not recognize them as signs of suicide.

In a study of 63 adult survivors of suicide to determine their needs (McMenamy, Jordan, & Mitchell, 2008), participants reported that several

LISTENING TO THE VOICES | A LETTER TO DAD: SOME TEN YEARS FOLLOWING HIS SUICIDE

Dear Dad,

It has been ten years, seven months, and eighteen days since you decided to take your life and leave your wife and three daughters all alone on this earth. It has gotten easier for all of us, but we all still miss you terribly: every Father's Day, every holiday, every time we listen to one of your plethora of Christmas music CDs and remember how much you loved it. No one in our family really likes classical music except for me, Dad. Guess what? I even like Wagner, the thundering composer that Mom could never stand, remember? Maybe if you had decided to stay with us for just a few more years, you and I could have gone to the opera together. But at the end of the day, I guess it does no good to be bitter.

I want you to know that I am thankful for the fact that you decided to stick around for as long as you did, for giving Mom "a great love" and for helping to give my two sisters and me life because I have to tell you, I have had a pretty nice one. We are all still really close, and we have travelled around the world, and are all at really great colleges. You said in your last letter that you left some money for us, and if that helped the three of us get here, thanks for that, too.

When I hear about people who grew up without ever knowing their fathers at all, I am thankful that I got to know you at all: that you were there to tuck me into bed at night, tickle me, grill us hamburgers on the barbeque, play soccer with me on our front lawn, answer pretty much whatever questions popped into our little heads, and play "airplane" with the three of us when you would drive us to school in your car. Did you know that we still call the person who sits in the shotgun seat the "co-pilot"? I think we always will too.

Thanks to you, we play Handel's *The Messiah* to death every Christmas and *Les Miserables* almost always gets played whenever we have an agonizingly long drive to take. I will try as hard as I can to hold onto all the memories that you left us for as long as I live because I know that the memory of coming into Mom's room and seeing her sitting on your bed, her face drenched with tears and having to tell us that you died will never go away. Every time I see a movie about a pair of lovers that have grown old together, I get angry at you, Dad. Every time I see a father walk his daughter down the aisle, I get angry at you. What was it, Pop? Why was all of that worth less than what made you decide to inhale all that carbon monoxide for? I do not want to keep being angry at you, but I never got the whole story about how sick you really were, and if you really tried to get help or not. I do not think I will ever ask, either. It does not really matter because I know that, no matter what, I will always love you and be thankful for what you gave me, Daddy. I just hope that one day I can see you again and tell you this myself.

Sincerely,
Your loving daughter

Source: The above letter was written by a college student in her introductory sociology class in response to the professor's assignment for students to write a thank-you letter to someone who had played a meaningful role in their lives. Permission was given by the student to include the letter here.

different types of resources were particularly helpful in their coping efforts. Suicide bereavement support groups were viewed by many as being as effective as individual counseling. Others reported that books and the Internet were also helpful. Survivor-to-survivor contact appeared to be particularly useful for several of the individuals. For formal support, those resources most often cited as a moderate-to-high level of helpfulness were mental health professionals (80 percent), funeral directors (70 percent), and clergy (63 percent). Most often named for informal support were close friends (87 percent), children (85 percent), spouses or partners (82 percent), neighbors or colleagues at work (74 percent), and brothers or sisters (70 percent). What might be helpful for one individual may not be for another; thus it is good to make various treatment options available.

ELDERLY SUICIDE

As noted earlier, Americans age 65 or older commit suicide at high rates. The high rates of suicide among the elderly are thought to be associated with depression, possibly linked to social isolation and negative attitudes toward elderly people (Howarth, 2009). The elderly use more lethal methods, predominately firearms, in their suicide efforts, and thus complete a higher proportion of their attempts. A major difference in the variables associated with suicide among younger age groups is the contribution of state-level public policies that have significant effects on the quality of life for many elderly individuals (Giles-Sims & Lockhart, 2006). The elderly can be affected by various state-level factors including both state-level (e.g., various degrees of material support for and regulation of long-term care) and private sector features (e.g., the extent of not-for-profit outreach programs). From Durkheim's analysis earlier in this chapter, elderly suicide seems more likely to arise from either high levels of social regulation (e.g., fatalistic suicide from nursing facility confinement) or low levels of social integration (e.g., egoistic suicide from relative social isolation), note Giles-Sims and Lockhart.

Suicide among the elderly may be even more prevalent than the data suggest. For example, intentional overdoses of prescribed medications may go unrecognized or unreported, or older persons may lose the will to live and simply stop caring for themselves by ceasing to take medications, stop drinking and eating, take increased health risks, or delay treatment for medical conditions.

Age by itself does not cause suicidal behavior. The person statistically most likely to commit suicide in the United States, however, is the white male older than 85. The traditional, assertive, achievement-oriented, middle-aged male may not be as adaptable to the retirement situation as the traditionally more passive, nurturant, middle-aged woman. Thus, the greater change in behavior and self-esteem necessitated in the older man may create more stress and depression if he is unable to find satisfaction in this new situation.

CAUSES OF SUICIDE The *threat* of suicide is uncommon in elderly persons—they simply kill themselves, and their attempts rarely fail. When an elderly person attempts suicide, he or she usually has a profound death wish and thus selects ways that are likely to fulfill that wish. Suicide is rarely an impulsive act unheralded by any warning signs but rather, as a rule, is an act that can be anticipated.

As noted earlier, a high correlation exists between suicide and depression. Elderly individuals often have much to be depressed about. They may have a variety of physical and mental afflictions, economic difficulties if on a fixed income, increased medical costs, loss of friends and family and thus limited social support, termination of work-related roles, and possible institutionalization. Urbanization in societies today may lead to the weakening of ties with family, friends, local religious institutions, and the original place of residence, thus perhaps contributing to depression and increased suicide rates among the elderly (Shah, 2008). Such an urbanization trend and suicide increase was suggested by Durkheim in the late 19th century.

A sense of belonging in the community can be a protective factor against depression, playing an important role. In order to have a sense of belonging, an individual must have energy, interest, and potential to develop a sense of

Michael R. Leming

Elderly persons are held in high esteem in many traditional societies because of the store of knowledge they have accumulated over the years. Perhaps not so coincidentally, suicides among the elderly in these cultures seem to be less common than in those of many modern industrialized nations.

belonging. Additionally, they have to feel valued, needed, and significant within their environment, note Suzanne McLaren and colleagues (2007).

Loneliness among the elderly may be a catalyst to depression. After each loss, an elderly person has to try to pull him or herself together and adjust in order to be able to retain psychological equilibrium. Losses for the elderly may include jobs, social status, health, independence, friends, and family. These losses may result in stress at a time when the individual is least able to cope. If there is no possibility of communication with other people, loneliness can become extremely distressing for the older individual.

The elderly have a high suicide rate, yet there are few suicide-prevention programs which target them, notes Sarah Skidmore (2007). In 2006, for example, 10 states passed laws intended to curb suicide among children and young adults, yet only the states of New Jersey and New Mexico passed laws addressing suicide among the elderly.

COPYCAT SUICIDES AND THE MEDIA

Friends of individuals who commit suicide may be at risk for copycat suicides. A **copycat suicide** occurs when one individual's suicide is at least partly caused by

PRACTICAL MATTERS | HOW YOU CAN HELP IN A SUICIDAL CRISIS

1. Recognize the clues to suicide. Look for signs of hopelessness and helplessness. Listen for suicide threats and words of warning. Notice if the person becomes withdrawn and isolated.
2. Trust your own judgment. If you believe that someone is in danger of suicide, act on your beliefs. Don't ignore the signs of suicide.
3. Tell others. Share knowledge with parents, friends, teachers, employers, or other people who might help. If you have to betray a secret to save a life, do it. Don't worry about breaking a confidence if someone's suicidal plans are revealed to you.
4. Stay with a suicidal person. Don't leave a suicidal person alone if you think there is immediate danger. Stay until help arrives or the crisis has passed.
5. Listen. Encourage a suicidal person to talk. Don't give false reassurances that "everything will be okay." Listen and sympathize with what the person says.
6. Urge professional help. Offer to make an appointment for and go with the person for professional help, if that is what it takes. Call your community hotline or crisis number for suggestions.
7. Be supportive. Show the person that you care. Help to make the person feel worthwhile and wanted.

From *Youth and suicide* (p. 96), by F. Klagsbrun, 1976, New York: Pocket Books.

exposure to another person's suicide. Using data from the National Longitudinal Survey of Adolescent Health, Feigelman and Gorman (2008) confirmed past evidence showing that a friend's suicide has an immediate impact on encouraging greater suicidal thinking, suicidal attempts, and higher depression for those experiencing this traumatic event. On the other hand, evidence from this analysis does not support the contention that enduring heightened suicidality remains an obstacle in the lives of those experiencing this type of traumatic loss.

Multiple suicides by people who meet on chat sites on the Internet appear to be increasing (Mishara & Weisstub, 2007). Meeting suicide companions online appears to be most prevalent in Japan. It is thought that the first wave of Internet suicide pacts occurred in 2000 in South Korea when there were three cases. In 2005 in Japan the number rose to 91, up from 55 in 2004 (McCurry, 2006). The number of suicide pacts has almost tripled since police began keeping records in 2003. The Japanese government plans to invest large sums of money into local services for people suffering from depression and other mental illnesses and will call on companies to improve pastoral care for employees.

Copycat suicides, particularly among teenagers, have been noted in the media, and imitation is generally cited as the cause. Research by Madelyn Gould and colleagues (2003) suggested a strong relationship between reports of suicide in the media and increased suicide rates. Teens who have already lost a friend to suicide see dying as glamorous and attention-getting. Studies of copycat suicides in clusters confirm that teens do not commit suicide without thought, but suicide by a classmate or acquaintance may give them the idea of a dramatic way to solve some of their problems. Studies by Katja Becker and colleagues (2004) give support that suicides increase proportionally to the amount, duration, and prominence of media coverage by illustrating the Internet's influence on adolescents by romanticizing or idealizing it as a heroic deed. Additionally, clusters of suicides around the same time or geographical area may be caused by this copycat effect.

John Van Hasselt/Sygma/Corbis

The suicide of Nirvana's Kurt Cobain in April 1994 shocked millions of young rock fans throughout the world and led to widespread grief and confusion. In Seattle, Washington, Cobain's hometown, a candlelight vigil was attended by 5,000 people, including these two fans.

With the suicide of Kurt Cobain, some individuals feared that large numbers of copycat teen suicides would occur (Science Daily, 2009). Such suicides that occur around the same time but across an entire geographic region are referred to as mass clusters. In recent years, satellite television and the Internet have increased the global range of the mass media, resulting in celebrities such as film actors and pop singers being assigned increasing importance. Additionally, reality television programs are increasing the number of celebrities within society (Science Daily, 2009). However, it has been noted that those apt to copy rock star Kurt Cobain were those who already had significant emotional problems. Those individuals with low self-esteem and self-identify tend to get hooked on heroes and are more likely to behave in the same self-destructive way.

The risk of copycat suicides after media coverage of celebrity suicides is well documented, however, as reported by Steven Stack (2000). In his meta-analysis of 42 studies concerning media impacts on suicide, Stack found that studies using entertainment or political celebrity suicides were 14.3 times more likely to find a copycat effect than studies that did not. If not well integrated into social life, copycat suicides may result, an example of Durkheim's egoistic suicide. A copycat suicide is particularly likely if the publicized stories deal with victims in similar life circumstances as the readers of the suicide accounts.

Though the preceding findings certainly suggest a correlation between media broadcasts on suicide and increased suicide clusters among adolescents, this is not to suggest that such media exposure causes an overall increased rate of suicides.

As Mishara and Weisstub (2007) conclude, there is no scientific proof that Internet sites and other media *cause* suicide, but a relationship may exist. Viewers previously thinking of suicide might be persuaded one way or the other after watching television programs on suicide. Those who do kill themselves are more likely to have made a previous attempt at suicide, to have lost a close friend or relative to violent death, or to have suffered a recent breakup with a girlfriend or boyfriend.

Adekola Alao and colleagues (2006) point out that the Internet can be a potential source of help for suicidal people of any age group and can be beneficial when suicidal people seek counseling online, in addition to helping individuals decide whether or not they should commit suicide. They report numerous websites that deal with suicide methods, with some being very graphic. Tam, Tang, and Fernando (2007) confirm how the Internet can serve a double purpose when it comes to suicide, both harming and helping. A suicidal person can use the Internet to learn methods to commit suicide. Rheana Murray (2008), in her study of suicide websites, interviewed individuals online who use the websites and also those who create and support the how-to-do-it websites. One of the websites offers a plethora of information about suicide, including the examination of the purpose of living, offering pro-life and pro-death viewpoints and information about euthanasia. The "Tying Loose Ends" page reminds viewers contemplating suicide to consider pets, bills, funeral plans, and more, giving links to such facilities, notes Murray. They even give instructions on how to write a suicide note. Additionally, information is available online about methods of suicide, the percentage of times such methods are lethal, the time it takes to actually die using particular methods, and the measure of agony with different methods, as pointed out in Murray's study.

For some individuals, as they travel down the road of life, they seemingly come to a dead end. They come "over the hill" and the road abruptly ends. The demands of life do not match their capacity, thus the stress becomes too much. To end it all appears to be a solution. As Mark Twain once said, "There are times in life when we would like to die temporarily." Unfortunately, completed suicide is a permanent answer to a temporary set of problems.

Some countries have attempted to block access by individuals to specific Internet content and sites. For example, the United States, Great Britain, and New Zealand passed laws to block certain Internet content, but those laws were overturned by the courts because of constitutional guarantees of freedom of expression (Mishara & Weisstub, 2007). Australia is the only country that currently has laws to specifically restrict sites that promote suicide or provide information on suicide methods. Control of Internet suicide promotion activities is certainly not easy. International jurisprudence makes it difficult to obtain jurisdiction over sites that originate outside the country, as the Internet allows for global access. Any attempt to control the Internet must be viewed along with the control and freedom of other media, note Mishara and Weisstub. Editors of newspapers, like Web masters, are free to publish what they please, even if it may incite suicides.

RATIONAL SUICIDE

There have been times in history when misery was so common and the outlook so bleak that suicide probably seemed a serious option. As Mark Twain once said, "there are times in life when we would like to die temporarily." The Judeo-Christian tradition, however, has generally advocated life and strongly urged that life not be disposed of, no matter what the temptations. For an elderly person, taking one's life tends to be a more rational or philosophical decision. Elderly individuals and others today seem to be asking if life is to be valued under all conditions because it has intrinsic value or if the value of life is relative to the circumstances. The "Listening to the Voices" box describing Mr. and Mrs. Saunders' final days suggests that the value of life for them was relative to the circumstances.

Although suicide is usually treated as the product of mental illness or as a desperate cry for help used by someone who does not really want to die, suicide can be considered a rational act. Indeed, suicide can be viewed as the purest form of control over dying (Kellehear, 2009). According to Charles McKhann (1999), **rational suicide** has two major foundations: the desire to avoid unnecessary suffering and the desire to exercise one's autonomy and self-determination. Stephen Ginn and colleagues (2011) suggest that for a suicide to be considered rational (therefore worthy of not being prevented) the person in question must have an unremittingly hopeless medical condition, should make the decision as a free choice, should have engaged in a sound decision-making process, and should have an assessment by a mental health professional.

In some ways, the Saunders' act in the Listening to the Voices box was a rational homicide/suicide. Jacques Choron's assertion (1972) that *rational* implies an absence of psychiatric disorder and a reasoning that is not impaired, with justifiable motives that are at least understandable by those in the same social or cultural group, seems to fit the case of the Saunders.

Richard Brandt (1986) stated that the person contemplating suicide is obviously making a choice between future world courses—the future world course that includes his or her demise an hour or so from now and several possible courses that include his or her demise at a later point. Although one cannot have precise knowledge about many features of the latter courses, it is certain that they all will end with death sooner or later. The basic question for the person to answer,

| LISTENING TO THE VOICES | AFTER 60 YEARS OF MARRIAGE, COUPLE DECIDES TO LEAVE WORLD TOGETHER |

Julia Saunders, 81, had her hair done. Her husband, Cecil, 85, collected the mail one final time and paused to chat with a neighbor. Inside their mobile home, they carefully laid out a navy blazer and a powder-blue dress.

After lunch, the Saunders drove to a rural corner of Lee County and parked. As cows grazed in the summer heat, the couple talked. Then Cecil Saunders shot his wife of 60 years in the heart and turned the gun on himself.

Near the clothes they had chosen to be buried in, the couple had left a note:

Dear children, this we know will be a terrible shock and embarrassment. But as we see it, it is one solution to the problem of growing old. We greatly appreciate your willingness to try to take care of us.

After being married for 60 years, it only makes sense for us to leave this world together because we loved each other so much.

On the floorboard of the car, Cecil and Julia Saunders had placed typewritten funeral instructions and the telephone numbers of their son and daughter.

In Philadelphia, a police officer stood by as the Saunders' son, Robert, 57, was told of his parents' death. His parents wanted no tears shed over their decision to die. The note they left for Robert and his sister, Evelyn, 51, ended with a wish: "Don't grieve because we had a very good life and saw our two children turn out to be such fine persons.

Love, Mother and Father."

From *Minneapolis Star and Tribune* (p. 12A), October 4, 1983.

to determine the best (or rational) choice, is which course would be chosen under conditions of optimal use of information, when all personal desires are taken into account. It is not a question just of what is preferred now.

Certainly one's desires, aversions, and preferences may change after a short while. When one is in a state of despair, nothing but the thing that he or she cannot have—a love that has been rejected or a job that has been lost—may seem desirable. The passage of time is likely to reverse all of this. So if one acts on the preferences of today alone, when the emotion of despair seems more than one can stand, death might seem preferable to life. But if one allows for the preferences of the weeks and years ahead, when many goals might be enjoyable and attractive, life might be found to be preferable to death. Thus, Brandt is suggesting that a future world course might well be different from that of the present.

Brandt further suggested that one must take into account the infirmities of one's sensing machinery. Knowing that the machinery is out of order will not tell one what results it would give if it were working; thus, the best recourse might be to refrain from making any decision when one is in a stressful frame of mind. If decisions have to be made, one must recall past reactions, in a normal frame of mind, to outcomes like those under assessment. The future world course obviously did not look enjoyable and attractive to the Saunders after 80 years of life.

OTHER SUDDEN, UNNATURAL, TRAUMATIC DEATHS

HOMICIDE

Homicide is defined as the intentional and sometimes unintentional or accidental killing of another person. Homicide is a tragic phenomenon that arouses great public and professional concern and the unlawful taking of human life. Violent dying via homicide is causally distinct from natural dying, as homicide follows an external human act. Homicide is a death that never should have happened, and the public demands a narration, thus complicating the private processing (Rynearson, 2012). Homicide is the third-leading cause of death of men in Latin America and ranks 20th worldwide. In the United States, homicide is the 21st cause of death in men and is the 57th cause in Western Europe (Cheng, 2012).

Personal experiences with death today are for some children too close to home, if they happen to live in communities besieged by chronic violence (Barrett, 1996). Daily exposure to episodes of violence and death increases children's risk of trauma and potentially undermines their psychological and emotional well-being. The cumulative effect of the exposure of young children to episodes of chronic violence and trauma cannot have anything but negative consequences for their learning potential and interpersonal conduct at school, at home, or in the community, noted Barrett (1996). Concepts of death are powerfully influenced by experience with death or threats of death. Mental health, anxiety management, and meaningfulness in life are the most powerful factors in developing concepts of death. Broad cultural-religious influences also enter significantly into each of these factors.

Children in urban environments are increasingly at risk of becoming victims of homicidal violence as they are likely to be both victims and perpetrators of lethal violence. Indeed, such an atmosphere of death is more realistic than that portrayed on television and in other media sources. In 2009, there were 16,799 deaths from homicide in the United States (3.7 deaths per 100,000 population), making homicide the 15th cause of death (CDC, 2012).

For children and adolescents living in certain inner cities in the United States, violent deaths are fairly routine in their neighborhoods. Especially in neighborhoods with heavy drug traffic, shootings and knifings may be all too common. For example, over 75 percent of young violent offenders interviewed in two poverty-stricken New York City neighborhoods had seen someone die in a violent incident (Science Daily, 2009). About half of them (51 percent) had been shot themselves and 78 percent said they had a close friend who died in a violent attack. Many inner-city children and adolescents show symptoms of post-traumatic stress. After they witness a violent death, such an image will probably be imprinted on their brains. Indeed, if this happened once, it could happen again, and perhaps the next time, the child or adolescent may be the victim—being in the wrong place at the wrong time. Images on the evening news and pictures in the newspapers of small children or adolescents being caught in cross-fire or being shot in their beds at night as retaliation by a juvenile gang or an older sibling are all too vivid for television news watchers or newspaper readers. Though they may not have been the object of the murder, nonetheless, if they are "in the vicinity," they may become a victim. Having to bury children and adolescents is typically not normal in the life cycle, but not uncommon in a subculture of violence.

The United States has a reputation, rightly so, of being a gun-totin' society. For example, on two separate occasions when George Dickinson was in England riding in a taxi, the driver said that he would love to move to the United States and drive a cab because he could make more money. However, both cab drivers said they were afraid that they would be murdered if driving a taxi in the United States. There is nearly one gun in the United States for every American, a figure that makes the United States by far the most heavily armed nation in the world (No. 2 is Yemen). The United States has the highest homicide rate by guns among the developed nations. In the wake of mass killings, other countries have adopted a range of new gun restrictions, yet this topic is very controversial in the United States, where gun control laws are among the most lax in the developed world (Spotlight: Arms Race, 2013). After the Sandy Hook Elementary School mass shooting in Newtown, Connecticut on December 14, 2012, President Obama called for a ban on the assault weapons and high-capacity magazines often used in such shootings, and created a working group led by Vice President Biden to propose ways to reduce gun violence.

Such mass shootings as occurred in Newtown, Connecticut, with 20 students and six staff members killed, are not new to the United States, however, as such events occurred back in 1949 when Howard Unruh used a Luger pistol to kill 13 individuals and wound three others in Camden, New Jersey (Pearce, 2012). The spectacular nature of such mass shootings are magnified by intense media coverage, thus giving a major impact to these happenings, though they make up a tiny fraction of overall gun crime.

Homicide has been the leading cause of death for African American males between the ages of 15 and 34 for the past 30 years and is the second-leading cause of death for Hispanic males between the ages of 15 and 34 (Science Daily, 2009). Most of these participants have seen violence used as a way to work through problems all their lives. As a result, their way of reacting when someone confronts them or shows disrespect often includes violence. Violence breeds violence.

Whether in a school setting or in a neighborhood, violence in recent years seems to have escalated. School-related violence has gained widespread media attention and concern, though most violent acts by adolescents take place off of school property (Diclemente & Sionean, 2012). Within these settings bullying and revenge appear to be common motives. Drive-by shootings by gang members

---

WORDS OF WISDOM | CAPITAL PUNISHMENT

In 2010 there were 36 states and the Federal Bureau of Prisons with 3,158 inmates under sentence of death (58 on death row in the federal system), and in 2011 13 states executed 43 inmates. Of those under sentence of death, 55 percent were white and 42 percent black; 98 percent were male.

The majority of countries in Western Europe and North and South America (more than 139 nations worldwide) have abandoned capital punishment in law or in practice. The United States remains in the same company as Iraq, Iran, and China as one of the major advocates and users of capital punishment.

There is no credible evidence that capital punishment deters crime, and innocent people have been convicted and executed. The death penalty is also very expensive, as noted in California, where, since 1978, over $4 billion has been spent in executing only 13 inmates.

The year 1999 was the apex for executions in the United States with 98 deaths. A recent switch in the method of execution has been from a single barbiturate to the sedative pentobarbital.

Taken from T. L. Snell (2011). Capital punishment, 2010—Statistical tables. U.S. Department of Justice; Death penalty focus (2009, March 31), Working for alternatives. http://www.deathpeanlty.org/section. Accessed on October 25, 2012; N. Welsh (2012, November/December). The death penalty is experiencing technical difficulties. *Pacific Standard*, 48–53.

are all too frequent in the 21st century. Violent teenagers seem to follow certain warning signs (Dennis, 2009): lack of connection, withdrawal, masking emotions, silence, trouble with friends, rage, and cruelty toward other children and animals.

Socialization in a milieu of violence may make it okay for violence to be considered an appropriate response in many circumstances. Such violence may not be strongly punished if it occurs under provocation and if the people toward whom it is directed are stigmatized or viewed as disreputable. Thus, violent deaths may become "normal" in a subculture of violence. At any given time in history, there are neighborhoods (even entire countries) warring with other groups, often for religious reasons. To live in circumstances in which terrorist bombings occasionally occur or sniper fire kills individuals must make for long nights and days for children and adolescents growing up in such an environment. They are all victims because they live in constant fear for their lives. Certainly, their perceptions of death would probably be different from those of individuals brought up in a more peaceful atmosphere.

Research on the effects of violent video games on children and adolescents demonstrated that exposure to violent media appears to increase aggressive behavior in young people, yet the impact of media violence continues to be underreported by news services (Anderson & Bushman, 2002). As early as the 1970s the U.S. Surgeon General warned of the potential negative effect of violent television. Thus, the media may be contributing to homicides and other crimes, though the correlation is not always easy to determine.

In 2009 the types of weapons used in homicides in the United States (U.S. Census Bureau, 2012) were firearms (handguns, rifles, shotguns), knives or cutting instruments, blunt objects (club, hammer), personal weapons (hands, fists, feet, pushed, etc.), poison, explosives, fire, narcotics, strangulation, and asphyxiation. Males were most frequently the victims (8.5 per 100,000 population to 2.3 for females) and blacks were most often the victims (19.6 per 100,000 population to 3.3 for whites).

PRACTICAL MATTERS | GUIDE TO SURVIVAL FOLLOWING A HOMICIDE OF A FRIEND/FAMILY MEMBER

The feelings one is having are likely similar to what others have felt, though these feelings are likely foreign. A feeling of numbness is common.

A family is not always pulled together by a homicide, thus communication is important in trying to reestablish bridges.

If the homicide is brought up, some individuals will not want to talk about or listen to details, though survivors need to talk.

Flashbacks of the moment will reoccur off and on, though the survivor may feel that she/he is almost "over it." Grief spasms may happen with crying and other outbursts. When the reality of death sinks in, depression often follows.

Holidays will likely be difficult, thus new traditions may need to be established.

Survivors may experience anger toward the murderer, society, or God. Revenge may also be a common emotion felt and the wish to kill the killer.

Most survivors slowly heal and get back to their daily activities. The memories of the decedent help with the healing process.

Taken from Guide to survival. *Justice for homicide victims* (2012). http://justiceforhomicidevictims.net/survival.html. Accessed on October 18, 2012.

ACCIDENTAL DEATHS

Deaths from accidents, not unlike deaths from suicides and homicides, are jarring experiences. Such deaths are unexpected with no chance to say goodbye. The most common accidental deaths in the United States are automobile (overwhelmingly the largest cause of accidental deaths), machinery-related, gases such as carbon monoxide, gunfire, food (e.g., suffocation by a stray piece of steak or something else), fires and smoke inhalation, drownings, drug overdoses, inhalation of toxins in cleaning products, and poisoning. Automobile accidents account for over three times the number of deaths as any of the other causes (DeathAnalysis.com, 2010).

DEATH ACROSS CULTURES | ROADSIDE MEMORIALS

As one drives across the United States, Canada, certain countries in Europe, New Zealand, or various other countries around the world, it is not unusual to occasionally see a roadside memorial such as a cross or a secular reminder that death occurred on or near that spot. Such memorials have become popular since the latter part of the 20th century. The religious symbol of a cross perhaps marks the sites as sacred. The memorial may be decorated from time to time with flowers, a teddy bear, a football jersey, a toy, photograph, or some other personal item of the deceased individual.

Memorials typically appear where there has been a sudden and violent death of a younger person. Such a memorial could be an enabling of a connection between the deceased's personal life and the impersonal site. Such memorials might serve as a memorialization or as a warning. As a relative or friend passes the memorial, there is the reminder of the life, and death, of that person, and thus symbolic representation of ongoing grief work. Perhaps this is supportive of the idea that the dead should live on in our memories; thus, through such memorials we strive to keep the deceased person "alive," enacting a symbolic immortality.

Traveling from state to state in the Untied States, there appears to be no uniformity as to the various roadside memorials. Federal laws prohibit placement on the right-of-way; thus an inconsistent patchwork of state policies controls these memorial sites. Twenty-one of the states have adopted a policy regarding the placement of roadside memorials along state highways. Some states, such as Illinois and Washington, have a DUI Memorial Sign Program. The sign may read "Please Don't Drink and Drive" and is posted "in memory of" the deceased individual. At least five sites have Adopt-a-Highway programs, recognized with a sign, for volunteer participation in roadside litter removal on designated roads. A "green movement" for roadside memorials is happening in 11 states where the Department of Transportation allows a tree to be planted near where a highway fatality occurred. Delaware was the first state to build a memorial garden dedicated to those who lost their lives in all types of traffic fatalities. Similarly, Maryland has started a Living Memorial Program in which a grove of trees will be planted each year to memorialize the individuals killed in highway crashes during the year.

The one universal theme among roadside memorials, at least in the United States, is that they are predominately religious in nature (73 percent estimated to be religious, 27 percent secular). These memorials can provide solace to grieving families, yet to others they are seen as distractions to motorists or as eyesores and a turnoff to some, for the religious symbols represented. Variation in these memorials by the general public is reflected in the diverse state policies and practices regarding roadside memorials. These memorials are relatively new to the scene on U.S. highways, yet they do not seem to be going away.

Adapted from G. E. Dickinson and H. C. Hoffmann. (2014). Roadside memorials: A 21st century development. In C. Staudt & J. H. Ellens (Eds.), *Our changing journey to the end: Reshaping death, dying, and grief in America*. Santa Barbara, CA: Praeger, 227–252.

Sudden, traumatic, or violent death, the young age of the deceased, and perceptions of death preventability are associated with post-traumatic stress disorder (PTSD). Road traffic fatalities, for example, contribute to the probability that the bereaved individual might experience a grief reaction very different from the "typical" grief reaction. Crash fatalities usually involve police investigations, insurance claims, court proceedings, hospital systems, and media attention, none of which is necessarily pleasant (Breen & O'Connor, 2010).

It is not uncommon for an individual receiving news of a death from an accident to go into a state of denial (Stage 1 of Kübler-Ross's stages of the dying process, as discussed in Chapter 5). One does not want to believe that such has happened and needs tangible evidence to prove it. An individual often wants to "see the body" in order to believe. Yet, others might shy away from seeing the body because of the fear of the truth being verified by the sight. Still others might be fearful of the unknown (death!), and may be afraid to see the dead body (de Bretagne, 2012).

It is important for those adjusting to an accidental death to establish a routine to help get through the next several days/weeks following the death. Support of family and friends is crucial as survivors adjust to the accidental death. Coping with death through accidents can go through three distinct time frames: immediately following the death, in the short term following the death, and in the long term following the death. Whether the death happened through a road traffic accident or an accident at work or elsewhere, accidental deaths are shocking and may require those working through bereavement to seek professional help (Morrisey, 2012). Bereaved individuals, following the death of a significant other in a car crash, tend to experience more depression, more psychiatric symptoms, greater mortality, and less future orientation than matched controls from a nonbereaved sample (Breen & O'Connor, 2010).

CONCLUSION

Any traumatic death is indeed a shock, whether from suicide, homicide, or an accident. Just what constitutes a suicide is not clear today. A rational suicide is typically thought out and planned, however, which is not necessarily true for other suicides, or for homicides or accidents. Many persons display suicidal behavior by constantly taking risks and gambling with their lives; yet if death results from such behavior, it is often classified as accidental or natural. Sociologists argue that suicidal persons construct their meanings of suicide and motivations for committing it out of collective values upon which the social structure rests. Meanings of suicide arise out of what people think, feel, and do.

Persons bereaved by a suicide may have a longer and more difficult grieving process due to their perceiving less social support than other bereaved individuals. Some grief reactions, such as shame or rejection, may account for this lack of support. Shame may occur because of the stigma surrounding death by suicide. Rejection may result because suicide implies that the deceased rejected not only life but also family and friends. Suicide survivors may also have more difficulties discussing the death with others; thus they may try to deny the cause of death.

Suicidal persons are more likely to be middle-aged adults or the elderly, male, and not married. Though age by itself does not cause suicidal behavior, a person most vulnerable to suicide is a white male older than age 85. Suicidal persons tend to talk about suicide before they act and often display observable signs of suicide. Males are perhaps more likely to complete suicide than are females because males tend to use a more lethal weapon—firearms. For suicidal persons, this act becomes an easy solution to their problems. Suicide can be a rational act, and one does not have to be insane to take one's own life.

How are survivors to cope with the death of a significant other from suicide? Religion appears to influence the coping process for suicide bereavement, although results are preliminary (Vandecreek & Mottram, 2009). A recent national poll (Harris Interactive Poll, 2007) found that 69 percent of individuals in the United States believe in an afterlife, 75 percent believe in heaven, and 62 percent believe in hell. With such large numbers of Americans believing in an afterlife, it would seem likely that religion would play a positive role in suicide bereavement, yet studies cited by Vandecreek and Mottram reveal varying results. Support groups, nonetheless, whether religious or otherwise, can be helpful for suicide bereavement. Individual therapy is also a most effective treatment for suicide bereavement, especially for children in the death of a parent from suicide (Webb, 2003). Combining both individual and group sessions may prove to be doubly effective.

Below are some suggestions for dealing with the aftermath of suicide:

1. Be available to the survivors by taking the initiative to help. Do not hesitate to visit them, even if you do not know what to say. Just being present shows that you care.
2. Discourage guilt feelings. Suicide tends to result from long-term feelings; thus, the grievers should not blame themselves with "if only" feelings.
3. Be truthful and honest at all times.
4. Discourage self-pity and encourage the individual to become involved in activities.
5. Encourage the griever to express his or her real feelings. Help guide the individual toward thinking ahead to the future.

Homicide, particularly in urban areas where the rates are higher, may be somewhat of a daily occurrence in some cities. Such trauma may be witnessed by children, who thus have an upbringing in a very unsafe environment. Mass shootings in schools, an event occurring too often in recent years, also constitute a most traumatic setting for both students and teachers. Though capital punishment is legal in some states in the United States, such action does not tend to serve as a deterrent for others. The United States seems to have an "eye for an eye" mentality regarding capital punishment, so unlike most of the rest of the world.

Accidental deaths are unexpected, with no chance to say goodbye. Such deaths are most common through automobile accidents. Roadside memorials are becoming commonplace in the United States and other countries around the world to "mark the spot" where an individual died in an auto crash. These markers, typically a cross, are especially found where a teenager or young adult died. If putting up such a marker is helpful in the grief process for some individuals, then so be it, assuming it is within the confines of the legal ramifications of the law in a particular state.

SUMMARY

1. Suicide has been around for as long as recorded history.
2. Suicide has historically been viewed in different ways in different places. Even today the general attitude toward suicide remains confused and sometimes contradictory.
3. The meaning of suicide is problematic. Suicide research centers have evolved in recent years to address the complex issues of suicide.
4. In the late 1800s, French sociologist Emile Durkheim noted that suicide is related to one's degree of social integration. Different types of society produce different suicide types: egoistic, altruistic, anomic, and fatalistic.
5. The dramaturgical approach to suicide study uses the drama or action to explain the behavior.
6. The existentialist perspective on suicide may view the choice of death as it views the choice of life—both the result of reflection—and not view death as the enemy.
7. If one feels alienated and powerless, suicide may appear to be a solution.
8. Though age by itself does not cause suicidal behavior, elderly males are most vulnerable to suicide in the United States.
9. People who commit suicide tend to talk about their attempt before they act.
10. The most frequent method of completed suicide in the United States is by firearms.
11. The concept of rational suicide suggests that a person does not have to be mentally ill to take his or her own life.
12. The media have contributed to copycat suicides by teenagers and college-age students, according to some researchers.
13. Suicidal deaths are virtually nonexistent before age five and are rare between five and nine years of age. Adults prefer to believe that children do not commit suicide.
14. Social factors tend to be associated with suicide: age, gender, marital status, and socioeconomic level.
15. Homicide is defined as the intentional and sometimes unintentional or accidental killing of another person.
16. Homicide is one of the leading causes of death for men in some countries.
17. In some urban environments, homicides occur frequently, making such a situation most traumatic for those living there.
18. The United States has a reputation for having a high homicide rate.
19. Capital punishment does not seem to serve as a deterrent to homicide, and research tends to support that position.
20. Accidental deaths are unexpected, with no chance to say goodbye.
21. Roadside memorials, marking the spot where an individual died in an automobile accident, have become rather frequent in numerous countries around the world.

DISCUSSION QUESTIONS

1. Discuss why the meaning of suicide continues to be problematic today.
2. Give examples of Durkheim's four types of suicide.
3. How does the dramaturgical approach view suicide?
4. How does an existentialist perspective explain suicide?

5. Describe signs of suicide. What should you do if you observe some of these signs in a friend?
6. Discuss why the suicide rate tends to be high among older persons.
7. The concept of rational suicide implies that one does not have to be mentally ill to take his or her own life. Discuss whether you feel that one can be sane and take his or her own life.
8. Discuss your reaction to the deaths of Julia and Cecil Saunders.
9. Why do you think that capital punishment does not serve as a deterrent to homicide and other violent crimes?
10. Discuss what you think it would have been like to have grown up in an urban environment where violent crimes such as homicide were common.
11. Discuss the pros and cons of allowing guns to be legal in the United States.
12. What is your opinion of roadside memorials? Should they be allowed, or are they simply a distraction on the highways?
13. Have you had personal experience with a friend or family member dying from an automobile accident? If yes, what was your initial gut reaction?

GLOSSARY

Alienation: Not feeling in control and not feeling a part of a situation. This often results in a lack of self-realization and a feeling of powerlessness and hopelessness.

Altruistic Suicide: Results from being overly integrated into society and willing to die for the group.

Anomic Suicide: Results from a lack of regulation of the individual when the norms governing existence no longer control that individual.

Completed Suicide: A suicide attempt in which the individual actually kills him- or herself; the act is completed with his or her death.

Copycat Suicide: Occurs when one person's suicide is at least partly caused by exposure to another person's suicide.

Dramaturgical Perspective: Highlights the construction of action by using the metaphor of drama to explain behavior.

Egoistic Suicide: Occurs when the person is inadequately integrated into the society.

Emic Approach: An analysis of behavior from the perspective of the participant (the one being studied). If an individual is studying suicide, for example, data could be gathered from one who has attempted suicide. What was this experience like from her or his point of view?

Etic Approach: An analysis of behavior from the perspective of the observer (the one doing the study, not the participant).

Fatalistic Suicide: Too much control over a person by society may lead to feelings of oppression under extremely strict rules.

General Strain Theory: Refers to strains on an individual and whether or not she or he can cope with this negative experience or relationship. Originates out of Robert Merton's idea of anomie and a state of normlessness where the individual does not know what to do. "What are the rules and regulations?" he or she may ask. An individual is unemployed at age 45 for the first time in her or his life and does not know where to begin in order to start over, for example. Suicide might be the "solution," if other options do not appear.

Homicide: Operationally defined as the intentional and sometimes unintentional or accidental killing of another person.

Karoshi: Death by overwork, a current phenomenon of concern in Japan.

Rational Suicide: Suicide in which the individual is not insane and is aware of what he or she is doing.

Suicidal Ideation: Refers to one who considers suicide, contemplates suicide, is prone to suicide, or has suicidal thoughts.

Suicide Gesture: Motivated by a need for aid and support from others. The intent for a completed suicide is not present, yet the individual wants to draw attention through a "call for help." Sometimes the gesture turns out to be a completed suicide, yet these are difficult to prove or disprove.

SUGGESTED READINGS

Coleman, L. (2004). *The copycat effect*. New York: Simon and Schuster. This book catalogues numerous tragedies by category, from mass suicides in ancient times to modern-day ones. Loren Coleman suggests a number of solutions to stem the problem of the copycat effect.

Colucci, E., & Lester, D. (Eds.). (2013). *Suicide and culture*. Ashland, OH: Hogrefe Publishing. The authors review the fundamental issues of why culture is of vital importance in understanding and preventing suicidal behavior.

De Leo, D., Cimitan, A., Dyregrov, K., Grad, O., & Andriessen, K. (Eds.). (2014). *Bereavement after traumatic death*. Ashland, OH: Hogrefe Publishing. Addresses the consequences of traumatic deaths after the sudden death of a significant other.

Guenther, L. (2013). *Solitary confinement: Social death and its afterlives*. Minneapolis, MN: University of Minnesota Press. The author examines the death-in-life experience of solitary confinement in America from the early 19th century to today's supermax prisons and documents how such confinement undermines prisoners' sense of identity and ability to understand the world.

Jordan, J. R., & McIntosh, J. L. (Eds.). (2011). *Grief after suicide: Understanding the consequences and caring for the survivors*. New York: Routledge. Written from a clinical and research perspective, the book takes a life cycle developmental approach to dealing with grief after a suicide.

Jurich, A. P. (2008). *Family therapy with suicidal adolescents*. New York: Routledge. A foundation book for all mental health practitioners working with adolescents.

Langford, J. M. (2013). *Consoling ghosts: Stories of medicine and mourning from Southeast Asians in exile*. Minneapolis, MN: University of Minnesota Press. The author looks at ways of facing death, conducting relationships with the dead and dying, and addressing the effects of violence that continue to reverberate in bodies and social worlds.

Leenaars, A. A. (2010). *Suicide and homicide-suicide among police*. Amityville, NY: Baywood. A good review of the classical and current studies of both suicide and homicide-suicide among police officers.

Leong, T. L., & Leach, M. M. (Eds.). (2008). *Suicide among racial and ethnic minority groups: Theory, research, and practice*. New York: Routledge. Provides a summary of available literature and suggests avenues for further research and intervention in the study of suicide.

Lester, D. (2003). *Fixin' to die: A compassionate guide to committing suicide*. Amityville, NY: Baywood. The author suggests that readers question and examine their own death and how they might prefer to die.

Marsh, I. (2010). *Suicide: Foucault, history and truth*. Cambridge: Cambridge University Press. Marsh, an allied health professional, illustrates that contemporary approaches to suicide have a history and draws on the work of Michel Foucault (1926–1984).

Miller, A. L., Rathus, J. H., & Linehan, M. M. (2006). *Dialectical behavior therapy with suicidal adolescents*. New York: Guilford Press. A manual on the why, how, and what of dialectical behavior therapy with multiproblem suicidal adolescents.

Minois, G. (1999). *History of suicide: Voluntary death in Western culture*. Baltimore, MD: John Hopkins University Press. Follows the religious, philosophical, literary, and judicial debate for and against self-murder from antiquity to the end of the Enlightenment.

Pat-Horenczyk, R., Brom, D., & Vogel, J. M. (Eds.). (2014). *Helping children cope with trauma: Individual, family and community perspectives*. New York: Routledge. The book emphasizes ways to build individual, family, and community resilience to help children cope with trauma.

Stack, S., & Bowman, B. (2012). *Suicide movies: Social patterns 1900–2009*. Ashland, OH: Hogrefe Publishing. A portrayal of suicide in the cinema and what it means for suicidal prevention.

Some children love to watch the cremations. The skull is usually the last thing to be burned. Sometimes it collapses with a loud pop, like a balloon bursting. When that happens, the children clap their hands.
—Alexander Campbell, The Heart of India

DIVERSITY IN DEATH RITUALS

CHAPTER 10

Skeeter Hagler

Death occurs in all societies, yet it evokes an incredible variety of responses. At the moment of death, survivors in some societies remain rather calm, others cry, and still others mutilate their bodies. Members of some societies officially mourn for months, whereas others complete the ritual within hours. In many societies families are involved in preparing the corpse for the funeral ritual; in others, families engage professional funeral directors to handle the job.

The variety of responses to death was further noted by Richard Huntington and Peter Metcalf in *Celebrations of Death: The Anthropology of Mortuary Ritual* (1992). They stated that corpses are burned or buried, with or without animal or human sacrifice; they are preserved by smoking, embalming, or pickling; they are eaten—raw, cooked, or rotten; they are ritually exposed as dead or decaying flesh or simply abandoned; or they are dismembered and treated in a variety of these ways. Funerals are times for avoiding people or holding parties, for weeping or laughing, or for fighting or participating in sexual orgies.

Whereas death rituals in the United States are generally subdued and rather gloomy affairs, some societies engage in rather spirited activities. The Bara of Madagascar, for example, engage in "drunken revelry" at a funeral—rum is consumed, sexual activities occur, dancing takes place, and contests involving cattle occur (Huntington & Metcalf, 1992). Among the Cubeo of South America, simulated and actual ritual coitus is part of the mourning ritual (Goldman, 1979). The dances, ritual, dramatic performances, and sexual license have the purpose of transforming grief and anger over a death into joy.

In most non-Western societies, death is not seen as one event but rather as a process whereby the deceased person is slowly transferred from the land of the living to the land of the dead (Helman, 1985). The process is illustrated by rituals marking biological death, followed by rituals of mourning, and then by rituals of social death. The deceased person is often viewed as a soul in limbo during rituals of mourning, though he or she is still a partial member of the society (Sweeting & Gilhooly, 1992).

For the Kota people of south India, for example, a person is not socially dead until after the dry funeral—the second funeral for the deceased, held annually and lasting 11 days (Mandelbaum, 1959).

UNDERSTANDING DEATH RITUALS

Ritual can be defined as the symbolic affirmation of values by means of culturally standardized utterances and actions (Taylor, 1988). People in all societies are inclined to symbolize culturally defined feelings in conventional ways. Ritual behavior is an effective means of expressing or reinforcing these important sentiments, and it helps make death less socially disruptive and less difficult for individuals to bear (Haviland, 1991). Rituals differ from other behaviors in that they are formal—stylized, repetitive, and stereotyped (Rappaport, 1974). Rituals are performed in special places, occur at set times, and include liturgical orders—words and actions set forth previously by someone. In cultures that deny death, rituals can make death a reality, normalize the grieving process, and introduce the possibilities for hope, imagination, and new life for survivors (Giblin & Hug, 2006).

| ALWAYS GO TO THE FUNERAL
by Deirdre Sullivan

I believe in always going to the funeral. My father taught me that.

The first time he said it directly to me, I was 16 and trying to get out of going to calling hours for Miss Emerson, my old fifth-grade math teacher. I did not want to go. My father was unequivocal. "Dee," he said, "you're going. Always go to the funeral. Do it for the family."

So my dad waited outside while I went in. It was worse than I thought it would be. I was the only kid there. When the condolence line deposited me in front of Miss Emerson's shell-shocked parents, I stammered out, "Sorry about all this," and stalked away. But for that deeply weird expression of sympathy delivered 20 years ago, Miss Emerson's mother still remembers my name and always says hello with tearing eyes.

That was the first time I went unchaperoned but my parents had been taking us kids to funerals and calling hours as a matter of course for years. By the time I was 16 I'd been to five or six funerals. I remember two things from the funeral circuit: bottomless dishes of free mints and my father saying on the ride home, "You can't come in without going out, kids. Always go to the funeral."

Sounds simple. When someone dies, get in your car and go to calling hours or the funeral. That I can do, but I think a personal philosophy of going to funerals means more than that.

"Always go to the funeral" means that I have to do the right thing when I really, really don't feel like it. I have to remind myself of it when I could make some small gesture, but I really don't have to and I definitely don't want to. I'm talking about those things that represent only inconvenience to me but the world to the other guy: you know, the painfully underattended birthday party, the hospital visit during happy hour, the shivah call for one of my ex's uncles. In my humdrum life, the daily battle hasn't been good vs. evil. It's hardly so epic. Most days my real battle is doing good vs. doing nothing.

In going to funerals I've come to believe that while I wait to make a grand, heroic gesture I should just stick to the small inconveniences that let me share in life's inevitable occasional calamity.

On a cold April night three years ago, my father died a quiet death from cancer. His funeral was on a Wednesday, middle of the workweek. I had been numb for days when, for some reason during the funeral, I turned and looked back at the folks in the church. The memory of it still takes my breath away. The most human, powerful and humbling thing I've ever seen was a church at 3 p.m. on a Wednesday full of inconvenienced people who believe in going to the funeral.

Source: http://www.npr.org/2005/08/08/4785079/always-go-to-the-funeral.

DEATH RITUALS AS A RITE OF PASSAGE

Throughout life we will occupy social positions within the societies of which we are a part. Whenever our social position changes, our identities change as well. These transitions in identity require of us, as well as of those who are significant in our life, the ability to adjust and adapt to the transitions. Human beings construct rituals as one means of acknowledging and adapting to change. Consider the following transitions: A child becomes an adult, a single person commits to another in marriage, a married couple become parents, a worker retires, and a person dies. In most societies, each of these transitions is marked by collective actions (or social rituals) that acknowledge a change in people's identities.

Death is a transition, but only the last in a long chain of transitions, or rites of passage, according to Richard Huntington and Peter Metcalf (1992). In many cultures and religions, being dead is another status change, replete with new roles and obligations. Often the dead are expected to give advice, cure illness, reward good deeds, and ensure a good harvest. In Ireland, the dead are called upon to cure the afflicted and to comfort the lonely (Enright, 1994). Thus, at funerals in Ireland, people never say farewell because they fully expect to hear from their friends and loved ones again.

The moment of death is related not only to the process of afterlife but also to the process of living, aging, and producing progeny. Death relates to life—to the recent life of the deceased and to the lives that he or she has procreated and now leaves behind. There is an eternity of sorts on either side of the line that divides the quick from the dead. Life continues generation after generation, and in many societies it is this continuity that is focused upon and enhanced during the rituals surrounding a death.

One of the best-known accounts of death as one of a series of such **rites of passage** through the life cycle comes from A. Van Gennep (1909/1960) in his treatment of funerals. Van Gennep compared funeral rituals with other rites of passage (such as marriage ceremonies and graduation exercises) and claimed that all rites of passage have three subrites—rites of separation, rites of transition, and rites of reincorporation.

In the rite of separation, those whose identity is undergoing change will be perceived by the group as different or "other." At the graduation or wedding, the graduates or the members of the bridal party dress differently. In the middle of the rite of passage the identity of the ones changing will experience some ambiguity—after the bride and groom have exchanged vows, are they married or single? Is the deceased person in the casket as dead as he or she will be after burial or cremation? Finally, for all rites of passage there will be a rite of reincorporation in which the community incorporates in a new way those who have taken on a new identity. These rites of reincorporation are often meals or feasts—graduation or wedding receptions and after-funeral meals.

Funeral or mortuary rites in many societies are critical because they ensure that the dead make the transition to the next stage of life or nonlife. Although many

In these two pictures we view French police officer Brigadier Benoît-Christophe Legendre (1975–2006). The first picture was taken a week before his death, and the second is of him lying in his bed at his father's house, near Marseille, in southern France, after he was prepared by the undertaker. Consider how difficult it must be psychologically for survivors to make the transition from alive to dead of their loved one. The social function of rites of passage is to ease this process of transition.

LISTENING TO THE VOICES

PARABLE OF IMMORTALITY
Henry Van Dyke 1852–1933

I am standing by the seashore.
A ship at my side spreads her white sails to the
morning breeze and starts for the blue ocean.

She is an object of beauty and strength,
and I stand and watch until at last she hangs
like a speck of white cloud
just where the sun and sky come down to mingle
with each other.
Then someone at my side says,
"There she goes!"

Gone where?

Gone from my sight . . . that is all.
She is just as large in mast and hull and spar
as she was when she left my side
and just as able to bear her load of living freight
to the place of destination.

Her diminished size is in me, not in her.

And just at the moment when someone at my
side says,
"There she goes!"
there are other eyes watching her coming . . .
and other voices ready to take up the glad
shout . . .

"Here she comes!"

cultures continue to have relationships with the dead, they only do so when the dead are safely in the next world. However, persons for whom no rites are performed

> ... are the most dangerous dead. They would like to be reincorporated into the world of the living and since they cannot be, they behave like hostile strangers toward it. They lack the means of subsistence which the other dead find in their own world, and consequently must obtain them at the expense of the living. (Van Gennep, 1909/1960, p. 160)

The view of death rituals as rites of passage is supported by a study of the similarities between the Irish funeral wake and the events staged to send immigrants to America in the 19th century (Metress, 1990). Both affairs involved public participation and an opportunity for the family and community to come together to grieve over their loss. Though sad, one could actually rejoice at the departed's rebirth to a new state: Death and immigration freed one forever from the stark hopelessness of poverty. Music, singing, dancing, and wake games were usually a part of the affair.

However, research by J. Hunter (2007–2008) suggests that funeral rituals in America often create an incomplete rite of passage. This is because, while funeral rituals may assist the bereavement in providing an immediate structure for coping with the death, funerary rituals are not effective in providing for the long-term emotional needs of survivors as they are attempting to reconstruct the meaning of their lives while they are grieving the loss of the loved one.

STRUCTURAL-FUNCTIONAL EXPLANATIONS

As we discussed in Chapter 1, the term *function* refers to the extent to which some part or process of a social system contributes to the maintenance of that system. Function means the extent to which an activity promotes or interferes with the maintenance of a system (Cuzzort & King, 1995). Rituals are a particularly effective means of promoting the maintenance of social systems.

The structural-functional perspective of sociologist Emile Durkheim (1915/1954) and anthropologists Bronislaw Malinowski (1925/1948) and A. R. Radcliffe-Brown (1964) reduced the ritual process to an equilibrium-producing system. The specific functions of rituals include validating and reinforcing values, providing reassurance and feelings of security in the face of psychological disturbances, reinforcing group ties, aiding status change by acquainting persons with their new roles, relieving psychological tensions, and reestablishing patterns of interaction disturbed by a crisis (Taylor, 1988). T. Oestigaard and J. Goldhahn (2006) argue that from the perspective of Norwegian society, death is more important for the living than the dead because funerals are one of the most important settings for re-creating society through the reestablishment of alliances in the society.

Similarly, burial rite functions can be enumerated as follows: First, they give meaning and sanction to the separation of the dead person from the living; second, burial rites help effect the transition of the soul to another, otherworldly realm; and third, they assist in the incorporation of the spirit into its new existence.

Finally, death rituals are emphasized as mechanisms that re-create social solidarity and reaffirm social structure. Death disrupts social networks, relationships, and patterns of interaction. Rituals performed at death help to restore order to that which has been disrupted. Individuals gain strength from the ritual affirmations of the community, and they are then better able to continue their lives.

The function of reaffirming social structure is often accomplished through the family reunion that occurs as an unintended consequence, or **latent function**, of the funeral ritual. The intended consequence, or **manifest function**, of a funeral is to pay respects to the deceased and support the survivors. The latent function is to bring family and community together. This latent function is as important as the manifest function.

Ronald Barrett (1992, p. 214) observed that African American funerals are indeed "a primary ritual and a focal occasion with a big social gathering after the funeral and the closest thing to a family reunion that might ever take place." This is a particularly important restorative function for African Americans, whose history of slavery included families being routinely, and arbitrarily, broken apart.

As I grow older, even though I am not an African American, I realize that funerals serve the function in my family as being a rarely held (almost obligatory) family reunion where our family gathers to remember (literally we "re-member," or reconstitute, the family and its memories of being a family) not only the one who has died but the meaning of our family.

However, a study by Melvin Williams (1981) in Pittsburgh, Pennsylvania, revealed class differences in the way that the restorative function of the funeral operates. For the middle class, a funeral may establish, validate, and reinforce social status, whereas for the lower class, funerals tend to be rites of intensification (where the feelings of the bereaved are experienced more intensely) and solidarity— as people put aside feuds and squabbles for the moment.

This function of reaffirming the social order is also evident in some egalitarian tribal cultures in which funerals involve elaborate feasting and gift-giving. The Death Across Cultures box illustrates how the egalitarianism of the Vanatinai is reinforced by funeral rituals.

For some, the burial ground can serve as a symbolic representation of the social order (Bloch & Parry, 1982). For the Chinese and Koreans the actual shape

DEATH ACROSS CULTURES | HONORING THE DEAD

Maria Lepowsky (1994) characterized the matrilineal society of the Vanatinai as being egalitarian; that is, there are no ideologies of male superiority or of female inferiority. This egalitarianism is reflected in the elaborate mortuary rituals. Mortuary ritual and the hosting of feasts and exchange of valuables that accompany it are the primary avenues to personal power and prestige in the society. On Vanatinai, the deaths of men and women are marked equally by elaborate mourning and feasting, and the burden of mourning obligations is the same for men and women. Women are expected to obtain and present ceremonial valuables, and both women and men may strive to enhance their reputations as "big men" (*giagia*, literally "givers") by first accumulating these valuables and subsequently giving them away and by hosting or contributing a great deal to feasts.

After the burial there will always be a series of feasts. The first feast, called *jivia*, occurs within a few days or weeks. During this feast valuables are exchanged and the deceased's kin ritually feed the mourning spouse or representatives of the father's lineage (Lepowsky, 1994). A *velaloga* feast two weeks or two months later releases a widow or widower from taboos against leaving the hamlet or bathing. There are a number of other feasts associated with mourning, but the final feast is the *zagaya*, which is the largest and most important of all. The *zagaya* is held after about three years of intensive preparations. After the *zagaya*, all taboos are lifted from people and places, and the bereaved spouses can again make themselves attractive, court, and remarry.

The relations of power revealed by ceremonial exchange on Vanatinai do not separate men from women. Different forms of wealth are not associated exclusively with one or the other. Strength, wisdom, and generosity are qualities looked for in both men and women. Only those with these qualities will be able to host a successful *zagaya* feast. The *giagia* are both male and female, and their power stems from their possession of wealth and their knowledge of magic.

From Honoring the dead, by D. G. Bates and E. M. Fratkin, 1999. In *Cultural anthropology* (2nd ed.). Boston, MA: Allyn & Bacon.

of the burial ground is similar to the human womb. The Hmong of Laos, after the birth of a baby, will bury the afterbirth in the dirt floor of the family home, and upon the death of that person, the preferred final disposition of the body is in that same place.

Among the Merina of central Madagascar, for example, after death one returns "home" to the tomb, representing a regrouping of the dead—a central symbol of the culture and an underlying joy of the second funeral. By the entry of the new corpses into the collective mausoleum, the tomb and the reunited dead within it represent the undivided and enduring descent group and become the source of blessings and the fertility of the future. The force of this symbol of the tomb as the representation of the eternal, undivided group can be sustained only by downplaying the individuality of the corpses that enter it.

C. Davis (2008), in her article "A Funeral Liturgy: Death Rituals as Symbolic Communication," claims that funerals are ultimate in expressing "final stories." These symbolic representations of a person's life provide the opportunity to bridge the liminal space between the living and the dead and the sacred and the human, while serving to bring together the past, present, and future of the deceased and their surviving loved ones. R. Marshall and P. Sutherland's (2008) anthropological research with Caribbean ethnic groups supports the research conclusions by Davis (2008), demonstrating that the social aspects of bereavement will have positive effects for intergroup relationships and social cohesiveness.

DEATH ACROSS CULTURES | NO PLACE LIKE HOME
Kevin Yang, 2013

"Spring Semester in Thailand" student from Hamline University

There is a belief in many Hmong traditions that after a person dies, their spirit will wander restlessly around the world unless it is guided back to its place of birth in an elaborate ritual. As the Hmong people have found themselves uprooted from their villages in northern Laos and Thailand and resettled throughout the world, the final journey home for many has become a long one. Although I identify as a Christian and I am strongly attached to Minnesota, I still find myself fascinated by this ritual. I ask how often have I felt similarly. How often have I felt as if I have been wandering in a land that does not always mirror me? How often have I felt that this land is not where I truly belong? One of the major reasons behind my decision to study abroad was the prospect of arriving in a part of the world that much of my family once called home and possibly find peace to some of these questions.

In older American cemeteries, family solidarity was preserved and symbolically communicated in the family plot, where many members of a single family were buried together in a space delineated by fences, headstones, and footstones. Shared tombstones for married couples often declared this family unity and solidarity with the epitaph "Together Forever."

Conversely, changes in rituals caused by changes in religious commitments can also lead to changes in the symbolic representation and experience of social unity. In other words, changes in religious meaning systems and patterns of funeralization can be dysfunctional for group maintenance. This perspective is illustrated by research in Papua New Guinea by R. Schram (2007), where he discovered that the latent consequence of conversion from traditional Papua religion to Christianity led to the new Christian converts emphasizing individualism over ties to the lineage group and extended relatives. When people begin to reject all reciprocal funeral feasts in favor of church meals, it is inevitable that the social fabric of society will also change.

MOURNING BEHAVIORS

The expression of emotions experienced by those who lose a loved one ranges from complete silence to amplified wailing, noted Effie Bendann in *Death Customs: An Analytical Study of Burial Rites* (1930). She stated that after a death the Aborigines of Australia and Melanesia indulge in the most exaggerated forms of weeping and wailing. They display other manifestations of emotional excitement seemingly because of grief for the departed. At the end of a certain designated time period, however, they cease with metronomic precision, and the would-be mourners indulge in laughter and other forms of amusement. Likewise, Hindus in India are encouraged to express their grief openly, even sometimes extravagantly, through shrieks of women mourners and floods of tears, yet no weeping is to occur during the cremation ceremony (Tully, 1994).

A study of **ethnographic** data from 78 societies (Rosenblatt, Walsh, & Jackson, 1976) to identify the essence of universal grief behavior showed that death is nearly universally associated with emotionality and that the most usual expression among the bereaved is crying. One could argue that people display the outward emotion of crying at the time of death because they are sorrowful. Perhaps it is not as simple as that. Anthropologist A. R. Radcliffe-Brown (1964) noted two types of weeping. There is reciprocal ritual weeping to affirm the existence of a social bond between two or more persons. It is an occasion for affirming social ties. Although participants may not actually feel these sentiments that bind them, participation in various rites will strengthen whatever positive feelings they do have. The second type of weeping—weeping over the remains of a significant other—expresses the continued sentiment of attachment despite the severing of this social bond.

Richard Huntington and Peter Metcalf (1992) observed that Radcliffe-Brown was strongly influenced by the French sociologist Emile Durkheim. Durkheim (1915/1954) argued that the emotions developed are feelings of sorrow and anger and are made stronger by participation in the burial rite, whereas Radcliffe-Brown argued that those participating in ceremonial weeping come to feel emotion that is not sorrow but rather togetherness. What Durkheim finds significant is the way that other members of society feel moral pressure to put their behavior into harmony with the feelings of the truly bereaved. Even if one feels no direct sorrow, weeping and suffering may result. Thus, to say that one weeps because of sadness may be too simplistic. From this perspective, death rituals become rites of intensification whereby feelings and emotional states are intensified by ritual participation. It is not uncommon for people to say "I was doing quite well in dealing with my emotions until the funeral began." Often the function of death rituals is to intensify feelings and emotions and then provide a means by which individuals can express their sentiments.

Even though expressions of sorrow seem to be pervasive in most cultures, there are other cultures in which excessive alcohol consumption, dancing, and spontaneous expressions of joy are encouraged as a socially acceptable response to death. Such experiences celebrate the new exalted status of the person who has died and proclaim that the deceased person is better off in this new situation. This is certainly evident in the Irish wake, where the merriment at wakes for the dead may seem disrespectful. From a cultural point of view, the wake actually demonstrates the strong belief that the physically or symbolically dead will continue to exist more happily elsewhere. The American wake reiterates the significance of transitional times for the Irish, whether changes in the seasons or major life changes such as birth, immigration, and death (Metress, 1990).

Expressions of joy and happiness during funerals or wakes may also offer a socially acceptable method for escaping feelings of sadness and anger. As discovered by T. Bordere (2008/2009), youth in New Orleans who lost friends through violent means found that funeral rites involving musical processions, dancing to good music, and expressing their grieving and emotions through t-shirt art, discovered, in their corporate grieving, reasons for celebration, remembrance, and community solidarity. Furthermore, during funerals for the Bara of Madagascar, rum is consumed, sexual activities occur, dancing takes place, and contests involving cattle are held (Huntington & Metcalf, 1992). In a functionally equivalent manner,

The "Day of the Dead" (*Día de los Muertos*) in Mexico demonstrates the belief in the continuity of relationships between the living and the dead.

the Cubeo of South America encourage simulated and actual ritual coitus as part of mourning rituals (Goldman, 1979). The Listening to the Voices box provides an example of how dancing can be used to transform cultural grief and sadness into joy and celebration.

As we have observed, societies encourage both the expressions of sadness and joy in their mourning rituals. In the former, participants are given the opportunity to collectively express their grief. In the latter, individuals will dance, participate in dramatic performances, and even experience sexual license, all with the purpose of transforming grief and anger over a death into joy.

| CUTTIN' THE BODY LOOSE: DANCING IN DEFIANCE OF DEATH IN NEW ORLEANS

"They gon cut the body loose!" One short brother with a mustache was running up and down the funeral procession explaining that they wasn't going to have to go all the way to the cemetery on account of they was going to cut the body loose. This meant that the hearse would keep on going and the band and the second liners and the rest of the procession was going to dance on back to some tavern not too far away. So we stood in a line and the second liners were shouting, "open it up, open it up," meaning for the people in front to get out the way so the hearse could pass with the body. After the hearse was gone we turned the corner and danced down to the bar.

Like a sudden urge to regurgitate and with the intensity of an ejaculation, an explosive sound erupted from the crowd under the hot New Orleans sun. People spontaneously answered the traditional call of the second line trumpet.

"Are you still alive?"
"YEAH!"
"Do we like to live?"
"YEAH!"
"Do you want to dance?"
"YEAH!"
"Well damn it, let's go!"

The trumpeter was taunting us now, and the older people were jumping from their front porches as we passed them, and they were answering that blaring hot high taunt with unmistakable fires blazing in their 60-year-old black eyes. They too danced as we passed them. They did the dances of their lives, the dances they used to celebrate how old they had become and what they had seen getting to their whatever number years. The dances they used to defy death.

Nowhere else in this country do people dance in the streets after someone has died. Nowhere else is the warm smell of cold beer on tap a fitting conclusion for the funeral of a friend. Nowhere else is death so pointedly belittled. One of us dying is only a small matter, an occasion for the rest of us to make music and dance. Nothing keeps us contained. With this spirit and this music in us, black people will never die, never die, never.

We were all ecstatic. We could see the bar. We knew it was ending, we knew we were almost there and defiantly we danced harder anyway. We hollered back even that much louder at the trumpeter as he squeezed out the last brassy blasts his lungs could throw forth. The end of the funeral was near, just as the end of life was near for some of us but it did not matter. When we get there, we'll get there.

From Cuttin' the body loose: Dancing in defiance of death in New Orleans, by K. Salaam, *Utne Reader* (1991, September–October), pp. 78–79.

CHANGES IN MOURNING BEHAVIORS IN THE UNITED STATES

In the days of the Puritans, elaborate and extended mourning was discouraged because it was thought to undermine the cheerful resignation to God's will that was essential to the Puritan experience. However, in the 19th century, when the Romantic influence's emphasis on emotions led to simple sentimentalism, mourning practices became much more dramatic.

For most of the 19th century, the main outlet for the grief of sentimentalism was the ritual of mourning, including the funeral but extending beyond that. This ritual differed markedly from the simple Puritan rite as mourners immersed themselves in grief to become, through their expressive (and often excessive) emotions, the central feature of the ritual. It allowed many members of the middle class (especially the women, who were supposed to be creatures of the "heart") to indulge in grief as therapeutic self-indulgence. Like other forms of sentimentality, the sentimental mourning ritual counterpointed the "real world" because it forced all mourners to consider the power of personal connections in their lives. It turned people from life to death, from the practical to the ceremonial, from the ordinary to the extraordinary, and from the banal to the beautiful (Taylor, 1980, pp. 39–48). This romantic emphasis led to the development of the beautiful funeral. Concerned with an etiquette of proper social relations, people used the funeral, which was designed to reflect taste and refinement, to preserve appearances among their middle-class peers and distinguish themselves from the common folk (Farrell, 1980, pp. 110–111). Especially around the mid-1880s, the tastefully refined middle-class funeral was a dark and formal affair. After death, which still generally occurred in the home, the family members either cleaned and dressed the corpse or, if possible, hired an undertaker to care for the corpse. If they had secured an undertaker, he would place a black badge over the door-bell or door knocker to indicate the presence of mourning and to isolate the family from the unwanted intrusions of everyday life. The family members would also close window shades and drapes. Sometimes they draped black crepe over pictures, mirrors, and other places throughout the house (Habenstein & Lamers, 1962, pp. 389–444).

By the time of the funeral, family members had swathed themselves in black mourning garb that symbolized their intimacy of relationship to the deceased and their depth of grief. After the funeral, custom encouraged the continued expression of grief, as widows were expected to spend a year in "deep" mourning and a year in "second" mourning. For the first year, a bereaved woman wore dull, black clothes, matched by appropriately somber accessories. In the second year, she gradually lightened her appearance by using a variety of materials in somewhat varying colors. Widowers and children were supposed to follow a similar regimen, but in practice, women bore the burden of the 19th-century mourning requirements. Social contact and correspondence followed similar rules, with widened social participation or narrowed black borders on stationery as indicators of different stages of mourning.

By the 20th century, the somber nature of the 19th-century mourning rituals had evolved along with attitudes toward death, which was now celebrated as a passage to eternal life, not a moment of judgment. Death was defined as "a normal change in an eternal process of growth" (Abbott, 1913). Mourning rituals were no longer somber dramatic affairs with elaborate displays of grief. Instead, restraint and passivity characterized 20th-century bereavement practices.

The funeral director used the culture of professionalism to become the authority on American death rituals. His middle-class clients, who feared death anyway, were all too happy to allow the funeral director to take control. This led to a scenario in which the funeral director was the stage manager and the family was the audience, responding to the drama in prescribed ways in the hopes of achieving a catharsis of death. Instead of the expressive grief of the 19th century, family members were expected to contain and control their emotions and to meet death stoically. At the turn of the century, some religious liberals saw grief as a lack of faith in the imminence of immortality. Others reacted to the central place of the mourners in the mournful Victorian funeral and charged that "over-much grief would seem mere selfishness" (Mayo, 1916, p. 6). Over and over again, writers proclaimed that "the deepest grief is the quiet kind" (Sargent, 1888, p. 51). An 1890s etiquette book suggested that "we can better show our affection to the dead

Library of Congress

During the latter part of the 19th century, mourners were expected to dress the part. In this picture, pallbearers wear mourning clothes, black sashes, badges, and dark hats.

by fulfilling our duties to the living, than by giving ourselves up to uncontrolled grief" (Pike & Armstrong, 1980, p. 125). Portraying grief as a selfish ploy to stop the ongoing business of life, mourners were persuaded to keep their grief controlled and private. In the long run, they predicted the modern practice of grief therapy in which grief is seen as a disorder (Mayo, 1916, p. 6; Pike & Armstrong, 1980, p. 125; Sargent, 1888, p. 51).

With this new ideal, Americans reduced their symbolic expressions of grief such as mourning wear because such expressions were thought to offend those who preferred to focus on life, not death. In concealing the uncouth and discordant expressions of grief, Americans reversed the 19th-century tradition of "good mourning," and isolated mourners were forced to discover their own private mourning rituals (Hillerman, 1980; Oxley, 1887).

The changes brought about by funeral directors early in the 20th century are still evident in mourning behaviors today. Private expression of grief in grief counseling sessions is preferred to public expression of grief. More recently, funeral services have become even briefer, often without the eulogy, which might be too upsetting. Psychologists are concerned about this decline in the expression of grief. Although many people do seek grief counseling, others do not, repressing their feelings of grief and becoming depressed or drowning their sorrow in alcohol.

The decision to seek grief counseling is clearly gender-related, with women being much more likely to receive professional therapy and participate in self-help groups for the bereaved.

GENDER DIFFERENCES IN MOURNING BEHAVIORS The Rosenblatt et al. (1976) study mentioned earlier noted that in the 78 cultures studied, there were significant gender differences in emotions during bereavement, with women tending to cry and self-mutilate more than men. The men tended to direct anger and aggression away from self. One of the traditional theories to explain gender differences in emotional expression is that it may be easier to socialize women than men to be overtly non-aggressive; thus, crying may represent a female expression of aggression. Another theory is that women are more affected by the loss from a death because of their stronger attachments through their role as mothers. On the other hand, women may not experience death more strongly, but they may simply be used as the persons symbolizing publicly, in burdensome or self-injuring ways, the loss that all experience. The analysis of data from these 78 societies concluded that the kind of data needed to explain the emotional difference by gender is lacking. An outgoing display of emotions is found among the Kapauku Papuans of west New Guinea (Pospisil, 1963). Their ritual at death requires female relatives of the deceased to give a formal expression of their grief as soon as the soul leaves the body. They weep, eat ashes, cut off their fingers, tear their garments and net carrying bags, and smear their faces and bodies with mud, ashes, or yellow clay. A loud singsong lamentation follows.

To express their grief, women among the Cheyenne Indians cut off their long hair and gash their foreheads causing blood to flow (Hoebel, 1960). If the deceased individual was killed by enemies, they slash their legs until caked with dried blood. Mourning gives the women their own masochistic outlet. Cheyenne men, on the other hand, simply let down their hair in mourning and do not bother to lacerate themselves. In traditional African funerals (Barrett, 1992) women tend to wail, whereas men sing and dance; men are not to cry in front of women because they would appear weak before the very group that they are to protect.

The Dinka women of the Sudan (Deng, 1972) cut their leather skirts and cover their bodies with dirt and ashes for as long as a year. Widows among the Swazi in Africa (Kuper, 1963) shave their heads and remain "in darkness" for three years before given the duty of continuing the lineage for the deceased through the required levirate marriage to her dead husband's brother. Mourning imposed on the Swazi husband is less conspicuous and shorter than that imposed on the wife.

Somewhat less dramatic than the previous examples, the Huicholes of Mexico (Weigand & Weigand, 1991) display a "great deal of crying and wailing" at the actual moment of death. This subsides during the preparation of the corpse but resumes again at the funeral site. However, Huichol women who have suffered the death of a mother or a child often express their loss by suicidal gestures—the stated goal of such behavior is to accompany the deceased. In reality, however, these gestures seldom result in suicide, according to C. G. Weigand and P. C. Weigand (1991).

In conclusion, cross-cultural studies of the ritual of mourning at death reveal that a double standard prevails among some cultures—different "scripts" for different genders. Women in many societies, including the United States, are expected to display more of their emotions than are men. Mourning rituals also tend to last for a set period of time in many societies, and women, in general, are expected to mourn longer than are men. One indication of a different duration of mourning by gender is that, when compared to women, men seem to remarry sooner without social sanction or ridicule, and women are more inclined to remain single after being widowed.

CUSTOMS AT DEATH

As will be discussed in Chapter 11, in the United States a professional is called upon to prepare the body for final **disposition** because we are a very specialized society with a high division of labor. The funeral director takes the body away and returns it later for viewing. The kin and friends in the United States normally play no significant role in handling the corpse. Compared with most societies, we are unique in the level of professional specialization relative to the preparation of the corpse and the actual disposition of the body. Most of the cultures discussed in this section encourage families and friends to become very involved in preparing the corpse for its final disposition.

There are two perspectives for viewing customs at death. From one viewpoint, it is possible to conclude that each society creates different practices unique to the culture in which they occur. An alternative point of view is to recognize that there are common human needs (e.g., to dispose of the body), and each society creates practices, rites, meanings, and rituals, which are functional equivalents to those found in other societies. The social-anthropological perspective, which employs the latter interpretation, is exemplified by a recent research study on the Muscogee Creek tribe by A. Walker and D. Balk (2007). This research gives an excellent example of the common rituals examined in the section below related to customs at death for the Muscogee Creek tribe.

NORMS PRIOR TO DEATH

Some societies have specific norms just before death. It is important in many societies, including the United States, that one be with the dying person at the time of death. So often it is said, "If only I had gotten there a few minutes earlier...." A visit before death allows one to say goodbye. Among the Dunsun of northern Borneo (Williams, 1965), relatives come to witness the death. The dying person is propped up and held from behind. When the body grows cold, the social fact of death is recognized by announcing "he exists no more" or "someone has gone far away."

The Salish Indians of the northwestern United States (Habenstein & Lamers, 1974) leave the dying person alone with an aged man who neither receives pay nor is expected to have any special qualifications for the task. One who is near death must confess his or her misdeeds to this man. The confession is to prevent a ghost from roaming the places that were frequented in life by the dying person.

The Ik of Uganda (Turnbull, 1972) place the dying person in the fetal position because death for them represents a "celestial rebirth." The Magars of Nepal

(Hitchcock, 1966) purify a dying person by giving him or her water that has been touched with gold.

The Maori of New Zealand (Mead, 1991) have a ceremony just before the person dies to send the spirit of the dying person away while the person is still alive and conscious of what is happening. After the spirit is gone, the individual may not be medically dead, but, in a Maori sense, is dead. Now that the person is a corpse, ceremonies begin and a space is set aside in the meeting house for the corpse to lie in an open coffin.

Having a cultural framework prescribing proper behavior at the time of death provides an established order and perhaps gives comfort to the bereaved. These behavioral norms give survivors something to do during the dying process and immediately thereafter, thus facilitating the coping abilities of the bereaved.

PREPARING THE CORPSE

It seems significant in most societies that the body be cleaned and steps be taken to assure a tolerable odor before final disposition. A reverence for the body also seems prevalent. As in the United States, it is important in many other societies that the deceased person look good to the mourners.

CLEANING, DECORATING, AND CLOTHING THE BODY   Ritual preparation of the body for final disposition, including cleaning and wrapping the body (**mummification**), is common in many societies, especially those that do not practice embalming. The material used for wrapping is often white cloth, which is considered natural and pure, but may also consist of other natural materials. For example, the Tiwi of

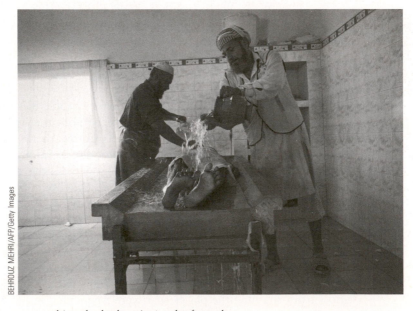

BEHROUZ MEHRI/AFP/Getty Images

Muslim men washing the body prior to the funeral.

Australia (Hart & Pilling, 1960) wrap the body in bark. The Qemant of Ethiopia (Gamst, 1969) wrap the body in a piece of white cloth and cover it with a mat of woven grass. After cleaning the body, the Ulithi in Micronesia (Lessa, 1966) cover it with a tuberous plant and decorate the head and hands with flower garlands.

Anthropologist Linda Connor (1995) noted that corpse washing in Bali is a pivotal point in the process of grieving for relatives. For the Bornu of Nigeria (Cohen, 1967), family members are required to wash the body, wrap it in a white cloth, place it onto a **bier**, and take it to the burial ground. Similarly, the Semai in Malaya (Dentan, 1968) have the housemates bathe the corpse and sprinkle it with perfume or sweet-smelling herbs to mask the odor of decay. They then wrap the body in swaddling clothes.

The Mapuche Indians of Chile (Faron, 1968) sometimes smoke the body, then wash and dress it in the person's best clothes, lay it out on a bier in the house, and place it into a pine coffin. Among some groups in southern Thailand (Fraser, 1966), the body is held and bathed with water specifically purified with herbs and clay. The corpse is then rinsed and dried, and all orifices are plugged with cotton.

In the French West Indies (Horowitz, 1967) the neighbors wash the body with rum and force a liter or more of strong rum down the throat as a temporary preservative before dressing the body and placing it onto a bed. Rather than use a strong drink like rum, the Zinacantecos of Mexico (Vogt, 1970) pour water into the mouth of the deceased about every half hour "to relieve thirst" while the grave is being dug.

The Tewa Indians in Arizona (Dozier, 1966) bury a woman in her wedding outfit and wrap a man in a blanket for burial. The Hopi Indians wrap a man in buckskin and a woman in her wedding blanket, bury them in the clothes that they were wearing at the time of death, and do not wash or prepare the body other than to wash and tie back the hair (Cox & Fundis, 1992). The Navajo in the southwestern United States (Cox & Fundis, 1992) bathe the corpse, then dress it in fine clothes, and put the right moccasin onto the left foot and the left moccasin onto the right foot.

Carrying Fish-like Casket in Ghana.

DEATH ACROSS CULTURES | CHRISTIAN KAREN FUNERAL OF MUSIKHEE

Michael R. Leming

In the Christian Karen funeral, a service is held in the home, and the casket is opened before carrying the body to the cemetery in the forest. While the body is being buried, the family burns the personal effects of the deceased. Following the committal service at the grave, members of the village return to the home of the deceased for a feast prepared by the family and members of the church.

Having lived in a Karen village for more than 15 years, on April 8, 2005, Michael Leming attended his first Karen funeral. The man had died of colon cancer on April 6 in the Karen village of Melaoop after a prolonged illness of more than one and a half years. While his family had converted to Christianity many years earlier, this man was a traditional Karen animist and infrequently worshipped the spirits found in the forest, land, and water. In the final two months of his life he decided to join his family and became a baptized Christian in the Karen Baptist tradition.

Prior to his death, he was visited by members of Melaoop village (a mixture of Christian and Buddhist Karen people). Melaoop has approximately 400 people living in it, and in the weeks before his death, many people came to the man's house and brought food, read the Bible, and attempted to comfort him in his disease of colon cancer. On the morning of Wednesday, April 6, he died in his home surrounded by his family, members of his church, and friends of the village.

After he died his oldest child (a man), helped by his other five children (males and females), placed his body on the floor in the corner of the living room in his house, took off his clothes, put a plastic sheet under his body, and washed the body with soap and water. His body was then dressed in traditional Karen clothes (including shoes, socks, pants, and Karen woven shirt). The family must dress the body in new or clean clothes before it is placed in a simple wooden box. Friends or members of the family typically build the burial box before or after the death.

In the two days prior to placing the body in the box, the family covered the dressed body in the corner of the living room, covered it with a Karen blanket, and poured whiskey down the man's mouth. It is thought that the whiskey provides comfort for the dead person, delays the process of decomposition, and minimizes the smell which mourners find unpleasant (temperatures at midday can reach the mid-90s in Melaoop in April). After preparations were made—and on the evening of the day of death—family, friends, and members of the village were called to come to the home to pray for the family and the deceased. This was a worship service lead by the pastor and leaders in the church. The service consisted of Christian Bible readings, songs, prayer, and words of encouragement for the family. During this worship service the family also announced plans for the funeral and burial service to be held in two days.

The service was lead by two Christian pastors, the village pastor and the father-in-law of the daughter, who lived in a village four miles away. The service began after a gong was rung to call the members of the village to the home of the deceased and the service of worship held in his honor. The service consisted of Christian prayers, Bible readings, songs, and a short sermon given by one of the pastors. Prior to the sermon, one of the members of the family (in this case a son, the eldest child) gave a eulogy that summarized the life of the man who died. The service ended with a prayer for the family, but there was no concluding song, which usually ends Christian Karen worship services.

The box, which had been closed during the service and covered with flowers from the village and from the forest, was then opened for all to see the man prior to his burial. The blanket covering his face was pulled back and he was then doused once again with whiskey, after which the box was closed again in preparation for the procession to the cemetery. At this point a four-inch-diameter bamboo pole, 12 feet in length, was tied to the box for the purpose of carrying the man to the cemetery, located in the jungle about one-half mile away from the village and up a very steep hill. There were no roads to the cemetery, only a dirt path. From all appearances, the cemetery was only a place dedicated for special use in the middle of the forest—old graves were overgrown with brush and weeds, and it was difficult to determine where the graves were located.

Some of the men at the service, perhaps 10 to 12 people, then carried the body to the grave that had been prepared the day before. All people attending the service followed the boxed body, which was preceded by a white cement cross on which the name and birth and death dates were written. This cross would be used to mark the grave. When the people reached the grave, the box was suspended over the grave and was hanging by the bamboo pole. At this time one of the pastors read a few Bible verses and said a prayer, and the 50 or more people at the grave concluded the service by singing the doxology (a Christian hymn or song). At this point one of the

family members cut the rope that tied the box to the bamboo pole, and the box dropped to the bottom of the grave. The pole was then removed from the grave and leaned against a tree that was near the grave.

At another site in the cemetery (forest), about 10 meters away, members of the family brought the personal effects of the man and burned them all. These included his mattress, blankets, sheets, some clothes, and a radio he played night and day before his death. The smell of the plastic and organic materials was not pleasant and caused many to cough. When I asked people why this burning took place, I was given two interpretations—one was that the man would want his personal effects in the afterlife, and the other interpretation was that this burning would prevent members of his family from arguing over their disposition. (These two interpretations may reflect animist and Christian interpretations of a traditional Karen practice created decades ago.)

Meanwhile, approximately 20 people remained at the gravesite while the other grievers were eating at the man's family home. A rectangular wooden form (approximately one meter by two meters) was placed over the grave and cement was mixed in the form to provide a covering for the grave. I saw no other grave in the cemetery that was covered with cement but was told that the man was wealthier than others who had been buried earlier. This 73-year-old man was buried next to his mother-in-law, who died at the age of 103.

After the cement cover was complete, the white cement cross marking the grave was installed; trees, flowers, and bushes were planted around the grave; and the members of the family posed for a picture next to the cross marking the grave. After the pictures were taken, these members of the family returned to their home to join other grievers in eating the meal of thanksgiving. Before eating, they too would wash their hands in the ritual water used by others.

I was told that at an unspecified time, usually less than one year, members of the family would have another service of remembrance and thanksgiving at their home, to which all members of their church and community would be invited. At Easter time (in late March or early April), members of the family would return to the cemetery, clean the grave site, and place flowers on the cement cover as they remembered their father, grandfather, and husband.

This funeral service is totally foreign to the traditional manner in which the Karen animists have disposed of their dead in the past. Normally they would

continued

have made animal sacrifices, chanted stories and incantations, and sung traditional songs and myths. The service would have been in the home, but it would have been overseen by the village priest. Furthermore, the body would have been burned and, depending upon the sex of the deceased, the home would have been moved; or in the case of a village priest, the entire village would have been relocated.

Today, by comparison, the financial effects upon the family are minimal, and the entire village assists the family as they grieve their loss and reestablish their lives without the deceased. It is an easier adjustment for the people, but it is different from everything they have known in the past.

From M. Leming, Professor and Director, Spring Semester in Thailand (www.SpringSemesterInThailand.com).

In Oregon, there is a Russian immigrant group whose body preparation practices date back to the 17th century (Morris, 1991). Death is a "village" affair, and all turn their attention to it when it happens. They rather rapidly do what they consider necessary to lay the body to rest, rarely taking more than 24 hours. When death occurs, the body is washed and dressed in a white, loosely based cloth. The body is then placed inside the casket with the arms crossed on the chest and the hands formed into the sign of the cross in the old style: The first two fingers are extended and the thumb is joined with the third and fourth fingers. The service occurs in the living room of the home, where the casket has been placed.

Body Preparation by Specialists.   Most of the practices cited have used family or friends to prepare the corpse for burial. However, this is not always the case. In many societies, there are designated individuals who perform body preparation rituals. For example, Muslims bring their dead to the mosque directly after death, where there are special rooms for washing the body. During this preparation, prayers and passages from the Qur'an are recited by a *hoca*—a lay holy man, not a priest—and care is taken that the body always faces Mecca.

The Huicholes of Mexico (Weigand & Weigand, 1991, p. 54) handle the corpse with a great deal of respect and care. A singer/curer who is usually closely related to the deceased supervises the preparation of the corpse for burial or placement into a cave. Clothing is changed, personal items are arranged to accompany the body, and the individual's hands and face are washed with water. Most adults aid in the preparation unless "emotional collapse prevents such activity." However, most people are rather calm during these preparations.

Anglo Canadians in and around Toronto, Canada (Ramsden, 1991) prefer to remove the corpse from the company of the living as soon as possible. The corpse is disposed of immediately after death, never to be seen again. This rapid removal of the corpse precludes any opportunity to note changes in the individual after death, thus reinforcing the perception of the Anglo Canadians that death is a static state. They wish to convey the idea that the dead are physically and socially gone. From the moment of death, the physical condition of the body is of no consequence, with the focus on permanently removing the body from the world of the living and banishing the dead person from social relations. The event of death is perceived as an instantaneous occurrence and marks a distinct boundary between a state of being alive and a state of being dead.

EMBALMING In contemporary American culture, the process of final disposition of the body should be studied within its cultural and historical context. As we have demonstrated earlier, it is wrong to believe that the funeral is either unique to Western culture or has been invented or created by it. To bury the dead is a common social practice. The methods to accomplish burial, and the meanings associated with it, are culturally determined.

The burial process often requires the involvement of a functionary, who may be a professional, tradesperson, religious leader, servant, or even a member of the family. The functionary in each society is closely associated with the folkways and mores of the society and its philosophical approach to life and death.

The Hebrew scripture reveals in Genesis 50:2 that physicians embalmed the body of Jacob, the father of Joseph. This description is followed by a detailed description of the funeral and burial. The historian Herodotus records embalming preparation as early as circa 484 B.C. These two sources and archaeological discoveries of earlier cultures give evidence of the disposition of the dead.

With the discovery of the circulatory system of the body, circa 1600, and the possibility of diffusion of preserving chemicals through that system, more sophisticated methods of embalming the body were developed. Dr. Hunter of England and Dr. Gannal of France, independently of each other, furthered the process in the early 1600s. In France, this work coincided with the advent of the bubonic plague. During this period of time, extensive attempts were made to preserve the bodies of the dead to protect the health of those who survived the plague.

Whereas the ancient Egyptian process of embalming required 70 days to perform, body preparation today is completed within a few hours and is more effective and acceptable. Body preparation may be as simple as bathing the body, closing the eyes and mouth, and dressing the body for final disposition. This procedure, though infrequently selected, is used by families who wish direct disposition, which we will discuss later in this chapter as a procedure that may be an acceptable and logical choice.

The most elaborate and famous burial customs were those of the Egyptians.

DEATH ACROSS CULTURES | DEATH IN ANCIENT EGYPT

The earliest burials known in Egypt date to a period well before 3000 B.C. and display evidence, through funerary gifts in the graves, of a belief of continued existence after death. During the middle and late predynastic period (before 3100 B.C.), the practice of wrapping the body in animal skins gradually gave way to other forms of protection, particularly the use of basket trays upon which the body was laid out.

Wrapped bodies of the first three dynasties were not truly mummified, since no treatment other than the use of linen bandages and resin was used. By the Fourth Dynasty, however, evidence was found of deliberate attempts to inhibit decomposition by removal of the soft internal organs from the body—accomplished by means of an incision in the side of the abdomen. Removal of the liver, intestines, and stomach improved the chances of securing good preservation because the emptied body cavity could be dried more rapidly. The removed organs were deposited in a safe place in the tomb in order for the body of the deceased to be complete once more in the netherworld.

Bandages soaked in resin were carefully molded to the shape of the body to reproduce the features, particularly in the face and the genital organs. As the resin dried, it consolidated the linen wrapping in position, preserving the appearance of the body for as long as it remained undisturbed. The corpse itself decomposed very rapidly within this linen shell, leaving the innermost wrappings in close contact with the skeleton.

An important factor in the development of the Egyptian tomb was the necessity to provide storage space for the items of funerary equipment considered essential for continued use by the deceased in the hereafter. A significant part of the material provided for the dead took the form of actual offerings of food and drink—required for the "very survival" of the deceased before enjoying all the other possessions in the tomb. Because of the need to provide offerings of food at the tomb, the tomb had to combine the function of a burial place with that of a mortuary chapel in which the priests could officiate.

From *Death in Ancient Egypt,* by A. J. Spencer, 1982, New York: Penguin Books.

In the United States, it is estimated that four out of five bodies are embalmed before final disposition. **Embalming,** by definition, is the replacement of normal body fluids with preserving chemicals. This process is accomplished by using the vascular system of the body to both remove the body fluids and to suffuse the body with preserving chemicals. The arterial system is used to introduce the chemicals into the body, and the venous system is used to remove the body fluids. This intravascular exchange is accomplished by using an embalming machine. The machine can best be described as an artificial heart outside of the body that produces the pressure necessary to accomplish the exchange of fluids. This, together with the filling of the chest and abdominal cavities with embalming fluid, constitutes the embalming procedure.

In addition to embalming and thorough bathing of the body, cosmetic procedures are used to restore a more normal color to the face and hands. When death occurs, the pigments of the skin, which give the body its normal tone and color, no longer function. Creams, liquids, or sprays are used to give the appearance of normal skin coloration.

The question "But why embalm or cosmeticize the dead body?" is based on the assumption that one of the needs of the family is the reality of death; thus, the body should be left in its most deathlike appearance. Those who have seen a person die (especially if the dying process was painful, prolonged, and emaciating) know that the condition of the body at the time of death can be very repulsive.

Many people cannot accept this condition. This is why contemporary funeral directors embalm and cosmeticize the body.

Another reason for embalming is the mobility of the American population. Viewing, which is practiced in more than 58 percent of the funerals today that involve earth burial and 42 percent of cremation services (Cremation Association of North America, 2011), often requires more than a bathing and dressing of the body. Embalming is necessary to accomplish a temporary preservation of the body to allow time for the family to gather—as many as two or three days may be needed. If the body were to remain unembalmed for this length of time, the distasteful effects of decomposition would create a significant problem for grievers.

Though arguments favoring embalming have been presented, it may not always be necessary or desired. Embalming is not required in all states. In many states, for example, if the body is disposed of within 72 hours, if it is not transported on a common carrier or across state lines, and/or if the person did not die of a contagious disease, embalming is not required. If a body is to be cremated and no public viewing is held, embalming would not be necessary. Many consumers just assume that embalming should or must occur.

A frequently asked question is, "If a body is embalmed, how long will it last?" There is no simple answer to this question. This is why funeral directors talk in terms of temporary preservation. Most families are interested in a preservation that will permit them to view the body, have a visitation, and allow the body to be present for the funeral. Beyond that, they are not concerned with the lasting effects of embalming. For an extended discussion of embalming, see P. Ashwood (2009).

FINAL DISPOSITION

Rituals and rites related to final disposition serve the functions of maintaining positive relationships with ancestral spirits, reaffirming social solidarity, and restoring group structures dismembered by death.

In many cultures and religions, individuals are considered to be composed of several elements, each of which may have a different fate after death (Palgi & Abramovitch, 1984). Thus, the actual destruction of the corpse, whether through cremation, burial, or decomposition, is thought to separate the elements—the various bodies and souls. For example, in India, when a son breaks the skull of his father on the funeral **pyre** (a combustible pile for burning a corpse), he is demonstrating that the body no longer has any value—because it is worn out—but the soul lives on (Tully, 1994).

As Gerry Cox and Ronald Fundis (1992, p. 193) stated in their discussion of Native American burial practices, "If nothing else is known, it is that tribal groups did not abandon their dead. They provided them with ceremony and disposal."

Indeed, many of the ceremonies and ways of disposing of the dead tend to go hand in hand with death among different cultures. These disposition rites not only reaffirm group structure but also enhance social cohesiveness.

There are two primary forms of final disposition, earth burial and **cremation**. As we will discuss, the dominant religions that practice earth burial are Judaism, Christianity, and Islam, whereas Buddhism and Hinduism practice cremation. Another practice, which used to be common among Plains Indians in the United States, is **above-ground burial** on scaffolds or in trees. This type of burial is often the first part of **secondary burial** rites, which allows for open-air decomposition of the body, followed by earth burial, and possibly subsequent reburials.

EARTH BURIAL Earth burial as a method of final disposition is by far the most widely used in the United States. It is used in approximately 58 percent of the 2.5 million American deaths annually. Almost without exception, earth burial takes place within established cemeteries. In some instances, earth burial can take place outside of a cemetery if the landowner where the interment is to be made and the health officer with jurisdiction grant their permission. In 1997, Bill Cosby acquired permission from local authorities to bury his son Ennis on the grounds of his estate. By law, cemeteries have the right to establish reasonable rules and regulations to be observed by those arranging for burial in them. A person does not purchase property within a cemetery but rather purchases the right to interment in a specific location within that cemetery. Most cemeteries require that the casket be placed into some kind of outer receptacle or burial vault. The cemetery will also control how the grave can be marked with monuments or grave markers. For an extended discussion of burial laws, see P. Elvig (2009).

Entombment, occurring in less than 5 percent of all final dispositions, might be considered as a special form of earth burial. It consists of placing the body (contained within a casket) into a building designed for this purpose. Cemeteries offer large buildings (**mausoleums**) as an alternative to earth burial or cremation. In some instances, families may purchase the right to interment in a cemetery and on the designated space build a private or family mausoleum that will hold as few as one or two bodies or as many as 12 to 16. Both types of mausoleums must be designed and constructed to provide lasting disposition for the body. Most states and/or cemeteries regulate the specifications and construction of the mausoleum.

Some societies keep the corpses in or near the home of the deceased. The Yoruba of Nigeria (Bascom, 1969), for example, dig the grave in the room of the deceased. In Uganda, a Lugbara male (Middleton, 1965) is buried in the center of the floor of his first wife's hut. The Swazi of Africa (Kuper, 1963) bury a woman on the outskirts of her husband's home.

To assure that the spirit will be reborn, some societies go to great lengths. For example, the Dunsun of northern Borneo (Williams, 1965) kill animals to accompany the deceased on the trip to the land of the dead. The Ulithi in Micronesia (Lessa, 1966) place a loincloth and a ginger-like plant in the right arm of the deceased so that gifts can be presented to the custodian at the entrance of the other world. For the Zinacantecos of Mexico (Vogt, 1970), a chicken head is put into a bowl of broth beside the head of the corpse. The chicken allegedly leads the "inner soul" of the deceased. A black dog then carries the soul across the river.

The Russian Orthodox community in Oregon, mentioned earlier (Morris, 1991), shares this concern for the rebirth of the spirit and constructs the grave site to accommodate that rebirth. All the graves are lined up facing east, with an Orthodox cross at the foot of each grave. At the Second Coming, the dead will rise from the grave and stand next to the cross, facing east, toward Christ.

As with the Zinacantecos of Mexico, a chicken is used in the burial rites of the Yoruba in southwestern Nigeria (Bascom, 1969). A man with a live chicken

Great Mausoleum at Forest Lawn in Glendale, California is the site of many internments of Hollywood "stars," including Michael Jackson, Nat King Cole, Red Skelton, Clark Gable, Carole Lombard, W. C. Fields, Jean Harlow, George Burns, Gracie Allen, and many more.

precedes the carrier of the corpse, plucking out feathers and leaving them along the trail for the soul of the deceased to follow back to town. Upon reaching the town gate, the chicken is killed by striking its head against the ground. The blood and feathers are then placed into the grave so that others will not die. A second chicken is killed, and its blood is put into the grave so that the soul of the deceased will not bother the surviving relatives.

For the Anangu in the northwestern part of south Australia, there have been many social changes in work and residential patterns. Recent burials have been shown to have stark contrasts with traditional mourning and burial practices for the Anangu. Their geographic mobility has forced them to some adapt some of their old values and practices and create new burial practices culturally compatible and supportive of their current geographic mobility (Edwards, 2013).

In the old days of the Wild West in the United States, one was buried "six feet under with his boots on." It is true that graves were six feet deep in earlier periods of U.S. history, but efficiency (and perhaps agnosticism regarding the belief in ghosts) dictates that today graves are less than six feet deep—typically around four-and-a-half feet, with 18 inches of dirt above the top of the casket or vault. With the sealed, heavier caskets of today—often placed within a steel or concrete vault—it is not necessary to place the body so deep in the ground as was the case with the unsealed pine box of an earlier period.

The Kalingas of the Philippines (Dozier, 1967) bury adults in graves six feet deep and three feet wide. The Mardudjara Aboriginal inhabitants of Australia (Tonkinson, 1978) dig a rectangular hole about three feet deep, line the bottom with leafy bushes and small logs, and then place the body inside. Similarly, the Semai of Malaya (Dentan, 1968) dig the grave two to three feet deep.

A less common form of burial is burial at sea or water burial. This form of burial can involve the submerging of the entire body, weighted down to prevent the body from resurfacing. However, more commonly, burial at sea will involve

The picture on the left was taken in Goa Lawah, Bali, Indonesia as the family says prayers to the ocean before they scatter the cremated remains of their loved one. The picture on the right was taken on February 14, 2014. On this occasion Michael Leming's older brother John's cremated remains scattered at sea in Oxnard, California where he used to scuba dive. Brother Doug gave the eulogy written by Michael (who was in Thailand) and was given the honor of returning John to the sea where he was re-united with his parents John and Eleanor and daughter Genny.

the scattering of cremated remains over (fresh or salt) water. For an extended discussion of burial at sea, see D. Stewart (2009).

CREMATION Cremation is the other method of final disposition. As mentioned before, this is the preferred method among Buddhists and Hindus. The practice of cremation among both of these religious groups will be discussed in detail later in the chapter. In the past, many Christians were adverse to cremation because it was associated with Asian and "pagan" religions. In 1886, the Vatican prohibited Catholics to use cremation as a way of body disposition; the ban lasted until 1963. Most current Christian denominations allow cremation even though Orthodox Judaism and Islam forbid the practice (S. Bolt, 2009).

Although earth burial is the dominant practice in the United States, the number of cremations has increased dramatically in the last two decades. In 1989, 16 percent of deaths in the United States involved cremation. In 2011, 42 percent of deaths in the United States involved cremation. Nevada leads the nation with 72 cremations per 100 deaths, followed by Washington, Oregon, Hawaii, and Montana (in descending order, from 70.9 percent in Washington to 65.7 percent in Montana).

States with the fewest cremations are Mississippi (13.8 percent), Alabama (17.2 percent), Kentucky (20.8 percent), Louisiana (21.2 percent), and West Virginia (23.2 percent). In Canada, 62.8 percent of all deaths involve cremation, with as many as 82 cremations per 100 deaths in the province of British Columbia (Cremation Association of North America, 2012). The Cremation Association of North America predicts that in the year 2017, 48.8 percent of all deaths in the United States and 68.2 percent of all deaths in Canada will involve cremation (Cremation Association of North America, 2012).

Asked why they were likely to choose cremation for themselves or a loved one, the respondents, in a random survey of 371 individuals by the Wirthlin Group in

2005 (cited by Cremation Association of North America, 2007), gave the following explanations:

1. Saves money (30 percent)
2. Saves land (13 percent)
3. Simpler (8 percent)
4. Body not in earth (6 percent)
5. Preference of the deceased (6 percent)

According to the Cremation Association of North America (2007), the following major trends are influencing preferences for cremation:

1. People are dying older.
2. Migration to retirement locations is increasing.
3. Cremation has become acceptable.
4. Environmental considerations are becoming more important.
5. Level of education is rising.
6. Ties to tradition are becoming weaker.
7. Regional differences are diminishing.
8. Religious restrictions are diminishing.
9. There is greater flexibility in memorialization services.

Until recently, the **crematory** was generally located within the cemetery. With the increase in cremation as an option for final disposition, however, many funeral homes have now installed crematories. There is a trend in the funeral industry to change the name of the establishment from "funeral home" to "funeral and cremation services." According to the Cremation Association of North America (CANA), by 2011 the percentage of American deaths involving cremation had increased to 42 percent. CANA predicts that number to go up to 58 percent by 2015 (Cremation Association of North America, 2011).

© Mike Baldwin / Cornered

"It's our newest cremation model.
Tank sold separately."

Baldwin, Mike/CSL, CartoonStock Ltd

Crematories generally require containment of the body in an appropriate casket or other acceptable rigid container. The containerized body is not removed or disturbed after it arrives at the crematory and is placed into a furnace or retort. Cremation is accomplished by the use of either extreme heat or direct flame. In either instance, reducing the casket (or alternative container) and the body to "ashes" takes two to three hours. Cremated remains do not have the appearance or chemical properties of ashes; they are primarily bone fragments. Some crematories process cremated remains to reduce the overall volume; others do not. Depending on the size of the body, cremation results in three to nine pounds of remains (National Funeral Directors Association, 1997).

Many people find that cremation opens up a wider range of memorialization options than traditional burials. Cremated remains can be collected, put into an **urn** or box, and then disposed of according to the wishes of the family—the **cremains** may be buried in a family plot or cemetery, placed into a **niche** in a **columbarium** (a special room in a cemetery), or kept in another place of personal significance, such as the home or church crypt. Subject to some restrictions, and with appropriate permission from property owners, cremated remains can be scattered by air, over the ground, or over water at a favorite location. Some cemeteries provide areas for scattering and may provide a space where families can place a commemorative plaque or other memorial. Recent alternatives involve having the cremains divided and incorporated into keepsake urns, jewelry, and other items so

Michael R. Leming

The columbarium provides families with an additional option for body disposition. For approximately 25 percent of the 2.43 million annual deaths in the United States, cremation is used as the method of final disposition, and approximately 10 percent choose to place the cremated remains in a columbarium.

TABLE 10.1	CREMATED REMAINS DISPOSITION AS REPORTED BY FUNERAL HOMES AND CREMATION SOCIETIES

Disposition	Percent
Taken home	38.7
Buried	36.6
Scattered on water/land	21.7
Placed in columbarium	3.1
TOTAL	**100**

Source: Cremation Association of North America. (2006) Disposition Survey. Chicago, IL: Author.

that multiple family members may keep a loved one close. Table 10.1 and Table 10.2 show a breakdown of the disposition of cremated remains in 2006 (current information is not available).

Many people choose to memorialize the site of disposition because they find consolation in knowing that there is a specific place to visit when they wish to remember and feel close to the person they have lost, regardless of whether the deceased person's remains are actually located at that place. Families should always choose an option that best fits their emotional needs.

One might assume that cremation would be the least expensive form of final disposition because the typical cost of a simple "no-frills" cremation is approximately 40 percent of the cost of the traditional funeral service with burial. However, a cremation service may be as simple or as elaborate as family members wish. Some people are surprised to learn that cremation does not preclude a funeral with all of the traditional aspects of the ceremony. Visitation and viewing with a funeral ceremony and church or memorial services are options to be considered. In some states, funeral homes are permitted to rent caskets for viewing and services. It is entirely possible to spend more money on a funeral involving cremation if plans include a traditional funeral that includes viewing a body in a casket and placing the cremated remains into an urn in a niche in a columbarium. This fact is support by the research findings of L. Kellaher, D. Pendergast, and J. Hockey (2005), which claimed that increasingly, families who select cremation as a means for disposition are burying cremative remains in cemeteries or placing them in columbaria—a much more costly alternative to scattering and disposal.

TABLE 10.2	CREMATED REMAINS DISPOSITION AS REPORTED BY CEMETERIES

Disposition	Percent
Scattered on dedicated property	1.7
Buried in cemetery	79.3
Placed in columbarium	10.6
Placed in common grave (ossuary)	8.4
TOTAL	**100**

Source: Cremation Association of North America. (2006) Disposition Survey. Chicago, IL: Author.

In Minnesota, Michael Leming's experience is noted thusly: Simple cremations (even by so-called cremation societies) may say that they cost $1,395, but the cremation providers know that that this price is just to get people to come through the doors. When some people realize how little they get from a "simple cremation" or "disposal service," they often upgrade to something more. I talked with my local funeral director, and he said, "When people come to my establishment and say, 'cremate my father and I will pick up the "ashes" in a week,' the cost for the family is $2,500, typically. If they want a full service at a church with the funeral home involved and perhaps a visitation at the chapel of the funeral home, the cost is $4,000. This compares with a traditional funeral at this funeral home, which is about $10,000.

So a "simple cremation" is about 25 percent of the traditional funeral service, while a memorial service directed by the funeral home is 40 percent of the traditional funeral service. As an analogy, if a local home improvement center advertises "solar clothes dryer $165," this sounds like a great deal until one learns that the product is actually an outdoor clothesline kit with clothes pins. As in most things, "one gets that which one pays for."

According to Thomas Lynch, an undertaker and poet (cited by Austerlitz, 2013):

> I think Americans figured out that if you can dispatch the body (through a thrifty disposal service), it is more convenient, emotionally, spiritually, and financially. It's easier in the near term. But I believe that my father had it right when he said, "You can pay the shrink, you can pay the bartender, or you can pay the funeral director." The dead are going to exact their pound of our spiritual and emotional flesh. And I think one of the things that makes it more bearable is if we do our duty to the dead.
>
> If you have a bunch of people gathered around an open piece of ground, and you lean [a shovel] towards someone, you don't have to give them the operating instructions. They know what to do with it! This is coded right in our genes as humans. I mean literally, we are part of the humus. This is us. We are people who dig holes in the ground and fill them.
>
> It's the hardest thing you're going to do, to carry and bury the body of someone you care deeply for, or are long associated with. Because when we bury someone like that, we're burying part of our history, too. But when you do it, you feel, "I did what I could. Someone will do it for me."

There are other forms of body disposition discussed by S. Bolt (2009), but the frequency of these alternatives is extremely limited in the United States. These alternatives include mummification, body donation to science, plastination, water resolution, cryonics, and cannibalism. Briefly, mummification involves dehydration of the body and is a permanent condition where the body may be placed on display. Examples of this would be a Roman Catholic saint who might be found in a glass sarcophagus under a church's altar or the body of Jeremy Bentham, which is kept in the University College of London.

> As requested in his will, his body was preserved and stored in a wooden cabinet called an "Auto-icon." Originally kept by his disciple Thomas Southwood Smith, it was acquired by University College London in 1850. It is normally kept on public display at the end of the South Cloisters in the main building of the college, but for the 100th and 150th anniversaries of the college, Bentham was brought to the meeting of the College Council, where he was listed as "present but not voting."

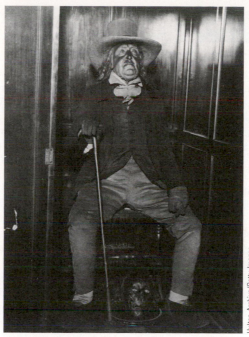

Jeremy Bentham: Present but not voting.

The Auto-icon has a wax head, as Bentham's head was badly damaged in the preservation process. The real head was displayed in the same case for many years between Bentham's feet, but became the target of repeated student pranks, including being stolen on more than one occasion. It is now locked away for security reasons.

Body donation to science, as the name implies, involves delivering to a medical or academic institution one's body by bequest. The reasons of such donations are for academic, medical, or scientific purposes. After the body has served its function it is cremated or buried, usually by the institution that has received the bequest. Typically family members are not involved in the final disposition, but this need not be the case. In Thailand, when a body is received by the medical school for the purpose of educating medical students, students refer to the cadaver as "Ajarn Yai" or "Big or respected teacher."

Plastination is a process developed by Gunther von Hagen in Heidelberg, Germany in 1977 to replace the natural body fluids with plastic to preserve human tissue. The outcome of this process allows the body to be preserved for an indefinite period of time. Gunther von Hagen's *Body Worlds* is a mobile museum of human anatomy that has traveled the world since 1995. Another similar show, created by Roy Glover and entitled *Bodies*, opened in Florida in 2005. Both exhibits have attracted millions of visitors and much controversy (Bolt, 2009).

Water Resolution is a relatively newer procedure by which the body is reduced to bone ash, similar to cremation, but involving chemicals rather than burning or intense heat.

| WATER RESOLUTION—A GREENER ALTERNATIVE TO CREMATION AND EARTH BURIAL

Water resolution is an accelerated version of the natural process of hydrolysis-driven decomposition after burial. It is accomplished by the application of alkaline hydrolysis to human cadavers to produce a pure ash bone shadow. The individual body is essentially placed into a horizontal pressure vessel, and then a fully automated process of pressure, high temperature, and alkalinity accelerates the natural process of tissue hydrolysis back into the building blocks of life. The product is sent back to the environment to be recycled.

The alkaline hydrolysis process has successfully been used worldwide for more than 12 years in laboratory and research applications and is a natural process. Bodies that are buried in the earth are degraded by alkaline hydrolysis, expedited by the soil bacteria. This is a very slow process. Food in the intestine is digested into usable nutrients by alkaline hydrolysis, expedited by enzymes that operate at a pH of 7 to 8 at body temperature. This is a moderately fast process for relatively small amounts of tissue. The alkaline hydrolysis process uses strong alkali (pH 14) to solubilize and hydrolyze tissue, expedited by heat at 150°C in a pressurized vessel. This process generates a solution of amino acids, peptides, sugars, and soap (salts of fatty acids) that is suitable for release to drain, applied to land as fertilizer, or recycled in many ways. Also produced are pure white bone shadows (ash), which may be easily powdered and given to the relatives as in cremation. The powder is 100 percent specific to that corpse.

Water Resolution is a "green" and genuine alternative to cremation and burial for the following reasons:

- Accelerated natural decomposition—returns organic elements to the ecosystem
- No mercury emissions, no scrubber abatement required, no contamination
- No burning of caskets—reduces CO_2 production
- Low carbon footprint—20 times less CO_2 emissions versus average cremation
- Energy efficient—one-tenth the energy per body versus cremation
- The water resolution process is three to five times less expensive than cremation
- Lower operating costs
- Embalming fluid neutralized and cytotoxic drugs destroyed
- Pacemakers can be left in, unlike cremation
- Compact units easily adapt to existing crematorium setup
- Titanium medical implants recovered, intact and sterile, for possible reuse in Third World countries
- Ashes may be returned as in cremation, but unlike cremation in that the ashes produced in the water resolution process are 100 percent separated from other bodies
- As nature intended, the body's organic building blocks of life are returned back to the earth

Presently, the major disadvantage to the water resolution process is that the cost of installation for the funeral professional is about double that of installing a traditional cremation retort chamber—approximately $200,000 compared with $100,000 (not including space).

Cryonics involves the preservation of the body through freezing to stop physical decay. Typically this process involves liquid nitrogen, and the body is cooled down to a temperature of –196°C (Bolt, 2009). Cryonic supporters hope that in the future advanced technology will be able to revive the "patient" and provide a cure for the individual who has been suspended in time. Like body donation, this form of body disposition is not thought to be permanent.

Cannibalism involves the eating of the human body as a form of body disposition. There are two forms of cannibalism. The first is exocannibalism, where the group eats the flesh of outsiders (or enemies), and the second is endocannibalism, where one consumes the body from one's own group. The latter involves a ritual that attempts to affirm group unity and maintain social solidarity that was threatened by the loss of one of the group's members. Exocannibalism is a ritual of humiliation or disrespect, whereas endocannibalism involves respect and honor for the deceased.

CASE STUDY: BURIAL AND MORTUARY PRACTICES OF NATIVE AMERICANS LIVING ON THE PLAINS

There is probably more generalized knowledge of Plains tribes than of most Native Americans, since they are the ones most often portrayed in movies and television. Unlike the more sedentary tribes of the Southwest, the Plains tribes were mobile on a large scale. With the great temperature variations of the Great Plains, the inhabitants needed to adapt to all kinds of climatic changes. They were primarily dependent upon the bison as a source of food, clothing, and shelter. Other animals and plant life were also major sources of food, but the bison offered the most dramatic picture of the life of the Plains tribes. There were sedentary tribes, but the various groups that spoke the Siouan language were the hunters and nomads of film and television. The coming of the horse, following the arrival of the Europeans, added greatly to their prowess as warriors and as hunters. A warlike tribe, the Sioux have made their name in history by fighting against the European Americans in famous battles such as Little Big Horn and suffering at their hands in the massacre of Wounded Knee. The former was recently portrayed in Kevin Costner's film *Dances with Wolves* and earlier in the film *Little Big Man*, which starred Dustin Hoffman.

The eight Sioux tribes (seven main tribes and one, the Assiniboin, that was outside of the loose confederation) had relatively similar burial and mortuary practices. Like many other western tribes, the Plains tribes believed that everything in the world around them was filled with spirits and powers that affected their lives, whether from the sun, the mountains, the buffalo, or the eagle (Capps, 1973).

The Sioux feared the dead and would burn the dwelling of the deceased, forbid using the individual's name, and bury personal goods with the corpse to keep the ghost of the deceased from coming along to live with friends and relatives (LaFarge, 1956). An individual was expected to pursue honored roles as he or she progressed through life and to learn how to play his or her roles well. The roles included a spirit of generosity, which meant giving to others from birth to death (Malan, 1958). This spirit meant that the Dakota cared for their dead with comradeship. Yet, death in old age was not feared nor were the ghosts of dead persons, who were often thought to remain for a time after their death (Spencer & Jennings, 1965). The Lakota practiced "ghost keeping" ceremonies to try to keep the soul of the deceased on earth to purify it and to ensure that it would return to its creator (Hirschfelder & Molin, 1992). To keep a ghost required that a family endure a great sacrifice and would ultimately give away all of their personal possessions to the needy in memory of the ghost (Powers, 1977).

The Sioux took the position that death will occur to all regardless of one's achievements, fame, wisdom, bravery, or whatever, and that the mortuary practices allowed the living a way of showing their reverent respect for the dead (Hassrick, 1964). The Dakota (Lakota or Sioux) tribes would prepare a tipi to honor the deceased. In front of the tipi, they would place a rack upon which robes and articles of clothing would be displayed, while inside the tipi mourners would prepare themselves for their bereavement (Hassrick, 1964). If the deceased was a young person, particularly a child, the mourners would gash their arms and legs and engage in ritual crying (Spencer & Jennings, 1965). When death occurred in the home, the burial would be delayed for a day and a half in hopes that the deceased might revive (Hassrick, 1964). The body would be dressed in the finest available clothes, provided by a relative if the deceased had none. The corpse would be wrapped tightly in robes, with the weapons, tools, medicines, and pipe included with the corpse. Then the bundle would be placed on a scaffold for air burial, with food and drink placed beneath the scaffold (Spencer & Jennings, 1965).

Some Dakota or Sioux groups used earth burial. There is evidence that in earlier times they used mound burial (Spencer & Jennings, 1965). During the winter, when scaffolds could not be built, trees were often used for burial (Hassrick, 1964). After the body was prepared and properly wrapped, the adult members of the family began *wacekiyapi*, or a worship rite for the deceased in which men might run pegs through their arms and legs, women might slash their limbs and cut off their little fingers at the first joint, and both men and women might cut their hair and express their grief by singing, wailing, or weeping (Hassrick, 1964). The favorite horse of the deceased would be killed beneath the scaffold of its owner and its tail tied to the scaffold, and the mourning would continue for as long as a year (Powers, 1977).

By placing the corpse in a tree or scaffold, the Dakotas and other similar tribes believed the soul would then be free to rise into the sky if the person died of natural causes. If the person died in battle, the Dakotas would often leave the person on the plains where he was slain to rise into the sky (Capps, 1973). For the Dakota or Sioux, the spirits of the dead are not gone and lost to humankind but rather continue to exist here and can be reached by the living for support and aid (DeMallie & Parks, 1987). The Lakota also have a memorial feast, which is held around the time of the anniversary of the death. This ceremony ends the mourner's duties, and the final giveaway will occur at this time (Hirschfelder & Molin, 1992). The ghosts of the dead appear at their will and communicate with the living (Powers, 1977).

There is evidence that Plains tribes used all known methods of disposal of the dead, including burial (both ground and air), cremation, and mummification. It is also probable that the cause of death, where the death occurred, and the age, sex, and social status of the deceased person had an impact on the mortuary and burial practices of the tribe, but information about how such factors influence burial practices is not conclusive.

Evidence suggests a general pattern that the tribes exhibit a fear of the dead. It is also likely that climate, weather, availability of materials to dispose of the body, and religious beliefs were major determinants in how bodies of the dead were disposed. Burial practices also seemed to remain stable for a remarkably long period of time among the tribes (Voegelin, 1944). Almost universally, tribes provide provisions for a spirit journey, whether for a single or for a group burial (Atkinson, 1935). If nothing else is known, it is clear that tribal groups did not abandon their dead. They provided them with ceremony and disposal.

(*Adapted from an article written by Gerry Cox and used with his permission.*)

INFLUENCE OF SOCIAL POSITION ON DEATH-RELATED BEHAVIORS AND RITUALS

A deceased person's social position and gender will often determine the way the body is prepared for burial and the final disposition of the corpse. For example, a deceased Buddhist common person in Thailand (Leming & Premchit, 1992) is cleaned, dressed, and placed into a casket. A high-status Buddhist individual, however, is embalmed and then bathed and dressed with new clothes. The face is covered with gold leaves before the body is placed into the casket.

In Greece (Brabant, 1994) the majority of individuals cannot afford permanent burial; thus, they rent a grave for three years, and then the remains are **exhumed** and the bones are placed into the "bone room." More prosperous individuals in Greece can afford to pay to have their remains stay buried permanently. The Death Across Cultures box about the Kapauku Papuans gives details about position in the community determining how and where burial or final disposition will occur.

Gender is also a factor. Among the Abkhasians near the Black Sea (Benet, 1974), women are buried 10 centimeters deeper than men. The Barabaig of Tanzania (Klima, 1970) place the bodies of women and children out into the surrounding bush, where they are consumed by hyenas. Only certain male and female elders will receive a burial.

In examining men's and women's 18th- and 19th-century gravestone epitaphs in cemeteries in the northeastern United States, Tarah Somers (1995) revealed major differences in the social expectations of men and women. Women were described in more passive and private terms ("meek and affectionate" and "joyfully

---

DEATH ACROSS CULTURES | BURIAL RITES AMONG THE KAPAUKU PAPUANS

Among the Kapauku Papuans of western New Guinea, burial rites are determined by the deceased's status and cause of the death. The simplest burial is given to a drowned man, whose body is laid flat on the bank of the river and protected by a fence erected around it. The body is then abandoned to the elements. Very young children and individuals not particularly liked and considered unimportant are completely interred. Children, women, and elderly persons who were unimportant but loved are tied with vines into a squatting position and semi-interred with the head above ground. A dome-shaped structure of branches and soil is then constructed to protect the head.

A respected and loved adult male among the Kapauku Papuans receives a tree burial. Tied in a squatting position, the corpse is placed in a tree house with a small window in front. Corpses of important individuals, who are feared by their relatives, and of women who died in childbirth require a special type of burial. Their bodies are placed in the squatting position on a special raised scaffold constructed in the house where death occurred. The house is then sealed and abandoned.

The most elaborate burial among the Kapauku Papuans is given to a rich headman. A special hut is built on high stilts, the body is tied in a squatting position, and a pointed pole is driven through the rectum, abdomen, chest cavity, and neck with its pointed end supporting the base of the skull. The body is then placed in the dead house with the face appearing in the front window of the structure. The body is pierced several times with arrows to allow the body fluids to drain away. Years later, the skull of the respected man may be cleaned and awarded a second honor of being placed on a pole driven into the ground near the house of the surviving relatives.

From *The Kapauku Papuans of West New Guinea*, by L. Pospisil, 1963, New York: Holt, Rinehart and Winston.

departed life"), whereas men were often memorialized with active and public terms ("skillful and valiant in truth" and "triumphant at the approach of death").

Thus, in death as in life, one's social status and gender determine how one is treated. Whether one is buried or left to the elements is often determined by gender, age, standing in the community, and cause of death. If one is buried in the ground, even the depth of burial may vary by one's social position. For an extended discussion of the relationship between social class and death-related behavior, see M. Kearl (2009).

DEATH RITUALS OF MAJOR RELIGIOUS GROUPS

Whatever their differences, religious rituals for the dead are always communal events. Feeding the survivors and the telling of stories cut across all traditions.

JEWISH CUSTOMS

Jewish mourning rituals focus more on the bereaved than on the body. By custom, Jews try to bury their dead within 24 hours—if possible, without embalming—in a plain, wooden coffin. Traditional Jews do not put the body on view or have it cremated.

PRACTICAL MATTERS | JEWISH GROUP BURIES ITS OWN

When a Jewish congregation here first began the practice of offering simple, inexpensive burials for its dead, some members were upset. But they now increasingly condone it. Rabbi Arnold M. Goodman, spiritual leader of Adath Jeshurun Congregation, says volunteers of its society to honor the dead—*Chevra Kevod Hamet*—now handle about half the funerals of members. The *Chevra* was formed in 1976 after Goodman, in a sermon, dealt with the impact of American values upon the funeral practices of Jewry. He suggested a committee study the requirements of the *Halacha*, or Jewish law, for responding to death.

Months of study convinced committee members that a simple wooden coffin should be used, the body should be washed in a ritual process called *tahara*, and because dust is to return to dust as quickly as possible, there should be no formaldehyde in the veins and no nails on the coffin. The society decided to offer traditional funerals free to Adath Jeshurun members. The congregation provided seed money. Memorial donations and voluntary contributions from the bereaved are accepted.

Here's how the *Chevra* functions: When death occurs, *chaverim* (friends) call on the family, aid in writing the obituary, explain death benefits, aid in other ways, and remain available for help. *Chevra Kadisha* (sacred society), people of the same sex as the deceased and usually five in number, wash the body at the mortuary while saying prayers. The body is dressed in a shroud sewn by *Chevra* members and placed in a wooden coffin.

Shomrin (guards) watch over the body, in blocks of two hours, until burial. The coffin with rope handles is light enough to be borne by pallbearers, including women. Spurning mechanical contrivances, the pallbearers lower the coffin into the grave. *Chaverim*, the rabbi, and cantor shovel in dirt. Family members may participate.

Judaism historically insists that the greatest commandment is to take personal involvement in burying the dead, Goodman says, but affluence enables people to pay surrogates to do it. Goodman says a funeral costs the *Chevra* less than one-third the normal price charged by a local funeral home director.

From Jewish group buries its own, June 25, 1982. Copyright 1982 Associated Press.

According to P. S. Knobel (1987), traditional Jewish burial customs require that the body be cleansed by members of the Jewish burial society (*hevra'qaddisha'* or "holy society") in a washing process called *tahorah*, or purification. Custom forbids embalming, cremation, and autopsy unless local laws require these procedures. The body is dressed in plain linen shrouds (*takhrikhim*); men are usually buried

Jewish Funeral Rituals.

with their prayer shawls (*tallit*). The body is then placed into a plain wooden casket and buried before sunset on the day of death, if at all possible. Reform Judaism allows for cremation and entombment, but burial is the most frequent form of body disposition. Throughout this process, it is considered inappropriate to use the funeral as a means for displaying one's social position and wealth.

For the bereaved, the Jewish mourning ritual begins by the rending (tearing) of garments. Survivors cut their clothing with a razor—on the left for a parent; on the right for a spouse, child, or sibling—to symbolize the tear in life that death has produced. For some, the ripping of a black ribbon, which is then attached to the clothing, has symbolically replaced the process of rending garments.

After a ritual healing meal, *shiva* begins. For the first week, men do not shave, survivors are not supposed to wash their whole bodies, and the entire family receives visitors while sitting on the floor or on low chairs. From the death until burial, mourners are exempt from normal religious obligations (e.g., morning prayers) and must not engage in the following activities: drinking wine, eating meat, attending parties, and engaging in sexual intercourse (Knobel, 1987).

The liturgy for the funeral consists of the recitation of psalms, a eulogy, and the following *El Male' Rahamin* memorial prayer:

> O God full of compassion, You who dwell on high! Grant perfect rest beneath the sheltering wings of Your presence, among the holy and pure who shine as the brightness of the heavens, unto the soul of [name of the deceased] who has entered eternity and in whose memory charity is offered. May his/her repose be in the Garden of Eden. May the Lord of Mercy bring him/her under the cover of His wings forever and may his/her soul be bound up in the bond of eternal life. May the Lord be his/her possession and may he/she rest in peace. Amen. (Knobel, 1987, p. 396)

During the interment service the body is lowered into the grave and covered with earth. The interment service consists of an acclamation of God's justice, a memorial prayer, and the recitation of *Qaddish*—a doxology reaffirming the mourner's faith in God despite the fact of death. After the burial service, the people in attendance form two lines between which the primary mourners pass. Those present comfort the mourners as they pass, saying, "May God comfort you among the rest of the mourners of Zion and Jerusalem" (Knobel, 1987, p. 397).

After the *shiva*, mourners continue to avoid social gatherings for 30 days after the death. When one is mourning the death of a parent, the restrictions are observed for one year. After one year, all ritual expressions of grief cease with the exception of the *Yahrzeit*—the yearly commemoration of the person's death. *Yahrzeit* is observed by lighting a memorial light, performing memorial acts of charity, and attending religious services to recite the *Qaddish* prayer (Carse, 1981; Knobel, 1987). A contemporary description of Jewish bereavement by M. Bukiet (2005) provides empirical evidence for the saliency of these issues, even though the article was written by a nonobservant Jew.

CHRISTIANITY

The Christian funeral service is primarily a worship service (or Mass of Christian burial for the Roman Catholic) that reflects the twin themes of victory and loss (described in Chapter 4). During the service, hymns are sung and scriptural

Christian Funeral Rituals.

passages are read to emphasize the resurrection of the dead and to provide consolation for the bereaved. It is common for a eulogy or biographical statement concerning the deceased to be read.

In the United States, Christian teaching does not discourage the process of embalming, nor does it prohibit autopsy, cremation, or any other form of final disposition (as is the case among traditional Jews)—provided that such practices do not indicate a rejection of a belief in the resurrection of the body. American Christian funerals are conducted by members of the clergy in a church, funeral chapel, and/or cemetery. Memorial services—religious services in which the dead body is not present—are becoming increasingly popular in many Protestant churches. At the occasion of death, the family of the deceased is expected to disengage from most normal social functioning until after the funeral. Funeral arrangements are typically made with the professional assistance of a funeral director and/or member of the clergy. For most Roman Catholics and many Protestants, on the day before the funeral, a wake or visitation service will be held at the funeral home. During this time (approximately five hours in duration), friends may view the

I was introduced to death early in life. In the Roman Catholic grade schools of my youth, funerals were part of the informal curriculum. When a classmate's parent died, we all assembled for the funeral mass, passing by the (usually) open casket and sharing—as best we could—the sorrow of the grieving family. Occasionally, it was a fellow pupil lying in the casket, snatched from life by an accident or—like my closest fifth-grade friend—from an illness, in his case a fatal epileptic seizure. What brought us together, young and old, was sacred ritual.

The liturgy that gathered and directed our emotions was familiar and color-coded. For funerals, the priest wore black vestments symbolizing death, just as he wore red for feast days of the martyrs and—on ordinary Sundays—green for the hope that all Christians have of life eternal. The music varied, too. Long before I learned a modern language, I knew by heart the somber Latin funeral hymn "Dies Irae" ("Day of Wrath"). I knew, too, that the clouds of incense billowing around the body were meant to honor flesh that was soon to turn to dust. By such appeals to the senses we children were inducted into the abstract mystery of death—our own as well as that of others. Sad? Yes. But never morbid. Death is real, the liturgy instructed us. But so is the promise of Resurrection.

From The ritual solution, by K. L. Woodward with A. Underwood, (1997, September 18), *Newsweek*, p. 62

body (if on view) and visit with the family of the deceased. Roman Catholic families may also have a rosary service and/or prayer service during the wake.

The funeral is typically held two to four days after the death. Occasionally, the funeral is delayed if family members are unable to make travel arrangements on such short notice (this would not be the case for a traditional Jewish or Islamic funeral). If final disposition involves cremation, the cremation can take place either after the funeral or before the memorial service. Burial and entombment dispositions are usually accompanied by rites of committal. On those occasions when there is no graveside service, words of committal will be read at the conclusion of the funeral service. At the conclusion of the funeral process, family members and others who have attended the services are often invited to share a meal together. This becomes a community rite of reincorporation.

HINDUISM

When a death takes place in Hindu society, the body is prepared for viewing by laying it out with the hands across the chest, closing the eyelids, anointing the body with oil, and placing flower garlands around it. This is done by individuals of the same gender as the deceased, and this process will be presided over in the home of the deceased by the dead person's successor and heir (Habenstein & Lamers, 1974).

Because Hindus believe that cremation is an act of sacrifice whereby one's body is offered to God through the funeral pyre, cremation is the preferred method of body disposition. In preparation for cremation, family members construct a bier consisting of a mat of woven coconut fronds stretched between two poles and supported by pieces of bamboo. The uncasketed body of the deceased is borne on the bier from the deceased's home to the place of cremation by close relatives. This funeral procession is led by the chief mourner—usually the eldest son—and

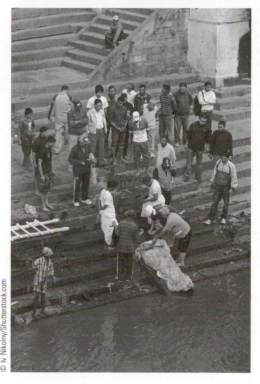

Hindu Funeral Rituals.

includes musicians, drum players, and other mourners. The wife of the deceased always remains behind in the home (Habenstein & Lamers, 1974).

    According to R. W. Habenstein and L. M. Lamers (1974), when the body reaches the place of cremation, usually a platform (*ghat*) located on the banks of a sacred river, the body will be removed from the bier and immersed in the holy waters of the river. During this process, a priest will perform a brief disposal ceremony. The body will then be smeared liberally with *ghi* (clarified

| DEATH ACROSS CULTURES | THE SATI |

Though not the common practice, in 1987 an 18-year-old woman in India rested the head of her dead husband on her lap, sat on his funeral pyre, and was burned alive (Salamat, 1987). No one could stop her from committing **sati**, the most noble act of loyalty that a widow can perform for her husband, who, in the Hindu religion, is supposed to be considered a god by his wife. A widow cannot remarry and is treated as an evil omen and an economic liability. If she chooses to live, she has to remain barefoot, sleep on the floor, and can never go out of the house because she would be slandered if seen talking to a man. Some would argue that she is better off dead. Before this widow's act, however, sati had not been performed in that community since early in the 20th century.

From A young widow burns in her bridal clothes, by A. Salamat, 1987, *Far Eastern Economic Review, 138*, pp. 54–55.

butter) and placed on the pyre for burning. At this point, the chief mourner, who has brought burning coals from the house of the deceased, lights the pyre, and a priest recites an invocation similar to the following:

> Fire, you were lighted by him, so may he be lighted from you, that he may gain the regions of celestial bliss. May this offering prove auspicious. (Habenstein & Lamers, 1974)

After the body has been consumed by the fire and only fragments of bones remain, the mourners will ritually wash themselves in the river in a rite of purification. They will then make offerings to the ancestral spirits of the deceased. Upon completion of this duty they will recite passages from the sacred texts.

Three days after this ritual, a few relatives of the deceased will return to the cremation site to gather the bones. A priest will again read from the sacred texts and sprinkle water onto the *ghat*, while any remains of the deceased will be placed in a vase and given to the chief mourner. It is then the obligation of the chief mourner to cast these remains into the Ganges or another sacred river (Habenstein & Lamers, 1974).

Between 10 and 31 days after the cremation, a *Shraddha* (elaborate ritual feast) is prepared for all mourners and priests who have taken part in the funeral rituals. During the *Shraddha*, gifts are given to the *guru* (religious teacher), the *purohita* (officiating priest), and other Brahmins (religious functionaries).

The social status of the family will determine how elaborate the *Shraddha* will be—for the poor this ritual will last eight to 10 hours, whereas the wealthy may give a *Shraddha* lasting several days. At the close of the *Shraddha*, the mourning period officially ends, even though later *Shraddhas* may be given as memorials (Habenstein & Lamers, 1974). Although it is believed that these ritual meals provide nourishment to the spirit of the deceased in its celestial abode, from a sociological perspective they serve as rites of family reincorporation and also differentiate the family regarding social status.

According to Habenstein and Lamers (1974), next to Hindu wedding ceremonies, funerals are the most important religious ceremonies. Although the funeral process is very costly to families, often resulting in their impoverishment, not providing the *Shraddha* creates more social problems for families than do the financial consequences of these rituals.

BUDDHISM

As the religious teachings of Buddha were disseminated throughout Asia, the beliefs and practices were adapted to indigenous cultural traditions dealing with death. Consequently, it is not possible to discuss Buddhist funeral rituals per se. Rather, there are Japanese, Korean, Chinese, and Thai funeral customs practiced within a Buddhist context.

In general, all Buddhist funeral ceremonies have some similarities. At the Buddhist temple, priests assist families as they engage in this important rite of passage. Prayers of the priests illustrate the "lesson of death"—that life is vanity. At funerals, Buddhist priests often read the following words from the Buddha:

> The body from which the soul has fled has no worth. Soon it will encumber the earth as a useless thing, like the trunk of a withered tree. Life lasts only for a moment. Birth and death follow one another in inescapable sequence. All that live must die. That man indeed is fortunate who achieves the nothingness of being. All animal creation is dying, or dead, or merits to be dead. All of us are dying. We cannot escape death. (cited by Habenstein & Lamers, 1974, p. 97)

For most Buddhists, cremation is the preferred form of body disposition, but earth burial is also frequently practiced. In Buddhism, unlike Hinduism, there is no "soul"—both the body and the idea of a soul distract from the proper meditation and attainment of nirvana. Cremation serves the function of promoting the process of liberation of the individual from the illusion of the present world.

During and after the funeral, family members will make offerings through the priests to the spirit of the deceased. They will also give ritual feasts for the priests and other mourners. As in other religious traditions, all of these funeral activities will emphasize the importance of the religious worldview, promote community cohesiveness, and reincorporate chief mourners into routine patterns of social life.

Michael Leming has lived for 25 years in Northern Thailand. During his residence, he has had a number of opportunities to observe the Thai Buddhist funeral. As in rituals of any culture, there are always some variations, yet the following description is generally accurate in most situations.

When a typical Buddhist person dies, the body is cleaned, dressed, and placed in a casket. The casket is kept either within the home or at the *wat* (temple) for a period of three days. During this period, monks come every evening to chant the Buddhist scriptures (*Abhidhamma*). Friends attend these services and offer gifts of floral tributes. On the fourth day, the body is taken to the charnel-ground (cremation site), which is usually a good distance from the *wat*.

The procession is a significant rite in the Buddhist funeral. The casket is put on a carriage and taken from the *wat* compound to the cremation site. During the procession, a long white cotton cord is attached to the casket and eight (the number is flexible) monks, together with lay devotees, carry the cord. In urban areas the carriage is motorized, whereas in traditional funerals and funerals in rural areas the carriage is pulled by walking members of the procession.

After arriving at the site of cremation, the relatives pose for pictures by the casket. Walking around the casket three times (which symbolizes the traveling in the cycle of death and rebirth), the relatives then place the casket in front of the crematorium. Many times young male family members are ordained as monks for a

Michael R. Leming

After arriving at the site of cremation, the relatives pose for pictures by the casket. Walking around the casket three times, they then place the casket in front of the crematorium. Many times, young male family members will be ordained as monks for a period of a few days during the funeral ceremonies. The merit earned by these young men can be dedicated to benefit the dead as well.

period of a few days during the funeral ceremonies. In addition, the merit earned by these young men can be dedicated to benefit the dead.

As a final merit-making rite, five or 10 important persons come forward one by one and place a set of yellow robes on the long white cotton cord, which is linked to the coffin. The most senior monk attending the service collects the robes after "contemplating symbolically the dead." According to religious practice in Buddhism, the contemplation of the corpse by Buddhist monks will bring merit to those who provide opportunity for the monks to do so. The merit earned can be dedicated further to the dead as well.

The stage is now set for the actual cremation. Just prior to the lighting of the fire, the biography of the deceased is read while *dok mai chan* is distributed to all in attendance. The *dok mai chan* is a sandalwood flower with one incense stick and two small candles attached. The mourners are all invited to come forward and deposit the *dok mai chan* before the casket. By so doing, mourners are deemed participants in the actual cremation.

The chairman of the ceremony then ignites the fire, and the casket is consumed. The next morning, the ashes (several pieces of bones) are gathered and made into a shape of a human being with the head facing east. Four monks attend this ritual, which culminates when the ashes are placed into a receptacle. Afterward, the ashes are enshrined in a reliquary built in the compound of the monastery.

Throughout the funeral ceremony, sorrow or lamentation is not emphasized. Rather, the focus is upon the impermanence of all things. The funeral is primarily a social event that affirms community values and group cohesiveness. Furthermore, it is a time when people are expected to enjoy the fellowship surrounding the rituals. Typically there is entertainment (dancing and musical performances) associated with funeral rituals for sorrow and loneliness to dissipate. It is believed that this assists the bereaved to conceptualize a happy and pleasant paradise in which the deceased will reside. For a contemporary study of Mahayana Buddhist rituals related to death, see N. Onishi (2008). This article demonstrates that when it

Michael R. Leming

On the day of the cremation, Buddhist mourners gather for a final ceremony. In the photo on the right lay participants in the service have placed robes in front of the casket to given to the monks and have dedicated the merit for this act to the person who has died. The monks then collect the robes and go back to their places.

comes to funerals, the Japanese have traditionally been inflexibly Buddhist—so much so that Buddhism in Japan is often called "funeral Buddhism," a reference to the religion's near-monopoly on the elaborate, and lucrative, ceremonies surrounding deaths and memorial services.

ISLAM

Islamic tradition requires that the funeral service take place without unnecessary delay and that burial rites be simple and austere (Rahman, 1987). When it is time for the funeral, the body is transported from the home or hospital to the mosque on the shoulders of the pallbearers. Within hours of death, the body is buried.

In preparation for burial, Islamic family members call into their home or the hospital a person of the same gender as the deceased who knows the prescribed ritual for washing and preparing the body. The eyes and mouth of the deceased are closed, the arms are straightened alongside the body, and the body is washed and wrapped in a white seamless cloth (shroud) similar to that worn for the pilgrimage to Mecca (Eickelman, 1987).

Muslims mourn their dead in mosques, never at funeral parlors. The body is washed in a special room (men prepare males; women prepare females), while the family and friends recite *suras* from the Qur'an as blessings for the deceased. At the mosque the body is placed onto a stone bier (*musalla*) in the outer courtyard. The funeral service is a part of one of the five regular daily religious services (usually the noon service). Because Muslims consider burying the dead a good deed, when worshippers leave the mosque and see the coffin in the courtyard, they participate in the procession to the cemetery even though they were not acquainted with the deceased (Habenstein & Lamers, 1974).

The body is then transported from the mosque to the cemetery on the shoulders of those male mourners in the procession. In most Islamic traditions, all the male members of the community walk in a procession to the grave site; women visit later during a 40-day mourning period (Habenstein and Lamers, 1974).

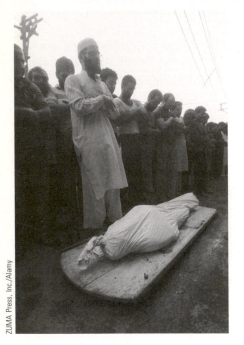

Islamic Funeral Rituals.

It is customary for every man in good health to carry the coffin on his shoulders for seven steps at least and for passers-by to accompany the procession for at least seven steps. When a new bearer pushes under the coffin, another steps away so that eight or 10 people are always under the load. These customs ensure that the remains have an escort, even though the dead person may have no living relatives. At a prearranged spot, a hearse and funeral cars await the procession. Where distances to the cemetery are short, the body is borne to the grave totally on foot.

At the cemetery the body is placed into the grave, and mourners place handfuls of dirt on top of it. A close friend of the deceased climbs into the grave to read final instructions to the dead in preparation for his or her meeting with Allah. Then, according to Islamic belief, the angels who accompany every believer in life enter the grave to question the departed soul on matters of faith and life: "Who is your Lord? Who is your prophet? What book do you follow?" The sexton then fills the remainder of the grave using a shovel.

At the grave site, rather than placing cut flowers, Muslim mourners plant flowers because they believe that every living plant utters the name of God. During this process, prayers are recited, and the service concludes with the preaching of a sermon (Habenstein & Lamers, 1974). After returning from the cemetery, all of the participants partake of a meal that is served at the home of the deceased. Occasionally, some food from this meal is placed over the grave for the first three days after the death. Mourning continues for another three days while family members receive social support and consolation from friends and members of the community.

Although Islamic women are allowed to openly express emotion in the bereavement process, Islamic men are encouraged to retain their composure as a sign that they are able to accept the will of Allah. A widow is required by the Qur'an to go into seclusion for four months and 10 days before she is allowed to remarry (Eickelman, 1987).

For two contemporary ethnographies related to Islamic bereavement, see studies by J. Millie (2008), dealing with Islamic funeral rituals in Indonesia, and F. Becker (2009), which documents the funerary practices of radical Muslims in Tanzania. These studies provide rich ethnographic data on Islamic funeral practices and their effects. For an extended discussion of all the religious groups discussed in this chapter, see C. Johnson (2009).

CONCLUSION

As discussed in Chapter 1, dying is more than a biological process. One does not die in a vacuum but rather in a social milieu. The act of dying has an influence on others because it is a shared experience. The sharing mechanism is death-related meanings composed of symbols.

Because death meanings are socially constructed, patterns of correct or incorrect behavior related to dying and death will largely be determined within the social setting in which they occur. Death-related behavior of the dying person and of those relating to him or her is in response to meaning relative to the audience and the situation. As noted, death-related behavior is shared, symboled, and situated.

Because death generally disrupts established interaction networks, shared "scripts" aid in providing socially acceptable behavior for the bereaved. Such scripts are essential because they prevent societal breakdowns while providing social continuity. Burial rituals are important in assuring social cohesion at the time of family dismemberment through death. Because death is a family crisis (Dickinson & Fritz, 1981), appropriate networks for coping must be culturally well grounded.

Death-related meanings are socially created and transmitted. Through participant observation, small children learn from others how to respond to death. If children are sheltered from such situations, their socialization will be thwarted.

As noted in this chapter, family involvement plays an important role in most societies as individuals prepare for death, as they prepare the corpse for final disposition, and as burial rituals that follow are performed. Therefore, whether death rituals involve killing a chicken, scraping the meat from the bones of the corpse, crying quietly, wailing loudly, mutilating one's own body, or burning or burying the corpse, all bereavement behavior has three interconnected characteristics—it is shared, symboled, and situated.

SUMMARY

1. Death occurs in all societies, yet it evokes an incredible variety of responses.
2. People in all societies are inclined to symbolize culturally defined feelings in conventional ways. Ritual behavior is an effective means of expressing or reinforcing these important sentiments.
3. Funeral or mortuary rites in many societies are critical because they ensure that the dead make the transition to the next stage of life or nonlife.
4. All rites of passage have three subrites—rites of separation, rites of transition, and rites of reincorporation.
5. Societies encourage the expression of both sadness and joy in their mourning rituals. In the former, participants are given the opportunity to collectively express their grief. In the latter, individuals engage in activities that enable them to transform grief and anger over a death into joy.
6. Cross-cultural studies of the ritual of mourning at death reveal that a double standard prevails among some cultures—different "scripts" for different genders.
7. There are common human needs (e.g., to dispose of the body), and each society or religious group creates practices, rites, meanings, and rituals, which are functional equivalents to those found in other societies or groups.
8. Customs for caring for the dying before death assist both the dying individual and the survivors in coping with the impending death.
9. For some, death is viewed as the end, whereas for others it is viewed as a continuation of life in a different form.
10. Burial rituals serve the functions of appeasing the ancestral spirits and the soul of the deceased, bringing the kin together, reinforcing social status, and restoring the social structure.
11. In most societies the body of the deceased is cleaned and prepared for burial. Some groups even keep parts of the body for ornamental or special purposes.
12. Anthropologists are interested in the meanings that different events, such as death, have for different cultures and religious groups.
13. Being familiar with cross-cultural death customs should help one to better understand the American concept of death.

DISCUSSION QUESTIONS

1. Why is it important to learn about bereavement patterns in other cultures? What are some of the manifest and latent functions of burial rites discussed in this chapter?
2. Drawing upon your own knowledge, discuss any U.S. behavior patterns for the dying just before death. How do these customs compare with those cited in this chapter?
3. Describe mourning rituals commonly found in the United States. Mourning rituals may differ by region of the country or ethnicity. Discuss these differences.

4. It is suggested that an explanation for crying when someone dies may be rather complex. Discuss the reasons for crying over a death.

5. Discuss the importance of the family being involved in some way related to the process of the final disposition of the deceased.

6. In death as in life, gender discrimination occurs. Discuss with a local funeral director the differences in funerals for males and females.

7. When you visit cemeteries in the United States, do you notice differences in graves (such as size of headstones and length of epitaphs) between males and females? How does social class or status make a difference in the treatment of the dead in U.S. cemeteries?

8. What are some of the functions of the burial rites discussed in this chapter?

GLOSSARY

Above-Ground Burial: The practice of placing the body on a scaffold or in trees for the purpose of allowing the body to decay or decompose. This type of burial is often the first step in the final disposition of the body. In the second stage the body is buried, which is referred to as **secondary burial**.

Bier: A framework upon which the corpse and/or casket is placed for viewing or carrying.

Columbarium: A building or wall for above-ground accommodation of cremated remains.

Cremains: That which is left after cremation.

Cremation: The reduction of a human body by means of heat or direct flame. The cremated remains are called cremains or ashes and weigh between three and nine pounds. *Ashes* is a very poor word to describe the cremated remains because they are actually processed bone fragments and calcium residue that have the appearance of crushed rock or pumice.

Crematory: An establishment in which cremation takes place.

Disposition: Final placement or disposal of a dead person.

Embalming: A process that temporarily preserves a deceased person by means of displacing body fluids with preserving chemicals.

Entombment: Opening and closing of a crypt, including placing and sealing of a casket within.

Ethnography: The systematic description of a culture based on firsthand observation.

Exhume: To remove a corpse from its place of burial.

Latent Function: Consequences of behavior that were not intended (e.g., a funeral brings the family together, usually in an amiable way).

Manifest Function: Consequences of behavior that are intended and overt (such as going to a funeral to pay respects to the deceased person).

Mausoleum: A building or wall for above-ground accommodation of a casket.

Mummification: The process of wrapping the body with cloth before its final disposition.

Niche: A chamber in a columbarium in which an urn is placed.

Pyre: A combustible pile (usually of wood) for burning a corpse at a funeral rite.

Rites of Passage: Ceremonies centering around transitions in life from one status to another (including baptism, the marriage ceremony, and the funeral).

Ritual: The symbolic affirmation of values by means of culturally standardized utterances and actions.

Sati: A controversial practice exercised rarely in some areas of India in which the widow throws herself onto the funeral pyre of her deceased husband. This "most noble act of loyalty" makes her a goddess who is worshipped at her cremation site.

Urn: A container for cremated remains.

SUGGESTED READINGS

Austerlitz, S. (2013). Why funerals demand a body: Undertaker Thomas Lynch on how American memorials went wrong. *Boston Globe*, October 20, 2013. This article provides a critique of contemporary funeral services, which may be more convenient (emotionally, spiritually, and financially) in the near term, but may be less personally satisfying in the long run.

Bolt, S. (2009). Body disposition. In C. D. Bryant and D. Peck (Eds.), *Encyclopedia of death and the human experience* (pp. 107–111). Thousand Oaks, CA: Sage Publications. This article provides an excellent discussion of modes of body disposition and the advantages and disadvantages of each.

Bryant, C. D. (2003). Funeralization in cross-cultural perspective. In C. D. Bryant (Ed.), *Handbook of death & dying* (pp. 611–693). Thousand Oaks, CA: Sage Publications. This is a series of essays dealing with funeralization, including articles on the Native American, Hindu, Muslim, Japanese, Chinese, European, and Jewish ways of death.

Cremation Association of North America. (2012). *2012 fact sheet*. Milwaukee: Author.

Edwards, B. (2013). Changes in Pitjantjatjara mourning and burial practices. *Australian Aboriginal Studies, 1*(Spring), 31.

Farrell, J. (1980). *Inventing the American way of death, 1830–1920*. Philadelphia, PA: Temple University Press. Examines the transformation from the Puritan way of death to the American way of death.

Garces-Foley, K. (Ed.). (2006). *Death and religion in a changing world*. Armonk, NY: M. E. Sharpe. This book is an excellent edited collection of articles on death from the perspective of many religious traditions.

Gowan, D. E. (2003). Christian beliefs concerning death and life after death. In C. D. Bryant (Ed.), *Handbook of death & dying* (pp. 126–136). Thousand Oaks, CA: Sage Publications. This article discusses the emergence of life-after-death beliefs in Judaism and Christianity.

Green, J. W. (2008). *Beyond the good death: The anthropology of modern dying*. Philadelphia, PA: University of Pennsylvania Press. An anthropologist provides an excellent interpretation of the ways in which Americans react when death is at hand for themselves or for those they care about.

Johnson, C. (2009). Funerals and funeralization in major religious traditions. In C. D. Bryant and D. Peck (Eds.), *Encyclopedia of death and the human experience* (pp. 499–503). Thousand Oaks, CA: Sage Publications. An extended discussion of the history, purpose, and process of funeralization from the perspective of the major religious traditions. Includes the following groups: Baha'i, Buddhist, Christian, Hindu, Jewish, Islamic, and Zoroastrian.

Kearl, M. (2009). Social class and death. In C. D. Bryant and D. Peck (Eds.), *Encyclopedia of death and the human experience* (pp. 875–878). Thousand Oaks, CA: Sage Publications. An extended discussion of the relationship between social class, race, and ethnicity and mortality and funeral practices.

Morgan, J. D., & Laungani, P. (Eds.). (2002). *Death and bereavement around the world*. Amityville, NY: Baywood Publishers. This edited work provides analysis of death-related behavior from the perspective of the major religious traditions. It also provides valuable bibliographic references and information.

Parkes, C. M., Laungani, P., & Young, B. (Eds.). (1997). *Death and bereavement across cultures*. New York: Brunner-Routledge. This book explores rich mourning traditions around the world, designed to meet the needs of those involved in the care of dying and the bereaved.

Rosen, S. J. (Ed.). (2008). *Ultimate journey: Death and dying in the world's major religions*. Westport, CT: Praeger Publishers. This book is written by different individuals who are experts on various religions. The focus is on the major world traditions and offers information about what death and dying means to those practicing these various faiths.

Rosenblatt, P. C. (2003). *Bereavement in cross-cultural perspective*. In C. D. Bryant (Ed.), *Handbook of death & dying* (pp. 855–861). Thousand Oaks, CA: Sage Publications. Comprehensive essay providing a comparative analysis of cross-cultural bereavement.

Spiro, H. M., Curnen, M. G. M., & Wandel, L. P. (Eds.). (1996). *Facing death: Where culture, religion, and medicine meet*. New Haven, CT: Yale University Press. Discussion of Christian, Judaic, Islamic, Hindu, and Chinese perspectives on death and rituals of mourning.

One could imagine a more pleasant means of livelihood, but almost any trade is bearable if the customers are sure.

—Clarence Darrow, discussing an ancestor reported to have been an undertaker

There is a major misconception about how much money funeral directors make. When I was at mortuary school, a teacher asked how many of us thought funeral directors made a lot of money. Several raised their hands. Then he told us how much funeral directors actually made. Two weeks later, about half those people were gone.

—John Everly (From H. Cox, *Insight*)

THE BUSINESS OF DYING

CHAPTER **11**

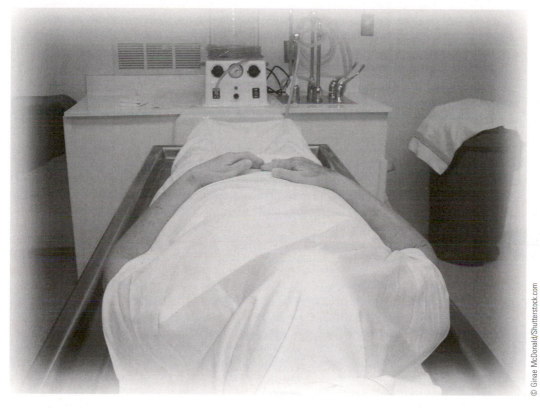

© Ginae McDonald/Shutterstock.com

THE BUSINESS OF PREPARING THE DEAD

A Minneapolis television station had a special "Dimension Report" on a new Twin Cities (Minneapolis and St. Paul, Minnesota) business called "Doggie-Do-Do Pick-Up." For a fee, these people would come to your home in the spring of the year (after the snow had melted) and remove a winter's supply of pet waste. Michael Leming and his son, who share an entrepreneurial spirit, concluded that there will always be money to be made by providing necessary services that no one wants to do. Furthermore, as the service becomes more necessary and the task more onerous, the price for the services rendered will escalate.

THE CHANGING AMERICAN FUNERAL

Funerals in the 21st century are obviously not the same as they were in colonial times in the United States. Indeed, "the times they are a-changin'," as singer Bob Dylan wailed in the 1960s. Like other aspects of our society, we have gone from being generalists, where a lot of people know how to do a lot of things, to a society of specialization and division of labor. The birth of the funeral industry in the late 19th century is a good example of this specialization.

THE PURITAN FUNERAL   For the Puritans, the tolling of the town bell announced the onset of a funeral, and participants gathered at the home for prayer and the procession. Death was feared by the Puritans, who prayed not for the soul of the deceased but rather for the comfort and instruction of the living. They believed that judgment occurs at death and that the dead are beyond human aid, so they prayed to reaffirm their faith and to glorify their God. In addition to prayers, Puritans might also read an **elegy**, which generally depicted the dead person as a saint freed from

---

DEATH ACROSS CULTURES | FUNERAL RITUALS IN A POLYNESIAN COMMUNITY

In a Polynesian village in Micronesia in the Pacific Ocean, anthropologist Michael D. Lieber observed funeral rituals. Funeral preparations tend to vary little by age or sex, except for more elaboration with age. The female head of the deceased's descent group assigns initial tasks for cooking food, preparing the body for viewing (washing, dressing, and closing orifices), and preparing the house for the wake. No weeping or other emotional expressions are allowed during this phase.

At the wake the body is viewed, women give forth a high-pitched wail, and weeping is permitted. Bodies of older people are buried under floors of the ancestral houses called grave houses. Others are weighted with stones and buried in the lagoon. Valuables are buried with the body. After the body is disposed of,

food is served to all guests. Chanting begins after people have eaten and continues until the next morning—the mood is more festive and entertaining.

Following these activities, all of the closest kin are given new names. The names are derived from chant lyrics and are usually names that ancestors used. Old names are "thrown away" to symbolize the rupture of a relationship, rather than simply as a sign of loss. To continue to use the old name would be a painful reminder of one's personal loss of an important relationship.

From Cutting your losses: Death and grieving in a Polynesian community (pp. 169–189), by M. D. Lieber, 1991. In D. R. Counts & D. A. Counts (Eds.), *Coping with the final tragedy: Cultural variation in dying and grieving*. Amityville, NY: Baywood Publishing.

the world and entering eternal bliss. Such elegies confirmed the new, separate status of the deceased, helping to bring grief under control, and provided a good example for structuring life after bereavement. Sometimes a copy of the elegy was pinned to the coffin or hearse for the funeral procession (Geddes, 1981).

The mourners walked from the home to the graveyard, where the men of the family, or the **sexton**, had opened a grave. They carried the coffin on a **bier**, covered with a black cloth ("pall")—both of which were the property of the town instead of the church, as was the custom in England. During the procession, mourners were supposed to "apply themselves to meditations and conferences suitable to the occasion." At the grave site, the pallbearers lowered the coffin into the earth, and the grave was refilled (Geddes, 1981, p. 111).

After the burial, the mourners returned home, where they shared food, drink, prayers, and comforting words. The family members thanked the pallbearers and participants and sometimes gave additional presents. They might also ask the minister to deliver a funeral sermon, which would occur not on the day of burial, but rather at the next regular meeting of the congregation. Sometimes families had the sermons and/or the elegies published and distributed to friends as a *memento mori*. At a later date, the family might also erect a marker over the grave to proclaim the imminence of death or God's promise of salvation. Such markers near the much-frequented meeting house were another way of maintaining the vitality of death in early American culture (Geddes, 1981; Ludwig, 1966).

The Puritan funeral was the primary social institution for channeling the grief of survivors. It provided for the disposition of the body and the acknowledgment of the absence of the deceased. It drew the community members together for mutual comfort, and it allowed mourners to honor the dead, to express their sorrow at separation, and to demonstrate their acceptance of God's will. After the funeral, Puritans expected mourners to return to their callings and to resume their life's work. They did not approve of elaborate or extended mourning of the sort that became customary in the 19th century because it undermined the cheerful resignation to God's will that was essential to the Puritan experience.

THE VICTORIAN FUNERAL   Victorian times brought a focus on aesthetics and soothing the mourner. The attitude was one of beautifying the whole funeral process and making the funeral more tolerable for bereaved individuals. To "get on" with the funeral and move on with life seemed to be emphasized. This reflected the "dying of death" movement discussed in Chapter 2.

The Focus on Aesthetics.   Between 1850 and 1920 funeral services changed significantly in the United States. Because "the growing wealth and prosperity of our country has caused people to demand something more in accordance with their surroundings" (Benjamin, 1882, p. 3) and because funeral directors cultivated a "steadily advancing appreciation of the aesthetics of society" (Funeral Directors, 1883), the new Victorian funeral was a work of art that attempted to restore order to the middle classes while hiding as much as possible the unpleasantries of death.

The demand for a new funeral service came from the American middle class, but it was created and supplied by casket manufacturers and funeral directors.

WORDS OF WISDOM | DISPLAY OF THE DEAD

Truly, we need to do away with some of the false ideas of Death which are shown in so many gruesome ways at funerals, and strive to give the young a different and truer idea of what the passing away of a soul means. The awfulness of some funerals is nothing short of criminal, especially as it affects the minds of the young. If there is work cut out for the minister of today, it is the enlightenment of his people on the subject of death and the funeral. But the minister must, first of all, imbibe a wholesome lesson of self-restraint for himself, and abolish the fulsome and tiresome eulogy which is the bane of so many funerals. He must learn for himself, and teach to his people, the beauty and solemnity of the brief service as prescribed by his church and attempt nothing more, and he must also relentlessly oppose the tendency which exists to turn the modern funeral, especially in the country, into a picnic. The present outpouring of a heterogeneous mass of folk from every point of the countryside is a farce that cannot be too soon abolished. A funeral is essentially a time for the meeting of the family and relatives and the closest friends, and the fewer the number of outsiders present the better. Nor is there anything quite so barbarous as the present custom at so many funerals of "viewing the remains" by a motley collection of folk, many of whom never even knew the dead in life, or, if they did, never thought enough of him to come and see him. The vulgar curiosity that prompts "a last look" at a loved one cannot be too severely denounced. Only second to it is the pretentious line of vehicles that "escort the remains to the grave" and the mental caliber of a community that bases the popularity of a man on the number of carriages that follow him to the grave! If there is a crying need of the gospel of simplicity it is in connection with funerals. It seems inconceivable that Death should be made the occasion for display, and yet this is true of scores of funerals. The flowers, including those fearful conceptions of the ignorant florist, such as "Gates Ajar"; the quality of casket and even of the raiment of the dead, the "crowd" at the "obsequies," the number of carriages in the "cortege"—oh, oh, "what fools these mortals be," to say naught of the wicked and wanton waste of much-needed money. It is difficult to conceive that a national love of display should have become so deep-rooted as to lead to the very edge of the grave!

From *Ladies Home Journal*, September 1903.

The National Funeral Directors Association (NFDA) was founded in 1882 in Rochester, New York, the home of the Stein Casket Manufacturing Company. The association's official journal was *The Casket*, founded and funded for several years by the Stein Company. As this name suggests, the first widespread innovation in funeral service was the casket, a stylish container for the corpse. Before 1850, most Americans were laid to rest in a coffin, a six-sided box that was constructed to order by the local cabinetmaker. By 1927 "the old wedge-shaped coffin (was) obsolete. A great variety of styles and grades of caskets (were) available in the trade, ranging from a cheap, cloth-covered pine box to the expensive cast-bronze sarcophagus" (Gebhart, 1927, p. 8). The rectangular shape of the new caskets complemented the artwork in concealing the uncouth corpse. In applying for a casket patent in 1849, A. C. Barstow explained:

> The burial cases formerly used were adapted in shape nearly to the form of the human body, that is they tapered from the shoulders to the head, and from the shoulders to the feet. Presently, in order to obviate in some degree the disagreeable sensations produced by a coffin on many minds, the casket, or square form has been adopted. (cited in Habenstein & Lamers, 1962, p. 270)

The adoption of the word *casket* also accelerated the "dying of death," as the word had previously denoted a container for something precious, like jewels. Accepting the associated idea of the preciousness of the body, Americans decided that a dead-looking corpse looked out of place in an elaborate silk-lined casket. Rather than remove the casket, they decided to stylize the body. Originally a way of preserving bodies for shipment home from Civil War battlefields or Western cities, embalming soon became a way of preserving appearances. Responding to the germ theory of disease and the public health movement, funeral directors attempted to gain professional status by emphasizing the disinfectant qualities of embalming. Most funeral directors, however, wanted simply "to retain and improve the complexion" so that the corpse would look "as natural as though it were alive" (Hohenschuh, 1921, p. 82). To do this, they began to cosmeticize the corpse and to clothe and position the body naturally. They replaced the traditional shroud with street clothes, and they tried "to lay out the body so that there will be as little suggestion of death as possible." By 1920, they succeeded so well that a Boston undertaker supposedly advertised:

For composing the features	$1
For giving the features a look of quiet resignation	$2
For giving the features the appearance of Christian hope and contentment	$5

(Dowd, 1921, p. 53)

Bereavement practices were affected by the change from coffin to casket and by the "restorative art" of the embalmer; they were also affected by the movement of the funeral from the domestic parlor to the funeral parlor. As people began banishing death from their homes to hospitals, they started moving the funeral from the family parlor to a specialized funeral parlor. After the Civil War, middle-class Americans began to exclude the formal parlor from their homes and to replace it with a "living room." At the same time, funeral directors wanted full control of the corpse and the funeral. The ease and efficiency of directing funerals in a funeral home made them more profitable. Despite these benefits, the transition to the funeral parlor was slow, extending well into the 20th century (Farrell, 1980).

The Focus on Soothing the Mourner.   Both in the domestic parlor and in the funeral parlor, the procedure of the turn-of-the-century funeral changed. In conjunction with the reform forces of religious liberalism, funeral directors began to redirect funerals to be shorter, more secular, and more soothing. They shortened the service by trying to revise the long sermon with its exhortations of repentance and renewal. Although some clerics resisted, funeral directors wanted the sermon redirected from theology to psychology, from preaching to grief therapy, and from the state of survivors' souls to the state of their emotions. The funeral director took care of all of the details of the funeral and performed as much as a stage manager as a mortician. "Really it is much the same," wrote one director, "I work for effect—for consoling and soothing effect" ("The Man Nobody Envies," 1914).

After 1880, funeral directors used their arts and the culture of professionalism to effect a massive change in the American way of bereavement. Professionalism

The practice of laying the deceased on a bier for viewing in the home persisted through the end of the 19th century.

was part of the middle-class strategy of specialization. It required education in an area of expertise and an ethic of service, and it provided autonomy and income for its practitioners. The American undertaker sought professional status because it would help to become "enough of an authority to convince his clients, without offense, that there are better methods than are prescribed by custom" (Hohenschuh, 1921, p. 9).

Etiquette books reinforced this culture of professionalism by advising readers that "the arrangements for the funeral are usually left to the undertaker, who best knows how to proceed" (Wells, 1887, p. 303). To the middle-class people who feared death anyway, this established a situation in which the public passively accepted changes in funeral services suggested by funeral directors (Hohenschuh, 1921; Wells, 1887).

The modern bereavement practices described proceeded from a simple desire to make death as painless for survivors as for the deceased. It came from a widespread cultural attempt "not to mention trouble or grief or sickness or sin, but to treat them as if they do not exist, and speak only of the sweet and pleasant things of life" (Strauss, "The Ideas of a Plain Country Woman," 1913). This "dying of death" came from the desire of the middle class for control—of self, society, and the environment. It ended exactly where de Tocqueville predicted:

> As they perceive that they succeed in resolving without resistance all the little
> difficulties their practical life presents, [the Americans] readily conclude that everything
> in the world can be explained, and that nothing in it transcends the limits of the
> understanding. Thus they fall to denying what they cannot comprehend. (1835/1945,
> pp. 2, 4) (Adapted from article written by James Farrell and used with his permission.)

Denying and disguising death, middle-class Americans achieved, on the surface at least, the dying of death.

THE CONTEMPORARY AMERICAN FUNERAL: MEETING THE NEEDS OF THE BEREAVED The modern thanatology movement, which was influenced by Jessica Mitford's *The American Way of Death* (1963) and Elisabeth Kübler-Ross's *On Dying and Death* (1969), effectively resurrected death in the consciousness of American life and thereby transformed the contemporary American funeral. Along with public education that attempts to provide some relevance for the American learner, funerals, too, attempt to be relevant to grievers and transform romanticism and sentimentality into an opportunity to honestly feel one's emotions and loss.

Mourners are encouraged to feel and experience the reality of death and then attempt to adapt to the environment in which the deceased is gone. As part of this process of combating the tendency to deny death, the modern funeral eschews euphemisms and allows mourners to personalize funeral rituals and work with professional funeral directors to create meaningful rituals that both reflect and celebrate the unique life of the person who had died. The process of personalization leads family members to deliver their eulogies and create memorial tables (consisting of pictures, artifacts, and mementos of the deceased) as attempts to combat the depersonalization and overprofessionalization that had characterized funerals of an earlier era.

Paul Irion (1956) described the following needs of the bereaved: reality, expression of grief, social support, and meaningful context for the death. For Irion, the funeral is an experience of significant personal value insofar as it meets the religious, social, and psychological needs of the mourners. Each of these must be met for bereaved individuals to return to everyday living and, in the process, resolve their grief.

The psychological focus of the funeral is based on the fact that grief is an emotion. Edgar Jackson (1963) indicated that grief is the other side of the coin of love. He contended that if a person has never loved the deceased—never had an emotional investment of some type and degree—he or she will not grieve upon death. As discussed in the opening pages of Chapter 2, evidence of this can easily be demonstrated by the number of deaths that we see, hear, or read about daily that do not have an impact on us unless we have some kind of emotional involvement with those deceased persons. We can read of 78 deaths in a plane crash and not grieve over any of them unless we personally knew the individuals killed. Exceptions to the preceding might include the death of a celebrity or other public figure, when people experience a sense of grief even though there has never been any personal contact.

In his original work on the symptomatology of grief, Erich Lindemann (1944) stressed this concept of grief and its importance as a step in the resolution of grief. He defined how the emotion of grief must support the reality and finality of death. As long as the finality of death is avoided, Lindemann believed, grief resolution is impeded. For this reason, he strongly recommended that the bereaved persons view the dead. When the living confront the dead, all of the intellectualization and avoidance techniques break down. When we can say, "He or she is dead, I am alone, and from this day forward my life will be forever different," we have broken through the devices of denial and avoidance and have accepted the reality of death. It is only at this point that we can begin to withdraw the emotional capital that we have invested in the deceased and seek to create new relationships with the living.

On the other hand, viewing the corpse can be very traumatic for some. Most people are not accustomed to seeing a cold body and a significant other stretched out with eyes closed. Indeed, for some this scene may remain in their memories for a lifetime. Thus, they remember the cold corpse, not the warm, responsive person.

Whether or not to view the body is not a cut-and-dried decision. Many factors should be taken into account when this decision is made.

Grief resolution is especially important for family members, but others are affected also—the neighbors, the business community in some instances, the religious community in most instances, the health care community, and the circle of friends and associates (many of whom may be unknown to the family). All of these groups will grieve to some extent over the death of their relationship with the deceased. Thus, many people are affected by the death. These affected persons will seek not only a means of expressing their grief over the death, but also a network of support to help cope with their grief.

Sociologically, the funeral is a social event that brings the chief mourners and the members of society into a confrontation with death. The funeral becomes a vehicle to bring persons of all walks of life and degrees of relationship to the deceased together for expression and support. For this reason, in our contemporary culture the funeral becomes an occasion to which no one is invited but all may come. This was not always the case, and some cultures make the funeral ceremony an "invitation only" experience. It is perhaps for this reason that private funerals (restricted to the family or a special list of persons) have all but disappeared in our culture. (The possible exception to this statement is a funeral for a celebrity—in which participation by the public may be limited to media coverage.)

At a time when emotions are strong, it is important that human interaction and social support become high priorities. A funeral can provide this atmosphere. To grieve alone can be devastating because it becomes necessary for that lone person to absorb all of the feelings into himself or herself. It has often been said that "joy shared is joy increased"; surely grief shared is grief diminished. People need each other at times when they have intense emotional experiences.

A funeral is in essence a one-time kind of support group to undergird and support the grieving persons. A funeral provides a conducive social environment for mourning. We may go to the funeral home either to visit with the bereaved family or to work through our own grief. Most of us have had the experience of finding it difficult to discuss a death with a member of the family. We seek the proper atmosphere, time, and place. It is during the funeral, the wake, the shivah, or the visitation with the bereaved family that we have the opportunity to express our condolences and sympathy comfortably.

Anger and guilt are often deeply felt at the time of death and will surface in words and actions. They are permitted within the funeral atmosphere as honest and candid expressions of grief, whereas at other times they might bring criticism and reprimand. The funeral atmosphere says in essence, "You are okay, I am okay; we have some strong feelings, and now is the time to express and share them for the benefit of all." Silence, talking, feeling, touching, and all means of sharing can be expressed without the fear of their being inappropriate.

Another function of the funeral is to provide a theological or philosophical perspective to facilitate grieving and to provide a context of meaning in which to place one of life's most significant experiences. For the majority of Americans, the funeral is a religious rite or ceremony (Pine, 1971). Those grievers who do not possess a religious creed or orientation will define or express death in the context of the values that the deceased and the grievers find important. Theologically or philosophically, the funeral functions as an attempt to bring meaning to the death and life of the deceased individual. For the religiously oriented person, the belief system will perhaps bring an understanding of the afterlife. Others may see only the end of biological life and the beginning of symbolic immortality created by the effects of one's life on the lives of others. The funeral should be planned to give meaning to whichever value context is significant for the bereaved.

"Why?" is one of the most often asked questions upon the moment of death or upon being told that someone we know has died. Though the funeral cannot provide the final answer to this question, it can place death within a context of meaning that is significant to those who mourn. If it is religious in context, the theology, creed, and articles of faith confessed by the mourners will give them comfort and assurance as to the meaning of death. Others who have developed a personally meaningful philosophy of life and death will seek to place the death in that philosophical context.

Cultural expectations typically require that we dispose of the dead with ceremony and dignity. The funeral can also ascribe importance to the remains of the dead. In keeping with the specialization found in most aspects of American life (e.g., the rise of professions), the funeral industry is doing for Americans that necessary task they no longer choose to do for themselves.

THE AMERICAN PRACTICE OF FUNERAL SERVICE

EDUCATION AND LICENSURE

With the evolution of the funeral, there likewise has been an evolution of the funeral functionary. Contemporary Americans refer to that functionary as a "funeral director," yet a hundred years ago this functionary was a "layer out of the dead," often a member of the family who physically and emotionally could perform the necessary tasks of bathing the body, closing the eyes and mouth, and dressing the body. It was not unusual for a midwife or other person who provided nursing-like services in the community to be called on to assist the family. Early advertisements indicate that nurses did offer such services.

As early as the 1890s, various states began to enact legislation to protect the public's health by licensing embalmers. Early licensing agencies directed their attention to the **embalming** process, and it was not until the 1920s and 1930s that licensing agencies began to regulate other aspects of the funeral and the operation of funeral homes.

The rationale for this licensure was based upon the need to protect the public, primarily in the financial area. However, regulations also addressed how a funeral should be conducted when the cause of death was a contagious disease. Another issue of public health protection was the transportation of the dead from the place of death to the location of final **disposition**. Public health authorities claim that regulating the treatment of the dead has significantly contributed to the advanced standard of health of our country.

Based on these concepts, the licensing agencies most often charged with responsibility to regulate the funeral industry have been the various state boards of health. In some states, special boards have been established to implement regulation and enforcement procedures.

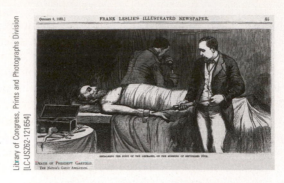

An artist's sketch of the embalming of the body of President Garfield on the morning of September 20, 1881, in preparation for the train ride from Elberon, New Jersey, to Washington, DC. Though originally a way of preserving bodies for shipment home from Civil War battlefields, embalming soon became a way of preserving appearances.

Licensure has been reserved for the individual states and includes three basic licenses. A license as an embalmer permits a person to legally remove the dead from the place of death and prepare the body through the process of embalming for viewing and holding the funeral. All states (with the exception of Colorado) require persons who perform these functions to be licensed. A second license, to practice as a funeral director, permits the holder to arrange the legal details of the funeral, including preparing the death certificate and counseling with the family to arrange and implement the kind of funeral desired for the deceased. A third license is one that permits the holder to practice mortuary science—an all-inclusive specialty that covers the practices of both embalming and funeral directing.

A few states issue a funeral director license, which may be held by only one person in each firm (usually the owner or manager) that serves to give the licensing agency control over all of the practitioners within that firm. A greater number of states have created a funeral home license or permit that is required for each funeral home and allows the state to close the funeral home by withdrawing the license, without taking action against the licensees employed by that firm. In addition to requiring embalmers to be licensed, with the exception of Colorado, all states and the District of Columbia require embalming practitioners to also be licensed as funeral directors. The exact number of states requiring funeral home licenses or permits is difficult to determine because some are required by law, some by regulation, and some by local ordinance. Approximately 25 states have some requirement that governs the operation of the funeral homes located in them.

According to the *Encyclopedia of Death and the Human Experience* (Tseng & Su, 2009), the American Board of Funeral Service Education was founded in the 1940s and is recognized (by the U.S. Department of Education and the Council on Higher Education Accreditation) as the sole national accrediting agency for academic programs that prepare funeral service professionals. However, the qualifications for licensure of funeral personnel differ by state regulations and requirements and chiefly concern age, citizenship, and specific education of the practitioner.

As of 2013, all states (with the exception of Colorado) require a high school education and some postsecondary education in mortuary science. Fourteen states require two years of college, including the one-year major in funeral service or mortuary science; 16 states require three years of college, including the one-year funeral service program; and four states require a bachelor's degree for licensure in funeral service. The typical funeral service student graduates with an associate degree in funeral service or mortuary science.

Following academic preparation, and in some cases before, all states (with the exception of Hawaii and Wyoming) require applicants to serve an apprenticeship that varies from one to three years, depending on individual state regulations. Upon the completion of academic and internship or apprenticeship requirements, applicants for licensure are required by all states to pass a qualifying examination before the issuance of the license to practice (NFDA, 2013).

Currently, 44 states mandate continuing education for renewal of the license to practice. The Academy of Professional Funeral Service Practice certifies and approves all continuing education programs and courses. The average number of continuing education hours per year required by states for relicensure is eight, but it varies from seven hours every two years (required by West Virginia) to 15 hours every two years, required by New Hampshire (NFDA, 2013).

According to NFDA (2013), there are 19,624 funeral homes in the United States with approximately 35,000 licensed personnel and 70,669 nonlicensed professionals. In the last decade the number of women entering colleges of funeral service education and becoming licensed has greatly increased. According to American Board of Funeral Service Education (ABFSE) (cited by NFDA, 2013), 45.6 percent of graduates were female in 2001, and by 2012 that percentage had increased to 53.4 percent (regarding salaries for funeral home employees, see Table 11.1).

TABLE 11.1 | NFDA's LATEST SALARY STATISTICS IN US DOLLARS, BASED ON 2012 DATA

	2012 Median Salary	2009 Median Salary	% change from 2009 to 2012
Professional			
Funeral Home Owner	$75,200	$74,000	1.6
Funeral Home Branch Manager	$63,000	$54,000	16.7
Non-owner Manager	$60,500	$64,000	−5.5
Staff			
Funeral Director/Embalmer	$49,549	$46,800	5.9
Funeral Director Only	$41,414	$38,000	9.0
Apprentice	$30,000	$28,000	7.1
Funeral Assistant	$24,960	$26,000	−4.0
Visitation Attendant	$20,800	$21,784	−4.5
Office Manager	$35,000	$32,925	6.3
Secretary/Clerical	$28,162	$25,765	9.3
Receptionist	$24,960	$22,460	11.1
Maintenance—Full Time	$30,000	$25,000	20.0

(NFDA, 2013).

THE ROLE OF THE FUNERAL DIRECTOR

Rabbi Earl Grollman (1972) described the role of the funeral director as that of a caretaker, caregiver, and gatekeeper. He indicated that the etymology of the word *undertaker* is based upon the activities of the early undertaker who "undertook" to do for people at the time of death those things that were crucial in meeting their bereavement needs. The funeral director, from the perspective of the community, was viewed as a secular gatekeeper between the living and the dead.

John Brantner (1973), elaborating upon the caregiver role, emphasized that the funeral director is a crisis intervener. This idea can be documented in the vast amount of literature on the counseling role of crisis interveners, who are not clinical practitioners by training but to whom the public turns in crisis.

The funeral director serves families by determining their needs and responding to them (Raether & Slater, 1974). This service includes, but is not limited to, the funeral (or its alternative) that the director and the family plan and implement together. As a licensee of the state, the funeral director handles details such as recording the death properly as required by law and files permits for transportation and final disposition of the body. The funeral director serves as a liaison with other professionals, working with the family's medical personnel, clergy, lawyers, cemetery personnel, and, when necessary, law enforcement officials.

According to the NFDA (2013), 88 percent of member funeral homes are family-owned and average 75 years of operation in the same community. According to NFDA, the association's average member handles 112 calls per year and has three full-time and four part-time employees; 74.7 percent of these funeral homes are located in small towns or rural areas, and 25.3 percent are located in urban or suburban areas (NFDA, 2013).

FUNERAL EXPENSES

Charges made by a funeral home ordinarily are for the services of the professional staff, the use of the funeral home facilities and equipment, transportation, and the casket or other container. In addition, most funeral homes provide burial **vaults** or other types of outer enclosures for the casket, and ancillary items that may be purchased from the funeral director—clothing, register books, acknowledgment cards, and stars of David, crucifixes, or crosses.

The other major cost of the funeral is the cemetery charge—either for the purchase of cemetery property for the right to interment therein, for **mausoleum** space, or for an **urn** for the **cremated** remains (in some instances, there will be a charge for providing a space in a **columbarium** for the urn). Most families will also select, in one form or another, a monument or marker to identify the grave or other place of final disposition.

A final category of expenses incurred by the family is money that the funeral home sometimes advances, at the request of the family, to other people involved in the final disposition. Such cash advances typically pay for the following: charges for opening and closing the grave, **crematory** costs, honoraria for clergy and musicians, **obituary** notices, certified copies of the death certificate, flowers, and transportation costs in addition to the transportation ordinarily furnished by the

| PRACTICAL MATTERS | SHOULD FUNERAL DIRECTORS PROFESSIONALIZE? |

Like many other occupations, funeral directors would like to be thought of as professionals. Should we as consumers support the attempts of funeral directors to become more professional?

Our first response to this question is, "Yes, of course." But let us not be so hasty. We must first consider what it is that makes an ordinary job a "profession." How do we distinguish a profession from a regular job? One way is to list the features or traits of occupations commonly accepted as professions. If asked which occupations are professions, most of us would answer: physician or lawyer. What sets these occupations apart? They are characterized by the following:

1. A specialized body of knowledge
2. A long period of training
3. An orientation toward service rather than profit
4. A commonly accepted code of ethics
5. Legal recognition (most often through licensure)
6. A professional association

Implicit in this definition is the assumption that professionals have the best interest of the public in mind. After all, they submit to a long period of training, look forward to serving others, abide by a code of ethics, and police themselves through their professional associations. All occupations should become professions!

But this is not the only way to define a profession. Others look more cynically on the role of professions in society. George Bernard Shaw said, "Professions are a conspiracy against the laity." What did he mean? Shaw's comment hints at an alternative definition of the professions, a definition that there is just one distinguishing characteristic of the professions: power. Professions are those occupations that have accumulated enough power to control the definition and substance of their work. For example, this view contends that physicians are professionals by virtue of their complete control of matters of health. Through their associations they control the number and training of doctors, they limit the practices of competitors (e.g., chiropractors or nurse practitioners), and they set rates of reimbursement for health care. Adherents to this view point out that professionals in fact incapacitate us: they limit our choices, make us feel unable to help ourselves, and encourage dependency.

Should funeral directors become more professional? Not all would agree. Funeral directors subscribe to the first definition and assert that professionals can better attend to the needs of the public. Followers of the second definition conclude that the move toward professionalization would limit competition, drive prices up, and, by promoting dependency, make us less able to deal with death.

From R. DeVries, professor of bioethics and sociology, University of Michigan Medical School, Ann Arbor, MI.

funeral home. Since the 1960s, NFDA has calculated the median cost of a funeral by totaling the costs of the following items: nondeclinable basic services fee, removal/transfer of remains to the funeral home, embalming, other preparation of the body, a metal casket, use of the funeral home and staff for viewing, use of the funeral home and staff for a funeral ceremony, use of a hearse, use of a service car/van, and a basic memorial printed package (e.g., memorial cards, register book, etc.). The national median cost of a funeral for calendar year 2012 was $7,045. If a vault is included, something that is typically required by a cemetery, the median cost is $8,343. The cost does not take into account cemetery, monument or marker costs, crematory fees (if cremation is selected instead of burial), or miscellaneous cash-advance items, such as flowers and obituaries. Many factors contribute to the final determination of how an individual funeral home prices its good and services, including the firm's business philosophy and the market in which it operates (NFDA, 2013) (see Table 11.2).

Table 11.2 | US Median Cost of an Adult Funeral: 2012 vs. 2009

Item	2012*	2009*	% Change
Nondeclinable basic services fee	$1,975	$1,817	$8.7
Removal/transfer of remains to funeral home	$285	$250	14.0
Embalming	$695	$628	10.7
Other preparation of the body	$225	$200	12.5
Use of facilities/staff for viewing	$400	$395	1.3
Use of facilities/staff for funeral ceremony	$495	$450	10.0
Hearse	$295	$275	7.3
Service car/van	$130	$125	4.0
Basic memorial printed package	$150	$125	20.0
Subtotal without casket	$4,650	$4,265	9.0
Metal casket (average charge for the most frequently purchased item)	$2,395	$2,295	4.4
Average Cost of a Funeral	**$7,045**	**$6,560**	**7.4**
Vault (average charge for the most frequently purchased item)	$1,298	$1,195	8.6
Total with vault	$8,343	$7,755	7.6

*Median Price = The amount at which half of the figures fall below and half are above. Prices have not been adjusted for inflation.

(NFDA, 2013).

Over the past decade, the median cost of an adult funeral in the United States has increased approximately 35.2 percent. This is similar to the 1991 to 1998 period, for which studies were completed, where the cost of a funeral increased by 34 percent. The percent increase during the 1980s (1980–1989) was significantly higher (87 percent), reflecting the significantly higher inflation rates of the 1980s (NFDA, 2013).

In a survey by the Federated Funeral Directors of America (2013) of the funerals conducted in 2012, it was determined that the average cost for a funeral for an adult in the United States was $6,631.94. This cost did not include grave or cremation expenses, the cost of a vault, clothing, or extra service requirements, such as the clergy, florist, and musicians. The approximate breakdowns for costs are as below:

Personnel	$2,424
Facilities	$1,385
Automotive	$394
Promotion	$264
Supplies	$143
Business Services	$187
After Sale	$331
Miscellaneous	$13
Casket Cost	$1,085
Profit	$406
Total Cost	**$6,632**

A survey of NFDA members determined that nationally, for every dollar taken in by affiliated funeral homes, money was distributed in the following manner:

32 cents for salaries and benefits

15 cents for other operating expenses

28 cents for merchandise (caskets, vaults, etc.)

15 cents for facilities

10 cents for before-tax profits (NFDA, 1999)

The 2013 NFDA General Price List surveys provide a number of other important findings. First, the data suggest that funeral service is a competitive industry. However, there is a difference of nearly $1,798 between the East/South Central region ($8,828) and the Mountain/Pacific region ($7,030) (NFDA, 2013). Second, approximately 63 percent of member firms operate only one funeral home facility, and 23.5 percent operate two facilities (NFDA, 2013).

Finally, the average profit per funeral in the past 25 years has declined by more than a third—as a percentage of sales, from 9.87 percent in 1988 to 6.12 percent in 2012 (Federated Funeral Directors of America, 2013). This relatively low return on equity makes funeral service (like other capital-intensive industries) a very difficult one to enter as a new entrepreneur. For the entrant, the price of land and buildings would result in higher operating expenses to perform the same service than those of a more established funeral home. With the percentage of immediate cremations increasing to 42 percent in recent years, the profit ratios will probably decrease even more and exacerbate this problem for young funeral directors.

PRICING SYSTEMS   Until 1984 a funeral home charged for its services and merchandise in basically three ways. The first was the unit method of pricing. In this method, all of the costs involved with the funeral home (including the staff, the facilities, the automotive equipment, and the casket) were included in a single charge. In making a selection under this method, the family looked at a "bottom line" figure to which only other charges paid by the funeral home (e.g., a vault and additional burial merchandise) might be added.

The second method was the biunit or triunit pricing system. The biunit system made separate charges for professional services and the casket, and the triunit

"He'd rest a lot better with thirty per cent off."

Donald Reilly/The New Yorker Collection/Cartoon Bank.Com

system made separate charges for professional services, the use of facilities, and the casket selected. This method enabled families to understand the charges for the three basic components making up funeral costs.

A third method of presenting costs was referred to as either functional, multi-unit, or **itemization**. In this method, each and every item of service, facility, and transportation was shown as a separate item together with its cost. In this method usually a minimum of eight to 10 items were listed, and the family decided in each instance whether or not that item would become a part of the funeral service. In 1984 the Federal Trade Commission (FTC) mandated that all funeral homes in the United States itemize their fees and that consumers have access to pricing information over the phone.

A major advantage of itemization is that family members have greater flexibility in arranging a funeral and controlling costs. The family members should have the freedom to decline those items that they do not want, and a proper allowance should be made for the items that are not used. For example, one may ask to see "the pine box"—usually a cloth-covered casket of wood or pressed wood. These caskets are relatively inexpensive and may or may not be in the display room. If the body is to be transported a great distance to the gravesite or crematorium, perhaps the funeral director's van or station wagon could be used rather than the expensive hearse. Similarly, the consumer should have the right to choose among types of **vault liners**. The greatest advantage of itemization is that one can look carefully at the itemized services and obtain the most adequate services at the best price (see Table 11.3).

There are two major disadvantages of the FTC policy requiring funeral homes to give prices over the phone and to itemize funeral expenses. The first is that when consumers receive prices over the phone it is difficult for them to make accurate price comparisons. One firm can say that it sells an oak casket for $3,600, and another will price a different oak casket at $6,300. It is entirely possible that the price/quality ratio is better for the more expensive casket. A helpful analogy might be to imagine calling two import automobile dealerships, Kia and Mercedes, to ask each, "How much do cars cost in your dealership?" The only way that one can make accurate price comparisons is to personally inspect both products. The same is true for the funeral industry. Families are better served when they can make careful comparisons between the costs of services provided by funeral firms—before the death of a significant other. Most people do not wait until a car breaks down before they shop for a new one. Furthermore, if they did find themselves in this situation, they would not purchase a car over the telephone.

Unlike other businesses, customers can provide many of their own products while they are purchasing services from a funeral establishment. It has always been possible to bring clothing to the funeral director that will be placed on the body of the deceased; however, increasingly, families are purchasing caskets online and from Costco or Wal-Mart. The FTC has now made this possible for consumers of funeral services. This would not be unlike restaurants being required to allow customers to bring alcoholic beverages into their establishments. However, one should remember that as the opportunity to make profit from the sales of products provided to the consumer decreases, the funeral industry will have to increase service fees in order to remain in business. In a similar manner, restaurants that

TABLE 11.3 | COST OF FUNERAL ITEMS AS PROVIDED BY US SURVEY OF NFDA MEMBERS

	*Median	Minimum	Maximum	# of Responses to Each Item
Funeral Items				
Nondeclinable basic services fee	$1,975	$550	$4,415	397
Embalming	$695	$230	$6,505	404
Other preparation (cosmetology, hairdressing)	$225	$30	$1,250	389
Removal/Transer of remains to FH	$285	$80	$4,050	395
Forwarding of remains to another FH	$2,095	$250	$5,295	390
Receiving of remains from another FH	$1,795	$250	$5,635	390
Use of facilities or staff for viewing	$400	$85	$1,285	393
Use of facilities/staff for funeral ceremony	$495	$45	$1,335	402
Use of facilities/staff for memorial service	$475	$100	$4,680	399
Use of equip/staff for graveside service	$350	$100	$4,500	388
Hearse	$295	$85	$650	403
Limousine	$250	$10	$2,754	244
Service car or van	$130	$30	$2,060	364
Printed materials (basic package)	$150	$25	$695	337
Disposition or Remains				
Direct cremation (container provided by family)	$2,115	$175	$5,140	395
Direct cremation (container provided by FH)	$2,245	$45	$5,140	395
Immediate burial (container provided by family)	$2,295	$695	$8,480	390
Immediate burial (container provided by FH)	$2,890	$795	$8,295	369
Packages Offered				
Casketed adult funeral with viewing and burial	$4,598	$1,150	$11,411	222
Memorial service and burial	$3,493	$400	$8,675	176
Casketed adult funeral with viewing and cremation	$4,715	$1,575	$8,500	222
Memorial service and cremation	$3,190	$595	$6,785	222
Caskets/Urn				
Metal burial casket	$2,395	$900	$5,300	371
Wood burial casket	$2,934	$970	$5,695	366
Cremation casket	$995	$60	$7,500	310
Urn	$275	$60	$2,500	364
Vault	$1,298	$300	$3,695	374

(NFDA, 2013).

BOB'S FUNERAL PARLOR

"ASK ABOUT OUR LAYAWAY PLAN"

Hawkins, Jonny/CSL, CartoonStock Ltd

allow customers the opportunity to bring wine with them while they dine usually charge a "corkage fee."

The disadvantage of the FTC ruling requiring itemization of funeral expenses is that itemization does not uniformly result in decreased expenses to families. Before the 1984 ruling, many funeral firms included some merchandise and services as part of the standard funeral. With itemization, funeral directors could provide justification to raise the cost for funeral services. The analogy of restaurant pricing may be appropriate at this point—it is often less expensive to order a complete meal than to order food a la carte.

ALTERNATIVES TO THE FUNERAL

People often ask if there are alternatives to the traditional funeral. There are three main alternatives: immediate disposition of the body of the deceased, the bequest of the body by him or her to a medical institution for anatomical study and research, and the memorial service. (See Chapter 10 for discussion of other low-incidence alternatives—mummification, plastination, cryonics, and cannibalism.)

Direct or immediate disposition occurs when the deceased is removed from the place of death to the place of cremation or earth burial without any ceremony; proper certificates are filed and permits received in the interim. In these instances, the family is not present, usually does not view the deceased after death, and is not concerned with any further type of memorializaton. Disposition is direct or immediate in that it is accomplished as quickly after death as is possible. In this situation the body probably will not be embalmed, and the only preparation will consist of bathing and washing. The 2002 survey of funeral directors found that 27.78 percent of all deaths involved immediate disposition (Cremation Association of North America, 2003). According to Cremation Association of North America (CANA), by 2011 the percentage of American deaths involving cremation had increased to

PRACTICAL MATTERS | THE HIGH COST OF SAYING GOODBYE

In the 1963 bestseller *The American Way of Death,* journalist Jessica Mitford was among the first to expose predatory practices in the funeral and cemetery business. Her work helped lead to the 1984 passage of regulations designed to curb abuses. Called the Funeral Rule, the regulations are enforced by the Federal Trade Commission.

Yet a three-month investigation by *Money* has found that consumers still face many of the old hazards—with new twists, as the industry faces revenue pressure caused by a growth in less expensive cremations.

The FTC's spot check of funeral homes last year found nearly one in four had serious violations of the Funeral Rule, involving their failure to properly disclose prices. Misleading buyers about federal and state law is another common problem. And consumer advocates say grieving families encounter many of the same issues with cemeteries.

Industry training tools obtained by *Money* reveal funeral pros sharing tips on how to hook grieving families into going over their budgets and to divert them from buying cheaper merchandise elsewhere.

And the magazine's survey of regulators in 48 states shows that financial disputes are growing over pre-need contracts, which allow you to buy funeral arrangements ahead of time and are now one of the top sources of complaints about the death care industry.

Funeral homes: Know your rights

It's the Funeral Rule's most basic requirement: Provide cost information over the phone and an itemized price list in person so that consumers can pick the specific goods and services they want and compare costs.

That's important given the wide disparity in funeral-home prices, even in the same area. An analysis of 10 major cities by Everest, a funeral-planning service, found that the difference in cost for a traditional funeral between the least and most expensive establishments within five miles of each other averages 164 percent.

Yet a surprising number of funeral homes still fail to follow the rules—27 percent of the 507 locations visited by the FTC in the past four years, up from an average noncompliance rate of 13 percent over the four years prior.

In New York City, one of the few cities or states with their own regs, 30 percent of facilities visited earlier this year did not have readily accessible price lists.

And when a city investigator, posing as the daughter-in-law of a terminally ill man, phoned funeral homes for pricing, 60 percent didn't provide it, as required by New York City law. "Stunning," says consumer affairs head Mintz. (The violators paid fines.)

The Funeral Rule also gives you the right to buy only the specific goods and services you want. Many families complain, though, that funeral directors pressure them to buy a package, says Lisa Carlson, executive director of the Funeral Ethics Organization, a nonprofit advocacy group.

"Often these packages include a lot of stuff you don't need, as a way to get you to pay more," she says. Extras such as grief counseling, digital tributes, and expensive floral displays can easily ratchet up the price.

A common tactic in selling such bundles, says Ed Markin, author of "An Affordable Funeral" is to present three price points. "Most people don't want to be considered cheap or overly extravagant," he says, so funeral directors know the family will typically choose the middle option and price accordingly.

A funeral home may also imply that state or local law requires a particular service, such as embalming, according to consumer advocates. The truth is most states almost never require embalming, and if a body does need to be preserved, refrigeration is just as effective, says Josh Slocum, executive director of the Funeral Consumers Alliance.

Because embalming brings in nearly $700 on average (and as much as $3,000), however, a funeral director might just forget to enlighten you about the facts, says Slocum.

Dan Rohling, a consumer advocate and former cemetery investigator from California, says homes are under pressure "to squeeze as much profit from every customer they can."

With 43% of funeral shoppers now choosing cremation over traditional burial, vs. just 26 percent in 2000, the profit per sale earned by independent funeral homes—the vast majority of the business—has fallen 27% in the past 10 years, the Federated Funeral Directors of America reports. Exacerbating the

continued

PRACTICAL MATTERS | THE HIGH COST OF SAYING GOODBYE *continued*

crunch: The industry has high fixed costs, including for embalming rooms, which are required by 37 states.

Financial pressures are also one reason the FTC has filed so few lawsuits against violators of the Funeral Rule—only six enforcement actions have been taken in the past 12 years. With limited resources to pursue costly litigation, the agency instead mostly diverts offenders into a three-year program that provides ongoing training and monitoring for compliance; they also pay fines (average: $6,000 per violator).

WHAT YOU SHOULD DO

Check prices at three homes, minimum. Four out of five consumers don't shop around when planning a funeral, AARP reports. Don't feel pressured to go with the first place you see or the one everyone you know used; you can compare costs by phone or online in less than a day.

Ask about both the mandatory basic-services fee, which covers the time and overhead of the funeral director and staff, and itemized charges for the additional services you select.

Get help with the search. If collecting multiple quotes feels too daunting, try efuneral.com or Everest's Price-Finder tool, good for pricing up to eight homes in your area ($29, everestfuneral.com).

You can also sometimes get discounted rates through your local branch of the Funeral Consumers Alliance (funerals.org).

Beating the high cost of caskets
With average prices ranging from $2,400 (for steel) to $3,500 (wood) and some copper and bronze models going for $10,000 or more, the casket is usually the most expensive item in a funeral.

The good news: Discount suppliers are growing in number, from independent online sellers to retailers like Costco and Wal-Mart, where coffins often go for less than half the price of comparable ones sold by funeral homes.

The bad news: Funeral homes don't make it easy for you to take your casket business elsewhere.

In some cases, funeral directors may refuse to accept outside merchandise or charge a handling fee, which violates the Funeral Rule. The practice resulted

in one of the three enforcement actions by the FTC this year: In June, Andrew Torregrossa & Sons Funeral Home in Brooklyn agreed to pay $32,000 to settle charges that it had refused services to two families (one was an FTC secret shopper) unless they bought a casket in-house.

Choosing cremation won't inure you to the casket push. A growing number of funeral homes offer casket rentals, usually for $800 or so, to view the body and expensive urns and containers for the remains.

Some also encourage buying a casket outright for the cremation, even though a cardboard box is sufficient.

Directors may try to get around that with questions that imply going without a casket is less than savory, says Markin. "They might say, 'Would you like a chance to say goodbye?' When the grieving party agrees, the funeral director might say, 'Well, you don't want to see your mom laid out in a cardboard box.' Suddenly you're back to buying a coffin."

WHAT YOU SHOULD DO

Look outside the home. Check out retailers like Costco.com, whose bestselling casket goes for $950, including shipping. Another direct supplier with a wide selection, BestPrice Caskets.com, has free ground shipping. Most caskets can be delivered within a day.

Ask for more options. Casket shoppers generally buy one of the first three models they're shown. Press for more choices and ask to see a catalogue, says Markin. Also request 20-gauge steel, not 18-gauge, which is thicker and costs $700 to $1,000 more. Markin says, "Both are suitable, and no one would know the difference."

And yet many homes do offer a slew of extras: memorial videos and websites, fancy vehicles for the funeral procession, keepsake jewelry (say, a pendant that holds ashes), even a white-dove release for the memorial service (six birds with music might run $200).

The family wanted a simple cremation and memorial. After being quoted $1,300 over the phone by one funeral home, Miller went with his wife and mother to finalize the arrangements. Once there, they were pitched a decorative box for the remains, a memorial video, and elaborate decorations.

"They insisted we buy a mandatory package with a floor of $2,600, even though we didn't need the components," says Miller. His family left, later finding a direct cremation service that charged a flat $895 fee.

The hard sell you may get at funeral homes can happen when you're arranging the burial too. There's a crucial difference: At the cemetery, there are no federal statutes like the Funeral Rule to protect you, and few states have regulations about sales practices either. Only six states require price lists, for example, says the Funeral Ethics Organization.

Yet equally big bucks are at stake. A single plot runs about $1,500 to $6,000, depending on where you live, according to IbisWorld, and you'll pay from $600 to $2,000 or more for cemetery workers to dig and re-cover the grave (you'll pay more if the burial is on a weekend).

Then there's the grave marker, which could mean $250 for a basic headstone or thousands for an elaborate bronzed, personalized monument.

To boost sales, cemeteries are also expanding offerings for cremation customers—only 15% buy merchandise, and when they do, they spend $700 on average, vs. $1,600 by families opting for a traditional burial, according to Janney Capital Markets.

Among the more common products: a niche, which is a recess in a wall where families can store an urn with cremated remains; and a bench ($3,000 at one Fort Lauderdale cemetery) or a "bench estate" for multiple family members ($15,000) in a "scattering garden," where relatives can spread the deceased's ashes and later visit.

Since cemeteries, unlike funeral homes, are not required to give you a price list, you also can't tell if you're being charged fairly, Jacobi says. Congressman Bobby Rush (D.-Ill.) has twice proposed legislation to extend the Funeral Rule to cemeteries and intends to try again next year.

Source: L. Gibbs and I. S. Mangla. (2012). The high cost of saying goodbye. *CNN Money Magazine*. November 9, 2012.

---

42 percent. CANA predicts that number to go up to 58 percent by 2015 (Cremation Association of North America, 2011).

Annual Growth Rate in U.S. Cremations

The annual growth rate is the difference between the yearly percentages of deaths cremated that are averaged over a five-year period. The table below shows the rates for the United States from 1996 through 2011. Further, this table illustrates that the cremation rate is increasing exponentially, with the current average growth at 1.64 percent annually.

TABLE 11.4 | PERCENT OF CREMATIONS OVER THE YEARS AND PERCENT OF GROWTH

Year	% Cremated
1996	21.8
2001	27.0
2006	33.8
2011	42.0

	% Change
% change 1996–2001	5.3
% change 2001–2006	6.8
% change 2006–2011	8.2
Average Annual % Growth Rate per Year over 2006–2011	1.64

Cremation Associations of North America, 2011.

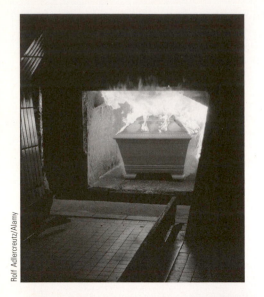

Rolf Adlercreutz/Alamy

AP Images/Michael Dwyer

CARLOS GARCIA RAWLINS/Reuters/Landov

Susan Van Etten/PhotoEdit

© Kzenon/Shutterstock.com

Johnny Green/PA Photos/Landov

Cremation involves burning the body, removing the bone-fragment remains, grinding them into pulverized fragments, placing them in an urn or some other receptical, and then burying or scattering them in a suitable place.

The frequency of this alternative differs by region of the country—it is performed most in the west and northwest part of the United States (approximately 70 percent of cases) and least in southern states (less than 19 percent of cases) (Cremation Association of North America, 2011). These percentages have gone up 10 percent in the west and northwest and almost doubled in the south since 2002. In a 2006 survey conducted by the SmithBucklin Corporation (cited by Cremation Association of North America, 2007), when **cremation** is involved, 62 percent of the cases involve immediate cremation and the body is not present during the funeral or memorial service.

Table 11.5 shows the five states with the highest and lowest percentage of cremations in 2010.

Reports from the Cremation Association of North America Statistics (CANA, 2008) on cremation are regarded as the most reliable and current ones available today. As an example, Figure 11.1 is just one summary from the 2008 CANA Statistics Report showing historical trends and predictions to the year 2025.

Body-bequest programs have become better known in the last four decades and permit the deceased (before death) or the family (after the death) to give the body to a medical institution. A compendium of body donation information (NFDA, 2003) indicates that, when the family desires, 75 percent of the donee institutions permit a funeral to be held before the delivery of the body to the institutions for study or research. Some medical schools will pay the cost of transporting the body to the medical school; others will not. This is the least expensive means of final disposition, especially if a memorial service is conducted without the body present. The compendium also indicates that in almost every instance the family may request that either the residue of the body or the cremated remains be returned when they are of no further benefit to the donee. In those instances when the

TABLE 11.5 | TOP FIVE STATES RANKED BY PERCENT OF CREMATIONS (2010)

State	Cremations	% Cremations
Nevada	14,697	72.2
Washington	34,019	70.9
Oregon	22,120	69.4
Hawaii	6,660	69.1
Montana	5,803	65.7

BOTTOM FIVE STATES RANKED BY PERCENT OF CREMATIONS (2010)

State	Cremations	% Cremations
Mississippi	3,918	13.8
Alabama	8,141	17.2
Kentucky	8,676	20.8
Louisiana	8,655	21.2
West Virginia	4,834	23.2

Cremation Association of North America, 2011.

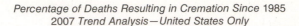

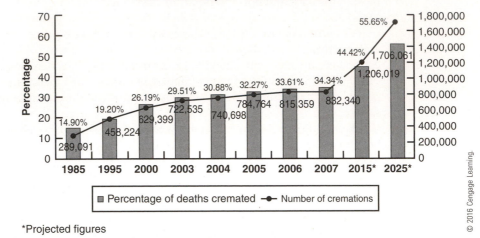

*Projected figures

FIGURE 11.1 | CREMATION DATA AND PREDICTIONS: DATA TRENDS
Cremation Association of North America, 2008.

family does not desire to have the body or the cremated remains returned, the donee institution will arrange for cremation and/or earth burial—often with an appropriate ceremony. People who are considering donating their bodies should be aware that at the time of death the donee institution may not need a body. If this does happen, the family will have to find another institution or make other arrangements for the disposition of the body. Furthermore, some families are surprised to find that they may be charged for transportation of the donated body. The 2007 NFDA survey found that gifts for anatomical research are made in less than 1 percent of all deaths (NFDA, 2007).

To some, body donation may not appear to be an alternative to the funeral (especially when a ceremony is held before delivery of the bequested body), but because the procedure is different from the most common methods of disposition, it may be considered as an alternative. According to the Centers for Disease Control, in 2010, approximately 15,197 such donations were made out of 2,468,435 deaths (CDC, 2010).

The memorial service is defined as a service without the body present. It is true that every funeral is a memorial service—because it is in memory of someone—but a memorial service, by our definition, is an alternative to the typical funeral. It may be conducted on the day of the death, within two or three days of the death, or sometimes weeks or months after the death. Those who wish to have a viewing can do so on the evening before the day of the memorial service. The memorial service typically places little or no emphasis on the death. Instead, it usually is a service of acclamation of philosophical concepts. Religious or nonreligious in content, such a service can meet the needs of the bereaved individuals. In cases involving cremation, less than 19 percent of the time is the body present (SmithBucklin Corporation, cited by Cremation Association of North America, 2007). In Minnesota and other states, it is becoming increasingly popular to have an urn present during the memorial service.

Organizations called **memorial societies** exist for consumers. An example is found in Ithaca, New York. The bylaws of this particular nonprofit and nonsectarian organization establish the following as purposes of this society:

1. To promote the dignity, simplicity, and spiritual values of funeral rites and memorial services
2. To facilitate simple disposal of deceased persons at reasonable costs, but with adequate allowances to funeral directors for high-quality services
3. To increase the opportunity for each person to determine the type of funeral or memorial service that he or she desires
4. To aid its members and promote their interests in achieving the foregoing purposes

Thus, such a memorial society would help educate consumers regarding death before the actual death of a significant other and present options for final disposition of the body. Likewise, many funeral directors today serve as valuable resource persons by sharing information about death with various community groups.

In Minnesota the largest memorial society is the Cremation Society of Minnesota (cremationsocietyofmn.com). This society is actually owned by the Waterston family, a private for-profit funeral service business with four locations—Minneapolis, Edina, Brooklyn Park, and Duluth, and a separate, high-volume crematorium (complete cremation can take place in less than 90 minutes) in Minneapolis. For $15 one can become a member of the Cremation Society of Minnesota and be able to access the services of the society in acquiring specialized economical cremation service.

At the time of one's death, the Minnesota Cremation Society is notified and the member's body is transported to the society's crematory, where it is held until proper medical authorization is secured, after which the body is cremated; no embalming is necessary. The cremated remains are handled according to the member's written instructions and may be picked up by the survivors or be delivered or mailed for a fee. The cremated remains are placed in a cardboard container, or permanent urns are available for as little as $12.

The cost of the basic cremation service, which includes removal of the body from the place of death, cremation, filing of the necessary papers, and cardboard container suitable for burial, is presently $1,395 for members (2013). This is payable at the time services are rendered; if the diseased is a nonmember, the cost is more. (This provides potential customers with an incentive to pay the $15 registration fee and become members of the society prior to death). If an individual dies out of state, the Minnesota Cremation Society will assist survivors in making all of the arrangements in the state in which the individual dies. If the death takes place in Minnesota, there is an additional transportation charge to the site of cremation from outside of the Twin City seven-county metropolitan area, or 25 miles from the Duluth facility. Presently the society charges $2 per mile one way.

In addition to transporting the body to the cremation, the Cremation Society's counselors will help survivors arrange a memorial service; write an obituary; prepare memorial folders; obtain certified copies of the death certificate; arrange cemetery services; select a permanent memorial urn, grave marker, or monument; notify the Social Security office; and file for veteran's benefits. There is an extra charge for these services. If families prefer to arrange a traditional funeral, they

PRACTICAL MATTERS | DIVERSITY IN DEATH: BODY DISPOSITION AND MEMORIALIZATON

Baby Boomers (individuals born between 1946 and 1964) are "coming of age" and beginning to reach the retirement years of their lives. With entry into this new cycle of life and with peers beginning to die, an interest in dying and death comes into focus. Yet, Baby Boomers tend to express their individuality regarding the final disposition of human remains and seek more personalized funerals: They are seeking alternatives to tradition. The TV shows *Six Feet Under* and *Family Plots* have given them some inside information about the funeral industry. They want to take control of their lives—and their deaths. Many are choosing ways to be memorialized that their parents and grandparents would not even have considered. They are beginning to rewrite the way America deals with the final chapter of life, thus becoming a major influence on the funeral industry.

Some recent personalized death trends involve cremation. Recent options for cremains include: (1) Placing them in paint for a painting, (2) putting them into orbit at a cost of from $1,000 for a less-than-10-year orbit of the earth in a one-gram capsule to $12,500 for a seven-gram capsule headed for the moon, or (3) inserting them in a concrete artificial reef (like a large wiffle ball) in the ocean at a cost ranging from $1,500 if in the reef with others' cremains or $5,000 if in a reef "alone," varying in weight from 400 to 4,000 pounds.

If a survivor is into diamonds, a diamond can be made from carbon in the brain of the deceased by a Chicago company called LifeGem. The process takes about 16 weeks. The cost of a quarter-carat diamond, not including cremation or setting, is $1,195, whereas a three-quarter-carat diamond costs $9,995. The diamonds are blue, due to the presence of boron, but LifeGem plans to offer red, yellow, and white diamonds in the future. Or if one is into gold, a ring can be made from the gold in the deceased individual's teeth.

If earth burial is to be the final form of disposition of the dead human remains, caskets that make a statement are available. Screen painting on metal caskets can be done. Caskets can be customized to fit the personality of the deceased. One can be frugal and buy a casket that also serves as a bookcase, a sofa, or an entertainment center, until needed for burial.

Lower-cost caskets can be purchased from a storefront business (Costco and Wal-Mart, for example, now offer caskets at discount prices). A funeral home is then obliged to complete the funeral process, though the casket was not bought from them.

Casket display rooms at many funeral homes are a thing of the past. Rather than walk into a room with 40 or 50 different caskets (a scene that can be overwhelming within 24 hours or so after a death), a products room may be found. This room might have less than a half dozen caskets, but will feature different linings for these caskets, cremation urns, miniature vaults, and numerous other funerary products.

Theme funerals are popular with some individuals. If the deceased was a Willie Nelson fan, for example, pallbearers can dress like the singer, as can the casketed body. If a beach person, ask the funeral home to bring in sand and cover the front of the funeral chapel. If a motorcycle enthusiast, bring Harleys into the funeral chapel, then ride off into the sunset behind the hearse. Music at the funeral may be that of the Beatles, rather than from a hymnbook. For those unable to attend the funeral, the service can be shown on Internet sites.

Other options include a butterfly or dove release at the funeral, a videotape of the funeral, photos of the deceased in the casket, a solar-powered video screen embedded in a headstone, sports attire placed on the deceased person's body, or a horse-drawn hearse for an old-timey funeral. Visitation at the funeral may include conversation stimulants such as memory boards (photos and other memorabilia). A hobby of the deceased can be brought to mind with her or his golf clubs, tennis rackets, needlepoint, or hunting or camping equipment.

For the environmentally concerned, a green funeral is an option—no embalming, no toxic chemicals, a biodegradable coffin, and no vaults, just whole bodies or ashes. Woodland burial sites are used. Memorial trees are often planted in lieu of a headstone. For a nongreen funeral, yet for one with environmental concerns, an Ecopad (casket made of recycled paper with cost around $500–$700) can be used for burial. If cremated, the earlier-mentioned artificial reef (Eternal Reef) provides "homes for the fish," thus an "environmentally correct" alternative.

The very expensive alternative ($28,000-plus) of cryonics (body freezing) has been around for a number of years. Another expensive option is mummification, done in Salt Lake City, Utah, at a cost of $35,000-plus. Though available, these two options are not popular and are infrequently utilized.

For keepsakes of the deceased, stuffed teddy bears can be given to the young survivors, with another bear placed in the casket, thus providing a "connection" with the dead person. Or "thumbies," fingerprint impression keepsakes, can remind one of the dead person.

Today's obituaries are often more personalized, giving particular traits of the deceased and making the reader feel more like she or he really knew the person after reading the obituary. Professional obituary writers are available to highlight the deceased individual to the public.

Just what sense can we make of the above? Back in the 1960s and 1970s, life in the United States began to change. Mainstream religious affiliations were declining, thus a more secular view of the world was emerging, making traditional religious rituals such as funerals and earth burials less attractive. A more heterogeneous population was evolving, with continued migration into the United States and with more internal migration from the farm to the city. With more mobility in today's society, there may no longer be the feeling of attachment to a particular locale by survivors of a deceased person, thus no need to bury a body to continually "visit." And for many individuals today, the idea of visiting a gravesite is appalling. With cremation, "visiting" is less necessary.

Baby Boomers seem concerned about the monetary aspects of final dispositions of dead human remains. Cost is likely a reason for the increase of cremation over earth burials, going from 10 percent of final dispositions in 1970 to nearly 30 percent in 2005 (price differences of $1,200–$1,500 for cremation and $5,000-plus for earth burials). In addition, cremation is quick, efficient, clean, does not waste land, and is less intrusive on the environment.

Beginning somewhere around the 1960s, death and sex were no longer taboo subjects. Somewhat of a deinstitutionalization in some areas began. For example, natural childbirth was reinvented, with less emphasis on hospitalized, controlled births. Baby Boomers are the generation that demanded public breastfeeding and homeschooling for their kids. Midwives are often preferred over medical doctors. Homeschooling is on the rise. Working at home became more popular in the 1980s than in previous years. Dying at home, away from the institutionalized settings of hospitals and nursing homes, via homebound hospice programs, emerged in the 1970s. Home funerals, without the involvement of a funeral home, are being talked about and practiced. In 45 of the 50 states an earth burial can take place without the aid of professional funeral services.

Baby Boomers are transforming an institutional death into a meaningful end-of-life experience. One could argue that death has been "resurrected," as evidenced through roadside memorials, Facebook, and "talking" gravestones in cemeteries. One's privacy in grief is being shared with the public online, thus "therapy" for one's loss. Traditional mourning is governed by conventions, but in the age of Facebook, selfhood is publically represented (Stone, 2010). Cremated remains of the dead person can be carried out in public via a necklace or bracelet or diamond made from the carbon in the dead person's body. What was a most institutionalized death is now being transformed into a meaningful end-of-life experience, an experience catering to the whims of those involved, not done merely out of tradition. As Bob Dylan wailed back in the 1960s, "The times they are a'changin'." Indeed, they are changing regarding body disposition and memorializaton trends.

Source: G. E. Dickinson. 2012. Diversity in death: Body disposition and memorializaton. *Illness, Crisis, and Loss,* 20(2), pp. 141–158.

may be held at one of the First Memorial Funeral Chapels (which performed 141 funerals in 2013) or in the church of your choice prior to cremation.

If one merely becomes a member of the society, all fees are paid at the time of the death of the member. However, members can also prearrange and prefund their funeral (also see section on "preneed funerals" later in this chapter) with the following two options for payment of services.

CREMATORIUM HOLDS OPEN HOUSE TO DEMYSTIFY PROCESS

Barbara Brotman, *Chicago Tribune*, November 13, 2012

In Illinois, Gerald Sullivan, president of the Cremation Society of Illinois, welcomed visitors to not just think about the cremation process, but tour the premises of the family-owned company's new crematory in Homewood, Illinois. "We want this whole thing to be transparent," said Sullivan, whose company has 10 locations in northern Illinois. "We like people to have information to make good decisions and show families that you can trust the people you're working with."

At the Cremation Society of Illinois' open house, visitors stood in the viewing area and looked through a glass wall at the auto loader, a raised track that pushes a container holding a body into the cremation chamber. "So we actually fired it up and showed what happened," he said. Through the open doors, people "could actually see the flames within the chamber." During a cremation, families can look through this glass to watch their loved one enter the chamber, Sullivan said; they can even turn a key to start the process if they want, or if their religious tradition requires it. But "people know the cemetery," Sullivan said. "No one says, 'Hey, let's go down and go to the crematory.'" But by the end of the day in Homewood, several dozen had.

Option 1: Funds can be placed in a trust account in a federally insured bank. The account is between the society and the member. In accordance with Minnesota law, this money is considered the member's asset, and he or she will receive a 1099 interest statement by January 31 of each year. If the member files income taxes, he or she will be required to pay the tax on the interest earned.

Option 2: Funds can be placed in a life insurance policy. With this option, members' increase in equity will not be taxed by the government.

Members can change any instructions related to their future funeral plans by contacting the society in writing. In reality, members and their families continue to control the funeral service until the family has delivered the service over to the Minnesota Cremation Society and other funeral service establishment (including the First Memorial Funeral Chapel). In 2013 the Minnesota Cremation Society performed 3,667 cremations in the Twin Cities area and 682 in Duluth, for a grand total of 4,349 for gross revenues of more than $6.1 million per year. Actually, most funeral homes in America provide services similar to those of the Minnesota Cremation Society, but the name of this society gives potential customers the impression that the business is a nonprofit organization, making this option more attractive to some consumers.

NEW TRENDS IN FUNERAL SERVICE

The newest trend in funeral service is to cultivate new business by courting the living with special services. Included in these services are working with the public in pre-need funeral planning, providing aftercare for survivors, and assisting community and religious organizations in providing death education for the public. All of these efforts can be thought of as an extension of past efforts to market and advertise funeral services, but the funeral industry has made concerted efforts to make these efforts effective in a very competitive industry where, in the past 32 years, the number of no-frills cremation services has increased significantly and corporate profits have declined by more than half—as a percentage of sales, from 13.2 percent in 1981 to 6.12 percent in 2012 (Federated Funeral Directors of America, 2013).

Finally, the average profit per funeral in the past 25 years has declined by more than a third—as a percentage of sales, from 9.87 percent in 1988 to 6.12 percent in 2012 (FFDA, 2013). This relatively low return on equity makes funeral service (like other capital-intensive industries) a very difficult one to enter as a new entrepreneur. For the entrant, the price of land and buildings would result in higher operating expenses to perform the same service than those of a more established funeral home. With the percentage of immediate cremations increasing to 42 percent in recent years, the profit ratios will probably decrease even more and exacerbate this problem even more for young funeral directors.

PRENEED FUNERALS: A "NEW" TREND IN PLANNING FUNERALS

One of the more controversial issues today in funeral service is the trend for **preneed funerals**. **Prearranging** is the process of arranging funerals in advance of need. This process can include selecting merchandise, planning the service, determining the method of viewing and final disposition, and selecting persons to be involved in the funeral. In addition to prearranging, **prefunding**, or making the legal commitment of money to pay for the funeral service, is also common. This is usually accomplished through insurance or a trust. *Preneed* is a generic term that refers to both processes of prearranging and prefunding.

Presently, approximately 1 million people a year prearrange their funerals, compared with 22,000 in 1960 (Anderson, 1997). A 1995 survey of its members conducted by the American Association of Retired Persons (AARP) determined that 7 million people had prearranged their funerals for a total cost exceeding 15 billion dollars (American Association of Retired Persons, cited by NFDA, 1997).

According to James Will (1988, p. 367) and William Hocker (1987, pp. 4–11), consumers prearrange and/or prefund funerals for these reasons:

1. To provide a forum for death-related discussions that is not profoundly affected by the grief that naturally accompanies death
2. To make one's funeral preferences known to one's survivors, thus assuring that survivors will not select a type of funeral that differs from the one desired
3. To provide an opportunity to personalize the funeral
4. To provide the dying with peace of mind—planning one's funeral can be one of the last pieces of unfinished business that one can accomplish for one's survivors
5. To give individuals an opportunity to get the most for their money through comparison shopping at a time when they are not faced with urgent need or overwhelmed by grief
6. To unburden loved ones of the obligation of planning and paying for a funeral
7. To protect an estate from funeral expenses in the future
8. To assure that funds are available in the future for the type of funeral that is desired

The disadvantage of preneed funerals from the perspective of consumers and funeral service professionals is the potential for consumer fraud. Some consumers have paid unethical funeral salespersons (some licensed and others not) for funerals and later found at the time of death that either the firm was no longer in business, the money had not been put into a trust account, and/or the deceased had moved to a location not served by the firm. For these reasons Thomas D. Bischoff, senior vice president of the Prearranged Funeral Division of Service Corporation International (the largest chain of funeral homes in the world), made the following recommendation:

> We, as funeral service professionals, must encourage that everything be done to minimize these eventualities. We would again suggest that an insurance-funded prearranged funeral program is the best safeguard that the funeral director and the consumer have. Insurance companies are very tightly and closely regulated and are required to maintain sufficient funds on hand to meet requirements. An insurance-funded prearranged funeral program has very little potential for mismanagement or fraud and as such should become the standard for the industry. (cited by Kelly, 1987, p. 32)

Ultimately, consumers must protect themselves against fraudulent entrepreneurs and dishonest "get-rich-quick" salespersons. Insurance-funded prearranged funeral programs are important potential safeguards, but the words of Robert W. Ninker,

---

LISTENING TO THE VOICES | MY FATHER'S COFFIN

My father, who is in his mid-80s, had his coffin built this past year. He has planned his service, down to the songs and the words. He wants it to be patriotic, positive, and irreverent, like him. The little family cemetery on their land in upstate New York, established by families long gone, has been cleaned up over the years my parents have lived there. He revived an old pink peony planted next to child Emily's grave, righted fallen stones, and dug away intrusive roots and vines. The surrounding stone wall was repaired and a new iron gate installed. Sons, daughters, and grandchildren helped with these projects, but mainly he did it himself, a little at a time.

The past winter, a carpenter friend built a coffin out of ash and maple cut and dried from my parents' woods. It is a beautiful object, with alternating light and dark stripes, carefully sanded, and shaped into the classic elongated triangle. We "viewed" it this Thanksgiving, wandering in small family groups from the cemetery to the woodshop where the coffin was drying in its latest coat of varnish. It is easier to look at old stones of unrelated people long dead than to ponder the very box your father will lie in when his life, as we know it, is over. One of my brothers would not come too close; he stood over in the corner by the tractor and looked at the coffin "sideways," as if unwilling to face the thought full on of what is to come.

Perhaps my father plays out in his mind the details of the ceremony—words and songs, pallbearers so familiar to him, sons and sons-in-law he raised to know how to work and how to stay loyal, the hole already dug, the shroud of lumber he knew as trees, the people standing about, especially my mother, all known and loved—he can picture it all and this is comforting. This is how he is assured his life continues because it is as if he can see it after he dies. It is as if he is still there.

If we think of God as a living quality that we transmit to each other in life, not as something or someone far away, but as the love and patience and sheer livingness of each other, then this may be my father's God. God will be there in those who carry the coffin, sing the songs, laugh about the old stories, say the prayers, share the food, stand under the sky, listen to the birds, and cry the tears. In that sense, my father will have eternal life.

From Martha Garrison, Spring 2005.

Matt Garrison

Today, many people prearrange their funerals. Bill Garrison is standing by his wooden coffin made by a friend from wood on Bill's farm in upstate New York. Bill died on May 5, 2013. As he had previously planned his funeral, the funeral went according to his arrangements, including his being buried in the coffin made by his friend.

(See Listening to the Voices on page 424.)

executive director of the Illinois Funeral Directors Association, should also be heeded:

> The best protection a buyer can have is to purchase from a funeral director with a long history of success, after asking and understanding that this person is licensed and, in fact, trusts his funeral funds…. The buyer should also check references with acquaintances. That is about as much certainty as there is in life. Of course, if the buyer responds to door-to-door sellers or boiler room operations, he is his own victim! No one can protect someone from his own fool-like actions. (cited by Kelly, 1987, p. 32)

AFTERCARE: THE FUTURE BUSINESS IS TO BE FOUND IN TODAY'S CUSTOMERS

The funeral industry has always known that 80 percent of its business is with families served in the past. However, aftercare is one method to increase the likelihood of getting repeat business. The newest trend in funeral service is to provide extensive aftercare services and products for widows and widowers. Among these services and products are grief therapy, bereavement support groups, video tributes, and even greeting cards sent to survivors to mark the anniversary of death or the deceased's birthday (Scott & Dolan, 1991).

In 1984 Accord Aftercare Services (now New Leaf Resourses; http://newleaf-resources.com/) began providing faith-based bereavement support materials to funeral homes, public and private organizations, hospices, hospitals, support groups, counselors, and individuals. Through its professional development seminars and grief programs, New Leaf Resources offers training in the areas of grief counseling, communications skills, and aftercare for funeral service personnel and the families they serve. New Leaf Resources designs materials, such as a self-study grief workbook, an online magazine (*New Leaf Magazine*), the videotape *Whitewater: The Positive Power of Grief*, and brochures for "forgotten grievers" that can be redistributed by local funeral homes to maintain contact with families after the funeral service. Annually, New Leaf Resources serves more than 50,000 bereaved families through its products, programs, and seminars.

Many other funeral homes in the United States are now offering grief counseling and support groups for their clients. In the 1970s the Carbon Funeral Home in Windsor Locks, Connecticut, created a support group called Begin Again. Begin Again meets weekly and sponsors seminars on a wide range of topics, including financial management, car maintenance, and occupational reentry. Begin Again even has meetings, the primary function of which is entertainment, featuring magicians and ventriloquists (Anderson, 1997).

Other funeral establishments are sponsoring tree-planting ceremonies and annual memorial observances as a service to the families of the deceased. In Chicago, the Blake-Lamb Funeral Home provides free limousine services for weddings in families of the deceased (Scott & Dolan, 1991). An optional service sponsored by the Fitzgerald and Son Funeral Home of Rockford, Illinois, is the annual Walk to Remember for families mourning stillbirths, miscarriages, and early infant death. During the walk, participants plant a tree in a public park, write their babies' names on paper, and put them next to the tree (Anderson, 1997).

Presently, the majority of funeral homes providing extensive systematic after-care are owned by individuals rather than large corporations. Funeral homes with local ownership have much more control over budget priorities and tend to be more responsive to community needs for involvement. These locally owned funeral homes also recognize that aftercare services attract more business and therefore can be good for "the bottom line," even though aftercare is not directly responsible for revenue enhancement.

Much of the growth, inspiration, and certification of aftercare workers come from membership and participation in the Association for Death Education and Counseling (ADEC, see http://www.adec.org). ADEC provides training and certification for individuals who provide aftercare. ADEC certifies individuals in thanatology: death, dying, and bereavement.

An example of such aftercare efforts can be found at Weeks's Funeral Home in Washington state. Catherine Johnson, a certified thanatologist and member of ADEC, is the bereavement services coordinator at Weeks' Funeral Home. As an employee of Weeks' Funeral Home she provides the following aftercare services for clients: extensive information on grief and local resources, individual grief counseling, a support group for bereaved parents, referrals to other support groups in the community, follow-up telephone calls two months after the funeral or memorial, special anniversary cards (on the birthday of a deceased child or the wedding anniversary of a widowed person), a "Handling the Holidays" seminar (which is open to the public), a library of books and tapes on grief-related topics, and the annual Christmas memorial service. In addition, Johnson speaks to local churches and service organizations, provides in-service education to hospice groups, and writes a twice-monthly article for the local newspapers on grief-related topics. Rather than using aftercare materials produced by Accord, Weeks' Funeral Homes produces most of its own aftercare materials given to client families (including the cards).

Duane Weeks (owner of the Weeks' Funeral Home) and Catherine Johnson have also edited a book on aftercare that provides several helpful chapters for those interested in funeral home aftercare. The book is *When All the Friends Have Gone: A Guide for Aftercare Providers*, and was published by Baywood Publishing in 2001.

Janet Blankenship, resource coordinator for the Ohio Funeral Directors Association, has done a survey of some of the aftercare services provided by Ohio funeral homes. To give an understanding of the range of services provided by funeral home aftercare services, following is a partial list of the more than 100 services offered:

- Grief Support Groups for various loss-related populations: pregnancy and infant loss, child loss, teen survivors, widows and widowers, suicide survivors, adult–child parental loss, grief groups for grandparents suffering loss, and grief programs in nursing homes.
- Special holiday services: Christmas, Thanksgiving, Valentine's Day, Memorial Day, Veteran's Day, and spring remembrances. Some of these services include ceremonies of candle lighting, tree planting, donating blood (gift of the heart), wearing black ribbons to remember loss, walks to remember, and potlucks.

- Grief counseling training programs for hospice workers, church workers, and community volunteers.
- Community-sponsored grief and loss seminars (including workshops, speakers, newspaper columns, newsletters, pamphlets, and statewide conferences on grief and loss) and facilitating men and women's groups in the community for the sharing of grief and loss.
- Cooperation with and support of community hospice programs and their work in aftercare.
- Sponsorship of death anniversary services of remembrances in the community.
- Maintenance of a grief and loss lending library as a community resource.

FUNERAL SERVICES ON THE INTERNET

Innovation and change are constants in any business, and death-related businesses are no exception. When the FTC in 1984 mandated that all funeral homes in the United States itemize their fees and that consumers have access to pricing information over the phone, and when e-commerce through the World Wide Web became accessible for all Americans at the end of the 20th century, the stage was set for death-related business decisions to be affected by high-technology communications.

In the United Kingdom, technology related to coping with death is now available as an application on one's iPad or iPhone. The application is called "What to Do When Someone Dies." This application covers a range of potentially complex processes, such as registering a death, making funeral arrangements, choosing burial or cremation, sorting out the deceased's affairs, dealing with probate, inheritance tax, and the administration of an estate. It can also help users to organize and keep track of arrangements by allowing them to make notes, and add and edit calendar events and contacts (Royal College of Nursing Publishing Company, 2012).

Today, in the United States and elsewhere, many funeral homes and cemeteries provide consumers and members of the community with obituaries, price information, and preneed and at-need funeral planning via the Internet. One can purchase funerals, caskets, vaults, cemetery plots, urns, flowers, obituary notices, cremation services, and scattering services entirely online and pay for these services with credit cards, without ever meeting face-to-face or talking with funeral personnel. Because the FTC has determined that the cost of funerals must be itemized and that consumers must only pay for those services they order from funeral establishments, it is possible to purchase caskets, vaults, and urns online from e-business vendors and have them delivered to funeral directors who are making the arrangements for the final disposition. Today's Internet consumer will find cost-comparison shopping possible when making funeral preparations.

Presently there are five major e-commerce full-service cooperative death-related businesses on the Internet—Legacy.com, Funeral.com, Thefuneraldirectory.com, Funeralnet.com, and Usafuneralhomesonline.com. These websites include direct links to websites of individual funeral homes, online obituaries, virtual cemetery visits, professional grief and religious counselors, a bereavement chat room, and an e-commerce component that allows the visitor to shop for and order services and products online. Consumers also have online access for sympathy gifts and cards, for making charitable contributions in a loved one's name, and for last-minute travel arrangements. A major advantage of these e-businesses is that they allow consumers to preplan their funerals (or a funeral for someone else who has recently died) and share these plans with relatives and/or significant others. Internet e-funeral websites are free to consumers because of fees paid by the funeral-related industries that make sales through the site.

For example, Legacy.com collaborates with more than 1,500 newspapers in the United States, Canada, Australia, New Zealand, United Kingdom, and Europe to provide ways for readers to express condolences and share remembrances of loved ones. Legacy.com draws more than 18 million unique visitors each month, making it one of the 50 most visited websites in the United States. In addition to hosting more than 20 million obituaries, Legacy.com also provides newspapers with online solutions for their celebration, pet, and public notice categories.

In recent years most funeral homes and providers of cremation and memorial services have their own websites to increase traffic to their own establishments. What promised to be a big Internet business at the turn of the century has been incorporated into local funeral homes and cremation services where they advertise professional services offered, merchandise sold, obituaries for pending or past funerals, and preplanning opportunities.

A related consumer service is the cyberfuneral, also known as funeralcasting or memorial webcasting. This type of Internet service allows families to broadcast the funeral service to a larger audience via the World Wide Web. This can be done live (especially for family members or friends who are unable to attend the actual service) or as an archived event to serve as a commemoration of the person's life and death. Some cyberfuneral companies attach the funeral webcast to personal web pages that include photos, video clips, or even PowerPoint presentations. For a broader discussion of cyberfunerals see C. Sofka (2009).

Another cyber service to the grieving is MyDeathSpace, where the details of deceased individuals' lives can be examined. For some, homes offer family photos and home movie footage for showing over the web, or the funeral can be viewed "live" over the Internet (Markowitz, 2005). Newspapers (through Legacy.com) offer cyber guestbooks where one can express condolences to the bereaved without going near the funeral home or church where the funeral is being held. CaringBridge.org gives families an opportunity to communicate with caring friends before and after the death.

A recent option for those grieving over the death of a significant other is Facebook.com. Facebook has a memorializing policy in regard to the pages of those who have died (Stone, 2010). Such is a place where individuals can save and share their memories of those who have died. Individuals all over the world can post messages, photos, and videos. A person's Facebook page can remain active in perpetuity, unless family members request that it be taken down.

Finally, a rather unique option on the Internet is an e-mail service called "my-last-email.com," which, on the day of the death of an individual, will e-mail prearranged messages to individuals from the deceased individual. Thus, if you wanted to say something positive (or negative) to someone, but could not do it to his or her face, the e-mail service would do that for you—after your death, with no chance for reprisal, if the message were negative.

"Counting you and the outpouring of comments on facebook, that makes three."

THE BUSINESS OF BURYING THE DEAD

Disposing of the dead in early American society meant burying the deceased in such a manner that respect was given to the body, the life that had been lived was memorialized, and the values of the living were reinforced; yet, the survivors were allowed to detach themselves from the deceased and return to their everyday lives. Like funeral functionaries, cemetery sextons and other grave-related workers assisted families as they dealt with the practical problems of disposing of the dead and returning to the task of living.

THE CHANGING AMERICAN CEMETERY

Like funeral services, contemporary American cemeteries have evolved from colonial landscapes that looked much different. Before the Industrial Revolution, cemeteries provided for the community a *memento mori*. For our 18th-century ancestors it was important to "remember death": The fact of our death should motivate us to work hard and do good works as a preparation for a final judgment by a righteous God.

THE PURITAN CEMETERY Before the development of the new cemeteries in the 19th century, Americans interred their dead in various types of places (Sloane, 1991). The earliest burials by pioneers were in unorganized, isolated places. The graves were typically unmarked because the grave would not be maintained. The European influence was present: The European custom was that the corpse was not highly regarded and dead bodies were relegated to trenches or places for the bones of the dead. In some cases a burial place on a family farm might be used.

Clusters of graves soon grew as the pioneers settled into small villages. These pioneers were breaking from European tradition in not burying their dead next to churchyards. Domestic burial grounds were found in all of the colonies, but were less popular in New England where the Puritans were more likely to follow English custom and bury their dead in a central burial ground beside the new churches, noted Sloane. Occasionally, a church member could be **entombed** beneath the church in public or in a private vault. African American slave burial places tended to be in community graveyards rather than in family plots. Religious graveyards, whether Protestant, Catholic, or Jewish, were remarkably similar in their layout, monuments, and management, stated Sloane.

Graves in these churchyards were designated with wood and stone markers, primarily using slate and sandstone. Carvings on gravestones appeared in the 1700s. Earlier stones had simply noted the deceased person's initials, age, and birth and death years. The appearance on gravestones of the cross-bones and skull and the soul faces symbolized the struggle between mortality and immortality of 17th-century Puritan New England's "almost obsessive concern with the oppressive inevitability of death and the necessity of reminding the living of the uncertain fate of the soul" (Combs, 1986, p. 18). Yet, the appearance of the soul figure embodies the Christian's confidence in a resurrection and offers a reassurance of salvation.

Most towns had a potter's field in which to bury the destitute or those not accepted into other burial places. Potter's fields were transient places and were often abandoned after a severe epidemic.

DEATH ACROSS CULTURES | CEMETERY ICONOGRAPHY

In the United States the evolution of cemetery practices and the shifting meaning of death can be traced to the early colonial period, when cemeteries took the form of church or town graveyards and iconography encompassed an age of sacred death. Prior to the mid-1600s there is no indication that the New England Puritans marked the graves of their deceased, but by the 1660s the practice of grave marking was widespread. James Deetz documents that stonecutters used three basic designs: death's heads, winged cherubs, and a willow tree overhanging a pedestaled urn. Death's heads were most common from the 1680s to the mid-1700s. In the same genre were carvings that emphasized the flight of time: the hourglass, a scythe in the hand of Death or Father Time, skeletons, crossbones, and death darts. Death's heads tended to become less severe during the early 18th century, metamorphosing into a pleasant cherub or angelic image. Some carvings began to use background designs of foliage, grapes, or hearts that softened the presentation of death. Verses of consolation and lines of poetry conveying a message of hope also began to appear. Increasingly, death was portrayed as benign sleep. The new gravestone art symbolized optimism, depersonalization of death, and a new interest and appreciation for nature. The optimism reflected changing religious beliefs from the Puritan doctrine of predestination to the notion that in Jesus Christ, individuals could find salvation that would ensure life after death. The willow tree and urn motif also marked the end of town graveyards of the colonial period and the rise of the modern cemetery.

From MacLean, V. M., & Williams, J. E. (2009). Cemeteries. In C. D. Bryant & D. Peck (Eds.), *Encyclopedia of death and the human experience* (pp. 169–173). Thousand Oaks, CA: Sage Publications.

Segregation by race occurred in life as well as in death. Both in the South and in the North, public and private graveyards were racially segregated, observed Sloane. Sometimes separate graveyards were established for African Americans and whites.

THE CITY GARDEN CEMETERY Nineteenth-century cemetery reform derived from religious liberalism, romantic naturalism, middle-class family sentimentality, and scientific concerns. In keeping with the middle-class philosophy of death with order, cemeteries were changed into small family gardens that hid the harsh reality of death in an attempt to create, instead, a forgetfulness of the certainty of death, and sentimental feelings about a peaceful afterlife.

In the early 19th century, embroidered or painted mourning pictures flourished as a form of memorialization. These pictures often showed stylized graveyard scenes, including standard features such as the weeping willow, the gravestone and **epitaph** (gravestone inscription) of the deceased, the mourners, and a *memento mori*. These remained popular until the 1830s, when printed memorials with spaces for names and dates came into vogue. During the middle third of the century, posthumous mourning portraiture also flourished, depicting the deceased with conventional symbols of mortality like the broken shaft or roses held with blooms downward. These paintings were drawings of the corpse, and they expressed the desire for the restoration of the dead through art. All of these forms of mourning art provided icons for the bereaved family to contemplate as a part of the extended mourning ritual (Lloyd, 1980).

LISTENING TO THE VOICES | GRAVE REMARKS: I'LL WRITE MY OWN EPITAPH BEFORE I LEAVE, THANK YOU

Epitaphs are footnotes chiseled on tombstones. They are parting shots taken at or by the deceased. They can be patriotic, poetic, profound, or pathetic. They can be wise, witty, or just weird. They can glorify, be grievous, or be gruesome. Few of them have summed up a person's attitude toward life as well as one found in a Georgia cemetery:

   I told you I was sick!

Perhaps the most famous epitaph is one credited to W. C. Fields, written for himself: *On the whole I'd rather be in Philadelphia.*

And in a Moultrie, Georgia, cemetery: *Here lies the father of twenty-nine He would have had more but he didn't have time.*

On the tombstone of a hanged sheep-stealer from Bletchley, Bucks, England: *Here lies the body of Thomas Kemp Who lived by wool and died by hemp.*

In a Thurmont, Maryland, cemetery: *Here lies an atheist—All dressed up and no place to go.*

In a Uniontown, Pennsylvania, cemetery: *Here lies the body of Jonathan Blake; Stepped on the gas instead of the brake.*

Written by a widow on her adulterous husband's tombstone in an Atlanta, Georgia, cemetery: *Gone. But not forgiven.*

Similarly, a Middlesex, England, widow put this on the gravestone of her wandering husband: *At last I know where he is at night!*

An English epitaph over the grave of Sir John Strange, a lawyer: *Here lies an honest lawyer, and that is Strange.*

An 1890 epitaph of Arthur C. Hormans of Cleveland, Ohio, puts things into startlingly clear perspective: *Once I wasn't. Then I was. Now I ain't again.*

From *The People's Almanac* (pp. 1312–1320), by D. Wallechinsky and I. Wallace, 1975, Garden City, NY: Doubleday; and *Lexington Herald*, Lexington, KY, September 21, 1979, pp. D5.

In 1895 a writer in *American Gardening* wrote that "the modern garden cemetery, like the modern religious impulse, seeks to assuage the cheerlessness and the sternness of life and to substitute the free and gracious charity of One who came to rob death of its hideousness" ("Extracts," 1895). By the 1830s many physicians had begun to worry about the possible health hazards of city graveyards. By the same time, commercial development had raised the price of land on which graveyards were located, and romantic ideas of landscape architecture had begun to affect the aesthetically oriented members of the middle class. Also, space limitations of city graveyards prevented the possibility of family plots, and many of the burial places had become overcrowded, unkempt, and unsightly. (*Adapted from an article written by James Farrell and used with his permission.*)

THE RURAL CEMETERY   The solution to these problems was the rural cemetery—a landscaped garden in a suburban setting. The rural cemetery movement sought to simplify the cemetery and provide an aesthetic environment in which grave markers did not detract from the beauty of the natural surroundings of grass, trees, flowers, and rolling hills. With this sense of romantic naturalism, cemeteries were a place to go and reflect upon the peace and beauty of the new world where the dearly departed had gone to reside.

In 1831 Mount Auburn Cemetery was founded four miles west of Boston, and its success stimulated the spread of such cemeteries all over the country. In rural

Lou Jones/Lonely Planet Images/Getty Images

Mount Auburn Cemetery in Massachusetts, founded in 1831, was the first rural landscaped cemetery in the United States. By moving cemeteries away from the cities, they were "out of sight," and thus "out of mind" (fitting the "dying of death" period from 1830 to 1945).

cemeteries, family plots averaging 300 square feet were nestled among trees and shrubs upon the slopes of soft hills or on the shores of little lakes. Paths curled throughout the grounds, passing lots enclosed by stone coping or wrought-iron fences and surmounted by a monument of some sort. Such cemeteries were established to bury bodies; to ease the grief of survivors; to bring people into communion with God, with nature, and with deceased family and friends; to teach important lessons of life; to surround bereavement with beauty and to divert the attention of survivors from death to the setting of burial; to display taste and refinement; and to reinforce the class stratification of the status quo (Bender, 1973; French, 1975; Rotundo, 1973).

The founders of Mount Auburn were liberal Unitarian reformers—progressive professionals and businesspeople who considered the cemetery in the context of social developments of their day. They equated the family with the garden and imagined both as a counterpoint to a society of accumulation. In the same way that upper-middle-class families moved to suburbs where curved roads and greenery contrasted with the grid-block plan of the cities and offered space to raise a family, they moved from the "cities of the dead" to rural cemeteries that offered space to "plant" a family.

This was also the period of the discovery of the asylum, when social deviants were located in restorative rural settings that would rehabilitate people away from the contaminating influence of urban life. Many reformers saw the cemetery, like the insane asylum, the orphanage, and the penitentiary, as an "asylum" from urban ills. In the cemetery, "the weary and worn citizen" was also rehabilitated. Cleaveland noted:

> Ever since he entered these greenwood shades, he has sensibly been getting farther and farther from strife, from business, and care.... A short half-hour ago, he was in the midst of a discordant Babel; he was one of the hurrying, jostling crowd; he was encompassed by the whirl and fever of artificial life. Now he stands alone in Nature's inner court—in her silent, solemn sanctuary. Her holiest influences are all around him. (1847, p. 13)

The rehabilitation of the rural cemetery paralleled the philosophy of education found in the common school. Like the new compulsory schools, rural cemeteries responded to middle-class fears of mobilization of the masses in the age of President Jackson. Educational reformers such as Horace Mann both reflected and affected cemetery proponents in their belief that "sentiment is the great conservative principle of society" and that "instincts of patriotism, local attachment, family affection, human sympathy, reverence for truth, age, valor, and wisdom ... constitute the latent force of civil society" (Tuckerman, 1856, p. 338).

Like these other reforms of antebellum society, rural cemeteries took the public mind by storm. Cities throughout the United States established rural cemeteries and people flocked to visit them. In New York, Baltimore, and Philadelphia, Andrew Jackson Downing estimated that more than 30,000 people a year toured the rural cemeteries. Consequently, he wondered whether they might not also visit landscaped gardens without graves. In articles such as "Public Cemeteries and Public Gardens" and "The New York Park," Downing (1921) advanced the idea that would result in New York's Central Park and a new direction for cemetery development.

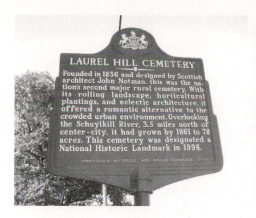

Michael R. Leming

Founded in 1836, Laurel Hill Cemetery in Philadelphia overlooks the Schuylkill River three-and-a-half miles north of the city.

THE LAWN PARK CEMETERY The new direction was the lawn park cemetery, emphasizing a new aesthetic—efficiency—and the absence of death. By the end of the century, members of the Association of American Cemetery Superintendents routinely wrote in journals like *Park and Cemetery* that "a cemetery should be a beautiful park. While there are still some who say 'a cemetery should be a cemetery,' ... the great majority have come to believe in the idea of beauty" (Simonds, 1919, p. 59). This new aesthetic emphasized the open meadows of the beautiful style over the irregular hill-and-dale outcroppings of the picturesque style and the irregular outcroppings of obelisks and monuments in the unregulated rural cemetery. This aesthetic coincided with considerations of efficiency, because the uncluttered landscape required less upkeep than the enclosures and elaborate monuments of the rural cemetery. Finally, this aesthetic buried death beneath the beauty of the design. Andrew Jackson Downing (1921, p. 59) had said that "the development of the beautiful is the end and aim of all other fine arts.... And we attain it by the removal or concealment of everything uncouth or discordant." The cemetery superintendents practiced what Downing preached. "Today cemetery making is an art," said one superintendent in 1910, "and gradually all things that suggest death, sorrow, or pain are being eliminated" (Hare, 1910, p. 41).

Cemetery superintendents eliminated suggestions of death by banning lot enclosures and grave mounds and by encouraging fewer gravestones and fewer inscriptions. They banned lot enclosures (fencing or stone coping) because these broke up the unified landscape, blocked the path of the lawn mower, and signified a "selfish and exclusive" possessive individualism. Whereas grave mounds obstructed the view and the lawn mower, they also reminded people of death. Without them, a cemetery lot would evoke "none of the gruesomeness which is invariably associated with cemetery lots.... *No grave mounds are used, so save the headstones, there is nothing to suggest the presence of Death*" (Smith, 1910, p. 539).

Some superintendents did not want to save the headstones either, proposing instead "a cemetery where there is no monument, only landscape" (Association of

American Cemetery Superintendents, 1889, p. 59). Most superintendents favored sunken stones at the site of the grave. Howard Evarts Weed (1912, p. 94), author of the influential publication *Modern Park Cemeteries*, argued that "with the head-stones showing above the surface we have the old graveyard scene, but buried in the ground they do not appear in the landscape picture and we then have a parklike effect." Some superintendents zoned the cemetery to permit monuments only on large "monument lots." This allowed the superintendent, like a real estate agent, to charge premium prices for such lots and for prime locations for corner lots or hillside or lakeside property. This **social stratification** of the cemetery allowed for the social mobility of the dead as ambitious dead people moved to better "neighborhoods" as their survivors saw fit. Where superintendents could not ban or limit the number of monuments, they tried to make them as unobtrusive as possible, preferring horizontal monuments to the earlier upright markers and preferring inexpressive epitaphs to the poetic epitaphs of earlier times. They wanted to replace the *memento mori* of earlier stones with "forgetfulness" as they buried death with the dead (Farrell, 1980).

Cemetery Services to Soothe the Mourners   Superintendents tried to structure cemetery services "to mitigate the harshness and cruelty of death and its attendant details and ceremonies" (Seavoy, 1906, p. 488) and to provide a sort of grief therapy for bereaved individuals. They encouraged private family funerals, and they tried to remove or conceal the uncouth and discordant aspects of interment. They carted the dirt away from the grave, or they hid it beneath cloth, flowers, or evergreens. They lined the grave with cloth to make it look like a little room. They suggested changes in religious services, and they escorted mourners away from the grave before filling it, to shield them from the finality of death. In all of these services, "everything that tends to remove the gloomy thoughts is done.... The friends cannot but leave the sacred spot with better, nobler thoughts, freed from the gloom and terror that otherwise would possess them" (Hay, 1900, p. 46). (*Adapted from an article written by James Farrell and used with his permission.*)

Graveyard Symbols of Death

When one goes to a cemetery, the purpose is usually to attend a burial service. At such times the individual reflects on the life of the recently deceased person and pays little attention to the surrounding gravestones. Try visiting a cemetery, especially an old one, under less stressful circumstances, to examine the gravestones.

If you have access to gravestones dating back to the early 18th century, you may see some markers with cross-bones and a skull, pointing out the stark reality of death—after death, muscle and skin deteriorate, and the skeleton remains. Other markers may have a soul face, usually a smiling, portrait-like face, often with wings. Unlike the cross-bones and skull, the soul face with wings symbolizes the idea of an afterlife. Some markers from the early 1700s may have a bust of the deceased engraved on the stone. These busts are usually reserved for someone of prominence, such as a minister. Eighteenth-century markers are typically made of slate or sandstone.

Walking through the cemetery, an individual might see various symbols of the passing of life and of death on the gravestones: an hourglass ("like sands in the hourglass, so are the days of our lives" (from the U.S. television show *Days of Our Lives*), babies and women weeping (men weeping are rarely seen), Greek and Roman urns (the open urn is symbolic of the separation of the soul from the body), broken columns (not whole and complete but broken), weeping willow trees (not just any willow tree), fallen trees (thus, will die), broken limbs (severed from life), the eternal flame (life continues), the Grim Reaper (who "cometh" in the night to "take one away"), and a lamb on children's markers (innocence and gentleness).

The idea that we do not die in the United States, but simply go to sleep, is evident in a cemetery. The word *cemetery* itself derives from the Greek and means "sleeping place." *R.I.P.* ("rest in peace") is sometimes seen on a gravestone. Bedshaped markers, complete with headboard, footboard, and sideboards surrounding the grave, are symbolic of resting. Double beds, single beds, and cribs for children can be found. One 19th-century marker in Columbia, South Carolina, gives the birth date of the deceased, but rather than saying "died" says "went to sleep" on a certain date.

The stroll through an old cemetery might also reveal "tabletop" markers with four legs and a top. These markers served a function during the Civil War, in that they were used as operating tables in some southern cemeteries. With a limited number of hospitals available and with a need for surgical facilities, these were available and about the right height. There are also false crypts that are "enclosed tables" with sides filled in between the "legs." False crypts are part of the English "table tombs," the preferred burial style of socially prominent people.

Epitaphs (inscriptions) on the markers were often rather lengthy in the 18th and 19th centuries and gave a brief history and personal traits of the persons buried there. Though worn by the passing of time, many of the epitaphs on these gravestones are still very legible. The viewer can sometimes note the history of when various plagues or diseases occurred, from the numerous members of a family who died within days of each other. An inscription of particular note is on a marker in Charleston, South Carolina, for a sea captain, aged 37, who went down with his ship in 1754. A reclining, smiling skeleton is displayed with its head

George Dickinson

Greek and Roman columns, false crypts, and tabletop markers are prominent in Charleston, South Carolina, where the oldest gravestone dates back to the 1690s.

resting on a winged hourglass. Above the smiling figure of death is the inscription "yesterday for me, today for thee."

In contrast to the earlier U.S. cemeteries, modern cemeteries are more uniform and less personalized and have limited gravestone variation in size and shape. Contemporary markers, typically made of granite or marble, generally have the name of the deceased and the birth and death dates. Rarely does a marker today have an epitaph other than perhaps an occasional quotation.

Thus, a walk through a cemetery, especially an older one, could prove to be a peaceful and enlightening experience for the observer. In a place for the dead, the cemetery seems to have life with all the flowers and trees. The venture can be like walking through a history book. In a *memento mori* field that reminds the individual of mortality, one is surrounded by immortality.

CEMETERIES TODAY

As one can see, there have been many changes in American cemeteries. Cemeteries constructed in the 20th and 21st centuries often have a great variety of displays to memorialize the dead. There may be some sections with large above-ground tombstones and even grave enclosures with head and foot stones to provide a **memento mori**. There might also be sections that only allow flat markers, and often these sections have a single above-ground memorial centerpiece to commemorate all those buried in these more simple surroundings.

Although the appearance of the modern cemetery will vary to some degree, in general, there are three basic types of cemeteries found in the United States: government cemeteries, nonprofit cemetery associations, and for-profit cemeteries.

The Government Cemetery

The federal government, through the Veterans Administration (http://www.cem.va .gov/), has established 131 national cemeteries in 39 states (and Puerto Rico) as well as 33 soldiers' lots and monument sites to honor individuals who have given their lives in the military service of their country. In addition the Department of the Army maintains two national cemeteries, the Arlington National Cemetery and the U.S. Soldiers' and Airmen's Home National Cemetery. Arlington National Cemetery, located in Arlington, Virginia, was the first and probably most famous of the national cemeteries. In this cemetery are more than 250,000 graves of not only soldiers but also of political leaders (e.g., John, Robert, and Edward Kennedy), 3,800 former slaves who were released as a result of the Civil War, and many war-correspondent journalists who died in combat. Annually, more than 4 million visitors to Arlington National Cemetery pay their respects at this 612-acre cemetery, which was formerly part of a plantation owned by Mary Anne Randolph Custis, wife of Robert E. Lee. Along with 14 other cemeteries, Arlington was set aside by President Abraham Lincoln in 1862 as a national cemetery. Today there are more than 2.5 million national cemetery gravesites, including full-casket gravesites, in-ground gravesites for cremated remains, and columbarium **niches**. The largest of the 130 national cemeteries is Calverton National Cemetery (1,045 acres), located in New York, and the smallest is Hampton National Cemetery (0.03 acre) in

George Dickinson

Arlington National Cemetery in Virginia is the most famous of U.S. national cemeteries. Note the precision with which the gravestones are lined up, very fitting for a military cemetery.

Virginia. Twenty-seven of the 130 national cemeteries are closed to new interments but may still accommodate family members in already-occupied graves.

The National Cemetery Act of 2003 authorized the Veterans Administration to establish new national cemeteries to serve veterans in the areas of Bakersfield, California; Birmingham, Alabama; Jacksonville, Florida; Sarasota County, Florida; southeastern Pennsylvania; and Columbia-Greenville, South Carolina. All six areas have veteran populations exceeding 170,000, which is the threshold the Veterans Administration has established for new national cemeteries. Ft. Jackson, Jacksonville, Sarasota, Alabama, and Bakersfield National Cemeteries opened for interments in 2009, and the Veterans Administration has just announced the names of two new VA national cemeteries planned to be built in Florida—the Cape Canaveral National Cemetery in Scottsmoor and the Tallahassee National Cemetery in Tallahassee. The Cape Canaveral Cemetery will have 318 acres of land and the Tallahassee Cemetery will have 250 acres.

"These two new cemeteries are a priority for the growing Veterans' community in Florida," said Secretary of Veterans Affairs Eric K. Shinseki. "As the VA moves forward with these new cemeteries, veterans, their spouses, and families can have peace of mind knowing that they have a final resting place in a nearby national shrine" (Department of Veterans Affairs, 2014).

Any honorably discharged veteran and his or her spouse and/or dependent children are eligible for burial within the 131 national cemeteries. With this free burial (including the cemetery plot, grave liner or vault, and grave marker), families may be provided with a dignified military funeral ceremony including folding and presenting the U.S. burial flag, the playing of "Taps," and a memorial certificate of appreciation signed by the president of the United States. The Veterans Administration will also furnish, upon request, a free grave marker (made of bronze, marble, or granite), which can be installed in a private nongovernment cemetery. This latter benefit is not provided to spouses and eligible dependents who choose not to be interred in national cemeteries.

At the time of a death, viewing and funeral services are performed, and then the deceased is transferred to the national cemetery for a committal service and/or burial. For reasons of safety, committal services are held at sites located away

from the gravesite, after which the casket is placed in the vault and then later taken
to the actual gravesite for burial. Committal services last approximately 15 minutes
and are performed with the assistance of two or more uniformed military persons
with at least one a member of the veteran's parent service of the armed forces.
Veterans' organizations may also assist in providing military funeral honors.
Although national cemeteries are opened seven days each week, burials are not
normally conducted on Saturdays or Sundays. For additional information on U.S.
national cemeteries go to http://www.cem.va.gov/.

Cities and states also build cemeteries as a service to their residents and as a
way of celebrating local history and honoring civil leaders. Government cemeteries
are not designed as profit-making ventures; rather they are administered as services
to the public and are paid for by fees charged to those receiving the benefits. A portion of these charges is set aside for endowments or trust funds, the interest from
which is used to pay for perpetual care of the cemetery.

THE NOT-FOR-PROFIT CEMETERY

The second type of cemetery is the not-for-profit cemetery, which is run by a cemetery association for the benefit of private members. Like a condominium association, a cemetery association consists of individuals and families who own lots in
the cemetery. Typically, meetings are called on a regular basis to employ workers
(sextons, grounds crew, and gravediggers), set lot prices and fees, and plan for
maintenance and improvements in the cemetery. For example, Michael Leming's
town of Northfield, Minnesota (population approximately 21,000), has three
not-for-profit cemeteries: one was created by the city of Northfield in the late 19th
century and deeded to the cemetery association after its creation; Mt. Calvary
Cemetery is owned and run by the Roman Catholic diocese for the benefit of
St. Dominic's Church members; and Oak Lawn Cemetery was established by private citizens as a nonprofit cemetery.

The normal way for nonprofit cemeteries to charge for their services is to
establish a fee for the lot and burial (opening and closing) and then take a portion
of these two charges (typically 10–15 percent) and place this money in an interest-
bearing trust fund or endowment. The annual income from the endowment should
be enough to pay for repairs, maintenance, and upkeep of the cemetery. When
annual income is not enough to pay for annual costs, the prices for lots and burials
are increased.

Unlike government cemeteries, lot owners and their families establish rules,
covenants, and ordinances to regulate the behaviors of the owners. Typical rules
include the following: All bodies will be buried in an east-west direction, lots
will be four feet by eight feet in size and four to five feet deep, a vault or grave
liner will be required for all earth burials to keep the ground from sinking,
monuments or markers will be placed on the east edge of the grave lot, plastic
flowers will be removed from all graves, the cemetery will be opened to visitors
at sunrise and closed at sundown, and there will be no more than one burial for
each full-casket gravesite and no more than two in-ground gravesite burials for
cremated remains.

FRED PROUSER/Reuters/Landov

FRED PROUSER/Reuters/Landov

Douglas Keister/Corbis

Allstar Picture Library/Alamy

Forest Lawn in California is considered by many to be the best example of the lawn park cemetery and is a special attraction for "tombstone tourists" as well as "home" for many Hollywood stars, including Michael Jackson, Nat King Cole, Red Skelton, Clark Gable, Carole Lombard, W. C. Fields, Jean Harlow, George Burns, Gracie Allen, and many more.

The following story is an example of local nonprofit cemetery politics. A few years ago, all three nonprofit cemeteries in Northfield, Minnesota (sometimes called *Mini-snowda*), contracted gravedigging to the same earth-moving firm. This firm provided a favorable price to the cemeteries, subject to the requirement that there would be no winter burials and that burials would only take place from April 1 through December 1. Because the cemetery associations were interested in

---

PRACTICAL MATTERS | CEMETERIES FEEL RECESSION'S CHILL

The recession is already difficult to live through, but it may also have everlasting effects. Cemeteries are having trouble expanding because of the high cost of real estate and a drop in revenue as strapped families increasingly turn to cheaper cremations.

Communities struggling with budget deficits have also had to curtail spending on public cemeteries, which means many people may find that their home-town cemeteries are full. "I know a lot of people living in Saugus would like to be buried in Saugus. But we're also thinking about the schools," said Donald Wong, chairman of the board of selectmen for Saugus, a town a few miles north of Boston. Wong said the town is thinking about creating new burial space on the same site as a new school.

Cemeteries are not only seeing a drop in money coming from municipalities. Customers are scaling back as well. Gary Brown, who oversees six cemeteries for the Catholic Diocese of Phoenix, said they've seen a 17% decrease in prepaid plots over the past two years. That leaves the diocese with a lot less money for expansion. "People are at this time reluctant to take their extra funds and spend it on items that are not necessary," Brown said.

Cemeteries also haven't been able to rely on investments to bail them out. David Heisterkamp, president of the Pennsylvania Cemetery, Cremation and Funeral Association, said most cemeteries invest part of their income into perpetual care funds for the maintenance of graves. The return on those investments is used to pay for daily operations. Heisterkamp said those investments have been clobbered by the recession, forcing cemeteries to cut back on maintenance and lay off workers.

"There's a myth that this industry is recession proof," he said. Brad Hansen, president of Hansen Mortuaries and Cemetery in Phoenix, said private cemeteries are getting squeezed from customers looking for deals. He said about 68% of Arizonans opted for cremations in the past year, compared with 60% the previous year. A funeral service can cost more than $6,300, compared to about $3,000 for a cremation service—that doesn't include the cost of a cemetery plot. "They see a value in cremation," Hansen said.

From Cemeteries feel recession's chill, by A. Gomez, *USA Today*, October 19, 2009.

---

keeping costs reasonable and their competition was following the same procedures, they were willing to make concessions to the gravedigger. A few years later, however, a prominent member of the community died in the dead of winter and his casket was to be kept in the cemetery crypt (actually the lawnmower maintenance building) until April 1. As a result of this important person's death and the holding of his body in the maintenance building until spring, a grassroots social movement (led by Michael Leming) took place soon thereafter to reform the cemetery's rules. Today, earth burials take place at any time of year in all three Northfield cemeteries.

THE FOR-PROFIT CEMETERY

The final type of cemetery is the for-profit cemetery. These are cemeteries whose goals are to provide a service for a fee to the consumer and a profit for the owners of the cemetery.

The most famous of all for-profit cemeteries is Forest Lawn Memorial-Parks & Mortuaries of California (http://www.forestlawn.com). This chain of 10 cemeteries, located in southern California, has served more than a million families since 1906. Forest Lawn is a full-service facility providing funeral and mortuary services,

| PRACTICAL MATTERS | CEMETERIES BREATHE LIFE INTO TOURISTS |

He's chipped his way through more than a foot of snow and ice to get to Ernest Hemingway. He's walked right up to Al Capone and Karl Marx. He's dragged his mom to visit the infamous cannibal Alferd Packer and just came back from seeing Farrah Fawcett. He is Jim Tipton, founder of Find a Grave (findagrave.com), a free online database of burial sites for the famous and otherwise around the globe.

"It does sound morbid and dark. But when you're actually visiting someone's grave, it's like visiting a relative; there's a closeness there," said Tipton, 37, of Salt Lake City, Utah. "And I've always liked the aesthetics of cemeteries. I've always called them parks for introverts because you don't have to worry about someone asking you to play a pick-up game."

At first glance, the idea of graveyard tourism may seem ghoulish. But for visitors who seek out head-stones, this sort of destination travel is about more than death and grief-seeking. It can be a form of entertainment and inspiration, a history and architecture lesson, a cultural appreciation course, a genealogical journey and a source of relaxation.

Providing solace and beauty for the living, in fact, is as important as honoring the dead at Mount Auburn Cemetery in Cambridge and Watertown, Massachusetts. This 175-acre institution, founded in 1831, made history as America's first garden-like, landscaped cemetery, inspiring similar sites across the country and even public parks like Central Park in New York.

After long days staring at a computer in Portland, Oregon, Scott Stanton caught this tourism bug when he'd unwind by strolling through neighborhoods, often cutting through graveyards. The self-described "frustrated rock 'n' roll star" soon started seeking out the burial sites of musicians, which took him to more than 550 plots throughout the world. From punk rocker Joey Ramone in New Jersey and composer Pyotr Ilyich Tchaikovsky in Russia to the Doors legend Jim Morrison in France and blues guitarist Stevie

Ray Vaughan in Texas, he spent more than 15 years compiling information and taking photographs for his book *The Tombstone Tourist: Musicians*. "I'm not a Carnival Cruise sort of guy," said Stanton, 50, who owns a software company. Doing the book "took a long, long time, because they kept dying. I could never figure out when to stop."

Many cities have cemeteries that have long attracted throngs of visitors. Père Lachaise in Paris, France, where Morrison is buried—along with Maria Callas, Frédéric Chopin and Oscar Wilde, to name a few—is one of them. After-death stargazers can stay busy in Los Angeles, California, where outfits bearing names like Dearly Departed Tours are dedicated to showing visitors the way. At Hollywood Forever Cemetery, about 2,000 people come out every Saturday night to spend the evening with Rudolph Valentino and Fay Wray, picnic and watch classic films.

Learning about strangers is part of what drives Cristina Lugo of New York, who with a club she calls the "Cemetery Girls" takes self-guided day trips to graveyards in the area. The headstones tell stories, such as the one in the Bronx that honors a family killed by a lightning strike, said Lugo, 37. She also sees visiting graves as a service to others—the departed and the descendants who can't get there on their own.

For the website Find a Grave, she often volunteers to track down and photograph the burial sites of people's ancestors. She recently ventured into one New York graveyard for a family in England, giving them a piece of their genealogical history. And perhaps more than anything, the cemetery tourism hobby brings Lugo a sense of peace she can't find in the urban, living jungle. "It's almost like church for me," she said. "It's a reminder that life is precious."

From Cemeteries breathe life into tourists, by J. Ravitz, CNN, October 30, 2009, http://www.cnn.com/2009/TRAVEL/10/30/graveyard.tourism/index.html.

cemetery plots, mausoleums and columbariums, florists, and the selling of death-related merchandise (monuments, vaults, caskets, clothing, and sundry items). Forest Lawn cemeteries have a huge endowment of more than 200 million dollars that pays for perpetual maintenance, upkeep, repairs, and improvements. The principal of the endowment funds can never be touched (by law); only the income generated from the endowment can be used. This is similar to nonprofit colleges and universities that have endowments that assist in funding operating budgets and scholarships for the learning enterprise, and tuition and fees charged to students cover the needed expenses not provided by the income from the endowment fund.

Present minimal costs at Forest Lawn (as of July 9, 2013) for cemetery plots and services are as follows:

- Property for burial beginning at $2,000. Contribution to the Endowment Fund—15 percent of lot price
- Charges for full-size interment (Saturday $325 extra)—$1,195, vault $725 and up
- Charges for cremation—$2,688, charges for cremated remains in niche or ground is $325, vault for cremated remains is $195 and up

Forest Lawn also provides funeral services and sells funeral merchandise and flowers. Forest Lawn Memorial-Parks have more than 1 million visitors each year who pay respect to the memories of loved ones, participate in Forest Lawn special events (such as Easter sunrise services and Memorial Day ceremonies) and educational programs, and enjoy the magnificent works of art, including exquisite stained glass windows, a collection of marble statuary—a rare combination of original works by modern masters and exact reproductions of Michelangelo's most timeless creations—a museum filled with original paintings and bronze statuary, and many historic artifacts. Check out the world famous Forest Lawn at forestlawn.com. As a child growing up in Los Angeles, Michael Leming's school classes and YMCA day-camp programs would visit Forest Lawn Memorial-Parks, as they would the other cultural and entertainment centers of the area—Disneyland, Knott's Berry Farm, Universal Studios, and Sea World.

There are many other for-profit cemeteries, such as those run by Service Corporation International (SCI), the world's largest provider of death care services. At the end of 2012, the company was North America's largest provider of death care products and services, with a network of funeral homes and cemeteries unequaled in geographic scale and reach. SCI operated 1,437 funeral service locations and 374 cemeteries (including 213 funeral service/cemetery combination locations) in North America, which are geographically diversified across 43 states, eight Canadian provinces, and the District of Columbia.

SCI's funeral service and cemetery operations consist of funeral service locations, cemeteries, funeral service/cemetery combination locations, crematoria, and related businesses. They also sell cemetery property and funeral and cemetery merchandise and services at the time of need and on a preneed basis.

On December 31, 2012, SCI employed 12,679 individuals on a full-time basis and 7,888 individuals on a part-time basis. As a result of such preneed sales, its backlog of unfulfilled preneed funeral and preneed cemetery contracts was $7.4 billion in December 31, 2012. In 2012 the gross revenues were $2.41 billion for all services performed. Investors can purchase stock in this and other for-profit corporations, in hopes of earning an income from corporate funeral homes, cemeteries, and floral businesses. Key financial developments in 2012 were as follows:

- Funeral gross profit increased $18.4 million, or 5.6 percent, due to higher case volume, primarily from preneed memorial merchandise sales of The Neptune

Society and higher General Agency revenues, partially offset by lower at-need revenues and higher selling-related expenses related to preneed sales initiatives.

- Cemetery gross profit increased $28.3 million, or 19.1 percent, due to an increase in preneed property sales production and trust fund income, partially offset by higher selling-related expenses and property costs.

If ever there was a business that specialized in providing a necessary service that no one wants to do, it is the death-related business of caring for and burying the dead.

CONCLUSION

In our discussion of the business of dying, we have described funerals and their alternatives within a cultural and historical perspective. In the United States an evolutionary, not a revolutionary, process has occurred. Americans did not invent the funeral or the funeral functionary. However, contemporary Americans have found an expression for their bereavement.

In this chapter we have also discussed the development of cemeteries and the building of funeral homes, and we have demonstrated how certain needs of Americans have been fulfilled. Security comes from the sheer orderliness and structure of these institutions. One should know what to expect from these services, and payment turns the responsibility over to the professionals. We Americans differ significantly from persons in nonliterate societies, in which such functions are completed within the kin network. However, paying someone else to perform a service fits middle-class Americans' ideas about specialization and division of labor. This historical perspective on changes in the funeral and the cemetery helps us understand our experience with contemporary American death customs. The consumer can be well informed about burial practices in our society by staying abreast of various funeral home practices and regulations.

SUMMARY

1. Death was feared by the Puritans, who prayed not for the soul of the deceased but rather for the comfort and instruction of the living. The funeral was the main social institution for channeling the grief of Puritan survivors.
2. The Victorian funeral brought a focus on aesthetics and soothing the mourner, reflecting the "dying of death" movement.
3. The contemporary American funeral focuses on meeting the needs of bereaved individuals in terms of social support and grief resolution.
4. Licensure in the funeral industry covers embalming, practicing as a funeral director, and practicing mortuary science. Education

requirements vary from state to state, but all states (except Colorado) require a one- to three-year apprenticeship.
5. The role of a funeral director is that of a caregiver, caretaker, and gatekeeper.
6. Funeral bills have been presented to customers using the following pricing systems: unit pricing, biunit or triunit pricing, and itemization. The latter is mandated in all states by the FTC.
7. In 2013 the cost of the average adult funeral was $7,045. If a vault is included, something that is typically required by a cemetery, the median cost is $8,343. The cost does not take into account cemetery, monument or marker costs, crematory fees (if cremation is

selected instead of burial), or miscellaneous cash-advance items, such as flowers and obituaries.

8. Alternatives to funerals include immediate disposition (burial or cremation), body donation, and memorial services.

9. New trends in funeral services include preneed funerals, aftercare services, and funeral services on the Internet.

10. The Puritans were buried in small, unadorned family plots, graveyards, and churchyards. The Victorian era ushered in the city garden cemetery, filled with elaborate headstones and epitaphs and designed to reflect the romantic naturalism and sentimentality of the period. Health concerns and crowding led to the development of the rural cemetery, designed to soothe the mourners by diverting attention from death to the rural surroundings. Whereas the rural cemeteries were characterized by elaborate monuments, the next phase of cemetery building, the lawn park cemetery, discouraged the use of gravestones, focusing on the aesthetics and efficiency of an uncluttered landscape of open meadows.

11. There are three types of cemeteries found in the United States: government cemeteries, nonprofit cemetery associations, and for-profit cemeteries.

DISCUSSION QUESTIONS

1. Describe and discuss the Puritan view of death. Describe the procedures and atmosphere surrounding the typical Puritan funeral.

2. How did the funeral reflect the values of the Victorian era?

3. What are the influences of the following occupations upon the "dying of death": life insurance agents, cemetery superintendents, and funeral directors?

4. How do funeral directors address the needs of the bereaved?

5. Describe the changes that have taken place with regard to the role of the family in the funeral process.

6. What are the requirements for becoming a funeral director, and how have they changed over the years?

7. Discuss the factors affecting costs after death and the expenses related to funerals and final disposition.

8. What would be your choice of final disposition of your body? Why would you choose this method, and what effects might this choice have upon your survivors?

9. What are the advantages and disadvantages of preneed funeral arrangements?

10. Describe and explain the reforms that have taken place over the years in the construction and maintenance of the cemetery.

GLOSSARY

Bier: A framework upon which the corpse and/or casket is placed for viewing and/or carrying.

Cremation: The reduction of a human body by means of heat or direct flame. The cremated remains are called cremains or ashes and weigh between three and nine pounds. *Ashes* is a very poor word to describe the cremated remains because they are actually processed bone fragments and calcium residue that have the appearance of crushed rock or pumice.

Crematory: An establishment in which cremation takes place.

Columbarium: A building or wall for above-ground accommodation of cremated remains.

Disposition: Final placement or disposal of a dead person.

Elegy: A song or poem expressing sorrow, especially for one who is dead.

Embalming: A process that temporarily preserves a deceased person by means of displacing body fluids with preserving chemicals.

Entombment: Opening and closing of a crypt, including placing and sealing of a casket within.

Epitaph: An inscription, often on a gravestone, in memory of a deceased person.

Itemization: A method of pricing a funeral in which every item of service, facility, and transportation is listed with its related cost. This is the method mandated by the Federal Trade Commission.

Mausoleum: A building or wall for above-ground accommodation of a casket.

Mechanical solidarity: Emile Durkheim's concept to describe the form of social cohesion that exits in small-scale societies that have minimal division of labor. In such societies there is little specialization or individuation. The "conscience collective" embraces individual awareness.

Memento mori: Any reminder of death.

Memorial Society: A group of people joined to obtain dignity, simplicity, and economy in funeral arrangements through planning.

Niche: A chamber in a columbarium into which an urn is placed.

Obituary: Notice of a death, usually with a brief biography.

Organic solidarity: Emile Durkheim's concept to describe the form of social cohesion that exits in complex societies with a high division of labor. Such societies have an organic character because of the necessary interdependence of their specialized and highly individuated members.

Prearranging: Arranging funerals in advance of need. This process can include selecting merchandise, planning the service, determining the method of viewing and final disposition, and selecting persons to be involved in the funeral.

Prefunding: Legally committing money to pay for the funeral service.

Preneed Funerals: A generic term that refers to both prearranged and prefunded funerals.

Probate: The judicial determination of the validity of a will.

Sexton: A church custodian charged with the upkeep of the church and parish buildings and grounds.

Social Stratification: A ranking of social status (position) in groups; upper, middle, and lower classes are basically distinguished in the U.S. social class system, for example, whereas India's stratification is a caste system.

Urn: A container for cremated remains.

Vault or Grave Liner: A concrete or metal container into which a casket or urn is placed for ground burial. Its function is to prevent the ground from settling.

SUGGESTED READINGS

Applegate, S. (2005). *Living among headstones: Life in a country cemetery*. New York: Thunder's Mouth Press. Filled with humor, singular events, pathos, original illustrations, and unexpected smiles, this book offers historical asides and moving personal stories. While the book is about rural cemeteries in contemporary America, it provides a broader context for understanding how we long for those we love, but who have died, to have a continuing place in our world.

Austerlitz, S. (2013). Why funerals demand a body: Undertaker Thomas Lynch on how American memorials went wrong. *Boston Globe*, October 20, 2013. This article provides a critique of contemporary funeral services that may be more convenient (emotionally, spiritually, and financially) in the near term, but may be less personally satisfying in the long run.

Benoit, T. (2003). *Where they are buried? How did they die?* New York: Black Dog and Leventhal Publishers. This entertaining book chronicles the "fitting ends and final resting places of the famous, infamous, and noteworthy." A good and interesting read for those who love trivia.

Bolt, S. (2009). Body disposition. In C. D. Bryant & D. Peck (Eds.), *Encyclopedia of death and the human experience* (pp. 107–111). Thousand Oaks, CA: Sage Publications. This article provides an excellent discussion of modes of body disposition and the advantages and disadvantages of each.

Carlson, L. (1997). *Final act of love: Caring for your own dead*. Hinesburg, VT: Upper Access Book Publishers. A consumer resource that assists bereaved individuals in bypassing the funeral director and arranging everything themselves, including burial at home, where it is feasible and permitted.

Davies, D., & Rumble, H. (2012). *Natural burial: Traditional/secular spiritualties and funeral innovation*. London: Continuum. With its theoretically informed analysis, Davies and Rumble's book makes a major contribution to our knowledge of natural burial. This is a very thorough exploration of this burial alternative.

Dickinson, G. E. (2012). Diversity in death: Body disposition and memorialization. *Illness, Crisis, and Loss, 20*(2), 141–158. This article provides an excellent description of changes in contemporary methods of body disposition and memorialization.

Farrell, J. (1980). *Inventing the American way of death, 1830–1920*. Philadelphia, PA: Temple University Press. Examines the transformation from the Puritan way of death to the American way of death.

Gibbs, L., & Mangla, I. S. (2012). The high cost of saying goodbye. *CNN Money Magazine*, November 9, 2012. This article provides an excellent critique of the high cost of dying in America. It is a nice follow-up to Jessica Mitford's famous 1963 book, *The American Way of Death*.

Gouin, M. (2010). *Tibetan rituals of death: Buddhist funerary practices*. Abingdon: Routledge. A comprehensive survey of the available literature on funerary practices in Tibetan societies.

Harris, M. (2007). Grave matters: A journey through the modern funeral industry to a natural way of burial. New York: Scribner; Tseng, T., & Su, C. (2009). Death care industry. In C. D. Bryant & D. Peck (Eds.), *Encyclopedia of death and the human experience* (pp. 305–309). Thousand Oaks, CA: Sage Publications. These resources provide a comprehensive discussion of the death care industry, including history, standards, practice, and current trends in funeral directing.

Holloway, K. F. C. (2002). *Passed on: African-American mourning stories*. Durham, NC: Duke University Press. Focuses on African American funeral homes and morticians, as well as the history of the profession and its practices.

Kopp, S. W., & Kemp, E. (2007, Winter). Consumer awareness of legal obligations of funeral providers. *Journal of Consumer Affairs, 41*(2), 326–340. Currently, all funeral home activities are regulated under the Funeral Industry Practices Rule of the Federal Trade Commission. The rule is premised on the assumption that it is difficult for consumers to make careful, informed purchase decisions in at-need situations because of emotional stress, time pressure, and lack of familiarity with available goods and services. This article provides an empirical assessment as to the efficacy of the FTC funeral rule, relative to consumer protection.

Laderman, G. (2003). *Rest in peace: A cultural history of death and the funeral home in twentieth-century America*. New York: Oxford University Press. Traces the origins of American funeral rituals from the evolution of embalming techniques during and after the Civil War, the shift from home funerals to funeral homes at the turn of the century, and the increasing subordination of priests, ministers, and other religious figures to the funeral director throughout the 20th century.

Leech, E., & Hull, R. T. (2009). *Funeral industry, unethical practices*. In C. D. Bryant & D. Peck (Eds.), *Encyclopedia of death and the human experience* (pp. 479–482). Thousand Oaks, CA: Sage Publications. This article provides a critical perspective on the American death industry.

MacLean, V. M., & Williams, J. E. (2009). Cemeteries. In C. D. Bryant & D. Peck (Eds.), *Encyclopedia of death and the human experience* (pp. 169–173). Thousand Oaks, CA: Sage Publications. Excellent discussion of the history of the cemetery, with special attention to differences between the cemeteries of the following ethnic groups: Jewish, Muslim, Spanish-Mexican, African American, and Native American.

Meyer, R. E. (Ed.). (1993). *Ethnicity and the American cemetery*. Bowling Green, OH: Bowling Green State University Popular Press. Anthology of eight essays exploring the manner in which representative ethnic groups in America have made their cemeteries. The book has an interdisciplinary focus, from folklore, cultural history, historical archaeology, landscape architecture, and philosophy.

Life's too short for worrying. Yes, that's what worries me.
—Origin Unknown

THE LEGAL ASPECTS OF DYING | CHAPTER 12

© Surgeon/Shutterstock.com

Most of the law that regulates the disposition of a person's body and property upon death is state law, not federal law, and considerable variation exists between the laws of different states. As a general rule, the law of the state where a person lives at the time of death controls disposition of the body and of property, although real estate (land and the structures thereon) is governed by the law of the state where the property is located. It is unlawful to dispose of a corpse until the cause of death has been satisfactorily ascertained (Mims, 1999). Several state laws, therefore, may have an impact on a person's death, in addition to some federal laws. Though an individual does not have to have legal counsel to handle some of the legal issues involved in death, it might behoove one to consider legal counsel, because lawyers are the experts on these topics. Perhaps "doing it one's self" and being one's own attorney can be likened to trying to fix the plumbing. After trying to fix the plumbing, the amateur often has to bring in a plumber to undo what he or she has done and usually at a much greater expense than would have originally been the case.

What does the law say regarding ownership of a dead body? It turns out that in English law, going back to 1614, no one owns a dead body (Mims, 1999). In other words, in England no one can be arrested for stealing a body, a fact that led to difficulties in the prosecution of "body snatchers." In England in 1614, a Mr. Haynes was accused of stealing the shrouds (burial sheets) from a corpse. In the court ruling, it was stated that the corpse could not own the sheets, and this has been misinterpreted as meaning that the corpse itself could not be owned. Thus, in 1856 in England, a son removed his mother's remains from a graveyard, but because the corpse was not anyone's property, he could only be charged with trespassing in the graveyard. Exactly who owns a human body even today is not clear, under British law, noted Mims (1999). A corpse, it seems, unlike a urine sample or a lock of hair, is not an item of property and cannot be stolen!

However, the same lack of ownership is *not* the case in the United States. As discussed in Chapter 8, various ethical questions revolve around ownership of one's body. Legally, however, the right to control the disposition of one's own body is more limited in the United States than the relatively broad power to dispose of property upon death. The decedent's body is not property in any conventional sense of the word and is not part of the decedent's estate. Therefore, the right to control one's own body, like other legal rights, ends at death, and any remaining rights pass to other individuals. Thus, any preferences for disposition of one's body expressed during life (e.g., funeral arrangements) will be considered but are not legally binding. Instead, the deceased person's next of kin have the right to arrange such details, because in the United States, unlike England, the next of kin have property rights of the corpse (Mims, 1999). For example, if an individual had requested (either orally or in writing) that the funeral director cremate upon death and the next of kin prefer earth burial, the funeral director may be in an awkward position, yet the wishes of the next of kin are legally binding.

As discussed in Chapter 1, social and biological deaths differ in various cultures. Our focus in this chapter is on biological deaths, legally determined.

ESTABLISHING THE CAUSE OF DEATH

If an individual is enrolled in a hospice program and dying of brain cancer, the cause of death is usually rather obvious. The process leading to death is probably related to the malignancy and can be tracked from there. The ultimate cause of death may be heart failure brought on by the weakened condition from the cancer. Yet, as discussed in Chapter 1, determining the cause of death is not always a clear-cut case. For example, an 88-year-old woman suffering from chronic heart disease, diabetes, and high blood pressure may fall and break her hip, forcing her to be bedridden and fitted with a catheter for urination purposes. She develops an infection from the catheter. Pneumonia sets in, and her lungs fill with fluid. She dies. What is the "cause" of her death? Such a question will not be easy to answer, yet the attending physician has to seek accuracy in filling out the death certificate.

DEATH CERTIFICATE

A **death certificate** is a permanent record of an individual's death that must be completed soon after death. The purpose of the death certificate is to obtain a simple description of the sequence or process leading to death, rather than a record

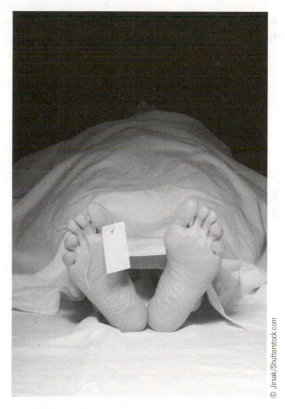

Dead body with toe-tag sent to the morgue for an autopsy.

describing all medical conditions present at death. Certified copies of a death certificate are needed to collect insurance proceeds and other death benefits. Copies can be obtained through the funeral home and added to the cost of the funeral bill (typically $15 for the first and $3 for each additional copy, up to $40 for the first and $3 for each additional copy, depending on the state—in Minnesota veterans get one free and pay for extras). Otherwise, copies can be obtained by writing to the vital statistics office or county health department in the county where the person died. If you wait for several weeks, copies can be obtained from the state's vital statistics office. In both cases there is a cost for each certificate.

Cause-of-death data are important for surveillance, research, design of public health and medical interventions, and funding decisions for research and development. The death certificate is a legal document used for legal, family, and insurance purposes. It is also used to compile mortality statistics. Although these data are overseen by the National Center for Health Statistics, physicians shape the content of mortality statistics by virtue of what they write on the death certificate and whether they obtain consent for an autopsy.

Causes of death on the death certificate represent a medical opinion that might vary with individual physicians. For situations in which the cause of death is difficult to certify, the certifier should select the causes that are suspected to have been involved and use words such as *probably* or *presumed* to indicate that the description provided is not completely certain (Hazlick, 1994). If the certifier really is not sure, he or she should state that the etiology (cause or origin) is unknown, undetermined, or unspecified; thus, it is clear that the certifier did not have enough information to provide even a qualified opinion about the etiology.

When preparing the death certificate for an elderly person, causes should not include terms such as *old age* because they have little value for public health or medical research. If malnutrition is noted, the certifier should consider if other medical conditions could have led to malnutrition. If several conditions resulted in death, the certifier should choose a single sequence to describe the process leading to death and list the other conditions in another section. *Multiple system failure* could be included as an "other significant condition." Finally, if the certifier cannot determine a descriptive sequence of causes of death, death may be reported as "unspecified natural causes" (Hazlick, 1994).

With the death of an infant, maternal conditions may have initiated or affected the sequence contributing to the death. Such conditions should be reported in addition to the causes related to the infant. When sudden infant death syndrome (SIDS) is suspected (discussed in Chapter 14), a complete investigation should be conducted. If the infant is younger than one year of age and no cause of death can be found, then the death can be reported as SIDS. If the investigation is not completed, the death may be reported as presumed to be SIDS.

Despite the importance of death certificates and the fact that physicians are the professionals who sign these documents, the majority of doctors in the United States have very little training in completing death certificates. According to a 2010 survey of the 122 U.S. medical schools, only 36 percent cover the topic of death certificates in the medical curriculum (Dickinson, 2011). To address this problem, the death certificate has been revised to include instructions for completing the form, along with material for additional information. For example, the

DEATH ACROSS CULTURES | INDIA'S LIVING DEAD FORM THEIR OWN CLUB

An association for dead people has been established in northern India by a group of men and women claiming that reports of their demise have been greatly exaggerated. The Union of Dead People was set up to represent the growing number of unfortunates who have been declared dead when they are actually very much alive.

The living dead are victims of a particularly nasty form of land-grabbing. They have been pronounced dead by relatives working in collusion with crooked bureaucrats to defraud them of their property. Once a death certificate has been issued and the person's land transferred into another person's name, it seems little can be done to rectify the error.

Ten "dead" people recently staged a sit-in protest against their plight outside local government offices. The range in ages of the protesters was from the teens to age 85. The 85-year-old woman said she was swindled out of her land so long ago that nobody now believes her claim. Apparently there are "thousands of people" like this. Many individuals are afraid to complain in case they are really killed by their relatives.

From India's living dead form their own club (p. 17), by D. Orr, June 18, 1999, *The London Times*.

primary cause of death is defined as "the disease or injury that initiated the train of morbid events leading directly to death, or the circumstances or violence which produced the fatal injury" (Kircher, 1992, p. 1268). It is stated that without the underlying cause (primary cause) death would not have occurred. The "immediate cause" of death is that which directly precedes death and is the ultimate consequence of the "underlying cause." The interval between the underlying and immediate causes of death may last for years or just seconds.

Though the importance of death certificates is pointed out above, death certificates are not universally used. Indeed, the World Health Organization, in reporting causes of death around the world, observes that death certificates are made out for only about 25 percent of all deaths (Mims, 1999). They are not used for deaths due to famines, massacres, wars, or natural disasters.

AUTOPSY

A dissection and examination of a just-deceased patient is an **autopsy**. The **pathologist** performing the autopsy assumes a mechanical, physical cause for the death. First, the pathologist gathers a history on how the death occurred and often obtains the past medical history of the deceased. The pathologist then examines the body externally and internally, taking biopsies of tissues to identify any diseases or abnormalities that may have contributed to the death. During the course of the autopsy, various laboratory tests may be performed, including x-rays, toxicological analysis of blood and urine, and cultures of body fluids and organs for evidence of infection.

Autopsies for cases of natural death or deaths occurring among patients under the care of a physician are usually performed at the hospital where the death occurred and with the permission of the next of kin. Local statutes may require an autopsy for cases of traumatic or sudden, unexpected deaths or for deaths due to external causes. Such an autopsy is requested by either a coroner or a medical examiner. A report summarizing the findings is prepared by the pathologist. Often, this medical specialist is then subpoenaed to testify in court about the findings.

U.S. STANDARD CERTIFICATE OF DEATH

LOCAL FILE NO.   STATE FILE NO.

NAME OF DECEDENT — For use by physician or institution

To Be Completed/ Verified By: FUNERAL DIRECTOR:

1. DECEDENT'S LEGAL NAME (Include AKA's if any) (First, Middle, Last)	2. SEX	3. SOCIAL SECURITY NUMBER

4a. AGE-Last Birthday (Years)	4b. UNDER 1 YEAR — Months / Days	4c. UNDER 1 DAY — Hours / Minutes	5. DATE OF BIRTH (Mo/Day/Yr)	6. BIRTHPLACE (City and State or Foreign Country)

7a. RESIDENCE-STATE	7b. COUNTY	7c. CITY OR TOWN

7d. STREET AND NUMBER	7e. APT. NO.	7f. ZIP CODE	7g. INSIDE CITY LIMITS? ☐ Yes ☐ No

8. EVER IN US ARMED FORCES? ☐ Yes ☐ No	9. MARITAL STATUS AT TIME OF DEATH ☐ Married ☐ Married, but separated ☐ Widowed ☐ Divorced ☐ Never Married ☐ Unknown	10. SURVIVING SPOUSE'S NAME (If wife, give name prior to first marriage)

11. FATHER'S NAME (First, Middle, Last)	12. MOTHER'S NAME PRIOR TO FIRST MARRIAGE (First, Middle, Last)

13a. INFORMANT'S NAME	13b. RELATIONSHIP TO DECEDENT	13c. MAILING ADDRESS (Street and Number, City, State, Zip Code)

14. PLACE OF DEATH (Check only one: see instructions)

IF DEATH OCCURRED IN A HOSPITAL:
☐ Inpatient ☐ Emergency Room/Outpatient ☐ Dead on Arrival

IF DEATH OCCURRED SOMEWHERE OTHER THAN A HOSPITAL:
☐ Hospice facility ☐ Nursing home/Long term care facility ☐ Decedent's home ☐ Other (Specify):

15. FACILITY NAME (If not institution, give street & number)	16. CITY OR TOWN , STATE, AND ZIP CODE	17. COUNTY OF DEATH

18. METHOD OF DISPOSITION: ☐ Burial ☐ Cremation ☐ Donation ☐ Entombment ☐ Removal from State ☐ Other (Specify):	19. PLACE OF DISPOSITION (Name of cemetery, crematory, other place)

20. LOCATION-CITY, TOWN, AND STATE	21. NAME AND COMPLETE ADDRESS OF FUNERAL FACILITY

22. SIGNATURE OF FUNERAL SERVICE LICENSEE OR OTHER AGENT	23. LICENSE NUMBER (Of Licensee)

ITEMS 24-28 MUST BE COMPLETED BY PERSON WHO PRONOUNCES OR CERTIFIES DEATH

24. DATE PRONOUNCED DEAD (Mo/Day/Yr)	25. TIME PRONOUNCED DEAD

26. SIGNATURE OF PERSON PRONOUNCING DEATH (Only when applicable)	27. LICENSE NUMBER	28. DATE SIGNED (Mo/Day/Yr)

29. ACTUAL OR PRESUMED DATE OF DEATH (Mo/Yr) (Spell Month)	30. ACTUAL OR PRESUMED TIME OF DEATH	31. WAS MEDICAL EXAMINER OR CORONER CONTACTED? ☐ Yes ☐ No

To Be Completed By: MEDICAL CERTIFIER

CAUSE OF DEATH (See instructions and examples)

32. **PART I.** Enter the chain of events--diseases, injuries, or complications--that directly caused the death. DO NOT enter terminal events such as cardiac arrest, respiratory arrest, or ventricular fibrillation without showing the etiology. DO NOT ABBREVIATE. Enter only one cause on a line. Add additional lines if necessary.

Approximate interval: Onset to death

IMMEDIATE CAUSE (Final disease or condition ----------> resulting in death)   a. _____
Due to (or as a consequence of): _____

Sequentially list conditions, if any, leading to the cause listed on line a. Enter the **UNDERLYING CAUSE** (disease or injury that initiated the events resulting in death) **LAST**
b. _____
Due to (or as a consequence of): _____

c. _____
Due to (or as a consequence of): _____

d. _____

PART II. Enter other significant conditions contributing to death but not resulting in the underlying cause given in PART I

33. WAS AN AUTOPSY PERFORMED? ☐ Yes ☐ No

34. WERE AUTOPSY FINDINGS AVAILABLE TO COMPLETE THE CAUSE OF DEATH? ☐ Yes ☐ No

35. DID TOBACCO USE CONTRIBUTE TO DEATH? ☐ Yes ☐ Probably ☐ No ☐ Unknown	36. IF FEMALE: ☐ Not pregnant within past year ☐ Pregnant at time of death ☐ Not pregnant, but pregnant within 42 days of death ☐ Not pregnant, but pregnant 43 days to 1 year before death ☐ Unknown if pregnant within the past year	37. MANNER OF DEATH ☐ Natural ☐ Homicide ☐ Accident ☐ Pending Investigation ☐ Suicide ☐ Could not be determined

38. DATE OF INJURY (Mo/Day/Yr) (Spell Month)	39. TIME OF INJURY	40. PLACE OF INJURY (e.g., Decedent's home; construction site; restaurant; wooded area)	41. INJURY AT WORK? ☐ Yes ☐ No

42. LOCATION OF INJURY: State: City or Town:

Street & Number: Apartment No.:   Zip Code:

43. DESCRIBE HOW INJURY OCCURRED:	44. IF TRANSPORTATION INJURY, SPECIFY: ☐ Driver/Operator ☐ Passenger ☐ Pedestrian ☐ Other (Specify)

45. CERTIFIER (Check only one):
☐ Certifying physician-To the best of my knowledge, death occurred due to the cause(s) and manner stated.
☐ Pronouncing & Certifying physician-To the best of my knowledge, death occurred at the time, date, and place, and due to the cause(s) and manner stated.
☐ Medical Examiner/Coroner-On the basis of examination, and/or investigation, in my opinion, death occurred at the time, date, and place, and due to the cause(s) and manner stated.

Signature of certifier: _____

46. NAME, ADDRESS, AND ZIP CODE OF PERSON COMPLETING CAUSE OF DEATH (Item 32)

47. TITLE OF CERTIFIER	48. LICENSE NUMBER	49. DATE CERTIFIED (Mo/Day/Yr)	50. FOR REGISTRAR ONLY- DATE FILED (Mo/Day/Yr)

To Be Completed By: FUNERAL DIRECTOR

51. DECEDENT'S EDUCATION-Check the box that best describes the highest degree or level of school completed at the time of death.	52. DECEDENT OF HISPANIC ORIGIN? Check the box that best describes whether the decedent is Spanish/Hispanic/Latino. Check the "No" box if decedent is not Spanish/Hispanic/Latino.	53. DECEDENT'S RACE (Check one or more races to indicate what the decedent considered himself or herself to be)
☐ 8th grade or less		☐ White
☐ 9th - 12th grade; no diploma		☐ Black or African American
☐ High school graduate or GED completed	☐ No, not Spanish/Hispanic/Latino	☐ American Indian or Alaska Native (Name of the enrolled or principal tribe) _____
☐ Some college credit, but no degree	☐ Yes, Mexican, Mexican American, Chicano	☐ Asian Indian / ☐ Chinese
☐ Associate degree (e.g., AA, AS)	☐ Yes, Puerto Rican	☐ Filipino / ☐ Japanese
☐ Bachelor's degree (e.g., BA, AB, BS)	☐ Yes, Cuban	☐ Korean / ☐ Vietnamese
☐ Master's degree (e.g., MA, MS, MEng, MEd, MSW, MBA)	☐ Yes, other Spanish/Hispanic/Latino (Specify) _____	☐ Other Asian (Specify)_____ ☐ Native Hawaiian ☐ Guamanian or Chamorro ☐ Samoan
☐ Doctorate (e.g., PhD, EdD) or Professional degree (e.g., MD, DDS, DVM, LLB, JD)		☐ Other Pacific Islander (Specify)_____ ☐ Other (Specify)_____

54. DECEDENT'S USUAL OCCUPATION (Indicate type of work done during most of working life. DO NOT USE RETIRED).

55. KIND OF BUSINESS/INDUSTRY

Cause-of-death – Background, Examples, and Common Problems

Accurate cause of death information is important
•to the public health community in evaluating and improving the health of all citizens, and
•often to the family, now and in the future, and to the person settling the decedent's estate.

The cause-of-death section consists of two parts. **Part I** is for reporting a chain of events leading directly to death, with the **immediate cause** of death (the final disease, injury, or complication directly causing death) on line a and the **underlying cause** of death (the disease or injury that initiated the chain of events that led directly and inevitably to death) on the lowest used line. **Part II** is for reporting all other significant diseases, conditions, or injuries that contributed to death but which did not result in the underlying cause of death given in Part I. **The cause-of-death information should be YOUR best medical OPINION.** A condition can be listed as "probable" even if it has not been definitively diagnosed.

Examples of properly completed medical certifications

CAUSE OF DEATH (See instructions and examples)

		Approximate interval: Onset to death
32. **PART I.** Enter the chain of events--diseases, injuries, or complications--that directly caused the death. DO NOT enter terminal events such as cardiac arrest, respiratory arrest, or ventricular fibrillation without showing the etiology. DO NOT ABBREVIATE. Enter only one cause on a line. Add additional lines if necessary.		
IMMEDIATE CAUSE (Final disease or condition resulting in death) ⟶	a. Rupture of myocardium Due to (or as a consequence of):	Minutes
Sequentially list conditions, if any, leading to the cause listed on line a. Enter the	b. Acute myocardial infarction Due to (or as a consequence of):	6 days
UNDERLYING CAUSE (disease or injury that initiated the events resulting	c. Coronary artery thrombosis Due to (or as a consequence of):	5 years
in death) **LAST**	d. Atherosclerotic coronary artery disease	7 years

PART II. Enter other significant conditions contributing to death but not resulting in the underlying cause given in PART I	33. WAS AN AUTOPSY PERFORMED? ■ Yes ☐ No
Diabetes, Chronic obstructive pulmonary disease, smoking	34. WERE AUTOPSY FINDINGS AVAILABLE TO COMPLETE THE CAUSE OF DEATH? ■ Yes ☐ No

35. DID TOBACCO USE CONTRIBUTE TO DEATH? ■ Yes ☐ Probably ☐ No ☐ Unknown	36. IF FEMALE: ■ Not pregnant within past year ☐ Pregnant at time of death ☐ Not pregnant, but pregnant within 42 days of death ☐ Not pregnant, but pregnant 43 days to 1 year before death ☐ Unknown if pregnant within the past year	37. MANNER OF DEATH ■ Natural ☐ Homicide ☐ Accident ☐ Pending Investigation ☐ Suicide ☐ Could not be determined

CAUSE OF DEATH (See instructions and examples)

		Approximate interval: Onset to death
32. **PART I.** Enter the chain of events--diseases, injuries, or complications--that directly caused the death. DO NOT enter terminal events such as cardiac arrest, respiratory arrest, or ventricular fibrillation without showing the etiology. DO NOT ABBREVIATE. Enter only one cause on a line. Add additional lines if necessary.		
IMMEDIATE CAUSE (Final disease or condition resulting in death) ⟶	a. Aspiration pneumonia Due to (or as a consequence of):	2 Days
Sequentially list conditions, if any, leading to the cause listed on line a. Enter the	b. Complications of coma Due to (or as a consequence of):	7 weeks
UNDERLYING CAUSE (disease or injury that initiated the events resulting	c. Blunt force injuries Due to (or as a consequence of):	7 weeks
in death) **LAST**	d. Motor vehicle accident	7 weeks

PART II. Enter other significant conditions contributing to death but not resulting in the underlying cause given in PART I	33. WAS AN AUTOPSY PERFORMED? ■ Yes ☐ No
	34. WERE AUTOPSY FINDINGS AVAILABLE TO COMPLETE THE CAUSE OF DEATH? ■ Yes ☐ No

35. DID TOBACCO USE CONTRIBUTE TO DEATH? ☐ Yes ☐ Probably ■ No ☐ Unknown	36. IF FEMALE: ☐ Not pregnant within past year ☐ Pregnant at time of death ☐ Not pregnant, but pregnant within 42 days of death ☐ Not pregnant, but pregnant 43 days to 1 year before death ☐ Unknown if pregnant within the past year	37. MANNER OF DEATH ☐ Natural ☐ Homicide ■ Accident ☐ Pending Investigation ☐ Suicide ☐ Could not be determined

38. DATE OF INJURY (Mo/Day/Yr) (Spell Month) August 15, 2003	39. TIME OF INJURY Approx. 2320	40. PLACE OF INJURY (e.g., Decedent's home; construction site; restaurant; wooded area) road side near state highway	41. INJURY AT WORK? ☐ Yes ■ No

42. LOCATION OF INJURY: State: Missouri		City or Town: near Alexandria
Street & Number: mile marker 17 on state route 46a	Apartment No.:	Zip Code:

43. DESCRIBE HOW INJURY OCCURRED: Decedent driver of van, ran off road into tree	44. IF TRANSPORTATION INJURY, SPECIFY: ■ Driver/Operator ☐ Passenger ☐ Pedestrian ☐ Other (Specify)

Common problems in death certification

The **elderly decedent** should have a clear and distinct etiological sequence for cause of death, if possible. Terms such as senescence, infirmity, old age, and advanced age have little value for public health or medical research. Age is recorded elsewhere on the certificate. When a number of conditions resulted in death, the physician should choose the single sequence that, in his or her opinion, best describes the process leading to death, and place any other pertinent conditions in Part II. If after careful consideration the physician cannot determine a sequence that ends in death, then the medical examiner or coroner should be consulted about conducting an investigation or providing assistance in completing the cause of death.

The **infant decedent** should have a clear and distinct etiological sequence for cause of death, if possible. "Prematurity" should not be entered without explaining the etiology of prematurity. Maternal conditions may have initiated or affected the sequence that resulted in infant death, and such maternal causes should be reported in addition to the infant causes on the infant's death certificate (e.g., Hyaline membrane disease **due to** prematurity, 28 weeks **due to** placental abruption **due to** blunt trauma to mother's abdomen).

When **SIDS** is suspected, a complete investigation should be conducted, typically by a medical examiner or coroner. If the infant is under 1 year of age, no cause of death is determined after scene investigation, clinical history is reviewed, and a complete autopsy is performed, then the death can be reported as Sudden Infant Death Syndrome.

When processes such as the following are reported, additional information about the etiology should be reported:

Abscess	Carcinomatosis	Disseminated intra vascular coagulopathy	Hyponatremia	Pulmonary arrest
Abdominal hemorrhage	Cardiac arrest		Hypotension	Pulmonary edema
Adhesions	Cardiac dysrhythmia	Dysrhythmia	Immunosuppression	Pulmonary embolism
Adult respiratory distress syndrome	Cardiomyopathy	End-stage liver disease	Increased intra cranial pressure	Pulmonary insufficiency
Acute myocardial infarction	Cardiopulmonary arrest	End-stage renal disease	Intra cranial hemorrhage	Renal failure
Altered mental status	Cellulitis	Epidural hematoma	Malnutrition	Respiratory arrest
Anemia	Cerebral edema	Exsanguination	Metabolic encephalopathy	Seizures
Anoxia	Cerebrovascular accident	Failure to thrive	Multi-organ failure	Sepsis
Anoxic encephalopathy	Cerebellar tonsillar herniation	Fracture	Multi-system organ failure	Septic shock
Arrhythmia	Chronic bedridden state	Gangrene	Myocardial infarction	Shock
Ascites	Cirrhosis	Gastrointestinal hemorrhage	Necrotizing soft-tissue infection	Starvation
Aspiration	Coagulopathy	Heart failure	Old age	Subdural hematoma
Atrial fibrillation	Compression fracture	Hemothorax	Open (or closed) head injury	Subarachnoid hemorrhage
Bacteremia	Congestive heart failure	Hepatic failure	Paralysis	Sudden death
Bedridden	Convulsions	Hepatitis	Pancytopenia	Thrombocytopenia
Biliary obstruction	Decubiti	Hepatorenal syndrome	Perforated gallbladder	Uncal herniation
Bowel obstruction	Dehydration	Hyperglycemia	Peritonitis	Urinary tract infection
Brain injury	Dementia (when not	Hyperkalemia	Pleural effusions	Ventricular fibrillation
Brain stem herniation	otherwise specified)	Hypovolemic shock	Pneumonia	Ventricular tachycardia
Carcinogenesis	Diarrhea			Volume depletion

If the certifier is unable to determine the etiology of a process such as those shown above, the process must be qualified as being of an unknown, undetermined, probable, presumed, or unspecified etiology so it is clear that a distinct etiology was not inadvertently or carelessly omitted.

The following conditions and types of death might seem to be specific or natural but when the medical history is examined further may be found to be complications of an injury or poisoning (possibly occurring long ago). Such cases should be reported to the medical examiner/coroner.

Asphyxia	Epidural hematoma	Hip fracture	Pulmonary emboli	Subdural hematoma
Bolus	Exsanguination	Hyperthermia	Seizure disorder	Surgery
Choking	Fall	Hypothermia	Sepsis	Thermal burns/chemical burns
Drug or alcohol overdose/drug or alcohol abuse	Fracture	Open reduction of fracture	Subarachnoid hemorrhage	

Source: http://www.cdc.gov/nchs/nvss/vital_certificate_revisions.htm

WORDS OF WISDOM | "VALUE" OF AN AUTOPSY

"Doctor," complained the patient, "all of the other physicians called in on my case seem to disagree with your diagnosis."

"Yes, I know they do," said the doctor, "but the autopsy will prove that I'm right."

A *complete autopsy* refers to an examination of the organs of the three major cavities of the body—the abdomen, chest, and head. The pathologist makes a Y-shaped incision extending from each armpit to the center of the lower abdomen. Various internal organs are removed and weighed, and blood, urine, and other fluids are sampled. Much can be determined from the size, color, and feel of various body parts. For example, the liver of an alcoholic patient may be pale and shriveled, and a diseased heart is likely to be flabby and grossly enlarged.

Other than seeking the cause of death in an "unknown" situation with possible criminal activity involved, there are other reasons for performing autopsies, as noted by the National Association of Medical Examiners: (1) The autopsy serves as a check on the accuracy of the clinical diagnoses and historical data. Did the physicians correctly identify the patient's problem? Such information could be helpful in the event of a lawsuit against the hospital and staff regarding the death. (2) Likewise, the autopsy can be a check on the appropriateness of medical and surgical therapy that followed the diagnoses. Were the doctors treating the patient in appropriate ways? (3) An autopsy can help gather data on new and old diseases and surgical procedures. Such gathered information could help in treating future patients and thus benefit many individuals. (4) Information from the autopsy can also be useful to the deceased's family. The autopsy could identify inherited diseases that constitute a risk for other family members (e.g., heart disease, cancer, and certain kinds of kidney disease).

The autopsy has ancient roots and has produced countless medical advances over time (Clark & Springen, 1986). The cumulative data gathered through autopsies have enabled physicians to treat or prevent a number of diseases. Autopsies confirmed the link between cigarette smoking and lung cancer. They helped in the discovery of abnormalities involved in congenital heart defects, multiple sclerosis, Alzheimer's disease, and viral infections in the brain that may cause the dementia shown by some people living with AIDS. The purpose of an autopsy is summed up by the statement on the entrance to the autopsy room at the Medical University of South Carolina in Charleston: "This place is where death rejoices to come to the aid of life."

Despite the usefulness of autopsies, there has been a sharp decline in the number of autopsies performed in the United States. In the mid-1940s, half of the patients who died in a hospital were dissected, whereas as of 2007 only 8.5 percent are subjected to a postmortem examination (Kircher, 1992 and CDC, 2011). The percentage of deaths for which an autopsy was performed declined more than 50 percent from 1972 through 2007, from 19.3 percent to 8.5 percent. More recently, in Great Britain autopsy numbers have fallen by up to 90 percent, making it harder to spot trends in disease development, misdiagnoses, and evidence of substandard care. Medical errors are picked up in nearly 40 percent of postmortem examinations,

© Picsfive/Shutterstock.com

Autopsy rooms like this one are being used less frequently than in the past, and valuable medical information is going undiscovered.

which offer a vital tool for improving patient care. Health officials then scrutinize death certificates to identify public health priorities (Lakhani, 2008).

There are two types of autopsy: those ordered by a coroner for sudden and suspicious deaths, and those in which doctors need consent from bereaved relatives for patients who die in the hospital. The current crisis is in the latter group. In Britain, consent autopsies were once carried out on 10 percent of people who died in the hospital; currently less than 1 percent of deaths now result in an autopsy (Lakhani, 2008).

Autopsy findings are believed to increase the reliability of death certificate data and therefore to provide an essential check on the accuracy of diagnoses—although there may be legitimate disagreement between the attending physician and the pathologist as to precisely what killed the patient. According to Lakhani (2008), autopsies should be seen as part of the whole care package. Autopsies can inform physicians of unknown complications and interactions, as well as problems that were missed before death, so that clinical practice can be improved. Accurate information about the cause of death also plays a vital role in identifying emerging problems such as superbug epidemics or industrial diseases.

Physicians, hospital administrators, and families of the deceased have shown reduced interest in autopsies. Physicians in particular fear that an autopsy will uncover an error and thus spur malpractice suits. Furthermore, autopsies are not covered under Medicare, Medicaid, or most insurance plans, though some hospitals—teaching hospitals in particular—do not charge for autopsies of individuals who passed away in the facility. A private autopsy by an outside expert can cost between $3,000 and $5,000. In some cases, there may be an additional charge

| DEATH ACROSS CULTURES | NATIVE CANADIAN BELIEFS REGARDING THE AUTOPSY |

An anthropological study of Native Canadian patients in a hospital setting found conflict between indigenous people and the medical establishment regarding postmortem care of the body. One case involved the death of a 10-month-old female Ojibway child from a reserve community.

The pediatric hospital contacted the parents for consent to perform an autopsy, because the cause of death was undetermined, though the child was being treated for a wide range of neurological problems. The parents refused to consent to the procedure, saying that it conflicted with their spiritual beliefs about the care of the body of the deceased. Though the parents continued to object, the medical examiner ordered the autopsy, nonetheless.

After this unpleasant experience, the medical staff of the pediatric hospital decided to hold an in-service training workshop dealing with the general issue of the cultural interpretation of autopsies among Native people. The meeting opened with a presentation of the biomedical perspective of the physicians, emphasizing the importance of an autopsy for improving the general level of medical knowledge. Native speakers argued that a primary value in Native culture was the corporal integrity of the body after death, a value linked to spiritual understandings about the length of time required for the transition of a person's soul to the afterworld.

A pathologist from the hospital stated that Native clients' objections to autopsies must be balanced against the wider community interest in assuring that deaths do not occur without determination of negligence or malevolent cause. The medical examiner also stressed that the postmortem examination was sometimes necessary to control contagious or environmentally caused diseases.

The director of the Native Services program was asked to provide an overview of the concerns of Native clients regarding current policies governing autopsies. She contrasted the analytical approach of the medical community with Ojibway beliefs about dying and postmortem care, emphasizing that traditional beliefs focused on the importance of understanding the reasons for a person's death in terms of violation of traditional beliefs or moral transgressions. She stressed that traditional believers were also concerned with identification of the cause of death, but that the Ojibway fostered an integrative approach, not a narrow understanding of causation, as in the biomedical culture. Funeral rituals provided mechanisms that transformed death into an integrative event for the community. The Ojibway beliefs emphasize any opposition to a procedure that would disturb corporal integrity, including amputation or tissue removed in biopsies. The Ojibway people believe that the spirit resides in the body for a defined period after death. To open the body cavity or to remove tissue would disrupt the departure of the spirit and provide the possibility for retribution.

The conference was followed up with continuing informal consultations between clinicians and Ojibway representatives. In some cases family acceptance of the autopsy resulted. In other situations the pathological examination was limited to minimally invasive procedures. Yet, in other instances the opposition to autopsy was pursued through legal channels. In the end, culturally sensitive support for dying Native patients and their families may require that individuals assert their prerogative to die in their homes, not in a hospital setting.

From Cultural mediation of dying and grieving among Native Canadian patients in urban hospitals (pp. 231–251), by J. M. Kaufert & J. D. O'Neil, 1991. In D. R. Counts & D. A. Counts (Eds.), *Coping with the final tragedy: Cultural variation in dying and grieving*. Amityville, NY: Baywood Publishing.

for the transportation of the body to and from the autopsy facility (Frontline, 2011). The average cost of a hospital autopsy in 2011 ranged from $500 to $1,000 for one organ, but for a full-body autopsy the cost ranged from $2,000 to $4,000 (Walker, 2011).

Additionally, physicians are reluctant to ask bereaved next of kin for permission to perform a procedure that many regard as upsetting and as an invasion of privacy. In fact, autopsies have been so reduced in numbers that the National

LISTENING TO THE VOICES | REFLECTIONS ON AN OBSERVED AUTOPSY

After observing an autopsy, health intern Elizabeth Maxwell wrote about her experience:

> We thanked the resident and autopsy specialist and left the room. It was nice to take off the gloves, masks, and aprons we had been wearing. Fresh air was welcome, too. But I began thinking about what I had seen that morning. I was upset by the whole situation: an innocent, older man suddenly dies, and with the loss of his life, he loses his identity and dignity. He was just a body for anatomical study. He was not a person, though he was still wearing his wedding ring.
>
> As I contemplated this paradox, I thought again of the whole autopsy process, and was dismayed again. It had been so brutal. It was nothing like the delicate surgery I had expected. Since an autopsy must show every detail, huge incisions must be made. Nothing needs replacement, so little care is taken to keep the parts intact. This carelessness upset me.
>
> I tried to sort everything out in my mind—the patient, the tragic accident, the brutal procedure, the stoic autopsy specialist, the helpful and concerned medical student, and the joking male medical students. All of this was too much for me to comprehend.
>
> All that day and night, I kept smelling the autopsy odor in various places I went. I could not escape it! The next morning I wrote down my feelings in my journal concerning what I had seen and felt. I also talked with people about my frustrated emotions for several weeks. With the passing of time I can now talk about it, but sometimes I think about that old man.

Source: From Elizabeth Maxwell, former undergraduate student intern at St. Olaf College, Northfield, MN.

Center for Health Statistics has stopped collecting autopsy statistics altogether ("Final Cut," 2001).

In the "Listening to the Voices" account, Elizabeth Maxwell, a former St. Olaf College student visiting a medical center during a health science internship, discusses her observation of an autopsy and her emotional reactions to it. Although it is difficult to generalize from one experience, it is possible that her first encounter with the autopsy is not unlike that of many first-year medical students in gross anatomy class.

CORONERS AND MEDICAL EXAMINERS

When a person dies of unnatural causes, by violence or accident, the body may be made available to the appropriate police authorities for evidentiary and record-keeping purposes (Wolf, 1995). The coroner or public medical examiner is authorized to view the body and conduct an autopsy. If criminal conduct or suicide is suspected or if public health may be affected, an autopsy may be performed in extreme circumstances over the objection of the deceased's family.

A **coroner** is a public official, appointed or elected, in a particular geographic jurisdiction, whose official duty is to make inquiry into deaths in certain categories (e.g., if the death is unexpected or unexplained or if there is the possibility of an injury or poisoning). The office of the coroner ("crowner") dates back to medieval times when the crowner was responsible for looking into deaths to be sure death duties were paid to the king (National Association of Medical Examiners, 1996). For example, it was the job of the coroner to maintain the private property of the

Hawkins, Jonny/CSL, CartoonStock Ltd

"Unfortunately, we won't know what's wrong with you untill we
do an autopsy."

British Crown. The coroner was responsible for investigating accidents such as shipwrecks to see what money could be obtained for the "royal purse" (Mims, 1999). The coroner's primary duties today, however, are to make inquiry into a death and complete the certificate of death. The coroner assigns a cause and manner of death and lists them on the certificate of death. The coroner also decides if a death occurred under natural circumstances or was due to accident, homicide, suicide, or undetermined means.

Coroners are called upon to decide if a death was due to foul play. Depending on the jurisdiction and the law, the coroner may or may not be trained in the medical sciences. The coroner in some states is a licensed physician, and in other states the coroner may have no more than a high school degree. The coroner often employs pathologists to perform autopsies when a questionable death occurs.

A **medical examiner** is a physician charged within a particular jurisdiction to investigate and examine persons dying sudden, unexpected, unexplained, or violent deaths. Unlike the coroner, the medical examiner is expected to bring medical expertise to the evaluation of the medical history and physical examination of the deceased. Though not required to have training in pathology, many part-time medical examiners are encouraged to take medical training to increase their ability to investigate death scenes.

ADVANCE DIRECTIVES

All states have legislation authorizing the use of legal documents such as durable power of attorney for health care (sometimes referred to as a health care proxy, surrogate, agent, or attorney-in-fact) and living wills, which together are known as **advance directives (ADs)**. Kelly Greene (2002) found that 77 percent of all Americans between the ages of 65 and 74 feel that having a living will is very important—the most important preparation you can make for later life—yet only between 20 and 30 percent actually have a living will to address end-of-life issues.

| PRACTICAL MATTERS | PERMIT FOR POSTMORTEM EXAMINATION (AUTOPSY) |

Date _____

Name of deceased _____ Hosp. No. _____

Date and time of death _____

Month   Day   Year   Hour   P.M.   A.M.

1. I hereby authorize any qualified member of the Staff of Physicians of the Hospital to perform an autopsy upon the body of the above named at such place as the hospital or physician may designate. My relation to the deceased being that of _____.

2. I know of no survivor of the deceased who is closer kin than I to authorize this postmortem examination.

3. The postmortem examination here authorized may be either complete or partial and such parts of the body may be removed as may be necessary for subsequent study in order to accomplish its

purpose. If the nature and extent of this examination or the right to remove parts of the body are to be limited in any way, these limitations must be clearly stated below. In the absence of any stated limitations, it is to be understood that the physician by whom the operation is performed is hereby permitted to examine any portion of the body that, in his or her opinion, should be examined, including organs within the chest, abdomen, skull, and extremities.

4. I further authorize him or her to have present at the postmortem examination such persons as he or she may deem proper.

Limitations (if any) _____

Signed _____

_____ (Witness)

_____ (Witness)

However, the Terri Schiavo case in Florida in 2004 created increased interest in the topic and may have been the reason why Americans changed their minds about living wills and ADs. According to the 2005 Gallup Poll, after the Schiavo case in 2004, the number of Americans with living wills doubled. However, it should be pointed out that the majority of Americans are still without a living will—59 percent. The purpose of these documents is to provide indications of the patient's treatment preferences to physicians and/or surrogate decision makers in the event of decisional incapacity. To complete them requires reflection on personal values surrounding disability (mental and physical), terminal care, and circumstances surrounding death (e.g., organ procurement).

In the absence of a legally binding AD, a court action may be necessary to decide whether treatment of an incompetent person or minor child may be terminated (Wolf, 1995), particularly if there is disagreement among decision makers or a disagreement between decision makers and the physician. Such judicial determinations in the recent past have focused on ascertaining what the patient would most likely have chosen had she or he been competent and whether the judgment of a third party (e.g., parent, guardian, or physician) may be substituted for the patient's consent. In some cases since 1976, a hospital ethics committee has become a forum for the analysis of such conflicts, and despite the fact that ethics committees do not usually make decisions, their involvement sometimes helps avoid an often prolonged process of litigation in court.

In 1992 the Joint Commission (formerly the Joint Commission on Accreditation of Healthcare Organizations) mandated that health care organizations come up with some way of addressing ethical concerns. By the year 2000, 95 percent of general hospitals offered ethics consultation or were starting up a consultation service (O'Reilly, 2008).

As noted in Chapter 8, since the implementation of the Patient Self-Determination Act in December 1991, health maintenance organizations (HMOs) and other health care institutions have been required to inform all adult patients about ADs, and hospitals have been required to ask all patients whether they have completed ADs. A study in California (Gordon & Shade, 1999) of more than 5,000 persons 65 or older regarding end-of-life care preferences revealed that less than one-third in this HMO group had completed and filed an AD, even though an all-out effort had occurred to encourage participation—waiting-room posters, articles about ADs in their health plan newsletter, and personal efforts on behalf of the health plan staff. Rates of completion were higher among the older persons who were in poorer health. A number of other educational strategies to increase AD completion rates have been tried, with varying amounts of success; however, most have been unsuccessful (Braun & Kayashima, 1999).

Researchers Braun and Kayashima (1999) decided to encourage ADs through churches and synagogues. The outcome of their 12-month project in Hawaii—using interviews and focus groups and distributing handbooks developed by their religious leaders to the congregations—was that the respondents saw real value in having the church or temple serve as a venue for death education. They suggested that an AD project would probably succeed in churches and temples where ministers encouraged these efforts.

More recently, in 2009 and beyond, in public discussions related to the health care reform bill in the U.S. Congress, commonly called the Affordable Care Act (ACA) or "Obamacare," one of the provisions for seniors was to be counseling for end-of-life provisions and for the creation of ADs. Those opposing the bill (mostly members of the Republican Party, especially "Tea Party" members) tried to convince seniors that the U.S. government was trying to kill the elderly in order to save on medical costs. Many seniors are suspicious of ADs because they fear that their options for health care may be limited.

Living Wills

As previously mentioned in Chapter 8, a **living will** is a document that states that an individual does not want medical intervention if the technology or treatment that keeps her or him alive cannot offer a reasonable quality of life or hope for recovery. Living wills were first introduced in the 1970s as the public came to fear dying while being connected to feeding tubes in the intensive care unit, even as they sought the cures hospitals and doctors offered (Plech, 2000). Despite a somewhat growing acceptance of living wills, the issue is still complicated by emotion and questions of when such a will should be invoked. Many persons are reluctant to take steps leading to a patient's death, and some are willing to withhold further medical treatment but do not want to disconnect existing life-support equipment. Others cannot distinguish living wills from "do not resuscitate" (DNR) orders (see Chapter 8).

In a 2001 study by Peter Ditto of the University of California at Irvine's School of Social Ecology, it was discovered that surrogate decision makers—most

| WORDS OF WISDOM | ADVANCE CARE PLANNING IN HEALTH CARE REFORM LEGISLATION |

There has been some confusion and misinformation regarding the advance care planning provision in the House's health care reform bill. Uninformed individuals have described this provision as a mandatory session that would dictate health care choices for older Americans. This is inaccurate and has resulted in confusion surrounding the value of advance care planning.

Section 1233 (pages 424–434) of the House Ways and Means Committee version of health care reform contains a new Medicare provision to allow coverage for an "Advance Care Planning Consultation." This provision is intended to offer Medicare beneficiaries an opportunity to engage in an informed and focused conversation with their health care practitioner about advance care planning options. This consultation would be in addition to the "Welcome to Medicare" physician consultation.

This consultation, like other consultations within the Medicare system, would be *voluntary* and would be reimbursable under Medicare when provided no more than once every five years, or whenever a patient undergoes a qualifying event, such as a life-threatening or terminal diagnosis; chronic disease diagnosis; or admission to a long-term care facility, a skilled nursing facility, or a hospice program.

Topics that are covered during the consultation, include:

- An explanation by a physician, nurse practitioner, or physician's assistant of advance care planning, including key questions and considerations, important steps, and suggested people to talk to.
- An explanation by the practitioner of advance directives, including living wills and durable powers of attorney, and their uses.
- An explanation by the practitioner of the role and responsibilities of a health care proxy.
- The provision, by the practitioner, of a list of national and state-specific resources to assist consumers and their families with advance care planning, including the national toll-free hotline, the advance care planning clearinghouses, and state legal services organization.
- An explanation of the end-of-life services and supports available, including palliative care and hospice.

The consultation is *not mandatory*. No one is required to undergo the consultation. NHPCO's Caring Connections program has provided a wide range of materials on advance care planning, including state-specific advance directives and information on discussing this important issue with loved ones and health care providers. People looking for more information, or who wish to download an advance directive, should visit Caring Connections at www.caringinfo.org/PlanningAhead.

From National Hospice and Palliative Care Organization, August 7, 2009.

often spouses and adult children—accurately predict patients' preferences regarding life-sustaining treatment only about 70 percent of the time. However, ADs do not significantly improve that level of accuracy, even when the surrogate discusses the instructions with the patient. Typically, the study shows, surrogates' predictions err on the side of overtreatment.

The results of Ditto's study (2001) showed that surrogates given the opportunity to read the patient's living will before making their predictions were no more accurate in predicting patients' life-sustaining treatment wishes than were surrogates who had not read the patient's living will. Even more surprising, allowing surrogates to discuss the living will with the patient prior to making their predictions also proved completely ineffective in improving the accuracy of surrogates' predictions. Predicting another person's wishes is no simple feat, even in mundane

PRACTICAL MATTERS | A LIVING WILL

To my family, my physician, my lawyer, and all others whom it may concern: Death is as much a reality as birth, growth, maturity, and old age. Death is the one certainty of life. If the time comes when I can no longer take part in decisions for my own future, let this statement stand as an expression of my wishes and directions, while I am still of sound mind.

If at such a time the situation should arise in which there is no reasonable expectation of my recovery from extreme physical or mental disability, I direct that I be allowed to die and not be kept alive by medications, artificial means or "heroic measures." I do, however, ask that medication be mercifully administered to me to alleviate suffering even though this may shorten my remaining life.

This statement is made after careful consideration and is in accordance with my strong convictions and beliefs. I want the wishes and directions here expressed carried out to the extent permitted by law. Insofar as they are not legally enforceable, I hope that those to whom this will is addressed will regard themselves as morally bound by these provisions.

Signed _____

Date _____

Witness _____

Witness _____

situations like predicting which movie your spouse wants to see. How much more difficult, then, to imagine whether your spouse would want to be tube-fed if he or she were in a coma.

This is not to say ADs are without value: "We found no evidence at all that having people talk and read about end-of-life decisions led them to understand each other any better. But discussion did seem to make both patients and surrogates feel better, particularly if the two had never dealt with end-of-life issues before," Ditto said (2001). Discussion of end-of-life issues often produced a sense of mutual understanding and comfort with end-of-life decision making, the study found (Ditto, 2001).

Perhaps even more problematic is the research by A. M. Cugliari and T. E. Miller (1994) that determined that 29 percent of New York state hospitals would not even honor a patient's request to withdraw or withhold treatment. In another study by G. Almgren (1993), in which 140 nursing home directors of nursing were interviewed, it was discovered that the competency of the patient and type of nursing home ownership were the primary factors influencing decisions to withdraw nutrition from terminally ill patients—the fact that a patient had signed a living will had little influence on these decisions. Furthermore, S. H. Miles and A. August (1990) demonstrated that the courts have a gender bias in the forced implementation of living wills. Because men were perceived as more rational and women as more unreflective, emotional, and immature, the courts accepted the treatment preference of 75 percent of the men but only 14 percent of the women. This pattern of gender bias had been discovered in many states and remains unchanged when one controls for the patient's age, condition, and treatment modality (Miles & August, 1990). Consistent with the research findings just cited, Donald Dilworth (1996) conducted his own research in five hospitals and reached the following conclusion:

> Living wills do virtually nothing to reduce patient suffering, because doctors and hospitals ignore them. Some 80 percent of doctors either misunderstood or set aside

PRACTICAL MATTERS	FIVE WISHES

The *Five Wishes* document helps you express how you want to be treated if you are seriously ill and unable to speak for yourself. It is unique among all other living will and health agent forms because it looks to all of a person's needs: medical, personal, emotional and spiritual. *Five Wishes* also encourages discussing your wishes with your family and physician.

Five Wishes lets your family and doctors know:

• Who you want to make health care decisions for you when you can't make them.
• The kind of medical treatment you want or don't want.
• How comfortable you want to be.
• How you want people to treat you.
• What you want your loved ones to know.

Five Wishes is changing the way America talks about and plans for care at the end of life.

More than 18 million copies of *Five Wishes* are in circulation across the nation, distributed by more than 35,000 organizations. *Five Wishes* meets the legal requirements in 42 states and is useful in all 50. With assistance from the United Health Foundation, *Five Wishes* is now available in 26 languages.

Five Wishes has become America's most popular living will because it is written in everyday language and helps start and structure important conversations about care in times of serious illness.

Five Wishes was introduced in 1997 and originally distributed with support from a grant by the Robert Wood Johnson Foundation, the nation's largest philanthropy devoted exclusively to health and health care. A copy of *Five Wishes* can be obtained for $5 (1-888-5-WISHES; www.aging-withdignity.org/5wishes.html).

Source: From http://www.agingwithdignity.org/5wishes.html

the dying requests, and a program to help patients avoid painful life-prolonging treatments had no effect at any of five hospitals I studied. The cause of the problem is the medical culture that rejects death and promotes technology.

Another problem with living wills is that they often address specific medical scenarios, yet possible treatments sometimes happen in "gray" areas (Chatzky, 2000). For example, many patients' living wills rule out some treatments if they are terminally ill. However, a profound stroke or end-stage Parkinson's disease may not be defined as a terminal condition.

A solution that forces one to think about the end-of-life gray areas ahead of time is called **Five Wishes**. *Five Wishes* was designed by lawyer Jim Towey, legal counselor to Mother Teresa. This "friendly-to-use" living will document guides people in making essential decisions about the care they want at the end of their life. *Five Wishes* speaks to people in their own language, not in "doctor speak" or "lawyer talk." It can be used in the living room instead of the emergency room. And it helps families talk with their physician about a subject that was too hard to face before. The idea of *Five Wishes* is to sit around the dinner table with your significant others and work through your wishes step by step.

Five Wishes is changing the way America talks about and plans for care at the end of life. More than 18 million copies of *Five Wishes* are in circulation across the nation, distributed by more than 35,000 organizations. *Five Wishes* meets the legal requirements in 42 states and is useful in all 50. *Five Wishes* is now available in 26 languages. The eight states where *Five Wishes* is not legally valid either require a specific state form or require that the person completing an AD be read a mandatory notice or warning. Residents of these states can still use *Five Wishes* to put their wishes in writing and communicate their wishes with their family and physician. Most health care professionals understand they have a duty to listen to the wishes of their patients, no matter how they are expressed.

If you complete a living will or *Five Wishes*, make several copies of it and share it with physicians and other health care providers, family members, and friends. A copy should also be given to your attorney to be kept in her or his office. This is not something that you should limit to a single copy, lock up in a safe deposit box, or keep as a big secret. Share the news!

A **durable power of attorney** for health care decisions allows an individual to designate someone else as a legal proxy to make treatment decisions if the person becomes unable to make them (Braun & Kayashima, 1999). *Durable* means that the document will be recognized in a court of law. There has been some disenchantment with the rather anemic power of living wills. Living wills seem to have no teeth, concluded Margaret Battin in *The Least Worst Death* (1994). However, states began in the 1990s to modify existing living will statutes or to adopt new ones to provide a durable power of attorney for health care. States provided some statutory mechanism for appointing a surrogate, so that a person could select someone—a friend, relative, or anyone she or he trusted—to make medical decisions authorizing the withholding or withdrawing of life-sustaining treatment if the individual were no longer able to do so. Durable power of attorney documents (and living wills) may not reliably represent a patient's true choices over time, noted Battin (1994). Indeed, in thorough research of this topic, Markson and colleagues (1997) concluded that those who commonly participate in surrogate decisions for incapacitated patients—family members and physicians—may place a different value on medical interventions than the patients they are supposed to represent and may not predict patient preferences accurately.

Durable Power of Attorney for Health Care

The durable power of attorney for health care can be structured to give an agent as little or as much authority as the patient chooses (Bennett, 1999). However, in general, the following are authorized by the agent: (1) using or withholding life support and other medical care; (2) placing the patient in, or taking her or him out of, a health facility such as a hospital; and (3) making decisions not specifically covered by a living will.

On the other hand, the durable power of attorney for health care does *not*: (1) give any authority to financial matters, administering involuntary psychiatric care, or sterilization; or (2) generally cover medical treatment that would provide comfort or relieve pain, which means that a health care proxy cannot refuse pain relief on your behalf (Bennett, 1999).

The durable power of attorney for health care generally only takes effect when two physicians, including the patient's physician of record, certify that the patient is not capable of understanding or communicating decisions about her or his health. Copies of durable health care powers of attorney should be distributed to the primary physician, the pharmacist, the individual's agent, and the hospital of choice, suggests Jarratt Bennett (1999).

PRACTICAL MATTERS | DURABLE POWER OF ATTORNEY FOR HEALTH CARE

PART 1 – POWER OF ATTORNEY FOR HEALTH CARE

DESIGNATION OF AGENT: I designate the following individual as my agent to make health care decisions for me:

Name of individual you choose as agent: _____

Address: _____

_____

Telephone: _____
     (home phone)   *(work phone)*   *(cell/pager)*

OPTIONAL: If I revoke my agent's authority or if my agent is not willing, able, or reasonably available to make a health care decision for me, I designate as my first alternate agent:

Name of individual you choose as first alternate agent: _____

Address: _____

_____

Telephone: _____
     (home phone)   *(work phone)*   *(cell/pager)*

OPTIONAL: If I revoke the authority of my agent and first alternate agent or if neither is willing, able, or reasonably available to make a health care decision for me, I designate as my second alternate agent:

Name of individual you choose as second alternate agent: _____

Address: _____

_____

Telephone: _____
     (home phone)   *(work phone)*   *(cell/pager)*

AGENT'S AUTHORITY: My agent is authorized to make all health care decisions for me, including decisions to provide, withhold, or withdraw artificial nutrition and hydration and all other forms of health care to keep me alive, except as I state here:

_____

_____

_____

(Add additional sheets if needed.)

continued

PRACTICAL MATTERS | DURABLE POWER OF ATTORNEY FOR HEALTH CARE *continued*

WHEN AGENT'S AUTHORITY BECOMES EFFECTIVE: My agent's authority becomes effective when my primary physician determines that I am unable to make my own health care decisions. _____
(Initial here)

OR

My agent's authority to make health care decisions for me takes effect immediately. _____
(Initial here)

AGENT'S OBLIGATION: My agent shall make health care decisions for me in accordance with this power of attorney for health care, any instructions I give in Part 2 of this form, and my other wishes to the extent known to my agent. To the extent my wishes are unknown, my agent shall make health care decisions for me in accordance with what my agent determines to be in my best interest. In determining my best interest, my agent shall consider my personal values to the extent known to my agent.

AGENT'S POSTDEATH AUTHORITY: My agent is authorized to make anatomical gifts, authorize an autopsy and direct disposition of my remains, except as I state here or in Part 3 of this form:

_____

_____

(Add additional sheets if needed.)

NOMINATION OF CONSERVATOR: If a conservator of my person needs to be appointed for me by a court, I nominate the agent designated in this form. If that agent is not willing, able or reasonably available to act as conservator, I nominate the alternate agents whom I have named, in the order designated.

PART 2 – INSTRUCTIONS FOR HEALTH CARE

If you fill out this part of the form, you may strike any wording you do not want.

END-OF-LIFE DECISIONS: I direct that my health care providers and others involved in my care provide, withhold, or withdraw treatment in accordance with the choice I have marked below:

Choice *Not To* Prolong Life:

_____ I do not want my life to be prolonged if (1) I have an incurable and irreversible
(Initial here) condition that will result in my death within a relatively short time, (2) I become unconscious and, to a reasonable degree of medical certainty, I will not regain consciousness, or (3) the likely risks and burdens of treatment would outweigh the expected benefits,

OR

Choice *To* Prolong Life:

_____ I want my life to be prolonged as long as possible within the limits of generally
(Initial here) accepted health care standards.

RELIEF FROM PAIN: Except as I state in the following space, I direct that treatment for alleviation of pain or discomfort be provided at all times, even if it hastens my death:

_____

_____

(Add additional sheets if needed.)

OTHER WISHES: (If you do not agree with any of the optional choices above and wish to write your own, or if you wish to add to the instructions you have given above, you may do so here.) I direct that:

_____

_____

_____

_____

(Add additional sheets if needed.)

PART 3 – DONATION OF ORGANS AT DEATH (OPTIONAL)

I. Upon my death:

I give any needed organs, tissues, or parts _____
                                             (Initial here)

OR

I give the following organs, tissues, or parts only: _____   _____
                                                                                (Initial here)

II. If you wish to donate organs, tissues, or parts, you must complete II and III.

My gift is for the following purposes:

Transplant _____   Research _____
           (Initial here)   *(Initial here)*

Therapy _____   Education _____
        (Initial here)   *(Initial here)*

III. I understand that tissue banks work with both nonprofit and for-profit tissue processors and distributors. It is possible that donated skin may be used for cosmetic or reconstructive surgery purposes. It is possible that donated tissue may be used for transplants outside of the United States.

1. My donated skin may be used for cosmetic surgery purposes.

    Yes _____   No _____
        (Initial here)   *(Initial here)*

2. My donated tissue may be used for applications outside of the United States.

    Yes _____   No _____
        (Initial here)   *(Initial here)*

continued

PRACTICAL MATTERS | DURABLE POWER OF ATTORNEY FOR HEALTH CARE *continued*

3. My donated tissue may be used by for-profit tissue processors and distributors.

Yes _____   No _____
        (Initial here)   *(Initial here)*

(Health and Safety Code Section 7158.3)

PART 4 – PRIMARY PHYSICIAN (OPTIONAL)

I designate the following physician as my primary physician:

Name of Physician: _____   Telephone: _____

Address: _____

_____

OPTIONAL: If the physician I have designated above is not willing, able, or reasonably available to act as my primary physician, I designate the following physician as my primary physician:

Name of Physician: _____   Telephone: _____

Address: _____

_____

PART 5 – SIGNATURE

The form must be signed by you and by two qualified witnesses, or acknowledged before a notary public.

SIGNATURE: Sign and date the form here:

Date: _____

Name: _____   _____
            (sign your name)   *(print your name)*

Address: _____

_____

STATEMENT OF WITNESSES: I declare under penalty of perjury under the laws of California (1) that the individual who signed or acknowledged this advance health care directive is personally known to me, or that the individual's identity was proven to me by convincing evidence (2) that the individual signed or acknowledged this advance directive in my presence, 3) that the individual appears to be of sound mind and under no duress, fraud, or undue influence, (4) that I am not a person appointed as agent by this advance directive, and (5) that I am not the individual's health care provider, an employee of the individual's health care provider, the operator of a community care facility, an employee of an operator of a community care facility,

Form 3-1 Advance Health Care Directive

the operator of a residential care facility for the elderly, nor an employee of an operator of a residential care facility for the elderly.

FIRST WITNESS

Name: _____   Telephone: _____

Address: _____

_____

Signature of Witness: _____   Date: _____

SECOND WITNESS

Name: _____   Telephone: _____

Address: _____

_____

Signature of Witness: _____   Date: _____

ADDITIONAL STATEMENT OF WITNESSES: At least one of the above witnesses must also sign the following declaration:

I further declare under penalty of perjury under the laws of California that I am not related to the individual executing this advance health care directive by blood, marriage, or adoption, and to the best of my knowledge, I am not entitled to any part of the individual's estate upon his or her death under a will now existing or by operation of law.

Signature of Witness: _____

LIFE INSURANCE

With that innovative institution, life insurance, man has tried to exorcise the anxiety and financial insecurity that would otherwise possess people contemplating death. Very early evidences of life insurance were seen in Barcelona in the 15th century, when the lives of slaves were insured, and in England in the late 17th century (Smith, 1941). The Aztecs in the 15th century and the ancient Babylonians in the Code of Hammurabi created elaborate systems of compensation for deaths (Greer, 1985). In the United States, as early as 1759, Presbyterian ministers established a company to provide a means whereby a minister of the Presbyterian faith could buy an annuity for his widow by making annual payments for a specified number of years (Presbyterian Ministers' Fund, 1938). By 1850, in the United States 48 companies held policies valued at $97 million; by 1920, 335 companies held 65 million policies worth $40 billion. Life insurance contributed to "the practical disappearance of the thought of death as an influence bearing upon practical life," and thus, it contributed to the "dying of death" (U.S. Bureau of the Census, 1975, p. 1050).

The tremendous growth of life insurance in the United States, after its beginnings in the mid-19th century, was due in part to our capitalistic economy (Clough, 1946). In earlier times society was organized on an essentially agricultural basis; many families were largely self-sufficient in an economic sense with the exchange of goods conducted mainly on a barter basis. The farm workers and the farmer had an equity in the land and, upon death, the children would then till the fields and carry on the tradition of the parents or continue as landholders. The farm was left to the heirs and the farm workers could typically continue working on the farm on which they grew up, if they so desired.

With the Industrial Revolution and the move to urban areas, however, the security of an agrarian society began to dissolve, observed Clough (1946). In a capitalistic economy, capital is a mobile surplus, which can be used for the accumulation of more surplus or profit, and the capitalist system is one in which the means of making a living are primarily based on the employment of capital. In the evolution of capitalism, the means of production have in part passed out of the possession or direct control of the worker, and the individual does not consume what he or she produces and must supply daily needs by going to the market. Livelihood is obtained either by money wages, by earnings on capital, or by the expenditure of surplus. Society has evolved away from economic self-sufficiency to economic interdependence, noted Clough. Such an evolution fits the description of French sociologist Emile Durkheim (1893/1946): We have moved from a society based on **mechanical solidarity** to one based on **organic solidarity**—from a fairly simple, homogeneous society in which individuals were generalists and could more or less take care of themselves by living off the land, to a more complex, diverse, heterogeneous society in which we are more dependent on each other. Organic solidarity fits the analysis of society from a structural-functional perspective, as discussed in Chapter 1.

Life insurance emerged from the same historical context as did funeral reform and rural cemeteries; it assumed the uniformity and continuity of death as natural occurrences. Life insurance depended on the science of statistics and on a species perspective of death and immortality that focused attention not on the life of the individual policyholder, but rather on the lives of beneficiaries. Over time, with experience in rates of mortality and the development of the laws of probability, life insurance on a scientific basis was possible. Yet, it was necessary to have enough people interested in sharing the risks of death so that there would be an average of expected losses before the scientific principles of insurance could operate correctly.

The value of life used by insurance agents, economists, legal experts, scientists, and agency administrators ranges from a few dollars to many millions of dollars, depending on the formulas used (Greer, 1985). One way to figure value is to break down the body into chemical elements—five pounds of calcium, 11.2 pounds of phosphorus, nine ounces of potassium, six ounces of sulfur, six ounces of sodium, a little more than one ounce of magnesium, and less than one ounce each of iron, copper, and iodine. On that basis a human life is worth $8.37. Another way to measure the monetary value of human life, noted Greer, is the growing price of contract murder. Such a killing might pay several thousand dollars to the killer, depending on how badly someone wants the person killed and how much the person wants to carry out the assignment. Insurance companies, however, do not figure the value of a person based on chemical elements in the body or on the price of a hired killing. The insurance company, in figuring what a person's life is worth, determines this by what one would have earned had he or she lived— earning power over the course of a working life.

Like the rural cemetery, life insurance was praised for its educational benefits, because it taught lessons of self-reliance, forethought, thrift, discipline, and (very) delayed gratification. For these reasons, clergymen such as Henry Ward Beecher endorsed the system of life insurance, responding to critics that, in effect, God helps those who help themselves. Also, life insurance accentuated the importance of the family, as did romantic sentimentalism and the family plot of the rural ceme- tery. Finally, life insurance provided families with money to pay for elaborate fun- erals, a fact that affected both the development of funeral service and the history of bereavement.

LIFE INSURANCE AS PROTECTION

The middle-class American virtue of seizing control and planning ahead for the future is exemplified by the fact that the majority of Americans have life insurance. The purpose of life insurance is to protect an individual's dependents and to give the insured person a feeling of security in knowing that his or her dependents will have some coverage in the event of death.

The amount of life insurance needed is related to the number of dependents a person has and the financial security of each. For example, if the person is the pri- mary provider in a family having three small children, he or she would have a greater need for life insurance than a single person with no dependents. If the pri- mary provider were to die, these dependents would have some financial coverage, at least for the time being. At the time of death, the added burden of the loss of income is not needed by the survivors.

The age of the individual and that person's place in the life cycle are also important considerations. If the individual is aged and there are no dependents or the individual's spouse has enough retirement income to provide for his or her living expenses, the primary need will be to cover death-related expenses. However, if one's only purpose for life insurance is to provide for one's dependents, term life insurance is probably the best choice. **Term life insurance** tends to have the lowest premiums for the greatest amount of coverage. Group policies, such as those provided by one's place of employment, tend to have lower premiums than policies obtained outside of a group. Term life insurance is for a specific number of years, such as five or 10, with premiums going up as one ages. "My suggestion is to start with term, but if there is a permanent insurance need, convert to a good **universal life** policy," said Andrew Gross, a financial planner in Washington, DC (Smith, 1988, p. 154).

LIFE INSURANCE AS AN INVESTMENT

Today, however, life insurance for many individuals is more than security for their dependents. Some policies, universal life policies for example, are a form of tax-deferred investment from which one can also borrow. Thus, one can have coverage in the event of death and can have the opportunity to build cash assets at the same time. This type of insurance policy is called **whole life** from which, for a set premium, one receives life insurance and a savings fund. **Universal life insurance** is more flexible, has investment opportunities, and allows one to raise or reduce premiums and the amount of coverage on one's life. **Variable life insurance** does not allow for the premium or minimum coverage on one's life to change, but one can switch the savings from among money markets or various forms of stock.

Another advantage of life insurance is that it is a way to pass assets to relatives and other survivors without having any estate tax assessed by the government. Proceeds from life insurance policies do not go through **probate** and therefore the tax-favored treatment of life insurance allows one to exceed the deductible allowances for estate tax and pass on more resources to one's survivors.

It is advisable to shop around and compare the benefits of the different life insurance programs. Premiums vary considerably for the same coverage. Be cautious of high-powered salespersons trying to sell coverage not needed. Become informed about life insurance by talking with knowledgeable consumers or by reading consumer magazines before purchasing any life insurance. Comparison shopping should pay off.

DISPOSITION OF PROPERTY

Upon death, a person's business affairs must be concluded (debts, taxes, and funeral expenses), and the property owned at death must be distributed to others. Real property, which includes land and what is attached to it, such as a house (as opposed to personal property such as clothes or jewelry), is transferred in three ways—through a deed, a will, or an inheritance. The latter two deal with property transfer at death. A will is important if one wishes to have any say in the transfer.

"And to my teenage grandchildren, I leave my unused
phone minutes."

Morgan, Ron/CSL, CartoonStock Ltd

INTESTATE AND INHERITANCE

If an individual dies without a will, that person dies **intestate**. In such a situation the property of the **decedent** (the deceased person) will pass through inheritance. To die intestate means that important decisions such as who gets your assets and raises your children may be made by a state judge in ways that you might not like. The court will appoint a person or institution (e.g., a bank), called an **administrator,** to administer the estate of a person who died intestate. Because only about 40 percent of American adults (and 25 percent of adults between 25 and 34) have a will (FindLaw Resources, 2007), dying intestate is the way most Americans die. Each of the 50 states has its own laws regarding wills; thus, what is true of one state may not necessarily be true of another.

If one dies intestate, classes of heirs (also called laws of descent) are established by the various states to determine how the property passes. Typically, if one is married at the time of death and has no children, the spouse receives the entire inheritance. If one has children, they also take from the inheritance (if the children are minors, the inheritance is to be held in trust until they reach a certain age). If one is married with children, the spouse takes half and the children take the other half. (E.g., if there is one child, she or he takes the entire half or 50 percent; if there are two children, they split the 50 percent, thus receiving 25 percent each—one half of one half; if there are three children, they each receive approximately 17 percent—one third of one half or one sixth, etc.). If the decedent has children but no spouse, then the children take the entire inheritance. (If there are two children, they each receive 50 percent of the inheritance, for example). If one dies intestate, the classification of heirs typically follows this pattern:

1. Spouse (no children), then spouse takes all.
2. Spouse and children, then the spouse takes half and the children split the other half.

3. Child or children and no spouse, then the child takes all or the children split equally.
4. If no spouse or children, then parents take from the inheritance.
5. If no spouse, children, or parents, typically siblings (brothers and sisters) take from the inheritance.
6. If "none of the above," then usually grandparents take from the inheritance, followed by aunts and uncles. If one dies intestate and without heirs, one's property goes to the state (the legal term for this is **escheat**).

Long-Term Care Insurance

Long-term care insurance helps provide for the cost of long-term care beyond a predetermined period. Long-term care insurance covers care generally not covered by health insurance, Medicare, or Medicaid. Individuals who require long-term care are generally not sick in the traditional sense, but instead, are unable to perform the basic **activities of daily living** (ADLs) such as dressing, bathing, eating, toileting, continence, transferring (getting in and out of a bed or chair), and walking.

About 60 percent of individuals over age 65 will require at least some type of long-term care services during their lifetime. About 40 percent of those receiving long-term care today are between 18 and 64. Once a change of health occurs, long-term care insurance may not be available. For example many people who are diagnosed with Alzheimer's and Parkinson's disease are under the age of 65. Presently only 10 percent of people who require long-term care are covered by long-term care insurance (National Bureau of Economic Research, 2014).

With long-term care in nursing homes costing as much as $7,000 per month ($84,000 per year), many people have realistic concerns that they will outlive their assets. I had a hospice patient who was confined to a nursing home (without long-term care insurance) and his wife was very concerned that when he finally died she would be left without any savings and only half of his social security benefit. Therefore, many people in their 50s will want to secure long-term care insurance before their health declines.

Michael Leming's wife was 58 when she secured such a policy at a guaranteed rate of $1,900 per year (all of it tax-deductible). At that time Lemings had two children in college and their resources were stretched. However, once a person purchases a policy, the language cannot be changed by the insurance company, the premiums cannot be increased, and the policy usually is guaranteed renewable for life. It can never be canceled by the insurance company for health reasons, but can be canceled for nonpayment. All payments for premiums are also deductible on one's federal taxes.

As in life insurance, the longer one waits to take out a long-term care insurance policy, the higher the premiums one pays and the greater the risk of poor health in qualifying for coverage. Unfortunately for Michael Leming, he had borderline diabetes and, unlike his wife, he was ineligible for coverage.

Some companies offering long-term care insurance are now going out of business because the cost of paying for benefits are going up at a rate faster than premiums for new customers, making this a poor investment opportunity for insurance companies. Insurance companies are dealing with this problem in three ways: 1) going out of business and selling their existing policies to other companies, 2) raising requirements for people applying for coverage (only insuring only the most healthy and unlikely to collect), or 3) raising premiums to cover expected payment benefits.

Long-term care insurance generally covers home care, assisted living, adult daycare, respite care, hospice care, nursing home and Alzheimer's facilities. If home care coverage is purchased, long-term care insurance can pay for home care, often from the first day it is needed. It will pay for a visiting or live-in caregiver, companion, housekeeper, therapist, or private duty nurse up to seven days a week, 24 hours a day—up to the policy benefit maximum. In Michael Leming's wife's policy, she can even hire a friend or family member to provide care for her in her home.

Most policies have an elimination period or waiting period similar to a deductible. This is the period of time that you pay for care before your benefits are paid. Elimination days may be from 20 to 120 days. The longer the elimination period, the lower the premium. Some policies require intended claimants to provide proof of 20 to 120 service days of paid care before any benefits will be paid. In some cases the option may be available to select zero elimination days when covered services are provided in the home in accordance with a "Plan of Care." Some policies require that the policy for long-term care be paid up to one year before becoming eligible to collect benefits.

WILLS

A **will** is essential if one does not want the state to determine the distribution of her or his property and decide who will raise the children (if children are involved). A well-executed will can provide orderly distribution of property, get the decedent's

Last Will and Testament.

assets to the people to whom she or he wants them to go, reduce the expense of probate court, and allow the **testator** (the one making the will) to name an individual of her or his choosing to administer the estate and to designate someone to be legally responsible for young children, rather than have the court appoint an administrator and place minor children (in the event of a single parent's death). To make out a will, the following criteria generally apply, though these may vary by state:

1. Must be at least 18 years of age or living in one of the few states that permit younger persons to make a will if they are married, in the military, or otherwise considered emancipated.
2. Must be of sound mind. One does not have to prove mental state in most cases. The testator may seem mentally "way out" and act "crazy," yet be of "sound mind" in a legal sense. A minimum mental capacity is imposed to ensure that persons writing wills have an appropriate appreciation of their actions, their property, and the parties to whom they wish their estates to pass.
3. The will must be in writing—typewritten or computer-printed.
4. The will must be signed by the testator.
5. The will must have two witnesses to the signing (Vermont requires three witnesses). The witnesses must indeed *witness* the testator signing the will. What one cannot do is sign the will, then rather casually ask two people to sign as witnesses. They would not have witnessed the signing, their sole function. The witnesses do not have to read the will. They are simply confirming that they saw the testator sign her or his will. Often the attorney and one of her or his employees will sign. Someone who is to benefit from the will should not be a witness.
6. The will must be dated.

A will must state who the **executor** (or "personal representative") of the estate will be (one responsible for supervising the distribution of property after death and for seeing that the debts and taxes are paid). The executor may also have the

Edgar Argo/CSL, CartoonStock Ltd

"MS. AVERY LEFT EACH ONE OF YOU, TEN MILLION DOLLARS... UNFORTUNATELY, HER ESTATE IS ONLY WORTH SEVEN DOLLARS AND SIXTEEN CENTS."

durable power of attorney, giving her or him the ability to handle finances (write checks and sell stock, for example), if the testator becomes incapacitated before death. The executor may perform in this capacity, though someone else could be chosen to have the durable power of attorney (Chatzky, 2000).

The will typically names one or more persons who will benefit from the will (called **beneficiaries**) and specifies the personal guardian for minor children. The will should be kept somewhere so that the executor can find it easily. Wills need to be updated occasionally, if one's situation changes (e.g., another child is born into the family). Generally, an individual is free to leave her or his property to whomever or whatever he or she desires.

If a testator does not wish to leave anything to a particular child, that child should be named and left something, say, one dollar, so that she or he cannot protest the will later and claim that the parent simply "forgot" her or him. If the child is named in the will, then the parent did not "forget."

It is not necessary to have an attorney make out a will, though as noted earlier, to hire a professional may be a good idea. If one has good self-help materials, it is not difficult to make a will that names who receives your earthly possessions and who will be guardians of your minor children (if applicable). For rather simple wills, which most individuals have, an attorney has a standard form in the computer and has the secretary type in the information, or one can seek such services as Legalzoom.com (at a minimum cost of $69). Unless a will is complicated, having a lawyer prepare the will should cost between $300 and $500. Attorneys usually charge by the hour, but some will draw up a simple will for a set fee.

A will can be changed by adding an amendment called a **codicil**. A codicil allows one to amend the will by adding new provisions without having to rewrite it entirely. For example, if additional property is acquired and the testator wishes to designate a certain person or organization to receive this, rather than write a completely new will, this information is simply added on as a codicil. Or, if a baby is born into the family or adopted, the testator will probably want to add her or him as a recipient from the estate.

There are several ways in which a will can be revoked (canceled or rescinded). Included among those ways are the following: The testator divorces or remarries, the testator writes another will (thus the importance of the date on the will), or the will is literally destroyed (by tearing, burning, or destroying in any way). If the will was accidentally destroyed, its terms may be proved by other competent evidence, usually a copy of the original will. For this reason, having two originally signed wills is recommended. The safest way to revoke a will is to execute a new will that expressly states that all previous wills are revoked. A will can be broken (shown to be useless) in various ways. For example, if the testator signed under coercion (someone forced her or him to sign against her or his wishes), and this can be proven in court, the will is then not legally binding. If it can be proven in court that the testator, at the time of signing the will, was incompetent, then the will can be broken. In addition, if fraud (deception or trickery) can be proven, the will can be broken.

A parent or parents with small children might prefer to set up a trust fund for each child, rather than have the assets go directly to minors. The trust does not usually require witnesses, but it should be notarized. The will can be used to name a trustee who will handle any property the child receives until the child reaches the

DEATH ACROSS CULTURES | KEEP THE DEAD CONTENT OR SUFFER THE CONSEQUENCES

The bodies of Zimbabwean soldiers killed in the Congolese civil war are being dug up to prevent their spirits from returning to haunt their families. It is believed that if the corpses of the soldiers killed fighting for the Democratic Republic of Congo are left in their graves or abandoned to decompose on the battlefield, their spirits will trouble their relatives. The number of soldiers estimated to have been killed ranges from a few dozen to a few hundred over a 10-month period. The spirit of the dead will worry the family as a way of punishing them, since the person was "thrown away in a foreign country." The spirit will cause problems "for many generations with sickness and death and other kinds of misfortune." To appease the spirits, the bodies will be exhumed for reburial at home. If the body itself cannot be located, soil from the area where the soldier died could be returned home.

From Zimbabwe to appease spirits by exhuming soldiers in Congo (p. 14), by J. Raath, May 31, 1999, *The London Times.*

---

age of majority, which in most states is age 18. Until the child reaches such an age, the trustee can spend the money for things such as health care, education, and general support of the child. Unless stated otherwise, when the child reaches the specified age, the trustee ends the trust and gives the child whatever is left. This arrangement affords wide flexibility in the management of the estate. The children have someone handle the money until the day when they are old enough to handle it themselves. An advantage of such a trust is that the child does not suddenly come into a lot of money and perhaps simply blow it all in a short period of time. A responsible adult is managing the assets for her or him.

HOLOGRAPHIC WILLS Though legal in only 25 states, **holographic wills** (written in one's own handwriting) differ from other wills. Such wills do not require witnesses. The will must be signed by the testator and should be dated. Even in the states where holographic wills are recognized, probate judges are often reluctant to accept them. Such wills are rather unreliable because proving their authenticity is often difficult. If, through handwriting comparisons, the holographic will can be authenticated, there is the additional problem of proving that the testator was in full possession of mental faculties when she or he signed the will. Also, the requirements for a holographic will vary among those states that accept them. Words that are part of a printed form or that have been typed and thus are not written in the testator's own hand may render the entire instrument invalid—it is then not wholly written in one's own handwriting.

George Dickinson recalls an attorney speaking to his death and dying class a few years ago in Kentucky, where holographic wills are legal. He said that an elderly woman came to him rather upset that her sister had recently died intestate. She said that she and her sister were very close. The sister's two adult children, however, paid no attention to their mother—never contacted her and did not send even a card on her birthday or Mother's Day. Yet, because the sister died without a will, from the classification of heirs, the two children would take 100 percent from her estate—50 percent each (the sister's husband had predeceased her).

The woman, rather casually, mentioned to the lawyer that her sister had written to her a few months before she died and stated that she wanted her to have everything she owned when she died. The attorney asked if she still had the letter. She produced the letter, written in her sister's own handwriting, signed, and dated by her sister. This letter in essence was a holographic will, which the lawyer took and successfully executed in court. The sister took all from the estate, and the two nonattentive children received nothing. Her sister had written a will—a holographic will—and indeed did *not* die intestate.

PROBATE

Probate ("to prove") is a legal process that takes place after someone dies. Probate involves proving in court that a deceased person's will is valid, identifying and inventorying the deceased person's property, having the property appraised, paying debts and taxes, and distributing the remaining property as the will directs (Probate FAQ, 1999). Probate generally involves paperwork and court appearances by attorneys. Probate can take up to one year to complete. A will must be probated to be effective. In many states, it is a crime to withhold a will from probate.

After a death, the executor of the estate or administrator appointed by a judge files papers in the local probate court. The executor proves the validity of the will, presents the court with lists of the deceased person's property, lists the debts, and names who is to benefit from the will. Creditors are notified of the death either directly or through an ad in the local newspaper, requesting anyone to come forward who is owed money by the deceased person's estate. This may take several months. The executor may have to make a decision as to whether or not to sell some liquid assets to pay off debts.

Not all property goes through probate court. Most states allow a certain amount of property to pass free of probate. In addition, money from life insurance policies goes directly to the person(s) named in the policy—the **beneficiary**.

If one wishes to avoid probate court to save time and legal expenses, there are several ways to do this (Randolph, 1999). Examples include setting up "payable-on-death accounts" so that one's liquid assets (like cash) go directly to a person(s) or institution. Or one could have a "transfer-on-death" to beneficiaries for stocks and bonds and vehicles. Or one could have property in joint tenancy with right of survivorship. Under this legal arrangement, two or more parties are given equal rights to the property during their lifetimes. When one party dies, that person's ownership rights dissolve, leaving nothing to pass through probate. The co-owner(s) becomes the complete owner of the property. If there are several co-owners, the last co-owner to survive becomes the complete owner.

ESTATE AND INHERITANCE TAXES

Death taxes are imposed in two general forms: estate tax and inheritance tax. Estate taxes are imposed on the decedent's estate and are paid from the estate before the remaining assets are distributed to beneficiaries and heirs. Inheritance taxes are levied against individuals who receive property through inheritance.

Before discussing the future of estate taxes, it is important to understand what the current federal estate tax rules provide for and to question if these rules are likely to be changed in the near future.

Under previous law, the federal estate tax exemption was supposed to drop significantly, from $5,120,000 in 2012 to $1,000,000 in 2013, and the estate tax rate was scheduled to jump from 35 percent to 55 percent. But Congress and President Obama acted very early in 2013 to pass the American Taxpayer Relief Act (ATRA), which, as mentioned above, has supposedly made the laws governing federal estate taxes, gift taxes, and generation-skipping transfer taxes permanent for 2013 and beyond (About.com, 2014).

Under ATRA, the federal estate tax, gift tax, and generation-skipping transfer tax exemptions were set at $5,000,000 but indexed for inflation beginning in 2011. This means that each exemption was $5,250,000 for 2013, increased to $5,340,000 in 2014, and continued to be increased according to inflation rates in 2015 and beyond. In addition, ATRA set the top estate tax, gift tax, and generation-skipping transfer tax at 40 percent and made portability of the estate tax and gift tax exemptions between married couples permanent for 2013 and future years (About.com, 2014).

The most popular method is used by married couples with children and is called an **AB trust** (sometimes called credit shelter trust, marital life estate trust, or marital bypass trust). Property is put in the trust. When one spouse dies, her or his half of the property goes to the children. The surviving spouse obtains the right to use the property for life and is entitled to any income generated from it. When the second spouse dies, the property goes to the children outright. Such an arrangement keeps the second spouse's taxable estate half the size it would otherwise have been, thus helping to avoid estate tax.

A similar method for tax savings is the **trust**. Trusts are legal entities that hold property for another and are often used in the estate planning process. A trust is a

---

| WORDS OF WISDOM | THE WILL |

Doug Smith is on his deathbed and knows the end is near.

His nurse, his wife, his daughter and 2 sons, are with him.

He asks for 2 witnesses to be present and a camcorder be in place to record his last wishes, and when all is ready he begins to speak:

- My daughter "Sybil, you take the apartments over in the east end."
- My son, "Bernie, I want you to take the Mayfair houses."

- My son, "Jamie, I want you to take the offices over in the City Centre."
- "Sarah, my dear wife, please take all the residential buildings on the banks of the river."

The nurse and witnesses are blown away as they did not realize his extensive holdings, and as Doug slips away, the nurse says, "Mrs. Smith, your husband must have been such a hard-working man to have accumulated all this property."

Sarah replies, "Property? …. the asshole had a paper route!"

way of transferring your property to an artificial legal entity or "person" (the trust) before your death, while still having the use and/or control of it during your lifetime. There are two kinds of trusts, revocable and irrevocable. If the trust is revocable, you can change it or decide to take the property back any time during your life. If the trust is irrevocable, you can't change it once you have set it up. If you name yourself as the sole trustee of your trust during your lifetime, you will be able to manage the trust while you are alive.

Another way to save on estate taxes is to make a sizable gift to a tax-exempt charity. Also, while still living, an individual can give away up to $14,000 (and $28,000 as a married couple) per year to any person or noncharitable institution without having to pay estate tax. That means a husband and wife can give their two grown children, and their kids' spouses, $112,000 a year. Cut in four grandchildren and $224,000 can be passed between generations every year. (Noncash gifts—say, a family vacation or cruise you treat the whole clan to—count against the $14,000.) In addition, the recipients would not have to pay any taxes on the $14,000 each year. Most states have either an inheritance tax or an estate tax. A few states have death taxes—inheritance and estate. Inheritance taxes are paid by those who inherit (the recipient). The taxes vary by state and according to the relationship of the two parties (e.g., formerly, Nebraska imposed a 15 percent tax if you leave money to a friend, but only 1 percent if left to a child. This tax has recently been repealed).

SAVING MONEY, TIME, AND PRIVACY: TRUSTS AND POUR-OVER WILLS

The trust owns the legal title to the property in it while you are still alive, and since a trust does not end at your death, it will still own the property when you die. You put instructions in the trust for how the trustee, or person controlling the trust, should distribute the trust property, and the trustee will carry out those directions.

Only property owned by the deceased at the time of death has to go through the court process called probate, so the property in the trust can be distributed without going through the probate process. Probate is the legal process that inherited property goes through in order to transfer the title of the property from the decedent to the beneficiary. If you have a large estate, or even a small estate with real property (i.e., real estate), it is often advantageous to set up a trust, as it is usually far less expensive for your heirs when you die.

A will is a document that transfers property to others after your death. Because you still own the property at the time you die, all the property transferred in the will must go through the probate process, which is often slow and costly. Even people with trusts sometimes have other property that is transferred by will and has to pass through probate.

WOULD A TRUST BE BETTER THAN A WILL?   The question of whether a trust is better than a will has no simple answer, since so many factors must be considered in estate planning. Your decision is personal and individual, and must suit your particular situation. What is right for one person might be very wrong for you. Briefly, a will is a legal document that gives your instructions for distributing your assets after you die. A will allows you to select an executor to manage the distribution of your assets,

© Mike Baldwin / Cornered

FUNERAL HOME

TISSUE

Mike Baldwin/CSL, CartoonStock Ltd

"Grief takes many forms. But don't worry, we're here to help you fill them out."

pay your debts, and handle other administrative duties. In a will, you may name a guardian to raise your minor children if you die before they turn 18, and you may choose a property guardian to oversee any assets you leave to your children. A will is not expensive to set up, but it must go through a sometimes costly, often lengthy legal process called probate before the assets may be distributed. During this process, your debts and any estate taxes will be paid out of your estate before your heirs receive their bequests. Fees your estate must pay to the executor, the attorney, and to the court add to the expense. Once filed, wills become part of the public record, accessible to everyone.

WHO SHOULD HAVE A TRUST?   There are many different reasons why you might want to have a trust. You might want to avoid probate; provide for minor children; provide for someone too young or who lacks the ability to manage money; avoid paying federal estate taxes; contribute to charity; distribute real property, particularly if it is located in another state; keep property separate; provide for yourself and your care if you become incapacitated and avoid a **conservatorship** proceeding; maintain privacy; and decrease the possibility of a legal challenge to the way you want your property distributed. If any of these situations apply to you, you should consider using a trust as part of your estate planning.

Probate is the legal process where a court oversees the payment of debts and distribution of property under a will. This can be a slow and costly process if the estate is even moderately large and complicated, and all details of the estate are made public as part of the court proceedings. Several states have summary procedures for small estates, but if your estate doesn't qualify for a summary procedure, you would be a good candidate for a trust. If you transfer assets to a trust while you are alive, then, when you die, the assets belong to the trust, not to you, so they are not included in probate, but will be distributed in the way you direct in the trust documents. The process of distribution will be private and confidential.

Minor children cannot inherit property, and therefore require an adult to manage property for them until they reach the legal age of majority in the state where they live. A parent can nominate a guardian for the child's financial matters in a will, but the probate court will have the final say on whether to approve the appointment. If you transfer assets to a trust that is for the benefit for your children while you are alive, you can name the trustee(s) and alternate trustees who will control the trust when you die. It is possible for a court to remove a trustee you have appointed, but only for misbehavior, so you have more control over who will control your children's assets.

If you leave children assets in a will, the children receive full control of the assets when they come of legal age. In most states this is 18, and if the assets are large, an 18-year-old may not be ready to manage that amount of money. If you set up a trust, you can control when the young person will receive full control of the assets, such as at age 25 or 30. Some older adults are also unable to manage money well. This may be because of a developmental disability or simply because the person lacks money skills. If you want to leave assets to someone like that, you might want to set up a trust that will control the assets throughout the life of the beneficiary. People also want to leave money for the care of pets that survive them, and a trust is a good way to do that.

If your estate is over the minimum amount for paying federal estate tax ($5.34 million minimum starting in 2014, unless the law is changed), trusts can be used to exempt some of your assets from your taxable estate. This can reduce your total estate to an amount below the estate tax minimum. You can also receive tax benefits during your lifetime by setting up a trust that makes a donation to charity when you die but keeps the income for you while you are alive.

Probate laws differ from state to state, so to avoid unforeseen results under different laws, if you own real property in different states, you can put that property in a trust while you are alive so there will be no change when you die. The trust will continue to own the property, and there will be no reason for the various states to be involved in how the property is distributed.

If you want to make sure that certain property is not divided between beneficiaries, which might result in the property being sold, you can put property in separate trusts and give separate instructions for distribution. That way, it will not be possible for a court to decide that the property can be divided and/or sold.

If you become incapacitated from old age or some accident or illness, someone will need to take care of your financial affairs for you. By setting up a trust where another trustee will take control of your assets if you become incapacitated, you can prevent someone from filing a petition to be named your conservator. A court will only appoint a conservator if no other arrangements have been made, so the court is not likely to grant a conservatorship to someone else if you have already made adequate arrangements for your own care. This allows you to choose who you want managing your affairs and allows you to exercise future control in the way you set up the trust.

In a conservatorship hearing, all of your affairs are made public, including details of your alleged incapacity. Most of us would not want the details of our private lives to be discussed by others in a public courtroom, so a trust avoiding that is useful. Trusts can also provide privacy about your assets. If you do not want all the details of your assets and the assets you are passing to others to be made public, you can use a trust to preserve privacy.

Finally, if you think someone might challenge a will because they do not like how you want to leave your property, you can make such a challenge much more difficult by using a trust instead of a will. To prove a will invalid, the challenger has to show that you were incompetent at the time the will was drafted and executed. A trust is not just drafted and executed at one point in time. Assets are often transferred to the trust over a long period during the person's lifetime, and the trust may manage those assets for many years before the death of the grantor. While you must be legally competent at the time you set up a trust, it is more difficult to prove such a trust invalid than a will, because the challenger has to claim that the person was incapacitated not only when the trust was set up, but during every transaction that the grantor carried on during the life of the trust.

CLOSING THE LOOPHOLES: THE "POUR-OVER" WILL A **pour-over will** is a particular type of will used in conjunction with a trust. This kind of will "pours" any property the deceased still owned at the time of death into the trust that the person set up during his or her life.

Some people intentionally choose not to put all their property into their trusts during their lifetimes. This may be to avoid inconvenience in dealing with certain kinds of property. For example, some states and insurance companies make it very difficult to buy, sell, or insure vehicles held in a trust. There may also be tax reasons for not transferring property to a trust. Subchapter S stock, for example, often does not fit well in a trust. The transfer of real estate from the individual to a trust may trigger a property tax reassessment.

In other situations people sometimes forget to put newly acquired property into a trust on an ongoing basis. For example, someone might acquire a rare book or a valuable work of art, but forget to transfer ownership to the trust.

If the person who intentionally or accidentally has left property out of the trust dies without a will, then the property that is not included in the trust, or transferred through some other estate planning device, will pass according to the state laws on what is called intestate succession (property inheritance when there is no will). This property will not pass according to the provisions of the trust, as the deceased probably wanted it to. To prevent the creation of an intestate estate, a pour-over will is created. It covers any property that was (intentionally or inadvertently) left out of the trust during the deceased's life. By the terms of the pour-over will, all the property the deceased owned at death is "caught" and is "poured over" into the existing trust. Though the property caught by a pour-over will has to go through probate, it will eventually be distributed according to the instructions of the deceased instead of being distributed under the state law.

Whenever a trust is used, it is essential to also have a pour-over will to catch your property that was not held by the trust or transferred in some other way. The information from this section "Saving Money, Time, and Privacy: Trusts and Pour-Over Wills" was used by permission from FreeAdvice.com. Advice Company maintains its headquarters along the shore of San Francisco Bay in Sausalito, CA. The mailing address is P.O. Box 1739, Sausalito, CA 94966-1739.

CONCLUSION

From the alpha to the omega of the life cycle—birth to death—a certificate is needed to legitimize one's existence and then to legitimize one's lack of existence. One can't just be born and just die; it all must be recognized within our legal system. Likewise, at death one's possessions go to other individuals, and this process is laid out within the laws of the various states. If one does not want the state to determine where the possessions go, a will is a must. Also, if an individual desires to have a say in end-of-life issues, she or he should have an AD (living will and legal power of attorney for health care). Americans sometimes joke about how many lawyers we have in our society, but, given all the legal requirements accompanying our various moves (birthing to dying), perhaps they are needed.

SUMMARY

1. Autopsies are sometimes performed after death to aid in determining the cause of death.

2. Though now legal throughout the United States, advance directives (ADs) are not popular, perhaps suggesting a reluctance to face death.

3. Medical examiners differ from coroners in that the latter usually do not have medical degrees. Both deal with trying to determine the causes of death in situations in which the cause of death is not known.

4. The death certificate states the reasons for the death of a person.

5. Life insurance emerged from the same historical context as did funeral reform and rural cemeteries; it assumed the uniformity and continuity of death as natural occurrences. Today, most Americans have life insurance, which is seen as a way to protect dependents as well as a tax-deferred investment from which one can borrow.

6. Long-term care insurance helps provide for the cost of long-term care beyond a prede-termined period. Long-term care insurance covers care generally not covered by health insurance, Medicare, or Medicaid.

7. If one wishes to have any say in the dispo-sition of one's property upon death, and in who would be legally responsible for the placement of minor children, he or she should have a will or a trust.

8. To die intestate (without a will) means that the state, via inheritance, will determine who receives one's property and also how minor children will be placed.

9. A holographic will (written in one's own handwriting) is very simple to make out, though sometimes difficult to process in court and not legal in half of the states.

10. Probate is a legal process that takes place after someone dies and involves proving in court that a deceased person's will is valid.

11. Under the American Taxpayer Relief Act (ATRA) of 2013, the federal estate tax, gift tax, and generation-skipping transfer tax exemptions were set at $5,000,000 but were indexed for inflation beginning in 2011. This means that each exemption was $5,250,000 for 2013, increased to $5,340,000 in 2014, and continued to increase according to inflation rates in 2015 and beyond.

12. ATRA set the top estate tax, gift tax, and generation-skipping transfer tax at 40 per-cent and made portability of the estate tax and gift tax exemptions between married couples permanent for 2013 and future years.

13. Individuals who want to make sure their wishes for medical treatment are honored should prepare an AD.

DISCUSSION QUESTIONS

1. Why are fewer autopsies occurring in the United States today?
2. What dangers are inherent in writing one's own will?
3. Would you be interested in being a coroner? Why or why not?
4. Discuss some of the problems with determining the causes of death.
5. Discuss the importance of a death certificate.
6. Discuss the advantages of an AD (living will and/or durable power of attorney for health care).
7. Discuss why most individuals do *not* have ADs.
8. What are the advantages and disadvantages of the following: term life insurance and whole life insurance? What factors should be taken into consideration when one is in the process of purchasing life insurance?
9. What are the advantages of having a will or trust versus dying intestate?
10. What are the criteria for making out a will?
11. Discuss who needs to make out a will and/or trust.
12. What are the advantages of trusts over wills? Why should one consider creating a trust rather than a will?

GLOSSARY

AB Trust: Half of a couple's property is placed in a trust. When one spouse dies, that half of the property goes to the children, but the surviving spouse gets the right to use the property for life. When the second spouse dies, the property goes to the children outright, thus reducing estate taxes by half.

Activities of daily living: Are the things we normally do such as feeding ourselves, bathing, dressing, grooming, working, homemaking, and leisure.

Administrator: A person or institution (e.g., a bank) appointed by the court either to administer the estate of a person who died intestate, or to administer the estate of a person who died with a will that does not properly name an executor or that names an executor who fails or refuses to qualify.

Advance Directives (ADs): Legal documents stating an individual's wishes regarding end-of-life decisions such as a living will and durable power of attorney for health care decisions.

Autopsy: A pathologist's medical examination of the organs of the dead body to determine the cause of death.

Beneficiary: One receiving benefit or advantage from a will, insurance policy, or trust.

Codicil: A supplemental modification of an existing will.

Conservatorship: If a person or persons become incapacitated or not fit to live according to their own wishes, relatives or the state may appoint someone to care for and be legally responsible for the care of that person or persons through a conservatorship.

Coroner: A public official, appointed or elected, in a particular geographic jurisdiction, whose official duty is to make inquiry into deaths in certain categories.

Death Certificate: A permanent record of an individual's death.

Decedent: The deceased person.

Durable Power of Attorney (for Health Care): A document that ascribes to a surrogate decision maker the legal authority to consent (or refuse to consent) to medical treatment should the patient lack competency to make such decisions.

Escheat: Property of a decedent goes to the state if not disposed of by a will and if the decedent has no heirs.

Executor: A person or institution (e.g., a bank) named in the will to administer an estate—to pay creditors and taxes and to distribute the estate to the beneficiaries. In some states the phrase *personal representative* is used.

Five Wishes: A living will in the form of an eight-page questionnaire that guides people in making essential decisions about their care at the end of life.

Holographic Will: A will written wholly in one's own handwriting that generally requires no witnesses.

Intestate: Describes a person who dies without a will.

Living Will: A document stating that an individual does not want medical intervention if the technology or treatment that keeps one alive cannot offer a reasonable quality of life or hope of recovery.

Long-term care insurance: Helps provide for the cost of long-term care beyond a predetermined period.

Medical Examiner: A physician charged to investigate and examine persons dying a sudden, unexpected, or violent death.

Pathologist: A physician who deals with the nature of disease, especially with the structural and functional changes caused by disease. A pathologist conducts autopsies to try to determine the cause(s) of death.

Pour-over will: A particular type of will used in conjunction with a trust. This kind of will "pours" any property the deceased still owned at the time of death into the trust that the person set up during his or her life.

Probate: A court proceeding resulting in a judgment that, when recorded, vests the title to property passing under a will to the beneficiaries the same way a deed does. The will then becomes effective. The procedure is the legal process of proving that a will is valid.

Term Life Insurance: A type of insurance policy covering the insured for a fixed period of time (e.g., 10 or 20 years). Premiums are usually lower for a greater amount of coverage than with other types of life insurance.

Testator: A person who makes a will.

Trust: A way of transferring your property to an artificial legal entity or "person" (the trust) before your death, while still having the use and/or control of it during your lifetime. Trusts minimize tax liabilities and make the transfer of assets more efficient than wills.

Universal Life Insurance: A type of insurance policy that is flexible and allows one to raise or reduce premiums and the amount of coverage on one's life.

Variable Life Insurance: A type of insurance policy that does not allow for the premium or minimum coverage on one's life to change but allows one to switch the savings from among money markets or various forms of stock.

Whole Life Insurance: A type of insurance policy with which, for a set annual premium, one receives life insurance coverage and, at the same time, invests one's money.

Will: A legal document in which a person states how he or she wants property and possessions distributed after death.

SUGGESTED READINGS

Block, S. (2009, October 23). 5 myths about wills—and what you should really do. *USA Today*. A simple and practical guide for people deciding how to manage their estate and whether or not to obtain legal counsel.

Bryant, C. D., & Snizek, W. E. (2003). The last will and testament: A neglected document in sociological research. In C. D. Bryant (Ed.), *Handbook of death and dying* (pp. 926–932). Thousand Oaks, CA: Sage Publications. Provides an overview of the role of the last will and testament in our society and its function in organizing and regulating inheritance.

Cassel, C. K. (Ed.). (1999). *The practical guide to aging: What everyone needs to know*. New York: New York University Press. Gives some very practical advice for estate planning and dealing with legal issues, whether one is younger or older than age 65.

Chozick, A. (2005, August 15). Transition plan: Putting your own affairs in order. *Wall Street Journal (Eastern Edition)*, 246 (31): R3. This article offers practical tips to parents on managing their personal finances.

Crowley, M. (2009, June 6). The "death tax" scam. *Rolling Stone, 1080*, pp. 49–53. The article discusses estate taxes in the United States and examines a bill sponsored by U.S. Senator Blanche Lincoln that attempted to reduce U.S. estate taxes. Details are provided about a movement among wealthy U.S. families to repeal the estate tax. According to the article, the tax was first introduced after the U.S. Gilded Age when families such as the Carnegies dominated the economic scene.

Downs, J. C. U. (2003). Coroner and medical examiner. In C. D. Bryant (Ed.), *Handbook of death and dying* (pp. 909–916). Thousand Oaks, CA: Sage Publications. Discusses the history, function, and role of the office of the coroner in the United States.

FreeAdvice.com. (http://law.freeadvice.com). *This is an incredible source for current legal information on finances, insurance, wills, trusts, and other financial matters related to estate planning. It is maintained by a large group of lawyers in California.*

Miller, R. K., Rosenfeld, J. P., & McNamee, S. J. (2003). The disposition of property: Transfers between the dead and the living. In C. D. Bryant (Ed.), *Handbook of death and dying* (pp. 917–925). Thousand Oaks, CA: Sage Publications. Discusses the processes whereby societies create patterns of inheritance—formal and informal rules for the disposition of accumulated property at death.

Peck, D. L. (2003). *The death certificate: Civil registration, medical certification, and social issues.* In C. D. Bryant (Ed.), *Handbook of death and dying* (pp. 899–908). Thousand Oaks, CA: Sage Publications. Discusses the evolution and function of the death certificate with implications for the future needs of society.

Pevey, C. (2003). Living wills and durable power of attorney for health care. In C. D. Bryant (Ed.), *Handbook of death and dying* (pp. 891–898). Thousand Oaks, CA: Sage Publications. A practical guide to living wills and durable power of attorney that contains a discussion of controversies and critiques of varying perspectives on legal issues related to the dying process.

Only people who avoid love can avoid grief. The point is to learn from it and remain vulnerable to love.
—John Brantner (from J. William Worden, *Grief Counseling and Grief Therapy*)

COPING WITH LOSS

CHAPTER 13

© Creatista/Shutterstock.com

If, at the conclusion of the funeral service, grief work were finished, the process of reintegration of the bereaved into society would be completed. The funeral service and the final disposition of the dead, however, mark only the end of public mourning; private mourning continues for some time.

THE BEREAVEMENT ROLE

In earlier chapters we discussed bereavement behavior within historical and cross-cultural perspectives. We have given a general description of the norms and cultural patterns that prescribe proper conduct for the bereaved within American society. When we apply these bereavement norms to particular persons occupying statuses within a group or social situation, we are talking about the bereavement role.

As discussed in Chapter 5, Talcott Parsons (1951) described the sick role as being composed of rights and obligations. The first right of the sick person is to be exempted from "normal" social responsibilities. The extent to which one is exempted is contingent upon the nature and severity of the illness. The second right is to be taken care of and to become dependent on others as one attempts to return to normal social functioning. In exchange for these rights, the sick person must express a desire to "want to get better" and must seek technically competent help.

J. D. Robson (1977) suggested that behavior related to the death of a significant other (e.g., a spouse or parent) is similar to illness behavior patterns. At the onset of death, bereaved persons are exempted from their normal social responsibilities. Depending on the nature and the degree of relationship with the deceased, the bereaved are awarded time away from employment in much the same way that they are given sick leave—spouses and children may be given a week, whereas close friends and relatives may be given time only to attend the funeral.

Bereaved individuals are also allowed to become dependent on others for social and emotional support and for assistance with tasks related to the requirements of normal daily living. In offering this type of support, neighbors and friends call on survivors with gifts of food, flowers, and other expressions of sympathy. This custom led Robert Kavanaugh's younger brother to ask if "dead people ate meatloaf and chocolate cake" (Kavanaugh, 1972).

In exchange for these privileges created by the death of a loved one, those adopting the bereavement role not only are required to seek technically competent help from funeral directors and clergy members but also are expected to return as soon as possible to normal social responsibilities. The bereavement role is considered a temporary one, and it is imperative that all role occupants do whatever is necessary to relinquish it within a reasonable period of time. Time extensions are usually granted to spouses and children, but there is a general American value judgment that normal grieving should be completed by the first anniversary of the death.

American folk wisdom contends that "time heals"—with the intensity of the grief experienced diminishing over time. According to Paul Rosenblatt (1983), however, a more accurate notion of mourning is that there are fewer periods of intensely felt grief with the passing of time. Furthermore, in analyzing 19th-century diaries, Rosenblatt discovered that it was quite common for a person to experience periods of mourning for losses that occurred many years earlier. What is abnormal behavior, from the perspective of the American bereavement role, is a person's being preoccupied with the death of the loved one and refusing to make attempts to return to normal social functioning. Examples of deviant behavior of this type would include the following:

- Malingering in the bereavement role and memorializing the deceased by refusing to dispose of articles of clothing and personal effects and living as if one expected the dead person to reappear.
- Rejecting attempts from others who offer social and emotional support, refusing to seek professional counseling, and taking up permanent residence in "Pity City."
- Rejecting public funeral rituals and requesting that the funeral functionaries merely pick up the body and dispose of it through cremation without any public acknowledgment of the death.
- Engaging in aberrant behavior such as heavy drug or alcohol usage.
- Rushing into major life changes such as hastily remarrying or moving to a new home. Behaviors such as these may be sanctioned by others through social avoidance, ostracism, and criticism. As a consequence, most people are not only encouraged but are also forced to move through the grieving process.

LISTENING TO THE VOICES | A WEDNESDAY AFTERNOON

The procession crept down Division Street. The neon sign at the Rock County Bank flashed 94 degrees and 3:06. Behind the plate glass of The River Inn, businessmen shook dice for their afternoon coffee. Lois Palmer came out of Hanson's Variety Store carrying a lamp shade.

Why did you die, Dan? Did you have to die when the children were so young? You were lean and fit, physically active, no apparent health problems. You were fine when you went to bed, but I heard your death rattle; a small cough, I thought. You were dead when I came to bed five minutes later.

Is there any money? Will Karen be able to finish her last year of college? How about Peter—he's only 17. And then Johanna and Tina, 13 and 11. At least Sig is out of school. Such problems. Such a loss. How will I ever manage?

Two things came back to me during those early days of worry, fear, and grief. I recalled 20 years of bridge games with Judy and Elmer and the late-night discussions of death and funerals as we ate brownies and drank coffee. We spoke of our own deaths and our personal wishes. Dan, always the most vocal, stated again and again, "Don't spend any money on me when I am dead," and "Give me a military

funeral." There was not a man in the United States more proud to be a Marine than Dan Soli. To hear him tell it, the Marines fought World War II single-handedly. Though he was 33 and the others in boot camp were 18, he was a member of the U.S. Marine Corps and never minded if the recruits called him "Grandpa."

About one hundred yards into the cemetery the procession came to a stop. I saw the fake grass surrounding the open grave over by the fence. Close by stood the caretaker's tool shed with a barrel for refuse in front. Someone had lettered a crude sign above the barrel. It said, "RUBBAGE." The grave diggers stood behind the tool shed waiting to complete their job.

As the casket moved from the hearse to the waiting pallbearers, the honor guard, smart in their black uniforms, aligned themselves on either side of the men carrying the casket and accompanied them to the grave site.

The children and I left the car to follow. Tina cried, "I'm too young to be without a father," and clung to me. I had vowed to remain dry-eyed. My silly little purse held a lawn hanky—no Kleenex, no billfold, no lipstick. I would not open that purse.

continued

Friends and relatives surrounded the grave as Pastor Jensen gave the committal service. Somehow the next part took me by surprise. Two of the honor guards stepped forward and together removed the American flag from the casket. With ceremony they folded it in the traditional way and presented it to me—the widow.

The two uniformed men rejoined the honor guard and they all stepped back from the mourners. At the command of the captain they fired a 21 gun salute.

At that point Peter moved away from the family, raised his trumpet, and blew *Taps* for his dad. Sweet,

melancholy notes rose plaintively over the gathered crowd. From the other side of the hill came the faraway echo. My heart ached with the pain of that moment.

Later, as we were leaving, Alice Overbeck came over to me, reached for my hand, and said, "Too bad your husband died, but you'll get over it."

She was wrong; I didn't get over it. I cry when I hear *Taps*.

Jane Soli was the retired secretary to the academic dean of Saint Olaf College, Northfield, Minnesota. The above account of her husband's death was written in 1989. Michael Leming read the above at Jane Soli's memorial service in 1999.

THE GRIEVING PROCESS

Grief is a very powerful emotion that is often triggered or stimulated by death. Thomas Attig makes an important distinction between grief and the grieving process. Although grief is an emotion that engenders feelings of helplessness and passivity, the process of grieving is a more complex coping process that presents challenges and opportunities for the griever and requires energy to be invested, tasks to be undertaken, and choices to be made (Attig, 1991).

Most people believe that grieving is a disease-like and debilitating process that renders the individual passive and helpless. According to Attig:

It is misleading and dangerous to mistake grief for the whole of the experience of the bereaved. It is misleading because the experience is far more complex, entailing diverse emotional, physical, intellectual, spiritual, and social impacts. It is dangerous because it is precisely this aspect of the experience of the bereaved that is potentially the most frustrating and debilitating. (1991, p. 389)

Death ascribes to the griever a passive social position in the bereavement role. Grief is an emotion over which the individual has no control. However, understanding that grieving is an active coping process can restore to the griever a sense of autonomy in which the process is permeated with choice and there are many areas over which the griever does have some control. James Miller provides several options for grievers involved in active coping (see the "Practical Matters" box).

NORMAL AND ABNORMAL GRIEF

Harvard psychiatrist Erich Lindemann (1944), over a half century ago, wrote an article in which he discussed "normal" and "abnormal" grief. Basically, he argued that normal (ordinary) grief will resolve on its own, whereas abnormal grief may require some help, either from a support group or a clinical psychologist or psychiatrist.

In a normal grieving process, bereaved individuals may have difficulty "letting go." They may have difficult and painful dreams about the deceased person.

PRACTICAL MATTERS | GRIEF TIPS—HELP FOR THOSE WHO MOURN

Following are ideas to help people who are mourning a loved one's death. Treat this list as a gathering of assorted suggestions that various people have tried with success.

Talk regularly with a friend—Talking with another about what you think and feel is one of the best things you can do for yourself.

Walk—Go for walks outside every day if you can. If you like, walk with another.

Visit the grave—Not all people prefer to do this. But if it feels right to you, then do so. Don't let others convince you this is a morbid thing to do. Spend whatever time feels right there.

Plant something living as a memorial—Plant a flower, a bush, or a tree in memory of the one who died.

If you're alone, and if you like animals, get a pet—The attention and affection a pet provides may help you adapt to the loss of the attention and affection you're experiencing after this significant person has died.

Plan ahead for special days—Birthdays, anniversaries, holidays, and other special events can be difficult times, especially for the first year or two. Give thought beforehand as to how you will handle those days. Do things a little differently than you used to, as a way of acknowledging this change in your life. But also be sure to invoke that person's presence and memory somehow during the day.

Allow yourself to laugh—Sometimes something funny will happen to you, just like it used to. You won't be desecrating your loved one's memory.

Allow yourself to cry—Crying goes naturally with grief. It may feel awkward to you, but this is not unusual for a person in your situation. A good rule of thumb is this: if you feel like crying, then cry.

Plan at least one thing you'll do each day—Even if your grief is very painful and your energy very low, plan to complete at least one thing each day, even if it's a small thing. Then follow through with your plan, day after day.

Keep a journal—Write out your thoughts and feelings. Do this whenever you feel the urge, but do it at least several times a week, if not several times a day.

Consider a support group—Spending time with a small group of people who have undergone a similar life experience can be very therapeutic. You can discover how natural your feelings are.

Vent your anger rather than hold it in—You may feel awkward being angry when you're grieving, but anger is a common reaction. Yell, if in an empty house. Cry. Hit something soft.

Give thanks every day—Whatever has happened to you, you still have things to be thankful for. Perhaps it's your memories, your remaining family, your support, your work, your own health—all sorts of things.

From *The caregivers book: Caring for another, caring for yourself,* by J. E. Miller, 1996, St. Paul, MN: Augsburg Fortress Press.

They may have delusions or assume habits of the individual. They may "see" the dead person or find themselves talking to her or him, turning to her or him while watching a television program and saying something, as they have done for perhaps decades. According to research by Glennys Howarth (2000), communication with a deceased person may occur during semiconsciousness or in dreams or may occur while one is fully conscious. Douglas Davies (1997, p. 154) reported that approximately 35 percent of the people contacted for his survey of 1,603 people in the United Kingdom "had gained some sense of the presence of the dead." Sensing the presence of a parent is the most common type of experience, followed by grandparents, then spouses. It is difficult to "let go" of something you have had for so much of your life.

It is also normal to try to avoid the pain of grief. Pain comes in waves. One may try to avoid reminders such as birthdays, religious holidays, or anniversaries. The survivor may avoid going to certain places where he or she often went before the individual died—a favorite restaurant, a walk on the beach, or a climb in the mountains. Tom Hanks in the movie *Sleepless in Seattle* moved from Chicago to Seattle to avoid those places he could not go since his wife had died. The opening scene of the movie has Hanks contemplating what to do. He decides to move—to avoid the pain of grief. He simply starts over in a new environment and thus will not be faced with trying to go to those places where he had memories of his deceased spouse.

A total absence of grief, on the other hand, is not normal. The survivors react as if nothing has happened. Life goes on. The television program *Benson* many years ago had an episode in which Benson's mother had come to visit him and died unexpectedly. Benson's reaction was to call the funeral home and have the body shipped back to her home for burial. Benson stayed there and went to work that day! His fellow employees did not know how to deal with him, because such behavior was not normal. He was displaying a total absence of grief.

Though some degree of depression is normal, attempting suicide or a prolonged inability to perform daily activities is a cause for concern. Most individuals are able to return to work and get out in the community within a few days. Though we may never get over the death of a significant other (like we get over the flu), we learn to live with the fact that the person is dead. We must accept the fact that that person will never again sit at her or his place at the breakfast table. Life goes on, however, and we must move on ourselves.

In addition to normal and abnormal grief, grief sometimes lingers and even intensifies—grief that dominates one's life, rather than receding into the background. Such unresolved grief has recently been identified as **complicated grief**, with an estimate that about 10 percent of bereaved individuals experience this kind of grief (Reynolds, 2013). Complicated grief is often accompanied by an inability to face the finality of an individual's death. Women are more susceptible to complicated grief than men. Additionally, other complicated grief situations might include the following: when an individual dies suddenly, parents' loss of a child, and multiple losses within a short timeframe—all can be overwhelming. "Complicated" refers to factors that interfere with the natural healing process. It is a long-lasting form of grief that takes over a person's life. Yet the good news is that by identifying this type of grief, appropriate treatments are being developed, according to M. Katherine Shear, Director of the Center for Complicated Grief at Columbia University School of Social Work in New York (Complicated Grief Program, 2013). For example, mourners revisit the moment they learned of the death of a significant other and begin to engage in an imaginary conversation with the decedent, perhaps explaining what they did or did not do, thus helping to relieve guilt, if that is an issue.

STAGES OF GRIEF

The grieving process, like the dying process, is essentially a series of behaviors and attitudes related to coping with the stressful situation of a change in the status of a relationship. As discussed in Chapter 5, many individuals have attempted to understand coping with dying as a series of universal, mutually exclusive, and linear stages. Not all people, however, will progress through the stages in the same manner. Seven behaviors and feelings that are part of the coping process were identified by Robert Kavanaugh (1972): shock and denial, disorganization,

WORDS OF WISDOM | THE BREVITY OF LIFE

How short is life. How soon comes death.   Epitaph on gravestone in Haworth, England.

---

volatile reactions, guilt, loss and loneliness, relief, and reestablishment. It is not difficult to see similarities between these behaviors and Kübler-Ross's five stages (denial, anger, bargaining, depression, and acceptance) of the dying process. According to Kavanaugh (1972, p. 23), "these seven stages do not subscribe to the logic of the head as much as to the irrational tugs of the heart—the logic of need and permission."

SHOCK AND DENIAL Even when a significant other is expected to die, at the time of death there is often a sense in which the death is not real. For most of us our first response is, "No, this can't be true." With time, our experience of shock diminishes, but we find new ways to deny the reality of death.

Some believe that denial is dysfunctional behavior for those in bereavement. However, denial not only is a common experience among the newly bereaved but

© Anneka/Shutterstock.com

Denial can be a temporary safe place for grievers who find it difficult to carry on with "normal" living in the face of extraordinary circumstances.

it also serves positive functions in the process of adaptation. The main function of denial is to provide the bereaved with a "temporary safe place" from the ugly realities of a social world that offers only loneliness and pain.

With time, the meaning of loss tends to expand, and it may be impossible for one to deal with all of the social meanings of death at once. For example, if a man's wife dies, not only does he lose his spouse, but also his best friend, his sexual partner, the mother of his children, a source of income, and so on. Denial can protect an individual from some of the magnitude of this social loss, which may seem unbearable at times. With denial, one can work through different aspects of loss over time.

DISORGANIZATION Disorganization is the stage in the bereavement process in which one may feel totally out of touch with the reality of everyday life. Some go through the two- to three-day time period just before the funeral as if on "automatic pilot" or "in a daze." Nothing normal "makes sense," and they may feel that life has no inherent meaning. For some, death is perceived as preferable to life, which appears to be devoid of meaning.

This emotional response is also a normal experience for the newly bereaved. Confusion is normal for those whose social world has been disorganized through death. When Michael Leming's father died, his mother lost not only all of those things that one loses with a death of a spouse, but also her caregiving role—a social role and master status that had defined her identity in the five years that her husband lived with cancer. It is only natural to experience confusion and social disorganization when one's social identity has been destroyed.

VOLATILE REACTIONS Whenever one's identity and social order face the possibility of destruction, there is a natural tendency to feel angry, frustrated, helpless, and/or hurt. The volatile reactions of terror, hatred, resentment, and jealousy are often experienced as emotional manifestations of these feelings. Grieving humans are sometimes more successful at masking their feelings in socially acceptable behaviors than other animals, whose instincts cause them to go into a fit of rage when their order is threatened by external forces. However apparently dissimilar, the internal emotional experience is similar.

In working with bereaved persons over the past 30 years, Michael Leming has observed that the following become objects of volatile grief reactions: God, medical personnel, funeral directors, other family members, in-laws, friends who have not experienced death in their families, and/or even the person who has died. Mild-mannered individuals may become raging and resentful persons when grieving. Some of these people have experienced physical symptoms such as migraine headaches, ulcers, neuropathy, and colitis as a result of living with these intense emotions.

WORDS OF WISDOM | MADE FOR LIFE!

It's funny the way most people love the dead. Once you are dead, you are made for life.

Jimi Hendrix, *Rolling Stone*, December 2, 1976.

The expression of anger seems more natural for men than expressing other feelings (Golden, 2000). Expressing anger requires taking a stand. This is quite different from the mechanics of sadness, where an open and vulnerable stance is more common. Men may find their grief through anger. Rage may suddenly become tears, as deep feelings trigger other deep feelings. This process is reversed with women, notes Tom Golden. Many times a woman will be in tears, crying and crying, and state that she is angry.

As noted earlier, a person's anger during grief can range from being angry with the person who died to being angry with God, and all points in between. Golden's mentor, Father William Wendt, shared the story of his visits with a widow and his working with her on her grief. He noticed that many times when he arrived she was driving her car up and down the driveway. One day he asked her what she was doing. She proceeded to tell him that she had a ritual she used in dealing with her grief. She would come home, go to the living room, and get her recently deceased husband's ashes out of the urn on the mantle. She would take a very small amount and place them on the driveway. She then said, "It helps me to run over the son of a bitch every day." He concluded the story by saying, "Now that is good grief." It was "good" grief because it was this woman's way of connecting to and expressing the anger component of her grief.

GUILT Feelings of guilt are similar to the emotional reactions discussed earlier. Guilt is anger and resentment turned in on one's self and often results in self-deprecation and depression. It typically manifests itself in statements like "If only I had … ," "I should have … ," "I could have done it differently," and "Maybe I did the wrong thing." Guilt is a normal part of the bereavement process.

From a sociological perspective, guilt can become a social mechanism to resolve the **dissonance** that people feel when unable to explain why someone else's loved one has died. Rather than viewing death as something that can happen at any time to anyone, people can **blame the victim** of bereavement and believe that the victim of bereavement was in some way responsible for the death—"If the individual had been a better parent, the child might not have been hit by the car," or "If I had been married to that person, I might also have committed suicide," or "No wonder that individual died of a heart attack, the spouse's cooking would give anyone high cholesterol." Therefore, bereaved persons are sometimes encouraged to feel guilt because they are subtly sanctioned by others' reactions.

LOSS AND LONELINESS Feelings of loss and loneliness creep in as denial subsides. The full experience of the loss does not hit all at once. It becomes more evident as bereaved individuals resume a social life without their loved one. They realize how much they needed and depended upon their significant other. Social situations in which we expected them always to be present seem different now that they are gone. Holiday celebrations are also diminished by their absence. In fact, for some, most of life takes on a "something's missing" feeling. This feeling was captured in the 1960s love song "End of the World":

Why does the world go on turning?
Why must the sea rush to shore?

WORDS OF WISDOM | GUILT OVER EUTHANASIA OF A PET

Some pet owners struggle with ambivalence about euthanasia; others do not. To "kill" one's pet is an awful thing to do, resulting in guilt feelings some 50% of the time. Yet others make peace with the procedure and realize that their behavior was out of love for the pet. Euthanasia compels pet owners to come to grips with what sociologists call the "caring-killing paradox." Giving themselves permission to aid in a pet's death allows owners to provide an alternative to suffering and pain. Such action fits the "quality of life" rather than the "quantity of life" philosophy.

Taken from George E. Dickinson (2014, summer). Household pet euthanasia and companion animal last rites. Phi Kappa Phi Forum, 94:4-6.

Don't they know it's the end of the world
Cause you don't love me anymore?
(*"End of The World," written by Arthur Keat and Sylvia Dee. Used with permission of the Keat Estate.*)

Loss and loneliness are often transformed into depression and sadness, fed by feelings of self-pity. According to Kavanaugh (1972, p. 118), this effect is magnified by the fact that the dead loved one grows out of focus in memory—"An elf becomes a giant, a sinner becomes a saint because the grieving heart needs giants and saints to fill an expanding void." Even a formerly undesirable spouse, such as an alcoholic, is missed in a way that few can understand unless their own hearts are involved. This is a time in the grieving process when anybody is better than nobody, and being alone only adds to the curse of loss and loneliness (Kavanaugh, 1972).

Those who try to escape this experience will either turn to denial in an attempt to reject their feelings of loss or try to find surrogates—new friends at a bar, a quick remarriage, or a new pet. This escape can never be permanent, however, because loss and loneliness are a necessary part of the bereavement experience. According to Kavanaugh (1972, p. 119), the "ultimate goal in conquering loneliness" is to build a new independence or to find a new and equally viable relationship.

RELIEF   The experience of relief in the midst of the bereavement process may seem odd for some and add to their feelings of guilt. Michael Leming observed a friend's relief six months after her husband died. This older friend was the wife of a minister, and her whole life before he died was his ministry. With time, as she built a new world of social involvements and relationships of which he was not a part, she discovered a new independent person in herself whom she perceived was a better person than she had ever before been.

Relief can give rise to feelings of guilt. However, according to Kavanaugh (1972, p. 121): "The feeling of relief does not imply any criticism for the love we lost. Instead, it is a reflection of our need for ever deeper love, our quest for someone or something always better, our search for the infinite, that best and perfect love religious people name as God."

"You're sure that's one of the stages of
grief?"

REESTABLISHMENT As one moves toward reestablishment of a life without the deceased, it is obvious that the process involves extensive adjustment and time, especially if the relationship was meaningful. It is likely that one may have feelings of loneliness, guilt, and disorganization at the same time, and that just when one may experience a sense of relief, something will happen to trigger a denial of the death. What facilitates bereavement and adjustment is fully experiencing each of these feelings as normal and realizing that it is hope (holding the grieving person together in fantasy at first) that will provide the promise of a new life filled with order, purpose, and meaning.

Reestablishment occurs gradually, and often we realize it has been achieved long after it has occurred. In some ways it is similar to Dorothy's realization at the end of *The Wizard of Oz*—she had always possessed the magic that could return her to Kansas. And, like Dorothy, we have to experience our loss before we really appreciate the joy of investing our lives again in new relationships.

Any time we experience a loss, especially a loss as significant as the death of someone close, it changes us. In most cases, we do not have control over the loss we have experienced. We do, however, have some control over the ways in which we respond to that loss. Some people grow through a grief experience while others seem to get stuck. How are such opposite results possible?

The answer to this question is closely related to how individuals grieve following their loss. Individuals who shut themselves off from the grief process also shut themselves off from the transformative experience that can result (Johnson, 1997). The transformative nature of the grief focuses not only on the process of "getting through" a time of sadness and loneliness immediately following the loss, but also on reconstructing one's life following the loss.

The transformative process that grief encourages includes three components that can best be understood by responding to the following three questions (Johnson, 1997): What have I lost? What do I have left? What may be possible for me?

Honestly answering these questions in response to one's situation may help facilitate a grieving process that shifts from limits to opportunities.

What have I lost? This question seeks to discover how extensive the loss is, that is, to clearly identify what has been lost. The grief process cannot begin or progress to completion until the loss and secondary losses that accompany it have been identified. Once the losses have been identified, and we believe we can go on, the healing process of grief has begun.

What do I have left? In responding to this question we allow the meaningful aspects of what remains in our lives to be recognized, remembered, and valued. Initially, and often depending upon our loss, it is possible to wonder if enough remains to even make our lives worthwhile. No matter how small what is left may seem, it is enough to build upon. It is important to remember to be appreciative of what one has and go forward from there, and not be ungrateful for what one does not have or has lost.

What may be possible for me? Once we have determined what is left, we are able to move on and begin to determine what is possible, in spite of the significant loss that has occurred. Success in responding to this question is based on our perspective. Rather than looking at our limitations, we view our opportunities instead. By focusing on what is possible, we allow ourselves to discover new ways to relate, understand, create, and commit to an ongoing process of renewal and discovery.

The process of transformation takes time, and each question mentioned above deserves attention. Honestly answering each of them is a significant step toward a healthy grief experience.

DISENFRANCHISED GRIEF

Like other aspects of death-related behavior, grief is socially constructed—people grieve only when they feel it is appropriate. Social scripts are provided for grievers, and social support is given to those who are recognized as having experienced loss and who act in accordance with the norms. However, not all loss is openly acknowledged, socially sanctioned, and publicly shared. Kenneth Doka (1989) used the term disenfranchised grief when referring to phenomena of this type.

According to Doka (1989) four types of situations lead to disenfranchised grief. The first situation is one in which **the relationship to the deceased is not socially recognized**. Examples of this type of disenfranchised grief include nontraditional relationships—such as extramarital affairs, heterosexual cohabitations, and homosexual relationships (Doka, 1987). If outsiders are unaware that a relationship exists and the bereaved individuals are unable to publicly acknowledge their loss, they will not receive social support for their grief, and their bereavement will be problematic.

A second type of disenfranchised grief occurs when the **loss is not acknowledged by others** as being a genuine loss. An abortion or miscarriage is often deemed to be of lesser significance because the mother never had the opportunity to develop a face-to-face relationship with the child (Thornton, Robertson, & Mlecko, 1991). In the case of induced abortion, it is assumed that because the pregnancy was unwanted, grieving is unnecessary. According to Idell Kesselman (1990, p. 241), whatever one's position on abortion, "we must acknowledge that at least one death occurs—in addition to the fetus, there is often the death of youth, of innocence, of dreams, and of illusion." Kesselman maintained that women who have had an abortion need to express "unresolved feelings of loss" and to deal with "issues of death, loss, and separation." Kesselman concluded that grief therapy is a necessary part of abortion counseling.

Other examples of unacknowledged losses are the death of a pet companion and the death of a former spouse. According to Avery Weisman (1990–1991), the loss of a pet companion is often accompanied by intense grief and mourning but is seldom recognized by others as an important and authentic occasion for bereavement. Likewise, the death of a former spouse is rarely thought of as a legitimate loss because most people believe that grief work should be completed shortly after the divorce.

Related to the unacknowledged loss is the third type of disenfranchised grief in which the **grievers are unrecognized**. The grief over the death of an adolescent peer or friend is rarely openly acknowledged and socially sanctioned. One of the reasons why wakes and visitation services often attract larger audiences than do funerals is that employers are increasingly unwilling to provide employees with time from work to attend a funeral of a person who is not a family member (Sklar & Hartley, 1990). Other unrecognized grievers are young children, mentally incompetent and/or retarded persons, and elderly adults. In each of these cases the bereavement needs of individuals are also often ignored by most social audiences (Kloeppel & Hollins, 1989; Sklar & Hartley, 1990).

Margaret Graham, Jane McCarthy, and Judy Ryan (2012) described their grief work in Ireland with people having intellectual disabilities ("learning disabilities" in the United Kingdom) living in residential homes. They give suggestions for working with unrecognized grievers with disabilities. Similar to the work of Graham et al., Community Living (2012) produced a film in which people with learning disabilities tell their stories and share their wishes to support other people with

"I FEEL RIDICULOUS!"

Farris, Joseph/CSL, CartoonStock Ltd

Michael R. Leming

Daniel died on the day he was to be born. The parents' loss is as real as that of parents who have lost a child after much time in caring for a child. This type of disenfranchised grief is not always acknowledged by others as being a genuine loss.

learning disabilities to become more comfortable talking about dying, death and bereavement.

The final type of disenfranchised grief occurs when the **death is not socially sanctioned,** such as a death occurring in the act of a crime or when death is caused by suicide or autoerotic asphyxia (Murphy & Perry, 1988; Thornton, Wittemore, & Robertson, 1989; Ness & Pheffer, 1990). When people feel ambivalent, awkward, and/or uncomfortable about the cause of the death, they are generally unable or unwilling to provide the social support needed by the bereaved.

According to Doka (1989), whenever disenfranchised grief occurs, the experience of grief is intensified and the normal sources of social support are lacking. Disenfranchised grievers are usually barred from contact with the deceased during the dying process. They are also frequently excluded from funeral rituals as well as from care and support systems that may assist them in their bereavement. Finally, they may often experience many practical and legal difficulties after the death of their loved one. All of these circumstances intensify the problematic nature of bereavement for disenfranchised grievers.

FOUR TASKS OF MOURNING

In 1982 J. William Worden published *Grief Counseling and Grief Therapy,* which summarized the research conclusions of a National Institute of Health study called the Omega Project (occasionally referred to as the Harvard Bereavement Study). Two of the more significant findings of this research—displaying the active nature of the grieving process—are that mourning is necessary for all persons who have experienced a loss through death and that four tasks of mourning must be accomplished before mourning can be completed and reestablishment can take place.

According to Worden (1982), unfinished grief tasks can impair further growth and development of the individual. Furthermore, the necessity of these tasks suggests that those in bereavement must attend to "grief work" because successful grief resolution is not automatic, as Kavanaugh's (1972) stages might imply.

The people in my newsroom got horrible news twice in the past few weeks.

Dan Freeborn, who wrote a business column for the *Star Tribune*, died of cancer in May.

Then, over the July 4th weekend, Michelle Foster, the wife of business reporter Jim Foster, died in a boating accident.

Dan was 48, married, with two sons, Michelle was 41. The Fosters have two sons, too.

Dan and Jim also have a workplace full of what I might call "second-row people"—all of us who are not part of their private lives but who really like that they were, or are, part of our work lives.

Everyone here would like to say and do the right thing.

It can be hard for second-row people to know what that is.

Tragedies like these come one time or another to every workplace. And when they do, the advice I've found encourages co-workers to err on the side of outreach, provide help according to the family's requests, and give the whole thing plenty of time.

Also, remember that second-row people can be very important to one another.

Here's more of what the experts advised:

Call or write the grieving family as soon as you hear the bad news.

"It might not be graceful or the best you could do if you had years to think, but just ask them how it's going and tell them you're thinking of them," said Paul Rosenblatt, a professor of family social science at the University of Minnesota.

"If you're saying something heartfelt, you can never say the wrong thing," said Jodi Aronson Prohofsky, senior vice president of clinical operations at Cigna Behavioral Health in Eden Prairie.

It works fine if one or two people become the conduits to convey sympathy and offers of help, but that's no replacement for individual messages.

"More is better," said Tom Ellis, executive director of the Center for Grief, Loss & Transition in the Twin Cities.

"When my father died, I got a condolence card signed by 30 people," Rosenblatt said. "That had meaning to me."

Watch out for co-workers, too, Ellis said. Tragic news can plunge them back into grief if they've recently lost someone important, or someone close to them is very sick.

A colleague's death stays hard for a while, said Marjorie Dyan Hirsch, a consultant in New York for Ceridian, the Bloomington-based benefits administrator. It's hard for anyone who walks past the empty desk, hard for bosses to think about hiring a replacement, and hard for whoever has to answer the lingering phone calls with, "I'm sorry to tell you that he's not here anymore."

That's why Hirsch thinks some kind of group counseling, or some kind of memorial time at work is important.

"It's something for the 'work-family' as opposed to the 'family-family,'" she said.

Workplaces have a certain code of composure, something grieving colleagues will hope to manage when they come back.

"Bereaved people have told me that they cry all the way into work, then they hold it together at work, then they cry all the way home," Rosenblatt said. "Sometimes they even value having the work keep them away from other thoughts."

That's why something like, "How are you doing?" is a good way to approach them, he said. If they just say, "Fine," you know they want to leave it at that for now.

Colleagues who are home dealing with a tragedy may call to talk about work. Your instinct is to want to stop them, but covering this can be important to them.

"You can tell them, 'If there are specific things you're concerned about let us know and we'll handle them for you,'" Ellis said.

Then there's the adjustment when they come back to work.

Rosenblatt has seen it go two ways: struggling to get work done or hyper-performance.

Some are numb, distracted and sleep-deprived, he said. Someone else may generate a 14-page memo for a small job.

Employers need to be flexible with the bereavement leave—one week, two weeks—because everyone is different.

Rosenblatt has also seen relationships grow or shrink through a tragedy.

"I've experienced that both as a bereaved person and as someone offering to help," he said. "Some

continued

relationships drift away, friends who just avoid talking. But others get closer, someone who says, 'Let me help you with the kids,' or something like that."

Co-workers immediately want to help. Some workplaces set up scholarship funds for the colleague's children. Some donate to a cause important to the family.

When a manager in her office lost her husband, Prohofsky said, everyone pitched in to handle the comings and goings of out-of-town family.

When another co-worker lost her husband, people from the office went to her house about two months later because she asked for help taking his clothes out of the closet. That kind of abiding offer of help is important, Hirsch said. But she also encourages people to go beyond a vague, all-inclusive offer to suggest something specific.

"You're better off making a suggestion because that way the person doesn't have to stop and think," she said, "because it's really hard to think when you're in grief."

Particularly helpful is anything to give the person some quiet time, a break from the weight of events. Rake their leaves. Dog-sit. Take the children to a movie.

Beyond all that guidance, this is a lot about making it up as you go along.

"You have to fly by the seat of your pants," Rosenblatt said. "What works one moment might not work the next. But caring about people is kinda why we're on the planet."

That includes second-row people.

Adapted from "Grief and the Second Row," by H. J. Cummins, Staff Writer, *Star Tribune*, July 7, 2005.

Each bereaved person must accomplish four necessary tasks: (1) accept the reality of the loss, (2) experience the pain of grief, (3) adjust to an environment in which the deceased person is missing, and (4) withdraw emotional energy and reinvest it in another relationship (Worden, 1982).

ACCEPT THE REALITY OF THE LOSS Especially in situations when death is unexpected and/or the deceased lived far away, it is difficult to conceptualize the reality of the loss. The first task of mourning is to overcome the natural denial response and realize that the person is dead and will not return. Bereaved persons can facilitate the actualization of death in many ways. The traditional ways are to view the body, attend the funeral and committal services, and visit the place of final disposition. The following is a partial list of additional activities that can assist in making death real for grieving persons.

- View the body at the place of death before preparation by the funeral director.
- Talk about the deceased person and the circumstances surrounding the death.
- View photographs and personal effects of the deceased person.
- Distribute the possessions of the deceased person among relatives and friends.

EXPERIENCE THE PAIN OF GRIEF Part of coming to grips with the reality of death is experiencing the emotional and physical pain caused by the loss. Many people in the denial stage of grieving attempt to avoid pain by choosing to reject the emotions and feelings that they are experiencing. As discussed by Erich Lindemann (1944), some do this by avoiding places and circumstances that remind them of the deceased. Michael Leming knows one widow who quit playing golf and quit eating at a particular restaurant because these were activities that she had enjoyed

| DON'T REDUCE MY GRIEF TO SOMETHING LOGICAL AND UNIVERSAL

Grief discriminates against no one. It kills. Maims. And cripples. It is the ashes from which the phoenix rises, and the mettle of rebirth. It returns life to the living dead. It teaches that there is nothing absolutely true, or untrue. It assures the living that we know nothing for certain. It humbles. It shrouds. It blackens. It enlightens.

Grief will make a new person out of you, if it doesn't kill you in the making. People say "How are you?" and they have that look in their faces again. And I look at them like they're crazy. I think, how do you think I am? I'm not strong enough for talk. A woman said to me, in a comforting tone of conspiracy, "My mother had breast cancer and lost her right breast. . . ." "Oh," I say, "that's awful . . ." and wonder if that is supposed to be some sort of comparison. A tit, a husband—same thing? Strange how people want to join you in your intensity, your epiphany, by trying to relate—grabbing at straws of a so-called "like" experience while simultaneously they are repulsed by the agony. They see you on the street, driving by and pretending they don't see you. We all have our stuff to deal with. I suppose it's hard enough dealing with your own.

I can't tell them to shut up, not because I'm not rude by nature (it takes too much energy to be polite) but because I am afraid they'll go away. One more abandonment. "I lost my father two years ago . . ." and I think, schmuck, that's supposed to happen, it's a natural part of becoming adult—children are supposed to outlive their parents. Did you lose the only other person in the world who would love your child the way you do? Did you lose the person you held all night, who slept next to you, warmed your bed so much you didn't need an extra blanket in the winter? Do you know how many blankets it takes to replace a husband? Did you lose the person who would worry about bills with you? Screw in a light bulb when you were busy with the baby? Don't reduce this experience to something logical, universal. Even if it is, I walk alone amongst the dead, it's my death, my pain. Don't pretend you know it, like you know batting averages. Don't sacrilege all over my crucifixion.

From The agony of grief (pp. 75–78), by S. Ericsson, 1991, September–October, *Utne Reader*.

with her husband. Another widow found it extremely painful to be with her dead husband's twin, even though he and her sister-in-law were her most supportive friends.

Worden cites the following case study to illustrate the performance of this task of mourning:

> One young woman minimized her loss by believing her brother was out of his dark place and into a better place after his suicide. This might not have been true, but it kept her from feeling her intense anger at him for leaving her. In treatment, when she first allowed herself to feel anger, she said, "I'm angry with his behavior and not him!" Finally she was able to acknowledge this anger directly. (1982, pp. 13–14)

The problem with the avoidance strategy is that people cannot escape the pain associated with mourning. According to Bowlby (cited by Worden, 1982, p. 14), "Sooner or later, some of those who avoid all conscious grieving, break down—usually with some form of depression." Tears can afford cleansing for wounds created by loss, and fully experiencing the pain ultimately provides wonderful relief to those who suffer while eliminating long-term chronic grief.

ASSUME NEW SOCIAL ROLES The third task, practical in nature, requires the griever to take on some of the social roles performed by the deceased person or to find others who will. According to Worden (1982), to abort this task is to become helpless by refusing to develop the skills necessary in daily living and by ultimately withdrawing from life.

An acquaintance of Michael Leming's refused to adjust to the social environment in which she found herself after the death of her husband. He was her business partner, as well as her best and only friend. After 30 years of marriage, they had no children, and she had no close relatives. She had never learned to drive a car. Her entire social world had been controlled by her former husband. Three weeks after his funeral she went into the basement and committed suicide.

The alternative to withdrawing is assuming new social roles by taking on additional responsibilities. Extended families who always gathered at Grandma's house for Thanksgiving will be tempted to have a number of small Thanksgiving dinners at different places after her death. The members of this family may believe that "no one can take Grandma's place." Although this may be true, members of the extended family will grieve better if someone else is willing to do Grandma's work, enabling the entire family to come together for Thanksgiving. Not to do so will cause double pain—the family will not gather, and Grandma will still be missed.

REINVEST IN NEW RELATIONSHIPS   The final task of mourning is a difficult one for many because they feel disloyal or unfaithful in withdrawing emotional energy from their dead loved one. One of Michael Leming's family members once said that she could never love another man after her husband died. His twice-widowed aunt responded, "I once felt like that, but I now consider myself to be fortunate to have been married to two of the best men in the world."

Other people find themselves unable to reinvest in new relationships because they are unwilling to experience again the pain caused by loss. The quotation from John Brantner at the beginning of this chapter provides perspective on this problem: "Only people who avoid love can avoid grief. The point is to learn from it and remain vulnerable to love."

Those who are able to withdraw emotional energy and reinvest it in other relationships find the possibility of a newly established social life. Kavanaugh depicts this situation well with the following description:

> At this point fantasies fade into constructive efforts to reach out and build anew. The phone is answered more quickly, the door as well, and meetings seem important, invitations are treasured and any social gathering becomes an opportunity rather than a curse. Mementos of the past are put away for occasional family gatherings. New clothes and new places promise dreams instead of only fears. Old friends are important for encouragement and permission to rebuild one's life. New friends can offer realistic opportunities for coming out from under the grieving mantle. With newly acquired friends, one is not a widow, widower, or survivor—just a person. Life begins again at the point of new friendships. All the rest is of yesterday, buried, unimportant to the now and tomorrow. (1972, pp. 122–123)

ASSISTING THE BEREAVED

In his book *Bereavement: Studies of Grief in Adult Life*, Colin Parkes (1972) noted that the funeral often precedes the "peak of the pangs" of grief that tends to be reached in the second week of bereavement. The face put on for the funeral can no longer be maintained, and the bereaved individuals have a need to be freed to grieve. The most valued person at this time is the one making few demands on the bereaved person, quietly completing household tasks, and accepting the bereaved person's vented anguish and anger—some of which may be directed against the helper.

It is important to recognize that the bereaved person has painful and difficult tasks to perform that cannot be avoided or rushed. The best assistance that one can offer to those in grieving is to encourage them to attend to the four tasks of mourning described earlier.

One can help bereaved individuals come to grips with the reality of the death by talking about the deceased person and encouraging them to conceptualize the loss. Parkes observed that it is often reassuring to the bereaved person when others show that they are not afraid to express feelings of sadness. Such expressions make the bereaved person feel understood, help reduce a sense of isolation, and help them express their own feelings of sadness.

It is not uncommon for one approaching a newly bereaved person to be unsure how to react. Parkes suggested that although a conventional expression of sympathy can probably not be avoided, pity is the last thing that the bereaved person wants. Pity makes one into an object—the bereaved person somehow becomes pitiful. Pity puts the bereaved person at a distance from, and in an inferior position to, the intended comforter. Parkes maintains that it is best to get conventional verbal expressions of sympathy over as soon as possible and to speak from the heart or not at all. There is not a proper thing to say at this time; a trite formula serves only to widen the gap between the two persons.

The encounter between the bereaved person and the helper may not seem satisfactory because the helper cannot bring back the deceased individual and the bereaved person cannot gratify the helper by seeming to be helped. Bereaved people do, however, appreciate the visits and expressions of sympathy paid by others.

"Sometimes a cute card just isn't
appropriate, Jane."

John Louthan

These tributes to the dead loved one confirm to the bereaved individual the belief that the deceased person is worth all of the pain. Bereaved individuals are also reassured that they are not alone and feel less insecure.

The Hallmark card company sells its products with the slogan "When you care enough to send the very best." Although sending a card is one way to validate the loss of the bereaved individuals and to assist them in their grieving, if you really care enough to send the very best, a personal visit or a handwritten letter is better. On the other hand, if you find it impossible to communicate a personal expression of sympathy, sending a card is preferable to doing nothing. In the "Practical Matters" box entitled "The Art of Condolence," seven suggestions are given for expressing condolence.

Although many bereaved people are frightened and surprised by the intensity of their emotions, reassurance that they are not going mad and that this is a perfectly natural behavior can be an important contribution of the helper (Parkes, 1972). On the other hand, absence of grief in a situation where expected, excessive guilt feelings or anger, or lasting physical symptoms should be taken as signs that not all is going well. These persons may require special help, and the caregiver should not hesitate to advise the bereaved person to get additional help if the caregiver is uncertain about the course of events.

Support groups empower persons to cope with social crisis and loss in a de-alienating environment. According to Louis LaGrand (1991, p. 212), the power of support groups lies in the "introduction of new ways of looking at one's problems and the development of belief systems that enhance the twin attributes of self-determination and interdependence, all of this accruing in a socially secure setting."

The grieving process often socially isolates the bereaved by marginalizing them from normal patterns of social interaction. Bereavement self-help groups, by their very nature, challenge the assumption that grievers are unique by confronting them with caring people who understand their experience of loss and are willing to practice patterns of coping that reduce feelings of despair and depression (LaGrand, 1991). For an extensive review of 100 different therapeutic and psychological techniques for counseling the grieving and bereaved, see Jones (2013) and Neimeyer (2012).

Some national support groups for grievers are Compassionate Friends, I Count Too, Candlelighters, Empty Arms, Begin Again, Widow to Widow Program, the Omega Project, Parents Without Partners, and various hospice programs and/or hospital grief therapy groups. Another resource is the Internet or World Wide Web. There are chat rooms for grievers, and one can do grief work by creating memorial tributes to the person who has died. John Buchanan offers some very practical advice in his article "Words for Grieving," which appeared in *The Christian Century* in July of 2013.

The most natural support base in this setting can be the family. Distress and grief associated with life events, such as a life-threatening illness and bereavement, can be helpfully shared with the family, who assist with processing and making sense of the event, provide mutual support and caregiving and, eventually, help with coming to terms with the losses and moving forward with life. Consequently, a significant group strategy for coping with loss is family-based grief therapy.

PRACTICAL MATTERS	The Art of Condolence

Based upon their study of thousands of condolence letters and analysis of their structure, Leonard and Hillary Zunin (1991) share seven components that provide the writer with a practical, simple, and clear outline for sharing his or her thoughts. Though all seven components need not be included in every letter of condolence, keeping them in mind will provide an effective guide for sharing thoughts in such a letter.

The components include the following:

1. **Acknowledge the loss.** Mention the deceased by name and indicate how you learned about the death. It is very acceptable to relate personal shock and dismay at hearing such news, and such an acknowledgment sets the tone for your letter.
2. **Express your sympathy.** Share your sorrow in an honest and sincere fashion. In so doing you are showing that you care and relate in some way to the difficult situation they are facing. Do not hesitate to use the words *died* or *death* in your comments.
3. **Note special qualities of the deceased.** As you reflect upon the individual who has died, think about those characteristics you valued most in that person and share them in your letter. These may be specific attributes, personality characteristics, or other qualities. Sharing these with the bereaved help them realize that their loved one was appreciated by others.
4. **Recount a memory about the deceased.** At the time of a death, memories we have of the deceased person are a most valued possession and something which can never be in too great a supply. Because the bereaved often have difficulty keeping those memories in the forefront of their thoughts, your sharing of memories will be very gratefully received. Feel free to recall humorous incidents as they can be very beneficial at this time.
5. **Note *special* qualities of the bereaved.** Grieving people also need to be reminded of their personal strengths and other positive qualities—those characteristics which will help them through this difficult time. By reminding them of the qualities you have observed in them, you will be encouraging them to use these qualities to their advantage at this time.
6. **Offer assistance.** Offering help need not be part of a condolence letter, but if help is offered, it should be for a specific thing. An open-ended offer of help places the burden for determining what that help will be on the bereaved, who have enough burdens already. Making an offer to do something specific and then doing it is a most welcome extension of yourself.
7. **Close with a thoughtful word or phrase.** The final words in your letter of condolence are especially important and should reflect your true feelings. Flowery or elaborate phrases do not help. Honest expressions of your thoughts and feelings communicate best.

From The art of condolence (pp. 3–4), by P. V. Johnson, 1991, *Caregivers Quarterly*, 6(3).

Tammy A. Schuler, Talia I. Zaider, and David W. Kissane (2012) have written an extensive discussion of family-based grief therapy in which they deal with many topics of significance: anticipatory grief, complicated grief, attachment theory, group adaptation, and efficacy of family grief.

COPING WITH VIOLENT DEATH

Deaths due to acute diseases (e.g., pneumonia), accidents, disasters, murder, war, and suicide provide special problems as well as advantages for survivors. For all quick deaths there is the problem of being unprepared for the death. The survivors did not have a support group in place and did not have the advantage of having experienced anticipatory grief. Some of the grieving that has preceded a death

DEATH ACROSS CULTURES | THE ONE FACE OF GRIEF

In most natural disasters, we glimpse the suffering of one region, one people, often a single city with victims of a single nationality. But in the aftermath of last Sunday's earthquake and tsunamis, we are witnessing something that hasn't been seen since the end of World War II: a nearly global catalog of woe. The islands and coastlines of the Indian Ocean and the South China Sea are home to an astonishing index of human diversity, made still more diverse by the tourists gathering for the holiday season. Before Sunday, it would have been hard to imagine a natural disaster that affected both aboriginal Andaman Islanders and northern Europeans. Now we know what it looks like.

Here, so far from the devastation, we naturally focus on the death count, now reported to be more than 80,000, and it will certainly keep rising. The victims include people of every age, but especially children. They include Indonesians, Thais, Indians, Sri Lankans, Burmese, Somalis, Swedes, Norwegians, Germans, British, Americans, and many others. Out of all that diversity, the tsunamis created a single, simple division, between the living and the dead.

At first, the concern of the living was naturally for the dead. But to avert a far worse disaster, the living must now look after themselves with all the aid the rest of the world can provide. The World Health Organization has stated that the greatest risk of disease doesn't come from the decaying of the numerous corpses. It comes from the way the survivors are now forced to live—without fresh water, without adequate sanitation and, in many places, without shelter or enough food.

It is not merely a symbol that the civil war going on in western Sumatra has been suspended in the wake of the disaster. It is a sober setting-aside of differences for a more important, more immediate cause. This relief effort is going to test all of us, all around the world. Right now, the problem is getting supplies into the region and figuring out where they are needed most, a task made all the harder by the geographical scope of the devastation and the fact that so many places are still out of contact.

There is only one face of grief—no matter how many languages and skin colors there are among the survivors—and there should be only a single face of determined, compassionate outreach, worn by the rest of us wherever we live.

"Tsunami Aftermath: The One Face of Grief," December 30, 2004. Editorial. *The New York Times*, 2004.

caused by a chronic disease cannot be expressed in deaths of this type. Consequently, grief is usually more intense when the dying takes place in a short period of time. Survivors may also experience more intense guilt: "If only I had done something, she wouldn't have died." Suicide creates special problems for survivors because they can become stigmatized when a relative commits suicide: "They drove him to it." Finally, when people die without warning, survivors often are troubled because they did not have a chance to mend a broken relationship or say goodbye.

Michael Leming recently gave a eulogy for a man who had died from a drug overdose. The man's wife wanted the cause of death to be mentioned, but the mother and brother of the man protested. The brother became so angry that he threatened Leming with bodily harm if he mentioned that the death was drug-related. The "compromise" was to have a second service for friends, when most of the family members would have gone, at which time Leming would deliver the eulogy.

The irate brother actually stayed in the wings of the church for the second service and listened. After the eulogy, he came up to Leming and apologized

PRACTICAL MATTERS | 21ST-CENTURY FORUM FOR GRIEF

In the weeks since 18-year-old Whitney Hendrickson's death in March in a fiery gas station crash, grief has taken a 21st-century turn. More than 100 people have flocked to Facebook to leave personal messages for Hendrickson on a public memorial group and semi-public profile page. Some signal a final farewell:

- "Hey Whitney. I hope alls well up there, we're gonna miss you an awful lot down here."
- "I won't say goodbye, but until we meet again."

Other messages offer updates to a friend who's no longer around:

- "I walked into church today and thought I saw you walking across the foyer. Just so you know, there's a girl that looks so much like you, it's insane."
- "Hey girl hey, so this morning I had the horror (or pleasure?) of waking up to a giant bushy squirrel. . . . Obviously I thought of you and our mutual squirrel obsession."

With more than 200 million active Facebook users, and half of all teenagers using some form of social networking, the practice of posting online messages to dead friends is a recent but increasing phenomenon.

Facebook does not delete profiles of the deceased, according to a spokesperson, but puts them in a "memorial state," which hides some information from view. A 2009 study published in the *Journal of Adolescent Research* on social networking and death found that users often continue a relationship with the profile of the deceased for months. The study looked at 20 deceased people's profiles on an undisclosed social-networking site and recorded 1,167 people posting 4,780 comments over an average 10-month period. "It may be changing how we see death," said Anya Salzgeber, a college student. "We may not see it as such a strong permanence because we still have a method to communicate with them."

Roger Sambrook, a professor who specializes in cyber-psychology, said the Internet's impact on the grief process has been significant. With the advent of social networking, "our lives are on display, and this is bringing the medium of death and mourning back into display," he said.

E. Findell, *Charleston Post and Courier*, May 17, 2009, p. 3G.

for his anger and inappropriate behavior. Such volatile emotions are a normal part of the grieving process, as Leming told him. Studies (Walter, 1999a) of bereavement clearly indicated that grieving people do not act rationally. Like falling in love and giving birth, grief makes people say and do irrational things temporarily.

Though various problems evolve from sudden deaths, survivors of sudden deaths are spared the following problems associated with **chronic diseases**:

1. Dying persons may not be willing to accept death and, when learning of their fates, may act in unacceptable ways.
2. Families may be unwilling to accept the death.
3. The dying process may be a long and painful process, not only for the dying patient, but also for the family.
4. The cost of dying from a chronic disease can be, and usually is, very expensive. The entire assets of a family can be wiped out by the medical bills of a chronically ill patient.

Because we have limited choices in the matter of how we die (unless we intervene), we will simply have to take whatever comes along. Who knows? Perhaps death will come when we are very old, at home, unexpectedly, in our own beds,

Factory of death is remembered 60 years later as survivors and world leaders gathered on January 27, 2005, to mark the day when the Red Army freed the Auschwitz death camp in Poland.

while sleeping—and with our full mental and physical capabilities. But we may not die the "all-American way" at home, as described in the previous sentence, or slowly in an institution. Death may be sudden and violent from an accident, a disaster, murder, or war.

ACCIDENTS

Accidents occur in disproportionately higher rates among younger individuals and are the leading cause of death among all persons from age 1 to age 24 years (ADAM, 2005). As Therese Rando (1993) observed, a high correlation exists between accidental death and violence, mutilation, and destruction. Such events leave survivors with losses that are often more shocking and difficult to assimilate. In addition, traumatic memories of the event or imagined scenes of what happened produce problems for the bereaved individuals. Guilt may be greater than usual as a result of the strong need to assign blame and responsibility in accidental deaths. For example, "*If only* I had not asked her to go to the store for the milk, this never would have happened," a parent may say. Such guilt can get in the way of getting on with bereavement, suggested Rando. There is often a need on the part of survivors to determine how much the individual suffered at the end; thus, they have a desire to talk to others who witnessed the event. The survivors want some reassurance that she or he did not suffer at the end.

The issue of unfairness and injustice often must be addressed with accidental deaths. The fact that the individual died because she or he was in the wrong place at the wrong time is difficult to accept. Such an injustice makes getting on with her or his own life difficult for the survivor and often grates on her or him day and night. Anger may be a reaction for survivors if the deceased person contributed in any way to her or his death (e.g., drinking and driving). The survivor may be angry because the deceased person caused the loss in part and contributed to a most painful experience for family and friends (i.e., the "How could you do this to us?" kind of reaction). Bereavement in accidental deaths can also be complicated by the media continually replaying the accident. This makes it difficult for bereaved individuals to "get away from" the accident, noted Rando.

What part do the media play in violent deaths? Friedrich Nietzsche, writing over a century ago in 1882, had this to say about the media and violent deaths:

> Our age is an agitated one, and precisely for that reason not an age of passion; it heats itself up constantly because it feels that it is not warm—basically it is freezing. . . . In our times it is merely by means of an echo that events acquire their "greatness"—the echo of a newspaper. (cited in Guignon, 1993, p. 290)

William Merrin (1999) suggested that Nietzsche provides a powerful image of a cold public looking to its media to fill an emptiness—an image of an era trying to warm up by overheating its events to feel alive, if only temporarily. Merrin looked at the automobile accident that took the life of Princess Diana and how the media responded.

The media is not a mirror reflecting the world but are themselves inseparable from the event. Today, the event and its broadcast have become a single phenomenon—the *media event*. The media may cover violent deaths for several days; then the coverage begins to wane. Such withdrawal does not mean that those grieving have completed their task, but the media's part in the situation is over. The media may return on the anniversary of certain tragic events, but the response may or may not be there. For example, on the first anniversary of the death of Princess Diana, a "Remembrance Walk" was planned and was expected to attract 15,000 people—300 came (Merrin, 1999). The accident that took Princess Diana's life attracted media attention to the extent that it did because of who she was. However, the extensive media coverage then attracted millions of viewers and readers. The media do contribute to highlighting violent deaths and to some extent playing them up beyond what they are. British theologian Douglas Davies (1999) observed that the "Diana event" was, primarily, a media event, yet with a difference. The difference consisted of two elements—the active and practical participation of millions of people, on the one hand, and the idea that the media were, maybe, responsible for the death, on the other hand. The media were reporting on something for which many thought the media responsible.

In trying to make some sense out of the "Diana event," Davies noted the loss of a highly charged symbol of social meaning—Diana was a condensed symbol of love, divorce, fragmented families, stardom and failure, beauty and eating disorders, and also of the marginalized groups of AIDS, leprosy, and land mine victims. Diana possessed the charisma that only certain royals have—a regality that sets them apart, along with the ability to connect with ordinary people. Through her the ordinary person could connect symbolically to society itself. Those who actually met and touched her could connect physically.

DISASTERS

Some common threads cut across deaths caused by disasters, according to B. Raphael (1986). Identification of the dead bodies may cause problems, contributing to complications for the bereaved. Dismemberment may make it difficult to put remains into a human form. When remains are not distinguishable (e.g., charred or mutilated), survivors are left with fear about the suffering that must have accompanied the death. Though one may feel traumatized at the thought of viewing the body, Raphael cites numerous studies that indicate that bereaved individuals who view the body do not regret it. Yet, family members are often advised not to see the body.

If unable to see the body of the deceased person, family members must deal with uncertainty. "Is that really her or him who was identified?" they may ask. Perhaps the individual is still alive. Thus, an element of hope may appear through the uncertainty. Such may have been the case, at least for a few days, for family members and friends with the sinking of the *Titanic* in April 1912. Nonetheless, a memorial service for the victims of the sinking took place at St. Paul's Cathedral, London, on April 19, 1912, with many thousands unable to get into the church (Davies, 1999). There were soldiers and sailors who had heard the playing of the "Dead March" for dead comrades, yet they too stood erect, with tears streaming down their faces, not ashamed of tears. The outpouring of grief was overwhelming, according to London papers.

Having lived through the tragedy of September 11, 2001; the tsunamis in Southeast Asia on December 26, 2004 and in Japan in March of 2011; the Madrid and London train bombings of 2004 and 2005; the 2013 Typhoon Haiyan in the Philippines where more than 10,000 people died; and most recently, the missing Malaysia Airlines Flight 370 in March of 2014 and the ferry carrying 459 people that capsized and sank off the coast of South Korea in April of 2014, we can all understand more fully the overwhelming nationwide grief experienced by Americans in earlier national disasters such as the Hindenburg fire, the attack on Pearl Harbor, the eruption of Mt. St. Helens, and the Oklahoma City Federal Reserve bombing. To have the 9/11 tragedy repeatedly enter our homes through live and rerun television coverage of the more than 3,000 people being killed by terrorists—will our nation ever be the same? Can any American ever feel safe again with the real possibility of nuclear power plant disasters, global terrorism directed toward the United States, and global climate change leading to repeated natural disasters such as tornadoes, hurricanes, floors, forest fires, and landslides?

With these tragic world events in the 21st century, not only are survivors suffering from several deaths, but also the number of significant others who could have offered social support is reduced. Thus, the whole issue is compounded. Furthermore, families were separated because of the disasters, adding to survivor

A natural disaster such as a hurricane destroys human lives, physical property, and personal possessions. Thus, there is grieving because of lost life but also because of lost treasures. A cemetery flooded by the storm surge from Hurricane Ike is shown, Saturday, September 13, 2008, in Galveston, Texas.

Michael R. Leming

The tsunami on Phuket Island took the Thai people by surprise. Another surprise was the way in which the people of the world came to their rescue. The international Rotary Clubs joined the efforts of the local Rotary Club to help rebuild many fishing boats. Without this help, the fishermen would not only have lost their homes and loved ones, but also their futures.

uncertainty. In some cases, bodies were never recovered and loved ones' grieving was complicated by the inability to visualize their loss.

In some disasters there are residual troubles created from the events. These include problems resulting from food and water shortages, contamination and unsanitary conditions, and poor medical care. In some disasters mobility may be impaired because of torn-up streets and highways; thus simply getting around may be difficult and add to a survivor's dilemma.

RECOVERING FROM TSUNAMI OR TYPHOON   Grieving any loss requires one to go through many stages of grief: denial, volatile emotions, depression, guilt, loneliness, and restoration.

Grief following a natural disaster is no different. From the perspectives of people who have experienced the death of one or more loved ones, there will be personal or private grief. From the perspective of the community that has experienced a natural disaster, there will also be corporate grief. The restoration of order created by a natural disaster such as the 2004 tsunami in Southeast Asia or the 2013 Typhoon Haiyan in the Philippines will take a long time, as the community adjusts to its loss and tries to move beyond it.

Ann and Mike Leming visited some of the beaches on Phuket Island in southern Thailand three months after the tsunami hit Southeast Asia on December 26, 2004, causing more than 300,000 deaths. The box entitled "Tsunami: Coping with a Natural Disaster" represents an e-mail they sent to friends.

MURDER

When the murder of a child is discussed, the child is described as *the victim*. Yet, the parents are also victims (Conrad, 1998). They are not dead, but they have

DEATH ACROSS CULTURES | TSUNAMI: COPING WITH A NATURAL DISASTER

Michael Leming and his wife were in Phuket, Thailand (an island off the coast of Thailand and one of the major sites of the Tsunami disaster on December 26, 2004) three months after the disaster. The following is an eyewitness account of some of the recovery in that part of the world.

While there were many places that experienced more devastation, the villages where we stayed (Karon, Patong, Kamala) were hit very hard. Most of our time was in Kamala. For Kamala beach, almost all of the small resorts and shops on the beach were destroyed at least on the first floor; a temple and school were also hit hard, and the small fishing boats that were in the harbor that morning were destroyed.

Today, three months after the disaster, about half of the repair work is complete. The island will probably be even more beautiful in a year than it was last tourist season. Of course the loss of life has created emptiness in the lives of the people from which they will never fully recover.

Today the biggest temporal problem is the lack of tourists and income to the local people. Taxi drivers, people who do laundry, hotel and resort workers, and so forth have lost income at a time when they are trying to rebuild. The government and others have tried to fill in the gaps but there is so much to do to restore normalcy. One of the problems is that Asian tourists (including Thais) are reluctant to visit a tourist place where so many have died and the potential for encountering angry ghosts exists.

A bright spot were the signs thanking people of the world for their support and aid. We saw a place on the beach where the local Rotary clubs (with the assistance of International Rotary) supported the building of ships. Every week another six new boats were placed back into service.

We wonder if all the money is getting to the people whose lives have been so profoundly affected. But still, there is so much being done.

March 2005

DEATH ACROSS CULTURES | THE ENGLISH WAY OF GRIEF

Noted British sociologist Tony Walter writes about the emotional reserve and the English way of grief. He says that in England the bereaved are not supposed to weep and hug each other, especially men—except on the soccer field. In England the "style of mourning" is really a matter of individual choice. However, when given total freedom, most individuals do not know how to express grief and look around anxiously to see how others are reacting.

The British feel most comfortable grieving in private. However, it is important that others know you are grieving, showing that you have cared for the deceased. Thus, one must provide clues of deep-seated feelings without actually breaking down in public and embarrassing others—not an easy task. The norm of private grief makes public appearances problematic; thus the funeral can be dreaded.

Over the past two to three decades, absent and private grief have been criticized by expressivists because it short-circuits the process of working through a range of feelings. However, the British have not immediately taken to this new way of grief, notes Walter. The English are now encouraged to cry, yet are aware that weeping spontaneously in the supermarket or outside the school gates will disturb others. Also, at the funeral they are encouraged to express their emotions today, yet uncontrollable emotion at funerals is frowned upon. Walter concludes that to talk of an English way of grief is not to claim that it is unchanging or unproblematic.

From Emotional reserve and the English way of grief, by T. Walter, 1997. In K. Charmaz, G. Howarth, & A. Kellehear (Eds.), *The unknown country: Death in Australia, Britain, and the USA.* London: Macmillan Press.

been injured—permanent changes have been brought about in their lives. Permanent voids are created in their lives. They think of the future with dread. They may lose interest in activities and events that once gave purpose to their lives. They may lose their faith in the "goodness of mankind" and their trust in the world—they feel betrayed.

When a child is murdered, the initial problem is to tell the parents, a job not cherished by police, yet it is their job. Joan Schmidt (1986) suggested that one giving such bad news be short and to the point: "I have bad news; your child has been hurt; your child is dead." Because the recipient of the news will probably be in shock and will disbelieve immediately after receiving the news, then is the time to add the simplest of details: The child was murdered, where, perhaps briefly how, and where the child is now.

If the murderer is known, the family may have divided alliances, conflict, and anger (Olson, 1997). Guilt is evident in most cases of family violence that end in homicide. If the killer is unknown, relatives may be in fear for their own lives. Coping for these survivors begins when they find out about the loss. They may need to spend time with the body. If it is impossible for them to see the body, just being in the same building or area where the body was may be helpful. These survivors need information about legal issues, such as an autopsy, and when results will be known.

As survivors of a murdered individual, there are certain enemies, observed Therese Rando (1993). Aside from the murderer(s), there are lawyers and courts, the media, and well-meaning "imbeciles." Swift justice does not often exist, and an attorney can many times find a loophole or a judge can grant a stay or overturn a conviction or give a delay. Friends may comment, "The trial is still going on?"

Life after the homicide of a significant other involves numerous practical and concrete changes developing out of the survivors' fear (Rando, 1993). Changes may be abstract (e.g., paralyzing fear) or concrete (e.g., keeping a gun for protection). Also, death imagery becomes intense and frightening for survivors who did not witness the trauma directly or view the body. Survivors are compelled to work through an internalized fantasy of grotesque dying that increases fears and complicates mourning by presenting the task of assimilating the violence and transgression implicit in this type of death, noted Rando.

The bereaved person's perception of the death as preventable, no matter the circumstance, is a high-risk factor for complicated mourning. If the death is perceived as having been preventable, anger becomes a central effect of this perception, stated Rando (1993). It is easy to assign blame and thus punish one's self or others. The outrage and indignation at the ultimate arrogance of someone who assumes the right to exterminate another human being contributes to a bereaved person's frustrations.

Assuming the murder is solved and the convicted murderer goes to prison (if not given the death penalty), the stress for the family is not over. "Life in prison" does not always literally mean that. For example, the convicted murderer may come up for parole within a few years. Such a situation gives the murderer of the family's child a chance to be free and creates fear that the murderer may strike again and cause grief to yet another family. A recent case in South Carolina has the murderer, though serving seven life sentences plus 355 years in prison, up for parole on an annual basis after serving only 17 of those years. Every year the 78-year-old mother and family of the murdered woman gather 8,000 signatures and take them to the parole board to encourage them to keep the convicted murderer behind bars. Thus, on an annual basis, the murder/rape of this elderly woman's daughter will surface. There are many victims of a single murder—the family continues to grieve and suffer.

MASS MURDERS

The rate of people killed by guns in the United States is 19.5 times higher than similar high-income countries in the world. In the last 30 years since 1982, America has mourned at least 61 mass murders (Shen, 2012).

On April 20, 1999, at Columbine High School in Littleton, Colorado, 13 students were killed and 21 wounded. Since Columbine there have been more than 20 events involving violent killings and mass murders in the United States. The most violent of these mass killings were 32 students killed at Virginia Tech University on April 16, 2007, and 20 children and six adults at Newtown, Connecticut, on December 14, 2012.

Furthermore, there have been 74 school shootings since Newtown, Connecticut (December 12, 2013) and the shooting in Troutdale, Oregon on June 10, 2014. That is almost one shooting per week (74 in 77 weeks) and most of these shooting have been in K-12 schools, not colleges or universities.

Using the most recent CDC estimates for yearly deaths by guns in the United States, it is likely that as of, June 11, 2014, roughly 51,258 people have died from guns in the U.S. since the Newtown shootings.

WAR

In war, an individual faces huge numbers of casualties, a threat to self and country, and the violence of modern weapons—all leading to a massive threat of death and the destruction of large numbers of individuals (Raphael, 1983). War deaths typically have all the risk factors and issues found in deaths caused by accidents, disasters, and homicides. Other factors include the serviceperson's and survivor's philosophical agreement with war in general, the type of death and circumstances, the stress of separation experienced by the survivor while the service person was deployed, and the extent of social support available (Rando, 1993).

For combat soldiers, external group demands and strong group cohesion keep them going. They interpret combat as a sequence of demands to be responded to by precise military performances (Bourne, 1970). Yet, some soldiers display long-lasting effects from combat (post-traumatic stress disorders) and may later have feelings of anger and guilt. A friend of George Dickinson's, a helicopter paramedic in Vietnam, was in a helicopter crash. He was the lone survivor out of a crew of eight. He felt guilty and often asked himself, "Why me? Why should I survive when everyone else died?"

Resul Cesur, Joseph J. Sabia, and Erdal Tekin (2013) conducted an empirical study of soldiers who were deployed overseas to estimate the effect of combat exposure on psychological well-being. Controlling for predeployment mental health, Cesur et al. (2013) found that active-duty soldiers deployed to combat zones are more likely to suffer from post-traumatic stress disorder (PTSD)

America honors its heroes by providing a special postmortem tribute during the military funeral.

than their counterparts deployed outside the United States in noncombat zones. Furthermore, among those deployed to combat zones and/or who witnessed allied or civilian deaths, it was discovered that there was an increased risk for suicidal ideation and PTSD relative to their active-duty counterparts deployed to combat zones without enemy combat. See also Griffith (2013) and Rice & Sher (2012).

Some bereaved individuals may experience postdeath reactions similar to those seen after the loss of a family member from a lingering illness, depending on the length and nature of the stress of the separation and the degree to which the serviceperson's death is anticipated (Rando, 1993). Nonetheless, such an encounter with death, violence, and mutilation often has profound and long-lasting effects: sights, sounds, and smells may linger for a lifetime. Without doubt, the death of a family member or friend in war is an unusual mixture of both sudden and anticipated loss. More military personal have committed suicide than have died in combat during the wars in Iraq and Afghanistan.

CONCLUSION

In this chapter we have discussed the grieving process, bereavement roles, normal adaptations to experiences of loss, the four tasks of mourning, and coping with violent deaths. We have attempted to demonstrate that bereaved individuals must attend to grief work because successful grief resolution is not automatic. This grief work refers to Worden's (1982) four necessary tasks of bereavement: accepting the reality of the loss, experiencing the pain of grief, adjusting to an environment in which the deceased is missing, and withdrawing emotional energy and reinvesting it in other relationships.

Grievers need support and assistance in the bereavement process. One can help grievers come to grips with the reality of the death by talking about the deceased person and encouraging them to conceptualize the loss that they are experiencing. It is often reassuring to the bereaved individual when others show that they are not afraid to express feelings of sadness. Such expressions make bereaved persons feel understood and reduce their sense of isolation. Many well-meaning friends find it difficult to find proper words to say. Unfortunately, their discomfort often causes them to do nothing. Even trite words are better than no words at all. A Hallmark card company slogan says, "When you care enough to send the very best" (send one of our cards). When comforting bereaved friends, however, a better personal slogan might be, "When you care enough to send the very best, send yourself." At this time, the very best that we have to offer is our own caring presence.

SUMMARY

1. The bereavement role is a temporary role that gives one the right to be exempted from normal social responsibilities and to become dependent upon others.

2. Abnormal bereavement behavior includes preoccupation with the death of the loved one and refusal to attempt to return to normal social functioning.

3. Disenfranchised grief involves loss that cannot be openly acknowledged, socially sanctioned, and publicly shared. There are four circumstances that lead to disenfranchised grief: (1) when the relationship to the deceased is not socially recognized; (2) when the loss is not acknowledged by others as being genuine loss; (3) when the grievers are unrecognized; and (4) when the death is not socially sanctioned.

4. The grieving process is similar to the dying process in that it is a series of behaviors and attitudes related to coping with the stressful situation of a change in the status of a relationship. The grieving process need not be passive, because bereaved individuals can make a number of decisions as they attempt to cope with their loss.

5. It is important in grieving to let feelings emerge into consciousness and not be afraid to express sadness.

6. It is not uncommon to be unsure of how to act around a newly bereaved person.

7. Rather than suggesting that "time heals," an accurate description of the mourning process suggests that the time intervals between intense experiences of grief increase with the passing of time.

8. Seven behaviors and feelings are part of the normal bereavement process: shock and denial, disorganization, volatile emotions, guilt, loss and loneliness, relief, and reestablishment.

9. All persons who have experienced a loss through death will need to attend to the four necessary tasks of grief work before mourning can be completed and reestablishment can take place. These tasks involve: accepting the reality of the loss, experiencing the pain of grief, adjusting to an environment in which the deceased person is missing, and withdrawing emotional energy and reinvesting it in other relationships.

10. Signs of abnormal grief suggest that an individual might need to seek professional help or at least affiliate with a support group.

11. Deaths from violent means create special bereavement problems for survivors.

DISCUSSION QUESTIONS

1. How can one avoid "deviant" or abnormal behavior regarding the bereavement role? What are some functions of defining bereavement roles as "deviant" or "abnormal"?

2. What is the relation between time and the feelings of grief experienced within the bereavement process?

3. Discuss how the seven stages of grieving over a death can also be applied to losses suffered from going through a divorce, moving from one place to another, or having an arm or a leg amputated.

4. If grief is a feeling that is imposed upon the individual through loss, how can grieving be conceived of as being an active process?

5. What are the unique problems faced by those whose grief is disenfranchised? What are the different types of disenfranchised grief?

6. Describe the four necessary tasks of mourning. What are some of the practical steps that one can take to accomplish each of these tasks?

7. What does Parkes mean by "The funeral often precedes the 'peak of the pangs'"? How can one assist friends in bereavement?

8. What could you do to assist people experiencing abnormal grief symptoms?

9. Distinguish between "normal" and "abnormal" grief.

10. How does grief from violent deaths differ from that for other deaths?

GLOSSARY

Acute Disease: A communicable disease caused by a number of microorganisms including viruses, fungi, and bacteria. Acute illnesses last for a relatively short period of time and result either in recovery or death. Examples of acute illnesses include smallpox, malaria, cholera, influenza, and pneumonia.

Bereavement Role: Behavioral expectations for the bereaved that are structured around specific rights and duties.

Blaming the Victim: A strategy asserted by individuals to relieve the dissonance experienced when innocent people suffer loss.

Chronic Disease: A noncommunicable self-limiting disease from which the individual rarely recovers, even though the symptoms of the disease can often be alleviated. Chronic illnesses usually result in deterioration of organs and tissues that makes the individual vulnerable to other diseases, often leading to serious impairment and even death. Examples of chronic illnesses include cancer, heart disease, arthritis, emphysema, and asthma.

Complicated Grief: An intense and long-lasting form of grief that takes over a person's life.

Disenfranchised Grief: Grief that cannot be openly acknowledged, socially sanctioned, and publicly shared.

Dissonance: An inconsistency in beliefs and values—relative to a particular social situation—that causes personal discomfort or tension for the individuals involved.

SUGGESTED READINGS

Balk, D. E. (2003). The evolution of mourning and the bereavement role in the United States. In C. D. Bryant (Ed.), *Handbook of death and dying* (pp. 829–837). Thousand Oaks, CA: Sage Publications. Provides a discussion of the evolution of mourning customs in the United States for both middle- and upper-class European Americans.

Bonanno, G. (2009). *The other side of sadness: What the new science of bereavement tells us about life after loss.* New York: Basic Books. We tend to understand grief as a predictable five-stage process of denial, anger, bargaining, depression, and acceptance. But in *The Other Side of Sadness*, Bonanno shows that our conventional model discounts our capacity for resilience. In fact, he reveals that we are already hardwired to deal with our losses efficiently—not by graduating through static phases. In the end, mourning is not predictable, but incredibly sophisticated. Combining personal anecdotes and original research, *The Other Side of Sadness* is a must-read for those going through the death of a loved one, mental health professionals, and readers interested in neuroscience and positive psychology.

Cahill, K. (Ed.). (1999). *A framework for survival: Health, human rights and human assistance in conflicts and disasters.* New York: Routledge. Focuses on mass death, destruction of the social order, and economic crisis.

Conrad, B. H. (1998). *When a child has been murdered: Ways you can help the grieving parents.* Amityville, NY: Baywood Publishing. Written to describe the painful experience of a child being murdered and to make individuals aware of the ways they can help the parents survive their grief.

Doka, K. J., & Breaux, J. (2002). *Living with grief: Loss later in life.* New York: Brunner-Routledge. Shows how older people come to terms with their mortality, how they are cared for, the fundamental realities of aging and dying, and the variety of particular losses experienced by older persons.

Henricks, J. H., Black, D., & Kaplan, T. (Eds.). (2000). *When father kills mother.* New York: Brunner-Routledge. Brings to public knowledge information about the effects of psychological trauma and bereavement on children.

Johns, A. (Ed.). (1999). *Dreadful visitations: Confronting natural catastrophe in the age of enlightenment.* New York: Routledge. A study of 18th-century European disasters and responses to them.

Jones, B. (2013). Large collection of practical techniques in grief therapy. *Journal of Psychology and Theology, 41*(3), 256. A review of Dr. Robert A. Neimeyer's prolific writings regarding nearly 100 different therapeutic techniques for counseling the grieving and bereaved, with the stated goal to "present a rich and representative smorgasbord of methods for engaging grief and its complications with greater creativity and awareness of alternatives." In addition to this primary goal, a secondary goal was the integration of research and practice, aiming to "foster an interchange between the field of bereavement research and scholarship ... and of clinical and counseling practice." To accomplish these goals, the edited work enlists contributions from 86 different authors, stemming from diverse clinical backgrounds including psychiatry, psychology, social work, nursing, and art therapy.

Neimeyer, R. A. (2012). *Techniques of grief therapy: Creative practices for counseling the bereaved.* New York: Routledge Publishers. *Techniques of Grief Therapy* is an indispensable guidebook to the most inventive and inspirational interventions in grief and bereavement counseling and therapy. Individually, each technique emphasizes creativity and practicality. As a whole, they capture the richness of practices in the field and the innovative approaches that clinicians in diverse settings have developed, in some cases, over decades, to effectively address the needs of the bereaved.

Neimeyer, R. A., Harris, D. L., Winokuer, H. R., & Thornton, G. F. (Eds.) (2011). *Grief and bereavement in contemporary society: Bridging research and practice.* New York: Routledge.An excellent job of integrating research and clinical practice related to grief and bereavement.

Parkes, C. M. (2001). *Bereavement: Studies of grief in adult life.* New York: Brunner-Routledge. Written to be accessible to bereaved people who want to understand the powerful and often frightening emotions and symptoms that they experience, while incorporating the latest scientific understandings of bereavements and its effects.

Pearson, C., & Stubbs, M. L. (1999). *Parting company: Understanding the loss of a loved one.*

Ontario, CA: Seal Press. Provides practical and emotional support for people confronting the death of a significant other through 14 firsthand accounts from caregivers reflecting on a wide range of experiences.

Rees, D. (2001). *Death and bereavement: The psychological, religious and cultural interfaces.* New York: Brunner-Routledge. Account of the psychology of death and bereavement, examining the thinking of Freud and Jung and of modern psychiatrists. This book also discusses aspects of death such as bereavement visions, euthanasia, grief for a pet, and suicide.

Roos, S. (2002). *Chronic sorrow: A living loss.* New York: Brunner-Routledge. Develops an analysis of chronic sorrow, a natural grief reaction to losses that are not final but continue to be present in the life of the griever, and looks at the effects of chronic sorrow on the griever and the people close to them.

Rosenblatt, P. C., & Wallace, B. R. (2005). *African-American grief.* New York: Routledge. This work considers the potential effects of slavery, racism, and white ignorance and oppression on the African American experience and conception of death and grief in America. The book is based on interviews with 26 African American experiences of grief.

Schuler, T. A., Zaider, T. I., & Kissane, D. W. (2012). Family grief therapy: A vital model in oncology, palliative care and bereavement. *Family Matters, 90*(Summer), 77ff. An extensive discussion of family-based grief therapy dealing with many topics of significance: anticipatory grief, complicated grief, attachment theory, group adaptation, and efficacy of family grief.

Smith, A. (2004). *Name all the animals: A memoir.* New York: Scribner. Alison Smith's close-knit Catholic family is the very picture of contentment—right up until the day her 18-year-old brother is killed in a car accident. In *Name All the Animals*, Smith walks readers through the breakdown and breakthroughs of her family in the days and years that follow.

Vera, M. I. (2003). *Social dimensions of grief.* In C. D. Bryant (Ed.), *Handbook of death and dying* (pp. 838–846). Thousand Oaks, CA: Sage Publications.

Now I lay me down to sleep; I pray the Lord my soul to keep. If I should die before I wake, I pray the Lord my soul to take.
—**New England Primer, 1781**

GRIEVING THROUGHOUT THE LIFE CYCLE

CHAPTER 14

George Dickinson

Perhaps no two individuals grieve in the exact same way. The way one grieves may depend on various situations, including past experiences with death, religious beliefs, personality, the particular person who died, the circumstances surrounding the death, and the age of the one grieving. It is the latter situation that we will address in this chapter—grieving throughout the life cycle.

GRIEVING PARENTS AND THE LOSS OF A CHILD

Adapting to the death of a family member is always difficult, but the death of a child is typically regarded as the most difficult of deaths. As discussed in Chapter 3, within the life cycle, an individual expects grandparents to die first, followed by parents, then self, then offspring. That is the "logical" way life is supposed to work. Unfortunately, that which is "supposed" to happen does not always occur. This is the case with children and death. Parents are not "supposed" to have to deal with a dying child, much less even think of having to attend their child's funeral. Yet, in the world of reality, not the ideal world, the life cycle is sometimes not played out as it is supposed to be, and parents must grieve over the loss of a child. Additionally, the death of a child symbolically threatens the family's hope for a future.

Research shows (Oliver, 1999) that spouses often grieve the loss of a child differently. The most consistent difference appears to be that, as a group, mothers tend to report more intense, long-lasting, and diverse grief reactions than do fathers. Gender differences have also been found in bereaved parents' coping styles. For example, bereaved mothers use more coping strategies as a whole and use more emotionally and verbally expressive forms of coping than men do. Mothers are more likely to feel a strong need to talk about the death and express their feelings, whereas fathers tend to try to resume regular activities and "keep busy," reported L. E. Oliver.

THE LOSS OF A FETUS OR AN INFANT

For Americans the loss of a fetus, or miscarriage, is considered not to be very common. However, it may be of surprise to learn that more than one million American women suffer the loss of a pregnancy annually (Ventura, Abma, Mosher, & Henshaw, 2009)—approximately the same number of total deaths in America resulting from heart disease and cancer combined (actually 1.2 million people, according to the Centers for Disease Control [www.cdc.gov/nchs/fastats/deaths.htm]).

To understand the nature of loss involved in the death of a child, we should remember that the relationship of a parent to a child begins long before birth (Raphael, 1983). For each parent, from the time of conception, the child becomes a source of fantasy—the imagined child whom he or she will become. These hoped-for extensions of self are very common among expectant parents. As the pregnancy progresses, the fantasy relationship with this imagined child intensifies with the selection of the name for the baby, the rehearsal for parenting, the fantasies shared with others, and finally the birth of the child. With a stillbirth or the death of a newborn, the bubble of one's fantasy world suddenly bursts.

Mourning the death of a fetus or **neonate** (infant in the first month of life) differs from mourning the death of another loved one (Furman, 1978). Mourning is a process of detachment from the loved one that is moderated by identification. The bereaved parents take into themselves aspects of the deceased, but because a fetus or newborn has not lived long enough as a separate person, the parents have little of their baby to take into themselves. Thus, they suffer detachment without identification. Parents must readjust their self-image with the knowledge that the baby will never again be part of them. In addition, bereaved parents are less likely to receive adequate validation for their loss—miscarriages, stillbirths, and infant deaths are often "discounted, minimized or negated," as are the deaths of adult children (Oliver, 1999).

In one of the first studies of **perinatal** loss (death occurring during the period closely surrounding the time of birth), J. H. Kennell, H. Slyter, and M. H. Klaus (1970) observed reactions of mothers who had lost a newborn infant and explored the strength of affectional ties between mothers and infants after first physical contact. The conclusion was that more intense mourning was associated with the previous loss of a baby either through miscarriage or death of a liveborn infant. Other studies (Hunfield, Wladimiroff, Verhage, & Passchier, 1995) also demonstrated more intense grief in mothers who had a history of prior perinatal loss. Perinatal mortality rates vary by race. For example, perinatal mortality data for African Americans demonstrate that the incidence of perinatal loss is greater than twice that of whites (Guyer et al., 1997).

The death of an infant places severe strains on parents and members of the family—a sense of loss that will probably persist over a number of years. Studies (Nicol, Tompkins, Campbell, & Syme, 1986) show that up to one-third of mothers experience a marked deterioration in their health and well-being after the loss of an infant. The death of a child also has dysfunctional consequences for family systems, like separation and divorce.

Longitudinal studies of grieving parents demonstrate that the experience of grief does not go away with time; rather, its focus changes. According to S. G. McClowry and colleagues (1987), even after nine years, parents are still experiencing significant pain and loss—which the researchers refer to as an **empty space**. This "empty space" feeling takes on three recurring patterns of grieving. The first is "getting over it." In this pattern the grievers do what they can to get back to life as usual and accept death as a matter of fact.

The second pattern of dealing with the empty space is to "fill the emptiness" by keeping busy. Some parents attend grief groups; others immerse themselves in work, become more religious, increase their food or alcohol intake, or become involved as volunteers in organizations such as Candlelighters (a support group for parents of children with cancer), or Compassionate Friends or Empty Arms (two support groups for bereaved parents). Other parents attempt to replace the dead child with another by getting pregnant or by adopting.

A contemporary trend is to have online support groups for bereaved parents. In researching this option for grievers, Elizabeth Pector (2012) discovered that most users are young, white, female, and well educated. According to Pector, "Although research about internet groups has not shown better health or grief outcomes, groups empower members to access information, resources and social support. Perinatal loss group participants appreciate convenient, safe online communities to validate grief and discuss emotions, experiences, and hopes."

The third pattern of dealing with the empty space is described as "keeping the connection." In this pattern of grieving, parents attempt to integrate this pain and loss into their lives:

> Although most of the grievers who were "keeping the connection" expressed satisfaction with their present lives, they continue to reserve a small part of themselves for the loss of a special relationship which they view as irreplaceable. (McClowry et al., 1987)

In a recent article published by Meghan Cholette (2012), in which the author does an extensive review of the research on perinatal loss, and based upon her personal professional and clinical experience as a nurse, she discusses differences in paternal and maternal grieving. Cholette points out that in most experiences of perinatal loss, fathers' grieving needs are often ignored and that most people are consciously unaware that when fathers experience a loss of this type, their lives, too, are changed forever. "By focusing the majority of time and care on the mothers' experience, the paternal experience of perinatal loss is not well understood and is often disregarded and overlooked." Cholette concludes that different grieving needs of fathers and mothers should be supported so that the healing process can be skillfully facilitated in families (fathers, mothers, and others) who have lost a child. It is her belief that, "When these unexpected tragedies occur, perinatal healthcare providers, childbirth educators and doulas are in a unique position to facilitate health, healing and a natural transcendence through creating positive memories."

FETAL DEATH   Fetal deaths include abortion. The abortion may be an induced abortion whereby the pregnancy is intentionally ended. This aspect of abortion continues to be very controversial in the United States, although it has been legal since the U.S. Supreme Court decision of *Roe v. Wade* in 1973. Abortion may also be in the form of a spontaneous abortion or miscarriage. Such an abortion is not intentional and usually occurs during the early part of the pregnancy.

Induced Abortion.   Grief after an induced abortion, particularly for an adolescent, may be especially difficult, because many teenagers do not want to talk about the experience or their feelings, noted psychologist William Worden (1982). Since 1980, the number of abortions per 1,000 women aged 15 to 44 has declined about 20 percent, falling from 25 per 1,000 women to 20 per 1,000 women in 1995 (Vobejda, 1997). Since 1973, the Supreme Court has consistently reaffirmed the basic principles underlying *Roe v. Wade* and has refused to overturn that decision. Public opinion polls show that the majority of Americans believe that abortion should be legal (25 percent in all cases, others, legal with some restrictions), with less than 21 percent favoring making abortion completely illegal (Gallup Poll, 2014).

In an induced abortion, a choice is made regarding life and death. Though the decision was made to abort the fetus, the individual(s) involved probably have very mixed emotions about this action. If they do not feel that the fetus is a human being, then the abortion is not killing but merely a medical procedure. The woman's right to terminate an unwanted pregnancy and have an abortion can be viewed as a service to women. The individual(s) involved may occasionally second-guess the decision, wonder what the fetus would have developed into, and periodically note how old that child would be at a given time. For anti-abortionists, to intentionally destroy a fetus is murder. Thus, the voices of those opposing abortion may echo in the ears of a woman who has had an induced abortion.

If the induced abortion is related to economic issues or is performed because the woman is unmarried, this is a very different situation from that of abortions because of rape or abortions because of evidence that the fetus has a serious illness. In the former cases, feelings of grief may indeed be strong—the decision to abort may have been especially difficult, and feelings of guilt may occur. The grief process may depend on the individual's views about abortion in general. In the latter cases (rape or illness of the fetus) the decision to abort would probably be more clear-cut. Though still a very difficult decision for most, no matter the reason, an individual's grieving could vary with the reasons for the choice.

Spontaneous Abortion.   When a fetus does not reach full term and life is terminated through a miscarriage or spontaneous abortion, the loss for the parents is particularly difficult. They had anticipated the birth of their child, and this is now not going to happen. Perhaps, they had been preparing the child's room and were eagerly making other plans—all to no avail. The child will not be coming home. The parents had also been looking forward to bonding with their child and this also will not happen.

A miscarriage fits the description of disenfranchised grief. In the past, the survivors of miscarriage were viewed as "illegitimate mourners" (Nichols, 1984). Though today their right to grieve is more acknowledged, such grief is still not socially sanctioned universally as "legitimate mourning." Not knowing what to say to the parents, some individuals may coldly say, "Don't worry, you can always have another baby!" But the parents wanted *this* baby and had anticipated such. It is like telling people when their pet dies that they can always get another pet. True, but that is not a very consoling statement to make. The bereaved parents probably feel cheated when a spontaneous abortion occurs and perhaps guilty—"Was it something from my sperm contribution?" the father may ask, or the mother may question her carrying of the fertilized egg (**zygote**) and its becoming a fetus. "My fault, perhaps?" she may wonder.

Rachel Evans (2012, p. 35), a nurse doing research in the United Kingdom, claims that sensitive, caring, and skilled nursing care for women experiencing miscarriage plays a crucial role in their long-term emotional recovery. She concludes:

> For some women, miscarriage is a traumatic life event and may even be regarded as the most painful form of bereavement. However, miscarriage is often not viewed by society as a valid reason for bereavement. Researchers and health care providers, who focus primarily on the physical aspects of miscarriage, often overlook the emotional effects of miscarriage. Nurses who work in gynecology and early pregnancy units should endeavor to provide sensitive and supportive care, while managing their own emotions. Some nurses may cope well with care in these specialized units, while others may

| BIRTH AND DEATH

Every time an earth mother smiles over the birth of a child, a spirit mother weeps over the loss of a child.

Ashanti saying

become emotionally overwhelmed. The stressful nature of providing care for women experiencing miscarriage should therefore be validated and recognized by those in nursing management and education.

In the 21st century, a movement is under way to mourn pregnancy losses, especially miscarriage. It seems that stillbirths have always been mourned. The movement is driven in part by the growing number of older women who, having put off childbearing, find themselves facing great difficulty trying to conceive healthy babies and carry them to term (Fein, 1998).

PERINATAL DEATH As noted earlier, death occurring during the period closely surrounding the time of birth produces grief for the parent(s). There is grief for that which did not occur—a bonding with the newborn. Such grief results from stillbirths, premature births, and deaths during the first month of life not related to sudden infant death syndrome.

Stillbirth. The stillbirth experience is unique within the context of perinatal bereavement, as death is anticipated before birth, observes S. H. Hutchins (1986). After months of feeling movement, the mother's first sign may be unexpected stillness. The death of a baby *in utero* during labor or shortly before birth more nearly parallels a sudden death, with surviving parents experiencing shock and fear. Parents experiencing stillbirth or other perinatal death are usually young and inexperienced in dealing with death. Subsequent attempts to have a child can be extremely stressful for them.

The shock of the death experience has the family going from the anticipation of joy to sudden heartbreak. Such grief was not anticipated, because the pregnancy was seemingly going along smoothly, and the day of birth was fast approaching. An intensive period of grief will follow such an unanticipated outcome. Holding the baby for as long as the parents like can be very therapeutic. Giving the baby a name helps to establish the identity of the offspring and makes it a "person" with

| EPITAPH UPON A CHILD THAT DIED

Here she lies, a pretty bud,
Lately made of flesh and blood:
Who as soon fell fast asleep
As her little eyes did peep.

Give her strewings, but not stir
The earth that lightly covers her.

Source: R. Herrick, Epitaph upon a child that died. From Washington, P. (Ed.). (1998), *Poems of mourning*. New York: Alfred A. Knopf.

George Dickinson

With a stillbirth, the parents do not have the opportunity of bonding. That baby that they so anticipated suddenly is not going to be. Nonetheless, for a brief moment, they had their child, though born dead, and must grieve over its loss.

---

whom the family can better identify. Having a memorial service may assist the family in coping with their loss.

For the people of southeast Tanzania in Africa, miscarriages and stillbirths are considered types of disease (Wehbah-Rashid, 1996). Yet these events are causes for bereavement to the couple and their family. Although there is no elaborate mourning, the couple receives assistance and consolation from their relatives and friends. They are encouraged to forget the loss and look ahead with courage and enthusiasm because of the potential of having other babies.

Preemies.   Babies born before full term (nine months) are referred to as being born premature (preemies). Typically a baby with a birth weight of less than five pounds is considered to be premature. Such a birth may result in numerous health problems for the baby, such as organs not being fully developed. Because the baby is born alive, the family has an opportunity to bond with him or her. Thus, they name and identify with their child. A preemie typically stays in the hospital for several weeks or months while the body continues to develop. Thus, the parents leave the hospital without their child. This separation in itself is cause for grief. The parents return home to the baby's room, minus the baby.

With the preemie still in the hospital, the parents spend much time going to and from there, to visit and determine the medical progress or lack thereof. The child is probably taped and wired to various pieces of medical equipment to enhance development and to keep the baby alive. If the prognosis is not good, the parents may begin to experience anticipatory grief, whereby they begin grieving for the child before he or she dies. Their "letting go" process perhaps begins with the initial negative prognosis. For an extended discussion of anticipatory grieving, see Cheng, Lo, and Woo (2013).

The parents may be faced with the decision at some point to have various life-support equipment removed, if death is inevitable. Such a situation presents a very stressful dilemma for the parents. To decide about passive euthanasia is a decision that the parents had probably not previously faced. Thus, their grief over the upcoming death of their child is made even more difficult because of the decision to allow the child to die, if such a situation occurs.

When the baby dies, preparing for and attending the service for final disposition can help the family move on with their grief. The support of others can be comforting. Just to know that others care can indeed help one with the grieving process.

Neonatal Death.   The death of a newborn baby during the first month of life can be especially devastating to the parents. Bonding is occurring, life is going on as expected, and quickly all is not well. As with preemies, the parents may be faced with tough decisions regarding the baby's health.

George Dickinson recalls one situation when the prognosis was not good, and the newborn's condition was rapidly deteriorating. The decision was to let the baby starve to death, which is indeed the way many people die in Third World countries. Thus, the process of passive euthanasia was begun. Various hospice volunteer workers held the baby around the clock for several days until she finally died. Such an ordeal is exhausting both physically and mentally for the family, hospice workers, and volunteers. Many individuals grieve in such a situation.

For parents to make their adjustment to perinatal deaths, whether stillbirths, preemies, or neonatal deaths, they need support from those around them. In a study of 130 parents who had experienced a perinatal death, Judith Murray and Victor Callan (1988) discovered that a consistent predictor of better adjustment was the parents' level of satisfaction with the comfort and support provided by physicians, nurses, and other hospital staff. Parents who were pleased with the level of support that they received from medical personnel were also less depressed and had higher levels of self-esteem and psychological well-being. Furthermore, emotional support from the other parent, family, and hospital staff was linked to fewer grief reactions and better overall adjustment. Therefore, support from others can go a long way to facilitate parental bereavement at the death of a child.

Hospital-based intervention is especially necessary and helpful to parents who are experiencing a perinatal death, and medical personnel are in a unique

George Dickinson

Though he lived only a few weeks, David Crawley indeed is missed by his family. He was "loved so much" but "is waiting" for his mother "just over the hill."

position to assist parents in their grief (Davis, Stewart, & Harmon, 1988). However, although parents seem to desire medical personnel to acknowledge their feelings of shock, guilt, and grief, such desires do not seem to be fulfilled in many instances.

Part of the emotional support needed by parents involves encouraging them to accept the reality of death and to express their feelings of loss, and then validating these feelings. Communication and understanding by the medical staff are great sources of support and comfort. By being available, medical personnel can help to reduce the isolation that parents often feel at this difficult time. Because grief is a necessary process, whenever bereavement is facilitated, the parents become subject to a lower risk for psychological and physical disturbances.

Procedures following a Perinatal Death.   The stillbirth or the death of a newborn infant is stressful not only for parents and siblings but also for all people who are involved with the child. Increasingly, parents are being included in the direct care of critically ill children until the time of death and after. Familiarity with the types of procedures and decisions that parents may face at the death of their child is essential. Most bereaved parents will learn about hospital procedures and how to share in decision making only when they are faced with such a situation.

The nursing procedures are fairly clear, and nurses are instructed to respond in ways that can help in socioemotional adjustments. Nurses notify a nursing/social worker counselor about the infant's trauma at the time of admission. Baptism is even offered by some hospitals, where it may be done by anyone in the absence of a chaplain. The pastoral care department of a hospital is notified, as is the communications department, so that accurate information is available to others.

In the case of stillbirths and infant deaths, the parents may be given the option to see and hold their infant, to learn the infant's gender, and to decide on an autopsy and funeral arrangements. The medical staff explains what the parents

© Angela Farley/Shutterstock.com

Though the bereaved parents take into themselves aspects of the deceased, because a fetus or newborn has not lived long enough as a separate person, the parents have little of their baby to take into themselves. Thus, they suffer detachment without identification.

can expect the infant to look like. Believing that the age of siblings is the major factor in determining their involvement in these settings, E. Furman noted:

> Adolescents should decide for themselves. Elementary grade children are helped by attending a service but not helped by seeing a malformed dead body. Children under school age are particularly not helped by seeing their dead brother or sister, but they are sometimes helped by being in the company of the parents at the time of the funeral. The extent of individual involvement depends on the preference of individual family members. (1978, p. 217)

The nurses' responsibilities include attaching identification bands to wrist and ankle; measuring the weight, length, and head circumference; taking footprints and possibly a handprint; and completing standard forms. These forms might include a fetal death certificate, an authorization for autopsy, and a record of the death for the receptionist. Full front and back view photographs as well as close-up shots of any abnormalities may be taken by the medical photographer. These are used by physicians to describe the infant's medical condition. Nurses are also encouraged to take nonmedical photographs that include the infant unclothed in a blanket, a close-up of the face, and the parents holding the child if they so desire. These photos are given to the counselor and later to the parents.

The dead infant is wrapped in a blanket, labeled, and taken to the morgue. In the case of a fetus, the body is sent to the pathology department with a surgical pathology laboratory slip. The physician may request that the placenta be included in the case of spontaneous abortions or fetal deaths. Genetic studies may also be requested: samples of cord blood, placenta, fresh tissue such as gonadal tissue or connective tissue around the kidneys, and skin may be sent to the state laboratory for a complete genetic study. A nurse then completes a checklist for assisting parents who are experiencing perinatal deaths, and she or he may, if procedure calls for it, place an identifier by the name tag at the entrance to the mother's room.

Hospitals with a special program to help the survivors of perinatal deaths provide nursing, medical, social work, and/or pastoral counselors who may assist the surviving parents and siblings. Time is taken to explain and help in the following matters: (1) autopsies and hospital procedures after death; (2) funeral or cremation options; (3) the nature and expression of grief and mourning; (4) coping with the reactions of friends and relatives; (5) helping siblings to deal with the death of their brother or sister; and (6) decisions regarding another pregnancy. Monthly meetings of bereaved parents can be established to provide a setting for sharing and learning about grief. These experiences of sharing with other parents allow for reality-based comparisons and for active support of other parents whose loss is also great.

This list of hospital procedures is by no means complete, and the attitudes and responses of physicians, nurses, and others may vary greatly. At times, for example, the helpers are in need of socioemotional support, along with parents and siblings. Many hospitals, on the other hand, have not dealt with the special needs of families experiencing perinatal deaths. Although these practices are becoming more common around the country, one should not be surprised if a nurse or physician appears stunned at the request of parents to spend some time with the body of their dead child. We can only hope that the human values of medical care will prevail over bureaucratic values as the welfare of the whole person is taken into account in the medical arena. This can be accomplished most effectively if parents are provided with accurate information, encouraged to ask questions, given plenty of time to make decisions, and given opportunities to share their experience in parental bereavement with others.

A recent study by Amanda Costin and Carol McMurrich (2013) discusses strategies on how to support the parents of a baby that is stillborn. They conclude that parents who have had the physical experience of meeting the child who has died are helped when medical personnel try to understand the dreams the parents had for the baby, become mindful of words used in describing the baby, encourage parents to create artifacts that represent the child who was lost, and encourage parents to talk about feelings of guilt, isolation, and, in many cases, the feeling of a loss of innocence.

Finally, one might think that families would be greatly helped by a leave of absence from work. However, research by two French scholars, Melanie Gagnon and Catherine Beaudry (2013), discovered that while it might be important to take a leave of absence following the death of an unborn child or infant, it is insufficient in and of itself, because adaptation to the loss is not yet complete when returning to work. The factor that appears to be more crucial to a successful return to work is the social support provided by clinical practitioners, organizations, self-help groups, and friends.

SUDDEN INFANT DEATH SYNDROME   There has been a continuing decrease in the rate of infant mortality in the United States, and the number of deaths attributed to the mechanism known as sudden infant death syndrome (SIDS) has been decreasing as well. In the Western world, SIDS is the most common cause of death of infants between one month and one year of age and accounts for approximately 50 percent of infant deaths between two and four months of age (Klonoff-Cohen et al., 1995). Much of the decrease in SIDS is attributed to an emphasis by the American Pediatric Association and other Western world pediatric groups on Back to Sleep, a program encouraging that the baby be placed on its back or side to sleep, not on its stomach.

Sudden, unexpected infant death is a major cause of death for infants between the ages of one week and one year in the United States—there were some 2,063 deaths from SIDS in the United States in 2010 (Centers for Disease Control and Prevention, 2014). According to a study by the Foundation for the Study of Infant Deaths, these babies do not cry out as if in pain, but simply die quietly in their sleep, after becoming unconscious. Sudden infant death is usually defined as the sudden, unexplained death of an infant younger than one year where no cause is found through a postmortem examination (Willinger, 1995).

Historically, unexpected, unexplainable deaths of infants were routinely attributed to the mother's lying on them because mothers often slept with infants. If a mother awoke and found her baby dead, she assumed that she had lain on the child, smothering or crushing her or him. Perhaps the earliest such death recorded is in the Bible in 1 Kings 3:19: "And this woman's child died in the night, because she overlaid it." It is believed that SIDS was occurring long before it was recognized and accepted as a diagnostic label (Beckwith, 1978).

The majority of parents whose infants have died of SIDS see it as the most devastating crisis that they have ever experienced. These parents show the strongest reactions among parents who experience death of their infants (Boyle, Vance,

Najman, & Thearle, 1996). Approximately one-fourth of parents bereaved by SIDS move from their homes to other communities in an effort to escape the pain of the baby's death (DeFrain, Jakub, & Mendoza, 1992). A study of 34 pairs of parents bereaved by SIDS (Carroll & Schafer, 1994) revealed that these parents sought support from within the family most often and from outside resources least often. Parents generally tend to "recover" from the death in approximately three years or more. However, a Norwegian study (Dyregrov & Dyregrov, 1999) of parents bereaved by SIDS reported that more mothers than fathers claimed to be affected by their grief 12 to 15 years after the death.

Some parents "turn to God," whereas others "turn away from God" as a response to their loss (DeFrain, Taylor, & Ernst, 1982). Their reaction may be reflected in church attendance. From 1985 through 1988 a study (Thearle et al., 1995) in Australia examining the emotional health of parents after SIDS, neonatal death, or stillbirth concluded that the bereaved parents who attend church regularly have less anxiety and depression compared with irregular church attenders and non-church attenders. Similarly, in interviews with parents who had lost children to SIDS, researchers Wortman and Silver (1992) found that parents' religious devotion and participation in religious activities were positively related to coping. Among these parents an important feeling was that they would someday see their children in heaven.

| WORDS OF WISDOM | *ESCHERICHIA COLI* INFECTION IMPLICATED IN SUDDEN INFANT CRIB DEATHS |

Sudden infant death syndrome (SIDS), in which apparently healthy babies die inexplicably in their cribs, may be linked to infection with a common bacterium, preliminary research suggests.

Researchers told in a conference on infectious diseases Thursday that a shock-producing by-product of *E. coli* was found in the blood of all SIDS-tested babies, but in none of the infants used as a comparison.

Experts not connected with the research said the toxic infection theory is plausible.

SIDS describes unexpected deaths that autopsies can't explain. Despite decades of research, scientists remain mystified by crib death, the top killer of babies ages one month to one year in the industrialized world.

Among the threats it has been tied to are sleeping position, passive smoke exposure, and genetic vulnerability. Infection is not a new idea, but this is the first time the specific *E. coli* protein has been implicated. Many researchers favor a theory that brain-stem birth defects somehow affect arousal reflexes, so that babies don't wake up when breathing, heart rate, blood pressure, or temperature problems arise.

However, some experts believe that such brain abnormalities may not be enough to cause death on their own. "Mainstream researchers have concentrated on respiratory obstruction as a possible mechanism, without any evidence that would support such a mode of death," said Dr. Paul Goldwater, who presented his study at the European Congress of Clinical Microbiology and Infectious Diseases in Milan.

"Those researchers ignored autopsy findings that consistently show wet, heavy lungs in SIDS babies. This is never seen in cases of suffocation," said Goldwater, a researcher at the Women's and Children's Hospital in North Adelaide, Australia.

"Such a lung condition is often seen in cases of infection. Autopsies also consistently show small hemorrhages on the heart and lungs, which is rare in suffocation, and the blood of SIDS babies is unclotted, which is something never seen in suffocation cases," he said.

"Also, SIDS deaths captured on medical monitors have shown that these babies died of a shock-like process," Goldwater said.

Source: Adapted from The Associated Press, 2005.

When an infant dies suddenly and unexpectedly, the sense of loss and grief may be overwhelming. When the sudden death is due to a known cause, the concrete character of the event can be incorporated into the normal rationalization of mourning. However, when death is due to an unknown mechanism, as in SIDS, feelings of inadequacy in caring for the child are reinforced for the medical staff and the parents (Mandell, McClain, & Reece, 1987).

Because there are still many unanswered questions about SIDS, parents have a tremendous guilt feeling and shoulder the responsibility for the infant's death. Marital conflict, difficulties with surviving children, and anxiety about future children becoming victims of SIDS are often experienced. With the cause of death in a SIDS case being questionable, the parents are likely to undergo a police interrogation, in addition to the stress of having lost their infant. In an era of child abuse, the parents may be suspected of smothering the child. Being questioned by the police is difficult enough, but the perceived inability of others, such as friends and relatives, to understand the depth of their despair is also very frustrating to these parents (DeFrain et al., 1992).

The Centers for Disease Control and Prevention ("Guidelines to Help Discern SIDS from Homicide," 1996) issued investigation guidelines to help coroners and police distinguish between SIDS and homicide in infants. The guidelines suggest noting aspects such as the position of the infant's body, any suspected injuries, and any evidence of drug use in the home. The death-scene investigation, the child's medical history, and an autopsy are necessary for a thorough investigation.

Young parents whose baby dies of SIDS have probably not experienced the death of a close relative; thus, they are not familiar with the social and emotional aspects of grief. The death of one's baby is traumatic under any conditions, but the sense of not knowing how to mourn adds to the difficulties of socially adjusting to the loss. The parents and siblings experience **anomic grief**—a grief without the traditional supports of family, church, and community.

A survey of newly trained SIDS counselors in North Carolina (Kotch & Cohen, 1985) reported that sharing the autopsy report with bereaved parents was a valuable part of the counseling process and removed some of the mystery surrounding the diagnosis of SIDS. The autopsy report, by documenting that the child died a natural death, may relieve some families of the feeling that they were somehow responsible for the death.

George Dickinson's own discussions with parents who have lost a child through SIDS suggest that friends may turn on them as if they are criminals. Parents tend to blame each other for the death—"If only you had ...." Parents become victims because SIDS is both personally traumatic and complicated by problems of social interaction. The uncertainty of the cause of death is frustrating to the parents and medical staff and clouds the whole issue from a societal perspective. A death due to SIDS must be one of the most traumatic experiences that parents can have in a lifetime.

The beginning of healing for parents of an infant who has died of SIDS is the closure of the relationship with the infant, advised Melodie Olson (1997). After the child has been pronounced dead, it is important for the parents to hold the child, perhaps several times between the time of death and the funeral. Parents' privacy needs to be respected. They should be allowed to take something from the child, such as a lock of hair or the baby blanket.

WORDS OF WISDOM	OH, DANNY BOY

... for you must go and I must die.
But come ye back when summer's in the meadow,
or when the valley's hushed and white with snow.

For I'll be here in sunshine and in shadow.
Oh Danny boy ... I love you so.

Irish folksong

THE LOSS OF A CHILD

C. M. Sanders (1980) noted that the death of a child is one of the most grievous of losses, significantly greater, on average, than that of a parent. Based on interviews with 155 families suffering the loss of a child ranging in age from one to 28 years, Ronald Knapp (1986) observed that the death of a child represents in a symbolic way the death of the self. Symbolically, a parent will die along with the child, only to survive in a damaged state with little or no desire to live today or to plan for tomorrow. In addition to losing part of themselves, parents lose some of their hope for the future, concluded Reiko Schwab (1992) from a study of 20 couples who had lost children.

Because children are "not supposed to die," especially before their parents do, the death of a child seems more tragic and traumatic than the death of an older person. We often take our children for granted. Although one can *imagine* the loss of a child, how often does one have such thoughts? Even so, such thoughts bear little resemblance to reality. For parents who have lost a child through death, however, the reality of the situation lingers forever. A friend whose son had died told George Dickinson that the stark reality hit him in the face every morning when he woke up and realized that this was not just a bad dream—his son was really dead.

Knapp (1986) noted that parents relive the child's death over and over in their imaginations as the end of the child's life is near. This imaginary scenario takes them from the moment of death through the funeral of the child. Parents may wish to keep the child at home so that he or she will not die in strange surroundings. When parents are able to make the decision to terminate all further treatment and let a disease take its course, a sense of tranquility results, and parents are ready to release their hold. Sometimes parents must take on the painfully hard task of telling the child that it is all right to let go. It is all right to stop fighting. It is all right to die! Giving permission to die is difficult. Sometimes it takes only gentle encouragement from parents—gentle persuasion that all has been done and that nothing more remains.

As mentioned earlier, available literature (Schwab, 1992), based on research and clinical observations, indicates that the death of a child strains the parents' marital relationship, sometimes resulting in separation and divorce. One study (Klass, 1987) concluded, however, that marriages did not end because of a child's death. Instead, parents in rocky marriages decided that their struggle with marital problems was no longer worth the effort. Nonetheless, even strong marriages often suffer after the death of a child.

DEATH ACROSS CULTURES | Coping with Death in the Philippines

The Philippines are some 600 miles southeast of the coast of mainland Asia. The people in the Philippines are predominately of Malay stock, and the religion of the majority is Roman Catholic. Death is viewed as a destiny or fate and part of the cycle of beginning and end—a natural occurrence. Dying of old age is seen as a positive occurrence. Death is viewed as more than sorrow and loss—happiness is also a part of death. Death is seen as the last cure for the incurably ill.

Though Filipinos are known culturally as shy people, they very openly discuss death. They believe that the more one shares of the situation, the easier the grieving will be. The socioeconomic background of the family may dictate how open they are when death occurs. Rich families, for example, are usually rather private, whereas those of lesser means will depend more on the community for support.

The death of a child is the most difficult death to face. Severe guilt is experienced by the parents. Children are perceived as "angels," innocent and "without sin"; thus they go directly to heaven after they die. They are expected to wait and meet their parents in heaven when their turn comes.

From "The Filipino perspective on death and dying" (pp. 215–220), by I. Weber, 1995. In J. K. Parry & A. S. Ryan (Eds.), *A cross-cultural look at death, dying, and religion*, Chicago, IL: Nelson-Hall Publishers.

Sometimes, the difficulties in the loss of a child are the result of gender differences in coping with grief, as noted earlier in a study by L. E. Oliver (1999). In another study composed of 145 bereaved parents, J. A. Cook (1983) reported that fathers find it difficult to grieve openly and thus keep their grief to themselves. Mothers, on the other hand, are more comfortable grieving openly and want to talk about their loss. They find their husbands' lack of expression of grief a barrier to communication. Cook (1988) noted that men are given little comfort and support in bereavement and are expected to be strong and to provide a source of support for others, yet their nonexpressiveness comes into conflict with their wives' needs for expression. Schwab (1992) concluded in a study of 20 couples who had lost children that husbands and wives appeared generally irritable and less tolerant of their spouse. On the other hand, Schwab noted that husbands and wives with a good marital relationship prior to a child's death appear to have come closer together through the tragedy that shattered their lives. Thus, a weak marriage may collapse with the death of a child, and a strong marriage may indeed be strengthened by the death. Whatever the result, the strain of the death of a child is certainly a test of the marital relationship.

THE LOSS OF AN ADULT CHILD

A description of the bereavement of parents who lose an adult child is provided by British anthropologist Goeffrey Gorer (1965), who noted that it is literally true that parents never get over it. As noted earlier, because it is "against the order of nature" for a child to die before her or his parents, the parents seem to interpret this as a "punishment for their own shortcomings," observed Gorer. Their self-image seems to be destroyed. For them, a reliance on the "orderliness of the universe" has been undermined.

PRACTICAL MATTERS | SUGGESTIONS FOR SURVIVORS OF A DEATH IN THE FAMILY

1. Don't judge the way people grieve—those who do not cry can be just as devastated as those who cannot stop crying.
2. Don't assume the death was for the best—"He was my dad, even if he was old."
3. Don't assume that because there are other children, the pain is any less—an amputated arm is still missed.

4. Don't say, "I know how you feel"—You are not that person.
5. Don't say, "Don't worry, you'll get married again" or "You'll have another baby" or "It's God's will."

From *Don't ask for the dead man's golf clubs: Advice for friends when someone dies*, by L. Kelly, 1998, New York: Kelly Communications.

Research reported by Therese Rando (1993) concluded that the death of an adult child is difficult for both fathers and mothers in that they see the adult child's responsibilities left unattended with death—fatherless or motherless children and unfinished tasks in life. The children probably feel shortchanged in that they will grow up without that parent. The surviving parents of the deceased adult child note this void in the lives of their grandchildren, observed Rando. They may see life as seeming unfair to their grandchildren.

The loss of an adult child leaves the parents with unfulfilled dreams for their offspring. Life may seem incomplete to them in that the chain in their generational lines has been broken—a loss of continuity in the life cycle occurs.

Michael R. Leming

"No farewell words were spoken, No time to say goodbye, you were gone before we knew it, and only God knows why."

GRIEVING CHILDREN AND ADOLESCENTS

A developmental approach to grieving the loss of a parent, from very young children through adolescence, reveals different manifestations of grief, concluded Grace Christ (2000). Children aged from three to five years tend to sleep with the surviving parent, suck their thumb, wet the bed, display clinging behavior, have bad dreams, and display various physical symptoms such as a stomach ache. Children aged six to eight years talk freely about the deceased parent, feel the parent's presence, and talk to the parent. Older school-aged children from nine to 11 years seem to be overwhelmed by their grief and do not like to talk about the deceased individual. Slightly older children, aged 12 to 14 years, avoid feelings and information about the illness and grieve alone. In addition, they grieve for the loss of the deceased's specific characteristics and special functions in the family. The grief of adolescents between the ages of 15 and 17 differs markedly from that of their younger peers in that they can become overwhelmed by their grief and not be able to control it.

Storytelling is a good way to keep memories alive for children and adolescents (Harvey, 1996). Writing down one's thoughts about the deceased person can be very insightful. Stories can both teach and heal. In telling stories about the deceased, even years later, one can help the young child—now grown—to better know her or his deceased parent. A close friend of George Dickinson's shared how he told the grown children of a mutual boyhood friend stories about their

Being a good observer and listener may help us to be a better caregiver of bereaved children.

now-deceased dad when he was growing up. This helped the children, whose dad committed suicide when they were small children, to better know what a great guy their dad had been.

Drawing is another way for children to express themselves. Pictures are a bridge to expression (Cox, 1998). The child can draw a picture, then explain what is meant by the picture. Drawing is an excellent way to bring a child's feelings out, especially those of a very young child whose verbal gifts may not be well developed.

In addition, music is a form of expression that can be used without oral fluency. Music is a therapeutic tool that can be used to impact on one's mood in a positive way. Music can also be used to promote spiritual development (Lowis & Hughes, 1997).

Grieving children and adolescents need to know that it is okay to laugh during a period of bereavement. George Dickinson, as noted earlier, has had college students tell him that when they were young and a grandparent had died it bothered them that people gathered at their home could be laughing "at a time like this." Humor, however, can be very healing to a grieving person. David Spiegel (1998) found that those grieving and dying lived an average of 18 months longer if they were happy—happiness is important to good health. Laughter gives the child and the adult something else to think about, observed Gerry Cox (1998). Given a reason to laugh, the individuals involved can forget their pain for a few minutes. Laughter promotes confidence and hope. Let us now take a closer look at grieving children in the loss of a parent, sibling, and grandparent.

LOSS OF A PARENT

How do children's grief reactions differ from those of other individuals? asked John Baker and Mary Sedney (1996). Children's grief reactions appear to last longer than those of adults. Their initial reactions appear to be less intense. Children have different ways of coping than adults—they are more likely to distract themselves and use fantasy to cope with the suddenness of the loss. They will tend to identify with the parent who has died, whereas adults are more likely to separate themselves from the characteristics of the person who died.

Parent death during these early years of attachment and dependency threatens the child deeply (Hatter, 1996). Basic survival needs may be met by another, but the unique qualities of the parental bond can never be replaced. The more positive and frequent the contact has been with the deceased parent, the more acutely a young child will be aware of the parent's absence (Norris-Shortle, Young, & Williams, 1993). Long-term implications of the death of a parent arise because children must grow up with the loss.

According to their self-report, 57 percent of children who lost a parent (Silverman, Nickman, & Worden, 1992) "spoke" to the deceased parent in some way, 55 percent dreamed of the parent who had died, and 81 percent felt the parent was watching them. Such an attachment to thoughts and memories of a dead parent may be a sign of positive adaptation rather than a sign of a pathological problem, noted Silverman and colleagues.

Preschool children up to five years of age most often show manifestations of anxiety or of aggressive behavior when a parent dies (Baker & Sedney, 1996).

"You couldn't put on a tie?"

School-aged children between six and 10 years of age may appear to deny that the death has occurred and strive to maintain an appearance of emotional control.

Adolescents show some trends similar to those in younger children, although their emotional reactions are often kept private at this age in an effort to appear "normal." In addition, adolescents are more likely than younger children either to become depressed or to try to escape from their emotions through acting-out behavior (e.g., running away, excessive drug use, and risk taking). Adolescents may also be preoccupied with the "unfairness" of death, noted Baker and Sedney.

LISTENING TO THE VOICES | DEATH OF A PARENT: EXPRESSION OF GRIEF BY AN ADULT CHILD

After being with his mother during her time of death, within six months, Don Bezanson—Michael Leming's colleague in the Library at St. Olaf College—also had to attend to the duties and sorrows related to the death of his father.

It is clear that Don is sad but at peace. It is also obvious where he got his eccentric sense of humor when reading the obituary he wrote for his parents. The obituary below follows his e-mail message to his colleagues at St. Olaf:

Juli and I will be attending my parents' burial at Ft. Snelling this afternoon. My mother, age 82, passed away July 7; my father, age 85, died October 4. I was their caretaker for 10 years; they spent the last two years at Three Links Nursing Facility.

The last eight years of increasing parental care were extraordinary, both in rewards and in demands. There were numerous times when people said, "I don't know how you do it"

(I knew all too well how much I could have done, but didn't). The answer? I would have burned out long ago without the support and concern of our library family.

Don

Stanley B. Bezanson
B. Aug. 16, 1919. Died Oct. 4, 2002. Preceded by his wife, Evelyn (July 7, 2002). Internment at Ft. Snelling.

There is no memorial service, but feel free to tell your boss you will be attending one. Say it's a Namnori funeral and takes all day. (Namnori is Ironman backwards.) Grab your camera and go for a long bike ride or walk. Clear your head. Take pictures of anything that catches your eye, especially weird mailboxes. When you get home, write a real, stamped letter recounting your thoughts and adventures to someone you care about. As long as they write back, enjoy corresponding for the rest of your life.

LOSS OF A SIBLING

Through the death of a sibling, bereaved brothers and sisters learn at an early age difficult lessons about the preciousness of human life, the importance of close personal relationships, the vulnerability of such relationships, the multiple impacts of loss on themselves and other family members, and the significance of the legacy left to them by the sibling who died (Stalman, 1996). Thus, the death of a sibling during childhood has immediate as well as long-term consequences for surviving siblings. In today's smaller modern family the surviving sibling may become the "only child" in the family.

In a study of 65 children (between the ages of four and 16) who were the siblings of deceased children, D. E. McCown and C. Pratt (1985) confirmed previous studies indicating that 30 to 50 percent of surviving children demonstrate increased behavior problems after the death of a sibling. Their studies showed that children in the middle age group, six to 11 years, developed more behavior problems than those in other age groups. Reasons cited for more problems in this age group are that the loss of a sibling at this phase may lead to feelings of vulnerability and inferiority and that for the child in this age group who is making the transition to concrete thought, the event and cause of sibling death may evoke confusion. The increased behavior problems may be a reflection of that confusion.

David Adams and Eleanor Deveau (1987) observed that, after a child's death, many siblings fear minor physical symptoms and worry about death occurring at the same age. Siblings resent parents for their preoccupation with the dead child and blame them for their inability to protect these siblings during illness. Parents are often so consumed by their own grief that they have little energy left to help surviving children. Problems also develop when parents expect surviving siblings to surpass or equal the achievement of a deceased child or to replace the deceased child.

Because of limited resources and energies in the home, peer support and special attention to the siblings of dying children are needed. Teachers, in particular, need to be alert to the special concerns and needs of these children. The personal worth of surviving siblings needs to be reinforced. As with siblings of a newborn coming into the family, siblings of a dying child have special needs. They, too, wish to be noticed and given some love and care.

LOSS OF A GRANDPARENT As noted in Chapter 3, many times a child's first experience with death is that of a grandparent. Grandparents often are important figures in the child's life, giving unconditional love and caregiving. Often a grandchild can do no wrong in the eyes of the grandparents. The positive reinforcement given to children by grandparents will no longer be a source of feedback for the bereaved grandchild.

On the other hand, in our mobile society grandchildren may not have bonded with their grandparents, owing to their living in a different geographical place; thus, the loss of a grandparent may not be so traumatic. Seeing their own parents upset and perhaps crying will probably bother the children, but their own personal loss may not seem severe. The child may feel that he or she is expected to grieve, but in fact may not feel like grieving.

Siblings of seriously ill children need attention also. Often, the siblings may feel neglected.

If this is the child's first experience with death, she or he may ask a lot of questions regarding death, because the event is new. Parents need to be receptive to questions to help the child cope with death. Indeed, for many children the death of a grandparent may be very traumatic or it may have a limited impact, depending on their degree of relationship with the deceased.

LISTENING TO THE VOICES | GOODBYE FOR NOW, GOOD FRIEND

"As you know, Gary's been really sick," I told the kids.
"Is he going to die?" Lola asked.
"Yes," I replied.
"Until he passes," the vet explained, before administering the first shot, a sedative, before Gary's lethal injection, "Gary's going to know you're here. He can hear you."
Jamison knelt on the floor, stroking Gary's face and body. Jamison spoke softly, repeating again and again what a good dog he is.
"I'm sorry for all those times I yelled at you," I said, "We love you, Gary""

"Bye, Gary!" Lola said. Noticing that Gary didn't look up like he normally does, Lola waved at him.
"Bye bye!" Solomon repeated, and waved too. You carved a place in our hearts, Gary, we will miss you. Goodbye for now, good friend.

Gary Lyngstad
2005–2014

Source: Written on June 12, 2014 by Caroline Lyngstad (St. Olaf College, Class of 1999), former student of Michael Leming.

GRIEVING ADULTS

For grieving adults, death is a reminder of our own mortality. The death of another adult highlights the fragility of life. As with any death, a void is left in one's life, especially if that person had been a part of the life *forever*, such as a parent, or for much of our adult life, such as a spouse or even a pet.

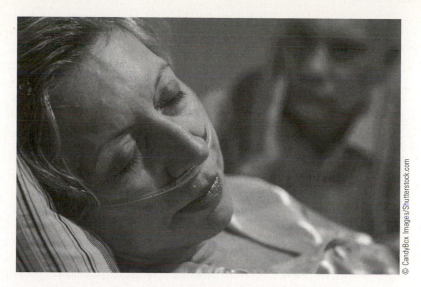

Even when expected, the death of a significant other always involves difficult adjustment patterns of behavior.

LOSS OF A SPOUSE

The death of a spouse, not unlike the death of a parent or child, is always unique and seems devastating. Two individuals have chosen to live together, and now that bond is broken. Life will be different, usually significantly so. Melodie Olson (1997) noted factors associated with the death of a spouse that may contribute to an especially prolonged adjustment: unexpected death of the spouse, high dependency on the deceased, and poor health of the surviving spouse before the death.

The dependency established in a marriage will leave voids in life for the surviving spouse. The division of labor found in a familial relationship will suffer when one member is taken away. Spouses are often best friends. Even in marriages far from perfect, important bonds are formed.

According to research reported by Catherine Sanders (1999), various situational variables affect bereavement in the loss of a spouse. If the surviving spouse has a good network of friends and family, such a support system helps tremendously in "making it," especially during the days immediately after the death. Eating alone can be a painful experience for many bereaved spouses. In fact, loneliness is a major problem of bereaved spouses. Sensing the presence of the deceased partner remains with the surviving spouse. Religion is a mainstay, providing hope to many when discouragement is almost overbearing.

LOSS OF A PARENT

Only 10 percent of adults have lost a parent by age 25, but by age 54, 50 percent of adults have lost both parents, and by age 72, 75 percent have lost both parents (Umberson, 2003). On average, about 13 years separates the death of one parent from the other. Most commonly, the death of a father precedes the death of a mother.

A mother's death is most likely to occur when her adult children are between the ages of 54 and 64 and a father's death is most likely to occur when his children are between the ages of 35 and 54.

The younger the adult child, the stronger the attachment to parents, because childhood memories of dependency are fresher in the minds of young adults than in those of older, more mature adults, observed Sanders (1999). In addition, the younger adult child probably has fewer significant others to draw from for support during grief.

The death of a parent can have a profound effect on the ways that adult siblings deal with each other, noted Lawrence Kutner (1990). Adult siblings may experience intense emotions as they reevaluate the meaning of family and their roles within it. The death of a parent, especially the second parent, often accentuates a pattern already existing—as siblings grow older, good relationships become better, and rotten relationships become worse. The death of a second parent often changes the focal point of family rituals such as organizing Thanksgiving or other significant holiday celebrations, observed Kutner. While the parents are alive, the relationships between siblings tend to be balanced compared with the unbalanced parent–child relationship; yet when parents die, one sibling tends to break out of this balanced relationship (Kutner, 1990).

Joan Douglas (1991), in studying reactions of middle-aged persons to the death of their parents, concluded that for many the integration of the loss of a parent involves confronting the loss of parental power and the reality of one's own mortality as part of the developmental process. The tension created by the need to both sever and maintain the parent–child bond and to face one's own death without giving way to despair forces many to move to a new perspective—a new level of integration.

For an extensive review of 100 different therapeutic and psychological techniques for counseling the grieving and bereaved, relative to all types of relationships, see Jones (2013) and Neimeyer (2012).

LOSS OF A PET

For many individuals, a pet is a significant member of the family. Examining the death of pets clarifies sociological issues, such as the process by which human traits are attributed to animals. We care for pets and talk to them *"as if"* they were members of the family. (Remember, W. I. Thomas taught us that if you define something as real, it becomes real in its consequences.) Pets often live with a family as many years as children live at home before leaving for college or emancipation. Pets can also make people feel needed, can relieve loneliness, and can serve as friends and companions. Indeed, are they not family members, "fictive kin" perhaps, yet part of the family? The death of a pet, therefore, is a traumatic experience for family members. As would occur for any other member of the family, the resulting death leaves a big void.

For non-pet lovers, such a death may not seem so significant, and they may find themselves somewhat cynical about another's outpouring of grief over a companion animal death. Such a reaction may contribute to disenfranchised grief, as discussed in Chapter 13. For this reason, John Homans (2012) notes in *What's a Dog For?* that to some non-pet owners caring for an animal at the end of its life and grieving after it is gone is in some ways more complicated than grieving for a person, as the question of what a companion animal *is* is far from settled—an honorary human or just an animal chattel.

Like other medical professionals relating to issues of dying and death, veterinarians have concerns about discussing euthanasia and giving bad news to pet owners. If the animal has a late-stage chronic illness, the dilemma for the pet owner may not just be "if" but "when" to terminate the life of the ailing animal. To take such action the owner relies heavily on the judgment of the veterinarian to give advice regarding the "right time" to euthanize—not too early, yet not too late. On average, veterinarians in the United States tend to euthanize around eight animals per month, according to one study (Dickinson, Roof & Roof, 2011). The veterinarian has assumed the "privilege" of helping decide on the life and death of animals (Frid & Perea, 2007).

The cost to euthanize a pet is between $50 and $100 in a veterinary office and $295 to $400 with at-home euthanasia (Costhelper, 2013). The cost following cremation (e.g., earth burial or cremation) ranges from $50 to $350 depending on the weight/size, or if it's a cat or dog. Few pets end up in a cemetery, yet the mere existence of a pet cemetery indicates a deep emotional connection to companion animals. Today, there are some 370 pet cemeteries and crematories in the United States, with earth burials costing $500 to $730, depending on the size of the animal, more expensive than having the veterinarian dispose of the animal (Pet Guardian, 2013).

When having a pet "put to sleep" is being contemplated, Herbert Nieburg and Arlene Fischer (1982) suggested asking the owner whether the pet can do the things that it once enjoyed, whether there is more pain or more pleasure in its life, whether the pet has become bad-tempered and snappish as a result of old age or illness, whether it has lost control of its bodily functions, and whether one can afford the expense and time involved in keeping a sick pet alive. Whatever the final decision, this is not an easy choice for an individual to make.

If the pet is euthanized, guilt feelings often follow, as 50 percent of pet owners feel such, according to one study (Fernandez-Mehler & colleagues, 2013), despite the consoling and suggestions of the veterinarian. Did I do the right thing? Should I have waited longer? Who am I to decide to have my pet (best friend?) "killed"? Am I playing God? An alternative to euthanasia is animal hospice, a relatively new offering in the United States. Such a service encourages an individual not to end the life abruptly of the animal, if the dying process can be achieved without intense or protracted suffering (Pierce, 2012). Hospice will promote attention to pain and palliation. As a good death can take many forms, hospice presents an option to help the pet die with relatively little pain.

Discussion of pet bereavement enables individuals to share their feelings of grief, fear, and loss. A. D. Weisman's work (1990–1991) with bereaved pet owners suggested that there were many similarities between grieving for a pet and grieving for a human—preoccupation was common; people reported mistaking shadows and sounds for their dead pet; guilt and ambivalence were experienced; and there were also corresponding feelings of loneliness and emptiness in the grieving process. According to Millie Cordaro (2012), counselors who acknowledge and validate the implications of pet loss will help to reenfranchise an undervalued grief for those who have experienced the loss of a companion animal.

PRACTICAL MATTERS ⎮ Veterinary Schools, Veterinarians, and End-of-Life Issues

The role of veterinarians regarding the death of animals is different from that of medical doctors and their dying and deceased patients, yet humans appear to have an emotional bond to their companion animals that is not unlike what is experienced with family and friends. Grieving for a pet and for a human have many similarities: guilt and ambivalence, preoccupation, mistaken shadows and sounds for the dead pet, and feelings of loneliness and emptiness.

Veterinarians, different from medical doctors, have the problem of helping the owner make the decision to end an animal's life through euthanasia. Whatever the final choice, this is not an easy decision for an individual to make. When euthanasia is decided upon, then the difficulty of this decision by the owner is often followed by disenfranchised grief, whereby the grief may not be openly acknowledged or socially sanctioned. Often the guilt felt by the pet owner at having the pet "put to sleep" involves the power of life and death held over the beloved companion animal. When the decision of euthanasia for a pet is pondered from the veterinarian's perspective, the most legitimate reasons for euthanizing revolves around the animal's quality of life. However looked upon, the whole scene is one of unpleasantness.

A survey of the 28 veterinary medicine schools in the United States in 2007 regarding end-of-life issues revealed that an average of 15 hours within the curriculum is devoted to dying and death, compared to 14 hours on this topic in U.S. baccalaureate nursing schools and 12 hours in U.S. medical schools. U.S. veterinary, medical, and nursing schools, therefore, devote about the same about of time in the curriculum to end-of-life issues. Topics most frequently covered in the veterinary curriculum were "communication with owners of dying animals" and "euthanasia," followed by "bereavement, analgesics for chronic pain, and attitudes toward death and dying."

As only the states of Oregon, Washington, Vermont and Montana allow a medical doctor to legally assist in a patient's death (physician-assisted suicide) and no state allows active voluntary euthanasia as veterinarians can do, the ethical issues are more complex for veterinarians than in human medicine. As opposed to human medicine, the cost of potential treatments is a primary consideration in veterinary decisions, with the euthanization of the animal as a viable final option if the client determines that the treatment expense outweighs the emotional and medical consequences.

When veterinarians were surveyed in 2007 in South Carolina (481 veterinarians with a 75 percent response rate), however, the majority did not think that their veterinary school training prepared them to relate to owners of terminally ill animals, though more recent graduates were more favorable than earlier graduates. Like other professionals dealing with end-of-life issues, veterinarians are concerned about discussing euthanasia and giving bad news to pet owners. These South Carolina veterinarians euthanized 7.53 animals per month on average. A striking feature of this demanding aspect of a veterinarian's job is the repetitive nature of the event. The issue of veterinarians' desire to help and bring comfort, when also confronted with the moral and emotional dilemma of euthanizing animals, puts them in a precarious position. Two-thirds of the veterinarians said that the owner stays during the euthanization procedure, with the majority (63 percent) preferring cremation (individual or group cremation) over earth burial. Of those pet owners who either leave prior to euthanization or wait outside the room, the majority are more likely to take the animal's body with them for final disposition (most often earth burial).

Pet hospice services are emerging in the 21st century, with some of these clinics focusing on teaching pet owners how to care for their terminally ill pets at home and others being handled at a pet facility. To help educate veterinarians about hospice care, the Nikki Hospice Foundation hosted the first international symposium on veterinary hospice care in 2007. For owners of dying/deceased pets, support groups and hotlines are available for grief support.

Source: Taken from G. E. Dickinson, P. D. Roof, and K. W. Roof (2010). End-of-life issues in US veterinary medicine schools. *Society and Animals*, 18(2), 152–162; G. E. Dickinson, P. D. Roof, and K. W. Roof (2011). A survey of veterinarians in the U.S.: Euthanasia and other end-of-life issues. *Anthrozoos*, 24(2), 167–174.

"Our Family"

DYING, DEATH, AND BEREAVEMENT IN THE 21ST CENTURY: A CHALLENGE

As we are now into the 21st century and living in an era when we grow to "very ripe old ages," the complexion of grief may take on a slightly different tone. Baby Boomers are reaching retirement, yet many still have one or two parents for whom they are caring. At the same time, Baby Boomers may fall into the "sandwich generation" and are taking care of their own young-adult offspring. In addition, with diseases such as Alzheimer's on the rise, these aged parents may be especially difficult to care for, complicating matters even more for the Baby Boomers.

These individuals in the "middle years" are being courted by funeral homes with the message that they are willing to provide whatever funeral a client might desire. Traditional funerals are not appealing to many today; thus, funeral homes seem receptive to change. The use of cremation is rapidly rising in the United States as land space becomes an issue in some areas and the whole idea of preservation of the body seems bizarre to many.

The "resurrection of death" period of bereavement, though beginning in 1945 with the dropping of the atomic bombs in Japan, seems to be alive and well today. Terrorist attacks on the United States, Spain, and England in the 21st century have been all-too-vivid reminders of the "resurrection of death" period. As Tony Walter (1999a) noted, the signs of death are becoming more visible with more public forms of mourning coming back into fashion—wearing a ribbon to support those with AIDS or having a moment of silence at the precise minute, one year or five years later, to recall an earlier disaster. The topic of death and grief is frequently seen in the media. End-of-life issues are addressed on television, in newspapers, in popular magazines, in bestselling books, on the radio, and on the Internet, as was evident by the Terri Schiavo case in Florida in 2004 and 2005, which involved intervention by all three branches of the federal government. With better mass communications today, a disaster anywhere in the world can be reported around the world practically as it is happening. We can now see wars being fought live on television from the comfort of our homes. Research money is available for topics related to dying and grief, rarely heard of until quite recently. College courses on dying, death, and bereavement are now almost ubiquitous on college campuses, yet were unknown in the 1960s. Euthanasia issues in medicine, many unknown a few years ago, are now daily news items. Individuals are more knowledgeable about end-of-life issues and thus are more demanding about their rights, whether regarding euthanasia or funeral practices.

The overwhelming focus in traditional medicine in the United States has been to prolong life. Yet, in the 21st century, palliative care is slowly creeping into American medicine. The focus previously has been on *curing* diseases, whereas now *caring* for those with diseases is gaining momentum. As bestselling author, Sherwin Nuland of Yale Medical School observed (Carlson, 2000), "Doctors today are filled with a kind of arrogance about their ability to defeat death." He noted that young physicians feel that if you cannot cure, you have failed. Nuland advocated a more humanistic bent to medicine. If the patient cannot be cured, then the doctor should try to make the patient comfortable—palliative care. Indeed, Jack Kevorkian alerted the medical world to the need for patients to have more care in controlling pain and in symptom control in general. The palliative care trend is there and probably will continue. Such an emphasis should enhance dying for the individual.

With chronic illnesses being prevalent in the United States today, rather than infectious diseases as in the past, an individual may simply live on and on—living with dying. Is this limited quality of life that a very elderly person often has really worth the thousands of dollars per month that is being paid to keep her or him alive in a nursing home? Will the soaring costs of drugs subside in the 21st century? How long will the family savings last to pay the bills? Will the aging parent outlive the coverage for living expenses during the last few years of life?

When family money is gone, will the government have enough money available from Medicaid and Medicare, and be willing to spend it, to cover the expenses? This issue of government responsibility for medical care costs for elderly persons is already a political football. As the population ages into the 21st century, will the problem only get worse before it gets better?

As a society, in an age when new technologies can miraculously prolong life, we duck questions about how to spend our health dollars on those for whom death is a close inevitability, said John Lantos (2000). As individuals, we also need to think about how we want to live as we are dying. Perhaps Daniel Callahan (1987) was correct in his book *Setting Limits: Medical Goals in an Aging Society* regarding governmental responsibility for the elderly—maybe we should "draw the line" regarding life-prolonging decisions with elderly persons. But then when "the elderly 'is' us," we might have a change of opinion! It seems that death can indeed be a relief in many situations. Should an individual have the right to say when and where she or he wishes to die? Given all that has happened in recent years, dying, death, and bereavement in the 21st century should be a challenge. Let's hope we are prepared.

CONCLUSION

No matter where in the life cycle a death occurs, it is not easy for the survivors to make the adjustment. The realization that the deceased individual will never again be in our presence is a shock that comes to us all throughout the life cycle. Such a loss is something that we do not simply "get over," but we learn to live with the fact that the individual is gone, never to return. Life will not be the same again because of the void left in our lives by the death of that significant other. Elisabeth Kübler-Ross stated in the title of one of her books (though referring to the one dying) that death is the final stage of growth. Experiencing the death of someone else is indeed a growth experience for the survivors, because we must adapt to life without that person—a major challenge indeed. With the numerous support systems in society today, however, we *can* learn to live without the deceased person. Life goes on, and we must go on with it.

SUMMARY

1. Parents often grieve the loss of a child differently.
2. The death of an infant places severe strains on parents and other family members.
3. Grief does not go away with time; rather, one adjusts to the absence of the deceased person.
4. A perinatal loss leaves the parents with an anticipation that will never happen.
5. Medical staff should be prepared to deal with the grief of parents after a perinatal death.
6. Sudden infant death syndrome (SIDS) is especially difficult for parents. Not only is their child dead, but they are initially suspected by law enforcement in the death.
7. Humor, art, storytelling, and music are good ways to help children cope with death.

8. In today's family, the surviving sibling may become the only child in the family.
9. The death of pets can be devastating for family members.
10. Younger adult children tend to have stronger attachments to parents than their older siblings.

11. Grief in the 21st century should be especially challenging with today's somewhat different perspectives toward death.

DISCUSSION QUESTIONS

1. Discuss particular stresses placed upon parents in the death of a child.
2. How does grief involving parental reactions to perinatal deaths differ from reactions to other deaths?
3. How can health care professionals give support to parents in the death of a child?
4. What nursing procedures follow perinatal deaths?
5. Why are deaths from SIDS especially difficult for parents?
6. How does the loss of older children and adolescents differ from other deaths?
7. How can storytelling, music, art, and humor serve as therapy for a grieving individual?

8. How do children manifest their feelings in the death of a parent?
9. What impact does the death of a sibling have on the surviving sibling(s)?
10. What might cause children to react differently to the death of a grandparent?
11. Why is grief for the loss of a pet so significant for families?
12. What losses are especially difficult in the death of a spouse?
13. How do you think that grief in the 21st century might differ from that in the 20th century?

GLOSSARY

Anomic Grief: A grief without the traditional support of others and no norms to help the bereaved deal with grief.

Back to Sleep: A program that encourages placing a baby on her or his back or side to sleep, not on her or his stomach.

Empty Space: Pain and loss still experienced years after the death of a child.

Induced Abortion: Intentional aborting of the fetus.

Neonate: Refers to an infant less than one-month-old.

Perinatal: Refers to the period closely surrounding the time of birth and includes miscarriages, stillbirths, and neonatal deaths.

Preemie: A baby born prematurely (pregnancy did not last the full term of nine months). The baby's birth weight is usually less than five pounds.

Spontaneous Abortion: Also called miscarriage—expulsion of a fetus from the womb before it is sufficiently developed to survive.

Sudden Infant Death Syndrome (SIDS): The sudden unexplained death of an infant when no cause is found in postmortem examination.

Zygote: Fertilized egg that is formed by the union of the sperm and the egg.

SUGGESTED READINGS

Bonanno, G. (2009). *The other side of sadness: What the new science of bereavement tells us about life after loss.* New York: Basic Books. We tend to understand grief as a predictable five-stage process of denial, anger, bargaining, depression, and acceptance. But in *The Other Side of Sadness,* Bonanno shows that our conventional model discounts our capacity for resilience. In fact, he reveals that we are already hardwired to deal with our losses efficiently—not by graduating through static phases. In the end, mourning is not predictable, but incredibly sophisticated. Combining personal anecdotes and original research, *The Other Side of Sadness* is a must-read for those going through the death of a loved one, mental health professionals, and readers interested in neuroscience and positive psychology.

Bryant, C. D. (2003). The social impact of survivorhood. In C. D. Bryant (Ed.). *Handbook of death and dying* (pp. 862–888). Thousand Oaks, CA: Sage Publications. Three essays edited by Dr. Bryant that discuss the loss of a spouse, child, and parent.

Cheng, J., Lo, R., & Woo, J. (2013). Anticipatory grief therapy for older persons nearing the end of life. *Aging Health*, 9(1), 103–114. An extended discussion of the concept of anticipatory grief and grief therapy.

Cholette, M. E. (2012). Through the eyes of a father: A perinatal loss. *International Journal of Childbirth Education*, 27(2), 33ff. Perinatal loss prompts profound grief and suffering. This article presents a review of nursing discourse on suffering and grief arising from perinatal loss.

Christ, G. H. (2000). *Healing children's grief: Surviving a parent's death from cancer*. New York: Oxford University Press. Relates the stories of 88 families through the terminal illness and death of one of the parents from cancer.

Felber, M. (1997). *Grief expressed: When a mate dies*. New York: Life Words. Helpful information to the bereaved in the death of a spouse from Marta Felber, a therapist whose husband died of cancer.

Holland, J. (2001). *Understanding children's experiences of parental bereavement*. New York: Brunner-Routledge. Explains how best to help and support a child whose parent or caregiver has died.

Jones, B. (2013). Large collection of practical techniques in grief therapy. *Journal of Psychology and Theology*, 41(3), 256. A review of the writings of Dr. Robert A. Neimeyer's prolific writing regarding nearly 100 different therapeutic techniques for counseling the grieving and bereaved, with the stated goal to "present a rich and representative smorgasbord of methods for engaging grief and its complications with greater creativity and awareness of alternatives." In addition to this primary goal, a secondary goal was the integration of research and practice, aiming to "foster an interchange between the field of bereavement research and scholarship ...

and of clinical and counseling practice." To accomplish these goals, the edited work enlists contributions from 86 different authors, stemming from diverse clinical backgrounds including psychiatry, psychology, social work, nursing, and art therapy.

Kelly, L. (1998). *Don't ask for the dead man's golf clubs: Advice for friends when someone dies*. New York: Kelly Communications. Practical advice for survivors of a death in the family from a 34-year-old widowed mother of three.

Kohn, I., Moffitt, P., & Wilkins, I. A. (2000). *A silent sorrow: Pregnancy loss—guidance and support for you and your family*. New York: Brunner-Routledge. A first step for bereaved parents and their families, providing support and guidance to help resolve the grief and enable them to look to the future with hope.

Kohner, N., & Henely, A. (2001). *When a baby dies: The experience of late miscarriage, stillbirth, and neonatal death*. New York: Brunner-Routledge. Offers understanding of what it means to lose a baby and the grief that follows, as well as information about why a baby dies, hospital practices, the process of grieving, sources of support, and the care parents need in future pregnancies.

Moulder, C. (2001). *Miscarriage: Women's experiences and needs*. New York: Brunner-Routledge. Explores the many different ways in which women physically experience miscarriage and emotionally react to it.

Neimeyer, Robert A. (2012). *Techniques of Grief Therapy: Creative Practices for Counseling the Bereaved*. New York: Routledge Publishers. *Techniques of Grief Therapy* is an indispensable guidebook to the most inventive and inspirational interventions in grief and bereavement counseling and therapy. Individually, each technique emphasizes creativity and practicality. As a whole, they capture the richness of practices in the field and the innovative approaches that clinicians in diverse settings have developed, in some cases over decades, to effectively address the needs of the bereaved.

Neimeyer, R. (2012). *Lessons of loss: A guide to coping*. New York: Routledge. *Lessons of Loss* explores how we react to loss, both physical and social, and adapt to it. This new edition has been updated and expanded to include all caregivers, not just the professional therapist.

Perschy, M. K. (2005). *Helping teens work through grief*. New York: Routledge. This book is a valuable guide in helping adults connect with grieving teens.

Pierce, J. (2012). *The last walk: Reflections on our pets at the end of their lives*. Chicago, IL: University of Chicago Press. A tribute to the complexity of human relations to companion animals and the range of issues and concerns that arise as their lives come to an end.

Riches, G., & Dawson, P. (2000). *An intimate loneliness: Supporting bereaved parents and siblings*. Buckingham: Open University Press. Through a sociological approach, an understanding that grief is not an individual matter, because individuals live in family networks that are part of larger communities.

Rosenblatt, P. (2000). *Parent grief: Narratives of loss and relationship*. New York: Brunner-Routledge. Explores what parents' stories say and do not say about the child's dying and death and about parent grief.

Ross, C. R., & Baron-Sorensen, J. (2007). *Pet Loss and Human Emotion*. New York: Routledge. Discusses the range of therapeutic interventions and pet loss for mental health professionals. Notes the need for mental health professionals to become increasingly more sensitive to the fact that their clients can experience real grief over the loss of a pet.

Rynearson, E. (2001). *Retelling violent death*. New York: Brunner-Routledge. Provides guidance for making the retelling of a violent death restorative and therapeutic, rather than entrenching the person in grief.

Schuler, T. A., Zaider, T. I., & Kissane, D. W. (2012). Family grief therapy: A vital model in oncology, palliative care and bereavement. *Family Matters*, 90(Summer), 77ff. An extensive discussion of family-based grief therapy dealing with many topics of significance: anticipatory grief, complicated grief, attachment theory, group adaptation, and efficacy of family grief.

Silverman, P. (2005). *Widow to widow: How the bereaved help one another*. New York: Routledge. The book shares the experiences of widows who have found comfort and continuity in mutual-help and community support programs.

Talbot, K. (2002). *What forever means after the death of a child: Transcending the trauma, living with the loss*. New York: Brunner-Routledge. Examination of the loss of an only child and the loss of the role of parent.

Umberson, D. (2003). *Death of a parent: Transition to a new adult identity*. New York: Cambridge University Press. A sociologist and former social worker, Debra Umberson sets out in a clear and comprehensive manner what the death of a parent means to most adults and how it functions as a turning point in the emotional, social, and personal aspects of our lives.

Walter, C. A. (2003). *The loss of a life partner: Narratives of the bereaved*. New York: Columbia University Press. A well-written, accessible, and compassionate book that provides an understanding of the experiences and needs of bereaved partners. The book discusses interviews with bereaved lesbian partners and grievers from other nontraditional relationships, giving insights into the universal bereavement challenges.

Walter, T. (Ed.). (1999). *The mourning for Diana*. Oxford: Berg. A sociological perspective on the various aspects of the aftermath of the death of Princess Diana.

Witt, D. D. (2003). Pet burial in the United States. In C. D. Bryant (Ed.), *Handbook of death and dying* (pp. 757–766). Thousand Oaks, CA: Sage Publications. A comprehensive essay on the treatment of dead companion animals including body disposition, pet cemeteries, pet funerals, grieving the death of a companion pet, and life after death issues related to pet loss.

Wurtele, M. (2003). *Touching the edge: A mother's spiritual journey from loss to life*. Hoboken, NJ: John Wiley & Sons, Inc. In this memoir Wurtele, co-founder of Ruminator Books, beautifully describes how the death of her only child darkened her life and tested her religious awakening.

REFERENCES

Abbott, L. (1913, August 30). There are no dead. *Outlook, 104,* 979–988.

About.com. (2014). Wills and estate planning. Accessed April 19, 2014. http://wills.about.com/od/understandingestatetaxes/a/future-of-estate-tax-2014-beyond.htm 2014).

ADAM. (2005). Death among children and adolescents. Healthcare Center. Retrieved July 13, 2010, from http://www.adam.about.com.

Adams, C. (1988). *More of the straight dope.* New York: Ballantine Books.

Adams, D. W., & Deveau, E. J. (1987). When a brother or sister is dying of cancer: The vulnerability of the adolescent sibling. *Death Studies, 11,* 279–295.

Alao, A., Soderberg, M., Pohl, E., & Alao, A. (2006). Cybersuicide: Review of the role of the Internet on suicide. *CyberPsychology & Behavior, 9,* 489–493.

Albom, M. (1997). *Tuesdays with Morrie.* New York: Doubleday.

Alcor Extension Foundation. (2009). *Cryonics at Alcor.* Retrieved July 13, 2010, from http://www.alcor.org.

Almgren, G. (1993). Living will legislation, nursing home care, and the rejection of artificial nutrition and hydration: An analysis of bedside decision-making in three states. *Journal of Health and Social Policy, 4*(3), 43–63.

Altman, D., & Levitt, L. (2003). *The sad history of health care cost containment as told in one chart.* Health Affairs Web Exclusive. Retrieved July 13, 2010, from www.healthaffairs.org.

Altman, L. K. (1989, November 14). Physicians endorse more humanities for premed students. *New York Times,* p. 22.

Alzheimer's Association. (2009). 2009 Alzheimer's disease facts and figures. *Alzheimer's & Dementia, 5*(3), 1–80.

Amella, E. J., Lawrence, J. F., & Gresle, S. O. (2005). Tube feeding: Prolonging life or death in vulnerable population? *Mortality, 10*(1), 69–81.

American Cancer Society. (2008). *Cancer statistics.* Retrieved July 13, 2010, from http://www.cancer.org.

American Foundation for Suicide Prevention. (2009). *International statistics.* Retrieved July 13, 2010, from http://www.afsp.org/index.cfm.

American Geriatrics Society Public Policy Committee. (1994). Voluntary active euthanasia. *Journal of the American Geriatrics Society, 39,* 826.

American Medical Association Council on Ethical and Judicial Affairs. (1992). *Current opinions.* Chicago, IL: American Medical Association.

American Psychological Association. (2008, August 17). *Suicidal thoughts among college students more common than expected.* Washington, DC: APA Office of Public Affairs.

American Psychological Association Commission on Violence and Youth. (1993). *Violence and youth: Psychology's response* (Vol. 1). Washington, DC: American Psychological Association.

Anderson, C. A., & Bushman, B. J. (2002). Human aggression. *Annual Review of Psychology, 53,* 27–51.

Anderson, J. (1997, May 27). Funeral industry, seeking new business, courts the living with special services. *New York Times,* p. B1.

Anderson, R. (1996). *Magic, science, and health.* Fort Worth, TX: Harcourt Brace College Publishers.

Anderson, W. G., Williams, J. E., Bost, J. E., & Barnard, D. (2008). Exposure to death is associated with positive attitudes and higher knowledge about end-of-life care in graduating medical students. *Journal of Palliative Medicine, 11*(9), 1227–1233.

Andriessen, K. (2006). On "intention" in the definition of suicide. *Suicide and Life-Threatening Behavior, 36,* 533–538.

Angell, M. (1997). The Supreme Court and physician-assisted suicide: The ultimate right. *New England Journal of Medicine, 337*, 50–53.

Apple, R. W., Jr. (2001, September 21). Bush presidency seems to gain legitimacy. *New York Times.*

Archives of the *Philadelphia Gazette* (1751–1752). Retrieved May 25, from CD-ROM [Folio II, Item 1841].

Aries, P. (1981). *The hour of our death.* New York: Oxford University Press.

Aristotle. (1941). Nicomachean ethics III. In R. McKeon (Ed. & Trans.), *The basic works of Aristotle.* New York: Random House.

Ashley-Cameron, S., & Dickinson, G. E. (1979, February). *Nurses' attitudes toward working with dying patients.* Paper presented at the Alpha Kappa Delta Research Symposium, Richmond, VA.

Ashwood, P. (2009). Embalming. In C. D. Bryant & D. Peck (Eds.), *Encyclopedia of death and the human experience* (pp. 404–406). Thousand Oaks, CA: Sage Publications.

Association of American Cemetery Superintendents. (1889). *3*, 59.

Atchley, R. C. (2004). *Social forces and aging.* Belmont, CA: Wadsworth.

Atkinson, M. J. (1935). *Indians of the Southwest.* San Antonio, TX: Naylor.

Attig, T. (1991). The importance of conceiving of grief as an active process. *Death Studies, 15,* 385–393.

Austerlitz, S. (2013). Why funerals demand a body: Undertaker Thomas Lynch on how American memorials went wrong. *Boston Globe*, October 20, 2013.

AVERT (2014). HIV and AIDS stigma discrimination. Retrieved April 6, 2014, from http://www.avert.org/hiv-aids-stigma-and-discrimination.htm.

Ayer, M. (1964). *Made in Thailand.* New York: Alfred A. Knopf.

Ayres, I., & Nalebuff, B. (2007, March 25). Do you have a better idea? *Parade Magazine*, 6–7.

Baby born without brain dies, but legal struggle will continue. (1992, March 31). *New York Times,* p. A8.

Baker, J. E., & Sedney, M. A. (1996). How bereaved children cope with loss: An overview. In C. A. Corr & D. M. Corr (Eds.), *Adolescence and death* (pp. 109–129). New York: Springer Publishing.

Balk, D. E. (2003). The evolution of mourning and the bereavement role in the United States. In C. D. Bryant (Ed.), *Handbook of death and dying* (pp. 829–837). Thousand Oaks, CA: Sage Publications.

Baring-Gould, W. S., & Baring-Gould, C. (1967). *The annotated Mother Goose.* New York: Random House Value Publishing.

Barrera, M., D'Agostino, N., Schneiderman, G., Tallet, S., Spencer, L., & Jovcevska, V. (2007). Patterns of parental bereavement following the loss of a child and related factors. *Omega: Journal of Death & Dying, 55*(2), 145–167.

Barrett, R. K. (1992). Psychocultural influences on African American attitudes toward death, dying and funeral rites. In J. Morgan (Ed.), *Personal care in an impersonal world* (pp. 213–230). Amityville, NY: Baywood Publishing.

Barrett, R. K. (1996). Young people as victims of violence. In R. G. Stevenson & E. P. Stevenson (Eds.), *Teaching students about death* (pp. 63–75). Philadelphia, PA: The Charles Press.

Barrett, W. (1958). *Irrational man: A study in existential philosophy.* New York: Doubleday.

Bascom, W. (1969). *The Yoruba of southwestern Nigeria.* New York: Holt, Rinehart and Winston.

Battin, M. P. (1994). *The least worst death: Essays in bioethics on the end of life.* New York: Oxford University Press.

Baudrillard, J. (1993). *Symbolic exchange and death.* London: Sage.

Becker, E. (1973). *The denial of death.* New York: The Free Press.

Becker, F. (2009). Islamic reform and historical change in the care of the dead: Conflicts over funerary practice among Tanzanian Muslims. *Africa, 79*(3), 416–434.

Becker, H. (1963). *Outsiders: Studies in the sociology of deviance.* Glencoe, IL: Free Press.

Becker, H. (1964, March). Personal change in adult life. *Sociometry, 27,* 40–53.

Becker, K., Mayer, M., Nagenborg, M., El-Faddagh, M., & Schmidt, M. (2004). Parasuicide online: Can suicide websites trigger suicidal behaviour in predisposed adolescents? *Nordic Journal of Psychiatry, 58*, 111–114.

Beckwith, J. B. (1978). *The sudden infant death syndrome* (DHEW Publication No. HSA 75-5137). Washington, DC: U.S. Government Printing Office.

Beecher, H. K. (1968, August). A definition of irreversible coma. *Journal of the American Medical Association, 205*, 85–88.

Bell, M. D. (1996, December). Magic time: Observations of a cancer casualty. *The Atlantic Monthly, 278*, 40–43.

Beloff, J. (1989). Do we have a right to die? In A. Berger, P. Badham, A. H. Kutscher, J. Berger, M. Perry, & J. Beloff (Eds.), *Perspectives on death and dying: Cross-cultural and multi-disciplinary view* (pp. 163–172). Philadelphia, PA: Charles Press.

Bendann, E. (1930). *Death customs: An analytical study of burial rites.* New York: Alfred A. Knopf.

Bender, T. (1973, June). The "rural cemetery" movement. *New England Quarterly, 47*, 196–211.

Benet, S. (1974). *Abkhasians: The long-living people of Caucasus.* New York: Holt, Rinehart and Winston.

Benjamin, C. L. (1882, February). Essay. *The Casket, 7*, 2.

Bennett, J. G. (1999). *Maximize your inheritance for widows, widowers & heirs.* Chicago, IL: Dearborn Financial Publishing.

Berger, P. L. (1969). *Sacred canopy: Elements of a sociological theory of religion.* New York: Doubleday.

Bhattacharya, P. (2013). Is there science behind the near-death experience: Does human consciousness survive after death? *Annals of Tropical Medicine and Public Health, 6*(2), 151.

Bibby, R. (2001). Canada's teens: *Today, yesterday, tomorrow.* Ontario, ON: Stoddart Publishing.

Bigelow, G. (1997, January 29). Letter from Dr. Gordon Bigelow, executive director of the American Board of Funeral Service Education, Brunswick, ME.

Billings, M. E., Engelberg, R., Curtis, J. R., Block, S., & Sullivan, A. M. (2010). Determinants of medical students' perceived preparation to perform end-of-life care in the medical curriculum: Students' opinions and knowledge. *Journal of Palliative Medicine, 13,* 319–326.

Blackhall, L. J., Murphy, S. T., Frank, G., Michel, V., & Azen, S. (1995). Ethnicity and attitudes toward patient autonomy. *Journal of the American Medical Association, 274,* 820–825.

Blank, R. H. (2001). Technology and death policy: Redefining death. *Mortality, 6,* 191–202.

Blau, P. M. (1964). *Exchange and power in social life.* New York: John Wiley & Sons.

Blauner, R. (1966). Death and social structure. *Psychiatry, 29,* 378–394.

Blendon, R. J., Szalay, U. S., & Knox, R. A. (1992). Should physicians aid their patients in dying? The public perspective. *Journal of the American Medical Association, 267,* 2658–2662.

Bloch, M., & Parry, J. (1982). *Death and the regeneration of life.* Cambridge, UK: Cambridge University Press.

Bluebond-Langner, M. (1978). *The private worlds of dying children.* Princeton, NJ: Princeton University Press.

Bluebond-Langner, M. (1989). Worlds of dying children and their well siblings. *Death Studies, 13,* 1–16.

Bolt, S. (2009). Body disposition. In C. D. Bryant & D. Peck (Eds.), *Encyclopedia of death and the human experience* (pp. 107–111). Thousand Oaks, CA: Sage Publications.

Bomanji, J. B., Britton, K. E., & Clarke, S. E. M. (1995). *Oncology.* London: British Nuclear Medicine Society.

Bordere, T. (2008/2009). To look at death another way: Black teenage males' perspectives on second-lines and regular funerals in New Orleans. *Omega, 50*(3), 213–232.

Bordewich, F. M. (1988, February). Mortal fears: Courses in "death education" get mixed reviews. *Atlantic Monthly, 261,* 30–34.

Borjigin, J., Lee, U., Liu, T., Pal, D., Huff, S., & Klarr, D. (2013). Surge of neurophysiological coherence and connectivity in the dying brain. *Proceedings of the National Academy of Sciences of the United States, 110*(35), 14432ff.

Bouma, H., Diekema, D., Langerak, E., Rottman, T., & Verhey, A. (1989). *Christian faith, health, and medical practice.* Grand Rapids, MI: Eerdmans.

Bourne, P. (1970). *Men, stress, and Vietnam.* Boston, MA: Little, Brown.

Boyd, M. (1997, November–December). Move forward with life. *Modern Maturity,* 71.

Boyle, F. M., Vance, J. C., Najman, J. M., & Thearle, M. J. (1996). The mental health impact of stillbirth, neonatal death or SIDS: Prevalence and patterns of distress among mothers. *Social Science and Medicine, 43,* 1273–1282.

Brabant, M. (1994, September). The high price of everlasting peace. *Worldwide report: Death.* London: BBC Worldwide.

Brabant, S. (2003). *Handbook of death and dying.* Volume 1: The presence of death. Thousand Oaks, CA: Sage Reference.

Brandt, R. B. (1986). The morality and rationality of suicide. In R. F. Weir (Ed.), *Ethical issues in death and dying* (2nd ed.) (pp. 330–344). New York: Columbia University Press.

Brantner, J. P. (1973, January). *Crisis intervener.* Paper presented at the Ninth Annual Funeral Service Management Seminar, National Funeral Directors Association, Scottsdale, AZ.

Braun, K. L., & Kayashima, R. (1999). Death education in churches and temples: Engaging religious leaders in the development of educational strategies. In B. de Vries (Ed.), *End of life issues: Interdisciplinary and multidimensional perspectives* (pp. 319–335). New York: Springer Publishing.

Breen, L. J., & O'Connor M. (2010). Acts of resistance: Breaking the silence of grief following traffic crash fatalities. *Death Studies, 34,* 30–53.

Breytspraak, L. (2008). Are older adults less anxious about death than are younger and middle-aged adults? *Aging.* University of Missouri Extension. Retrieved March 31, 2014, from http://missourifamilies .org/quick/agingqa/agingqa32.htm.

Brody, J. E. (2007, September 4). For living donors, many risks to weigh. *The New York Times,* p. D7.

Brotman, B. (2012). Crematorium holds open house to demystify process. *Chicago Tribune* (November 13, 2012).

Brown, R. (1996, February). *A free market in human organs.* Fairfax, VA: The Future of Freedom Foundation.

Bruno, M., Ledoux, D., & Laureys, S. (2009). The dying human: A perspective from biomedicine. In A. Kellehear (Ed.), *The study of dying* (pp. 51–75). Cambridge: Cambridge University Press.

Buchanan, J. (2013). Words for grieving. *The Christian Century, 130*(15), 3.

Bukiet, M. J. (2005). Custom and law: After the death of his father, a not-notably observant Jew turns to the mourning rituals of his faith. *American Scholar, 74*(4), 100–113.

Bursack, C. B. (2012). Have "the talk" with elders: End-of-life issue conversations. *VITAS: Innovative Hospice Care.* http://www.agingcare .com/articles/having-connversations-with-elderly-about-end-of-life. Accessed on October 8, 2012.

Byock, I. (1997). *Dying well: The prospect for growth at the end of life.* New York: Riverhead Books.

Cacciatore, J. (2007). Effects of support groups on post traumatic stress responses in women experiencing stillbirth. *Omega: Journal of Death & Dying, 55*(1), 71–90.

Cairney, T. H. (2011). Key themes in children's books. Literacy, families and learning. http://trevorcairney .blogspot.com. Accessed on October 8, 2012.

Califano, J. A., Jr. (1986). *America's health care revolution: Who lives? Who dies? Who pays?* New York: Random House.

Callahan, D. (1987). *Setting limits: Medical goals in an aging society.* New York: Simon & Schuster.

Callahan, D. (2000). *The troubled dream of life: Living with mortality.* Washington, DC: Georgetown University Press.

Callender, C. O. (1987). Organ donation in blacks: A community approach. *Transplantation Proceedings, 19,* 1551–1554.

Campbell, T. W., Abernethy, V., & Waterhouse, G. J. (1984). Do death attitudes of nurses and physicians differ? *Omega, 14,* 43–49.

CancerNet. (2000, February). *Fever, chills, and sweats.* Bethesda,

MD: National Cancer Institute. Retrieved July 13, 2010, from http://www.cancernet.nci.nih.gov.

Cann, C. (2014). *Virtual afterlives: Grieving the dead in the twenty-first century*. Lexington, KY: University of Kentucky Press.

Canning, R. R. (1965). Mormon return-from-the-dead stories: Fact or folklore? *Utah Academy Proceedings*, 42, I.

Cant, R., Cooper, S., Chung, C., & O'Connor, M. (2012). The divided self: Near death experiences of resuscitated patients—A review of literature. *International Emergency Nursing*, 20(2), 88.

Cantor, J., Bushman, B. J., Huesmann, L. R., Grobel, J., Malamuth, N. M., Impett, E. A., et al. (2001). Some hazards of television viewing: Fears, aggressions, and sexual attitudes. In D. G. Singer & J. L. Singer (Eds.), *Handbook of children and the media* (pp. 207–307). Thousand Oaks, CA: Sage Publications.

Capps, B. (1973). *The Indians*. New York: Time-Life Books.

Carlson, E. (2000, September). Taking back the end of life. *AARP Bulletin*, 43, 18–20.

Carroll, J. (2006). Public continues to support right-to-die for terminally ill patients. At least 6 in 10 Americans support euthanasia, doctor-assisted suicide. *Gallup News Service*, June 19.

Carroll, R. M., & Schafer, S. S. (1994, May). Similarities and differences in spouses coping with SIDS. *Omega*, 28, 273–284.

Carron, A., Lynn, J., & Keaney, P. (1999). End-of-life care in medical textbooks. *Annals of Internal Medicine*, 130, 82–86.

Carse, J. (1981). Death. In K. Crim, R. A. Bullard, & L. D. Shinn (Eds.), *Abingdon dictionary of living religions*. Nashville, TN: Abingdon Press.

Carter, G. T., Flanagan, A. M., Earleywine, M., Abrams, D. I., Aggarwal, S. K., & Grinspoon, L. (2011). Cannabis in palliative medicine: Improving care and reducing opioid-related morbidity. *American Journal of Hospice & Palliative Medicine*, 28(5), 297–303.

Center for Advanced Palliative Care. (2002). *CAPC Manual*. New York: Beth Israel Medical Center.

Centers for Disease Control and Prevention (2014). CDC-INFO. Atlanta, GA.

Centers for Disease Control and Prevention. (2007a). HIV-AIDS statistics (2007). *NIAID Fact Sheet*. Bethesda, MD: National Institutes of Health.

Centers for Disease Control and Prevention. (2011). The changing profile of autopsied deaths in the United States, 1972–2007. Number 67 (August). http://www.cdc.gov/nchs/data/databriefs/db67.htm.

Centers for Disease Control and Prevention. (2007b). *National center for injury prevention and control*. Retrieved July 13, 2010, from http://www.cdc.gov/ncipc/wisqars.

Centers for Disease Control and Prevention. (2008). *National center for injury prevention and control*. Web-based injury statistics query and reporting system. Retrieved July 13, 2010, from http://www.cdc.gov/ncipc/wisqars.

Centers for Disease Control and Prevention. (2009a). *Suicide and self-inflicted injury*. Retrieved July 13, 2010, from www.cdc.gov.

Centers for Disease Control and Prevention. (2009b). *National center for injury prevention and control*. Retrieved July 13, 2010, from http:// www.cdc.gov/ncipc/wisqars.

Centers for Disease Control and Prevention (2012). *Suicide and self-inflicted injury*. http://www.cdc.gove.nchs/fastats/suicide.htm. Accessed on October 16, 2012.

Centers for Medicare and Medicaid Services (2009). *2007 National health care expenditures data*. Baltimore, MD: Office of the Actuary, National Health Statistics Group.

Cesur., R., Sabia, J., & Tekin, E. (2013). The psychological costs of war: Military combat and mental health. *Journal of Health Economics*, 32(1), 51ff.

Chambliss, D. F. (1996). *Beyond caring: Hospitals, nurses, and the social organization of ethics*. Chicago, IL: University of Chicago Press.

Chang, A. (2009, October 15). Studies: Elderly care often futile (dialysis, dementia). *Associated Press* release.

Charmaz, K. (1975). The announcement of death by the coroner's deputy. *Urban Life*, 4(3), 296–316.

Charmaz, K. (1980). *The social reality of death*. Reading, MA: Addison-Wesley.

Chase, R. (2009, April 28). *Return of fallen soldiers: Most families consent to coverage*. New York: *Associated Press*.

Chatzky, J. S. (2000, September). The last word: Making your health-care wishes clear—and comprehensive. *Money*, p. 172.

Chen, P. (2007). *Final exam: A surgeon's reflections on mortality*. New York: Alfred A. Knopf.

Cheng, J., Lo, R., & Woo, J. (2013). Anticipatory grief therapy for older persons nearing the end of life. *Aging Health*, 9(1), 103.

Cheng, M. (2012, December 14). Study: People living longer, but are sicker. *Associated Press. Charleston Post & Courier*, Charleston, SC, p. A16.

Chochinov, H. M., Wilson, K. G., & Ennis, M. (1995). Desire for death in the terminally ill. *American Journal of Psychiatry*, 152, 1185–1191.

Choron, J. (1972). *Suicide*. New York: Charles Scribner's Sons.

Christ, G. H. (2000). *Healing children's grief: Surviving a parent's death from cancer*. New York: Oxford University Press.

Christakis, N. A. (1999). *Death foretold: Prophecy and prognosis in medical care*. Chicago, IL: University of Chicago Press.

Clark, D. (2007). End-of-life care around the world: Achievements to date and challenges remaining. *Omega*, 56, 101–110.

Clark, D., Dickinson, G., Lancaster, C., Noble, T., Ahmedzai, S., & Philip, I. (2000, March 27–29). UK geriatricians' attitudes to active voluntary euthanasia and physician-assisted death. *Age and Ageing*, 30, 395–398.

Clark, M., & Springen, K. (1986, November 17). The demise of autopsies. *Newsweek*, p. 61.

Claxton-Oldfield, S., Gosselin, N., & Claxton-Oldfield, J. (2009). Imagine you are dying: Would you be interested in having a hospice palliative care volunteer? *American Journal of Hospice & Palliative Medicine*, 26, 47–51.

Cleaveland, N. (1847). *Green-wood illustrated*. New York: R. Martin.

Clifford, D., & Jordan, C. (1999). *Plan your estate*. Berkeley, CA: Nolo Press.

Clifton, E. (2012, June 8). Suicide rate in military at highest level in ten years. *ThinkProgress*. http://thinkprogress .org/security. Accessed on October 18, 2012.

Cholette, M. (2012). Through the eyes of a father: A perinatal loss. *International Journal of Childbirth Education, 27*(2), 33ff.

Christian Century. (2012). Heaven above or below? *Christian Century, 129*(14), 9.

Clough, S. B. (1946). *A century of American life insurance*. New York: Columbia University Press.

CNN. (2009). U.S. and coalition casualties in Afghanistan and Iraq. Retrieved July 13, 2010, from http://www.cnn.com/SPECIALS/ 2003/ iraq/forces/casualties/index .html.

CNN. (2013). 25 deadliest mass shootings in U.S. history fast facts. CNN Library. http://www.cnn.com/ 2013/09/16/us/20-deadliest-mass-shootings-in-u-s-history-fast-facts/. Accessed on November 13, 2013.

Cockerham, W. C. (1991). *The aging society*. Englewood Cliffs, NJ: Prentice-Hall.

Cockerham, W. C. (2009). *Medical sociology*. Upper Saddle River, NJ: Prentice Hall.

Cockerham, W. C. (2012). *Medical sociology*. Upper Saddle River, NJ: Prentice Hall.

Codex juris canonici. (1918). Rome, Italy: Polyglot Press.

Coe, R. M. (1970). *Sociology of medicine*. New York: McGraw-Hill.

Cohen, E., & Kass, L. R. (2006, January). Cast me not off in old age. *Commentary, 121*(1) 32–39.

Cohen, J. S., Fihn, S. D., Boyko, E. J., Jonsen, A. R., & Wood, R. W. (1994). Attitudes toward assisted suicide and euthanasia among physicians in Washington state. *New England Journal of Medicine, 331,* 89–94.

Cohen, R. (1967). *The Kanuri of Bornu*. New York: Holt, Rinehart and Winston.

Cohn, F., Harrold, J., & Lynn, J. (1997, May 30). Medical education must deal with end-of-life care. *Chronicle of Higher Education*, A56.

Colon, M. (2012). The experience of physicians who refer Latinos to hospice. *American Journal of Hospice and Palliative Medicine, 29*(4), 254–259.

Colucci, E., & Lester, D. (Eds.) (2013). *Suicide and culture*. Ashland, OH: Hogrefe Publishing.

Colucci, E., & Martin, G. (2008). Religion and spirituality along the suicidal path. *Suicide and Life-Threatening Behavior, 38,* 229–244.

Combs, D. W. (1986). *Early gravestone art in Georgia and South Carolina*. Athens, GA: University of Georgia Press.

Community Living. (2012). Talking about dying, death and bereavement. *Community Living, 26*(2).

Complicated Grief Program. (2013). Center for Complicated Grief, Columbia University School of Social Work, New York. http:// www.complicatedgrief.org/ bereavement. Accessed on October 29, 2013.

Conley, H. N. (1979). *Living and dying gracefully*. New York: Paulist Press.

Conner, N. E. (2012). Predictive factors of hospice use among blacks: Applying Andersen's behavioral model. *American Journal of Hospice & Palliative Medicine, 29*(5), 368–374.

Connor, L. H. (1995, September). The action of the body on society: Washing a corpse in Bali. *Journal of the Royal Anthropological Institute, 1,* 537–560.

Connor, S. R. (2009). *Hospice and palliative care: The essential guide*. New York: Routledge.

Conrad, B. H. (1998). *When a child has been murdered: Ways you can help the grieving parents*. Amityville, NY: Baywood Publishing.

Cook, J. A. (1983). A death in the family: Parental bereavement in the first year. *Suicide and Life-Threatening Behavior, 13,* 42–61.

Cook, J. A. (1988). Dad's double binds: Rethinking fathers' bereavement from a men's studies perspective. *Journal of Contemporary Ethnography, 17,* 285–308.

Cooney, W. (1998). The death poetry of Emily Dickinson. *Omega, 37,* 241–249.

Cope, L. (1978, June 22). Is death, like pregnancy, an all-or-nothing thing? *Minneapolis Tribune*, pp. 1, 6A.

Copp, G. (1998). A review of current theories of death and dying. *Journal of Advanced Nursing, 28,* 382–390.

Copp, G., Richardson, A., McDaid, P., & Marshall-Searson, D. A. (1998).

A telephone survey of the provision of palliative day care services. *Palliative Medicine, 12,* 161–170.

Cornies, L. (2012). Emergency responders put their life on the line. http:// sweeneyalliance.org'grievingbehindthe badge/cornies-ptsd. Accessed on November 2, 2012.

Corr, C. A., & Corr, D. M. (2002). Stage theory. *Encyclopedia of death and dying*. New York: Macmillan.

Council on Ethical and Judicial Affairs. (1991, July). Report 32: Decisions to forgo life-sustaining treatment for incompetent patients. In *Code of medical ethics*. Chicago, IL: American Medical Association.

Cordaro, M. (2012). Pet loss and disenfranchised grief: Implications for mental health counseling practice. *Journal of Mental Health Counseling, 34*(4, October), 283ff.

Costhelper. (2013). Pet euthanasia cost, http://pets.costhelper.com/dog-euthanasia. Accessed on July 17, 2013.

Costin, A., & McMurrich, C. (2013). A discussion of parental validation following stillbirth using transforming counseling and education. *International Journal of Childbirth Education, 28*(3), 71ff.

Counts, D. A., & Counts, D. R. (1991). Loss and anger: Death and the expression of grief in Kaliai. In D. R. Counts & D. A. Counts (Eds.), *Coping with the final tragedy: Cultural variation in dying and grieving* (pp. 191–212). Amityville, NY: Baywood Publishing.

Cowell, D. D., Farrell, C., Campbell, N. A., & Canady, B. E. (2002). Management of terminal illness. *Academic Psychiatry, 26*(2), 76–81.

Cox, G. R. (1998, May). *Using humor, art, and music with dying and bereaved children*. Paper presented at the International Conference on Death and Bereavement in Ontario, Canada.

Cox, G. R., & Fundis, R. J. (1992). Native American burial practices. In J. Morgan (Ed.), *Personal care in an impersonal world* (pp. 191–204). Amityville, NY: Baywood Publishing.

Cox, H. G. (1996b). *Later life: The realities of aging* (4th ed.). Upper Saddle River, NJ: Prentice-Hall.

Cox, M., Garrett, E., & Graham, J. A. (2004–2005). Death in Disney films: Implications for children's understanding of death. *Omega, 50*, 267–280.

Crandall, R. C. (1991). *Gerontology: A behavioral approach*. New York: McGraw-Hill.

Cremation Association of North America. (1997). *1997 fact sheet*. Milwaukee: Author.

Cremation Association of North America. (2007). *2007 fact sheet*. Milwaukee: Author.

Cremation Association of North America. (2012). *2012 fact sheet*. Milwaukee: Author.

Crimmins, E. M., & Beltran-Sanchez, H. (2010). Mortality and morbidity trends: Is there compression of morbidity? *Journal of Gerontology: Social Sciences, 66*(1), 75–86.

Crissman, J. K. (1994). *Death and dying in central Appalachia: Changing attitudes and practices*. Urbana, IL: University of Illinois Press.

Cugliari, A. M., & Miller, T. E. (1994, April). Moral and religious objections by hospitals to withholding and withdrawing life-sustaining treatment. *Journal of Community Health, 19*(2), 87–100.

Culver, C. M. (1990). *Ethics at the bedside*. Hanover, NH: University Press of New England.

Cummins, R., Oranto, J. P., & Thies, W. H. (1991). Improving survival from sudden cardiac arrest: The 'chain of survival' concept. *Circulation, 83*, 1832–1847.

Cupit, I. N., Sofka, C. J., & Gilbert, K. R. (2012). Death education. In C. J. Sofka, K. N. Cupit, & K. R. Gilbert (Eds.), *Dying, death, and grief in an online universe* (pp. 163–182). New York: Springer Publishing Company.

Cuzzort, R. P., & King, E. W. (2002). *Social thought into the twenty-first century*. Fort Worth, TX: Harcourt College Publishers.

Daniel, S. S., & Goldston, D. B. (2009). Interventions for suicidal youth: A review of the literature and developmental considerations. *Suicide and Life-Threatening Behavior, 39*, 252–268.

Dattel, A. R., & Neimeyer, R. A. (1990). Sex differences in death anxiety: Testing the emotional expressiveness hypothesis. *Death Studies, 14*, 1–11.

Daugherty, C. K., & Hlubocky, F. J. (2008). What are terminally ill cancer patients told about their expected deaths? A study of cancer physicians' self-reports of prognosis disclosure. *Journal of Clinical Oncology, 26*(36), 5988–5993.

Davies, D. (1999). The week of mourning. In T. Walter (Ed.), *The mourning for Diana* (pp. 3–18). Oxford: Berg.

Davies, D. J. (1997). *Death, ritual and belief: The rhetoric of funerary rites*. London: Cassell.

Davis, C. S. (2008, September/October). A funeral liturgy: Death rituals as symbolic communication. *Journal of Loss and Trauma, 13*(5), 406–421.

Davis, D. L., Stewart, M., & Harmon, R. J. (1988, December). Perinatal loss: Providing emotional support for bereaved parents. *Birth, 14*, 242–246.

Dawson, G. D., Santos, J. F., & Burdick, D. C. (1990). Differences in final arrangements between burial and cremation as the method of body disposition. *Omega, 21*, 129–146.

Day, J. (1999, January). Alleviating bone pain using strontium therapy. *Professional Nurse, 14*(4), 1–4.

Deathanalysis (2012). Death analysis—Coping with death of love[d] ones. http://www.deathanalysis.com. Accessed on October 18, 2012.

Hategan, E. (2007, August 20). Death in children's movies: the loss of innocence as a subversive agenda in Hollywood. Incognito Press. http://incognitopress.wordpress.com/2007/08/20. Accessed on October 5, 2012.

de Beauvoir, S. (1965). *A very easy death* (P. O'Brien, Trans.). New York: G. P. Putnam.

de Bretagne, R. (2012). How to deal with grief following accidental death. Helium. http://www.helium.com/items/2182637-how-to-deal-with-grief-followiing-accidental-death. Accessed on October 18, 2012.

DeFrain, J. D., Jakub, D. K., & Mendoza, B. L. (1992). The psychological effects of sudden infant death on grandmothers and grandfathers. *Omega, 24*, 165–182.

DeFrain, J. D., Taylor, J., & Ernst, L. (1982). *Coping with sudden infant death*. Lexington, MA: Lexington Books.

De Leo, D., Cimitan, A., Dyregrov, K., Grad, O., & Andriessen, K. (Eds.) (2014). *Bereavement after traumatic death*. Ashland, OH: Hogrefe Publishing.

De Leo, D., Conforti, D., & Carollo, G. (1997, Fall). A century of suicide in Italy: A comparison between the old and the young. *Suicide and Life-Threatening Behavior, 27*(3), 239–249.

DeMallie, R. J., & Parks, D. R. (1987). *Sioux Indian religion: Tradition and innovation*. Norman: University of Oklahoma Press.

Democratization of death. (February 1, 2002). *The Chronicle of Higher Education*, B4, B6.

Deng, F. M. (1972). *The Dinka of the Sudan*. New York: Holt, Rinehart and Winston.

Dennis, D. (2009). *Living, dying, grieving*. Boston, MA: Jones and Bartlett Publishers.

Dennis, M. R. (2009). Condolences. In C. D. Bryant & D. Peck (Eds.), *Encyclopedia of death and the human experience* (pp. 219–221). Thousand Oaks, CA: Sage Publications.

Dentan, R. K. (1968). *The Semai: A nonviolent people of Malaya*. New York: Holt, Rinehart and Winston.

Department of Veterans Affairs. (2014). http://www.cem.va.gov/cems/newcem.asp.

Derry, S. (1997). Dying for palliative care. *European Journal of Palliative Care, 4*(3), 66–71.

de Tocqueville, A. (1945). *Democracy in America* (Phillips Bradley, Trans.). New York: Vintage Books. (Original work published 1835.)

DeVries, R., & Subedi, J. (1998). *Of bioethics and society: Constructing the ethical enterprise*. Upper Saddle River, NJ: Prentice Hall.

Dickinson, G. (2012). Diversity in death: Body disposition and memorialization. *Illness, Crisis, and Loss, 20*(2), 141–158.

Dickinson, G. E. (1988, January). Death education for physicians. *Journal of Medical Education, 63*, 412.

Dickinson, G. E. (1992). First childhood death experiences. *Omega, 25*, 169–182.

Dickinson, G. E. (2005, September 15–18). *Baby boomers and personalized death trends*. Paper presented at the Dying, Death, and Disposal Conference, Bath, England.

Dickinson, G. E. (2006, July). Teaching end-of-life issues in U.S. medical schools: 1975–2005. *American Journal of Hospice and Palliative Medicine*, 23(3), 197–204.

Dickinson, G. E. (2007a, September). End-of-life and palliative care issues in medical and nursing schools in the United States. *Death Studies*, 31(8), 713–726.

Dickinson, G. E. (2007b). A deathly education. *Cancer Nursing Practice*, 6, 24–25.

Dickinson, G. E. (2011a). Shared grief is good grief. *Phi Kappa Phi Forum*, 91(3), 10–11.

Dickinson, G. E. (2011b). Thirty-five years of end-of-life issues in US medical schools. *American Journal of Hospice and Palliative Medicine*, 28(6), 412–417.

Dickinson, G. E. (2012a). Twenty-first century end-of-life issues in selected US professional schools. *Illness, Crisis & Loss*, 20(1), 19–32.

Dickinson, G. E. (2012b). End-of-life and palliative care education in US pharmacy schools. *American Journal of Hospice and Palliative Medicine*. Accepted on July 13 for publication in a forthcoming issue.

Dickinson, G. E. (2012c). Diversity in death: Body disposition and memorialization. *Illness, Crisis and Loss*, 20(2), 141–158.

Dickinson, G. E., & Ashley-Cameron, S. (1986, April 4–6). *Sex role socialization versus occupational role socialization: A comparison of female physicians' and female nurses' attitudes toward dying patients*. Paper presented at the Eastern Sociological Society's annual meeting, New York City.

Dickinson, G. E., Clark, D., & Sque, M. (2008). Palliative care and end of life issues in UK pre-registration, undergraduate nursing programmes. *Nurse Education Today*, 28, 163–170.

Dickinson, G. E., Clark, D., Winslow, M., & Marples, R. (2005). U.S. physicians' attitudes concerning euthanasia and physician-assisted death: A systematic literature review. *Mortality*, 10(1), 43–52.

Dickinson, G. E., & Field, D. (2002). Teaching end-of-life issues: Current status in United Kingdom and United States medical schools. *American Journal of Hospice and Palliative Care*, 19, 181–186.

Dickinson, G. E., & Fritz, J. L. (1981, September 2). Death in the family. *Journal of Family Issues*, 379–384.

Dickinson, G. E., & Hoffmann, H. C. (2014). Roadside memorials: A 21st century development. In C. Staudt & J. H. Ellens (Eds.), *Our changing journey to the end: Reshaping death, dying, and grief in America*. Santa Barbara, CA: Praeger, 227–252.

Dickinson, G. E., & Hoffmann, H. C. (2009, September 9–12). *Roadside memorial policies in the United States*. Paper presented at the Social Context on Dying, Death & Disposal Conference, Durham, England.

Dickinson, G. E., & Hoffmann, H. C. (2010). Roadside memorial policies in the United States. *Mortality*, 15 (2), 152–165.

Dickinson, G. E., Lancaster, C. J., Summer, E. D., & Cohen, J. S. (1997). Attitudes toward assisted suicide and euthanasia among physicians in South Carolina and Washington. *Omega*, 36, 201–218.

Dickinson, G. E., Lancaster, C. J., Winfield, I. C., Reece, E. F., & Colthorpe, C. A. (1997). Detached concern and death anxiety of first-year medical students: Before and after the gross anatomy course. *Clinical Anatomy*, 10, 201–207.

Dickinson, G. E., & Paul, E. S. (2014). End-of-life issues in UK medical schools. *American Journal of Hospice & Palliative Medicine*. Accepted for publication on March 5, 2014 in a forthcoming issue.

Dickinson, G. E., & Pearson, A. A. (1979a). Differences in attitudes toward terminal patients among selected medical specialties of physicians. *Medical Care*, 17, 682–685.

Dickinson, G. E., & Pearson, A. A. (1979b). Sex differences of physicians in relating to dying patients. *Journal of the American Medical Women's Association*, 34, 45–47.

Dickinson, G. E., & Pearson, A. A. (1980–1981). Death education and physicians' attitudes toward dying patients. *Omega*, 11, 167–174.

Dickinson, G., Roof, P., & Roof, K. (2011). A survey of veterinarians in the US: Euthanasia and other end-of-life issues. *Anthrozoos*, 24(2), 167–174.

Dickinson, G. E., Roof, P., & Roof, K. (2009, September 9–12). *South Carolina veterinarians: Euthanasia and other end-of-life issues*. Paper presented at the Conference on the Social Context of Dying, Death and Disposal, Durham, England.

Dickinson, G. E., Sumner, E. D., & Frederick, L. M. (1992, May–June). Death education in selected health professions. *Death Studies*, 16, 281–289.

Dickinson, G. E., & Tournier, R. E. (1993, January–February). A longitudinal study of sex differences in how physicians relate to dying patients. *Journal of the American Medical Women's Association*, 48, 19–22.

Dickinson, G. E., & Tournier, R. E. (1994). A decade beyond medical school: A longitudinal study of physicians' attitudes toward death and terminally-ill patients. *Social Science and Medicine*, 38, 1397–1400.

Dickinson, G. E., Tournier, R. E., & Still, B. J. (1999). Twenty years beyond medical school: Physicians' attitudes toward death and terminally ill patients. *Archives of Internal Medicine*, 159, 1741–1744.

Diclemente, R., & Sionean, C. (2012). Street violence. eNotes.com. Accessed on October 5, 2012.

Dignitas: Swiss suicide helpers. (2003, January 20). *BBC News*.

Dilworth, D. C. (1996, February). Dying wishes are ignored by hospitals and doctors. *Trial*, 32(2), 79–81.

Ditto, P. (2001). But living will, or advance directive, may help ER, critical-care physicians who don't know patients. *Healthcare News Releases*. Irvine, CA: University of California at Irvine. Retrieved July 13, 2010, from http://www.healthcare.uci.edu/news_releases.asp?filename=LivingWills.htm.

Doctors "kill" patient to save her life. (1988, September 22). Charleston, SC, *News and Courier*, p. 1A.

Doka, K. J. (1987). Silent sorrow: Grief and the loss of significant others. *Death Studies*, 11, 441–449.

Doka, K. J. (1989). *Disenfranchised grief*. Lexington, MA: Lexington Books.

Doka, K. J. (1993). *Living with life-threatening illness: A guide for patients, their families, and caregivers*. New York: Lexington Books.

Doka, K. J. (2003). What makes a tragedy public. In M. Lattanzi-Licht & K. J. Doka (Eds.), *Living with grief: Coping with public tragedy* (pp. 3–13). Washington, DC: Hospice Foundation of America.

Dollimore, J. (2001). *Death, desire and loss in Western culture*. New York: Routledge.

Donne, J. (1930). *Biathanatos*. New York: Facsimile Text Society. (Original work published 1644.)

Donnison, D., & Bryson, C. (1995). Matters of life and death: Attitudes to euthanasia. In R. Jowell, J. Curtice, A. Park, L. Brook, & D. Thomson (Eds.), *British social attitudes: The 13th report* (pp. 161–184). Aldershot, UK: Dartmouth Publishing.

Dore, M. (2013, June 25). Washington state's annual assisted suicide report: What it doesn't say. Retrieved April 6, 2014, from http://www.lifenews.com/2013/06/25/washington-states-annual-assisted-suicide-report-what....

Doughty, C. (2014). *Smoke gets in your eyes: And other lessons from the crematory*. New York: W. W. Norton.

Douglas, J. D. (1991). Patterns of change following parent death in midlife adults. *Omega, 22*, 123–137.

Dowd, Q. L. (1921). *Funeral management and costs: A world survey of burial and cremation*. Chicago, IL: University of Chicago Press.

Downing, A. J. (1921). *Landscape gardening* (10th ed.). New York: John Wiley & Sons.

Dozier, E. P. (1966). *Hano: A Tewa Indian community in Arizona*. New York: Holt, Rinehart and Winston.

Dozier, E. P. (1967). *The Kalinga of Northern Luzon, Philippines*. New York: Holt, Rinehart and Winston.

Duff, R. W., & Hong, L. K. (1995). Age density, religiosity and death anxiety in retirement communities. *Review of Religious Research, 37*, 19–32.

Dumont, R. G., & Foss, D. C. (1972). *The American view of death: Acceptance or denial?* Cambridge, MA: Schenkman.

Dunham, W. (2008, December 15). U.S. stroke, heart disease death rates down sharply. *Reuters*. Retrieved July 13, 2010, from http://www.reuters.com/ article/idUSTRE4BE6FU20081215.

Durand, R. P., Dickinson, G. E., Sumner, E. D., & Lancaster, C. J. (1990, Spring–Summer). Family physicians' attitudes toward death and the terminally-ill patient. *Family Practice Research Journal, 9*(2), 123–129.

Durbin, K. F. (2003). Death, dying, and the dead in popular culture. In C. D. Bryant (Ed.) *Handbook of death and dying* (pp. 43–49). Los Angeles: Sage Publications.

Durkheim, E. (1915). *The elementary forms of religious life*. New York: George Allen and Unwin.

Durkheim, E. (1946). *The division of labor in society* (G. Simpson, Trans.). New York: Free Press. (Original work published 1893.)

Durkheim, E. (1951). *Suicide: A study in sociology*. New York: The Free Press. (Original work published 1897.)

Durkheim, E. (1954). *Elementary forms of the religious life*. New York: The Free Press. (Original work published 1915.)

Durkheim, E. (1961). *Moral education*. Glencoe, IL: Free Press.

Durkheim, E. (1964). *The rules of the sociological method*. New York: The Free Press.

Durkheim, E. (1966). *Suicide*. New York: Free Press.

Durkin, K. F. (2003). Death, dying, and the dead in popular culture. In C. D. Bryant (Ed.) *Handbook of death and dying*. Thousand Oaks, CA: Sage Publications.

The dying of death. (1899, September). In Albert Shaw (Ed.), *Review of Reviews, 20*, 364–365.

Dyregrov, A., & Dyregrov, K. (1999). Long-term impact of sudden infant death: A 12- to 15-year follow-up. *Death Studies, 23*, 635–661.

Eben, A. (2012). *Proof of heaven: A neurosurgeon's journey into the afterlife*. New York: Simon and Schuster.

Eckerd, L. M. (2009). Death and dying course offerings in psychology: A survey of nine Midwestern states. *Death Studies, 33*(8), 762–770.

Edney, A. T. B. (1988). Breaking the news: The problems and some answers. In W. J. Kay et al. (Eds.), *Euthanasia of the companion animal*. Philadelphia, PA: The Charles Press.

Edwards, B. (2013). Changes in Pitjantjatjara mourning and burial practices. Australian Aboriginal Studies, 1(Spring), 31.

Edwards, G., & Mazzuca, J. (1999, March 24). Three quarters of Canadians support doctor-assisted suicide. *Gallup News Service*, Ontario, ON.

Eggertson, L. (2012). Organ donation's 'silver bullet'? *Canadian Medical Association Journal, 184*(16, November 6), E835ff.

Eickelman, D. F. (1987). Rites of passage: Muslim rites. In M. Eliade (Ed.), *The encyclopedia of religion* (12th ed.). New York: Macmillan.

Elias, N. (1985). *On the loneliness of dying*. New York: Blackwell Publishers.

Ellis, J. B., & Range, L. M. (1989). Characteristics of suicidal individuals: A review. *Death Studies, 13*, 485–500.

Elvig, P. (2009). Burial laws. In C. D. Bryant & D. Peck (Eds.), *Encyclopedia of death and the human experience* (pp. 127–130). Thousand Oaks, CA: Sage Publications.

Emanuel, E., & Emanuel, L. (1998). The promise of a good death. *Lancet, 351*, 21–29.

Emmons, N. (1842). Death without order. In J. Ede (Ed.), *The works of Nathaniel Emmons* (Vol. 3). Boston, MA: Crocker and Brewster.

Enck, R. E. (2010). Physician-assisted suicide. *American Journal of Hospice & Palliative Medicine, 27*(7), 441–443.

Encyclopaedia Britannica (1768), Vol. 2. Edinburgh, Scotland: Encyclopaedia Britannica, Inc.

Enright, L. (1994, September). Keeping the corpse company with a whiskey. *Worldwide report: Death*. London: BBC Worldwide.

Ens, C., & Bond, J. B. (2007). Death anxiety in adolescents: The contributions of bereavement and religiosity. *Omega, 55*, 169–184.

Epstein, J. A., & Spirito, A. (2009). Risk factors for suicidality among a nationally representative sample of high school students. *Suicide and Life-Threatening Behavior, 39*, 241–251.

Erickson, K. A. (2013). *How we die now: Intimacy and the work of dying*. Philadelphia, PA: Temple University Press.

Erikson, E. (1959). Identity and the life cycle: Selected papers. *Psychological Issues*, 1(1), 1–171.

Erikson, E. (1963). *Childhood and society*. New York: Norton.

Euthanasia.ProCon.org. (2013). State-by-state guide to physician-assisted suicide. http://euthanasia.procon.org/view.resource.php?resourceID=000132 (Last updated on: 5/28/2013 3:49:15 PM PST). Accessed on November 6, 2013.

Evans, R. (2012). Emotional care for women who experience miscarriage. *Nursing Standard*, 26(42), 35ff.

EXIT. (1980). *A guide to self-deliverance*. London: Author.

Extracts. (1895, August). *Park and Cemetery*, 5, 108.

Faber-Langendoen, K., & Bartels, D. M. (1992, May). Process of forgoing life-sustaining treatment in a university hospital: An empirical study. *Critical Care Medicine*, 20(5), 570–577.

Facts About SIDS. (2009). *What everyone needs to know*. Retrieved July 13, 2010, from http://www.sids-network.org/facts.htm.

Fallis, J. (2012). Stem cell donations. *CMAJ: Canadian Medical Association Journal*, 10(January 10), E13ff.

Faron, L. C. (1968). *The Mapuche Indians of Chile*. New York: Holt, Rinehart and Winston.

Farrell, J. J. (1980). *Inventing the American way of death*, 1830–1920. Philadelphia, PA: Temple University Press.

Faulkner, J., & DeJong, G. F. (1966). Religiosity in 5-D: An empirical analysis. *Social Forces*, 45, 246–254.

Federated Funeral Directors of America. (2013). *Survey of funeral homes*. Springfield, IL: Author.

Feifel, H. (1959). *The meaning of death*. New York: McGraw-Hill.

Feigelman, W., & Gorman, B. S. (2008). Assessing the effects of peer suicide on youth suicide. *Suicide and Life-Threatening Behavior*, 38, 181–194.

Feigelman, W., Gorman, B. S., & Jordan, J. R. (2009). Stigmatization and suicide bereavement. *Death Studies*, 33, 591–608.

Fein, E. B. (1998, January 25). For lost pregnancies, new rites of mourning. *New York Times*, p. 30.

Fenigsen, R., & Fenigsen, R. (2012). Dutch government-ordered surveys of euthanasia. *Issues in Law & Medicine*, 28(2, Fall), 237ff.

Fenigsen, R., & Fenigsen, R. (2012). Who is leading us there? *Issues in Law & Medicine*, 28(2, Fall), 333f.

Fernandez-Mehler, P., Gloor, P., Sager, E., Lewis, F. I., & Glaus, T. M. (2013). Veterinarians' role for pet owners facing pet loss. *Veterinary Record*, 10, 1–7.

Final cut—Medical arrogance and the decline of the autopsy, by A. Gawande, *The New Yorker*, March 19, 2001. http://www.newyorker.com/archive/2001/03/19/010319fa_gwnd_dept_fact.

FindLaw Resources. (2007, June 30). Most Americans don't have a will, says new FindLaw.com survey. Mountain View, CA: Author. Retrieved July 13, 2010, from http://www.findlaw.com.

Fingarette, H. (1996). *Death: Philosophical soundings*. Peru, IL: Open Court Publishing.

Fingerhut, L. A., & Warner, M. (1997). *Injury chartbook. Health, United States*, 1996–97. Hyattsville, MD: National Center for Health Statistics.

Fishman, T. C. (2010). *Shock of gray: The aging of the world's population and how it pits young against old, child against parent, worker against boss, company against rival, and nation against nation*. New York: Simon and Schuster.

Fletcher, J. (1966). *Situational ethics: The new morality*. Philadelphia, PA: Westminster.

Flynn, C. P. (1986). *After the beyond: Human transformation and the near-death experience*. Englewood Cliffs, NJ: Prentice-Hall.

Foderaro, L. W. (1994, April 7). Death no longer a taboo subject. *Palm Beach Post*.

Fortner, B. V., & Neimeyer, R. A. (1999). Death anxiety in older adults: A quantitative review. *Death Studies*, 23, 387–411.

Frank, A. W. (1991). *At the will of the body: Reflections on illness*. Boston, MA: Houghton Mifflin.

Fraser, H. C., Kutner, J. S., & Pfeifer, M. P. (2001). Senior medical students' perceptions of the adequacy of education on end-of-life issues. *Journal of Palliative Medicine*, 4(3), 337–343.

Fraser, T. M., Jr. (1966). *Fishermen of south Thailand: The Malay villagers*. New York: Holt, Rinehart and Winston.

Freidson, E. (1972). *Profession of medicine*. New York: Dodd Mead.

French, S. (1975). The establishment of Mount Auburn and the "rural cemetery" movement. In D. E. Stannard (Ed.), *Death in America*. Philadelphia, PA: University of Pennsylvania Press.

Freund, P. E. S., & McGuire, M. B. (1995). *Health, illness, and the social body: A critical sociology* (2nd ed.). Englewood Cliffs, NJ: Prentice-Hall.

Freund, P. E. S., & McGuire, M. B. (1999). *Health, illness and the social body: A critical sociology*. Englewood Cliffs, NJ: Prentice-Hall.

Frid, M. H., & Perea, A. T. (2007). Euthanasia & thanatology in small animals. *Journal of Veterinary Behavior*, 2(2), 35–39.

Friedman, H., & Kohn, R. (2008). Mortality, or probability of death, from a suicidal act in the United States. *Suicide and Life-Threatening Behavior*, 38, 287–301.

Frontline. (2011). Post mortem: Death investigation in America. http://www.pbs.org/wgbh/pages/frontline/post-mortem/things-to-know/autopsy-101.html, Posted February 1, 2011.

Frosch, D. (2007, August 30). Colorado police link rise in violence to music. *The New York Times*.

Fulton, G. B., & Metress, E. K. (1995). *Perspectives on death and dying*. Boston, MA: Jones and Bartlett Publishers.

Fulton, R., & Owen, G. (1988). Death and society in twentieth century America. *Omega*, 18, 379–394.

Funeral directors. (1883, June). *The Casket*, 8.

Furman, E. (1978). The death of a newborn: Care of the parents. *Birth Family Journal*, 5, 214.

Gabriel, T. (1991, December 8). A fight to the death. *New York Times Magazine*, 46–48.

Gagnon, M., & Beaudry, C. (2013). Return to work during perinatal mourning: The case for organizational support. *Relations Industrielles/Industrial Relations*, 68(3), 457ff.

Gallup, A. M., & Newport, F. (2014, May 18). *The Gallup Poll: Public Opinion 2005*. Lanham, MD: Rowman & Littlefield Publishers. http://www.gallup.com/poll/1576/abortion.aspx.

Gallup Poll. (2005, May 17). *3 in 4 Americans back euthanasia*. Retrieved July 13, 2010, from http://www.editorandpublisher.com Gallup Poll. (2005, July 13–14). Reported on CBS News.

Gallup Survey. (1993, April). Physicians need to detect suicide warning signs. *Geriatrics*, 48, 16.

Gamst, F. C. (1969). *The Qemant: A pagan-Hebraic peasantry of Ethiopia*. New York: Holt, Rinehart and Winston.

Ganley, E. (2008, March 21). French revive euthanasia debate. Charleston, SC, *Post and Courier*, p. 9AA.

Gasperson, K. R. (1996). *Delivering bad news in the clinical context: Current recommendations and student perspectives*. Unpublished master's thesis, University of Kentucky, Lexington.

Gawande, A. (2001). Final cut—Medical arrogance and the decline of the autopsy. *The New Yorker*, March 19, 2001.

Gebhart, J. C. (1927). *The reasons for present-day funeral costs*. Unpublished article.

Geddes, G. (1981). *Welcome joy: Death in Puritan New England*. Ann Arbor, MI: U.M.I. Research Press.

Gellene, D. (2014, March 4). Sherwin B. Nuland obituary. *The New York Times*.

Gentile, M., & Fello, M. (1990, November). Hospice care for the 1990s: A concept coming of age. *The Journal of Home Health Care Practice*, 1–15.

Georges, J., Onwuteaka-Philipsen, B., Muller, M., Van der Wal, G., Van Der Heide, A., & Van Der Maas, P. (2007). Relatives' perspective on the terminally ill patients who died after euthanasia or physician-assisted suicide: A retrospective cross-sectional interview study in the Netherlands. *Death Studies*, 31(1), 1–15.

Gervais, K. G. (1987). *Redefining death*. New Haven, CT: Yale University Press.

Gibbs, J. T. (1997, Spring). African-American suicide: A cultural

paradox. *Suicide and Life-Threatening Behavior*, 27(1), 68–79.

Gibbs, L., & Mangla, I. S. (2012). The high cost of saying goodbye. *CNN Money Magazine* (November 9).

Giblin, P., & Hug, A. (2006). The psychology of funeral rituals. *Liturgy*, 21(1), 11–19.

Gibson, R. (2012). The case for euthanasia and physician-assisted suicide. *ISAA Review: Journal of the Independent Scholars Association of Australia*, 11(1, April), 55ff.

Gilbert, S. M. (2006). *Death's door: Modern dying and the ways we grieve*. New York: W. W. Norton & Company.

Giles-Sims, J., & Lockhart, C. (2006). Explaining cross-state differences in elderly suicide rates and identifying state-level public policy responses that reduce rates. *Suicide and Life-Threatening Behavior*, 36, 694–708.

Gill, S. D. (1981). *Sacred words: A study of Navajo religion and prayer*. Westport, CT: Greenwood.

Gillick, M. (2000). *Lifelines: Living longer, growing frail, taking heart*. New York: W. W. Norton & Company.

Gillon, R. (1990). Editorial: Death. *Journal of Medical Ethics*, 16(1), 3–4.

Gilman, C. P. (1935). The living of *Charlotte Perkins Gilman*. Madison, WI: University of Wisconsin Press.

Ginn, S., Price, A., Rayner, L., Owen, G. S., Hayes, R. D., Hotopf, M., & Lee, W. (2011). Senior doctors' opinions of rational suicide. *Journal of Medical Ethics*, 37, 723–726.

Ginsberg, E. (1991). Access to health care for Hispanics. *Journal of the American Medical Association*, 165, 238–241.

Ginsberg, H., & Opper, S. (1988). *Piaget's theory of intellectual development* (3rd ed.). Englewood Cliffs, NJ: Prentice-Hall.

Glaser, B., & Strauss, A. (1965). *Awareness of dying*. Chicago, IL: Aldine.

Glock, C., & Stark, R. (1966). *Christian beliefs and anti-Semitism*. New York: Harper and Row.

Goffman, E. (1959). *The presentation of self in everyday life*. New York: Doubleday.

Goffman, E. (1963). *Stigma*. Englewood Cliffs, NJ: Prentice-Hall.

Golden, T. R. (2000). *Swallowed by a snake: The gift of the masculine side of healing*. New York: McDonald and Woodward Company.

Goldenberg, J., Pyszczynski, T., Greenberg, J., & Solomon, S. (2000). Fleeing the body: A terror management perspective on the problem of human corporeality. *Personality and Social Psychology Review*, 4, 200–218.

Goldenberg, S. (2000, April 11). Many in America are resigned to pain. *New York Times*, p. D8.

Goldman, I. (1979). *The Cubeo: Indians of the northwest Amazon*. Urbana: University of Illinois Press.

Goleman, D. (1989, December 5). Fear of death intensifies moral code, scientists find. *New York Times*, p. 19.

Gomez, A. (2009, October 19). Cemeteries feel recession's chill. *USA Today*.

Goode, E. (2003, October 28). And still, echoes of a death long past. *New York Times*, p. D1.

Goodwin, D. M., Higginson, I. J., Myers, K., Douglas, H. R., & Normand, C. E. (2000, March 27–29). *Methodological issue in evaluating palliative day care: A multi-centre study*. Paper presented at the Palliative Care Congress, Coventry, England.

Gordon, E. (2013, December 20). By the numbers: HIV/AIDS stigma changing, but infection still spreads. *Newsworks*. Retrieved April 6, 2014, from http://www.newsworks.org/index.php/local/the-pulse/63104-hiv-aids-stigma-changing-but-i....

Gordon, N. P., & Shade, S. B. (1999, April 12). Advance directives are more likely among seniors asked about end-of-life care preferences. *Archives of Internal Medicine*, 159, 701–704.

Gorenstein, D. (2013, December 19). How doctors die: Showing others the way. *New York Times*, p. 2.

Gorer, G. (1955, October). The pornography of death. *Encounter*, 5, 49–53.

Gorer, G. (1965). *Death, grief, and mourning*. Garden City, NY: Doubleday.

Gottlieb, B. (1993). *The family in the Western world: From the black death to the industrial age*. Oxford: Oxford University Press.

Gould, M. S., Greenberg, T., Munfakh, J. L. H., Kleinman, M., & Lubell, K. (2006). Teenagers' attitudes about seeking help from telephone crisis services (hotlines). *Suicide and Life-Threatening Behavior, 36,* 601–613.

Gould, M., Jamieson, P., & Romer, D. (2003). Media contagion and suicide among the young. *American Behavioral Scientist, 46,* 1269–1280.

Grady, D. (2000a, May 30). Charting a course of comfort and treatment at the end of life. *The New York Times,* p. D7.

Grady, D. (2000b, May 29). At life's end, many patients are denied peaceful passing. *The New York Times,* p. A13.

Graham, M., McCarthy, J., & Ryan, J. (2012). How clients cope with the death of a parent: *Learning Disability Practice, 15*(4, May), 14–18.

Granda-Cameron, C., & Houldin, A. (2012). Concept analysis of good death in terminally ill patients. *American Journal of Hospice & Palliative Medicine, 29*(8), 632–639.

Grassi, L., Magnani, K., & Ercolani, M. (1999). Attitudes toward euthanasia and physician-assisted suicide among Italian primary care physicians. *Journal of Pain and Symptom Management, 17,* 188–196.

Greenberg, M. (2003). Good grief: The different ways to cope after loss. *Psychology Today, 36,* 44.

Greene, K. (2002). Many consider a living will important, but wait too long. *The Wall Street Journal,* and for *Encore,* the journal's guide to life after 55. www.uslivingwillregistry.com/wsj.shtm. Accessed on November 6, 2013.

Greening, L., Stoppelbein, L., Dhossche, D., Erath, S., Brown, J., Cramer, R., et al. (2008). Pathways to suicidal behaviors in childhood. *Suicide and Life-Threatening Behavior, 38,* 35–45.

Greer, W. R. (1985, June 30). Putting a price on human life. Louisville, KY, *Courier-Journal,* p. 1D.

Greyson, B. (1997). The near-death experience as a focus of clinical attention. *Journal of Nervous and Mental Disease, 185*(5), 327–334.

Griffith, J. (2013). Suicide and war: The mediating effects of negative mood, posttraumatic stress disorder symptoms, and social support among Army National Guard soldiers. *Suicide and Life-Threatening Behavior, 42*(4), 453–469.

Grollman, E. A. (1972, May). Commencement address. Minneapolis, MN: Department of Mortuary Science, University of Minnesota.

Groopman, J. (2002, October 28). Dying words: How should doctors deliver bad news? *The New Yorker,* pp. 62–70.

Gubrium, J. B. (1975). *Living and dying at Murray Manor.* New York: St. Martin's Press.

Guidelines to help discern SIDS from homicide. (June 21, 1996). Charleston, SC, *Post and Courier,* p. 4A.

Guignon, C. (1993). *The Cambridge companion to Heidegger.* Cambridge: Cambridge University Press.

Guillon, C., & LeBonniec, Y. (1982). *Suicide, its use, history, technique and current interest.* Paris: Editions Alain Moreais.

Guterman, L. (2000, June 2). The dope on medical marijuana. *Chronicle of Higher Education, 46,* A21–A22.

Guyer, B., Martin, J. A., MacDorman, M. F., Anderson, R. N., & Strobino, D. M. (1997). Annual summary of vital statistics—1996. *Pediatrics, 100*(6), 905–918.

Guyer, R. L. (1998, February 6–8). When decisions are life-and-death. *USA Weekend,* 26.

Habenstein, R. W., & Lamers, W. M. (1962). The pattern of late 19th century funerals. In *The history of American funeral directing.* Milwaukee, WI: Bulfin.

Habenstein, R. W., & Lamers, W. M. (1974). *Funeral customs the world over* (Rev. ed.). Milwaukee, WI: Bulfin Printers.

Hafferty, F. W. (1991). *Into the valley: Death and the socialization of medical students.* New Haven, CT: Yale University Press.

Haight, B. K. (1992). Long-term effects of a structured life review process. *Journal of Gerontology: Psychological Sciences, 47,* 312–315.

Hale, N. G. (1971). *The origin and foundations of the psychoanalytic movement in the United States, 1876–1918.* New York: Oxford University Press.

Hall, G. S. (1949). *Youth: Its education, regimen and hygiene.* New York: D. Appleton.

Hammond, P., & Mosley, M. (1999). *Trust me (I'm a doctor).* London: Metro Books.

Hancock, D. R., Williams, M. M., Taylor, A. J. W., & Dawson, B. (2004). Impact of dissection on medical students. *New Zealand Journal of Psychology, 22,* 17–25.

Hanks, G. W., deConno, F., Ripamonti, C., Hanna, M., McQuay, H. J., Mercadante, S., et al. (1996). Morphine in cancer pain: Modes of administration. *British Medical Journal, 312,* 823–826.

Hare, S. J. (1910). The cemetery beautiful. *Association of American Cemetery Superintendents, 24,* 41.

Harper, S. (2009). Advertising six feet under. *Mortality, 14,* 203–225.

Harris Interactive Poll. (2007, November 29). The religious and other beliefs of Americans. Harris Poll #119. Retrieved July 13, 2010, from http://www.harrisinteractive.com/harris_poll.

Hart, C. W. M., & Pilling, A. R. (1960). *The Tiwi of North Australia.* New York: Holt, Rinehart and Winston.

Harvey, J. H. (1996). *Embracing their memory: Loss and the social psychology of storytelling.* Needham Heights, MA: Allyn & Bacon.

Hassan, R. (1996). The euthanasia debate. *Medical Journal of Australia, 165,* 164–165.

Hassrick, R. B. (1964). *The Sioux: Life and customs of a warrior society.* Norman: University of Oklahoma.

Hastings Center, 2011. Bioethics Graduate Programs. *TheHastingsCenter.org* http://www.thehastingscenter.org/BioethicsWire/BioethicsGraduatePrograms/Default.aspx. Accessed on November 8, 2013.

Hatter, B. S. (1996). Children and the death of a parent or grandparent. In C. A. Corr & D. M. Corr (Eds.), *Handbook of childhood death and bereavement* (pp. 131–148). New York: Springer Publishing.

Haviland, W. A. (1991). *Anthropology* (6th ed.). Fort Worth, TX: Holt, Rinehart and Winston.

Hawkins, A. H. (1990). Constructing death: Three pathographies about dying. *Omega, 22,* 301–317.

Hay, E. E. (1900). Influence of our surroundings. *Association of American Cemetery Superintendents, 14,* 46.

Hays, J. C., Gold, D. T., Flint, E. P., & Winer, E. P. (1999). Patient preference for place of death: A qualitative approach. In B. de Vries (Ed.), *End of life issues* (pp. 3–21). New York: Springer Publishing.

Hazlick, R. (Ed.). (1994). *The medical cause of death manual.* Northfield, IL: College of American Pathologists.

Heaven above or below? (2012.) *The Christian Century, 129*(14), 9.

Heidegger, M. (1962). *Being and time.* San Francisco, CA: Harper. (Original work published 1927.)

Helman, C. (1985). *Culture, health, and illness.* Bristol: Wright.

Hendin, H. (1997). Seduced by death: *Doctors, patients and the Dutch cure.* New York: W. W. Norton & Co.

Henneman, E. A., & Karras, G. E. (2004). Determining brain death in adults: A guideline for use in critical care. *Critical Care Nurse, 24,* 50–56.

Hertz, R. (1960). The collective representation of death. In R. Needham & C. Needham (Trans.), *Death and the right hand* (pp. 84–86). Aberdeen: Cohen and West.

Hesselink, B. A. M., Pasman, H. R. W., Was, G., Soethout, M. M. M., & Onwuteaka-Philipsen, B. D. (2010). Education on end-of-life care in the medical curriculum: Students' opinions and knowledge. *Journal of Palliative Medicine, 13*(4), 381–387.

Higginson, I. J. (1999, November/December). Evidence-based palliative care. *European Journal of Palliative Care, 6,* 188–193.

Hill, P. T. (1992). Individual rights vs. state interests: Ethical concerns in thanatology. *Loss, Grief and Care, 6*(1), 51–59.

Hillerman, B. (1980). Chrysalis of gloom: Nineteenth century mourning costume. In M. V. Pike & J. V. Armstrong (Eds.), *A time to mourn: Expressions of grief in nineteenth century America.* Stony Brook, NY: The Museums at Stony Brook.

Hinton, J. (1979). Comparison of places and policies for terminal care. *Lancet, 1*(8106), 29–32.

Hirschfelder, A., & Molin, P. (1992). *The encyclopedia of Native American religions.* New York: Facts on File.

Hitchcock, J. T. (1966). *The Magars of Manyan Hill.* New York: Holt, Rinehart and Winston.

Hocker, W. V. (1987). *Financial and psychosocial aspects of planning and funding funeral services in advance as related to estate planning and life-threatening illness.* Unpublished article distributed by the National Funeral Directors Association.

Hoebel, E. A. (1960). *The Cheyennes: Indians of the Great Plains.* New York: Holt, Rinehart and Winston.

Hoefler, J. M. (2000). Making decisions about tube feeding for severely demented patients at the end of life: Clinical, legal, and ethical considerations. *Death Studies, 24,* 233–254.

Hohenschuh, W. P. (1921). *The modern funeral: Its management.* Chicago, IL: Trade Periodical Company.

Holden, J. E. (2009). Near-death experiences. In C. D. Bryant & D. Peck (Eds.), *Encyclopedia of death and the human experience* (pp. 773–776). Thousand Oaks, CA: Sage Publications.

Holland, H. S. (1919). King of terrors. In C. Cheshire (Ed.), *Fact of the faith: Being a collection of sermons not hitherto published in book form by Henry Scott Holland.* London: Longmans, Green & Co., pp. 125–134 (first preached in 1910).

Hollander, C. (2013). What's the fairest way to dispense donated organs? *National Journal* (July 25). http://www.nationaljournal.com/njonline/. Accessed on November 8, 2013.

Holloway, K. F. C. (2002). *Passed on: African American mourning stories.* Durham, NC: Duke University Press.

Holson, L. (2011). For funerals too far, mourners gather on the Web. *The New York Times,* January 24, p. A1.

Holy Bible (Revised Standard Version) (1962). New York: Oxford University Press.

Homans, G. C. (1965). Anxiety and ritual: The theories of B. Malinowski and R. A. Radcliffe-Brown. In W. A. Lessa & E. Z. Vogt (Eds.), *Reader in comparative religion: An anthropological approach.* New York: Harper and Row.

Homans, J. (2012). *What's a dog for? The surprising history, science, philosophy, and politics of man's best friend.* New York: The Penguin Press.

Hooyman, N. R., & Kiyak, H. A. (1988). *Social gerontology: A multidisciplinary perspective.* Boston, MA: Allyn & Bacon.

Horowitz, M. M. (1967). *Morne-Paysan: Peasant village in Martinique.* New York: Holt, Rinehart and Winston.

Horowitz, R., Gramling, R., & Quill, T. (2013). Palliative care education in US medical schools. *Medical Education, 48*(1), 59–66.

Howarth, G. (2000). Dismantling the boundaries between life and death. *Mortality, 5*(2), 127–138.

Howarth, G. (2009). The demography of death. In A. Kellehear (Ed.), *The study of dying* (pp. 99–122). Cambridge: Cambridge University Press.

Howe, D. W. (1970). *The Unitarian conscience: Harvard University Press, 1805–1861.* Cambridge, MA: Harvard University Press.

Hsu, C. Y., O'Connor, M., & Lee, S. (2009). Understandings of death and dying for people of Chinese origin. *Death Studies, 33*(2), 153–174.

Huang, Z., & Ahronheim, J. C. (2000). Nutrition and hydration in terminally ill patients. *Clinics in Geriatric Medicine, 16*(2), 313–325.

Huber, R., Meade-Cos, V., & Edelen, W. B. (1992). Right to die responses from a random sample of 200. *Hospice Journal, 8*(3), 1–19.

Hughes, J. A., Martin, P. J., & Sharrock, W. W. (1995). *Understanding classical sociology: Marx, Weber, and Durkheim.* London: Sage Publications.

Hughes, T., Schumacher, M., Jacobs-Lawson, J. M., & Arnold, S. (2008). Confronting death: Perceptions of a good death in adults with lung cancer. *American Journal of Hospice & Palliative Medicine, 25,* 39–44.

Humphry, D. (1981). *Let me die before I wake: Hemlock's book of self-deliverance for the dying*. Los Angeles, CA: The Hemlock Society.

Humphry, D. (1991). *Final exit: The practicalities of self-deliverance and assisted suicide for the dying*. Eugene, OR: The Hemlock Society.

Hunfield, J. A. M., Wladimiroff, J. W., Verhage, F., & Passchier, J. (1995). Previous stress and acute psychological defense as predictors of perinatal grief: An exploratory study. *Social Science and Medicine, 40*, 829–835.

Hung, N. C., & Rabin, L. A. (2009). Comprehending childhood bereavement by parental suicide: A critical review of research on outcomes, grief processes, and interventions. *Death Studies, 33*, 781–814.

Hunter, J. (2007–2008). Bereavement: An incomplete rite of passage. *Omega, 56*(2), 153–173.

Hunter, S. B., & Smith, D. E. (2008). Predictors of children's understandings of death: Age, cognitive ability, death experience and maternal communicative competence. *Omega, 57*, 143–162.

Huntington, R., & Metcalf, P. (1992). *Celebrations of death: The anthropology of mortuary ritual*. Cambridge: Cambridge University Press.

Hutchins, S. H. (1986). Stillbirth. In T. A. Rando (Ed.), *Parental loss of a child* (pp. 129–144). Champaign, IL: Research Press Company.

Iglehart, J. K. (2009). A new era of for-profit hospice care—The Medicare benefit. *The New England Journal of Medicine, 360*, 2701–2703.

Illich, I. (1976). *Medical nemesis: The expropriation of health*. New York: Pantheon.

Institute of Medicine. (1997). *Approaching death: Improving care at the end of life*. Washington, DC: National Academy Press.

Ireland, J. (2010, April 3). Children coping with death. *IComfort*. http://www.livestrong.com/article/101614-children-coping-death. Accessed on October 8, 2012.

Irion, P. E. (1956). *The funeral: An experience of value*. Milwaukee, WI: National Funeral Directors Association.

Irvine, M. (2009, October 9). Topic of suicide prevention coming out of closet. *Associated Press* release. *Post and Courier*, Charleston, SC, p. 4A.

Is life expectancy now stretched to its limit? (1990, November 2). *New York Times*, p. A13.

Jackson, E. N. (1963). *For the living*. Des Moines, IA: Channel Press.

Jacoby, S. (2012, March 30). Taking responsibility for death. *New York Times*.

Jacques, E. (1965). Death and the mid-life crisis. *International Journal of Psychoanalysis, 46*, 502–514.

Jauhar, S. (2000, January 4). When decisions can mean life or death. *New York Times*, p. D8.

Jecker, N. S. (1994). Physician-assisted death in the Netherlands and the United States: Ethical and cultural aspects of health policy development. *Journal of the American Geriatric Society, 42*, 672–678.

Jeffrey, D. (1993). *"There is nothing more I can do!": An introduction to the ethics of palliative care*. Cornwell: The Patten Press.

Jenkins, R. (1999, February 18). Expert wants trade in live body parts. London, *The Times*, p. 8.

Joffe, P. (2008). An empirically supported program to prevent suicide in a college student population. *Suicide and Life-Threatening Behavior, 38*, 87–103.

Johnson, C. (2009). Funerals and funeralization in major religious traditions. In C. D. Bryant & D. Peck (Eds.), *Encyclopedia of death and the human experience* (pp. 499–503). Thousand Oaks, CA: Sage Publications.

Johnson, G. R., Krug, E. G., & Potter, L. B. (2000, Spring). Suicide among adolescents and young adults: A cross-national comparison of 34 countries. *Suicide and Life-Threatening Behavior, 30*(1), 74–82.

Johnson, P. V. (1997, April). Creating meaningful events that celebrate life. *Bradshaw Quarterly*.

Johnson, S. (1990, August 26). Near-death experiences almost always change lives. Charleston, SC, *News and Courier/The Evening Post*, p. 13I.

Jonas, E., & Fischer, P. (2006). Terror management and religion: Evidence that intrinsic religiousness mitigates worldview defense following mortality salience. *Journal of Personality and Social Psychology, 91*, 553–567.

Jones, B. (2013). Large collection of practical techniques in grief therapy. *Journal of Psychology and Theology, 41*(3), 256.

Jones, L. S., Paulman, L. E., Thadani, R., & Terracio, L. (2001). Medical student dissection of cadavers improves performance on practical exams but not on NBME anatomy subject exam. *Medical Education Online, 6*(2), np. Retrieved July 13, 2010, from http://www.med-ed-online.org.

Jones, M. (June 9, 2008). Graveyard shifting. *Newsweek*.

Joralemon, D. (1995). Organ wars: The battle for body parts. *Medical Anthropology Quarterly, 9*(3), 335–356.

Joralemon, D. (1999). *Exploring medical anthropology*. Boston, MA: Allyn & Bacon.

Jung, C. (1923). *Psychological types*. London: Pantheon Books.

Jung, C. (1933). *Modern man in search of a soul*. New York: Harcourt and Brace.

Jung, C. (1971). The stages of life. In J. Campbell (Ed.), *The portable Jung*. New York: Viking Press.

Kagan, S. (2012). *Death*. New Haven, CT: Yale University Press.

Kaiser Family Foundation and Health Research and Educational Trust. (2008). *Employer health benefits 2008 annual survey*. Retrieved July 13, 2010, from http://www.kaiseredu.org.

Kaldjian, L. C., Wu, B. J., Jekel, J. F., Kaldjian, E. P., & Duffy, T. P. (1999). Insertion of femoral vein catheters for practice by medical house officers during cardiopulmonary resuscitation. *The New England Journal of Medicine, 341*, 2088–2091.

Kalish, R., & Reynolds, D. (1976). *Death and ethnicity: A psychocultural study*. Los Angeles, CA: University of Southern California.

Kamerman, J. B. (1988). *Death in the midst of life: Social and cultural influences on death, grief and mourning*. Englewood Cliffs, NJ: Prentice-Hall.

Kaplan, A. (2008, November 11). Interview with Neal Conan (host) on National Public Radio. Medical advances complicate definition of death.

Kass, L. R. (1971). Death as an event: A commentary on Robert Morison. *Science*, 173, 698–702.

Kastenbaum, R. J. (1986). Death in the world of adolescence. In C. A. Corr & J. N. McNeil (Eds.), *Adolescence and death* (pp. 4–15). New York: Springer Publishing.

Kastenbaum, R. J. (2000). *The psychology of death* (3rd ed.). New York: Springer Publishing.

Kastenbaum, R. J. (2001). *Death, society, and human experience* (7th ed.). Boston, MA: Allyn & Bacon.

Kastenbaum, R. J. (2009). *Death, society, and human experience*. Boston, MA: Allyn & Bacon.

Kaufman, S. R. (2000, March). In the shadow of death and dying: Medicine and cultural quandaries of the vegetative state. *American Anthropologist*, 102, 69–83.

Kavanaugh, R. E. (1972). *Facing death*. Baltimore, MD: Penguin Books.

Kaveny, C. (2011). Dignity & the end of life: How not to talk about assisted suicide. *Commonweal*, 138(13, July 15), 6.

Kearl, M. (2009). Social class and death. In C. D. Bryant & D. Peck (Eds.), *Encyclopedia of death and the human experience* (pp. 875–878). Thousand Oaks, CA: Sage Publications.

Kellaher, L., Pendergast, D., & Hockey, J. (2005, November). In the shadow of the traditional grave. *Mortality*, 10(4), 237–250.

Kelland, K. (2012, September 20). Middle-aged men have higher suicide risk. *Reuters. NBC News*. http://vitals.nbc.news.com/.

Kellehear, A. (1984). Are we a "death-denying" society? A sociological review. *Social Science and Medicine*, 18, 713–723.

Kellehear, A. (1990). *Dying of cancer: the final year of life*. Chur, Switzerland: Harwood Academic Publishers.

Kellehear, A. (1996). *Experiences near death: Beyond medicine and religion*. New York: Oxford University Press.

Kellehear, A. (2009). *The study of dying: from autonomy to transformation*. Cambridge: Cambridge University Press.

Kelly, T. E. (1987, February). Predict preneed vital to financial future. *The American Funeral Director*, 31–69.

Kelner, N. J., & Bourgeault, I. L. (1993). Patient control over dying: Responses of health care professionals. *Social Science and Medicine*, 36, 757–765.

Kennell, J. H., Slyter, H., & Klaus, M. H. (1970). The mourning response of parents to the death of a newborn infant. *The New England Journal of Medicine*, 283, 344–349.

Kesselman, I. (1990). Grief and loss: Issues for abortion. *Omega*, 21(3), 241–247.

Kidd, S., Henrich, C. C., Brookmeyer, K. A., Davidson, L., King, R. A., & Shahar, G. (2006). The social context of adolescent suicide attempts: Interactive effects of parent, peer, and school social relations. *Suicide and Life-Threatening Behavior*, 36, 386–395.

Kircher, L. T. (1992). Autopsy and mortality statistics: Making a difference. *Journal of the American Medical Association*, 267, 1264–1270.

Kirchoff, K. (2003). Analysis of end-of-life content in critical care nursing textbooks. *Journal of Professional Nursing*, 19, 372–381.

Klass, D. (1987). Marriage and divorce among bereaved parents in a self-help group. *Omega*, 17, 237–249.

Kleinfield, N. R. (September 11, 2002). Still New York in all its pain and glory. *New York Times*.

Klima, G. J. (1970). *The Barabaig: East African cattle-herders*. New York: Holt, Rinehart and Winston.

Kloeppel, D. A., & Hollins, S. (1989). Double handicap: Mental retardation and death in the family. *Death Studies*, 13, 31–38.

Klomek, A. B., Marrocoo, F., Kleinman, M., Schonfeld, S., & Gould, M. S. (2008). Peer victimization, depression, and suicidality in adolescents. *Suicide and Life-Threatening Behavior*, 38(2), 166–180.

Klonoff-Cohen, H., Edelstein, S. L., Leftkowitz, E. S., Srinivasan, I. P., Kaegi, D., Chang, J., et al. (1995). The effect of smoking and tobacco exposure through breast milk on sudden infant death syndrome. *Journal of the American Medical Association*, 273, 795–798.

Knapp, R. J. (1986). *Beyond endurance: When a child dies*. New York: Schocken Books.

Knobel, P. S. (1987). Rites of passage: Jewish rites. In M. Eliade (Ed.), *The encyclopedia of religion* (12th ed.). New York: Macmillan.

Koenig, H. G. (1993). Legalizing physician-assisted suicide: Some thoughts and concerns. *Journal of Family Practice*, 37, 171–179.

Kolata, G. (1995, November 2). After 80, Americans live longer than others. *The New York Times*, p. A13.

Koshal, A. (1994, July). Ethical issues in xenotransplantation. *Bioethics Bulletin*, 6(3).

Kotarba, J. A. (1983). Perceptions of death, belief systems and the process of coping with chronic pain. *Social Science and Medicine*, 17, 681–689.

Kotch, J. B., & Cohen, S. R. (1985). SIDS counselors' reports of own and parents' reactions to reviewing the autopsy report. *Omega*, 16, 129–139.

Kowalczyk, L., & Heisel, W. (1999, September 26). Cadavers becoming commodities? *The Orange County [CA] Register*.

Kristof, N. D. (1996, September 29). For rural Japanese, death doesn't break family ties. *New York Times*, p. 10.

Kübler-Ross, E. (1969). *On death and dying*. New York: Macmillan.

Kübler-Ross, E. (1975). *Death: The final stage of growth*. Englewood Cliffs, NJ: Prentice-Hall.

Kuper, H. (1963). *The Swazi: A South African kingdom*. New York: Holt, Rinehart and Winston.

Kushner, H. S. (1981). *When bad things happen to good people*. New York: Schocken Books.

Kushner, H. S. (1985, October). Lecture given in Charleston, SC.

Kutner, L. (1990, December 6). The death of a parent can profoundly alter the relationships of adult siblings. *New York Times*, p. B7.

LaFarge, O. (1956). *A pictorial history of the American Indian*. New York: Crown.

LaGrand, L. E. (1991). United we cope: Support groups for the dying and bereaved. *Death Studies*, 15, 207–230.

Lakhani, N. (2008). Sharp fall in number of autopsies puts patients' lives at risk. *The Independent*, (August 3).

Landberg, J. (2009). Per capita alcohol consumption and suicide rates in the U.S., 1950–2002. *Suicide and Life-Threatening Behavior, 39,* 452–459.

Lantos, J. (2000, September 8). How to live as we are dying. *The Chronicle of Higher Education, 47,* B18–B19.

Leake, J. (2009, September 13). Heart attacks plummet after smoking ban. *The Sunday Times,* 1–2.

Lemelle, A. J., Harrington, C., & Leblanc, A. J. (2000). *Readings in the sociology of AIDS.* Upper Saddle River, NJ: Prentice Hall.

Leming, M. R. (1979–1980). Religion and death: A test of Homans' thesis. *Omega, 10*(4), 347–364.

Leming, M. R., & Premchit, S. (1992). Funeral customs in Thailand. In J. Morgan (Ed.), *Personal care in an impersonal world.* Amityville, NY: Baywood Publishing.

Leming, M. R., Vernon, G. M., & Gray, R. M. (1977, July). The dying patient: A symbolic analysis. *International Journal of Symbology, 8,* 77–86.

Lepowsky, Maria. (1994). *Fruit of the motherland: Gender in an egalitarian society.* Madison, WI: University of Wisconsin Press.

Lerner, M. (1970). When, why and where people die. In O. G. Brim, H. E. Freeman, & N. A. Scotch (Eds.), *The dying patient* (p. 14). New York: Russell Sage Foundation.

Lessa, W. A. (1966). *Ulithi: A Micronesian design for living.* New York: Holt, Rinehart and Winston.

Lester, D. (2009). The use of the Internet for counseling the suicidal individual: Possibilities and drawbacks. *Omega, 58,* 233–250.

Lewis, C. S. (1961). *A grief observed.* London: Faber and Faber.

Lewis, R. (2013, January). A cure for what ails you. *Scientific American, 308*(1), 32.

Liaison Committee on Medical Education (LCME). (2010, June). Functions and structure of a medical school. Standards for accreditation of medical education programs leading to the M.D. degree (ED-13). www.lcme.org/functions2010march.pdf, p. 9. Accessed on December 13, 2010.

Lifton, R. J., & Olson, E. (1974). *Living and dying.* New York: Praeger.

Lin, D. (1992). Hospice helps make death a time of dignity. *Free China Journal,* 5.

Lindemann, E. (1944, September). Symptomatology and management of acute grief. *American Journal of Psychiatry, 101,* 141–148.

Linehan, M. M. (2000). Behavioral treatments of suicidal behavior: Definitional obfuscation and treatment outcomes. In R. W. Maris, S. S. Canetto, J. L. McIntosh, & M. M. Silverman (Eds.), *Review of suicidology* (pp. 84–111). New York: Guilford.

Lino, M. (1990, July). The $3,800 farewell. *American Demographics,* 8.

Liu, Ka-Yuet. (2009). Suicide rates in the world: 1950–2004. *Suicide and Life-Threatening Behavior, 29,* 204–213.

Lloyd, P. (1980). Posthumous mourning portraiture. In M. V. Pike & J. G. Armstrong (Eds.), *A time to mourn: Expressions of grief in nineteenth century America.* Stony Brook, NY: The Museums at Stony Brook.

Lock, M. (1995). Contesting the natural in Japan: Moral dilemmas and technologies of dying. *Culture, Medicine and Psychiatry, 19,* 1–38.

Logue, B. J. (1994). When hospice fails: The limits of palliative care. *Omega, 29,* 291–301.

Long, J. B. (1987). Underworld. In M. Eliade (Ed.), *The encyclopedia of religion* (12th ed.). New York: Macmillan.

A look at statistics (2012). Online Schools. http://www.onlineschools.org/visual-academy/suicide-stats. Accessed on October 15, 2012.

Lowis, M. J., & Hughes, J. (1997). A comparison of the effects of sacred and secular music on elderly people. *Journal of Psychology, 131*(1), 45–55.

Ludwig, A. (1966). *Graven images: New England stonecarving and its symbols.* Middletown, CT: Wesleyan University Press.

Lyness, D. (2012). Helping your child deal with death. *KidsHealth.* http://kidshealth.org. Accessed on October 7, 2012.

Maciejewski, P. K., Zang, B., Block, S. D., & Prigerson, H. G. (2007). An empirical examination of the stage theory of grief. *Journal of the American Medical Association, 297,* 716–723.

Madwar, S. (2011). United States officials propose further retreat from first-come, first-served organ donation. *CMAJ: Canadian Medical Association Journal, 12*(July), 639ff.

Magill, L. (2009). The meaning of the music: The role of music in palliative care music therapy as perceived by bereaved caregivers of advanced cancer patients. *American Journal of Hospice & Palliative Medicine, 26,* 33–39.

Maguire, P., & Faulkner, A. (1992). Communicating with cancer patients: Handling bad news and difficult questions. *British Medical Journal, 297,* 907–909.

Mails, T. E. (1991). *The people called Apache.* Englewood Cliffs, NJ: Prentice Hall.

Maimon, D., Browning, C. R., & Brooks-Gunn, J. (2010). Collective efficacy, family attachment, and urban adolescent suicide attempts. *Journal of Health and Social Behavior, 51*(3), 307–324.

Malan, V. D. (1958). *The Dakota Indian family.* Bulletin No. 470. Brookings, SD: South Dakota State College.

Malinowski, B. (1929). *The sexual life of savages.* New York: Harcourt, Brace and World.

Malinowski, B. (1948). *Magic, science and religion, and other essays.* Boston, MA: Beacon Press. (Original work published in 1925.)

Mandelbaum, D. (1959). Social uses of funeral rites. In H. Feifel (Ed.), *The meaning of death.* New York: McGraw-Hill.

Mandell, F., McClain, M., & Reece, R. M. (1987). Sudden and unexpected death. *American Journal of Diseases of Children, 141,* 748–750.

Mander, J. (1991). *Absence of the sacred: Failure of technology and the survival of the Indian Nations.* San Francisco, CA: Sierra Club Books.

Mandrusiak, M., Rudd, M. D., Joiner, T. E., Berman, A. L., Orden, K. A. V., & Witte, T. (2006). Warning signs for suicide on the Internet: A descriptive study. *Suicide and Life-Threatening Behavior, 36,* 263–271.

Manning, M. (1998). *Euthanasia and physician-assisted suicide.* Mahwah, NJ: Paulist Press.

The man nobody envies: An account of the experiences of an undertaker. (1914, June). *American Magazine, 77,* 68–71.

Mansnerus, L. (1995, November 26). Dying writer Leary wants creative ending to his story. Charleston, SC, *Post and Courier*, p. 8A.

Marchione, M. (2008, December 17). Nation's 1st face transplant done. *Associated Press*.

Marcu, O. (2007). Meaning making and coping: Making sense of death. *Cognition, Brain, Behavior, 11*, 397–416.

Markowitz, J. (2005, April 17). Funerals have come a long way, baby. *Pittsburgh Tribune Review*, Pittsburgh, PA.

Marks, S. C., & Bertman, S. L. (1980). Experiences with learning about death and dying in the undergraduate anatomy curriculum. *Journal of Medical Education, 55*, 844–850.

Markson, L., Clark, J., Glantz, L., Lamberton, V., Kern, D., & Stollerman, G. (1997). The doctor's role in discussing advance preferences for end-of-life care: Perceptions of physicians practicing in the VA. *Journal of the American Geriatrics Society, 45*, 399–406.

Marshall, R., Sutherland, P. (2008). The social relations of bereavement in the Caribbean. *Omega, 57*(1), 21–34.

Martin, S. C., Arnold, R. M., & Parker, R. M. (1988, December). Gender and socialization. *Journal of Health and Social Behavior, 29*, 333–343.

Martini, D. (2009). *Helping children cope with chronic illness.* Chicago, IL: American Academy of Child & Adolescent Psychiatry. Retrieved August 7, 2009 from http://www .aacap.org.

Martinson, I. M., & Campos, R. G. (1991). Long-term responses to a sibling's death from cancer. *Journal of Adolescent Research, 6*, 54–69.

Marx, K. (1967). *Capital: A critique of political economy* (Vol. 1). New York: International Publishers.

Matcha, D. A. (2000). *Medical sociology.* Boston, MA: Allyn & Bacon.

Mayo, W. R. (1916). Address. *Association of American Cemetery Superintendents, 2*, 51.

Mazzarino-Willett, A. (2010). Deathbed phenomena: Its role in peaceful death and terminal restlessness. *American Journal of Hospice & Palliative Medicine, 27*(2), 127–133.

McBrien, R. P. (1987). Roman Catholicism. In M. Eliade (Ed.), *The encyclopedia of religion* (12th ed.). New York: Macmillan.

McCarthy, A. (1991, September 13). The country of the old. *Commonweal, 118*, 505–506.

McClowry, S. G., Davies, E. B., May, K. A., Kulenkamp, E. J., & Martinson, I. M. (1987). The empty space phenomenon: The process of grief in the bereaved family. *Death Studies, 11*, 361–374.

McCown, D. E., & Pratt, C. (1985). Impact of sibling death on children's behavior. *Death Studies, 9*, 323–335.

McCurry, J. (2006, June 2). Internet suicides rise in Japan. *The Guardian* (London), p. 18.

McGrath, P., Pun, P., & Holewa, H. (2012). Decision-making for living kidney donors: An instinctual response to suffering and death. *Mortality, 17*(3), 201–220.

McGuire, D. J., & Ely, M. (1984, January–February). Childhood suicide. *Child Welfare, 63*, 17–26.

McGuire, D. (1988). Medical student, fourth year. In I. Yalof (Ed.), *Life and death: The story of a hospital* (pp. 339–344). New York: Random House.

McKhann, C. F. (1999). *A time to die: The place for physician assistance.* New Haven, CT: Yale University Press.

McLaren, S., Gomez, R., Bailey, M., & Van der Horst, R. K. (2007). The association of depression and sense of belonging with suicidal ideation among older adults: Applicability of resiliency models. *Suicide and Life-Threatening Behavior, 37*, 89–102.

McMenamy, J. M., Jordan, J. R., & Mitchell, A. M. (2008). What do suicide survivors tell us they need? Results of a pilot study. *Suicide and Life-Threatening Behavior, 38*, 375–389.

McNamara, B. (1997, April 4–6). *A good enough death?* Paper presented at the Social Context of Dying, Death and Disposal Third International Conference, Cardiff University, Wales.

McNamara, B., Waddell, C., & Colvin, M. (1994). The institutionalization of the good death, *Social Science and Medicine, 39*, 1501–1508.

McNeil, J. N. (1986). In talking about death: Adolescents, parents, and peers. In C. A. Corr & J. N. McNeil (Eds.), *Adolescence and death* (pp. 185–199). New York: Springer Publishing.

Mead, H. M. (1991). Sleep, sleep, sleep; Farewell, farewell, farewell: Maori ideas about death. In D. R. Counts & D. A. Counts (Eds.), *Coping with the final tragedy: Cultural variation in dying and grieving* (pp. 43–51). Amityville, NY: Baywood Publishing.

Mebane, E. W., Oman, R. F., Kroonen, L. T., & Goldstein, M. K. (1999). The influence of physician race, age, and gender on physician attitudes toward advance care directives and preferences for end-of-life decision-making. *Journal of the American Geriatrics Society, 47*, 579–591.

MEC/PAC (2012, June). A data book: Health care spending and the Medicare program. Medicare payment advisory committee, p. 5.

Medhanandi, Sister. (1996). *The joy hidden in sorrow* [Online]. Retrieved July 16, 2010, from http://www.buddhanet.net/ joydeath.htm.

Medhanandi, Sister. (1998). *The way of the mystic* [Online]. Retrieved July 16, 2010, from http://www .buddhanet.net/mystic.htm.

Medical marijuana. (2009). [Online] Retrieved July 16, 2010, from http://medicalmarijuana.procon .org.

Medicare payment policy (2011, March). Report to the Congress. Hospice, Chapter 11. Washington, DC.

Meier, D. E., Emmons, C. A., Wallenstein, S., Quill, T., Morrison, R. S., & Casel, C. K. (1998). A national survey of physician-assisted suicide and euthanasia in the United States. *The New England Journal of Medicine, 338*, 1193–1201.

Meij, L. W., Stroebe, M., Stroebe, W., Schut, H., Van Den Bout, J., Van Der Heijden, P., et al. (2008, March). The impact of circumstances surrounding the death of a child on parents' grief. *Death Studies, 32*(3), 237–252.

Melo, C. G., & Oliver, D. (2011). Can addressing death anxiety reduce health care workers' burnout and improve patient care? *Journal of Palliative Care, 27*(4), 287–295.

Merchant, C., Kramer, A., Joe, S., Venkataraman, S., & King, C. A. (2009). Predictors of multiple suicide attempts among suicidal black adolescents. *Suicide and Life-Threatening Behavior, 39,* 115–124.

Mermann, A. C. (1997, July 11). Preparing medical students to provide care for patients at the end of life. *Chronicle of Higher Education,* p. B3.

Mermann, A. C., Gunn, D. B., & Dickinson, G. E. (1991, January). Learning to care for the dying: A survey of medical schools and a model course. *Academic Medicine, 66,* 35–38.

Merrin, W. (1999). Crash, bang, wallop! What a picture! The death of Diana and the media. *Mortality, 4*(1), 41–62.

Merton, R. K. (1949). The bearing of sociological theory on empirical research. In *Social structure and social theory.* New York: The Free Press.

Merton, R. K. (1968). The self-fulfilling prophecy. In *Social theory and social structure.* New York: Free Press.

Metress, E. (1990). The American wake of Ireland: Symbolic death ritual. *Omega, 21,* 147–153.

Middleton, J. (1965). *The Lugbara of Uganda.* New York: Holt, Rinehart and Winston.

Miles, S. H., & August, A. (1990, Spring–Summer). Courts, gender and "the right to die." *Law, Medicine and Health Care, 18*(1, 2), 85–95.

Miller, F. G. (2011). Death and organ donation: Back to the future. *Issues in Law & Medicine, 27*(1, Summer), 88ff.

Millie, J. (2008, February). Supplicating, naming, offering: Tawassul in West Java. *Journal of Southeast Asian Studies, 39*(1), 107–122.

Mills, E. (2013, March 10). So tell me: how do you want to die? *The Sunday Times,* London, p. 4.

Mims, C. (1999). *When we die: The science, culture, and rituals of death.* New York: St. Martin's Press.

Mishara, B. L., & Weisstub, D. N. (2007). Ethical, legal, and practical issues in the control and regulation of suicide promotion and assistance over the Internet. *Suicide and Life-Threatening Behavior, 37,* 58–65.

Missler, M., Stroebe, M., Geurtsen, L., Mastenbroek, M., Chmoun, S., & Van der Houwen, K. (2012). Exploring death anxiety among elderly people: A literature review and empirical investigation. *Omega, 64*(4), 357–379.

Mitford, J. (1963). *The American way of death,* Greenwich, CT: Fawcett Publications.

Moller, D. W. (1996). *Confronting death: Values, institutions, and human mortality.* New York: Oxford University Press.

Moller, D. W. (2000). *Life's end: Technocratic dying in an age of spiritual yearning.* Amityville, NY: Baywood Publishing.

Monaghan, P. (2002, February 22). The unsettled question of brain death. *Chronicle of Higher Education,* 14–16.

Montagu, A. (1968). *The natural superiority of women* (Rev. ed.). New York: Collier Books.

Montgomery, L. (1996, December 12). AMA to teach physicians how to aid the dying patient. Charleston, SC, *Post and Courier,* p. 4A.

Montross, C. (2007). *Body of work: Meditations on mortality from the human anatomy lab.* New York: The Penguin Press.

Moody, R. A., Jr. (1975). Life after life: The investigation of a phenomenon—Survival of bodily death. Boston, MA: G. K. Hall.

Morris, R. A. (1991). Po starykovsky (the old people's way): End of life attitudes and customs in two traditional Russian communities. In D. R. Counts & D. A. Counts (Eds.), *Coping with the final tragedy: Cultural variation in dying and grieving* (pp. 91–112). Amityville, NY: Baywood Publishing.

Morrisey, B. (2012). Coping with death through accidents. Facing bereavement. http://www .facingbereavement.co.uk/coping-with-death-though-accidents.html. Accessed on October 19, 2012.

Morrison, R. S., & Morris, J. (1995, July). When there is no cure: Palliative care for the dying patient. *Geriatrics, 50,* 45–50.

Morriss, F. (1987). Euthanasia is never justified. In J. Rohr (Ed.), *Death and dying: Opposing viewpoints.* St. Paul, MN: Greenhaven Press.

Morrissey, M. B. (2011). Phenomenology of pain and suffering at the end of life: a humanistic perspective in gerontological health and social work. *Journal of Social Work in End-of-Life & Palliative Care, 7*(1), 14–38.

Mulkay, M. (1993). Social death in Britain. In D. Clark (Ed.), *The sociology of death: Theory, culture, practice* (pp. 31–49). Oxford, UK: Blackwell Publishers.

Mumford, L. (1947, March). Atom bomb: Social effects. *Air Affairs, 1,* 370–382. Reprinted as Mumford, L. (1954). *In the name of sanity* (pp. 10–33). New York: Norton.

Murphy, P., & Perry, K. (1988). Hidden grievers. *Death Studies, 12,* 451–462.

Murray, J., & Callan, V. J. (1988). Predicting adjustment to perinatal death. *British Journal of Medical Psychology, 61,* 237–244.

Murray, R. (2008). A search for death: How the Internet is used as a suicide cookbook. *Chrestomathy, 7,* 142–156.

Nagamine, T. (1988). Attitudes toward death in rural areas of Japan. *Death Studies, 12,* 61–68.

Nagi, M. H., Puch, M. D., & Lazerine, N. G. (1978). Attitudes of Catholic and Protestant clergy toward euthanasia. *Omega, 8,* 153–164.

NAHIC (National Adolescent Health Information Center). (2007). *Fact sheet on violence: Adolescents and young adults.* San Francisco, CA: Author, University of California at San Francisco.

Nano, S. (2008, August 15). When can a donor be declared dead? *Associated Press.*

Nardi, P. (1990). AIDS and obituaries: The perpetuation of stigma in the press. In D. A. Feldman (Ed.), *Culture and AIDS* (pp. 159–168). New York: Praeger.

National Association of Medical Examiners. (1996). *So you want to be a medical detective?* St. Louis, MO: Author.

National Bureau of Economic Research. (2014). Long-term care insurance and nursing home use. http://www .nber.org/digest/mar04/w9957.html. Accessed on April 18, 2014.

National Cancer Institute. (2000, May 14). *Annual report to the nation on the status of cancer, 1973–1997.* Bethesda, MD: Author.

National Center for Children Exposed to Violence (2005). *Media violence.* http://nccev.org/violence/media.html. Accessed on October 5, 2012.

National Center for Health Statistics. (1995). *Health risk behaviors among our nation's youth* (PHS Publication No. 95-1520). Hyattsville, MD: Author.

National Center for Health Statistics (2012). *National Vital Statistics Reports 2010, 60*(4), Atlanta, GA, Author.

National Funeral Directors Association. (1997). *Fact sheet for 1996.* Milwaukee, WI: Author.

National Funeral Directors Association. (1999). *Fact sheet for 1998.* Milwaukee, WI: Author.

National Funeral Directors Association. (2003). *Fact Sheet for 2002.* Milwaukee, WI: Author.

National Funeral Directors Association. (2009). *Fact Sheet for 2008.* Milwaukee, WI: Author.

National Funeral Directors Association. (2013). *Fact Sheet for 2008.* Milwaukee, WI: Author.

National Hospice and Palliative Care Organization (NHPCO). (2012). *NHPCO facts and figures: Hospice care in America*, 2011 edition.

National Institute of Mental Health. (2003). *Suicide facts.* Bethesda, MD: Author.

National Television Violence Study (2003). Kaiser Family Foundation, Key Facts: TV Violence. http://nccev.org/violence/media.html. Accessed on October 5, 2012.

Neale, R. E. (1973). *The art of dying.* New York: Harper and Row.

Neergaard, L. (2013, January 8). Cancer death rates falling, report says. *Associated Press.*

Neimeyer, R. (2012). Techniques of grief therapy: Creative practices for counseling the bereaved. New York: Routledge Publishers.

Nelson, B. (2000, June 11). Research use of embryos sets off clash. Charleston, SC: *The Post and Courier*, 15-A [Newsday release].

Ness, D. E., & Pheffer, C. R. (1990, March). Sequelae of bereavement resulting from suicide. *American Journal of Psychiatry, 147*(3), 279–285.

Newport, F. (2007). Questions and answers about Americans' religion. Gallup Organization. Retrieved July 13, 2010, from http://www.gallup.com/poll/103459/Questions-Answers-About-Americans-Religion.aspx#3.

New York Organ Donor Network (2009). Online. Retrieved from http://www.donatelifeny.org.

Nichols, J. (1984, November). Illegitimate mourners. In *Children and death: Perspectives and challenges.* Symposium sponsored by Children's Hospital Medical Center of Akron, Akron, OH.

Nicol, M. T., Tompkins, J. R., Campbell, N. A., & Syme, G. J. (1986). Maternal grieving response after perinatal death. *The Medical Journal of Australia, 144*, 287–289.

Nieburg, H. A., & Fischer, A. (1982). *Pet loss: A thoughtful guide for adults and children.* New York: Harper & Row.

Niemiec, R. M., & Schulenberg, S. E. (2011). Understanding death attitudes: The integration of movies, positive psychology, and meaning management. *Death Studies, 35*, 387–401.

Noonan, D. (2008, April 28). Doctors who kill themselves. *Newsweek*, p. 16.

Noppe, I. C., Noppe, L. D., & Bartell, D. (2006). Terrorism and resilience: Adolescents' and teachers' responses to September 11, 2001. *Death Studies, 30*, 41–60.

Nordheim, D. V. (1993). Vision of death in rock music and musicians. *Popular Music and Society, 17*, 21–31.

Norris-Shortle, C., Young, P. A., & Williams, M. A. (1993). Understanding death and grief for children three and younger. *Social Work, 38*, 736–742.

Novak, M. (2009). *Issues in aging.* Boston, MA: Pearson.

Nuland, S. B. (1994). *How we die: Reflections on life's final chapter.* New York: Alfred A. Knopf.

Nuland, S. B. (1998, November 2). The right to die. *The New Republic, 219*(18), 29–35.

Nursing Standard. (2011). The gift of life—at a cost: Should the health service pay for the funerals of organ donors? We asked our readers panel. *Nursing Standard, 26*(14, December 7), 28.

O'Dea, T. (1966). *The sociology of religion.* Englewood Cliffs, NJ: Prentice-Hall.

Oestigaard, T., & Goldhahn, J. (2006). From the dead to the living: Death as transitions and re-negotiations. *Norwegian Archeological Review, 39*(1), 27–48.

Ogden, R. D. (1995, December). *The right to die: A rejoinder to Bruce Wilkinson's critique. Canadian Public Policy, 21*(4), 456–460.

O'Halloran, C. M., & Altmaier, E. M. (1996). Awareness of death among children: Does a life-threatening illness alter the process of discovery? *Journal of Counseling and Development, 74*, 259–262.

Oken, D. (1961). What to tell cancer patients. *Journal of the American Medical Association, 175*(13), 1120–1128.

Oliver, L. E. (1999). Effects of a child's death on the marital relationship: A review. *Omega, 39*, 197–227.

Olson, L. K. (2003). *The not so golden years: Caregiving, the frail elderly, and the long-term care establishment.* New York: Rowman & Littlefield.

Olson, M. (1997). *Healing the dying.* Albany, NY: Delmar Publishers.

O'Mathuna, D. P. (1996). Medical ethics and what it means to be human. *Irish Bible School Journal*, 12–19.

O'Meara, K. P. (1999). Harvesting fetal body parts. *Insight*, News World Communications.

On death as a constant companion. (1965, November 12). *Time Magazine, 86*(20), 52–53.

Onishi, N. (2008, July 14). In Japan, Buddhism, long the religion of funerals, may itself be dying out. *New York Times*, 6.

Onstad, E. (2000, November 28). Dutch approve law on mercy killings, protests start. *Associated Press* release.

Orbach, I., Gross, Y., Glaubman, H., & Berman, D. (1985). Children's perception of death in humans and animals as a function of age, anxiety and cognitive ability. *Journal of Child Psychology and Psychiatry, 26*, 453–463.

O'Reilly, K. (2008). Willing, but waiting: Hospital ethics committees. *American Medical News* (amednews.com). Posted Jan. 28, 2008. http://www.amednews.com/article/20080128/profession/301289970/4/.

Oswalt, W. H. (1986). *Life cycles and lifeways*. Palo Alto, CA: Mayfield Publishing.

Oxley, J. M. (1887, February). The reproach of mourning. *Forum, 2*, 608–614.

Palgi, P., & Abramovitch, H. (1984). Death: A cross-cultural perspective. *Annual Review of Anthropology, 13*, 385–417.

Pardue, P. (1968). Buddhism. In *International encyclopedia of the social sciences* (2nd ed.) (pp. 165–184). New York: Macmillan.

Parents Television Council (2011). TV bloodbath: Violence on prime time broadcast TV. Retrieved September 28, 2012, from http://www .parentstv.org/PTC/publications/ reports/stateindustryviolence/ exsummary.asp.

Parker, G. D., Smith, T., Corzine, M., Mitchell, G., Schrader, S., Hayslip, B., & Fanning, L. (2012). Assessing attitudinal barriers toward end-of-life care. *American Journal of Hospice & Palliative Medicine, 29*(6), 438–442.

Parkes, C. M. (1972). *Bereavement: Studies of grief in adult life*. New York: International Universities Press.

Parkes, C. M. (2009). *Bereavement: Studies of grief in adult life*. New York: International Universities Press.

Parsons, T. (1951). *The social system*. New York: Free Press.

Parsons, T. (1958). The definitions of health and illness in light of American values and social structure. In E. E. Jaco (Ed.), *Patients, physicians, and illness*. New York: Free Press.

Parsons, T. (1978). Death in the Western world. In Parsons, T., *Action theory and the human condition* (pp. 331–351). New York: Free Press.

Parsons, T., & Fox, R. (1952). Illness, therapy and the modern urban American family. *Journal of Social Issues, 8*, 31–44.

Parvin, K. V., & Dickinson, G. E. (2010). End-of-life issues in US child life specialist programs. *Child Youth Care Forum, 39*, 1–9.

Pear, R. (2002, February 18). 9 in 10 nursing homes lack adequate staff, study finds. *New York Times*, p. A11.

Pearce, M. (2012, December 24). Mass shootings not new phenomenon in U.S. *Los Angeles Times*.

Pector, E. (2012). Sharing losses online: Do Internet support groups benefit the bereaved? *International Journal of Childbirth Education, 27*(2), 19.

Peluso, P. (2012). Chicago firefighter suicide report seeks answers. Firehouse.com. http://www .firefighterclosecalls.com/news; fullstory/news/Chicago. Accessed on November 2, 2012.

Perry, B. L., Pullen, E. L., & Oser, C. B. (2012). Too much of a good thing? Psychosocial resources, gendered racism, and suicidal ideation among low socioeconomic status African American women. *Social Psychology Quarterly, 75*(4), 334–359.

Perry, P. (1988, September). Brushes with death. *Psychology Today, 22*, 14–17.

Pet Guardian (2013). Pet burial options. Best Friends Animal Society. http:// www.petfuardian.com/common .php. Accessed on July 18, 2013.

Piaget, J. (1958). *The growth of logical thinking from childhood to adolescence*. New York: Basic Books.

Pierce, J. (2012). *The last walk: Reflections on our pets at the end of their lives*. Chicago, IL: University of Chicago Press.

Pike, M. V., & Armstrong, J. V. (Eds.). (1980). *A time to mourn: Expressions of grief in nineteenth century America*. Stony Brook, NY: The Museums at Stony Brook.

Pine, V. R. (1971, June). *Findings of the professional census*. Milwaukee, WI: National Funeral Directors Association.

Pirkis, J., Burgess, P., & Dunt, D. (2000). Suicidal ideation and suicide attempts among Australian adults. *Crisis, 21*(1), 16–25.

Planalp, S., & Trost, M. (2009). Reasons for starting and continuing to volunteer for hospice. *American Journal of Hospice & Palliative Medicine, 26*, 288–294.

Plech, E. H. (2000). *Celebrating the family: Ethnicity, consumer culture, and family rituals*. Cambridge, MA: Harvard University Press.

Pomerantz, D. (2013). Michael Jackson leads our list of the top-earning dead celebrities. *Forbes Magazine* (online at http://www.forbes.com/ sites/dorothypomerantz/2013/10/ 23/michael-jackson-leads-our-list- of-the-top-earning-dead-celebrities/). Accessed on October 23, 2013.

Pospisil, L. (1963). *The Kapauku Papuans of West New Guinea*. New York: Holt, Rinehart and Winston.

Potter, A. C. (1993). Will the "right to die" become a license to kill? The growth of euthanasia in America. *Journal of Legislation, 19*(1), 31–62.

Powers, W. K. (1977). *Oglala religion*. Lincoln, NE: University of Nebraska Press.

Powner, D. J., Ackerman, B. M., & Grevnik, A. (1996, November 2). Medical diagnosis of death in adults: Historical contributions to current controversies. *The Lancet, 348*, 1219–1224.

Presbyterian Ministers' Fund. (1938). *Presbyterian Synod minutes*. Philadelphia, PA: Author.

Pridmore, S., & Pasha, M. I. (2004). Psychiatry and Islam. *Australasian Psychiatry, 12*, 380–385.

Prior, L. (1989). *The social organization of death: Medical discourse and social practices in Belfast*. New York: St. Martin's Press.

Probate FAQ. (1999). *Nolo's legal encyclopedia*. Berkeley, CA: Nolo Press.

Proctor, R. N. (1995). *Cancer wars: How politics shapes what we know and don't know about cancer*. New York: Basic Books.

Pyenson, B., Conner, S., Fitch, K., & Kinzbrunner, B. (2004). Medicare cost in matched hospice and nonhospice cohorts. *Journal of Pain and Symptom Management, 28*(3), 200–210.

Pyszczynski, T., Solomon, S., & Greenberg, J. (2003). *In the wake of 9/11: The psychology of terror*. Washington, DC: American Psychological Association.

Quill, T. E. (1991). Bad news: Delivery, dialogue, and dilemmas. *Archives of Internal Medicine, 151*, 463–468.

Rachels, J. (1975). Active and passive euthanasia. *New England Journal of Medicine, 292*, 78–80.

Radcliffe-Brown, A. R. (1964). The *Andaman Islanders*. New York: The Free Press.

Raether, H. C., & Slater, R. C. (1974). *Facing death as an experience of life*. Milwaukee, WI: National Funeral Directors Association.

Rahman, F. (1987). *Health and medicine in the Islamic tradition*. New York: Crossroad Press.

Rahman, F. (1989, February 27). Routine CPR can abuse the old and sick. *Minneapolis Star and Tribune*, p. 9A.

Ramsden, P. G. (1991). Alice in the afterlife: A glimpse in the mirror. In D. R. Counts & D. A. Counts (Eds.), *Coping with the final tragedy: Cultural variation in dying and grieving* (pp. 27–41). Amityville, NY: Baywood Publishing.

Ramsey, P. (1970). *The patient as person: Explorations in medical ethics*. New Haven, CT: Yale University Press.

Rando, T. A. (1993). *Treatment of complicated mourning*. Champaign, IL: Research Press.

Randolph, M. (1999). *Eight ways to avoid probate*. Berkeley, CA: Nolo Press.

Raphael, B. (1983). *The anatomy of bereavement*. New York: Basic Books.

Raphael, B. (1986). *When disaster strikes: How individuals and communities cope with catastrophe*. New York: Basic Books.

Raphael, S. (2003). Richness of collaboration for children's response to disaster. *Journal of Children and Adolescent Psychiatric Nursing*, 16(1), 35–36.

Rappaport, R. A. (1974). Obvious aspects of ritual. *Cambridge Anthropology*, 2, 2–60.

Rasmussen, C. A., & Brems, C. (1996, March). The relationship of death anxiety with age and psychosocial maturity. *The Journal of Psychology*, 130, 141–144.

Ratner, E. R., & Song, J. Y. (2002, June 7). Education for the end of life. *The Chronicle of Higher Education*, p. B12.

Reader Supported News. (2013). How many people have been killed by guns since Newtown? http://readersupportednews.org/off-site-news-section/428-foreclosure/15298-focus-how-many-people-have-been-killed-by-guns-since-newtown. Accessed on November 13, 2013.

Report of the Committee on Physician-Assisted Suicide and Euthanasia. (1996). *Suicide and Life-Threatening Behavior*, 26(Suppl., 1–19).

Reynolds, G. (2013). A grief that won't heal. *Parade*, 20 (October 27).

Rhymes, J. A. (1996, May–June). Barriers to palliative care. *Cancer Control Journal*, 3(3), 1–9.

Rice, T., & Sher. L. (2012). Suicidal behavior in war veterans. *Expert Review of Neurotherapeutics*, 12(5), 611ff.

Richardson, W. C. (1992, June 3). Educating leaders who can resolve the health-care crisis. *Chronicle of Higher Education*, 39, B1.

Rifkind, H. (2006, June 10–16). The screen stars. Available at http://timesonline.co.uk/the knowledge.

Ring, K. (1980). *Life at death: A scientific investigation of the near-death experience*. New York: Coward, McCann and Geoghegan.

Robbins, B. D., Tomaka, A., Innus, C., Patterson, J., & Styn, G. (2008). Lessons from the dead: The experiences of undergraduates working with cadavers. *Omega*, 58, 177–192.

Robins, A., & Fiske, A. (2009). Explaining the relation between religiousness and reduced suicidal behavior: Social support rather than specific beliefs. *Suicide and Life-Threatening Behavior*, 39, 386–395.

Robinson, L., & Mahon, M. M. (1997). Sibling bereavement: A conceptual analysis. *Death Studies*, 21, 477–499.

Robson, J. D. (1977). Sick role and bereavement role: Toward a theoretical synthesis of two ideal types. In G. M. Vernon (Ed.), *A time to die*. Washington, DC: University Press of America.

Rocco, J., & Rocco, H. (2007, October 29). Should death education be part of the public schools curriculum? New York Teachers. http://nyteachers.wordpress.com/2007/10/29/should-death-education-be-part-of-the-public-schools-curriculum? Accessed on September 28, 2012.

Rodabough, T. (1985). Near-death experiences: An examination of the supporting data and alternative explanations. *Death Studies*, 9, 95–113.

Rodabough, T., & Cole, K. (2003). Near-death experiences as secular eschatology. In C. D. Bryant (Ed.), *Handbook of death and dying* (pp. 611–693). Thousand Oaks, CA: Sage Publications.

Rogers, R. G., Hummer, R. A., & Nam, C. B. (2000). *Living and dying in the USA*. San Diego, CA: Academic Press.

Rosenberg, C. E. (1973, May). Sexuality, class, and role in nineteenth century America. *American Quarterly*, 25, 137.

Rosenberg, J. F. (1983). *Thinking clearly about death*. Englewood Cliffs, NJ: Prentice-Hall.

Rosenberg, J. F. (2000, May 15). Art of prognosis becoming an increasingly valued skill. *American Medical News*, 1–4.

Rosenblatt, P. (1983). *Bitter, bitter tears*. Minneapolis, MN: University of Minnesota Press.

Rosenblatt, P. C., Walsh, R., & Jackson, A. (1976). *Grief and mourning in cross-cultural perspective*. New Haven, CT: Human Relations Area Files Press.

Rotundo, B. (1973, July). The rural cemetery movement. *Essex Institute Historical Collections*, 109, 231–242.

Roy, H., & Russell, C. (2006). Attitudes toward death. *The Encyclopedia of aging & the elderly*. http://www.medrounds.org/encyclopedia-of-aging. Accessed on October 8, 2012.

Royal College of Nursing Publishing Company. (2012). When someone dies. *Emergency Nurse*, 20(2, May), 12.

Rumbelow, H. (1999, April 8). Transplant boom raises prospect of divorce haggling. London: *The Times*, p. 13.

Russac, R. J., Gatliff, C., Reece, M., & Spottswood, D. (2007). Death anxiety across the adult years: An examination of age and gender. *Death Studies*, 31, 549–561.

Ryff, C. D. (1991). Possible selves in adulthood and old age: A tale of shifting horizons. *Psychology and Aging*, 6, 286–295.

Rynearson, E. K. (2012). The narrative dynamics of grief after homicide. *Omega*, 65(3), 239–249.

Rzhevsky, L. (1976). Attitudes toward death. *Survey*, 22, 38–56.

Saad, L. (2013). U.S. support for euthanasia hinges on how it's described: Support is at low ebb on the basis of wording that mentions "suicide." *Gallup Politics*. May 29, 2013. http://www.gallup.com/poll/162815/support-euthanasia-hinges-described.aspx. Accessed November 6, 2013.

Sade, R. M. (1999, March 8). Cadaveric organ donation: Rethinking donor motivation. *Archives of Internal Medicine, 159*, 438–442.

Sagan, L. (1987). *The health of nations.* New York: Basic Books.

Sahler, O. J. Z. (1978). *The child and death.* St. Louis, MN: C. V. Mosby.

Salahi, L. (2011, June 21). FDA cigarette warning labels include tracheotomy hole and rotting teeth. ABCNews .go.com.

Salamat, A. (1987). A young widow burns in her bridal clothes. *Far Eastern Economic Review, 138*, 54–55.

Sandeen, P. (2012). Massachusetts voters deny rights to terminally-ill people. Death with Dignity National Center. http://www .deathwithdignity.org/2012/11/07/ massachusetts-voters-deny-rights- terminally-ill-people. Accessed on November 7, 2013.

Sanders, B. S., Burkett, T. L., Dickinson, G. E., & Tournier, R. E. (2004). Hospice referral decisions: The role of physicians. *American Journal of Hospice and Palliative Medicine, 21*(3), 196–202.

Sanders, C. M. (1980). A comparison of adult bereavement in the death of a spouse, child, and parent. *Omega, 10*, 303–322.

Sanders, C. M. (1999). *Grief the mourning after: Dealing with adult bereavement.* New York: John Wiley & Sons.

Sanders, S., Mackin, M. L., Reyes, J., Herr, K., Titler, M., Fine, P., & Forcucci, C. (2010). Implementing evidence-based practices: Considerations for the hospice setting. *American Journal of Hospice & Palliative Medicine, 27*(6), 369–376.

Sargent, A. H. (1888). Country cemeteries. *Association of American Cemetery Superintendents, 2*, 51.

Saunders, C. (1992, Winter). The evolution of the hospices. *Free Inquiry*, 19–23.

Sawchik, T. (2012, October 20). Starting to heal with new team. *The Post and Courier*, A1, A6.

Schadenberg, A. (2013, January 30). Oregon assisted suicide deaths hit record high in 2012. Retrieved April 6, 2014, from http://www.lifenews. com/2013/01/30/oregon-assisted- suicide-deaths-hit-record-high-in- 2012/.

Schatz, H. (1976, March 1). The ill in America's armchairs. *The Minneapolis Tribune.*

Schecter, H., & Everitt, D. (1997). *The A-Z encyclopedia of serial killers.* New York: Pocket Books.

Schiedermayer, D. (1994). *Putting the soul back in medicine: Reflections on compassion and ethics.* Grand Rapids, MI: Baker Books.

Schim, S. M., Briller, S. H., Thurston, C. S., & Meert, K. L. (2007). Life as death scholars: Passion, personality, and professional perspectives. *Death Studies, 31*, 165–172.

Schmidt, J. D. (1986). Murder of a child. In T. A. Rando (Ed.), *Parental loss of a child* (pp. 213–220). Champaign, IL: Research Press Company.

Schneider, J. W. (1993). Family care work and duty in a "modern" Chinese hospital. In P. Conrad & E. B. Gallagher (Eds.), *Health and health care in developing countries: Sociological perspectives* (pp. 154–179). Philadelphia, PA: Temple University Press.

Schram, R. (2007, July). Sit, cook, eat, full stop: Religion and the rejection of ritual in Auhelawa (Papua New Guinea). *Oceania, 77*(2), 172–190.

Schroepfer, T. (1999). Facilitating perceived control in the dying process. In B. de Vries (Ed.), *End of life issues* (pp. 57–76). New York: Spring Publishing.

Schuler, T., Zaider, T., & Kissane, D. (2012). Family grief therapy: A vital model in oncology, palliative care and bereavement. *Family Matters, 90*(Summer), 77ff.

Schwab, R. (1992). Effects of a child's death on the marital relationship: A preliminary study. *Death Studies, 16*, 141–154.

Science Daily (2009, January 15). Most young violent offenders in two NYC neighborhoods have seen someone killed. http://www .sciencedaily.com/releases/2009/01.

Science Daily. (2009, October 1). Celebrities spawn copycat suicides, study confirms. http://www .sciencedaily.com/releases/ 2009/09.

Scott, C. R., & Dolan, C. (1991, April 11). Funeral homes hope to attract business by offering services after the service. *Wall Street Journal*, p. B1.

Seale, C. (1991). Communication and awareness about death: A study of a random sample of dying people. *Social Science and Medicine, 32*, 943–952.

Seale, C. (1997). Social and ethical aspects of euthanasia: A review. *Progress in Palliative Care, 5*, 1–6.

Seale, C. (1998). *Constructing death: The sociology of dying and bereavement.* Cambridge: Cambridge University Press.

Seavoy, M. R. (1906, February). Twentieth century methods. *Park and Cemetery, 15*, 488.

Secomb, L. (1999). Philosophical deaths and feminine finitude. *Mortality, 4*(2), 111–125.

Seligson, H. (2014, March 21). An online generation redefines mourning. *The New York Times.* http://www.nytimes.com/2014/03/ 23/fashion/an-online-generation- redefines-mourning.html?_r=0.

Sell, L., Devlin, B., Bourke, S. J., Munro, N. C., Corris, P. A., & Gibson, G. J. (1993). Communicating the diagnosis of lung cancer. *Respiratory Medicine, 87*, 61–63.

Selvin, P. (1990, July 17). Cryonics goes cold: People just aren't dying to be frozen. Charleston, SC, *News and Courier*, p. 5C.

Seymour, J. (2012). Looking back, looking ahead: the evolution of palliative and end-of-life care in England. *Mortality, 17*(1), 1–17.

Shah, A. (2008). A cross-national study of the relationship between elderly suicide rates and suicide rates and urbanization. *Suicide and Life-Threatening Behavior, 38*, 714–719.

Shalit, R. (1997). When we were philosopher kings. *The New Republic, 216*(17), 24–28.

Sharlet, J. (2000, January 28). The truth can comfort the dying, a physician argues. *The Chronicle of Higher Education, 47*, A20–A21.

Shen, A. (2012). A timeline of mass shootings in the US since Columbine. *Think Progress.* http:// thinkprogress.org/justice/2012/12/ 14/1337221/a-timeline-of-mass- shootings-in-the-us-since- columbine/December 14, 2012 at 3:01 pm. Accessed on November 13, 2013.

Shepherd, D. M., & Barraclough, B. M. (1976). The aftermath of parental suicide for children. *British Journal of Psychiatry, 129*, 267–276.

Shine, T. M. (2000) *Fathers aren't supposed to die: Five brothers reunite to say good-bye.* New York: Simon and Schuster.

Shneidman, E. (1973). Megadeath: Children of the nuclear family. In *Deaths of man.* Baltimore, MD: Penguin Books.

Shneidman, E. (1980). *Voices of death.* New York: Harper and Row.

Silverman, M. M. (2006). The language of suicidology. *Suicide and Life-Threatening Behavior, 36,* 519–532.

Silverman, P. R. (2000). *Never too young to know: Death in children's lives.* New York: Oxford University Press.

Silverman, P. R., Nickman, S., & Worden, J. W. (1992). Detachment revisited: The child's reconstruction of a dead parent. *American Journal of Orthopsychiatry, 62,* 494–503.

Simonds, O. C. (1919). Review of progress in cemetery design and development with suggestions for the future. *Association of American Cemetery Superintendents, 24,* 41.

Simonson, R. H. (2008). Religiousness and non-hopeless suicide ideation. *Death Studies, 32,* 951–960.

Singg, S. (2009). Types of grief. In C. D. Bryant & D. Peck (Eds.), *Encyclopedia of death and the human experience* (pp. 538–542). Thousand Oaks, CA: Sage Publications.

Sirmons, K. L., Dickinson, G. E., & Burkett, T. L. (2010). End-of-life issues: Dental schools and dentists. *Journal of Dental Education, 74*(1), 43–49.

Skidmore, S. (2007, September 19). Elderly at highest risk for suicide. *Associated Press* release. *Post and Courier,* Charleston, SC, p. 14A.

Skinner, B. F. (1990, August 7). Skinner to have last word. *Harvest Personnel.*

Sklar, F., & Hartley, S. F. (1990). Close friends and survivors: Bereavement patterns in a "hidden" population. *Omega, 21*(2), 103–112.

Slater, P. (1974). *Earthwalk.* Garden City, NY: Doubleday.

Sloane, D. C. (1991). *The last great necessity: Cemeteries in American history.* Baltimore, MD: Johns Hopkins University Press.

Smith, B. (1910, March). An outdoor room on a cemetery lot. *Country Life in America, 17,* 539.

Smith, B. W., Pargament, K. I., Brant, C., & Oliver, J. M. (2000). Noah

revisited: Religious coping by church members and the impact of the 1993 Midwest flood. *Journal of Community Psychology, 28,* 169–186.

Smith, D. C., & Maher, M. F. (1993). Achieving a healthy death: The dying person's attitudinal contributions. *The Hospice Journal, 9,* 21–32.

Smith, J. M. (1996). *AIDS and society.* Upper Saddle River, NJ: Prentice Hall.

Smith, M. T. (1988, December). Why you might go for a cash-value policy. *Money,* 153–161.

Smith, R. S. (1941, May). Life insurance in fifteenth century Barcelona. *The Journal of Economic History.*

Smith, T. L., & Walz, B. J. (1995). Death education in paramedic programs: A nationwide assessment. *Death Studies, 19,* 257–267.

Smith, T. W., Rasinski, K. A., & Toce, M. (2001). *America rebounds: A national study of public response to the September 11th terrorist attacks, preliminary findings.* National Opinion Research Center, University of Chicago. Retrieved August 19, 2009, from http://www .norc.uchicago.edu/projects/reaction/ pubresp.pdf.

Smith, W. (2014). At the bottom of the slippery slope. In G. E. Dickinson & M. R. Leming (Eds.), *Annual Editions: Dying, death, and bereavement.* New York: McGraw-Hill Companies.

Snell, C. (2005). *Peddling Poison.* Westport, CT: Praeger.

Sofka, C. (2009). Cyberfunerals. In C. D. Bryant & D. Peck (Eds.), *Encyclopedia of death and the human experience* (pp. 249–251). Thousand Oaks, CA: Sage Publications.

Somers, T. S. (1995, June 22–25). *Relict, consort, wife: The use of Connecticut Valley gravestones to understand concepts of gender in the late eighteenth and early nineteenth centuries.* Paper presented at the annual meeting of the Association for Gravestone Studies, Westfield, MA.

Span, P. (2014, January 7). Bounced from hospice. *The New York Times.*

Spencer, R. F., & Jennings, J. D. (1965). *The Native Americans.* New York: Harper and Row.

Spiegel, D. (1998). Getting there is half the fun: Relating happiness to health. *Psychological Inquiry, 9*(1), 66–68.

Spinney, L. (2011). Battling the body brokers: A hard-hitting book calls for greater transparency to deter the illegal trade in human blood, organs and eggs. *Nature, 474*(7350, June 9), 156–158.

Spotlight: Arms race (2013, January 14). *Time,* p. 14.

Springer, K. W., & Mouzon, D. (2009, August 8–11). *Masculinity and healthcare seeking among midlife men: Variation by adult SES.* Presented at the American Sociological Association annual meeting, San Francisco, CA.

Srinivas, M. N., & Shah, A. M. (1968). Hinduism. In *International encyclopedia of the social sciences* (Vol. 6, pp. 358–366). New York: Macmillan.

Stack, S., & Bowman, B. (2012). *Suicide movies: Social patterns 1900–2009.* Ashland, OH: Hogrefe Publishing.

Stack, S. (2000). Media impacts on suicide: A quantitative review of 293 findings. *Social Sciences Quarterly, 81,* 957–971.

Stack, S., & Wasserman, I. (2007). Economic strain and suicide risk: A qualitative analysis. *Suicide and Life-Threatening Behavior, 37,* 103–112.

Stack, S., & Wasserman, I. (2009). Gender and suicide risk: The role of wound site. *Suicide and Life-Threatening Behavior, 39,* 13–20.

Stalman, S. D. (1996). Children and the death of a sibling. In C. A. Corr & D. M. Corr (Eds.), *Handbook of childhood death and bereavement* (pp. 149–164). New York: Springer Publishing.

Steiger, B. (1974). *Medicine power: The American Indian's revival of his spiritual heritage.* Garden City, NY: Doubleday.

Stevenson, I., Cook, E. W., & McClean-Rice, N. (1989). Are persons reporting "near-death experiences" really near death? A study of medical records. *Omega, 20*(4), 45–54.

Stewart, D. (2009). Burial at sea. In C. D. Bryant & D. Peck (Eds.), *Encyclopedia of death and the human experience* (pp. 123–125). Thousand Oaks, CA: Sage Publications.

Stillion, J. M. (1985). *Death and the sexes*. Washington, DC: Hemisphere/McGraw-Hill.

Stobbe, M. (2008, July 1). Gun use in suicides tops homicides. *Associated Press* release, *Post and Courier*, Charleston, SC.

Stockwell, E. G. (1976). *The methods and materials of demography*. New York: Academic Press.

Stoeckel, H. H. (1993). *Survival of the spirit*. Reno, NV: University of Las Vegas.

Stolberg, S. G. (2002, October 3). War, murder and suicide: A year's toll is 1.6 million. *New York Times*, p. A12.

Stone, E. (2010). Grief in the age of Facebook. *The Chronicle Review* (March 5), B20.

Strauss, J. V. (1913, April). The ideas of a plain country woman. *Ladies Home Journal*, 30, 42.

Sturgill, B. (Producer). (1995). *Death on my terms: Right or privilege* [Videotape]. Chicago, IL: Terra Nova Films.

Sudnow, D. (1967). *Passing on: The social organization of dying*. Englewood Cliffs, NJ: Prentice Hall.

Sullivan, A. D., Hedberg, K., & Fleming, D. W. (2000, February). Legalized physician-assisted suicide in Oregon—The second year. *The New England Journal of Medicine*, 342(8), 598–604.

Sullivan, A. M., Warren, A. G., Lakoma, M. D., Liaw, K. R., Hwang, D., & Block, S. D. (2004). End-of-life care in the curriculum: A national study of medical education deans. *Academic Medicine*, 79, 760–768.

Sweeting, H. N., & Gilhooly, M. L. M. (1992). Doctor, am I dead? A review of social death in modern societies. *Omega*, 24, 251–269.

Tam, J., Tang, W. S., & Fernando, D. J. S. (2007). The Internet and suicide: A double-edged tool. *European Journal of Internal Medicine*, 18, 453–455.

Tanner, J. G. (1995). Death, dying, and grief in the Chinese-American culture. In J. K. Perry & A. S. Ryan (Eds.), *A cross-cultural look at death, dying, and religion* (pp. 183–192). Chicago, IL: Nelson-Hall Publishers.

Taylor, L. (1980). Symbolic death: An anthropological view of mourning ritual in the nineteenth century.

In M. V. Pike & J. V. Armstrong (Eds.), *A time to mourn: Expressions of grief in nineteenth century America*. Stony Brook, NY: The Museums at Stony Brook.

Taylor, R. B. (1988). *Cultural ways* (3rd ed.). Boston, MA: Allyn & Bacon.

Thearle, M. J., Vance, F. C., Najman, J. M., Embelton, G., & Foster, W. J. (1995). Church attendance, religious affiliation and parental responses to sudden infant death, neonatal death and stillbirth. *Omega*, 31, 51–58.

Thomas, E. (2009, September 21). The case for killing Granny. *Newsweek*, 34–40.

Thomas, W. I. (1923). *The unadjusted girl*. New York: Little, Brown & Company.

Thomson, G. (November 14, 2008). The art of dying. *The Chronicle of Higher Education*, B16–17.

Thornton, G., Robertson, D. U., & Mlecko, M. L. (1991). Disenfranchised grief and evaluations of social support by college students. *Death Studies*, 15, 355–362.

Thornton, G., Wittemore, K. D., & Robertson, D. U. (1989). Evaluation of people bereaved by suicide. *Death Studies*, 13, 119–126.

Timmermans, S. (1999). *Sudden death and the myth of CPR*. Philadelphia, PA: Temple University Press.

Titus, S. L., Rosenblatt, P. C., & Anderson, R. M. (July, 1979). Family conflict over inheritance of property. *The Family Coordinator*, 337–338.

Tobin, S. (1972). The earliest memory as data for research in aging. In D. P. Kent, R. Kastenbaum, & S. Sherwood (Eds.), *Research, planning and action for the elderly*. New York: Behavioral Publications.

Tonkinson, R. (1978). *The Mardudjara aborigines*. New York: Holt, Rinehart and Winston.

Townsend, E. (2007). Suicide terrorists: Are they suicidal? *Suicide and Life-Threatening Behavior*, 37(1), 35–49.

Trammell, R. L. (1975). Saving life and taking life. *Journal of Philosophy*, 72, 131–137.

Truog, R. D. (1997). Is it time to abandon brain death? *Hastings Center Report*, 27(1), 29–37.

Tseng, T., & Su, C. (2009). Death care industry. In C. D. Bryant & D. Peck (Eds.), *Encyclopedia of death and the human experience* (pp. 305–309). Thousand Oaks, CA: Sage Publications.

Tuckerman, H. (1856, November). The law of burial and the sentiment of death. *Christian Examiner*, 61, 338–342.

Tully, M. (1994, September). When body and soul go their separate ways. *Worldwide report: Death*. London: BBC Worldwide.

Turnbull, C. M. (1972). *The mountain people*. New York: Simon and Schuster.

Turnbull, C. M. (1983). *The human cycle*. New York: Simon and Schuster.

Turner, J. H. (1985). *Sociology: A student handbook*. New York: Random House.

Twenty-five Facts About Organ Donation and Transplantation. (2009). National Kidney Foundation. Retrieved July 13, 2010, from http://www.kidney.org.

Twycross, R. G. (1996, February). Euthanasia: Going Dutch? *Journal of the Royal Society of Medicine*, 89, 61–63.

Tylor, E. B. (1873). *Primitive culture*. London: John Murray.

Uchendu, V. C. (1965). *The Igbo of southeast Nigeria*. New York: Holt, Rinehart and Winston.

Umberson, D. (2003). *Death of a parent: Transition to a new adult identity*. New York: Cambridge University Press.

UMM (2011). Death among children and adolescents. University of Maryland Medical Center website at www.umm.edu. Accessed on October 5, 2012.

UNAIDS. (2013). Global Report: UNAIDS report on the global AIDS epidemic 2013. Retrieved April 6, 2014, from http://www.unaids.org/en/resources/documents/2013/name.85053.en.asp.

Ungureanu, I., & Sandberg, J. G. (2008). Caring for dying children and their families: MFTs working at the gates of the Elysian Fields. *Contemporary Family Therapy*, 30, 75–91.

United Nations. (1953, August). *Principles for a vital statistics system* (Statistical Papers, Series M, No. 19, p. 6).

United Network for Organ Sharing. (2009). *UNOS facts and statistics about transplantations* [Online]. http://204.127.237.11/stats.htm.

U.S. Bureau of the Census. (1975). *Historical statistics of the United States, colonial times to 1970* (pp. 1050–1059). Washington, DC: Department of Commerce, Bureau of the Census.

U.S. Census Bureau. (2012). Homicide victims by race and sex. Law enforcement, courts, and prisons. Statistical abstract of the United States. Washington, DC: Department of Commerce.

U.S. Health Care Costs. (2009). http://www.KaiserEDU.org.

Vandecreek, L., & Mottram, K. (2009). The religious life during suicide bereavement: A description. *Death Studies, 33,* 741–761.

Vander Veer, J. B. (1999, May). Euthanasia in the Netherlands. *Journal of the American College of Surgeons, 188,* 532–537.

van Eys, J. (1988, Summer). In my opinion … normalization while dying. *Children's Health Care, 17,* 18–21.

Van Gennep, A. (1960). *The rites of passage* (M. B. Vizedom & G. L. Caffee, Trans.), Chicago, IL: University of Chicago Press. (Original work published 1909.)

Vargens, O. M. C., & Bertero, C. (2012). The phantom of death improving quality of life: You live until you die. *American Journal of Hospice & Palliative Medicine, 29*(7), 555–562.

Vaughn-Cole, B. (2000, September 28–October 2). *Suicide: Psychological and physiological assessment of grief.* Abstracts of the American College of Nurse Practitioners National Clinical Symposium, Salt Lake City, UT.

Veatch, R. M. (1995). The definition of death: Problems for public policy. In H. Wass & R. A. Neimeyer (Eds.), *Dying: Facing the facts* (3rd ed.), pp. 405–432. Washington, DC: Taylor and Francis.

Ventura, S. J., Abma, J. C., Mosher, W. D., & Henshaw, S. K. (2009). Estimated pregnancy rates for the United States, 1990–2005: An update. *National Vital Statistics Reports, 58*(4). http://www.cdc.gov/nchs/data/nvsr/nvsr58/nvsr58_04.pdf.

Verhoef, M. J., & Kinsella, T. D. (1996). Alberta euthanasia survey: Three-year follow-up. *Canadian Medical Association Journal, 155,* 885–890.

Vernon, G. M. (1970). *Sociology of death: An analysis of death-related behavior.* New York: Ronald Press.

Vernon, G. M. (1972). *Human interaction.* New York: Ronald Press.

Vernon, G. M., & Cardwell, J. D. (1981). *Social psychology: Shared, symboled, and situated behavior.* Washington, DC: University Press of America.

Viner, R. M., Coffey, C., Mathers, C., Bloem, P., Costello, A., Santelli, J., & Patton, G. C. (2011). 50-year mortality trends in children and young people, *The Lancet, 377,* 1162–1174.

Vobejda, B. (1997, December 5). Abortion rate in U.S. off sharply. *Washington Post,* p. A1.

Voegelin, E. W. (1944). *Mortuary customs of the Shawnee and other Eastern tribes.* Indianapolis: Indiana Historical Society.

Vogt, E. Z. (1970). *The Zinacantecos of Mexico: A modern Mayan way of life.* New York: Holt, Rinehart and Winston.

Vovelle, M. (1976). La redécouverte de la mort. *Pensée, 189,* 3–18.

Waldrop, D. P., & Kusmaul, N. (2011). The living-dying interval in nursing home-based end-of-life care: family caregivers' experiences. *Journal of Gerontological Social Work, 54*(8), 768–787.

Walker, B. A. (1989). Health care professionals and the near-death experience. *Death Studies, 13,* 63–71.

Walker, A., & Balk, D. (2007, August). Bereavement rituals in the Muscogee Creek tribe. *Death Studies, 31,* 633–652.

Walker, M. (2011). Average cost of an autopsy. *KayCircle Everyday References.* http:www.kaycircle.com/Average-Cost-of-an-Autopsy. Accessed on November 4, 2013.

Walker, R. L., & Flowers, K. C. (2011). Effects of race and precipitating events on suicide versus nonsuicide death classification in a college sample. *Suicide and Life-Threatening Behavior, 41*(1), 12–20.

Walsh, K. (1974). *Sometimes I weep.* Valley Forge, PA: Judson Press.

Walter, T. (1995). Natural death and the noble savage. *Omega, 30,* 237–248.

Walter, T. (1999a). And the consequence was. … In T. Walter (Ed.), *The mourning for Diana* (pp. 271–278). Oxford: Berg.

Walter, T. (1999b). The questions people asked. In T. Walter (Ed.), *The mourning for Diana* (pp. 19–47). Oxford: Berg.

Walter, T. (2008). The sociology of death. *Sociology Compass, 2*(1), 317–336.

Wanzer, S. H., Federman, D. D., Adlestein, S. J., Cassel, C. K., Cassem, E. H., Cranford, R. E., et al. (1989). The physician's responsibility toward hopelessly ill patients. *The New England Journal of Medicine, 320,* 844–849.

Ward, B. (2013, July 24). The Death Cafe discussion group steers its members on how to live. *StarTribune,* Minneapolis, MN.

Wass, H. (1979). Death and the elderly. In H. Wass (Ed.), *Dying: Facing the facts.* Washington, DC: Hemisphere.

Wass, H. (1995). Death in the lives of children and adolescents. In H. Wass & R. A. Neimeyer (Eds.), *Dying: Facing the facts* (pp. 269–301). Washington, DC: Taylor and Francis.

Wass, H. (2004). A perspective on the current state of death education. *Death Studies, 28,* 289–308.

Wass, H., Berardo, F. M., & Niedermeyer. R. A. (Eds.), *Dying: Facing the facts.* Washington, DC: Hemisphere Publishing Corporation.

Wass, H., Miller, D., & Redditt, C. A. (1991). Adolescents and destructive themes in rock music: A follow-up. *Omega, 23,* 199–206.

Wass, H., Miller, M. D., & Thornton, G. (1990). Death education and grief/suicide intervention in the public schools. *Death Studies, 14,* 253–268.

Wass, H., & Sisler, S. (1978, January). *Death concern and views on various aspects of dying among elderly persons.* Paper presented at the International Symposium on the Dying Human, Tel Aviv, Israel.

Watts, D. T., Howell, T., & Priefer, B. A. (1992). Geriatricians' attitudes toward assisting suicide of dementia patients. *Journal of the American Geriatrics Society, 40,* 878–885.

Weaver, R. R., & Rivello, R. (2006–2007). The distribution of mortality in the United States: The effects of income (inequality), and social capital, and race. *Omega, 54,* 19–39.

Webb, N. B. (2002). September 11, 2001. In N. B. Webb (Ed.), *Helping bereaved children* (pp. 365–384). New York: Guilford Press.

Webb, N. B. (2003). Play and expressive therapies to help bereaved children: Individual, family, and group treatment. *Smith College Studies in Social Work, 73,* 405–422.

Weber, L. J. (1981). The case against euthanasia. In D. Bender (Ed.), *Problems of death: Opposing viewpoints.* St. Paul, MN: Greenhaven Press.

Weber, M. (1966). *The theory of social and economic organization.* New York: The Free Press.

Weber, M. (1968). Bureaucracy. In H. Gerth & C. C. Mills (Trans.), *Max Weber.* New York: Free Press.

Wechter, E., O'Gorman, D. C., Singh, M. K., Spanos, P., & Daly, B. J. (2013). The effects of an early observational experience on medical students' attitudes toward end-of-life care (published online November 6, 2013). *American Journal of Hospice and Palliative Medicine.*

Weed, H. E. (1912). *Modern park cemeteries.* Chicago, IL: R. J. Haight.

The Week. (2014, April 11). Obamacare crosses its first hurdle, p. 2.

Weeks, O. D., & Johnson, C. A. (2001). *When all the friends have gone: A guide for aftercare providers.* Amityville, NY: Baywood Publishing.

Wehbah-Rashid, J. A. R. (1996). Explaining pregnancy loss in matrilineal southeast Tanzania. In R. Cecil (Ed.), *The anthropology of pregnancy loss: Comparative studies in miscarriage, stillbirth and neonatal death* (pp. 75–93). Oxford: Berg.

Weidner, N. J., Cameron, M., Lee, R. C., McBride, J., Mathias, D. M., & Byczkowski, T. L. (2011). End-of-life care for the dying child: What matters most to parents. *Journal of Palliative Care, 27*(4), 279–286.

Weigand, C. G., & Weigand, P. C. (1991). Death and mourning among the Huicholes of western Mexico. In D. R. Counts & D. A. Counts (Eds.), *Coping with the final tragedy: Cultural variation in dying and grieving* (pp. 53–68). Amityville, NY: Baywood Publishing.

Weijer, C. (1995, Spring). Learning from the Dutch: Physician-assisted death, slippery slopes and the Nazi analogy. *Health Law Review, 4*(1), 23–29.

Weisman, A. D. (1990–1991). Bereavement and companion animals. *Omega, 22,* 241–248.

Weisman, A. D. (1993). *The vulnerable self: Confronting the ultimate questions.* New York: Insight Books.

Weiss, G. L., & Lonnquist, L. E. (2003). *The sociology of health, healing, and illness.* Englewood Cliffs, NJ: Prentice Hall.

Weiss, G. L., & Lonnquist, L. E. (2009). *The sociology of health, healing, and illness.* Upper Saddle River, NJ: Prentice Hall.

Wells, R. A. (1887). *Decorum: A practical treatise on etiquette and dress of the best American society.* Springfield, MA: King, Richardson.

Wells, R. V. (2000). *Facing the "King of Terror": Death and society in an American community, 1970–1990.* Cambridge: Cambridge University Press.

Wertenbaker, L. T. (1957). *The death of a man.* New York: Random House.

Wholey, D. R., & Burns, L. R. (2000). Tides of change: The evolution of managed care in the United States. In C. C. Bird, P. Conrad, & A. A. Fremont (Eds.), *Handbook of medical sociology* (pp. 217–237). Upper Saddle River, NJ: Prentice Hall.

Wilkins, R. (1996). *Death: A history of man's obsessions and fears.* New York: Barnes and Noble Books.

Wilkinson, B. W. (1995, December). "The right to die" by Russel Ogden: A commentary. *Canadian Public Policy, 21*(4), 449–455.

Will, J. (1988). Preneed: The trend toward prearranged funerals. In H. Raether (Ed.), *The funeral director's practice management handbook.* Englewood Cliffs, NJ: Prentice-Hall.

Williams, M. D. (1981). *On the street where I lived.* New York: Holt, Rinehart and Winston.

Williams, T. R. (1965). *The Dunsun: A North Borneo society.* New York: Holt, Rinehart and Winston.

Willinger, M. (1995). Sleep position and sudden infant death syndrome. *Journal of the American Medical Association, 273,* 818–819.

Willis, C. A. (2002). The grieving process in children: Strategies for understanding, educating, and reconciling children's perceptions of death. *Early Childhood Education Journal, 29,* 221–226.

Wilson, S. (2009). Obama reverses Bush policy on stem cell research. *Washington Post* (March 10).

Wolf, S. S. (1995). Legal perspectives on planning for death. In H. Wass & R. A. Neimeyer (Eds.), *Dying: Facing the facts* (pp. 163–184). Washington, DC: Taylor and Francis.

Wong, P. W. C., Yeung, A. W. M., Chan, W. S. C., Yip, P. S. F., & Tang, A. K. H. (2009). Suicide notes in Hong Kong in 2000. *Death Studies, 33,* 372–381.

Wong, P. T. P., & Tomer, T. (2011). Beyond terror and denial: the positive psychology of death acceptance. *Death Studies, 35,* 99–106.

Worden, J. W. (1982). *Grief counseling and grief therapy: A handbook for the mental health practitioner.* New York: Springer Publishing.

Wortman, C. B., & Silver, R. C. (1992). Reconsidering assumptions about coping with loss: An overview of current research. In L. Montada, S. Filipp, & M. J. Lerner (Eds.), *Life crises and experiences of loss in adulthood* (pp. 341–365). Hillsdale, NJ: Lawrence Erlbaum Associates.

Wyman, L. C. (1970). *Sandpaintings of the Navaho Shootingway and the Walcott Collection.* Washington, DC: Smithsonian Institution Press.

Yalom, I. D. (2008). *Staring at the Sun: Overcoming the terror of death.* San Francisco, CA: Jossey-Bass.

Yancu, C. N., Farmer, D. F., & Leahman, D. (2010). Barriers to hospice use and palliative care services use by African American adults. *American Journal of Hospice and Palliative Medicine, 27*(4), 248–253.

Yang, B., Lester, D., & Yang, C. (1992). Sociological and economic theories of suicide: A comparison of the U.S.A. and Taiwan. *Social Science and Medicine, 34,* 333–334.

Yang, C. S., & Chen, S. (2006). Content analysis of free-response narratives to personal meanings of death among Chinese children and adolescents. *Death Studies, 30,* 217–241.

Yapp, K. A. (2012). Culture and end-of-life care: An epidemiological evaluation of physicians. *American Journal of Hospice & Palliative Medicine, 29*(2), 106–111.

Ying, Y., & Chang, K. (2009). A study of suicide and socioeconomic factors. *Suicide and Life-Threatening Behavior, 39,* 214–226.

Yokota, F., & Thompson, K. M. (2000). Violence in G-rated animated films. *Journal of the American Medical Association, 283,* 2716–2720.

Young, E., Bury, M., & Elston, M. A. (1999, November). Live and/or let die: Modes of social dying among women and their friends. *Mortality, 4,* 269–289.

Youngner, S. J., Landefeld, C. S., Coulton, C. J., Juknialis, B. W., & Leary, M. (1989). Brain death and organ retrieval: A cross-sectional survey of knowledge and concepts among health professionals. *Journal of the American Medical Association, 261,* 2205–2210.

Zarkowski, P., & Avery, D. (2006). Hotel room suicide. *Suicide and Life-Threatening Behavior, 36,* 578–581.

Zeyrek, E. Y., Gencoz, F., Bergman, Y., & Lester, D. (2009). Suicidality, problem-solving skills, attachment style, and hopelessness in Turkish students. *Death Studies, 33,* 815–827.

Zucker, A. (2000). Assisted suicide. *Death Studies, 24,* 359–361.

Zuger, A. (2004, March 23). Anatomy lessons, a vanishing rite for young doctors. *New York Times,* D1, D6.

Zunin, L. M., & Zunin, H. S. (1991). *The art of condolence.* New York: HarperCollins.

CREDITS

These pages constitute an extension of the copyright page. We have made every effort to trace the ownership of all copyrighted material and to secure permission from copyright holders. In the event of any question arising as to the use of any material, we will be pleased to make the necessary corrections in future printings. Thanks are due to the following authors, publishers, and agents for permission to use the material indicated.

Literary Credits

Chapter 1. 16: Source: H. Philpot. (January 2010). Still Life. *Skirt*, p. 24. Heather Philpot is a wife, mother, and freelance writer who lives in Travelers Rest, South Carolina. **19:** Sources: *Abstract of the Twelfth Census of the United States, 1900.* Table 93. Washington, DC: U.S. Government Printing Office, 1902; National Center for Health Statistics, *National Vital Statistics Reports*, 2010, 60(4), Centers for Disease Control and Prevention. Atlanta, GA, January 11, 2012. **30:** George E. Dickinson **34:** George E. Dickinson **37:** From *A Bag of Noodles* (p. 8), by W. A. Armbruster, 1972, St. Louis: Concordia. Copyright 1973 by Concordia Publishing House. Reprinted by permission.

Chapter 2. 45: Reprinted by permission of the publishers and the Trustees of Amherst College from *The Poems of Emily Dickinson*, Thomas H. Johnson, ed., Cambridge, MA: The Belknap Press of Harvard University Press, Copyright © 1951, 1955, 1979, 1983 by the President and Fellows of Harvard College. **48:** Doctor explains the "persistent vegetative state." S. Russell, March 23, 2005. *San Francisco Chronicle.* Copyright 2005 by San Francisco Chronicle. Reproduced with permission of San Francisco Chronicle. **57:** Marcus Aurelius (121 A.D.–180 A.D.). Taken from *Death: Philosophical soundings* (p. 162), by H. Fingarette, 1996, Chicago, IL: Open Court. **62:** Taken from W. Marx (2012). "Requiem sempiternam"? Death and the musical requiem in the twentieth century. *Mortality*, 17(2), 119–129. **68:** Taken from *Death: Philosophical soundings* (p. 123), by H. Fingarette, 1996, Chicago, IL: Open Court. **71:** © Cengage Learning. **72–73:** Developed by Michael Leming (1979–1980). Religion and death: A test of Homans' thesis. *Omega*, 10(4), 347–364.

Chapter 3. 85: From First childhood death experiences (pp. 176–177), by G. Dickinson, 1992, *Omega*, 25. **86:** Taken from *Death among children and adolescents.* (2011). University of Maryland Medical Center web site at www.umm.edu. Accessed on October 5, 2012. **91:** From "First Childhood Death Experiences" (p. 173)

by G. Dickinson, 1992, *Omega*, 25. **93:** Sources: S. B. Hunter and D. E. Smith. (2008). Predictors of children's understandings of death: Age, cognitive ability, death experience and maternal communicative competence. *Omega*, 57, 143–162; H. Ginsberg and S. Opper. (1988). *Piaget's theory of intellectual development* (3rd ed.). Englewood Cliffs, NJ: Prentice-Hall; D. Martini. (2009). *Helping children cope with chronic illness.* Chicago, IL: American Academy of Child & Adolescent Psychiatry. Retrieved August 7, 2009 from www.aacap.org. **95:** Sources: E. Erikson. (1968). *Identity: Youth and crisis.* New York: W. W. Norton & Company; E. Erikson. (1950). *Childhood and society.* New York: W. W. Norton & Company; E. Erikson. (1997). *The life cycle completed.* New York: W. W. Norton & Company. **106:** Source: Adapted from J. N. McNeil. (1986). In talking about death: Adolescents, parents, and peers (pp. 197–198). In C. A. Corr & J. N. McNeil (Eds.), *Adolescence and Death*, New York: Springer Publishing Company. **107:** Source: Ecclesiastes 3:1. The Holy Bible (Revised Standard Version). (1962). New York: Oxford University Press. **114:** J. Joseph. (1987). Warning. In S. Martz (Ed.), *When I am an old woman I shall wear purple: An anthology of short stories and poetry.* Watsonville, CA: Papier-Mache Press. **116:** Source: Adapted from Benet, Sula. (1974). *Abkhasians: The longliving people of the Caucasus.* New York: Holt, Rinehart, & Winston.

Chapter 4. 136: Source: Annant Rambachan, Department of Religion, St. Olaf College. By permission of the author. **137:** Source: From The way of the mystic, by Sister Medhanandi, 1998. Available online at http://www.buddhanet.net/mystic.htm **140:** Source: Damon Runyon. **143:** Source: *60 Minutes*, September 27, 2009 (CBS News). Available online at http://www.cbsnews.com/video/watch/?id=5345034n&tag=contentMain;contentBody. **144:** Source: Percy Bysshe Shelley.

Chapter 5. 157: Adapted from *Tuesdays with Morrie*, by M. Albom, 1997, New York: Doubleday. **158:** Dialectic on dying (pp. 182–189), by J. M. Boyle, 1981. In M. M. Newell et al. (Eds.), *The role of the volunteer in the care of the terminal patient and the family.* New York: The Foundation of Thanatology/Arno Press. **159:** Adapted from *I don't know what to say: How to help and support someone who is dying*, by R. Buckman, 1992, New York: Viking Press; R. S. Morrison & J. Morris (1995, July), When there is no cure: Palliative care for the dying patient. *Geriatrics*, 50, 45–50. **160:** A physician's comment from a survey of 1,093 physicians, "Death education and physicians' attitudes toward dying patients"

585

(pp. 167–174), by G. E. Dickinson and A. A. Pearson, 1980–1981, *Omega*, 11(2). **166:** From E. Kübler-Ross. (1969). *On death and dying*. New York: Macmillan Publishing Company. **170:** From *Facing death* (p. 6), by R. E. Kavanaugh, 1972, Baltimore: Penguin Books. **173:** From *The hour of our death*, by P. Aires, 1981, New York: Knopf. **178:** Source: Hospice in the Home Program, Visiting Nurse Association of Los Angeles. **182:** From *A Protestant legacy: Attitudes to death and illness among older Aberdonians* (pp. 98–100) by R. Williams, 1990, Oxford: Clarendon Press. **183:** Source: T. Hughes, M. Schumacher, J. M. Jacobs-Lawson, and S. Arnold. (2008). Confronting death: Perceptions of a good death in adults with lung cancer. *The American Journal of Hospice and Palliative Medicine*, 25, 39–44. **184:** From *Cancer care nursing* (p. 33), by M. Donovan and S. Pierce, 1976, New York: Appleton-Century-Crofts. **185:** From *Facing death* (pp. 142–143), by R. E. Kavanaugh, 1972, Baltimore, MD: Penguin Press. **187:** Taken from J. Gold (2012, July 23). Child life specialists help sick kids be kids. *Kaiser Health News*; Child life: Empowering children and families (2012). Rockville, MD: Child Life Council. www.childlife.org. **190:** Taken from K. A. Erickson (2013). *How we die now: Intimacy and the work of the dying*. Philadelphia, PA: Temple University Press.

Chapter 6. 197: From *The grace in dying*, by K. D. Singh, 1999, Dublin, Ireland: Newleaf. **200:** Source: American Cancer Society. (2008). www.cancer.org **202:** Adapted from *Cancer wars*, by R. T. Proctor, 1995, New York: Basic Books. **204:** M. Marchione (2009, November 2). Alternatives gain ground. Associated Press. **207:** From *Magic, science, and health: The aims and achievements of medical anthropology*, by R. Anderson, 1996, Fort Worth, TX: Harcourt Brace College Publishers. **208:** Told to George Dickinson by a nurse in the spring of 1999 in Sheffield, England. **209:** Taken from J. Silberner (2012, December 7). Public Radio International, Cancer Series, Part V: Dispensing Comfort. **211:** Source: New York Organ Donor Network, 2009. **214:** From An old man's friend, by W. A. Hensel, 2000, *Journal of the American Medical Association*, 283(14), 1793–1794. **216:** Source: *NHPCO Facts and Figures: Hospice Care in America*, 2011 Edition (2012). National Hospice and Palliative Care Organization. **217:** Source: *NHPCO Facts and Figures: Hospice Care in America*, 2011 Edition (2012). National Hospice and Palliative Care Organization. **218:** Source: K. Stone, I. Papadopoulos, and D. Kelly (2012), Establishing hospice care for prison populations: An integrative review assessing the UK and USA perspective, *Palliative Medicine*, 26(8), 969–978. **224:** From O. Kelly, *Make Each Day Count* newsletter. **224:** Source: *NHPCO Facts and Figures: Hospice Care in America*, 2011 Edition (2012). National Hospice and Palliative Care Organization. **225:** Taken from J. Pierce's *The last walk: Reflections on our pets at the end of their lives* (2012). Chicago, IL: University of Chicago Press. **228:** A. Parker. (2009, October 11). Grit and grace: Military man battles health issues. *The Post and Courier*, Charleston, SC.

Chapter 7. 238: From C. Middlebrook. 1996. *Seeing the crab* (p. 80). New York: Basic Books. **241:** From How the poor die, by G. Orwell, 1950. In *Shooting an elephant and other essays*, New York: Harcourt Brace. **245:** Source: From a letter received on November 12, 1997, from Courtney Schomp, a former student in George Dickinson's class entitled "Death and Dying." **247:** Sources: G. E. Dickinson and D. Field. (2002). Teaching end-of-life issues: Current status in United Kingdom and United States medical schools. *American Journal of Hospice and Palliative Care*, 19, 181–186; G. E. Dickinson. (2007). End-of-life and palliative care issues in medical and nursing schools in the United States. *Death Studies*, 31, 713–726; G. E. Dickinson. (2011). Thirty-five years of end-of-life issues in US medical schools. *American Journal of Hospice and Palliative Medicine*, 28(6), 412–417; G. E. Dickinson, D. Clark, and M. Sque. (2008). Palliative care and end of life issues in UK pre-registration, undergraduate nursing programmes. *Nurse Education Today*, 28, 163–170. **250:** Adapted from Medical education must deal with end-of-life care (pp. A56–A57), by F. Cohn, J. Harrold, and J. Lynn, May 30, 1997, *The Chronicle of Higher Education*. **254:** Taken in part from *Medical sociology*, by W. C. Cockerham, 2012, Upper Saddle River, NJ: Prentice Hall.

Chapter 8. 262: Adapted from When decisions are life-and-death, by R. L. Guyer (1998, February 6–8), *USA Weekend*, p. 26. **269:**

From Teaching surgery without a patient, by K. S. Mangan (2000, February 25), *The Chronicle of Higher Education*, 45, pp. A49–A50. **278:** British court permits man to help wife die. November 30, 2004. Copyright © 2004 Reuters Limited. All rights reserved. **286:** Adapted from Routine CPR can abuse the old and sick by F. Rahman (1989, February 27), *Minneapolis Star and Tribune*, p. 9A. **287–288:** Adapted from Euthanasia in the Netherlands, by J. B. Vander Veer (1999, May), *Journal of the American College of Surgeons*, 188(5), 532–537; Slippery slopes in flat countries—A response, by J. J. M. van Dalden (1999, February), *Journal of Medical Ethics*, 25, 22–24; Dutch becoming first nation to legalize assisted suicide, by M. Simons (2000, November 28), *New York Times*, A3. **292–293:** Adapted from Death and dignity: A case of individualized decision making, by T. Quill (1991, May), *Harper's Magazine*, pp. 32–34; originally published in *The New England Journal of Medicine*, March 7, 1991, 324. **294:** Petrarch (1304–1374) **294:** Euthanasia .ProCon.org. (2013). http://euthanasia.procon.org/view.resource .php?resourceID=000132.

Chapter 9. 311: *A look at statistics* (2012). Online Schools. **314:** Taken from M. J. Goldblatt and J. T. Maltsberger. (2011). Self-harming behavior and suicidality: Suicide risk assessment. *Suicide and Life-Threatening Behavior*, 41(2), 227–233. **320:** Source: Figures from the National Center for Health Statistics for the year 2006. Taken from the American Foundation for Suicide Prevention, 2009. **321:** Source: Bullying and suicide. (2012). OLWEUS Bullying Prevention Program. http://www.bullyingstatistics.org/content/bullyingand-suicide.html. Accessed on October 16, 2012. **323:** From Helping suicidal adolescents: Needs and responses, by A. L. Berman, 1986. In C. A. Corr & J. N. McNeil (Eds.), *Adolescence and death* (pp. 151–166) [Microfilm]. New York: Springer Publishing Company. **324:** Sources: E. Clifton (2012, June 8). Suicide rate in military at highest level in ten years. *ThinkProgress*; T. Sawchik (2012, October 20). Starting to heal with new team. *Post and Courier*, Charleston, SC, pp. A1, A6. **325:** From M. Silverman (2004, March). College student suicide prevention. *Student Health Spectrum*. **328:** Source: The above letter was written by a college student in her introductory sociology class in response to the professor's assignment for students to write a thank-you letter to someone who had played a meaningful role in their lives. Permission was given by the student to include the letter here. **331:** From *Youth and suicide* (p. 96), by F. Klagsbrun, 1976, New York: Pocket Books. **335:** From *Minneapolis Star and Tribune* (p. 12A), October 4, 1983. **337:** Taken from T. L. Snell (2011). Capital punishment, 2010—Statistical tables. U.S. Department of Justice; Death penalty focus (2009, March 31), Working for alternatives. http://www.deathpenalty.org/section. Accessed on October 25, 2012; N. Welsh (2012, November/December). The death penalty is experiencing technical difficulties. *Pacific Standard*, 48–53. **338:** Taken from Guide to survival. *Justice for homicide victims* (2012). http://justiceforhomicidevictims.net/survival.html. Accessed on October 18, 2012. **339:** Adapted from G. E. Dickinson and H. C. Hoffmann. (2014). Roadside memorials: A 21st century development. In C. Staudt & J. H. Ellens (Eds.), *Our changing journey to the end: Reshaping death, dying, and grief in America*. Santa Barbara, CA: Praeger, 227–252.

Chapter 10. 347: Source: http://www.npr.org/2005/08/08/4785079 /always-go-to-the-funeral. **351:** From Honoring the dead, by D. G. Bates and E. M. Fratkin, 1999. In *Cultural anthropology* (2nd ed.). Boston, MA: Allyn & Bacon. **355:** From Cuttin' the body loose: Dancing in defiance of death in New Orleans, by K. Salaam, *Utne Reader* (1991, September–October), pp. 78–79. **366:** From *Death in Ancient Egypt*, by A. J. Spencer, 1982, New York: Penguin Books. **373:** Source: Cremation Association of North America. (2006) Disposition Survey. Chicago, IL: Author. **373:** Source: Cremation Association of North America. (2006) Disposition Survey. Chicago, IL: Author. **379:** From *The Kapauku Papuans of West New Guinea*, by L. Pospisil, 1963, New York: Holt, Rinehart and Winston. **380:** From Jewish group buries its own, June 25, 1982. Copyright 1982 Associated Press. **384:** From The ritual solution, by K. L. Woodward with A. Underwood, (1997, September 18), *Newsweek*, p. 62. **386:** From A young widow burns in her bridal clothes, by A. Salamat, 1987, *Far Eastern Economic Review*, 138, pp. 54–55.

Chapter 11. 396: From *Cutting your losses: Death and grieving in a Polynesian community* (pp. 169–189), by M. D. Lieber, 1991. In D. R. Counts & D. A. Counts (Eds.), *Coping with the final tragedy: Cultural variation in dying and grieving*. Amityville, NY: Baywood Publishing. **398:** From *Ladies Home Journal*, September 1903. **405:** (NFDA, 2013). **407:** From R. DeVries, professor of bioethics and sociology, University of Michigan Medical School, Ann Arbor, MI. **408:** (NFDA, 2013). **411:** (NFDA, 2013). **413–415:** Source: L. Gibbs and I. S. Mangla. (2012). The high cost of saying goodbye. *CNN Money Magazine*. November 9, 2012. **415:** Cremation Associations of North America, 2011. **417:** Cremation Association of North America, 2011. **418:** Cremation Association of North America, 2008. **420–421:** Source: G. E. Dickinson. 2012. Diversity in death: Body disposition and memorializaton. *Illness, Crisis, and Loss*, 20(2), pp. 141–158. **424:** From Martha Garrison, Spring 2005. **431:** From MacLean, V. M., & Williams, J. E. (2009). Cemeteries. In C. D. Bryant & D. Peck (Eds.), *Encyclopedia of death and the human experience* (pp. 169–173). Thousand Oaks, CA: Sage Publications. **432:** From *The People's Almanac* (pp. 1312–1320), by D. Wallechinsky and I. Wallace, 1975, Garden City, NY: Doubleday; and *Lexington Herald*, Lexington, KY, September 21, 1979, pp. D5. **442:** From Cemeteries feel recession's chill, by A. Gomez, *USA Today*, October 19, 2009. **443:** From Cemeteries breathe life into tourists, by J. Ravitz, CNN, October 30, 2009, http://www.cnn.com/2009/TRAVEL/10 /30/graveyard.tourism/index.html.

Chapter 12. 453: From India's living dead form their own club (p. 17), by D. Orr, June 18, 1999, *The London Times*. **454–455:** Source: http://www.cdc.gov/nchs/nvss/vital_certificate_revisions .htm **458:** From Cultural mediation of dying and grieving among Native Canadian patients in urban hospitals (pp. 231–251), by J. M. Kaufert & J. D. O'Neil, 1991. In D. R. Counts & D. A. Counts (Eds.), *Coping with the final tragedy: Cultural variation in dying and grieving*. Amityville, NY: Baywood Publishing. **459:** Source: From Elizabeth Maxwell, former undergraduate student intern at St. Olaf College, Northfield, MN. **463:** From National Hospice and Palliative Care Organization, August 7, 2009. **465:** Source: From http://www.agingwithdignity.org/5wishes.html **480:** From Zimbabwe to appease spirits by exhuming soldiers in Congo (p. 14), by J. Raath, May 31, 1999, *The London Times*. **482:** Reprinted by permission of the publishers and the Trustees of Amherst College from THE POEMS OF EMILY DICKINSON, Thomas H. Johnson, ed., Cambridge, MA: The Belknap Press of Harvard University Press, Copyright © 1951, 1955, 1979, 1983 by the President and Fellows of Harvard College.

Chapter 13. 493–494: Jane Soli was the retired secretary to the academic dean of Saint Olaf College, Northfield, Minnesota. The above account of her husband's death was written in 1989. Michael Leming read the above at Jane Soli's memorial service in 1999. **495:** From *The caregivers book: Caring for another, caring for yourself*, by J. E. Miller, 1996, St. Paul, MN: Augsburg Fortress Press. **497:** Epitaph on gravestone in Haworth, England. **498:** Jimi Hendrix, *Rolling Stone*, December 2, 1976. **500:** Taken from George E. Dickinson (2014, summer). Household pet euthanasia and companion animal last rites. Phi Kappa Phi Forum, 94:4–6. **505–506:** Adapted from "Grief and the Second Row," by H. J. Cummins, *Staff Writer, Star Tribune*, July 7, 2005. **507:** From The agony of grief (pp. 75–78), by S. Ericsson, 1991, September–October, *Utne Reader*. **511:** From The art of condolence (pp. 3–4), by P. V. Johnson, 1991, *Caregivers Quarterly*, 6(3). **512:** "Tsunami Aftermath: The One Face of Grief," December 30, 2004. Editorial. *The New York Times*, 2004. **513:** E. Findell, *Charleston Post and Courier*, May 17, 2009, p. 3G. **518:** From Emotional reserve and the English way of grief, by T. Walter, 1997. In K. Charmaz, G. Howarth, & A. Kellehear (Eds.), *The unknown country: Death in Australia, Britain, and the USA*. London: Macmillan Press.

Chapter 14. 532: Ashanti saying **532:** Source: R. Herrick, Epitaph upon a child that died. From Washington, P. (Ed.). (1998), *Poems of mourning*. New York: Alfred A. Knopf. **539:** Source: Adapted from The Associated Press, 2005. **541:** Irish folksong **542:** From "The Filipino perspective on death and dying" (pp. 215–220), by

I. Weber, 1995. In J. K. Parry & A. S. Ryan (Eds.), *A crosscultural look at death, dying, and religion*, Chicago, IL: Nelson-Hall Publishers. **543:** From *Don't ask for the dead man's golf clubs: Advice for friends when someone dies*, by L. Kelly, 1998, New York: Kelly Communications. **546:** Stanley B. Bezanson, B. Aug. 16, 1919. Died Oct. 4, 2002. Preceded by his wife, Evelyn (July 7, 2002). Internment at Ft. Snelling. **548:** Source: Written on June 12, 2014 by Caroline Lyngstad (St. Olaf College, Class of 1999), former student of Michael Leming. **552:** Source: Taken from G. E. Dickinson, P. D. Roof, and K. W. Roof (2010). End-of-life issues in US veterinary medicine schools. *Society and Animals*, 18(2), 152–162; G. E. Dickinson, P. D. Roof, and K. W. Roof (2011). A survey of veterinarians in the U.S.: Euthanasia and other end-of-life issues. *Anthrozoos*, 24(2), 167–174.

Cartoon Credits

13 Campbell, Martha cartoon—"I think you'll be interested in the next patient. He's ninety two years old and accompanied by his parents." Campbell, Martha/CSL, CartoonStock Ltd

52 Guy and Rodd cartoon—"And finally, I would like to be buried with an elephant bone…Just to confuse future archaeologists." Guy and Rodd/Distributed by Universal Uclick for UFS/CSL, CartoonStock Ltd

70 Kamagurka cartoon—"Death doesn't frighten me, but the night nurse does…" Kamagurka/CSL, CartoonStock Ltd

96 Calvin and Hobbes cartoon—"THIS IS WHERE DAD BURIED THE LITTLE RACCOON." "I DIDN'T EVEN KNOW HE EXISTED A FEW DAYS AGO AND NOW HE'S GONE FOREVER. ITS'S LIKE I FOUND HIM FOR NO REASON. I HAD TO SAY GOOD BYE AS SOON AS I SAID HELLO." "STILL…IN A SAD. AWFUL, TERRIBLE WAY. I'AM HAPPY I MET HIM." "WHAT A STUPID WORLD." Calvin and Hobbes © Watterson. Used by permission of Universal Uclick. All rights reserved.

149 Kes cartoon—"… And you say you first noticed them just after you had a 'near death experience'?" Kes/CSL, CartoonStock Ltd

171 Reynolds cartoon—"The bad news is chemo can kill you before the cancer does. The good news is the medical bills and health insurance can kill you before the chemo does." Reynolds, Dan/CSL, CartoonStock Ltd

205 Oliphant cartoon—"Cartoon on Patient's Bill of Rights" Oliphant © Universal Uclick. Used by permission. All rights reserved.

213 Judd, Phil cartoon—"He's our new Pallitative Specialist!" Judd, Phil/CSL, CartoonStock Ltd

256 Bacall, Aaron cartoon—"Take As Directed or as Budget Allows" Bacall, Aaron/CSL, CartoonStock Ltd

266 Harris, S cartoon—"Ethics ethics" Harris, S/CSL, CartoonStock Ltd

274 Fran cartoon—"Your insurance only covered the removal of the damaged organ…you'll have to put the transplant in yourself." Fran/CSL, CartoonStock Ltd

289 Kes cartoon—"After reviewing how much your treatment is costing this hospital. I'm going to recommend assisted suicide." Kes/CSL, CartoonStock Ltd

290 Andy White cartoon—"Would you also like to purchase the extended warranty" Andy White/CSL, CartoonStock Ltd

299 Fischer, Ed cartoon—"10 Commandments Exceptions" Fischer, Ed/CSL, CartoonStock Ltd

310 Doonesbury Cartoon—"Doctor standing over dead body with two friends near him. Doctor says the man must have died of an overdose but won't know of what until they do an autopsy. The male friend passionately denied that his friend was using drugs or alcohol despite the bottle on the bed stand." Doonesbury © 1986 G. B. Trudeau. Reprinted by permission from Universal Click. All rights reserved.

326 Reynolds cartoon—"Buy Grimbalto for depression. (side effects include headaches, paranoia, thoughts of suicide and depression." Reynolds, Dan/CSL, CartoonStock Ltd

371 Baldwin cartoon—"It's our newest cremation model. Tank sold separately." Baldwin, Mike/CSL, CartoonStock Ltd

409 Donald Reilly cartoon—"He'd rest a lot better with thirty per cent off." Donald Reilly/The New Yorker Collection/Cartoon Bank.Com

412 Hawkins, Jonny cartoon—Man and woman at funeral parlor see sign, 'Ask about our Layaway Plan'. Hawkins, Jonny/CSL, CartoonStock Ltd

430 Martha Gradisher cartoon—"Counting you and the outpouring of comments on facebook, that makes three." Martha Gradisher/CSL, CartoonStock Ltd

460 Jonny Hawkins cartoon—"Unfortunately, we won't know what's wrong with you until we do an autopsy." Jonny Hawkins/CSL, CartoonStock Ltd

475 Morgan, Ron cartoon—"And to my teenage grandchildren, I leave my unused phone minutes." Morgan, Ron/CSL, CartoonStock Ltd

478 Edgar Argo cartoon—"Ms. Avery left each one of you, ten million dollars...unfortunately, her estate is only worth seven dollars and sixteen cents." Edgar Argo/CSL, CartoonStock Ltd

484 Mike Baldwin Cartoon—"Grief takes many forms. But don't worry, we're here to help you fill them out." Mike Baldwin/CSL, CartoonStock Ltd

501 Nick Downes cartoon—"You're sure that's one of the stages of grief?" Downes, Nick/CSL, CartoonStock Ltd

503 Joseph Farris cartoon—"I feel ridiculous!" Farris, Joseph/CSL, CartoonStock Ltd

509 John Louthan cartoon—"Sometimes a cute card just isn't appropriate, Jane" (Woman showing card to deceased who is in the casket) John Louthan

546 P. Byrnes cartoon—"You couldn't put on a tie?" Pat Byrnes/The New Yorker/Cartoon Bank.Com

Photo Credits

1 DECEMBER 12, 2013 - BERLIN: mourning for Nelson Mandela: flowers, candles and images at the South African Embassy in Berlin.—© 360b/Shutterstock.com

11 Roadside Memorial Photo. This is a picture with two crosses with artificial flowers. Captions on the crosses read: Sgt. Mundy and Deputy Musice.—George Dickinson

15 Appalachian family with dead child—Tennessee Tech University Department of Archives

21 Photo of UK cigarette cartons regarding the ills of smoking—George Dickinson

22 Photo of infants' graves in Switzerland. These multiple white crosses have year of birth/death on them.—George Dickinson

32 Karl Marx's grave—George Dickinson

35 Saint Christopher's Hospice entrance—George Dickinson

36 Saint Christopher's Hospice exit—George Dickinson

43 Gravestones through archway—George Dickinson

51 Danger of Death Keep Off here. Danger of Death Keep Off with figure (inside triangle) being struck by lightning.—George Dickinson

55 Gravedigger—George Dickinson

65 Photo of gravestone of Olivia Clarke with "fell asleep". Photo of gravestone of Olivia Clarke (1995–2005).—George Dickinson

67 Photo of portion of poem "In Memoriam" which is on marble Photo of marble slab on brick wall at Highgate Cemetery in London. A portion of the poem "In Memoriam" by Tennyson: "There sat the Shadow feared of man."—George Dickinson

74 A bloody and scary looking zombie girl—© Andreas Gradin/Shutterstock.com

83 Young girl praying in cemetery—© Kim Reinick/Shutterstock.com

97 Children in a small rural elementary school in southeastern United States were instructed by their teacher to draw a picture of death. Above is a sample of their depictions. Death may be viewed as the taker of life, like the Grim Reaper.—George Dickinson

102 Happiness teenagers enjoying in lunch at restaurant—© Lucky Business/Shutterstock.com

111 Rear view of a senior man and woman couple walking arms around each other on a deserted tropical beach with bright clear blue sky—© Darren Baker/Shutterstock.com

121 Officials Pay Respect To Pope John Paul II—Pool/Getty Images News/Getty Images

126 This Japanese funeral emphasizes the social status of the man who has died. The living honor the dead and thereby create a symbolic community consisting of their ancestors and living members of their communities.—Bernard Pierre Wolff/Science Source

130 The wailing wall prayer.—© iStockphoto.com/Boryak

132 Priest Anointing an Ill Patient.—Fuse/Jupiter images

133 Praying on cemetery.—© Sinan Isakovic/Shutterstock.com

135 Devoted Hindus in Bali make pilgrimage to temple, the home of the Hindu gods. One third of a woman's life in Bali will be spent either preparing for or performing Hindu religious rituals.—Michael R. Leming

138 The central object of worship in Buddhism is the Buddhist "trinity."—the Buddha, the Darma (teachings of the Buddha), and the Sangha (the disciples of the Buddha).—Michael R. Leming

142 A mourner places a garland of flowers on the body of Mother Teresa, at a church in Calcutta Sunday, September 7, 1997. The funeral of the "sister of mercy" will be held on Saturday.—AP Images/Sherwin Crasto

155 Doctor talking to elderly patient lying in bed in hospital—© AlexanderRaths/Shutterstock.com

168 A pretty indian doctor bonding with her elderly patient. Isolated on white—© Lisa F. Young/Shutterstock.com

195 Senior woman with seriously ill husband in hospital—© Monkey Business Images/Shutterstock.com

201 Heads of surgeons with masks looking down obviously doing surgery—George Dickinson

211 Kidney Arrives for Transplant Surgery A kidney arrives by plane in Rome for a transplant operation at the Umberto I hospital. Location: Rome, Italy—Vittoriano Rastelli/Historical/Corbis

215 Photo of Saint Christopher's Hospice in London (a building)—George Dickinson

220 Photo of Cicely Saunders and George Dickinson sitting behind a desk with lot of books behind them—George Dickinson

223 Happy nurse holding hands of elderly patient sitting side by side at home, laughing.?—© StockLite/Shutterstock.com

228 Photo of a man with an oxygen tube with his wife standing behind him—Grace Beahm/The Post and Courier

233 Photo of a street sign saying "Elderly People" and shows two persons helping each other across the street—George Dickinson

239 Surgeons in operative room—© Brasiliao/Shutterstock.com

239 A cardiologist implants a heart defibrillator in a patient. Focus is on the fluoroscopy monitors. (14MP camera, real operation).—© Carolina K. Smith MD/Shutterstock.com

243 Connecticut Hospice—www.hospice.com

251 Health care workers and elderly woman in wheelchair—© Alexander Raths/Shutterstock.com

261 UK. Senior woman heart attack patient being monitored in Resuscitation unit in Accident and Emergency department of NHS hospital—Realimage/Alamy

272 Donor Heart Procurement for Cardiac Transplantation—© kalewa/Shutterstock.com

279 Dr. Jack Kevorkian Assisted suicide advocate Dr. Jack Kevorkian poses with his "suicide machine" in Michigan, in this Feb. 6, 1991, file photo. Kevorkian is expected to leave prison Friday, June 1, 2007 after serving more than eight years of a 10- to 25-year sentence in the death of a Michigan man. (AP Photo/Richard Sheinwald, file)—AP Images/Richard Sheinwald

305 Take an overdose—© Photogrape/Shutterstock.com

308 People carry a coffin of a victim of a suicide bomb attack in Baqouba, capital of Iraq's Diyala province, 60 kilometers (35 miles) northeast of Baghdad, Iraq, Friday, Jan. 18, 2008. A suicide bomber struck Shiites Thursday, as worshippers prepared for their most important holiday, killing at least 11 at a mosque in violent Diyala province, one day after a similar attack by a woman in a nearby village.—AP Images/STR

317 Office workers walk beneath a stock quotation board flashing a drop of benchmark Nikkei average in a Tokyo business district September 24. The Nikkei 225 average tumbled more than three percent on Friday morning, piercing the psychological 17,000 barrier. The Nikkei average closed out the morning session at 16,743.83, down 581.93 points or 3.36 percent from Wednesday's close. ta/Photo by Toshiyuki Aizawa REUTERS—Reuters/Corbis

330 Karen edler singing with young man at a wedding—Michael R. Leming

332 RALLY IN COMMEMORATION OF KURT COBAIN Location: Seattle, Washington State, USA—John Van Hasselt/Sygma/Corbis

333 Dead End Road Sign Photo—George Dickinson

INDEX